Green Approach to Alternative Fuel for a Sustainable Future

Green Approach to Alternative Fuel for a Sustainable Future

Edited by

Maulin P. Shah

ELSEVIER

Elsevier
Radarweg 29, PO Box 211, 1000 AE Amsterdam, Netherlands
The Boulevard, Langford Lane, Kidlington, Oxford OX5 1GB, United Kingdom
50 Hampshire Street, 5th Floor, Cambridge, MA 02139, United States

Copyright © 2023 Elsevier Inc. All rights reserved.

No part of this publication may be reproduced or transmitted in any form or by any means, electronic or mechanical, including photocopying, recording, or any information storage and retrieval system, without permission in writing from the publisher. Details on how to seek permission, further information about the Publisher's permissions policies and our arrangements with organizations such as the Copyright Clearance Center and the Copyright Licensing Agency, can be found at our website: www.elsevier.com/permissions.

This book and the individual contributions contained in it are protected under copyright by the Publisher (other than as may be noted herein).

Notices
Knowledge and best practice in this field are constantly changing. As new research and experience broaden our understanding, changes in research methods, professional practices, or medical treatment may become necessary.

Practitioners and researchers must always rely on their own experience and knowledge in evaluating and using any information, methods, compounds, or experiments described herein. In using such information or methods they should be mindful of their own safety and the safety of others, including parties for whom they have a professional responsibility.

To the fullest extent of the law, neither the Publisher nor the authors, contributors, or editors, assume any liability for any injury and/or damage to persons or property as a matter of products liability, negligence or otherwise, or from any use or operation of any methods, products, instructions, or ideas contained in the material herein.

ISBN: 978-0-12-824318-3

For information on all Elsevier publications visit our website at https://www.elsevier.com/books-and-journals

Publisher: Candice Janco
Acquisitions Editor: Anita Koch
Editorial Project Manager: Maria Elaine Desamero
Production Project Manager: Sruthi Satheesh
Cover Designer: Greg Harris

Typeset by TNQ Technologies

Contents

List of contributors   xv

1. Policies of biofuel for commercialization
Prangan Duarah, Dibyajyoti Haldar and Mihir Kumar Purkait

1. Overview of policies for biofuel production and commercialization   1
2. Factors affecting the commercialization of cellulosic biofuel   2
    2.1 Drivers   2
    2.2 Barriers   3
3. Latest policies undertaken by USA to commercialize cellulosic biofuels   3
4. Brazil's biofuel national policies implemented for its commercialization   4
5. Recent progress in policies taken by the government of China to drive biofuel generation   5
6. The national policy of India to promote its biofuel commercialization program   6
7. Different policies for algal biofuel commercialization   7
8. Challenges with the policies and future recommendations   8
9. Conclusions   9
Acknowledgment   9
References   9

2. Life cycle assessment of biofuels derived from Mahua (*Madhuca* species) flowers and seeds: key issues and perspectives
Urmimala Das, Bishnupriya Hansdah, Sudhanshu S. Behera and Ramesh C. Ray

1. Introduction   13
    1.1 Geographical and climatic resolution   13
    1.2 *Madhuca* species as a green approach for biofuel production   14
2. Biomass (mahua flowers) for conversion to bioethanol   14
    2.1 Bioethanol production process   15
    2.2 Bioethanol refinery   16
3. Biomass (mahua seeds) for conversion to biodiesel   16
    3.1 Seed oil and oil yield   16
    3.2 Biodiesel production process and recent advances   17
    3.3 Biodiesel refinery   20
4. Life cycle assessment   20
    4.1 Goal, scope, and functional unit   20
    4.2 System boundaries and reference system   20
    4.3 Inventory analysis   21
    4.4 Coproduct, byproduct, residue   21
    4.5 Impact assessment   21
5. Conclusion and future trends   21
References   21
Further reading   23

3. Generation of biofuels from rice straw and its future perspectives
Protha Biswas, Sujata Mandal, Tuyelee Das, Satarupa Dey, Mimosa Ghorai, Sayan Bhattacharya, Arabinda Ghosh, Potshangbam Nongdam, Vineet Kumar, Abdel Rahman Al-Tawaha, Ercan Bursal and Abhijit Dey

1. Introduction   25
2. Rice crop residue   26
3. Composition and properties   26
4. Emissions formed from crop residues and associated health hazard   27
5. Management of rice straw   27
    5.1 Management of rice straw in the field   27
    5.2 Utilization of crop residues off-field   28
6. Biofuel production and its processes   29
7. Life cycle assessment (LCA) and SWOT analysis for biofuels production from rice straw   30

8. Advantages and disadvantages of producing biofuels from rice straw and its socioeconomic evaluation	31
9. Conclusion and future perspectives	31
References	32
Further reading	33

4. Prospects of rice straw for sustainable production of biofuels: Current trends and challenges

Priyanka Kamilya, Omar Aweiss Ali, Subrat Kumar, Sandeep Kumar Panda and Ritesh Pattnaik

1. Introduction	35
2. Rice production in Asian countries	35
3. Composition of rice by-products: rice straw and rice husks	36
4. Chemical components of different types of lignocellulosic biomass	36
5. Potential of rice residue in biorefineries	37
5.1 Physical pretreatment	37
5.2 Chemical pretreatment	37
5.3 Biological pretreatment	39
5.4 Thermochemical pretreatment	39
6. Conversion of rice straw into biofuels	40
6.1 Bioethanol	40
6.2 Biogas	41
6.3 Bio-oil	42
6.4 Other value-based products	42
7. Challenges and future perspectives	42
8. Conclusion	42
Acknowledgments	42
References	43

5. Biofuel production from algal biomass: Technologies and opportunities addressing the global energy crisis

Shibam Dey, Ishanee Mazumder, Chandrashish Roy, Omar Aweiss Ali, Subrat Kumar and Ritesh Pattnaik

1. Introduction	45
2. Algae: a potential source of biofuels	46
2.1 Microalgae as a source of biofuel	46
2.2 Macroalgae as a source of biofuel	46
3. Microalgal-based biofuel: technological perspectives	47
4. Technologies for microalgal cultivation	47
4.1 Open ponds	47
4.2 Photobioreactors	47
4.3 Two-stage hybrid system	48
4.4 Lipid induction technique	48
5. Technologies for macroalgal cultivation	48
6. Technologies for algal harvesting	48
6.1 Filtration	48
6.2 Sedimentation	48
6.3 Flocculation	48
6.4 Flotation	49
7. Physicochemical parameters for biofuel production from algae	49
7.1 Effect of temperature	49
7.2 Effect of light	49
7.3 Effect of pH	49
7.4 Effect of salinity	49
7.5 Effect of nutrients	49
8. Genetically engineered algae for biofuel production	50
8.1 Lipid augmentation by genetic alterations	50
8.2 Genetic engineering to steer hydrogen production	50
9. Bioconversion technologies of algal biomass to biofuel	52
9.1 Biodiesel	53
9.2 Bioethanol	53
9.3 Biogas	54
9.4 Biohydrogen production	54
9.5 Bioelectricity	54
9.6 Biomass production and phycoremediation	54
9.7 Value-added products	55
10. Conclusion and future perspectives	56
Acknowledgments	57
References	57

6. Bioenergy production from algae: biomass sources and applications

Mostafa M. El-Sheekh, Ghadir Aly El-Chaghaby and Sayed Rashad

1. Introduction	59
2. Algae cultivation conditions	59
3. Macro and microalgae biomass production	60
4. Algal biomass chemical composition	60
5. Macro and microalgae for bioenergy production	61
6. Methods for algal biomass conversion to bioenergy products	63
7. Benefits and limitations of using algae for bioenergy production purposes	64
8. Economics of bioenergy production from algal biomass	65
9. Conclusion and future prospects	65
References	66

7. Microbial bioprospecting of biodiesel industry-derived crude glycerol waste conversion into value-added products

Hiren K. Patel, Nidhi P. Patel and Maulin P. Shah

1. Introduction 71
2. Biodiesel 72
   2.1 Chemistry of biodiesel transesterification of oils 72
   2.2 Chemistry of biodiesel transesterification of oils 72
   2.3 The transition to a fuel: 19th-century internal combustion engine use and fuel choices 73
   2.4 Problems brought about by biodiesel 73
3. Application of biodiesel 74
   3.1 Biodiesel reduces greenhouse gas emissions 75
   3.2 Biodiesel reduces tailpipe emissions 75
   3.3 Biodiesel and human health 75
   3.4 Biodiesel improves engine operation 75
   3.5 Biodiesel is easy to use 75
4. Production of biodiesel 75
   4.1 Biodiesel sources 76
5. Crude glycerol waste from biodiesel as a by-product 77
   5.1 Value-added opportunities for crude glycerol 77
6. Bioconversion of crude glycerol waste into value-added products 78
7. Microbial bioprospecting 83
8. Application of microbial prospecting 83
9. Conclusion 84
10. Future prospects 84
References 84

8. Microalgae-based biofuel synthesis

Mohamed Gomaa, Mustafa A. Fawzy and Mostafa M. El-Sheekh

List of abbreviations 89
1. Introduction 89
2. Upstream processes for biodiesel synthesis 90
   2.1 Strain selection 90
   2.2 Algae cultivation and its effects on biodiesel production 92
3. Downstream processes for biodiesel synthesis 94
   3.1 Microalgae harvesting techniques 94
   3.2 Technologies for effective lipid extraction from algae 96
   3.3 Techniques for biofuel production from algal biomass 98
4. Properties of microalgal biodiesel 100
   4.1 Iodine value 100
   4.2 Cetane number 100
   4.3 Saponification value 101
   4.4 Oxidation stability 101
   4.5 Kinematic viscosity 102
   4.6 Density 102
   4.7 Higher heating value 102
   4.8 Cold flow properties 102
5. Future perspectives: challenges and opportunities 103
6. Conclusion 103
References 103

9. A plausible scenario for the third generation of biofuels from microalgae

Rohit Saxena, Rosa M. Rodríguez-Jasso, Mónica L. Chávez-González, Cristóbal N. Aguilar, Cristina González-Fernández, Guillermo Quijano and Héctor A. Ruiz

1. Introduction 107
2. Third-generation biorefinery 108
3. Technologic applications of microalgae 108
   3.1 Spirulina platensis 108
   3.2 Cultivation system 110
   3.3 Harvesting 113
   3.4 Challenges in cultivation and harvesting 115
   3.5 Bioethanol production 115
   3.6 Biodiesel 118
   3.7 Biohydrogen from microalgae 118
   3.8 High value-added by-product (phycocyanin) 119
4. Future perspective 119
5. Conclusion 119
Acknowledgments 119
References 119

10. Microalgae as a promising feedstock for biofuel production

Sanaa M.M. Shanab, Mostafa M. El-Sheekh and Emad A. Shalaby

1. Introduction 123
2. Microalgal biomass production 124
   2.1 Photoautotrophic algal production 124
   2.2 Heterotrophic algal production 125
   2.3 Mixotrophic algal production 125

3. Harvesting of microalgal biomass ... 126
4. Extraction and purification of microalgal biomass ... 126
5. Conversion techniques/methods of algal biomass to biofuels ... 126
   5.1 Thermochemical conversion ... 126
   5.2 Biochemical conversion ... 127
6. Algal biofuel types ... 128
   6.1 Biogas (biomethane production) ... 128
   6.2 Bioethanol ... 128
   6.3 Biodiesel ... 128
7. Genetically modified algae ... 130
8. Algal fuel properties ... 131
9. Economic importance of algae as biofuel sources ... 131
10. Applications of biomass ... 132
11. Conclusion ... 132
References ... 132
Further reading ... 135

11. Thermochemical conversion of biomass into valuable products and its modeling studies

R. Saravanathamizhan, V.T. Perarasu and K. Vetriselvan

1. Introduction ... 137
   1.1 Biomass ... 137
   1.2 Classification of biomass ... 137
   1.3 Characteristics of biomass ... 138
2. Biomass conversion techniques ... 138
3. Thermochemical conversion ... 138
   3.1 Combustion ... 139
   3.2 Gasification ... 141
   3.3 Pyrolysis ... 143
   3.4 Liquefaction ... 145
4. Modeling studies ... 146
   4.1 Equilibrium models ... 146
   4.2 Kinetic models ... 147
   4.3 ASPEN Plus models ... 147
   4.4 CFD models ... 148
   4.5 ANN models ... 148
5. Conclusion ... 149
References ... 149

12. Rice straw for biofuel production

Pranjal P. Das, Ankush D. Sontakke and Mihir Kumar Purkait

1. Introduction ... 153
2. Crop residues ... 154
3. Concept of rice straw ... 155
   3.1 Composition of rice straw ... 155
   3.2 Field management of rice straw ... 155

4. Life cycle assessment (LCA) and SWOT analysis for biofuels production from rice straw ... 156
5. Processes to improve the biofuel yield from rice straw ... 157
   5.1 Pretreatment processes ... 157
   5.2 Anaerobic codigestion process ... 161
   5.3 Innovative process based on additives ... 161
6. Mathematical models and tools to scrutinize the sustainability of biofuels ... 163
7. Conclusion ... 164
References ... 164

13. Biofuel production from algal biomass

Pranjal P. Das, Niladri S. Samanta, Simons Dhara and Mihir Kumar Purkait

1. Introduction ... 167
2. Categories of biofuel generation ... 168
3. Algal biomass ... 168
   3.1 Algal ... 168
   3.2 Cultivation ... 169
4. Techniques involved in transformation of third-generation biofuels from algal biomass ... 170
   4.1 Extraction of oil or lipid from algal biomass ... 170
   4.2 Lipids or oil transesterification ... 171
   4.3 Saccharification and pretreatment of algal biomass ... 172
   4.4 Microbial fermentation of algal biomass ... 173
   4.5 Anaerobic digestion of microalgal biomass ... 173
5. Biofuels obtained from the biovalorization of algal biomass ... 174
   5.1 Biomethane ... 174
   5.2 Biodiesel ... 174
   5.3 Bioethanol ... 175
   5.4 Biochar (BC) ... 175
6. Opportunities and challenges in the production of biofuels from algal biomass ... 176
7. Conclusion ... 176
References ... 177

14. Microalgae biofuels: a promising substitute and renewable energy

Indu Chauhan, Vivek Sharma, Pawan Rekha and Lovjeet Singh

1. Introduction ... 181
2. Benefits of using microalgae for biofuel productions ... 181

3. Cultivation and harvesting of microalgae for biofuel production — 181
   3.1 Growth factors for microalgae cultivation — 182
   3.2 Bioreactors used for culturing — 183
4. Harvesting of algal biomass — 184
   4.1 Physical methods — 184
   4.2 Chemical methods — 185
   4.3 Biologic method — 185
5. Biofuel production using microalgae — 185
   5.1 Biohydrogen — 185
   5.2 Bioethanol — 186
   5.3 Biodiesel — 186
   5.4 Biogas — 186
6. Applications — 186
   6.1 Transport — 186
   6.2 Aviation — 188
   6.3 Lubricants — 188
   6.4 Power/electricity generation — 188
References — 188

15. Biofuel production by catalysis

Vivek Sharma, Prashnasa Tiwari, Indu Chauhan and Pawan Rekha

1. Introduction — 191
2. Different approaches for biofuel production — 191
3. Catalysts used so far for biofuel production — 192
   3.1 Metal/alloy based nanocatalyst — 192
   3.2 Metal oxide nanoparticles — 194
   3.3 Biomass-based catalysts — 194
4. Comparative highlights of catalytic efficiency of developed catalysts for biofuel production — 196
5. Conclusion — 197
References — 197

16. Potential of lignin as biofuel substrate

Sagarjyoti Pathak and Hitesh S. Pawar

1. Introduction — 201
   1.1 Current energy status globally — 201
   1.2 Climate change—a serious threat worldwide — 201
   1.3 Potential biomass feedstocks for energy production — 202
   1.4 Availability and abundance of biomass — 202
2. Generations of biofuel — 203
   2.1 First-generation biofuels — 203
   2.2 Second-generation biofuels — 203
   2.3 Third- and fourth-generation biofuels — 204
3. What is lignin chemically? — 204
   3.1 Role of lignin in the biological living world — 204
   3.2 Synthesis of lignin in the living system — 205
   3.3 Lignin carbohydrate complex (LCC) — 206
4. Isolation and extraction of lignin from lignocellulosic biomass — 206
   4.1 Pretreatment technologies to separate and depolymerize lignin — 206
   4.2 Commercially available lignin — 210
5. Elucidation and characterization of the extracted lignin — 212
6. Routes of lignin conversion — 213
   6.1 Pyrolysis method to convert lignin — 213
   6.2 Lignin conversion by acid catalysis — 215
   6.3 Lignin conversion through base catalysis — 215
   6.4 Lignin conversion through reductive catalysis — 215
   6.5 Oxidative conversion of lignin — 216
7. Challenges in lignin conversion — 217
8. Conclusions and future outlook — 218
List of abbreviations — 219
References — 219

17. Biohydrogen production from dark fermentation of lignocellulosic biomass

Prajwal P. Dongare and Hitesh S. Pawar

1. Introduction — 223
   1.1 Lignocellulosic biomass for dark fermentation — 225
   1.2 Sources of lignocellulosic biomass — 225
   1.3 Characteristics of lignocellulosic biomass — 226
   1.4 Compositional variation in cellulose, hemicellulose, and lignin content in lignocellulosic biomass — 226
2. Role of pretreatment — 227
3. Pretreatment methods for dark fermentation — 228
   3.1 Physical pretreatment method for dark fermentation — 228
   3.2 Chemical pretreatment methods for dark fermentation — 228
   3.3 Biological pretreatment methods — 230
   3.4 Combination of pretreatment methods — 230
4. Inhibitor formation and solution strategies to overcome inhibitor formation for dark fermentation — 231
5. Strategies for dark fermentation — 232

    5.1 Pretreatment followed by enzyme hydrolysis 232
    5.2 Natural strategy relevant to dark fermentation 233
    5.3 Separation 233
6. Comparative study with dark fermentation 234
7. Dark fermentation advantages and stoichiometry 234
8. Factor affecting dark fermentation 235
9. Cultivation techniques of bioreactor 236
10. Yield enhancement strategies 236
11. Conclusion and future perspectives 238
List of abbreviations 239
References 239

18. Theoretical insight into methanol sono-conversion for hydrogen production

Aissa Dehane, Leila Nemdili, Slimane Merouani and Atef Chibani

1. Introduction 243
2. The sonochemical process 243
3. Some literature data 245
4. Methanol pyrolysis model 246
5. Optimal gas conditions 246
6. Ultrasound frequency effect 249
7. Acoustic intensity effect 250
Acknowledgements 252
8. Conclusion 252
References 252

19. Upscaling of $LaNi_5$-based metal hydride reactor for solid-state hydrogen storage: numerical simulation of the absorption—desorption cyclic processes

Atef Chibani, Aissa Dehane, Leila Nemdili and Slimane Merouani

1. Introduction 257
2. Previous works 258
3. Physical description of the unit 258
4. Model 259
5. Analysis of the absorption process (hydrogen storage) 259
6. Analysis of the desorption process (hydrogen discharging) 262
7. Conclusion and future perspectives 266
Nomenclature 267
Greek symbols 267
References 267

20. Oleaginous microbes for biodiesel production using lignocellulosic biomass as feedstock

Falak Shaheen, Palvi Ravinder, Rahul Jadhav, Navanath Valekar, Sangchul Hwang, Ranjit Gurav and Jyoti Jadhav

1. Introduction 271
2. Prospective oleaginous microorganisms for biodiesel production 272
    2.1 Oleaginous yeast and filamentous fungi 273
    2.2 Microalgae 274
    2.3 Oleaginous bacteria 274
    2.4 Cocultivation of microorganisms 274
3. Lignocellulosic biomass as a carbon source 274
    3.1 Food crops 275
    3.2 Nonfood/energy crops 275
    3.3 Forest trees and residues 275
    3.4 Industrial process residues 275
4. Structural features of lignocellulosic biomass and recalcitrance 276
5. Pretreatment 277
    5.1 Mechanical pretreatment 277
    5.2 Nonmechanical pretreatment 279
    5.3 Importance of catalyst recovery 282
6. Downstream processing of the biomass 283
    6.1 Hydrolysis/saccharification 283
    6.2 Fermentation 284
    6.3 Transesterification 285
    6.4 Purification of biodiesel 286
7. Environmental issues 286
8. Opportunities and challenges 286
    8.1 Challenges 287
9. Recent advancement 287
    9.1 Genetic and metabolic engineering 287
    9.2 Metabolic modeling 289
    9.3 Biorefineries 289
    9.4 Bioflocculation 289
    9.5 Biocatalyst 289
    9.6 Cocultivation 289
    9.7 Osmotic stress 290
10. Applications of biodiesel 290
    10.1 On-road and off-road vehicles 290
    10.2 Marine vessels 290
    10.3 Solvents 290
    10.4 Automotive diesel engine 290
    10.5 Drilling fluids 291
    10.6 Stationary power generation 291
11. Future prospects and scope 292
References 293

21. An insight into rice straw–based biofuel production

Manswama Boro and Anil Kumar Verma

1. Introduction 297
2. Rice straw 297
   2.1 Rice ecosystem 297
   2.2 RS availability and management 298
   2.3 RS constituents and characteristics 299
3. Biofuel production from RS 300
   3.1 Types of biofuels that can be produced from RS 300
   3.2 Pretreatment of RS 302
   3.3 The biofuel production process 304
4. Economic evaluation of biofuel produced from rice straw 305
5. Conclusion and future prospects 307
6. Conflict of interest statement 307
7. List of abbreviations 307
Acknowledgments 307
References 307

22. Sources and techniques for biofuel generation

S.A. Aransiola, M.O. Victor-Ekwebelem, S.S. Leh-Togi Zobeashia and Naga Raju Maddela

1. Introduction 311
2. Biofuel and its sources 312
   2.1 Bioethanol 312
   2.2 Biodiesel 313
   2.3 Biogas 314
3. Environmental benefits of biofuel: global perspectives 314
4. Biomass feedstocks in the production of biofuel 315
   4.1 Biomass with some of their application 315
5. Extraction and determination of different biofuels 316
   5.1 Fast pyrolysis 316
   5.2 Biodiesel extraction methods 319
6. Conclusion 320
References 321

23. Biomass waste and feedstock as a source of renewable energy

Kondakindi Venkateswar Reddy, Nalam Renuka Satya Sree, Pabbati Ranjit and Naga Raju Maddela

1. Introduction 325
2. Biomass 326
   2.1 Biomass—various forms available 326
   2.2 Biomass composition and their energy content 327
   2.3 Measuring and analysis of biomass 329
3. Available biomass conversion technologies 329
   3.1 Pyrolysis 330
   3.2 Torrefaction 331
   3.3 Gasification 331
   3.4 Combustion 331
   3.5 Biological conversion 331
4. Products and their applications 332
   4.1 Biochar 332
   4.2 Nanocellulose 332
   4.3 Biofuel 332
5. Implementation 333
6. Recent advances 333
7. Environmental impacts 333
8. Challenges 333
9. Conclusion 334
References 334

24. Bioenergy and beyond: biorefinery process sources, research, and advances

Thamarys Scapini, Sérgio Luiz Alves Júnior, Aline Viancelli, William Michelon, Aline Frumi Camargo, Angela Alves dos Santos, Laura Helena dos Santos and Helen Treichel

1. Introduction 335
2. Biomass sources for integrated biorefineries: bioenergy and beyond 336
   2.1 Lignocellulosic biomass 339
   2.2 Starchy biomass 340
   2.3 Pectin biomass 340
   2.4 Microalgae 341
3. Genetically engineered microorganisms for CO_2 fixation 343
4. Key barriers to converting biomass into bioenergy 344
5. Final considerations 345
References 345

25. Endemic microalgae biomass for biorefinery concept and valorization

Samanta Machado-Cepeda, Rosa M. Rodríguez-Jasso and Héctor A. Ruiz

1. Introduction 349
2. Methodologies to extract and isolate microalgae 350
3. Review of morphological microalgae identification 352

4. Optimal culture media for laboratory-level photobioreactor biomass production, depending on microalgae species ... 357
   4.1 Closed system production ... 358
   4.2 Recovery of microalgal biomass ... 358
   4.3 Production of lipids ... 363
   4.4 Production of carbohydrates ... 363
5. Biomass fractionation ... 365
6. Techniques to identify and quantify lipids and carbohydrates ... 365
7. Conclusions ... 365
Acknowledgments ... 366
References ... 366

26. Microalgae biomass: a model of a sustainable third-generation of biorefinery concept

Alejandra Cabello-Galindo, Rosa M. Rodríguez-Jasso and Héctor A. Ruiz

1. Introduction ... 369
2. Current status of third-generation bio-refineries with microalgae ... 370
   2.1 Biorefinery classification ... 370
   2.2 First-generation biorefineries ... 370
   2.3 Second-generation biorefineries ... 370
   2.4 Third-generation biorefineries ... 370
3. Microalgae and sustainability ... 370
4. Bioremediation capacity of microalgae ... 372
   4.1 Environmental toxicant monitoring ... 373
   4.2 Use of microalgae for wastewater treatment ... 374
   4.3 Microalgae mitigation capacity of CO_2 ... 375
5. Integration and reuse of industrial waste for microalgae cultivation ... 375
6. Co-production of microalgae-derived products for circular bio-economy-based bio-refinery ... 376
7. Future perspectives ... 376
8. Conclusion ... 377
Acknowledgments ... 377
References ... 377

27. Green hydrogen production: a critical review

Gilver Rosero-Chasoy, Rosa M. Rodríguez-Jasso, Cristóbal N. Aguilar, Germán Buitrón, Isaac Chairez and Héctor A. Ruiz

1. Introduction ... 381
2. Hydrogen production from renewable energy ... 381
3. Hydrogen production from biofuels obtained by thermochemical processes of microalgae biomass ... 383
   3.1 Direct combustion ... 383
   3.2 Liquefaction ... 383
   3.3 Gasification ... 384
   3.4 Pyrolysis ... 385
4. Hydrogen production by microorganisms ... 385
   4.1 Direct and indirect bio-photolysis ... 385
   4.2 Dark fermentation ... 386
   4.3 Photo-fermentation ... 386
5. Benefits and challenges of hydrogen production ... 387
6. Conclusions and perspectives ... 388
Acknowledgments ... 389
References ... 389

28. Algal biofuel production using wastewater: a sustainable approach

Sougata Ghosh, Bishwarup Sarkar and Sirikanjana Thongmee

1. Introduction ... 391
2. Algal biofuel production using wastewater ... 391
3. Conclusion and future perspectives ... 397
Acknowledgment ... 397
References ... 397

29. Rice straw: a potential substrate for bioethanol production

Quratulain, Ali Nawaz, Hamid Mukhtar, Ikram ul Haq and Vasudeo Zambare

1. Introduction ... 399
2. Rice straw potential for bioethanol production ... 400
3. Production of bioethanol from rice straw ... 400
   3.1 Basic concepts of rice straw ... 400
   3.2 Cellulolytic strain culture and enzyme activity testing ... 401
   3.3 Pretreatment techniques ... 401
   3.4 Fermentation of rice straw hydrolysates ... 403
   3.5 Bioethanol production's techno-economic worthiness ... 404
4. Conclusions and future recommendations ... 405
References ... 405

30. Algal bioenergy: the fuel for tomorrow

Elham M. Ali, Mostafa Elshobary and Mostafa M. El-Sheekh

1. Introduction ... 409
   1.1 Climatic change consequences on energy ... 409

		1.2 Climate impacts on mediterranean countries	410
		1.3 Potentiality of algae as a biologic source of energy	411
		1.4 Types of microalgae	411
		1.5 Size classification	411
2.	Species		411
3.	Potential strains		411
		3.1 History of biofuel research	411
4.	What is biofuel?		412
		4.1 First-generation biofuel	412
		4.2 Second-generation biofuel	412
		4.3 Third-generation biofuel	412
5.	Algal biofuel		413
		5.1 Advantages of algae as a biofuel source	413
		5.2 Sources of algal biomass	414
		5.3 Artificial cultured biomass	414
6.	Lipids yield of microalgae		415
7.	Growth requirements for optimal microalgal growth		416
		7.1 Temperature	416
		7.2 Light	416
		7.3 Nutrient provision	416
		7.4 Mixing	416
8.	Closed system		417
		8.1 Horizontal tube photobioreactor	417
		8.2 Vertical tube photobioreactor	417
		8.3 Stirring photobioreactor	418
		8.4 Flat-panel photobioreactor	418
9.	Open system		418
		9.1 Static pond	418
		9.2 Raceway pond	418
		9.3 Circular ponds	419
		9.4 Hybrid photobioreactors	419
10.	Macroalgal cultures		419
11.	Types of algal bioenergy		419
		11.1 Bioethanol production	419
		11.2 Biodiesel production	419
12.	Difference between diesel and biodiesel		420
13.	Microalga species for biodiesel		421
14.	Cultivation systems		421
		14.1 Photoautotrophic versus heterotrophic	421
		14.2 Biohydrogen production	421
15.	Methodologies of production		422
		15.1 Steps of bioethanol production from algae biomass	422
		15.2 Pretreatment process	422
		15.3 Fermentation process	423
		15.4 Bioethanol recovery	424
16.	Steps of biodiesel production from algae biomass		424
		16.1 Steps of biohydrogen production from algae biomass	424
17.	Algal resources in Egypt		426
18.	Future perspective in Egypt: initiatives toward a safe environment with clean energy		427
References			427
Further reading			432

31. Plant-based biofuels: an overview

Soumya Singh and Shalini Singh

1.	Introduction		433
	1.1 Biofuels and their economics		433
	1.2 Generations of biofuels		434
	1.3 Plant biofuels: an introduction		434
	1.4 The renewable fuel standards (RFS) II		435
	1.5 Feedstocks for biofuel production		435
2.	Switchgrass		436
3.	Feedstock processing technologies		437
	3.1 Deconstruction		437
	3.2 Upgrading		437
4.	Production strategies		437
	4.1 Bioethanol production		437
	4.2 Biodiesel production		440
	4.3 Biomethane production		440
5.	Future and scope of biofuels		441
References			442

32. Bioenergy: biomass sources, production, and applications

Tanvi Taneja, Muskaan Chopra and Indu Sharma

1.	Introduction		443
	1.1 Bioenergy		443
	1.2 Biomass sources		443
	1.3 Bioenergy production techniques		445
2.	Biomass conversion techniques to produce electricity and heat		445
	2.1 Gasification techniques		446
	2.2 Pyrolysis		446
	2.3 Fermentation		446
	2.4 Bioethanol production from biomass created by agricultural waste		447
3.	Enzymatic hydrolysis		447
	3.1 Fermentation		447
References			450
Further reading			451

33. Bioprospecting microalgae for biofuel synthesis: a gateway to sustainable energy

Nahid Akhtar, Atif Khurshid Wani, Reena Singh, Chirag Chopra, Sikandar I. Mulla, Farooq Sher and Juliana Heloisa Pinê Américo-Pinheiro

1. Introduction — 453
2. Modification of growth and physiologic conditions for enhancing microalgae biomass and biofuel components — 454
3. Modification of growth and physiologic conditions for enhancing lipid content of microalgae — 455
4. Modification of growth and physiologic conditions for enhancing carbohydrate content of microalgae — 456
5. Genetic engineering for enhancing microalgae biomass and biofuel components — 457
6. Future prospects — 458
7. Conclusion — 459
References — 459

34. Technoeconomic analysis of biofuel production from agricultural residues through pyrolysis

Kondragunta Prasanna Kumar and Neelancherry Remya

1. Introduction — 463
2. Properties of agricultural residues — 463
3. Thermochemical conversion process for agricultural residue — 465
   3.1 Combustion — 465
   3.2 Gasification — 465
   3.3 Pyrolysis — 465
4. Technoeconomic analysis — 466
5. Technoeconomic analysis of pyrolysis process of agricultural residue — 466
6. Summary — 467
References — 468

35. Different methods to synthesize biodiesel

José Manuel Martínez Gil, Ricardo Vivas Reyes, Marlón José Bastidas Barranco, Liliana Giraldo and Juan Carlos Moreno-Piraján

1. Biodiesel: definition, variations, properties, and comparison with the diesel — 471
   1.1 Biodiesel production by transesterification — 474
   1.2 Enzyme-catalyzed transesterification for biodiesel production — 478
2. Sources for the synthesis of biodiesel — 479
   2.1 Vegetable oils for biodiesel — 481
   2.2 Animal fat for biodiesel — 482
   2.3 Biodiesel from used cooking oil — 482
3. Biodiesel from microalgae oils — 482
4. Production of biodiesel from oleaginous microorganisms with organic waste as raw materials — 483
5. Application of nanotechnologies in the production of biodiesel — 483
6. Conclusions — 484
7. Perspectives and challenges for the future of biodiesel — 484
Acknowledgments — 484
References — 485
Further reading — 490

Index — 491

List of contributors

Cristóbal N. Aguilar, Biorefinery Group, Food Research Department, School of Chemistry, Autonomous University of Coahuila, Saltillo, Coahuila, Mexico

Nahid Akhtar, Department of Biotechnology, School of Bioengineering and Biosciences, Lovely Professional University, Phagwara, Punjab, India

Abdel Rahman Al-Tawaha, Department of Biological Sciences, Al-Hussein Bin Talal University, Maan, Jordon

Elham M. Ali, Department of Aquatic Environment, Faculty of Fish Resources, Suez University, Suez, Egypt

Omar Aweiss Ali, School of Biotechnology, Kalinga Institute of Industrial Technology (Deemed-to-be University), Bhubaneswar, Odisha, India

Angela Alves dos Santos, Department of Biochemistry, Federal University of Santa Catarina, Florianópolis, Santa Catarina, Brazil

Sérgio Luiz Alves Júnior, Laboratory of Biochemistry and Genetics, Federal University of Fronteira Sul, Chapecó, Santa Catarina, Brazil; Department of Biological Science, Graduate Program in Biotechnology and Bioscience, Federal University of Santa Catarina, Florianópolis, Santa Catarina, Brazil

Juliana Heloisa Pinê Américo-Pinheiro, Department of Forest Science, Soils and Environment, School of Agronomic Sciences, São Paulo State University (UNESP), Ave. Universitária, Botucatu, SP, Brazil; Graduate Program in Environmental Sciences, Brazil University, São Paulo, SP, Brazil

S.A. Aransiola, Bioresources Development Centre, National Biotechnology Development Agency, Ogbomoso, Nigeria

Marlón José Bastidas Barranco, Grupo de Investigación Desarrollo de Estudios y Tecnologías Ambientales del Carbono (DESTACAR), Facultad de Ingeniería, Universidad de La Guajira, Colombia

Sudhanshu S. Behera, Centre for Food Biology & Environment Studies, Bhubaneswar, Odisha, India

Sayan Bhattacharya, School of Ecology and Environment Studies, Nalanda University, Bihar, India

Protha Biswas, Department of Life Sciences, Presidency University, Kolkata, West Bengal, India

Manswama Boro, Department of Microbiology, Sikkim University, Gangtok, Sikkim, India

Germán Buitrón, Laboratory for Research on Advanced Processes for Water Treatment, Unidad Académica Juriquilla, Instituto de Ingeniería, Universidad Nacional Autónoma de Mexico, Queretaro, Mexico

Ercan Bursal, Department of Biochemistry, Mus Alparslan University, Turkey

Alejandra Cabello-Galindo, Biorefinery Group, Food Research Department, School of Chemistry, Autonomous University of Coahuila, Saltillo, Coahuila, Mexico

Aline Frumi Camargo, Laboratory of Microbiology and Bioprocess (LAMIBI), Federal University of Fronteira Sul, Erechim, Rio Grande do Sul, Brazil; Department of Biological Science, Graduate Program in Biotechnology and Bioscience, Federal University of Santa Catarina, Florianópolis, Santa Catarina, Brazil

Isaac Chairez, Unidad Profesional Interdisciplinaria de Biotecnología, UPIBI, Instituto Politécnico Nacional, Mexico City, Mexico

Indu Chauhan, Department of Biotechnology, Dr. B. R. Ambedkar National Institute of Technology Jalandhar, Jalandhar, Punjab, India

Atef Chibani, Laboratory of Environmental Process Engineering, Department of Chemical Engineering, Faculty of Process Engineering, University Salah Boubnider Constantine 3, El Khroub, Algeria

Muskaan Chopra, Department of Bio-Sciences and Technology, Maharishi Markendeshwar (Deemed-to-be University) Mullana, Ambala, Haryana, India

Chirag Chopra, Department of Biotechnology, School of Bioengineering and Biosciences, Lovely Professional University, Phagwara, Punjab, India

Mónica L. Chávez-González, Biorefinery Group, Food Research Department, School of Chemistry, Autonomous University of Coahuila, Saltillo, Coahuila, Mexico

Pranjal P. Das, Department of Chemical Engineering, Indian Institute of Technology Guwahati, Guwahati, Assam, India

Urmimala Das, Maharaja Sriram Chandra Bhanja Deo University, Sri Ram Chandra Vihar, Baripada, Odisha, India

Tuyelee Das, Department of Life Sciences, Presidency University, Kolkata, West Bengal, India

Aissa Dehane, Laboratory of Environmental Process Engineering, Department of Chemical Engineering, Faculty of Process Engineering, University Salah Boubnider Constantine 3, El Khroub, Algeria

Shibam Dey, School of Biotechnology, Kalinga Institute of Industrial Technology (Deemed-to-be University), Bhubaneswar, Odisha, India

Abhijit Dey, Department of Life Sciences, Presidency University, Kolkata, West Bengal, India

Satarupa Dey, Department of Botany, Shyampur Siddheswari Mahavidyalaya, Howrah, West Bengal, India

Simons Dhara, Department of Chemical Engineering, Indian Institute of Technology Guwahati, Guwahati, Assam, India

Prajwal P. Dongare, DBT-ICT Centre for Energy Biosciences, Institute of Chemical Technology, Mumbai, Maharashtra, India

Prangan Duarah, Centre for the Environment, Indian Institute of Technology Guwahati, Assam, India

Ghadir Aly El-Chaghaby, BioAnalysis Lab., Regional Center for Food and Feed, Agricultural Research Center, Giza, Egypt

Mostafa M. El-Sheekh, Department of Botany, Faculty of Science, Tanta University, Tanta, Egypt

Mostafa Elshobary, Department of Botany, Faculty of Science, Tanta University, Tanta, Egypt

Mustafa A. Fawzy, Botany and Microbiology Department, Faculty of Science, Assiut University, Assiut, Egypt; Biology Department, Faculty of Science, Taif University, Taif, Saudi Arabia

Mimosa Ghorai, Department of Life Sciences, Presidency University, Kolkata, West Bengal, India

Sougata Ghosh, Department of Physics, Faculty of Science, Kasetsart University, Bangkok, Thailand; Department of Microbiology, School of Science, RK University, Rajkot, Gujarat, India

Arabinda Ghosh, Department of Botany, Gauhati University, Guwahati, Assam, India

Liliana Giraldo, Facultad de Ciencias, Departamento de Química, Grupo de Calorimetría Universidad Nacional de Colombia, Sede Bogotá, Bogotá, Colombia

Mohamed Gomaa, Botany and Microbiology Department, Faculty of Science, Assiut University, Assiut, Egypt

Cristina González-Fernández, Biotechnological Processes Unit, IMDEA Energy, Madrid, Spain

Ranjit Gurav, Ingram School of Engineering, Texas State University, San Marcos, TX, United States

Dibyajyoti Haldar, Department of Biotechnology, Karunya Institute of Technology and Sciences, Coimbatore, Tamil Nadu, India

Bishnupriya Hansdah, Maharaja Sriram Chandra Bhanja Deo University, Sri Ram Chandra Vihar, Baripada, Odisha, India

Ikram ul Haq, Institute of Industrial Biotechnology, Government College University Lahore, Lahore, Punjab, Pakistan

Laura Helena dos Santos, Laboratory of Microbiology and Bioprocess (LAMIBI), Federal University of Fronteira Sul, Erechim, Rio Grande do Sul, Brazil

Sangchul Hwang, Ingram School of Engineering, Texas State University, San Marcos, TX, United States

Rahul Jadhav, Department of Biotechnology, Shivaji University, Kolhapur, Maharashtra, India

Jyoti Jadhav, Department of Biochemistry, Shivaji University, Kolhapur, Maharashtra, India; Department of Biotechnology, Shivaji University, Kolhapur, Maharashtra, India

Priyanka Kamilya, School of Biotechnology, Kalinga Institute of Industrial Technology (Deemed-to-be University), Bhubaneswar, Odisha, India

Kondragunta Prasanna Kumar, Indian Institute of Technology, Bhubaneswar, Odisha, India

Vineet Kumar, Department of Basic and Applied Sciences, School of Engineering and Sciences, GD Goenka University, Gurugram, Haryana, India

Subrat Kumar, School of Biotechnology, Kalinga Institute of Industrial Technology (Deemed-to-be University), Bhubaneswar, Odisha, India

S.S. Leh-Togi Zobeashia, National Biotechnology Development Agency, Abuja, Nigeria

Samanta Machado-Cepeda, Biorefinery Group, Food Research Department, School of Chemistry, Autonomous University of Coahuila, Saltillo, Coahuila, Mexico

Naga Raju Maddela, Departamento de Ciencias Biológicas, Facultad la Ciencias de la Salud, Universidad Técnica de Manabí, Portoviejo, Manabí, Ecuador; Instituto de Investigación, Universidad Técnica de Manabí, Portoviejo, Manabí, Ecuador

Sujata Mandal, Department of Life Sciences, Presidency University, Kolkata, West Bengal, India

José Manuel Martínez Gil, Grupo de Investigación Catálisis y Materiales, Facultad de Ciencias Básicas y Aplicadas, Universidad de La Guajira, Colombia; Facultad de Ciencias, Departamento de Química, Grupo de Investigación en Sólidos Porosos y Calorimetría, Universidad de los Andes, Sede Bogotá, Bogotá, Colombia

Ishanee Mazumder, School of Biotechnology, Kalinga Institute of Industrial Technology (Deemed-to-be University), Bhubaneswar, Odisha, India

Slimane Merouani, Laboratory of Environmental Process Engineering, Department of Chemical Engineering, Faculty of Process Engineering, University Salah Boubnider Constantine 3, El Khroub, Algeria

William Michelon, University of Contestado, Concórdia, Santa Catarina, Brazil

Juan Carlos Moreno-Piraján, Facultad de Ciencias, Departamento de Química, Grupo de Investigación en Sólidos Porosos y Calorimetría, Universidad de los Andes, Sede Bogotá, Bogotá, Colombia

Hamid Mukhtar, Institute of Industrial Biotechnology, Government College University Lahore, Lahore, Punjab, Pakistan

Sikandar I. Mulla, Department of Biochemistry, School of Allied Health Sciences, REVA University, Bengaluru, Karnataka, India

Ali Nawaz, Institute of Industrial Biotechnology, Government College University Lahore, Lahore, Punjab, Pakistan

Leila Nemdili, Laboratory of Environmental Process Engineering, Department of Chemical Engineering, Faculty of Process Engineering, University Salah Boubnider Constantine 3, El Khroub, Algeria

Potshangbam Nongdam, Department of Biotechnology, Manipur University, Imphal, Manipur, India

Sandeep Kumar Panda, School of Biotechnology, Kalinga Institute of Industrial Technology (Deemed-to-be University), Bhubaneswar, Odisha, India

Hiren K. Patel, School of Science, P. P. Savani University, Surat, Gujarat, India; School of Agriculture, P. P. Savani University, Surat, Gujarat, India

Nidhi P. Patel, School of Science, P. P. Savani University, Surat, Gujarat, India

Sagarjyoti Pathak, DBT-ICT Centre for Energy Biosciences, Institute of Chemical Technology, Mumbai, Maharashtra, India

Ritesh Pattnaik, School of Biotechnology, Kalinga Institute of Industrial Technology (Deemed-to-be University), Bhubaneswar, Odisha, India

Hitesh S. Pawar, DBT-ICT Centre for Energy Biosciences, Institute of Chemical Technology, Mumbai, Maharashtra, India

V.T. Perarasu, Department of Chemical Engineering, A C Tech, Anna University, Chennai, Tamil Nadu, India

Mihir Kumar Purkait, Department of Chemical Engineering, Indian Institute of Technology Guwahati, Guwahati, Assam, India

Guillermo Quijano, Laboratory for Research on Advanced Processes for Water Treatment, Instituto de Ingeniería, Unidad Académica Juriquilla, Universidad Nacional Autónoma de México, Querétaro, Mexico

Quratulain, Institute of Industrial Biotechnology, Government College University Lahore, Lahore, Punjab, Pakistan

Pabbati Ranjit, Center for Biotechnology, Institute of Science and Technology, JNTUH, Hyderabad, Telangana, India

Sayed Rashad, BioAnalysis Lab., Regional Center for Food and Feed, Agricultural Research Center, Giza, Egypt

Palvi Ravinder, Department of Biochemistry, Shivaji University, Kolhapur, Maharashtra, India

Ramesh C. Ray, Centre for Food Biology & Environment Studies, Bhubaneswar, Odisha, India

Kondakindi Venkateswar Reddy, Center for Biotechnology, Institute of Science and Technology, JNTUH, Hyderabad, Telangana, India

Pawan Rekha, Department of Chemistry, Malaviya National Institute of Technology Jaipur, Jaipur, Rajasthan, India

Neelancherry Remya, Indian Institute of Technology, Bhubaneswar, Odisha, India

Ricardo Vivas Reyes, Grupo de Investigación Química Cuántica y Teórica, Facultad de Ciencias Exactas y Naturales, Universidad de Cartagena, Colombia

Rosa M. Rodríguez-Jasso, Biorefinery Group, Food Research Department, School of Chemistry, Autonomous University of Coahuila, Saltillo, Coahuila, Mexico

Gilver Rosero-Chasoy, Biorefinery Group, Food Research Department, School of Chemistry, Autonomous University of Coahuila, Saltillo, Coahuila, Mexico

Chandrashish Roy, School of Biotechnology, Kalinga Institute of Industrial Technology (Deemed-to-be University), Bhubaneswar, Odisha, India

Héctor A. Ruiz, Biorefinery Group, Food Research Department, School of Chemistry, Autonomous University of Coahuila, Saltillo, Coahuila, Mexico

Niladri S. Samanta, Centre for the Environment, Indian Institute of Technology Guwahati, Guwahati, Assam, India

R. Saravanathamizhan, Department of Chemical Engineering, A C Tech, Anna University, Chennai, Tamil Nadu, India

Bishwarup Sarkar, College of Science, Northeastern University, Boston, MA, United States

Nalam Renuka Satya Sree, Center for Biotechnology, Institute of Science and Technology, JNTUH, Hyderabad, Telangana, India

Rohit Saxena, Biorefinery Group, Food Research Department, School of Chemistry, Autonomous University of Coahuila, Saltillo, Coahuila, Mexico

Thamarys Scapini, Bioprocess Engineering and Biotechnology Department, Federal University of Paraná, Curitiba, Parana, Brazil; Laboratory of Microbiology and Bioprocess (LAMIBI), Federal University of Fronteira Sul, Erechim, Rio Grande do Sul, Brazil

Maulin P. Shah, Industrial Waste Water Research Lab, Environ Technology Limited, Ankleshwar, Gujarat, India

Falak Shaheen, Department of Environmental Biotechnology, Shivaji University, Kolhapur, Maharashtra, India

Emad A. Shalaby, Department of Biochemistry, Faculty of Agriculture, Cairo University, Cairo, Egypt

Sanaa M.M. Shanab, Department of Botany and Microbiology, Faculty of Science, Cairo University, Cairo, Egypt

Vivek Sharma, Department of Chemistry, Banasthali Vidyapith, Newai, Rajasthan, India

Indu Sharma, Nims Institute of Allied Medical Science and Technology, NIMS University Rajasthan, Jaipur, Rajasthan, India

Farooq Sher, Department of Engineering, School of Science and Technology, Nottingham Trent University, Nottingham, United Kingdom

Soumya Singh, Shri Ramswaroop Memorial University, Lucknow, Uttar Pradesh, India

Shalini Singh, Amity Institute of Microbial Technology, Amity University, Jaipur, Rajasthan, India

Reena Singh, Department of Biotechnology, School of Bioengineering and Biosciences, Lovely Professional University, Phagwara, Punjab, India

Lovjeet Singh, Department of Chemical Engineering, Malaviya National Institute of Technology Jaipur, Jaipur, Rajasthan, India

Ankush D. Sontakke, Department of Chemical Engineering, Indian Institute of Technology Guwahati, Guwahati, Assam, India

Tanvi Taneja, Department of Bio-Sciences and Technology, Maharishi Markendeshwar (Deemed-to-be University) Mullana, Ambala, Haryana, India

Sirikanjana Thongmee, Department of Physics, Faculty of Science, Kasetsart University, Bangkok, Thailand

Prashnasa Tiwari, Department of Chemistry, Banasthali Vidyapith, Newai, Rajasthan, India

Helen Treichel, Laboratory of Microbiology and Bioprocess (LAMIBI), Federal University of Fronteira Sul, Erechim, Rio Grande do Sul, Brazil; Department of Biological Science, Graduate Program in Biotechnology and Bioscience, Federal University of Santa Catarina, Florianópolis, Santa Catarina, Brazil

Navanath Valekar, Department of Chemistry, Shivaji University, Kolhapur, Maharashtra, India

Anil Kumar Verma, Department of Microbiology, Sikkim University, Gangtok, Sikkim, India

K. Vetriselvan, Department of Chemical Engineering, A C Tech, Anna University, Chennai, Tamil Nadu, India

Aline Viancelli, University of Contestado, Concórdia, Santa Catarina, Brazil

M.O. Victor-Ekwebelem, Department of Microbiology, Alex Ekwueme Federal University, Ndufu-Alike, Abakaliki, Nigeria

Atif Khurshid Wani, Department of Biotechnology, School of Bioengineering and Biosciences, Lovely Professional University, Phagwara, Punjab, India

Vasudeo Zambare, Om Biotechnologies, Nashik, Maharashtra, India; Aesthetika Eco Research Pvt Ltd, Nashik, Maharashtra, India

Chapter 1

Policies of biofuel for commercialization

Prangan Duarah[1], Dibyajyoti Haldar[2] and Mihir Kumar Purkait[3]

[1]*Centre for the Environment, Indian Institute of Technology Guwahati, Assam, India;* [2]*Department of Biotechnology, Karunya Institute of Technology and Sciences, Coimbatore, Tamil Nadu, India;* [3]*Department of Chemical Engineering, Indian Institute of Technology Guwahati, Guwahati, Assam, India*

1. Overview of policies for biofuel production and commercialization

As shown in Fig. 1.1, global energy consumption is concentrated in a few major sectors: buildings, transportation, manufacturing, agriculture, forestry, and fisheries. Every sector's consumption grew to 9.1 Gtoe in 2019, although their proportions to total consumption remained stable. The most energy-intensive sectors were construction, industry, and transportation. Since 2000, building consumption has increased by 1.2% per year as a result of population growth, constructed area expansion, greater building services, and comfort levels, and an increase in time spent within buildings. Even at times of crisis, such as the 2008 economic downturn or the COVID-19, this rising tendency has been maintained. Without more rigorous rules, energy consumption in buildings as well as in the other sectors is expected to rise in the future as demand rises in emerging nations [1,2].

Biofuels have shown to be one of the most effective strategies to decarbonize the transportation industry, with global production and consumption rising from more than 64 billion liters in 2007 to more than 145 billion liters in 2017 [3,4]. Total biofuel consumption is expected to reach 2019 levels in 2021, following a historic fall in 2020 due to the COVID-19 outbreak, which disrupted worldwide transportation. According to the International Energy Agency (IEA), an increase of 28% in annual worldwide demand for biofuels is expected by 2026. Even though the United States has seen the most significant gain in volume, a large part of this increase is due to a rebound from the decline caused by the epidemic. Over the course of the projection period, Asia will overtake European biofuel output, accounting for about 30% of new production. A combination of strong domestic policy, rising demand for liquid fuels, and export-driven production have

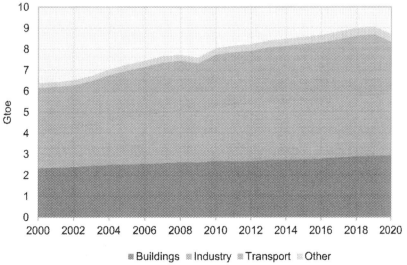

FIGURE 1.1 Sectorwise global energy consumption data from 2000–20.

resulted in this growth. Most of the rise in Asia is due to the recent Indian ethanol policy and biodiesel blending objectives in Indonesia and Malaysia. By 2026, India is expected to become the world's third-largest ethanol market [5].

To boost the development and use of biofuels, biofuel policies have been devised and implemented. Blending regulations, excise tax discounts and incentives, sustainable or low-carbon fuel mandates, government subsidies, and public finance are all examples of various measures of biofuel policies. These policy instruments have an impact on various phases of biofuel production and consumption. There are two basic types of actions that governments may take: market-pull policies and technology-push policies. Research and development (R&D), demonstration, and commercialization of biofuels are often supported through technology-push policies. In order to lower the cost of R&D, propel new ideas, and support early stage innovations over the financial "valley of death" that lies between original development and commercialization, these tools are utilized [6]. Policymakers are also depending on market-pull policies, such as those that try to promote the use of existing technologies like ethanol or biodiesel. For example, ethanol has been shown to significantly reduce greenhouse gas emissions [7]. Policies that employ technology-neutral financing to promote competition for cash among various technologies and creative initiatives are known as "technology push policies" Different governments use loan guarantees, governmental investments in technology firms, and subsidies or grants to support the development and implementation of biofuel technologies. The majority of the biofuels technology initiatives that have been sponsored through loan guarantee programs are those that are in their final stages of research and demonstration. On the other hand, grants have been the principal means for financing the development and pre-deployment of biofuels technology during the pilot and demonstration phases.

More established technologies have profited in a complementary manner from market-pull regulations such as biofuel blending mandates, fuel/CO_2 excise tax exemptions/reductions, and renewable or environmental friendly fuel standards [8]. These policies intend to shield biofuels from direct competition with fossil fuels, encourage investment, make use of economies of scale, and further technological and organizational education by fostering their widespread use. It's the objective of this chapter to look at the current literature on bioenergy policy efforts that have been implemented by various federal agencies in a number of different countries. In conclusion, a variety of points of view and problems connected to the generation of bioenergy are summed up, and an attempt is made to look into the future for potential solutions.

2. Factors affecting the commercialization of cellulosic biofuel

The techno-economic analysis is crucial for commercializing cellulosic-derived biofuels. The commercialization of biofuels is hampered by a number of factors, some of which have been recognized in the last few decades. Energy security, rural development, and pricing uncertainty are the main factors driving this change. However, the commercialization of cellulosic biofuels is hindered by several factors, including feedstock costs, pretreatment costs, production costs, and profitability [9].

2.1 Drivers

Chen et al. (2017) have found out the drivers of the commercialization of biofuels. The nine drivers that have been identified include added value from nonfuel coproducts, decrease in carbon emissions, dependency on fossil fuels, energy security, the food vs. fuel controversy, government policies, rural economic growth, and fluctuating oil prices. Discussions on land usage often bring up the issue of food vs. fuel, and this can have an effect on policy and/or incentives [10]. It was determined that government policies were far more important than any other component in scaling up cellulosic biofuel production. Other studies show that incentives and supportive policies are crucial for the growth of nascent businesses like biofuels because of the high initial production costs and the requirement for infrastructure development. It has also been shown that rules and regulations that support the energy market are vital [11,12].

Many studies have shown that the value added to nonfuel coproducts is the second-most important factor in scaling up production, demonstrating the need for coproducts for additional profit in order to make cellulosic biofuels more affordable than fuels derived from petroleum. In comparison to fossil fuels, cellulosic biofuels produce fewer greenhouse gases [13,14]. Throughout the study, this has been a constant source of motivation. Sustainability analysts place a high value on carbon emissions due to public and political debates over the environmental implications of climate change. Climate change is seen as a major concern by 87% of expert respondents, according to a study by Ref. [15].

According to this study's survey, US oil prices fluctuated between $100 in September 2014 and $45 a barrel in September 2015, which may justify the fourth most rated scaling-up reason uncertain price of oil. Biofuels have been suggested as a potential solution to the price instability that is prevalent in the petroleum business [14]. There were also "relatively major" reasons for the expansion of the biofuels business, with many researchers calling for research and development of renewable biofuels to reduce demand for fossil fuels and to boost national security. Rural economic growth is the sixth most significant driver driving the cellulosic biofuels sector [16,17].

At the P = 10 significance level, the research found that there were significant variations in the group mean values for two of the eight drivers that were discovered. Scaling up cellulosic biofuels is more critical to feedstock and sustainability specialists than it is to economics experts, according to these experts. Carbon emission reduction was seen by sustainability researchers as a far more critical incentive than it was by the other three expert groups [18].

2.2 Barriers

Chen et al. (2017) identify nine barriers that potentially affect the commercialization of cellulosic biofuels. The nine potential barriers include capital availability, cellulosic biofuel logistics, consistent feedstock supply, competition vs. corn-grain ethanol, competition vs. petro-fuels, policy uncertainty, feedstock costs, high production costs, and technology availability. The fact that feedstock costs make up around one-third of total biofuel production costs brings attention to the fact that this is the fourth most significant scale-up barrier [18]. According to the research that has been conducted thus far, achieving cost competitiveness with fuels derived from petroleum requires addressing two key aspects of overall biofuels costs: feedstock supply costs as well as storage costs [19,20]. Further analysis revealed that competition vs. petrofuels was the number one commercialization hurdle. High manufacturing costs, followed by policy uncertainty, were the second and third most cited barriers to commercialization. Competition vs. petrofuels accounted for the biggest percentage of Feedstock, Processing, and Sustainability specialists' group ratings, corroborating the rating scale results.

Commercialization of cellulosic biofuels is further hampered by a lack of readily available money, technology, cellulosic biofuels logistics, and a reliable supply of feedstock. Lignocelluloses are a complex mixture of cellulose, hemicellulose, and lignin. Ethanol may be produced biologically via delignification and depolymerization of carbohydrate polymers, both of which are required for the delignification process to separate cellulose and hemicellulose from lignin. For the lignocellulosic biomass sector to attain large-scale production, effective technology to densify, manage, and store this biomass must be available. Finally, the literature has recorded the availability of funding for these early stage technologies, developing worries about agricultural production changes and harvested area swings, and logistical issues.

Cellulosic biofuels logistics were evaluated much higher by feedstock and processing professionals than by sustainability participants. High production costs were seen by processing specialists as a far greater impediment than feedstock participants. These findings demonstrate the significance of looking at problems from several angles [18].

Because of the high cost of production, regulatory uncertainty, and competition from petrofuels, analysts claim cellulosic biofuels pose the greatest challenges to commercialization. Several studies have shown that cellulosic biofuels have greater production costs than petroleum-based fuels or first-generation biofuels, including pretreatment expenses. According to current estimates of the expenses associated with producing biofuels, the cost of producing cellulosic ethanol is much higher than that of producing gasoline on an energy equivalent basis.

3. Latest policies undertaken by USA to commercialize cellulosic biofuels

As per a study that was published by the Energy Information Administration of the United States Department of Energy (DOE/EIA), domestic consumption of renewable energy increased by 32% between the years 2006 and 2012. Renewable sources of energy were responsible for roughly 9% of the world's total energy consumption in 2012. The Arab oil embargo of 1973 brought to light the degree to which the United States is reliant on oil from other countries, which ultimately led to a crisis with fossil fuels. Fuel economy laws and alternative fuel assistance were directly impacted by the energy crisis. As a result, the Energy Policy and Conservation Act (EPCA) was passed by the US government in 1975 [21]. The Energy Policy and Conservation Act's primary goals were to cut energy use, increase energy security, and enhance energy efficiency. The government of the US approved the Energy Tax Act of 1978 in order to provide a tax exemption of $0.40 per gallon of ethanol and a gasoline excise tax of $0.04 per gallon for gasoline that contains ethanol. In an effort to promote the utilization of alternative fuels, this action was taken. The Tax Reform Act of 1984 extended the ethanol tax exemption from US$0.50 per gallon to US$0.60 per gallon under the Surface Transportation Assistance Act of 1982. The 1988 Alternative Motor Fuels Act (AMFA) recommended using methanol, ethanol, or natural gas as a major energy source or as a complement to diesel or gasoline. In addition, the AMFA also encouraged the use of biodiesel. By virtue of the Omnibus Budget Reconciliation Act of 1990, the tax exemption for ethanol was prolonged until the year 2000 (although at a decreased price of $0.54 per gallon). US$0.51 per gallon was reduced to US$0.51 by 2005 as a result of the 1998 Transportation Efficiency Act of the 21st century. The Jobs Creation Act of 2004 retained the exemption until 2010.

To put it another way, for the first time in American history, biofuel's importance was expressly acknowledged in the Energy Policy Act of 1992. A number of guidelines set by EPCA 1975 and AMFA 1988 were updated by this legislation. The Energy Policy Act of 2005 included significant measures to encourage use of ethanol as an alternative fuel source. According to this law, efforts were concentrated on commercializing biofuels through research and development in areas such as power generation system development and production facilities for lignocellulosic biomass-to-biofuel conversion technology integration. In a similar vein, the Energy Independence and Security Act (EISA) of 2007, which is sometimes commonly referred to as the "Clean Energy Act of 2007," supported the research and development of biofuels for the next generation. The act's primary goals were to increase the production of renewable fuels, primarily corn-based ethanol from new plants, as well as biodiesel, cellulosic, and advanced biofuels. EISA 2007 predicts biofuel production would rise from 4.7 billion US gallons (18,000,000 m^3) in 2007 to 36 billion US gallons (140,000,000 m^3) in 2022. In 2012, ethanol refineries in the United States generated 13.3 billion gallons of the fuel. The ethanol industry was expected to add $43.4 billion to the US GDP in 2012 and $29.9 billion to average profits [22]. Biodiesel demand has grown significantly in the United States in recent years. According to the EIA 2021 study, demand for biomass-based fuel in the United States climbed by 12% in 2020. In 2021 and 2022, biodiesel imports in the United States are predicted to rise by 43% and 49%, respectively. More than half of the biodiesel produced in the United States was produced in Midwest states, including Iowa, Missouri, and Illinois. However, biodiesel output in the United States is predicted to decline by 2% in 2020, or 49 million gallons. Biodiesel shipments from the United States have increased by 29% over the previous year [3,23].

4. Brazil's biofuel national policies implemented for its commercialization

To meet geopolitical and energy security objectives, a national alcohol program was launched by Brazil's military administration in 1975. Since then, several biofuel efforts in the country have gone through a number of stages and are still ongoing. The promotion of ethanol as an alternative to fossil fuels is one example of this initiative. Sugarcane ethanol was made to improve trade balance and energy independence amid the 1970s international oil crisis [24]. In addition, the National Program for Biodiesel Production and Use (PNPB), which was introduced in 2005, follows in the footsteps of Proálcool in focusing on biodiesel supply while also emphasizing social and regional development [25]. Notably, both initiatives were founded with the primary purpose of ensuring supply and energy security, with the environmental benefits of biofuels being identified afterward [26].

In 2019, the transportation sector in Brazil consumed 33% of total energy and produced 190.5 MtCO$_2$eq, accounting for 45.4% of total energy emissions [27]. Furthermore, transportation relies heavily on oil as a fuel source, which contributes significantly to pollution [27]. A system for allocating prices on externalities to meet environmental goals, as well as incentives for expanding biofuel production and usage in transportation, was introduced by the National Biofuels Policy (RenovaBio), which was developed in December 2017 [27,28].

In conclusion, the objective of Brazil fulfilling its commitments under the Paris Agreement served as the driving force for the establishment of RenovaBio. This strategy provides procedures for Life Cycle Assessment (LCA), commercialization, and predictability of the fuel market. In addition to promoting the utilization of biofuels, improving national energy security, and lowering GHG emissions, these goals will also be accomplished by implementing this strategy. The major goal is to enhance biofuel use and expansion in the Brazilian energy grid through a strong decarbonization policy, producer incentives, and the creation of a market for carbon-reduction credits known as biofuel Decarbonization Credit (CBIO) [28].

However, COVID-19 distribution is taking place in a global context, and its long-term effects are yet to be discovered. Because of the initial drop in demand for fuel, the agribusiness sector in Brazil was hit by a number of challenges, including low prices for ethanol and sugar; contract cancellations between producers and distributors; shifts in the production mix; operational constraints in plants, ports, and distribution centers; and monetary challenges, particularly in regard to production costs [27]. Oil and gasoline prices are expected to remain volatile in Brazil through 2021, and this will have a direct impact on domestic fuel prices for Brazilian consumers due to rising exports of commodities and a lack of precipitation, which has led to one of Brazil's worst droughts, which has harmed crop and livestock production. In terms of the economy, these effects extend beyond the gasoline market, resulting in increased food and electricity costs, deprivation of essential goods, and inflation. An emphasis on system efficiency rather than cost consequences and negative externalities is recommended by the government, as has been noted [29].

There is a need to reassess Brazil's biofuel policies, including RenovaBio, the Biodiesel Auction, and decisions from the Brazilian Petroleum Natural Gas, and Biofuels Agency, in light of multiple uncertainties, changes in market conditions, and socioeconomic factors. Despite the hurdles that this new public policy in the energy sector faces, RenovaBio promotes the use of biofuels in a country that is a leader in the field and helps Brazil meet its NDC targets. Even though the transportation industry averted 14.5 million tonnes of CO$_2$ as a consequence of the 2019/2020 objective, the program has yet to be fully implemented. However, the government is expected to fill up some of RenovaBio's shortcomings to meet Brazil's climate goals [27].

5. Recent progress in policies taken by the government of China to drive biofuel generation

In order to achieve both its low-carbon transition goals and its goals for environmental sustainability, China is making a concerted effort to advance the development of renewable energy sources. This has led to a slew of government initiatives aimed at encouraging the utilization of bioenergy. Following the issuance of an urgent notice by the National Development and Reform Commission (NDRC), the People's Republic of China Law on Renewable Energy went into effect on January 1, 2006. At the same time, a variety of associated protocols, technical benchmarks, and implementation plans were finally enacted and implemented. Provisional Measures for Administration of the Subsidy Capital for Crop Straw Energy Utilization were published by the Ministry of Finance of the People's Republic of China on October 30, 2008. These initiatives included specific financial aid for the briquettes gasification and carbonization sectors of the economy [30].

Among China's bioenergy plans, biogas is a key component. This technique was pioneered by a methane digester located in the south-east of China in the 1980s that used drains to store and consume exhaust gas [31]. The Hydraulic Cylinder Digester, Meandering stream fabric digester, spheroidal digester, separated floating bell-type digester, and prefabricated block digester are only a few of the efficient biogas digesters produced by China in recent years. Consequently, by the end of 2007, 2.65×10^7 residential biogas digesters had built 2.66×10^4 poultry-and-livestock-farm biogas projects across China, producing 1.08×10^{10} m^3 of biogas. Announcing plans to spend RMB 3,000,000 on rural home biogas digesters, rural biogas service systems, and biogas in intensive agricultural projects, China's Ministry of Agriculture announced this intention on November 14th, 2008. This was done as part of an attempt to encourage post-earthing rehabilitation and economic recovery in the aftermath of the financial crisis, both of which have been beneficial to China's biogas sector. However, operating a methane digester in rural China comes with a number of drawbacks that should be considered. It is anticipated that China's biogas generation would decrease in tandem with the falling number of domestic digesters. As per a report, biogas output from home digesters would fall to 10×10^{27} m^3 by 2020, whereas biogas generation from production units will rise to $3-5 \times 10^{27}$ m^3. Because China is such a large country with many different elevations, weather conditions vary greatly, which has a detrimental influence on biogas generation in rural houses. In China's western provinces, such as Sichuan and Guangxi, biogas production and consumption are promoted. Domestic digesters perform poorly in cold areas, including Heilongjiang and regions of Inner Mongolia; as a result, household biogas plants are not ideal for harsh and snowy locations. As a consequence of this, the authorities of the provincial government need to design thorough plans that take into account the advantages offered by certain municipalities [32].

The current status of China's biofuel production is detailed in Table 1.1, which presents data derived from a wide range of indigenous production techniques and a variety of biomasses. As seen in the chart, bioethanol plays a critical part in China's bioenergy future. It's also worth mentioning that corn accounts for more than half of all biofuel production. Heilongjiang, Jilin, Liaoling, Hen'an, Jingshu, Anhui, Hubei, Hebei, Guangxi, and Shandong are among the 10 provinces where bioethanol transportation infrastructure has been built [30]. China has also created ethanol-sweet species for 6% carbonate alkalinity wasteland in sugar-based ethanol sectors. Fermentation in solid form and rapid fermentation in liquid form were used to reduce stalks of sweet sorghum to 10% of the feedstock. A demonstration plant capable of producing 3000 tonnes of sweet sorghum-based bioethanol per year has been set up in Dongtai, which is located in Jiangsu Province. Various lignocellulose-based ethanol production systems can produce up to 10,000 tonnes of gasoline-ethanol per year with feedstock conversion rates of more than 18%. Bioethanol taxes have also changed in China. All bioethanol production value-added taxes were abolished in 2007 by the government to aid cellulose-based bioethanol companies. In 2010, the government reduced the duty on ethanol imports from 30% to 5%, a significant reduction [33].

Until the early 21st century, biodiesel production on a large scale had not begun. Several private enterprises, huge corporations, and international energy organizations began investing in China's biodiesel industry in 2002 as a result of the NDRC's promotional efforts. The "Medium and Long-Term Development Plan for Renewable Energy," the "Renewable Energy Development of the 11th Five-Year Plan," and the "Renewable Energy Development of the 12th Five-Year Plan" have resulted in significant changes to the country's biodiesel policy [34]. The COVID-19 epidemic has a detrimental influence on biofuel production's predicted consequences. The production of biodiesel in China is anticipated to increase to 1.45 billion liters by the year 2020, increasing from 0.939 billion liters in 2019, as a result of the scheduled completion of 42 biodiesel plants by the end of the year 2020. The Chinese government had expected to have nationwide coverage of E10 ethanol gasoline for car usage by the end of 2020. However, only 15 of mainland China's 31 provinces have fully or partially adopted ethanol-gasoline thus far. China's real blend rate is expected to be 2.1% in 2021, which is a little uptick from 2020 but still well below the 10 year peak of 2.8%. The amount of biodiesel produced in China in 2021 was projected to rise by more than 54% from 2020 as a result of increasing exports [35].

TABLE 1.1 The current state of bioenergy production in China, which makes use of both traditional and cutting-edge methods and a variety of biomass.

Location	Biomass used	Production company	Type of bioenergy generated	Bioenergy generation capacity
Jilin Province	Corn	Jilin Fuel Alcohol Limited Company	Bioethanol	0.6 million ton/year
Heilongjiang	Corn	Heilongjiang Huarun Alcohol Limited Company	Bioethanol	0.10 million ton/year
Henan	Wheat	Henan Tianguan Enterprise Group Corporation	Bioethanol	0.6 million ton/year
Anhui	Corn	Anhui Fengyuan Group Corporation	Bioethanol	0.32 million ton/year
Heilongjiang	Corn	COFCO Bio-energy co. Ltd.	Bioethanol	0.28 million ton/year
Guangxi	Cassava	Guangxi COFCO Bio-energy co. Ltd.	Bioethanol	0.20 million ton/year
Inner Mongolia	Sorghum	ZTE Energy Ltd.	Bioethanol	0.05 million ton/year
Shandong	Corncob	Longlive Biotechnology co. Ltd	Bioethanol	0.05 million ton/year
Shanxi province	Maize stover silage, dairy cattle manure	Shanxi Shenmu New Energy co. Ltd.	Biogas	1570 Nm^3

Reproduced with permission from P. Duarah et al., A review on global perspectives of sustainable development in bioenergy generation, Bioresour. Technol. (2022) 348 p. 126791.

6. The national policy of India to promote its biofuel commercialization program

In India's rural areas, biomass has long been used as a significant source of home energy for a variety of purposes, including lighting and cooking. Animal manure, waste products from agriculture (particularly rice husk and bagasse), and wood fuels are all instances of forms of biomass energy. Wood fuels make up a bigger portion of the country's total biomass energy. Punjab has the most bioenergy potential (16,860 MJ), followed by Haryana, Gujarat, and Uttar Pradesh. Aside from these states, India's northeastern states have the ability to generate bioenergy [36].

A 2013 survey found that India was one of the top markets for bioenergy, with an installed capacity of roughly 0.4 GW. Over 4.4 GW of bioelectricity was generated last year from bagasse-based combined heat and power (CHP) plants. India's bioenergy strategy in the 1980s was focused on boosting the efficiency of conventional biomass use and supply, using current technology. Cooking stoves known as "Chulhas" (cooking stoves) were developed in 1985–1986 as a national effort to replace inefficient mud-based cooking devices and portable iron stoves with chimneys. However, in 2002, MNRE made the decision to halt the program. The new National Biofuels Policy, the New National Biogas and Organic Manure Program (NNBOMP), and the Program on Energy from Urban, Industrial, and Agricultural Waste/Residues were introduced in 2018 in the bioenergy industry. The expected outcomes form the national biofuel policy 2018 is illustrated in Fig. 1.2. These initiatives aim to modernize current energy policies and programs to meet the problems of the future. Additional initiatives have been taken by the Indian government to increase the country's output of biogas. Biogas power generation (off-grid) and thermal energy application (BPGTP) programs have recently been developed to support the establishment of off-grid biogas generating units by the government [37].

It is anticipated that the national strategy on biofuels that will be implemented in 2018 would reduce the costs of production while simultaneously improving affordability for end-users. The technological infrastructure, government subsidies, pricing, and institutional activities are brought together to bring the vision, objectives, and strategy for developing biofuels to fruition. From sugarcane juice, B-molasses may be extracted, and C-molasses can be used to extract C-molasses. C-molasses are the primary source of ethanol in India, which operates 25 million tonnes of sugar per year. Revised sugarcane control orders issued in 1966 have been amended so that ethanol can be produced directly from juice or B-molasses. Ethanol production was able to take advantage of sugarcane surpluses while lowering sugar prices. Oil marketing businesses buy ethanol from sugar processing facilities in order to take part in the ethanol blending scheme. By 2030, it is envisioned that a 20% ethanol/gasoline blend (E20) and a 5% biodiesel/diesel blend may be achieved using this strategy.

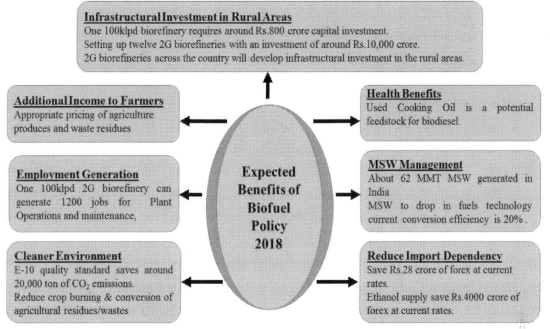

FIGURE 1.2 Benefits from Indian biofuel policy 2018. *Reproduced with permission from R. Kothari et al., Assessment of Indian bioenergy policy for sustainable environment and its impact for rural India: strategic implementation and challenges, Environ. Technol. Innovat. (2020) 20 p. 101078.*

Ethanol pledges from producers, however, fall well short of the necessary demand for blending. Between 2007 and 2009, however, ethanol mixes were limited to 2% [38]. Ethanol blends in 2015−16 and 2016−17 were just 3.5% and 2.07%, respectively. In the biofuels business, the government encouraged joint ventures and investment. Additionally, the policy enables for 100% foreign direct investment in biofuel technology through the automated method; however, this is only permitted if the ethanol that is generated is utilized solely for reasons within the country [39]. The program continues to give financial incentives, subsidies, and grants to enterprises that produce biofuels. It suggests that the government establish a "National Biofuel Fund" to offer a variety of financial incentives for "advanced biofuels," such as tax credits as well as other price structures. In addition, the proposal includes a viability gap funding plan for advanced biofuel refineries. If implemented, this plan would provide an investment of rupees five thousand crores (approximately $750 million) spread out over a period of 6 years, along with the tax benefits and a higher purchase price than for first-generation biofuels.

Lawmakers have mandated that public sector banks like NABARD and others give loans at a better interest rate to farmers and businesses working on renewable energy projects. This new regulation encourages the use of a different pricing model as a means to expand the consumption of advanced biofuels across the country. As a means of bolstering the local biofuel economy and meeting the demands of our nation's government with regard to biofuel, it is imperative that emphasize the policy's severe prohibition on the import and export of biomass. Biodiesel feedstock can be imported, notwithstanding a ban on imports without prior clearance from the National Biofuel Coordination Committee [39].

For the biofuels program to be a success, state governments must be actively engaged, as well as numerous ministries must be coordinated. According to biofuel policy, a Biofuel Development Board should be created to assist biofuel efforts throughout the state. In India's Uttar Pradesh, Karnataka, Chhattisgarh, Rajasthan, and Uttarakhand, five similar bodies are now in operation. In order to guarantee that the policy is properly executed, the government should provide a single point of contact for the approval of biofuel initiatives. In line with the National Policy on Biofuel 2018, the National Biofuel Coordination Committee has given its approval for the conversion of surplus agricultural grains into ethanol. It is essential to have the committee's prior consent before moving further. Currently, there is no official entity in place that can sanction the conversion of excess food grains into biofuel. In addition, the policy does not effectively address the challenges of technical feasibility [40].

7. Different policies for algal biofuel commercialization

Since the 1950s, scientists have been investigating the potential of microalgae as a source of renewable energy. Several national research initiatives on large-scale culture and process technologies for marine algal energy exploration were begun by the United States in the 1970s after the outbreak of the oil crisis. For example, the Marine Biomass Program (MBP) in 1974 and the Aquatic Species Program (ASP) from 1978 to 1996 are examples of these programs. Oil prices have remained high since 2005, which has led to renewed interest in microalgae energy. After Green Fuel Company's industrial demonstration of microalgae cultivation using power plant flue gas, numerous other microalgae firms followed suit. These businesses began making preparations for demonstrations and commercial production one after the other. The governments, huge corporations, and venture capital firms all put up a substantial amount of money. Algae biofuels are expected to replace 17% of US transportation petroleum fuels, according to official estimates. It was in 2007 that the United States relaunched its mini-Manhattan microalgae energy research project. Priorities and aims for developing algae energy technology were outlined in 2008 in a National Algal Biofuels Technology Roadmap, which served as a useful example in furthering the development of the algae energy sector. By 2019, the US Department of Energy's Bioenergy Technologies Office (BETO) aimed to cut the cost of algal biofuels and coproducts to $7/gge (gasoline gallon equivalent), a 50% reduction.

Several microalgae initiatives have been funded by the US Department of Energy (DOE) and the US Department of Defense (DOD) since 2007. Seaweed energy research in the United States has received approximately 450 million dollars in funding from several federal agencies since 2008. Algae biofuel has been supported by the US government in a variety of ways in recent years. The majority of microalgae oil and biomass research was funded before 2010. However, in 2010 the focus turned to the commercialization of algal biomass refinement and the large-scale production of military aviation fuel. European Union (EU) has a well-developed liquid biofuel sector, and algal biofuel development is given high priority. Microalgae biofuel research and development costs were hoped to be reduced by the Carbon Trust in 2008 through a public investment program involving £20–30 million. The BioMara project, which the European Union funded with €6 million in 2009 with the help of England and Ireland, aims to find the most desired algae strain and put it into commercial production. Algae biofuel generation is the focus of the year's largest scientific research project, FP7, in 2010. The InteSusAI Project, the All-Gas Project, and the BIOFAT Project each received a €20 million investment to help them get their ideas off the ground. EU began a 4 year algal energy development project EnAlage with project funds up to €14 million in 2011, with the goal of solving today's issue of lack of knowledge on algae production in the Northwest. Despite its late start, China's microalgae energy industry has grown significantly. There was a significant investment in marine energy research by the Ministry of Science and Technology, the State Oceanic Administration, and the Shanghai Municipal People's Government, among other organizations. When the Chinese Academy of Sciences launched its "Solar Energy Initiative" in January 2009, they were looking at a number of essential approaches for producing biodiesel from high-efficiency microalgae. There are a number of critical phases involved in these investigations that include genetic alteration, breed selection, oil collection, photosynthesis, large-scale production, and refining of oil-producing microalgae strains. The Chinese Academy of Sciences and Sinopec Group joined forces in the following month to work on the Microalgae Biodiesel Series Techniques project, aiming for outdoor pilot research by 2015 and long-term industrial demonstrations of 10,000-scale. In 2010 and 2012, National High Technology Research and Development Programs of the Ministry of Science and Technology invested heavily in microalgae energy technology to build a polygenerational system for microalgae growing and carbon fixing.

Algae biofuel was supported by governments throughout the world, which adopted regulations to encourage its development. An algal bio-oil production process developed by the US Department of Energy's Pacific Northwest National Laboratory resulted in high yields of algae bio-oil production in a shorter period and at a lower cost [41]. An important advance has been made in algal metabolic engineering at the Scripps Institution of Oceanography (SIO), increasing oil output from algae. Supported by the United States, the Department of Energy (DOE) also accomplished a great deal. Algenol Biotech, Sapphire Energy, Solazyme, Inc., and BioProcess Algae have received BETO funding. An algal biomass full refining process has been completed by Algenol Company with the help of BETO. Nine thousand gallons of ethanol and 1100 gallons of hydrocarbon fuel per year were exceeded, according to Algenol, which validated the company's production rates [42]. There are four major issues that need to be addressed before algae may be commercialized: culture management and stability, scalability of the system designs, long-term water use, and nutrient supplies.

8. Challenges with the policies and future recommendations

Throughout the decades, numerous efforts have been made to create and implement biofuels, which has earned these endeavors a high priority of research interests among academics all over the world. At the business level, however, it is still vital to have a comprehensive understanding of process optimization. The security of the food supply is another significant issue that arises with the development of bioenergy. The manufacture of biofuels requires the use of a variety of food crops, such as sugarcane, maize, corn, coconut, sunflower, and soya. This practice is coming under greater scrutiny since it is believed to encourage competition between food crops and biofuel raw materials. Cellulosic materials are used in the production of second-generation biofuels as a direct consequence of this. However, the distillation process that is used in the production of biofuel is a time, money, and energy-consuming one. Other problems related to the production of bioenergy include high initial investment costs and poor overall efficiency of the equipment used in commercial production [43]. Lignocellulosic biomass pretreatment is critical to the biomass conversion process. It helps to break down the complex structure of biomass. Various pretreatment technologies are being developed to enhance the efficiency of the biomass in the biofuel conversion process. To boost pretreatment efficiency, a cost-effective industrial-scale approach must be developed [43].

It has emerged as perhaps the most efficient photosynthetic process, using just sunshine and simple supplements to produce liquid and gaseous biofuels, such as biodiesel, bioethanol, biomethane, biohydrogen, as well as biobutanol [44]. Ethanol and other biofuels derived from this process are prohibitively expensive due to the high cost of harvesting. As reported, a liter of algal biofuel can cost between $1 and $2, but a liter of gasoline can be as low as $0.63 to $1. Therefore, it is necessary to optimize the microalgae-based manufacturing technology to reduce the overall production costs at a commercial level [45]. Electroporation and particle bombardment, two of the earliest forms of genetic engineering, have given way to more advanced techniques such as zinc-finger nucleases (ZFN) and RNA interference, which contribute significantly to the development of more efficient methods for the production of microalgae. The commercial production of bioproducts from a wide range of genetically modified organisms is possible if these approaches achieve popular adoption. The resources available for genetic manipulation, the formation of a synthetic gene network, and the construction of genome-scale microalgae models are limited. It is important not to undervalue their dependability when it comes to keeping the genetic integrity of newly produced algae strains [46-49].

9. Conclusions

It is believed that the use of sustainable biofuel would improve the social, economic, and environmental challenges that humanity faces. As a result of this, governments all over the country have come up with different strategies to encourage the development of bioenergy and ensure its ability to compete economically. The policies that are put into place by various countries encourage the production of domestic feedstock as well as the development of innovative technology for biofuel feedstock. Feedstock expenses account for around one-third of biofuel production costs, highlighting the fourth largest scale-up barrier feedstock costs. While rushing to reach biofuel quotas and blending targets, most governments overlook serious sustainability issues. Cellulosic biofuel logistics, constant feedstock supply, competition vs. corn-grain ethanol, competition vs. petro-fuels, feedstock costs, high production costs, regulatory uncertainty, and technological availability are all possible impediments. One-third of biofuel production costs are accounted for by feedstock costs, the fourth most significant scale-up barrier. A good policy intervention might aid in the resolution of these issues. The proper use of feedstock and the identification of acceptable feedstock based on the country's topography are critical. More regional collaboration is also necessary to help with logistics, management, sustainability, and project completion, which can only be achieved through diverse regulations. Furthermore, energy policy research is in its infancy, and several stakeholders, including technical scientists, policymakers, companies, and the general public, must step in to build a framework for policies.

Acknowledgment

The study is supported by Indian National Academy of Engineering (INAE/121/AKF/22), Gurgaon, India. The authors are solely responsible for all of the opinions, results, and conclusions expressed in this study; INAE's viewpoints are not necessarily reflected in any of these aspects.

References

[1] M. González-Torres, et al., A review on buildings energy information: trends, end-uses, fuels and drivers, Energy Rep. 8 (2022) 626—637.
[2] P. Duarah, et al., Progress in the electrochemical reduction of CO2 to formic acid: a review on current trends and future prospects, J. Environ. Chem. Eng. 9 (6) (2021) 106394.
[3] P. Duarah, et al., A review on global perspectives of sustainable development in bioenergy generation, Bioresour. Technol. 348 (2022) 126791.
[4] P. Duarah, D. Haldar, M.K. Purkait, Technological advancement in the synthesis and applications of lignin-based nanoparticles derived from agro-industrial waste residues: a review, Int. J. Biol. Macromol. 163 (2020) 1828—1843.
[5] IEA, Bioenergy, 2022 [cited 2022 12.05.2022]; Available from: https://www.iea.org/fuels-and-technologies/bioenergy.
[6] S.M. Jordaan, et al., The role of energy technology innovation in reducing greenhouse gas emissions: a case study of Canada, Renew. Sustain. Energy Rev. 78 (2017) 1397—1409.

[7] V. Costantini, et al., Demand-pull and technology-push public support for eco-innovation: the case of the biofuels sector, Res. Pol. 44 (3) (2015) 577–595.
[8] J. Hoppmann, et al., The two faces of market support—how deployment policies affect technological exploration and exploitation in the solar photovoltaic industry, Res. Pol. 42 (4) (2013) 989–1003.
[9] M. Gunasekaran, et al., Lignocellulosic biomass as an optimistic feedstock for the production of biofuels as valuable energy source: techno-economic analysis, environmental impact analysis, breakthrough and perspectives, Environ. Technol. Innovat. 24 (2021) 102080.
[10] M. Carus, L. Dammer, Food or non-food: which agricultural feedstocks are best for industrial uses? Ind. Biotechnol. 9 (4) (2013) 171–176.
[11] D. Yue, F. You, S.W. Snyder, Biomass-to-bioenergy and biofuel supply chain optimization: overview, key issues and challenges, Comput. Chem. Eng. 66 (2014) 36–56.
[12] B.E. Dale, et al., Take a Closer Look: Biofuels Can Support Environmental, Economic and Social Goals, ACS Publications, 2014.
[13] J.J. Bozell, G.R. Petersen, Technology development for the production of biobased products from biorefinery carbohydrates—the US Department of Energy's "Top 10" revisited, Green Chem. 12 (4) (2010) 539–554.
[14] G. Fiorese, et al., Advanced biofuels: future perspectives from an expert elicitation survey, Energy Pol. 56 (2013) 293–311.
[15] E. Hodgson, et al., Horizon scanning the European bio-based economy: a novel approach to the identification of barriers and key policy interventions from stakeholders in multiple sectors and regions, Biofuels Bioprod. Biorefin. 10 (5) (2016) 508–522.
[16] M. Ebadian, et al., Modeling and analysing storage systems in agricultural biomass supply chain for cellulosic ethanol production, Appl. Energy 102 (2013) 840–849.
[17] V. Balan, D. Chiaramonti, S. Kumar, Review of US and EU initiatives toward development, demonstration, and commercialization of lignocellulosic biofuels, Biofuels Bioprod. Biorefin. 7 (6) (2013) 732–759.
[18] M. Chen, P.M. Smith, The U.S. cellulosic biofuels industry: expert views on commercialization drivers and barriers, Biomass Bioenergy 102 (2017) 52–61.
[19] V. Balan, Current Challenges in Commercially Producing Biofuels from Lignocellulosic Biomass, International Scholarly Research Notices, 2014, 2014.
[20] R.E. Sims, et al., An overview of second generation biofuel technologies, Bioresour. Technol. 101 (6) (2010) 1570–1580.
[21] R.K. Dixon, et al., US energy conservation and efficiency policies: challenges and opportunities, Energy Pol. 38 (11) (2010) 6398–6408.
[22] R.D. Schnepf, B.D. Yacobucci, Renewable Fuel Standard (RFS): Overview and Issues, vol 40155, Congressional Research Service, Washington, DC, 2010.
[23] Z. Zhongming, et al., US Imports of Biomass-Based Diesel Increased 12% in 2020, 2021.
[24] L. Benites-Lazaro, N. Mello-Théry, M. Lahsen, Business storytelling about energy and climate change: the case of Brazil's ethanol industry, Energy Res. Social Sci. 31 (2017) 77–85.
[25] S.L. Stattman, O. Hospes, A.P. Mol, Governing biofuels in Brazil: a comparison of ethanol and biodiesel policies, Energy Pol. 61 (2013) 22–30.
[26] M.A.F.D. de Moraes, D. Zilberman, Production of Ethanol from Sugarcane in Brazil: From State Intervention to a Free Market, vol 43, Springer Science & Business Media, 2014.
[27] C. Grangeia, L. Santos, L.L.B. Lazaro, The Brazilian biofuel policy (RenovaBio) and its uncertainties: an assessment of technical, socioeconomic and institutional aspects, Energy Convers. Manag. X 13 (2022) 100156.
[28] F. Gonçalves, et al., Strategies to improve the environmental efficiency and the profitability of sugarcane mills, Biomass Bioenergy 148 (2021) 106052.
[29] L.L.B. Lazaro, et al., Policy and governance dynamics in the water-energy-food-land nexus of biofuels: proposing a qualitative analysis model, Renew. Sustain. Energy Rev. 149 (2021) 111384.
[30] Z. Peidong, et al., Bioenergy industries development in China: dilemma and solution, Renew. Sustain. Energy Rev. 13 (9) (2009) 2571–2579.
[31] D. Liu, L. Zhu, Current situation and development strategy of biogas industry in Guangxi, Guangxi Agric. Sci. 39 (4) (2008) 445–448.
[32] A.S. Giwa, et al., Prospects of China's biogas: fundamentals, challenges and considerations, Energy Rep. 6 (2020) 2973–2987.
[33] J. Zhao, Development of China's biofuel industry and policy making in comparison with international practices, Sci. Bull. 60 (11) (2015) 1049–1054.
[34] Y.-J. Xu, G.-X. Li, Z.-Y. Sun, Development of biodiesel industry in China: upon the terms of production and consumption, Renew. Sustain. Energy Rev. 54 (2016) 318–330.
[35] USDA, China: Biofuels Annual, 2021 [cited 2022 12.05.2022]; Available from: https://www.fas.usda.gov/data/china-biofuels-annual-7.
[36] M. Hiloidhari, D. Das, D. Baruah, Bioenergy potential from crop residue biomass in India, Renew. Sustain. Energy Rev. 32 (2014) 504–512.
[37] MNRE, Bio Energy, 2021 [cited 2022 5.05.2022]; Available from: https://mnre.gov.in/bio-energy/current-status.
[38] S. Das, The national policy of biofuels of India—a perspective, Energy Pol. 143 (2020) 111595.
[39] NPB, National Policy on Biofuels, 2018 cited 2022 5.05.2022]; Available from: http://164.100.94.214/sites/default/files/uploads/biofuel_policy_0.pdf.
[40] A.P. Saravanan, A. Pugazhendhi, T. Mathimani, A comprehensive assessment of biofuel policies in the BRICS nations: implementation, blending target and gaps, Fuel 272 (2020) 117635.
[41] A. Sukenik, G. Shelef, Algal autoflocculation—verification and proposed mechanism, Biotechnol. Bioeng. 26 (2) (1984) 142–147.
[42] J. Lane, Algenol hits 9K gallons/acre mark for algae-to-ethanol process, Biofuels Digest. (2013). Available from: https://www.iea.org/fuels-and-technologies/bioenergy. (Accessed 5 May 2022).
[43] D. Haldar, M.K. Purkait, A review on the environment-friendly emerging techniques for pretreatment of lignocellulosic biomass: mechanistic insight and advancements, Chemosphere 264 (2021) 128523.

[44] J.N. Rogers, et al., A critical analysis of paddlewheel-driven raceway ponds for algal biofuel production at commercial scales, Algal Res. 4 (2014) 76–88.
[45] C.N. Ogbonna, E.G. Nwoba, Bio-based flocculants for sustainable harvesting of microalgae for biofuel production a review, Renew. Sustain. Energy Rev. 139 (2021) 110690.
[46] V. Godbole, M.K. Pal, P. Gautam, A critical perspective on the scope of interdisciplinary approaches used in fourth generation biofuel production, Algal Res. 58 (2021) 102436.
[47] R. Kothari, et al., Assessment of Indian bioenergy policy for sustainable environment and its impact for rural India: strategic implementation and challenges, Environ. Technol. Innovat. 20 (2020) 101078.
[48] M. Shah (Ed.), Microbial Bioremediation & Biodegradation, Springer, Singapore, 2020.
[49] M. Shah (Ed.), Removal of Refractory Pollutants from Wastewater Treatment Plants, CRC Press, USA, 2021.

Chapter 2

Life cycle assessment of biofuels derived from Mahua (*Madhuca* species) flowers and seeds: key issues and perspectives

Urmimala Das[1], Bishnupriya Hansdah[1], Sudhanshu S. Behera[2] and Ramesh C. Ray[2]

[1]Maharaja Sriram Chandra Bhanja Deo University, Sri Ram Chandra Vihar, Baripada, Odisha, India; [2]Centre for Food Biology & Environment Studies, Bhubaneswar, Odisha, India

1. Introduction

In the past decade, carbon dioxide emissions have reached an all-time high for the second time. However, last year's emissions rise was only 0.5%, less than half of the 10 year average [1]. Global carbon dioxide emissions have increased by 50% since the 1997 Kyoto Protocol was negotiated to reduce emissions (International Energy Agency, IEA). Nonrenewable energy sources, on the other hand, have long-term detrimental effects on human health and environmental issues. Exhaust gases such as NOx and CO are produced as a result of the growing demand for and usage of diesel engines in a variety of areas, causing major environmental pollution and hazards such as global warming and respiratory difficulties [1].

Because fossil fuels are rapidly depleting, and the rise in greenhouse gas emissions, there has been an increasing interest in the search for alternative energy sources over the last 2 decades [2,3]. Bioethanol produced from renewable resources could serve as a partial alternative for fossil fuels, reducing 32% of worldwide gasoline consumption in transportation while also lowering greenhouse gas emissions [4]. Sugars and starches such as sugarcane, cheese whey, lactose, maize, cassava, corn, mahua (*Madhuca* spp.) flowers, and lignocelluloses such as agroindustrial waste, pretreated rice straw, and bagasse, as well as microalgae, are available in nature for preparing biofuels that can be classified as first, second, and third generation biomass [4].

The search for new, inexpensive, nonfood biomass sources for ethanol generation is gaining traction around the world. Mahua flowers and seeds for bioenergy production can help communities who rely on the forest to attain energy independence; jobs are also being created the economy is strengthened while lowering anthropogenic carbon emissions. Using a Life Cycle Assessment (LCA) approach, this chapter examines the possible use of mahua flowers and seeds as primary raw materials for the manufacturing of two key biofuels, bioethanol and biodiesel.

1.1 Geographical and climatic resolution

According to a study, mahua is an important economic tree that grows in the Indo-Pak subcontinent's subtropical region, which includes the northern, central, and southern regions of peninsular India, Sri Lanka, and Burma [5–7]. *Madhuca latifolia* L. grows in mixed deciduous forests on sandy and rocky soils throughout Asia and Australia [8–10]. Several other species are found in Malaysia, Sri Lanka, Australia, New Guinea, and Thailand [11]. In a survey conducted by the Indian Central Oilseeds Committee, there are approximately 8.5 million mahua trees in India [12–14]. *M. latifolia* and *M. longifolia* are the two *Madhuca* species that are found almost anywhere in India. These two Malaysian species have merged and are now known as *M. longifolia* var. *longifolia* and *M. longifolia* var. *latifolia* [15,13]. Mahua trees are fast-growing trees with evergreen or semi-evergreen leaves that can reach a height of 20 m with a temperature range of 2–46°C. These trees may grow in a variety of soil types, but they favor alluvial soil. The optimal soil for increased development and productivity is deep loamy or sandy-loam soil with sufficient drainage. The tree's shallow root system is broad and spread out, which keeps the soil together and prevents erosion [16,11].

1.2 *Madhuca* species as a green approach for biofuel production

Biodiesel made from edible oils has a detrimental influence on society, the agriculture problem, and the decline of food supplies, resulting in economic imbalance. Food competition and difficulties related with food vs. fuel are eliminated with nonedible oil. Nonedible food crops are seen as a promising replacement for the cultivation of traditional edible food crops. Mahua is a nontraditional forest tree with a significant annual output potential of roughly 60 million tons in India. Mahua oil, extracted from the kernel, contains 21% free fatty acids (such as stearic acid, linoleic acid, and palmitic acid). The choice of feedstock is critical in the manufacture of biodiesel since it directly affects the cost of manufacturing. Screening of various accessible feedstocks, ranging from edible to nonedible oil, has a significant impact on manufacturing costs [17]. Biodiesel is gaining popularity these days due to its good environmental impact. Biodiesel is made mostly from vegetable oils. Among the tree-borne nonedible oilseed, one of the most appropriate plants is jatropha (*Jatropha curcas*). Tree-borne nonedible oilseeds are used for the biodiesel program in India. However, it has a significant flaw in terms of oxidation stability. *Madhuca indica* is a nonedible oilseed that grows on trees and has a production capacity of 60 MT per year in India. It has 35%—40% oil content and a calorific value that is comparable to mineral diesel [18,19]. There is a need to find alternate fuels because fuels derived from petroleum degrade every now and again. One option is to use biofuels instead of existing fossil fuels because they have desired features that will fulfill the needs of automobiles. The topic of this report is mahua-based biodiesel, which is made from a nonedible oil source. Straight vegetable oils cannot be used due to their carbon deposition, excessive viscosity, poor atomization of fuel, and other factors. Transesterification is used to make biodiesels in order to circumvent these restrictions. These transesterified oils are mixed in the right proportions to assure compatibility with current diesel engines. This chapter will offer a case study on increasing the engine's performance by altering the combustion chamber shape and fuel additions such as oxygenated additives and metal-based additives in the design parameter. Also discussed are the results of two alternative combustion chamber designs, hemispherical and toroidal. The effects of an oxygenated ethanol addition and a metal-based alumina additive on performance, combustion, and emission trends are discussed [20].

2. Biomass (mahua flowers) for conversion to bioethanol

In India's tribal regions, the mahua tree is a largely unexploited species that may adapt to thrive in arid climates and provide a cheaper supply of biomass, food, and fuel [21,4]. Mahua flowers are dense fascicles near the end of branches that are 1.5 cm long and droopy fleshy off-white in color. They contain primarily reducing sugars (36.3%—50.62%, w/w) (Table 2.1) [4]. Mahua flowers are traditionally used to make the country liquor "mahuli," which is readily available across India's central and northern plains [21,4].

TABLE 2.1 Proximate composition (%) of mahua flower.

Constituent	Mahua flower	Dry flower
Moisture	73.6—79.82 (%, d.b.)	11.61—19.8 (%, w.b)
pH	4.6	—
Starch (g/100 g^{-1})	0.94	—
Ash (%)	1.5	1.4—4.36
Total sugars (g/100 g^{-1})	47.35—54.06	41.62
Total inverts (%)	54.24	
Cane sugars (%)	3.43	
Reducing sugars (g/100 g^{-1})	36.3—50.62	28.12
Proteins (%)	6.05—6.37	5.62
Crude fat	0.8—1	

T. Banerjee, A. Samanta, Improvement over traditional brewing techniques for production of bioethanol from mahua flowers (Madhuca indica), Int. J. Energy Sect. Manag. 7 (4) (2018a) 12—21. and R. Kumar, A.K. Ghosh, P. Pal, Fermentative ethanol production from Madhuca indica flowers using immobilized yeast cells coupled with solar driven direct contact membrane distillation with commercial hydrophobic membranes, Energy Convers. Manag. 181 (2019) 593—607.

2.1 Bioethanol production process

The hunt for sustainable fuel alternatives is fueled by rising petroleum prices, rising greenhouse gas emissions, and dwindling fossil fuel supplies. Biofuel production, particularly bioethanol, has enormous potential to meet present energy demand [22,4]. Bioethanol production systems now rely heavily on sugarcane and starch-based food supplies like biomass, putting major strain on food security [22,23]. Nonedible biomass resources, such as mahua flowers, could be regarded suitable for ethanol production in this case. The corolla of mahua flowers is fleshy and contains a lot of reducing sugar [24]. Ref. [25] used immobilized *Saccharomyces cerevisiae*-3044 cells to study the generation of ethanol from mahua flower (*M. latifolia* L.) in submerged fermentation (SmF). It has been investigated how specific parameters change during fermentation and how to keep them stable while producing ethanol. The most ethanol is produced when the pH is between 5 and 5.5, the inoculum level is 1.5 g 100 mL^{-1}, the inoculum age is 48 h, the temperature is 30–32°C, the nitrogen source is 0.05%–0.06%, the sodium potassium tartrate is 1.2 g L^{-1}, and the fermentation period is 2–4 days. For a 1:5 slurry composition, the maximum ethanol yield was found to be 338 mL^{-1}, and this technology and design can be used to produce ethanol on a wide scale.

Because of the rise in fuel prices, the recent development of bioethanol as an alternative energy source through microbial fermentation has reignited scientific interest. The most extensively employed microorganisms for ethanol production are *S. cerevisiae* and *Zymomonas mobilis*. Ref. [26] compared the ability of the yeast *S. cerevisiae* (CTCRI strain) and the bacterium *Z. mobilis* (MTCC 92) to ferment ethanol from mahua flowers in a study [27]. Using free cells of *S. cerevisiae* and *Z. mobilis*, the ethanol production after 96 h was 149.0 and 122.9 g kg^{-1} flowers, respectively. In comparison to *Z. mobilis*, the *S. cerevisiae* strain produced 21.2% more final ethanol. After 96 h of fermentation, *S. cerevisiae*'s ethanol yield (Yx/s), volumetric product productivity (Qp), sugar to ethanol conversion rate (percent), and microbial biomass concentration (X) were found to be 5.2%, 21.1%, 5.27%, and 134% higher than *Z. mobilis*, respectively.

In another investigation, Ref. [28] used *Z. mobilis* MTCC 92 free cells and cells immobilised in a calcium alginate matrix to perform batch fermentation of mahula flowers. After 96 h of fermentation, free and immobilized cells produced 122.9 ± 0.972 and 134.6 ± 0.104 g kg^{-1} flowers, respectively, on a dry weight basis, indicating that cells entrapped in calcium alginate matrix produced 8.7% more ethanol than free cells. Furthermore, the immobilized cells were physiologically active for three more fermentation cycles, yielding 132.7 ± 0.095, 130.5 ± 0.09, and 128.7 ± 0.056 g ethanol per kg flower, respectively, in the first, second, and third cycles.

On application of different immobilized substrates (agar agar and calcium alginate), Ref. [21] investigated batch flower fermentation employing immobilized and free *S. cerevisiae* cells. Using immobilized (in agar agar and calcium alginate) and free cells, the ethanol yields were 151.2, 154.5, and 149.1 g kg^{-1} flowers, respectively. In terms of ethanol output from mahula flowers, cell entrapment in calcium alginate was shown to be marginally superior than those in agar agar (2.2% more) and over-free cell (3.5% more). Furthermore, the immobilized cells remained physiologically active for at least three cycles of ethanol fermentation [150.6, 148.5, and 146.5 g kg^{-1} (agar agar) and 152.8, 151.5, and 149.5 g kg^{-1} flowers (calcium alginate) for the first, second, and third cycles, respectively] without lowering productivity.

Ref. [21] reported the production of bioethanol from mahua flowers using *S. cerevisiae* in the batch fermentation process. Various models for optimization of process parameters and also three different initial substrate concentrations at 30 C for 96 h (pH 5) were investigated. Hinshelwood model was found to be fitted the concentration-time profile most satisfactorily and predicted the bioethanol yield within the acceptable error range.

Non-Timber Forest Produce (NTFP) such as mahua flowers are abundant and inexpensive, making them a suitable substrate for sustainable bioethanol production in bioindustries without compromising the nation's food supply. Mahua flowers could be used as a low-cost, plentiful biomass for bioethanol production in a cost-effective manner [4,22,26,29]. Ref. [24] used a commercial yeast strain, *S. cerevisiae* MTCC 4780, and a bacterium strain, *Z. mobilis* MTCC 92, to produce ethanol from mahua flowers. The fermentation medium was sterilized mahua flower and water, and the ethanol content was evaluated using a specific gravity method. *S. cerevisiae* and *Z. mobilis* produced 9.11% (v/v) and 9.96% (v/v) ethanol after 72 and 120 h of incubation at 30 ± 1°C, respectively.

In a study, Ref. [26] reported using *S. cerevisiae* to produce bioethanol from *M. longifolia* species (mahua flowers were harvested in Eastern Ghats region of Andhra Pradesh, India). In test-1 (*S. cerevisiae* + mahua flowers + media), test-2 (mahua flowers + media), and control (media), 1000 mL^{-1} acidic fermented medium at pH-4, 5, and 5.7 generated 170.03 mL^{-1}, 142.3 mL^{-1}, and 127.7 mL^{-1} of bioethanol, respectively. The percentage of bioethanol was validated using an alcohol meter, with findings showing 46.6%, 24.6%, and 0% in test-1, test-2, and control, respectively. A spectrophotometer set to 204–240 nm was used to confirm the presence of bioethanol.

Ref. [29] conducted a comparative study on the production of bioalcohol from wastewater of potato chips industry and tap water using mahua flower (*M. longifolia*) and baker's yeast. The study indicated that the potato industry wastewater could be used for the production of bioalcohol. Application of potato industry wastewater not only saves the treatment cost of wastewater, but it will also save our precious drinking/tap water resource. Mahua flowers and potato industry wastewater in the ratio of 1:5 found the yield of 400 mL^{-1} bioalcohol/kg of mahua flower in a fermentation period of 14 days.

Bioethanol generated from mahua flowers having a high concentration of fermentable carbohydrates was reported by Ref. [20] as a first-generation biofuel. Mahua flower extracts are fermented using yeast strain *S. cerevisiae* 3078 culture in two different ways (batch and fed-batch). With aging and prolonged fermentation, the concentration of generated alcohol grew until it reached a constant level. After 14 days at 33°C (pH 5.7), the maximum yields of ethanol from fresh and 6 month-stored mahua flowers were 18% and 15% (using batch) and 22% and 16% (using fed-batch fermentation), respectively.

Ref. [4] used immobilized yeast cells to evaluate the fermentation of reducing sugar derived from low-cost mahua flowers and compared it to free cells for ethanol production. After fermentation, the fermenter was connected to a solar-driven membrane distillation (SDMD) system for ethanol separation and concentration. In continuous fermentation using immobilized yeast cells, increased ethanol output (0.48 g g^{-1}) and productivity (28 g L^{-1} h^{-1}) were achieved at a dilution rate of 0.4 h^{-1} compared to conventional procedures.

Ref. [30] studied the synthesis of ethanol from mahua flowers using a yeast strain (*Pichia kudriavzevii*) obtained from milk whey. The isolated yeast strain was found using a basic local alignment search method and molecular phylogenetic analysis to sequence the D1/D2 domain of the rRNA gene. Optimizing operational parameters (fermentation period of 48 h, temperature of 25°C, and pH of 5.0) resulted in maximum ethanol output of 371 g kg^{-1} from the flowers. The effect of nitrogen supplements, enzymatic treatment, and coculture on bioethanol production under optimised circumstances revealed that the best condition produced the most ethanol without any supplementing or treatment. The production of ethanol was increased when the *P. kudriavzevii* strain was cocultured with *P. stipitis* NCIM 3497.

Ref. [23] discovered *Meyerozyma caribbica* M72, a new yeast strain from the mahua flower. For ethanol synthesis from rice straw, an efficient procedure was devised employing mild sulfuric acid (0.8%, w/v) treatment at 160°C for 10 min, followed by saccharification with the identified new yeast. The identified yeast can use both the pentose and hexose sugars present in rice straw. *M. caribbica* M72 from mahua flower was discovered to produce 24.36 g L^{-1} of ethanol from saccharified hydrolysate of rice straw in 48 h at 32°C and pH 7 with an ethanol tolerance of 8%.

2.2 Bioethanol refinery

If favorable policies and investments are in place, biofuels have the potential to meet more than a quarter of global transportation fuel demand by 2050, according to the International Energy Agency (IEA). Currently, various governments support biofuel incentives to encourage local biofuel production and use include blending mandates or targets, subsidies, tax exemptions and credits (exemptions from excise and pollution taxes, corporate tax breaks for biofuel producers), reduced import duties, support for research and development and direct involvement in biofuel production, and other incentives (https://promfgmedia.com).

3. Biomass (mahua seeds) for conversion to biodiesel

Mahula seeds, a nonfood-grade inexpensive fat substrate from a nonagricultural setting such as a forest can be used to make biodiesel instead of food-grade sugar/starchy crops like maize and sugarcane.

3.1 Seed oil and oil yield

The mahua (*M. indica*) is a nonedible oil tree that can be found in abundance throughout India. Mahua fruit kernels yield 35%–40% oil and the seed contains maximum up to 51.5% of oils/fatty acids (palmitic, stearic, oleic, linoleic, linolenic, arachidic) (Table 2.2) making them an excellent source of biodiesel [31]. The thermal pyrolysis of mahua seed for biofuel production was reported by Ref. [32]. Pyrolysis was carried out in a semi-batch reactor at different temperatures (450–600°C) with a nitrogen flow rate of 30 mL^{-1} min^{-1} and a continuous heating rate of 20°C min^{-1}. The maximum bio-oil output of 49% and 18% biochar were produced at an optimal temperature of 525°C. Bio-oil was discovered to have a calorific value of 39.02 MJ^{-1}kg^{-1}, which is similar to the calorific values of conventional petroleum fuels.

TABLE 2.2 Proximate composition (%) of mahua seed.

Constituent	Mahua seed
Protein (N × 6.25)	16.9
Oil/Fatty acids composition (palmitic, stearic, oleic, linoleic, linolenic, arachidic)	51.5
Fiber	3.2
Carbohydrates	22.0
Ash	3.4
Saponins	2.5
Tannins	0.5

S.K. Nayak, B.P. Pattanaik, Experimental investigation on performance and emission characteristics of a diesel engine fuelled with mahua biodiesel using additive, Energy Proc. 54 (2014) 569–579. and A. Singh, I.S. Singh, Chemical evaluation of mahua (Madhuca indica) seed, Food Chem. 40 (2) (1991) 221–228.

3.2 Biodiesel production process and recent advances

Biodiesel is an alkyl ester of fatty acids that may be produced by transesterification from any vegetable oil and is a renewable, biodegradable, and nontoxic fuel [33]. The biodiesel from mahua oil has high free fatty acids through the transesterification process [34]. Using the transesterification process, Ref. [35,36] investigated the production of biodiesel from raw mahua oil and also studied the different properties. The study revealed that the physicochemical properties of biodiesel were found within the range of different specification standards (ASTM D-6751 of USA, EN-14,214 of Europe, and IS-15607 of India) and claimed that biodiesel composited with mixed (41.2% saturated and 58.8% unsaturated) fatty acids. In a study, Ref. [37] reported the preparation of biodiesel from the mahua oil by esterifications followed by the transesterifications. The most favorable condition for the production of the biodiesel was sodium methoxide (2.2%), alcohol-oil ratio (0.1%), 1.15 h reactions time, and at 60 C of temperature.

Through the transesterification process, blends of dual oils/biofuels improve the performance of biodiesel [38]. Ref. [34] used a mechanical stirrer to optimize biodiesel synthesis from neem and mahua oil through the transesterification process utilizing KOH (0.5 wt%) as a catalyst. Blends of neem–mahua–based diesel improve the performance of biodiesel. Two-step transesterification was used by Ref. [33] to produce biodiesel from mahua and jatropha oil. The dual biofuel samples met the requirements of Europe, the United States, and India, according to the findings. The results showed that the pretreatment technique developed lowered the free fatty acid value of biodiesels effectively (1%). The sample blends with 15%–20% biodiesel had improved features and were tested further on a diesel engine for fuel combustion characteristics.

3.2.1 Improvement of performance

Mahua biodiesel has been used as an alternative fuel for diesel engines [39,40]. The effect of injection timing, compression ratio on the performance of mahua biodiesel has been reported by several authors [39,41,42]. Mahua biodiesel is generally blended/mixed with diesel fuel up to 15%–20% (at any compression ratio) which increases the performance [42], and even more its performance increases if an additive is added [43].

Ref. [40] found that when a biodiesel blend with coated engine was compared to a base engine running on straight diesel, the specific fuel consumption was lowered by 8.5% and the brake thermal efficiency (BTE) was increased by 6.2% [40]. Ref. [39] used diesel, mahua methyl ester, and methanol additive mixes to investigate the indirect injection diesel (IDI) engine. The best combination of load and fuel was found to be 20 kg of mahua methyl ester + 3% methanol.

Biodiesel generation mediated by heterogeneous solid catalysts has emerged as a favored technique because it has the potential to be cost-effective. The Mn-doped ZnO heterogeneous nanocatalyst has demonstrated effective catalytic activity in the conversion of mahua oil to biodiesel and is appropriate for large-scale biodiesel production (yield of 97%) [17].

When compared to KOH as a catalyst, red mud has a higher calorific value (10,601 $kcal^{-1}kg^{-1}$). Ref. [44] examined biodiesel generation from mahua oil using two different catalysts: KOH and activated red mud by catalytic cracking (waste from the aluminum industry) in different diesel fuel blends. The study indicated that red mud (as a catalyst) not only improved most of the qualities of the fuel, but it also lowered environmental stress by reducing emissions and fuel consumption.

Metal oxides incorporated in biodiesel blends have been found to increase the working efficiency and emission behavior of diesel engines by a few researches [42,45]. The addition of aluminum oxide nanoparticles to a mahua biodiesel blend in various quantities resulted in a significant increase in brake thermal efficiency [32]. Ref. [42] investigated the combustion and performance characteristics of an injection diesel engine using cerium oxide nanoparticles (Nanocera) added water emulsified mahua biodiesel blend (NWEB) in comparison to water emulsified mahua biodiesel blend (WEB), B20, and diesel. The maximum braking thermal efficiency for NWEB was found to be 30.41%, compared to 28.25% for diesel. When compared to B20 and WEB, the mean brake-specific fuel consumption was 3.76% and 3.93% lower with NWEB. The effects of silicon dioxide (SiO_2) nanoparticles added to the mahua methyl ester mix on the diesel engine were explored by Ref. [46]. When compared to other blends previously evaluated, the brake thermal efficiency of SiO_2 mixed brakes showed a modest rise, but BSFC (brake specific fuel consumption) exhibited a declining trend.

The performance parameters at a rated load revealed that as the biodiesel concentration was increased, the BTE declined [47]. Because of its higher viscosity and lower calorific content, BTE for mahua biodiesel (B100) was 3.5% lower than diesel. However, mahua biodiesel (B100) has a 36% greater brake-specific energy consumption than diesel [47].

Most biodiesel have a low oxidation stability, which is a big disadvantage. The oxidation process degrades the quality of biodiesel and causes the engine fuel pump to wear out faster [19]. Ref. [19] compared mahua biodiesel's oxidation and storage stability to those of jatropha biodiesel and mineral diesel in a paper. The biodiesels were blended with mineral diesel to improve oxidation stability, and it was discovered that 20% blended jatropha biodiesel (JB20) and 30% blended mahua biodiesel (MB30) met the EN-590 standard (20 h).

Ref. [48] used three fuel samples to test the performance of four-stroke diesel engines: biodiesel (B100), diesel-biodiesel (B40), and diesel-biodiesel-nanoparticles (B40T105MWCNT45). The findings of the experiment revealed that adding nanoparticles, such as TiO_2 and multi-walled carbon nanotubes (MWCNT) to a diesel−biodiesel fuel mixture (i.e., B40T105MWCNT45) improved the brake thermal efficiency (BTE) significantly when compared to B100 and diesel.

Ref. [49] studied the performance of dual biodiesel (rapeseed (RA) and mahua biodiesel (MU)) compression ignition (CI) engines. Its performance was also compared to that of the CI diesel engine. Biodiesel (RA and MU) and biodiesel mixes with diesel have higher brake-specific fuel consumption than diesel. However, the BL20 (20%RM + 80% diesel) demonstrated the closest brake-specific fuel consumption (BSFC) to diesel of all the biodiesel blends tested. BL20 had an 8.18% greater BSFC than diesel. When compared to other biodiesel blends, the BL20 blend had the highest braking thermal efficiency but was closest to diesel (i.e., 2.79% lower than diesel).

Ref. [43] recently found that using mahua oil with di-tert-butyl peroxide (DTBP) (as an additive) improved the performance qualities of mahua biodiesel compared to mahua biodiesel without additive [43].

3.2.2 Reduction of exhaust emission

Mahua biodiesel oil offers the advantage of lowering greenhouse gas emissions (without requiring any engine modifications) (Table 2.3) and has the potential to become a long-term fuel source like biodiesel [16,49]. Ref. [32] showed a minor reduction in hazardous pollutants (e.g., CO, HC, and smoke) for aluminum oxide (Al_2O_3) nanoparticles combined mahua biodiesel. Ref. [16] discovered that mahua biodiesel has the lowest residual oxygen, CO, HC, and NO emissions when compared to diesel. For B20 (20% mahua oil + 80% diesel oil), the percentage of carbon dioxide reduced by up to 20.84%, the percentage of HC decreased by up to 50%, and the percentage of nitrogen oxides decreased by up to 22.1%. Similarly, with B20, the fraction of residual oxygen dropped to 24.58%. The results of the experiments showed that using mahua oil biodiesel as a diesel engine fuel is a feasible alternative to diesel.

In a stationary diesel engine, Ref. [33] investigated the emission characteristics of neat mahua oil biodiesel combined with different quantities of octanol (10% and 20%, v/v). The use of mahua oil biodiesel resulted in considerable reductions in all emissions (7.5% CO, 5.7% HC, 5.4% NOx, and 2.9% smoke).

Ref. [47] found that biodiesel and its mixes emit less smoke than diesel due to higher oxygen content and full combustion. Because biodiesel (obtained from mahua seeds) has more oxygen available, it has a shorter ignition delay than diesel. When compared to diesel at rated load, unburned hydrocarbon (UBHC) and smoke emissions were reduced by 22% and 23%, respectively. The amount of nitric oxide (NO) released rose by 6% as the biodiesel concentration increased.

Ref. [49] studied the emission characteristics of a dual biodiesel (rapeseed (RA) and mahua biodiesel (MU)) variable compression ratio diesel engine (VCRDE). CO and HC emissions were found to be lower in all biodiesel and blends tested than diesel. CO and HC emissions were 20.66% and 8.56% lower in the BL20 (20% RM + 80% diesel) blend than in diesel. Furthermore, all of the biodiesel and blends tested emitted more NOx than diesel, with the B20 blend emitting 3.77% more NOx than diesel. The smoke opacity of all tested biodiesel and blends was lower than that of diesel. The BL20 mix had a smoke opacity that was closer to diesel (i.e., 6.97% lower than diesel).

TABLE 2.3 Comparison of suitable blending conditions and atmospheric emission characteristics of mahua biodiesel with diesel (petroleum fuel).

Mahua biodiesel	Suitable blending composition/Preparation	Emission characteristics with respect to diesel	References
Dual biodiesel BL20	Rapeseed and mahua biodiesel (1:1)	CO emission: 20.66% ↓ HC emission: 8.56% ↓ Smoke opacity: 6.9% ↓ NO_x emission: 3.79% ↑	[49]
Biodiesel blends	10% biodiesel and 90% diesel	CO_2 emission 46.91% ↓	[38]
Blended biodiesel	80% diesel and 20% (MME20 + AONP120)	Smoke 5.38% ↓, HC 6.39% ↓, CO 10.24% ↓, NO_x ↓, and BTE 8.8% ↑	[50]
Mahua biodiesel	—	UBHC emission 22% ↓ and smoke emissions 23% ↓	[47]
Mahua biodiesel	20% mahua oil + 80% diesel oil	CO_2 emission: 20.84% ↓ HC emission: 50% ↓ NO emission: 22.1% ↓	[16]
Mahua oil biodiesel	Mahua oil + octanol (10% and 20%, v/v)	CO emission 7.5% ↓, HC 5.7% ↓, 5.4% NO_x ↓, and 2.9% smoke ↓	[33]

Abbreviation: *Mme*, mahua methyl ester; *Bte*, brake thermal efficiency; *AONP*, Al_2O_3 nanoparticle; *Ubhc*, unburned hydrocarbon. ↓: reduced; ↑: higher.

Ref. [38] investigated the potential of dual biodiesel blends made from nonedible bio-oils such as jatropha (*Jatropha Curcas*), karanja (*Pongamia pinnata*), mahua (*Madhuca indica*), and neem (*Azadirachta indica*). The effects of mixes on CO, CO_2, HC, and NOx emissions were studied and compared to straight diesel. The blend D90 + JB5 + NB5 (i.e., 10% biodiesel and 90% diesel) shows the greatest reduction in CO_2 (46.91%). When compared to straight diesel, all of the blends under consideration emit more CO_2.

Ref. [50] evaluated the effect of fuel-borne additives when applied to a mahua methyl ester (MME) blend used in a CRDI diesel engine. MME was combined with clean diesel fuel and 40, 80, and 120 ppm mass fractions of Al_2O_3 and Fe_2O_3 nanoparticles, and biodiesel were made by combining 80% diesel and 20% MME. The results showed that when the Al_2O_3 nanoparticle additives' blended biodiesel (MME20 + AONP120) was used, the number of harmful pollutants such as smoke (5.38%), HC (6.39%), carbon monoxide (10.24%), NOx, and other pollutants were significantly reduced, and the BTE was improved by 8.8% when compared to MME20 [50].

3.3 Biodiesel refinery

In India, there is a vast potential for the production of biodiesel from *J. curcas* and karanj (*P. pinnata*) as they occur in plenty in forests and wastelands. However, the biodiesel biorefinery from mahua has not been reported.

4. Life cycle assessment

The LCA is a method of compiling and evaluating inputs (energy and materials) and outputs (products, byproducts, pollutants, and emissions) in order to estimate the potential environmental impacts of a product system over its entire life cycle [51]. In the case of biofuels production, the system's boundaries are typically defined as "well to wheel" (WTW), with the impacts assessed based on the extraction or production of raw materials from the "well," as well as their transformation, distribution, and end use: the "wheels." Other approaches, such as "well to tank" (WTT), focus solely on raw material production and the bioethanol production process, and "tank to wheel" (TTW), measure improvements in impact factors while using bioethanol in engines [51]. The focus of the analysis has been on upstream activities, such as agriculture or forestry, and downstream activities, such as engine use, with little attention paid to the bioethanol manufacturing process [51].

4.1 Goal, scope, and functional unit

The goal of the study represents the environmental and social implications of producing biofuels from mahua flowers and seeds from a life cycle perspective. The study's scope includes feedstock planting and harvesting, feedstock processing, biofuel generation, and related transportation from cradle to gate. The feedstock specified to the mahua flowers and seeds. The functional unit remains to assess the potential life cycle environmental impacts and resources consumptions in producing mahua as biofuel feedstocks [52] and followed by documentation and analysis of the environmental emissions and resource consumptions of producing mahua [52].

4.2 System boundaries and reference system

The system boundaries/reference system depicts significant environmental and social issues associated with biofuels production from both feedstocks (mahua flowers and seeds) in comparison with conventional gasoline and suggestions to improve the environmental and social performance of biofuels production [36]. The mahua flowers and seeds are considered to be promising feedstock for several purposes [32,53,54]. They can be used as raw materials to produce high-value products, e.g., for ethnomedicinal applications [53], a food source for wild animals [54], and suitable substrate for biofuel production [32]. Other research attempting to identify the cradle-to-gate with different technologies can also benefit from this assessment. The biochemical study revealed that mahua flowers contained moisture (10%–15%), sugar (64%–68%), reducing sugars (50%–55%), invert sugars (10%–14%), ash (2%–4%), crude protein (4%–5%), crude fat (0.8%–1%), Fe^{2+}, and Ca^{2+} [20]. Significant variations in oil content were found between agroclimatic zones (50.07%–53.85%) [6]. Oil parameters such as kernel oil percent, palmitic acid, stearic acid, oleic acid, and linoleic acid showed significant differences. The saturated fatty acids, palmitic acid, and stearic acid ranged from 11.7%–25.9% to 19.1%–32.2% in kernel oil, respectively [6,55]. The hemicellulose/cellulose extraction technology is not used for ethanol production from mahua flowers, and the life cycle environmental impact of the downstream processes in system boundary is just a linear deletion. Consequential life cycle assessment (cLCA) linked to the market mechanism, i.e., economic responses/market implications of the system are taken into account by expanding the system boundary [36].

4.3 Inventory analysis

A life cycle inventory (LCI) is a process of quantifying energy and raw material requirements, environmental pollution for the entire life cycle of a product, process, or activity [51]. The main issue of inventory analysis includes data collection and estimations, validation of data, and relating data to the specific processes within the system boundaries [51].

4.4 Coproduct, byproduct, residue

The production of renewable biomass often involves the generation of coproducts, by-products, or wastes [56]. According to Clean Development Mechanism [29], coproducts are defined as items with revenues similar to the primary product, byproducts are products with lower revenues than the main product, and wastes have little or no revenue [57]. According to alternative definitions, coproducts are all output streams other than the principal product that is not wasted and are not employed as raw materials elsewhere in the system [56]. Several byproducts/coproducts/residues, such as seed oil cake (made after oil extraction from mahua seed) for fish feed and bovine feed from mahua biomass (flowers and seeds) through biofuels production [28] have recently gained attention.

4.5 Impact assessment

Impact assessment creates a link between a product or process and its possible negative effects on human health, the environment, and the depletion of natural resources [56]. Despite the fact that mahua biofuels have a lower GHG footprint and a positive energy balance when compared to fossil fuel alternatives, the assessment of other impact categories such as acidification and eutrophication has been underreported, owing to the fact that they become relevant only when intensive mahua tree planting is required in the feedstock production, with fertilizers being the main contributors to those categories [51].

5. Conclusion and future trends

Mahua biomass-based biofuels (bioethanol and biodiesel) is one of the most promising alternative fuels and occupies a significant position among the alternatives to traditional petroleum-based fuels owing to various technological and economic aspects as well as effective analytical procedures for quality control. In this chapter, the key issues/difficulties and methodological assumptions responsible for large ranges and uncertainties in mahua bioenergy LCA have been highlighted and discussed. However, some critical elements (for example, indirect effects) are unknown and highly dependent on local and climate variables. Although policymakers are claiming for methodological standards, scientific research for estimating indirect effects is still at a preliminary stage. Moreover, a more consistent and accessible characterization of the essential metrics, such as boundary conditions, functional units, and effect categories, can facilitate the comparison of LCA results on mahua biofuels.

References

[1] P.V. Elumalai, M. Parthasarathy, J.S.C.I.J. Lalvani, H. Mehboob, O.D. Samuel, C.C. Enweremadu, A. Afzal, Effect of injection timing in reducing the harmful pollutants emitted from CI engine using N-butanol antioxidant blended eco-friendly Mahua biodiesel, Energy Rep. 7 (2021) 6205–6221.

[2] W. Gao, L.G. Tabil, T. Dumonceaux, S.E. Ríos, R. Zhao, Optimization of biological pretreatment to enhance the quality of wheat straw pellets, Biomass Bioenergy 97 (2017) 77–89.

[3] P. Shah Maulin, Microbial Bioremediation & Biodegradation, Springer, 2020.

[4] R. Kumar, A.K. Ghosh, P. Pal, Fermentative ethanol production from Madhuca indica flowers using immobilized yeast cells coupled with solar driven direct contact membrane distillation with commercial hydrophobic membranes, Energy Convers. Manag. 181 (2019) 593–607.

[5] S. Behera, R.C. Mohanty, R.C. Ray, Biochemistry of post-harvest spoilage of mahula (Madhuca latifolia L.) flowers: changes in total sugar, ascorbic acid, phenol and phenylalanine ammonia-lyase activity, Arch. Phytopathol. Plant Protect. 45 (7) (2012) 846–855.

[6] M. Munasinghe, J. Wansapala, Study on variation in seed morphology, oil content and fatty acid profile of Madhuca longifolia grown in different agro-climatic zones in Sri Lanka, Sci. Res. 3 (3) (2015) 105–109, https://doi.org/10.11648/j.sr.20150303.20.

[7] P. Shah Maulin, Removal of Refractory Pollutants from Wastewater Treatment Plants, CRC Press, 2021.

[8] T. Banerjee, A. Samanta, Improvement over traditional brewing techniques for production of bioethanol from mahua flowers (Madhuca indica), Int. J. Energy Sect. Manag. 7 (4) (2018) 12–21.

[9] G. Baskar, A. Gurugulladevi, T. Nishanthini, R. Aiswarya, K.J.R.E. Tamilarasan, Optimization and kinetics of biodiesel production from Mahua oil using manganese doped zinc oxide nanocatalyst, Renew. Energy 103 (2017) 641–646.

[10] S.S. Behera, R.C. Ray, Forest bioresources for bioethanol and biodiesel production with emphasis on mohua (*Madhuca latifolia* L.) flowers and seeds, in: Bioethanol Production from Food Crops, Academic Press, 2019, pp. 233–247.

[11] M. Sunita, P. Sarojini, Madhuca lonigfolia (Sapotaceae): a review of its traditional uses and nutritional properties, Int. J. Human. Soc. Sci. Invent. 2 (5) (2013) 30–36.

[12] S. Behera, S. Kar, R.C. Mohanty, R.C. Ray, Comparative study of bio-ethanol production from mahula (Madhuca latifolia L.) flowers by *Saccharomyces cerevisiae* cells immobilized in agar agar and Ca-alginate matrices, Appl. Energy 87 (1) (2010) 96–100.

[13] J. Sinha, V. Singh, J. Singh, A.K. Rai, Phytochemistry, ethnomedical uses and future prospects of Mahua (*Madhuca longifolia*) as a food: a review, J. Nutr. Food Sci. 7 (2017) 573.

[14] S.V. Ghadge, H. Raheman, Process optimization for biodiesel production from mahua (*Madhuca indica*) oil using response surface methodology, Bioresour. Technol. 97 (3) (2006) 379–384.

[15] Y.C. Awasthi, S.C. Bhatnagar, C.R. Mitra, Chemurgy of sapotaceous plants: Madhuca species of India, Econ. Bot. (1975) 380–389.

[16] S. Kumar, S. Kumar, A. Kumar, S. Maurya, V. Deswal, Experimental investigation of the influence of blending on engine emissions of the diesel engine fueled by mahua biodiesel oil, Energy Sour. Part A Recover. Util. Environ. Eff. 40 (8) (2018) 994–998.

[17] S. Behera, R.C. Mohanty, R.C. Ray, Ethanol fermentation of mahula (Madhuca latifolia L.) flowers using free and immobilized bacteria Zymomonas mobilis MTCC 92, Biol., Sect.Cell. Mol.Biol. 65 (3) (2010) 416–421.

[18] N. Acharya, P. Nanda, S. Panda, S. Acharya, A comparative study of stability characteristics of mahua and jatropha biodiesel and their blends, J. King Saud Univ. Eng. Sci. 31 (2) (2019) 184–190.

[19] S. Behera, R.C. Ray, M.R. Swain, R.C. Mohanty, A.K. Biswal, Traditional and current knowledge on the utilization of Mahula (*Madhuca latifolia* L.) flowers by Santhal tribe in Similipal biosphere reserve, Orissa, India, Ann. Trop. Res. 38 (2016) 94–104.

[20] U.S. Jyothi, S. Santhi, J. Jhansi, Performance enhancement approaches for Mahua biodiesel blend on diesel Engine, in: Sustainable Manufacturing and Design, Woodhead Publishing, 2021, pp. 201–222.

[21] G. Halder, S.H. Dhawane, D. Dutta, S. Dey, S. Banerjee, S. Mukherjee, M. Mondal, Computational simulation and statistical analysis of bioethanol production from *Madhuca indica* by batch fermentation process using *Saccharomyces cerevisiae*, Sustain. Energy Technol. Assess. 18 (2016) 16–33.

[22] T. Agrawal, S.K. Jadhav, A. Quraishi, A sustainable approach for bioethanol production from mahua flowers by *Saccharomyces cerevisiae* MTCC 4780 & *Zymomonas mobilis* MTCC 92, J. Ravishankar Univ. 29 (1) (2016).

[23] A. Purohit, S. Kaur, S.K. Yadav, Identification of a yeast Meyerozyma caribbica M72 from mahua flower for efficient transformation of rice straw into ethanol, Biomass Convers. Biorefin. (2021) 1–13.

[24] N.S. Dugala, G.S. Goindi, A. Sharma, Evaluation of physicochemical characteristics of Mahua (Madhuca indica) and Jatropha (Jatropha curcas) dual biodiesel blends with diesel, J. King Saud Univ.-Eng. Sci. 33 (6) (2021) 424–436.

[25] P. Mandal, N. Kathale, Production of ethanol from mahua flower (Madhuca Latifolia L.) using saccharomyces cerevisiae-3044 and study of parameters while fermentation, ABHINAV national monthly refereed, J. Res. Sci. Technol. 1 (9) (2009).

[26] R. Gedela, R.T. Naidu, S. Rachakonda, A. Naidu, Madhuca longifolia flowers for high yields of bio-ethanol feedstock production, Int. J. Appl. Sci. Biotechnol. 4 (4) (2016) 525–528.

[27] S. Behera, R.C. Mohanty, R.C. Ray, Comparative study of bio-ethanol production from mahula (Madhuca latifolia L.) flowers by *Saccharomyces cerevisiae* and Zymomonas mobilis, Appl. Energy 87 (2010) 2352–2355.

[28] A. Gupta, R. Chaudhary, S. Sharma, Potential applications of mahua (Madhuca indica) biomass, Waste and Biomass Valoriz. 3 (2) (2012) 175–189.

[29] R.N. Singh, A. Mishra, D. Asha, Optimization of operational parameters for production of Bio-ethanol using mahua flower and food processing waste water, Invertis J. Renew. Energy 7 (4) (2017) 218–223.

[30] T. Agrawal, A. Quraishi, S.K. Jadhav, Bioethanol production from Madhuca latifolia L. flowers by a newly isolated strain of Pichia kudriavzevii, Energy Environ. 30 (8) (2019) 1477–1490.

[31] S.K. Nayak, B.P. Pattanaik, Experimental investigation on performance and emission characteristics of a diesel engine fuelled with mahua biodiesel using additive, Energy Proc. 54 (2014) 569–579.

[32] D. Pradhan, R.K. Singh, H. Bendu, R. Mund, Pyrolysis of Mahua seed (*Madhuca indica*)—Production of biofuel and its characterization, Energy Convers. Manag. 108 (2016) 529–538.

[33] A. Mahalingam, Y. Devarajan, S. Radhakrishnan, S. Vellaiyan, B. Nagappan, Emissions analysis on mahua oil biodiesel and higher alcohol blends in diesel engine, Alex. Eng. J. 57 (4) (2018) 2627–2631.

[34] H.C. Joshi, M. Negi, Study the production and characterization of Neem and Mahua based biodiesel and its blends with diesel fuel: an optimum blended fuel for Asia, Energy Sour. Part A Recover. Util. Environ. Eff. 39 (17) (2017) 1894–1900.

[35] N. Acharya, P. Nanda, S. Panda, S. Acharya, Analysis of properties and estimation of optimum blending ratio of blended mahua biodiesel, Eng. Sci. Technol., Int. J. 20 (2) (2017) 511–517.

[36] M. Martín-Gamboa, D. Iribarren, D. García-Gusano, J. Dufour, A review of life-cycle approaches coupled with data envelopment analysis within multi-criteria decision analysis for sustainability assessment of energy systems, J. Clean. Prod. 150 (2017) 164–174.

[37] S. Ramasubramanian, S.S. Kumar, L. Karikalan, S. Baskar, Performances emissions behaviors of Compression ignition engine by mahua oil, Mater. Today Proc. 37 (2021) 982–985.

[38] S. Sayyed, R.K. Das, K. Kulkarni, Experimental investigation for evaluating the performance and emission characteristics of DICI engine fueled with dual biodiesel-diesel blends of Jatropha, Karanja, Mahua, and Neem, Energy 238 (2022) 121787.

[39] K.P. Rao, B.A. Rao, Parametric optimization for performance and emissions of an IDI engine with Mahua biodiesel, Egypt. J. Petrol. 26 (3) (2017) 733–743.

[40] R.T. Sarath Babu, M. Kannan, P. Lawrence, Performance analysis of low heat rejection diesel engine, using Mahua oil bio fuel, Int. J. Ambient Energy 38 (8) (2017) 844–848.

[41] N.R. Banapurmath, A.S. Chavan, S.B. Bansode, S. Patil, G. Naveen, S. Tonannavar, M.S. Tandale, Effect of combustion chamber shapes on the performance of Mahua and Neem biodiesel operated diesel engines, J. Petrol Environ. Biotechnol. 6 (4) (2015) 1.

[42] N. Kumar, H. Raheman, Combustion, performance, and emission characteristics of diesel engine with nanocera added water emulsified Mahua biodiesel blend, Environ. Prog. Sustain. Energy 40 (4) (2021) e13572.

[43] C.A. Harch, M.G. Rasul, N.M.S. Hassan, M.M.K. Bhuiya, Modelling of engine performance fuelled with second generation biodiesel, Procedia Eng. 90 (2014) 459–465.

[44] M. Senthil, K. Visagavel, C.G. Saravanan, K. Rajendran, Investigations of red mud as a catalyst in Mahua oil biodiesel production and its engine performance, Fuel Process. Technol. 149 (2016) 7–14.

[45] C.S. Aalam, C.G. Saravanan, Effects of nano metal oxide blended Mahua biodiesel on CRDI diesel engine, Ain Shams Eng. J. 8 (4) (2017) 689–696.

[46] P.K. Nutakki, S.K. Gugulothu, J. Ramachander, Effect of metal-based SiO2 nanoparticles blended concentration on performance, combustion and emission characteristics of CRDI diesel engine running on mahua methyl ester biodiesel, Silicon (2021) 1–15.

[47] A. SanthoshKumar, V. Thangarasu, R. Anand, Performance, combustion, and emission characteristics of DI diesel engine using mahua biodiesel, in: Advanced Biofuels, Woodhead Publishing, 2019, pp. 291–327.

[48] A. Jasrotia, A.K. Shukla, N. Kumar, Impact of nanoparticles on the performance and emissions of diesel engine using mahua biodiesel, in: Proceedings of International Conference in Mechanical and Energy Technology, Springer, Singapore, 2020, pp. 49–59.

[49] A. Saravanan, M. Murugan, M.S. Reddy, S. Parida, Performance and emission characteristics of variable compression ratio CI engine fueled with dual biodiesel blends of Rapeseed and Mahua, Fuel 263 (2020) 116751.

[50] P.K. Nutakki, S.K. Gugulothu, Influence of the effect of nanoparticle additives blended with mahua methyl ester on performance, combustion, and emission characteristics of CRDI diesel engine, Environ. Sci. Pollut. Control Ser. (2021) 1–12.

[51] M. Morales, J. Quintero, R. Conejeros, G. Aroca, Life cycle assessment of lignocellulosic bioethanol: environmental impacts and energy balance, Renew. Sustain. Energy Rev. 42 (2015) 1349–1361.

[52] B. Neupane, A. Halog, S. Dhungel, Attributional life cycle assessment of woodchips for bioethanol production, J. Clean. Prod. 19 (6–7) (2011) 733–741.

[53] A.V. Dambhare, P.S. Patil, R.H. Khetade, M.J. Umekar, A review on: phytochemical screening and pharmacological activity on Madhuca longifolia, J. Med. Plants 8 (2) (2020) 54–60.

[54] D.J. Pinakin, V. Kumar, S. Kumar, S. Kaur, R. Prasad, B.R. Sharma, Influence of pre-drying treatments on physico-chemical and phytochemical potential of dried mahua flowers, Plant Foods Hum. Nutr. 75 (4) (2020) 576–582.

[55] S. Yadav, P. Suneja, Z. Hussain, Z. Abraham, S.K. Mishra, Genetic variability and divergence studies in seed and oil parameters of mahua (*Madhuca longifolia* Koenig) JF Macribide accessions, Biomass Bioenerg. 35 (5) (2011) 1773–1778.

[56] A. Singh, D. Pant, N.E. Korres, A.S. Nizami, S. Prasad, J.D. Murphy, Key issues in life cycle assessment of ethanol production from lignocellulosic biomass: challenges and perspectives, Bioresour. Technol. 101 (13) (2010) 5003–5012.

[57] CDM (Clean Development Mechanism), Draft guidance on apportioning of project emissions to co-products and by-products in biofuel production, in: CDM Meth Panel, 31st Meeting Report-Annex 7, 2007.

[58] A. Singh, I.S. Singh, Chemical evaluation of mahua (Madhuca indica) seed, Food Chem. 40 (2) (1991) 221–228.

Further reading

[1] S.A. Edrisi, P.C. Abhilash, Exploring marginal and degraded lands for biomass and bioenergy production: an Indian scenario, Renew. Sustain. Energy Rev. 54 (2016) 1537–1551.

[2] S.V. Ghadge, H. Raheman, Biodiesel production from mahua (*Madhuca indica*) oil having high free fatty acids, Biomass Bioenergy 28 (6) (2005) 601–605.

[3] D. Kumar, V.K. Chhibber, A. Singh, Adding ZnO nanoparticle in mahua oil methyl ester (MoME) biodiesel for eco-friendly and better performance in DI engine, Natl. Acad. Sci. Lett. (2021) 1–4.

[4] P. Roy, A. Dutta, Life cycle assessment (LCA) of bioethanol produced from different food crops: economic and environmental impacts, in: Bioethanol Production from Food Crops, Academic Press, 2019, pp. 385–399.

[5] V. Singh, J. Singh, R. Kushwaha, M. Singh, S. Kumar, A.K. Rai, Assessment of antioxidant activity, minerals and chemical constituents of edible mahua (*Madhuca longifolia*) flower and fruit of using principal component analysis, Nutr. Food Sci. (2020). ISSN: 0034-6659.

[6] P. Yadav, D. Singh, A. Mallik, S. Nayak, Madhuca longifolia (Sapotaceae), a review of its traditional uses, phytochemistry and pharmacology, Int. J. Biomed. Res. 3 (7) (2012) 291–305.

Chapter 3

Generation of biofuels from rice straw and its future perspectives

Protha Biswas[1], Sujata Mandal[1], Tuyelee Das[1], Satarupa Dey[2], Mimosa Ghorai[1], Sayan Bhattacharya[3], Arabinda Ghosh[4], Potshangbam Nongdam[5], Vineet Kumar[6], Abdel Rahman Al-Tawaha[7], Ercan Bursal[8] and Abhijit Dey[1]

[1]*Department of Life Sciences, Presidency University, Kolkata, West Bengal, India;* [2]*Department of Botany, Shyampur Siddheswari Mahavidyalaya, Howrah, West Bengal, India;* [3]*School of Ecology and Environment Studies, Nalanda University, Bihar, India;* [4]*Department of Botany, Gauhati University, Guwahati, Assam, India;* [5]*Department of Biotechnology, Manipur University, Imphal, Manipur, India;* [6]*Department of Basic and Applied Sciences, School of Engineering and Sciences, GD Goenka University, Gurugram, Haryana, India;* [7]*Department of Biological Sciences, Al-Hussein Bin Talal University, Maan, Jordon;* [8]*Department of Biochemistry, Mus Alparslan University, Turkey*

1. Introduction

Rice straw is one of the world's most abundant waste materials made up of lignocellulose. There is a surplus of rice straw worldwide, making it a valuable feedstock for energy production. Among all cereal crops, rice is the third most important after corn and wheat based on the total production. In 2007, the FAO reported that world rice production was approximately 650 million tons and each kilogram of harvested grain yields 1–1.5 kg of straw [1,2]. The increased rice production has contributed to an increase in nonedible rice residues such as rice straw, rice husk, etc., with the bulk increase of rice [3]. Due to their underuse from the beginning of rice production, such residues negatively impact the economy and the environment. The rice residues contain cellulose, hemicellulose, and lignin, as well as starch, extractives, and inorganic matter [4]. These residues are used as fertilizer additives, as a source of heat and electricity, etc. and have no nutritional value making them unsuitable for feeding animals. In light of these characteristics, researchers worldwide are seeking sustainable ways to use lignocellulosic biomass as a source of bioenergy [5,6]. Rice straw is estimated to amount to between 650 million and 975 million tons annually globally, of which a substantial amount is used as cattle feed and the remainder as waste. The low bulk density, slow soil degradation, presence of rice stem diseases, and high mineral content of rice straw limit the types of residues that can be disposed of with rice straw [7]. Rice straw is largely burned in the field itself, which results in air pollution, as well as the release of particulate matter causing greenhouse effect. As a result of open field burning, India produces surplus rice straw residue and emits 0.05% of greenhouse gases [1]. Considering that climate change is widely recognized as a threat to world, there is increasing interest in the alternative use of agricultural residue for energy purposes.

As one of the renewable sources of biofuels, biomass contributes about 45 ± 10 EJ to global energy consumption [8]. Biofuels have been converted from biomass in numerous ways over the past few decades, taking their transportation, storage, and energy density into consideration [9]. Presently, biofuels are an attractive option for lowering greenhouse gas emissions and managing biodiversity loss, as well as facilitating the concept of sustainable development, including improved utilization of water and other resources, and increased production of food using biomass. Several factors affect the yield of biofuels, including the composition of the biomass, the optimal conditions for reaction, including temperatures, rates of reaction, heating, and so on [10]. As a consequence of globalization, fossil fuel reserves are depleting and threatening the environment, therefore it is more important to focus on renewable sources of energy like biomass exploitation. In the process of fermentation of wheat, corn, sugarbeet, crop residues, etc, biomass can produce biofuels such as bioethanol, biosyngas, biodiesel, biogas, biochar, and biohydrogen [11].

Rice straw has alluring characteristics due to its abundance throughout the world that make it an excellent material for the production of bioenergy. The crop is one of the greatest producers of silica-rich biomass [12]. Developing or designing cost-effective bioenergy conversion processes for rice straw is a need of the hour. The silica present in rice has a major role in facilitating mechanical strength, polymer properties, redesigning of the network of the cell wall, and other functions. Though there are very little information on its effects on the process of saccharification and fermentation. As a result, rice straw exploitation is comparatively difficult, and improving the economics of such conversion technologies is an urgent necessity [3]. A pretreatment method is required for rice straw due to its chemical composition, consisting of lignin which shields the carbohydrates, decreasing their availability. Furthermore, rice straw also contains highly polymerized cellulose and hemicellulose making it highly resistant to biodegradation [13].

This article provides information on rice straw's composition, adverse effects on the environment and human health, rice straw management schedules, biofuel production from rice straw, life cycle assessments, and SWOT analyses, which give a better understanding of how to utilize rice straw for biofuel production with limited effect on the human being and environment.

2. Rice crop residue

A variety of crop residues, such as rice straw, rice husk, etc., can be used as feedstocks for bioenergy production, sustainably, and can ensure short-term energy security without interfering with the production of human food [14,15]. Straw is a product that remains after harvest of rice in the field and its biomass is determined by a host of different factors such as rice varieties, soil management practices, nutrient management practices, and weather. There are many variables that determine the amount of rice straw gathered from the field, including the cutting height, which is the height of the stubble which remains after harvest [16]. In terms of bulk rice straw production, Asia holds the first place [17]. There are four main parts to rice straw: rachis, leaves, and leaves sheaths [18]. Traditionally, rice straw harvesting was done manually, and the raw material was used for industries such as pulp and papermaking, fertilizer manufacturing, and animal feed. However, with the advent of advanced technologies, the collection has become too expensive and time-consuming [1]. Developing an alternative approach to valorize rice straw with a view to combating the issue of burning rice straw and rendering farmers profit from collecting rice straw is an urgent necessity [19].

3. Composition and properties

More than half of the world's population depends on rice as a major source of energy and protein. It is the third most important cereal crop in the world after maize (corn) and wheat. With a production of 731 million tons annually, rice is one of the most important crops cultivated around the globe. It is grown in a number of Asian countries including Africa, Asia, Europe and even the United States of America (USA) [5]. Upon harvesting rice grains from the paddy, rice straw is the byproduct from panicle rachis, leaf blades, leaf sheaths, and stems.

It is important to consider the chemical composition of biomass, as it changes with the seasons and with variety. Variation affects consistency and yield of final products, so it is considered a disadvantage. Rice straw must therefore be characterized structurally [20]. The cell wall of rice straw is tough and recalcitrant by nature, which has evolved to accommodate a number of growing demands including protection from harsh environmental conditions, insects and pathogens. Hexoses (which include glucose, galactose, and mannose), hemicelluloses (such as xylose and arabinose), lignin (both acid-soluble and insoluble), ash, silica, and extractives are some of the main components in the biomass. The extracts (nonstructural components) of plants are mainly made up of proteins (about 30%) and pectins (about 28%) with smaller amounts of free sugars, chlorophyll, fats, oils, waxes, and chlorophyll. Among the major components of the cell wall, lignin is an amorphous polymer composed of three aromatic alcohols—p-coumaryl, coniferyl, and sinapyl alcohol [21]. The macromolecule cellulose is composed of monomeric D-glucose molecules linked through glycosidic bonds. Based on the type of organization of the structure, cellulose can be further subdivided into crystalline (more organized) and amorphous (less organized) forms. Hemicellulose is a carbohydrate that has an intricate structure consisting of numerous small polymeric units such as pentose, hexose, uronic acids, etc [22]. The hemicellulose molecules provide the structure with tensile strength as lignin and cellulose bind together. Monolignols consist of p-coumaryl, coniferyl, and sinapyl alcohols and are shielded by lignin [23]. There is no independent lignin in plant tissue, but instead binds to other polymers like cellulose and hemicelluloses via covalent bonds. In rice straw lignin and carbohydrate complexes, 63.9% of the content is carbohydrates, 2.8% is uronic acid, 0.8% is trans-ferulic acid, 27.7% klason lignin, 4.2% is acetyl content, and 4% is trans-p-coumaric acid [12]. There are p-coumaric acid and ferulic acids in rice straw in both esterified and etherified forms, and the ester linkage is lower while the ether linkage is higher for p-coumaric acid. Rice straw has a high ash content (10%—17%), high inorganic compound and a high ferulic acid content but has poor inorganic content and alkali content [24].

Rice straw generated after harvest is currently used or disposed of in a variety of ways, including ploughing into fields, composting, animal feed, serving as cattle house flooring, making straw crafts, and covering fields with it [25]. Currently, rice straw is extensively used for various bioenergy production processes such as the production of biodiesel, bioethanol, biobutanol, and biohydrogen [22]. By pyrolyzing rice straw liquid (bio-oil), solid (biochar), and a mixture of light gases (syngases) can be produced. Furans, phenols, phenolics, aromatics, alcohols, ketones, acids, as well as furanoglucosides are present in bio-oil derived from rice straw. Hemicellulose pyrolysis produces ethanol and pyranoglucose, while cellulose pyrolysis produces ketones. The use of rice straw as a second-generation biofuel has been also mentioned in several reports. Rice straw can be anaerobically digested to produce biogas. By substituting fossil fuels for rice straws, anaerobic digestion of rice straws provides a green technology that generates better waste utilization and reduces greenhouse gas emissions [26]. The effective use of rice straw for composite materials has been documented in several reports. In the presence of 5% microfibrilized rice straw, polypropylene composites with rice straw reinforcement showed improved properties [27].

4. Emissions formed from crop residues and associated health hazard

Because of the abundant energy resources from fossil fuels, traditional biofuels are being tapped less in China as its economy develops rapidly. As a result, there was less demand for crop residues for bioenergy production, which forced farmers to burn crop residues, causing environmental hazards such as air pollution. This pollution causes respiratory diseases and also releases a huge amount of greenhouse gas which cause depletion of ozone layer in the troposphere [28]. It is also responsible for the negative impact of regional climate and crop yield due to the accumulation of atmospheric pollutants, such as sulfur oxides, CO_2, CO, and hydrocarbons, caused by the burning of straw. Particles of very small sizes (particulate matters with size of particles ~ 2.5 and ~ 10 μm) are formed in the air when straw is burned, and these particles can remain in the air for a long time and cause difficulty in breathing [29]. Methane is the primary greenhouse gas produced by burning crop residues and it absorbs infrared radiation 60 times more efficiently than carbon dioxide, making it one of the most potent greenhouse gases. Carbonyl sulfide has a greater impact on the environment than methane and carbon dioxide due to its prolonged stay in the troposphere, where it ensures maximum saturation of sulfur compounds while hydroxyl radicals eventually react with it, creating tropospheric sulfate to counterbalance global warming [23]. In addition to removing soil nutrients about 80% of nitrogen, 25% of phosphorus, 21% of potassium, and destroying the majority of beneficial insects and microorganisms in the soil when rice straw is burned, burning also diminishes soil organic matter. As a result of burning rice straw, a leveled bed is created which makes next crops grow more easily and also removes phyto-mass, which can cause various diseases and pathogens and reduce yields [23]. Modern rice cultivars almost produce 6–7 tonnes per hectare of straw which is basically not necessary to deal with the issue of soil and water erosion. A better-quality fertilizer, the implementation of plausible recycling of plant-based nutrients, the reduction of greenhouse gas emissions, regulation of rice burning are all factors that improve crop productivity. These factors can be further enhanced by field management [23].

5. Management of rice straw

Due to the use of high-yielding, short-duration rice varieties, as well as the shorter turnaround time between crops, rice crops have widespread production and use. Due to short time for the decomposition of rice straw, and poor fertilization qualities of leftover rice straw, incorporating rice straw into the soil has numerous obstacles, especially in intensive agriculture systems with two to three rounds of cropping per year. As a result of such factors, the burning of rice straw in an open field has increased extensively in recent decades despite a ban, which was enforced in major rice-producing nations because of air pollution problems. It is the major challenge for researchers, government organizations, and farmers to solve the problem of in situ burning of rice straws in different states [30]. Therefore, it is essential to search for sustainable alternatives and advanced technologies that will decrease the environmental impact following value-added rice products by maximizing rice revenue generation.

5.1 Management of rice straw in the field

Previously, these rice straw materials were regarded as wastes that had to be disposed of, but it has become increasingly apparent that they are natural resources and not wastes. Following rice crops, straw may be used as mulch, removed from the field, piled or spread in the field, incorporated into the soil, or burned in situ. This combination of activities affects soil in a different way, whether it's nutrient balance or long-term soil fertility. Rice straw is commonly burned in open fields, but this practice has a negative impact on the environment and agriculture due to greenhouse gas emissions, air pollution, and degradation in soil quality. Based on the investigations reported, almost 98% fewer greenhouse gases are emitted when rice straw is burned than when the straw is incorporated into the soil because methane is reduced when it decomposes. However, the loss of carbon during combustion if included, results in a reduction in soil-carbon sequestration of newly incorporated rice straw due to a sudden reduction of carbon content (90%) during the combustion process [31].

5.1.1 Retention of surface as mulch

Mulching flooded rice usually is not feasible because the traditional practice of puddling incorporates residual crop residue on the surface of the soil. For mulching straw, the biomass needs to be removed from the field before soil preparation, and then returned after the land has been prepared. The amount of soil organic carbon (SOC) and total nitrogen increased with mulching, but yields were sometimes lower than when rice was grown under flooded conditions. The South Chinese government has developed a ground covering rice production system for conserving water savings and improved nitrogen efficiency. Numerous reports indicate that rice can be grown under nonflooded conditions and less water can be consumed by mulching the rice with crop residues. Although nonflooded mulched rice often yields comparable yields to conventionally flooded rice, its water-use efficiency is often higher (water use is 3.3 and 2.4 times less than flooded rice) [32,33].

5.1.2 Incorporation of residue in soil

Considering residue incorporation into flooded soil as a viable alternative to open field burning has continued to grow in popularity. The short interval between rice crops makes residue incorporation most challenging in intensive monocropping systems. If suitable machinery is not available, incorporating large quantities of fresh residue involves considerable labor. Rice can be harvested using a combine harvester that leaves straw on the field, and then the residue is ploughed or disked into the soil. Rice—rice cropping systems with two or 3 months between rice crops have been proposed as a way to speed up residue decomposition and increase plant nutrients by incorporating rice residue into aerobic soil as soon as possible after harvest rather than waiting until flooded soil is prepared before rice crop establishment. The earlier residues are incorporated into the soil, the better is the congruence between soil N supply and crop demand, although the amount and quality of residues incorporated greatly affect the size of this effect. In rice—rice systems without applying N or with moderate rates of applied N, early residue incorporation increased grain yields by 13%—20%, compared to late residue incorporation. Rice—rice rotations accumulated 11%—12% more C and 5%—12% more N than rice—maize rotations, and N-fertilized treatments sequestered greater amounts of C.

5.1.3 The burning of crop residue

In south Asia, rice straw is used as cattle feed and for other purposes after being taken from fields. Currently, farmers are burning in-situ large amounts of agricultural residues left in the field after mechanized harvesting. North-West India, Thailand, and China have mechanized harvesting, so all straw remain in the field and are burned rapidly on the spot. After harvest, straw is piled on threshing sites in Indonesia and the Philippines and burned to make planting easier the following season. This practice destroys soil pests and pathogens as well as causing 80% nitrogen loss, 25% phosphorous loss, and 21% phosphorus loss and 4%—60% potassium loss. Upon burning, the residue produces 13 tons of CO_2 ha^{-1}, resulting in significant air pollution and killing of beneficial soil organisms. As straw burns, it emits gaseous pollutants such as CO_2, CH_4, CO, N_2O, NOx, SO_2, Ozone, THC, TC, and BTX, which adversely affect health. During the straw burning season in Punjab and Haryana, suspended particulate matter (PM25, PM10) reaches 100—200 μg m^{-3} in the air. As a result, burning provides the land with a quick clean-up of residues before the next crop is planted, which enhances seed germination and establishment [34].

5.2 Utilization of crop residues off-field

5.2.1 Forage for livestock

In terms of all feeding stuffs, rice straws have the lowest protein content and are the rich in crude fiber, but also have low levels of P, Ca, and trace elements, but are high in silica. Almost no crude protein is digestible while 40% of the total nitrogen is digestible in rice straw. In China, North Vietnam, Eastern India, Bangladesh and Nepal, rice straw is collected from fields and used as fodder, thus there is a great demand for straw as fodder. Approximately 70% of rice straw's dry weight is carbohydrates, which makes it a potential source of energy for ruminants.

5.2.2 The composting of residues

Using rice straw in anaerobic composting with manure reduces 23% in-field methane emissions as well as nitrogen oxide (77%) and methane emissions from rice straw management. Comparing aerobic composting with anaerobic composting, aerobic composting can reduce methane emissions by about 90% [35]. Compostable crop residue can be removed from the field, mixed with other organic matter originating at the farm, like animal waste, and then returned to the soil as rice crop manure. Traditional passive composting in India involves simply accumulating crop residue in piles or pits, where they decompose gradually over a long period. The rural composting methods in China use turnings and aeration holes for passive aeration, and they produce compost within 2 to 3 months [36]. Composting requires labor input, but not capital investments or sophisticated infrastructures or machinery. Composter technology is most likely to benefit small farmers with little to no manual labor requirements. A hectare of rice land yields about 3.2 tons of nutrient-rich manure as FYM. Preparation costs for the enriched compost are USD 154; manure costs USD 48; and enriched straw manure yields 206 kg NPK at a cost of about USD 150 per ton [31].

5.2.3 The production of biofuel and biogas

It is possible to use rice residues as biofuel in a variety of ways, including direct use as a fuel source for earthen ovens, which is a traditional rural method of cooking rice. Ethanol is the most common biofuel produced from crop residues. An option for generating energy from bio-based sources is combustion burning straw alone or in conjunction with coal to produce a stream to turn a turbine to generate electricity. Households and small communities could use it to produce low-cost energy. Anaerobic digestion, or biogas, is a second bio-power method that decomposes straw and other wastes, like manure, in an aerobic environment [31,37]. In China, anaerobic digestion is becoming more and more popular, with 7.64 million households using it at the end of 2000 [38]. Biochar is also capable of reducing methane emissions associated with incorporating rice, and it has a better potential to reduce emissions than composting since it can enhance soil-carbon sequestration (50%) by converting rice straw into a better and more stable form of carbon [39,40]. The life cycle assessment (LCA) of rice produced from open fields and burning biochar indicates an approximately 50%–70% decrease in total carbon footprint and methane emissions [41].

5.2.4 Roof thatching

A rice farmer with little resources covers the bamboo roofs in her home with long rice straws (90–130 cm). Thatch generally lasts for 2 years, though it also depends on rainfall amount, the slope of the roof, cultivar, and silicon content of the straw. Thatch made from rice straws of slender grain variety has a long lifespan. Using straw as a roof thatch keeps the roof cool in summer and warm in winter, and is really economical for farmers, but it is relatively flammable if a fire breaks out [31].

5.2.5 Mushroom grown in rice straw

Rice straw is an excellent substrate for growing mushrooms, and there are many edible varieties such as *Agaricus volvacea*, *Volvariella volvacea*, *Vaginata virgata* and *Amanita virgata* are cultivated in South-East Asia on this material. As straw mushrooms are rich in amino acids and can supplement protein levels that are lacking in Asian diets, they have a very short cropping cycle (30 days) and yield a few 10% to 15% percent of the dry substrate. It is easy and inexpensive to grow rice straw mushrooms, and they require very little space and investment [31].

5.2.6 Transportation packaging

Paddy straw has also become a common packaging material for fruits and furniture in metropolitan areas of India, due to its low cost. The packaging materials should be dry; clean; free of insects, soil, feces, prohibited or restricted seeds, other plant matter, animal carcasses, or any other materials that could constitute a quarantine risk.

6. Biofuel production and its processes

As a result of climatic changes, global warming, and environmental degradation due to burning fossil fuels extensively, researchers and engineers are now investigating alternatives to fossil fuels, like biofuels, with an aim to combat global energy security. Biofuels derived from food crops, such as corn, sugarcane, rapeseed, etc., are considered as first-generation. These are primarily used for making bioethanol and biobutanol by fermenting starch/sugar, as well as for producing biodiesel by transesterifying oil crops.

Rice straw can be used for the production of biofuels and bioenergy in numerous ways. In order to make rice straw suitable for use on a biological platform, a pretreatment step is required to modify the interactions between cellulose, hemicelluloses, and lignins, improving cellulose accessibility, removing carbohydrate-lignin complexes, and decreasing crystallinity of cellulose. Hence, it increases biofuel yields from enzymatic hydrolysis [42]. There are many approaches for generating biofuels, including pyrolysis, gasification, and torrefaction or roasting [20]. Pyrolysis is the process of rapidly heating biomass or other feedstocks to a maximum temperature without oxygen or anaerobic condition, or by holding them for a short time period, to produce noncondensable gases (CO, Co_2, H_2, CH_4, etc.), solid char, and liquid product, respectively. During pyrolysis, variables, such as temperature and holding time, play a major role in determining the type of products obtained. The reaction formula is-

$$C_nH_mO_p \text{ (Biomass)} \longrightarrow \sum C_xH_yO_z(\text{liquid}) + \sum C_aH_bO_c(\text{gas})H_2O + C(\text{Char})\ldots\ldots\ldots$$

During pyrolysis, large molecules are broken down into small molecules of varying sizes. In pyrolysis, the products consist of liquid products such as tar, heavier hydrocarbons (HCNs), and water, solid products include char, and gases such as carbon dioxide, carbon monoxide, water, ethylene, benzene, ethane, and ethene [20].

The hydrothermal gasification process involves the gasification of an aqueous medium at extremely high temperatures (t) and pressures (P) exceeding or approaching the critical value. Utilizing hydrothermal gasification as a means to convert biomass into energy, there are three primary routes. The first pathway is in liquefaction, in which, liquid fuels are formed above critical pressure (22.1 MPa) but near critical temperature between 300 and 4000°C. In the second pathway, under low temperatures (350–5000°C) in the presence of a catalyst, methane (CH_4) is formed or gasified. In the third pathway hydrogen (H_2) is produced when high temperatures (over 6000°C) are present either with a catalyst or without one. The formula of the reaction is

$$mC_6H_{12}O_6 + nH_2O \longrightarrow wH2þxCH4þyCOþzCO2\ldots$$

Torrefaction or roasting is a process involving cofiring biomass in a power plant running on fossil fuels along with a small quantity of fuel. Compared to complete biomass conversion, complete biomass switching has various advantages, including greater cost-effectiveness, reduction of greenhouse gases, and higher efficiency. Other advantages include reduced technical risks, Large-scale implementation within a short timeframe, lower cost of CO_2 abatement when cofiring has been implemented, synergistic effect when corrosion is dealt with, clean conversion process with proper pollution control, etc. Cofiring biomass can be accomplished through three approaches: direct cofiring, indirect cofiring, and parallel cofiring. When coal and biomass are cofired directly in a boiler, the fuels can be fed directly into each other through mills and burners that are connected or segregated, making this more viable, cost-effective, and simple. The improved viability is attributed to the intrinsic properties of torrefied biomass, such as their more brittle and less fibrous nature and the absence of pulverization mills. The direct feeding of biomass along with coal thus improves the grinding process of biomass, thereby requiring less or no energy to grind it, and provides a biomass that is the same particle size and quantity required for the boiler's furnace heat input. Torrefaction reduces the heating value and the varied combustion properties of the biomass and eliminates the need for expensive fuel storage containers [34].

7. Life cycle assessment (LCA) and SWOT analysis for biofuels production from rice straw

Producing biofuel from rice straw involves various cultivation methods, crop management practices, harvesting procedures, and postharvesting techniques which are the part of life cycle [30]. Life cycle assessment (LCA) examines the impact on the environment and the energy balance from raw materials to the end product with waste disposal. Using this assessment, one can calculate or simulate energy balance and its impact on various aspects of the environment, such as climate change, ozone depletion, acidification of the land, and eutrophication of freshwater and marine environments. Additionally, crop production and numerous agricultural systems can also be evaluated using LCAs for various purposes including comparative analysis of best practices based on various production methods, activities, technologies, etc. Several precise economic and environmental factors influence LA, including product development, product marketing, strategic planning, and decision support. As part of the framework, specific objectives, scope, and functional units are defined, inventories are scrutinized, impact evaluations conducted, and statistics analyzed.

Rice straw is used to produce low-cost cellulosic ethanol and also to produce other valuable byproducts that can be used for commercial purposes. As part of the process, the biomass is fractionated into its constituent parts, such as cellulose, hemicellulose, and lignin. This process starts with size reduction to 200–500 mm, pretreatment to remove xylose, then enzymatic hydrolysis polysaccharides into monosaccharides, cofermentation, ethanol distillation and ends with water treatment. This process reduces global warming potential and by utilizing biofuel as renewable energy resources like solar or hydroelectric power. In addition, a large number of value-added products are generated, such as methanol, silica, food-grade carbon dioxide, inorganic compounds, etc [43].

A SWOT (strengths, weaknesses, opportunities, threats) analysis is a set of scientific techniques that can provide the opportunity to assess all possible factors or approaches that can be used to prioritize development strategies in a specific sector with the intention of identifying bottlenecks or favorable conditions. The transition from traditional to sustainable policymaking is characterized by high levels of uncertainty, among other factors, leading to a failure of traditional policies. Thus, understanding how structural and functional aspects of a particular system change at various phases, such as policies, industries, commercialization, expenditure, etc., is of paramount importance [43]. SWOT analyses seek to identify characteristics of a system both internally (internal factors strengths and weaknesses) and externally (external factors—opportunities) in order to support decision-making processes [44,45]. A SWOT analysis helps in strategizing how to better manage wastes for the production of bioethanol, thereby enhancing the effectiveness and boosting competitiveness [46]. The SWOT analysis has been completed in order to evaluate better strategies for rice straw mushroom value chains which will further help in exploring the possibilities of enhancing the subsector of value chains to bring awareness about the use of rice straw beyond the burning stage to value-added commercial products [28].

8. Advantages and disadvantages of producing biofuels from rice straw and its socioeconomic evaluation

The advantages and disadvantages of biofuels as an alternative to fossil fuels which can decrease the CO_2 emissions have been debated. Biofuels have better net positive energy or better efficacy, a higher loss of biodiversity, etc. There has been evidence and documents that the use of crop residues, for the production of biofuels, is less detrimental to the environment, ecosystem, the economy of the nation, and public health, as they produce renewable energy with no carbon footprint, which reduces fossil fuel dependence and cause fewer respiratory illnesses due to less air pollution. Research found that crop residues can be a potential source of biofuel or power generation, but their impact on food production is limited. Furthermore, if crop residues aren't used, they degrade and emit CO_2, which accelerates global warming and other environmental impacts. There are some negative impacts including the collection, transportation energy, and other related resources, but these are comparatively smaller than the impacts that arise from the production of biofuels from energy crops. The energy crisis is partially solved because there are sufficient resources to meet all demand. The energy crisis is not fully solved though there are sufficient resources to meet all demand. Biofuel production needs huge water consumption thus requires good management practices. For biofuels to be economically viable and technologically advanced, extensive financial incentives are needed. As the rice crop residue production is abundant it is a security for energy supply (sustainable renewable energy) replacing fossil fuel and it can improve rural economy by giving opportunity of more jobs [47,48].

Biofuel economy has significant effects on the economy and social aspects such as job creation, land use, water use, improved regional productivity, and more [49]. One of the most common misconceptions about biofuels is the debate between food and fuel. By utilizing improved farming techniques, it is possible to increase crop yields and generate food and fuel sustainably. It is expected that the policies connected to biofuels will boost the economy of biofuels, curbing fossil fuel dependence and promoting a socioeconomically combined approach [27].

9. Conclusion and future perspectives

The abundance of rice straw around the world makes it a highly competitive energy feedstock. Rice is the most residue producing crop and the use of its straw as feedstock in biotechnological production processes shows great potential since it is a lignocellulosic agricultural residue. Straw from rice fields is traditionally used as cattle feed and for other uses but recently, mechanized harvesting has led farmers to burn the rice straw in situ in large amount which is left in the field, leading to reduced nutrient levels and soil organic matter (SOM). In order to reduce the adverse health (respiratory diseases) and environmental (air pollution, green-house effect, harmful effect on soil etc.) impacts, researchers have developed innovative bioconversion technologies for producing biofuels and bioenergy and also rice straw management strategies. On-field residue management is beneficial to water conservation, soil health, and soil productivity, as well as the environment. Utilization of rice crop residues for off-field purposes includes feeding livestock, roof thatching, composting rural residues, growing mushrooms on paddy straw, and packaging for transportation. The global economy is highly dependent on fossil fuels, so alternative means of producing fuels and other byproduct chemicals are urgently needed. By developing technologies to produce different fuels and chemicals from straw, various products will be produced that are best suited to the local market conditions and circumstances. Rice straw's composition is the main challenge in using it for producing bioenergy; from it in bioenergy because it consists of lignin, cellulose, and hemicellulose, which makes biorefineries use appropriate pretreatment methods, adding to the cost and hampering the national economy. Thus, researchers must invent novel and feasible technologies to produce biofuel from rice straw which will be cost-effective, efficient, and will not produce any harmful pollutant and will be environmentally safe as well.

References

[1] P. Binod, R. Sindhu, R.R. Singhania, S. Vikram, L. Devi, S. Nagalakshmi, N. Kurien, R.K. Sukumaran, A. Pandey, Bioethanol production from rice straw: an overview, Bioresour. Technol. 101 (13) (2010) 4767–4774.

[2] B.L. Maiorella, Ethanol fermentation, in: M. Young (Ed.), Comprehensive Biotechnol, vol. III, Pergamon Press, Oxford, 1985, pp. 861–914.

[3] Y.F. Huang, S.L. Lo, Utilization of rice hull and straw, in: Rice, AACC International Press, 2019, pp. 627–661.

[4] R. Potumarthi, R.R. Baadhe, P. Nayak, A. Jetty, Simultaneous pretreatment and sacchariffication of rice husk by Phanerochete chrysosporium for improved production of reducing sugars, Bioresour. Technol. 128 (2013) 113–117.

[5] E.R. Abaide, M.V. Tres, G.L. Zabot, M.A. Mazutti, Reasons for processing of rice coproducts: reality and expectations, Biomass Bioenergy 120 (2019) 240–256.

[6] N. Kaur, G. Singh, M. Khatri, S.K. Arya, Review on neoteric biorefinery systems from detritus lignocellulosic biomass: a profitable approach, J. Clean. Prod. 256 (2020) 120607.

[7] S.I. Mussatto, I.C. Roberto, Optimal experimental condition for hemicellulosic hydrolyzate treatment with activated charcoal for xylitol production, Biotechnol. Prog. 20 (1) (2004) 134–139.

[8] International Energy Agency, (IEA), World Energy Outlook 2019, 2019.

[9] D. Vamvuka, Bio-oil, solid and gaseous biofuels from biomass pyrolysis processes—an overview, Int. J. Energy Res. 35 (10) (2011) 835–862.

[10] R. Sindhu, P. Binod, A. Pandey, S. Ankaram, Y. Duan, M.K. Awasthi, Biofuel production from biomass: toward sustainable development, in: Current Developments in Biotechnology and Bioengineering, Elsevier, 2019, pp. 79–92.

[11] M. Demont, T.T.T. Ngo, N.V. Hung, G.P. Duong, T.M. Dương, N.T. Hoang, M.C. Custodio, R. Quilloy, M. Gummert, Rice straw value chains and case study on straw mushroom in Vietnam's Mekong River Delta, in: Sustainable Rice Straw Management, Springer, Cham, 2020, pp. 175–192.

[12] F. Li, G. Xie, J. Huang, R. Zhang, Y. Li, M. Zhang, Y. Wang, A. Li, X. Li, T. Xia, C. Qu, Os CESA 9 conserved-site mutation leads to largely enhanced plant lodging resistance and biomass enzymatic saccharification by reducing cellulose DP and crystallinity in rice, Plant Biotechnol. J. 15 (9) (2017) 1093–1104.

[13] B. Gadde, S. Bonnet, C. Menke, S. Garivait, Air pollutant emissions from rice straw open field burning in India, Thailand and the Philippines, Environ. Pollut. 157 (5) (2009) 1554–1558.

[14] M. Ali, M. Saleem, Z. Khan, I.A. Watson, The use of crop residues for biofuel production, in: Biomass, Biopolymer-Based Materials, and Bioenergy, Woodhead Publishing, 2019, pp. 369–395.

[15] L. Yang, X.Y. Wang, L.P. Han, H. Spiertz, S.H. Liao, M.G. Wei, G.H. Xie, A quantitative assessment of crop residue feedstocks for biofuel in North and Northeast China, GCB Bioenergy 7 (1) (2015) 100–111.

[16] M. Gummert, N.V. Hung, P. Chivenge, B. Douthwaite, Sustainable Rice Straw Management, Springer Nature, 2020, p. 192.

[17] J.K. Saini, R. Saini, L. Tewari, Lignocellulosic agriculture wastes as biomass feedstocks for second-generation bioethanol production: concepts and recent developments, 3 Biotech 5 (4) (2015) 337–353.

[18] Y. Oladosu, M.Y. Rafii, N. Abdullah, U. Magaji, G. Hussin, A. Ramli, G. Miah, Fermentation Quality and Additives: A Case of Rice Straw Silage, BioMed Research International, 2016, 2016.

[19] A. Satlewal, R. Agrawal, S. Bhagia, P. Das, A.J. Ragauskas, Rice straw as a feedstock for biofuels: availability, recalcitrance, and chemical properties, Biofuels Bioproduc. Biorefin. 12 (1) (2018) 83–107.

[20] D. She, X.N. Nie, F. Xu, Z.C. Geng, H.T. Jia, G.L. Jones, M.S. Baird, Physico-chemical characterization of different alcohol-soluble lignins from rice straw, Cellul. Chem. Technol. 46 (3) (2012) 207.

[21] N.G. Lewis, E. Yamamoto, Lignin: occurrence, biogenesis and biodegradation, Annu. Rev. Plant Biol. 41 (1) (1990) 455–496.

[22] H.V. Scheller, P. Ulvskov, Hemicelluloses, Annu. Rev. Plant Biol. 61 (2010) 263–289.

[23] J.H. Grabber, How do lignin composition, structure, and cross-linking affect degradability? A review of cell wall model studies, Crop Sci. 45 (3) (2005) 820–831.

[24] M. Zevenhoven, The prediction of deposit formation in combustion and gasification of biomass fuels, Rapp. Tech. (2000) 1–2.

[25] Y. Matsumura, T. Minowa, H. Yamamoto, Amount, availability, and potential use of rice straw (agricultural residue) biomass as an energy resource in Japan, Biomass Bioenergy 29 (5) (2005) 347–354.
[26] J.D. Murphy, N.M. Power, An argument for using biomethane generated from grass as a biofuel in Ireland, Biomass Bioenergy 33 (3) (2009) 504–512.
[27] Y. Wu, D.G. Zhou, S.Q. Wang, Y. Zhang, Polypropylene composites reinforced with rice straw micro/nano fibrils isolated by high intensity ultrasonication, Bioresourcs 4 (4) (2009) 1487–1497.
[28] S. Yang, H. He, S. Lu, D. Chen, J. Zhu, Quantification of crop residue burning in the field and its influence on ambient air quality in Suqian, China, Atmos. Environ. 42 (9) (2008) 1961–1969.
[29] S.K. Lohan, H.S. Jat, A.K. Yadav, H.S. Sidhu, M.L. Jat, M. Choudhary, J.K. Peter, P.C. Sharma, Burning issues of paddy residue management in north-west states of India, Renew. Sustain. Energy Rev. 81 (2018) 693–706.
[30] N. Van Hung, M.C. Maguyon-Detras, M.V. Migo, R. Quilloy, C. Balingbing, P. Chivenge, M. Gummert, Rice Straw Overview: Availability, Properties, and Management Practices, vol. 1, Sustainable Rice Straw Management, 2020.
[31] S.B. Goswami, R. Mondal, S.K. Mandi, Crop residue management options in rice–rice system: a review, Arch. Agron. Soil Sci. 66 (9) (2020) 1218–1234.
[32] X. Fan, J. Zhang, P. Wu, Water and nitrogen use efficiency of lowland rice in ground covering rice production system in south China, J. Plant Nutr. 25 (9) (2002) 1855–1862.
[33] J. Qin, F. Hu, B. Zhang, Z. Wei, H. Li, Role of straw mulching in non-continuously flooded rice cultivation, Agric. Water Manag. 83 (3) (2006) 252–260.
[34] B.S. Sidhu, V. Beri, Rice residue management: farmer's perspective, Indian J. Air Pollut. Control 8 (1) (2008) 61–67.
[35] S.O. Petersen, M. Blanchard, D. Chadwick, A. Del Prado, N. Edouard, J. Mosquera, S.G. Sommer, Manure management for greenhouse gas mitigation, Animal 7 (s2) (2013) 266–282.
[36] M.A. Guang-Ting, Ecological organic fertilizers and sustainable development of agriculture, Chin. J. Eco. Agric. 12 (3) (2004) 191–193.
[37] P. Shah Maulin, Microbial Bioremediation & Biodegradation, Springer, 2020.
[38] X. Zeng, Y. Ma, L. Ma, Utilization of straw in biomass energy in China, Renew. Sustain. Energy Rev. 11 (5) (2007) 976–987.
[39] J. Lehmann, J. Gaunt, M. Rondon, Bio-char sequestration in terrestrial ecosystems—a review, Mitig. Adapt. Strateg. Glob. Change 11 (2) (2006) 403–427.
[40] Y.F. Yin, X.H. He, G.A.O. Ren, Y.S. Yang, Effects of rice straw and its biochar addition on soil labile carbon and soil organic carbon, J. Integr. Agric. 13 (3) (2014) 491–498.
[41] A. Mohammadi, A. Cowie, T.L.A. Mai, R.A. de la Rosa, P. Kristiansen, M. Brandao, S. Joseph, Biochar use for climate-change mitigation in rice cropping systems, J. Clean. Prod. 116 (2016) 61–70.
[42] P. Shah Maulin, Removal of Refractory Pollutants from Wastewater Treatment Plants, CRC Press, 2021.
[43] A. Sreekumar, Y. Shastri, P. Wadekar, M. Patil, A. Lali, Life cycle assessment of ethanol production in a rice-straw-based biorefinery in India, Clean Technol. Environ. Policy 22 (2) (2020) 409–422.
[44] T. Baycheva-Merger, B. Wolfslehner, Evaluating the implementation of the Pan-European Criteria and indicators for sustainable forest management—A SWOT analysis, Ecol. Indicat. 60 (2016) 1192–1199.
[45] M. Kurttila, M. Pesonen, J. Kangas, M. Kajanus, Utilizing the analytic hierarchy process (AHP) in SWOT analysis—a hybrid method and its application to a forest-certification case, For. Pol. Econ. 1 (1) (2000) 41–52.
[46] P. Rauch, U.J. Wolfsmayr, S.A. Borz, M. Triplat, N. Krajnc, M. Kolck, R. Oberwimmer, C. Ketikidis, A. Vasiljevic, M. Stauder, C. Mühlberg, SWOT analysis and strategy development for forest fuel supply chains in South East Europe, For. Pol. Econ. 61 (2015) 87–94.
[47] E.C. Petrou, C.P. Pappis, Biofuels: a survey on pros and cons, Energy Fuels 23 (2) (2009) 1055–1066.
[48] G.A. Reinhardt, E. Von Falkenstein, Environmental assessment of biofuels for transport and the aspects of land use competition, Biomass Bioenergy 35 (6) (2011) 2315–2322.
[49] C. Hunsberger, L. German, A. Goetz, "Unbundling" the biofuel promise: querying the ability of liquid biofuels to deliver on socio-economic policy expectations, Energy Pol. 108 (2017) 791–805.

Further reading

[1] J.B. Guinée, M. Gorrée, R. Heijungs, G. Huppes, R. Kleijn, A. De Koning, L. Van Oers, A. Wegener Sleeswijk, S. Suh, H. Udo de Haes, G. de Bruijn, Handbook on LCA, Operational Guide to the ISO Standards, 2002.
[2] L.K. Singh, G. Chaudhary (Eds.), Advances in Biofeedstocks and Biofuels, Biofeedstocks and Their Processing, John Wiley & Sons, 2016.
[3] A. Song, P. Li, F. Fan, Z. Li, Y. Liang, The effect of silicon on photosynthesis and expression of its relevant genes in rice (Oryza sativa L.) under high-zinc stress, PLoS One 9 (11) (2014) e113782.
[4] R.C. Sun, J. Tomkinson, P.L. Ma, S.F. Liang, Comparative study of hemicelluloses from rice straw by alkali and hydrogen peroxide treatments, Carbohydr. Polym. 42 (2) (2000) 111–122.

Chapter 4

Prospects of rice straw for sustainable production of biofuels: Current trends and challenges

Priyanka Kamilya, Omar Aweiss Ali, Subrat Kumar, Sandeep Kumar Panda and Ritesh Pattnaik

School of Biotechnology, Kalinga Institute of Industrial Technology (Deemed-to-be University), Bhubaneswar, Odisha, India

1. Introduction

Rice straw is produced in excess all over the world. It has an annual production of 731 million tons, of which Asia contributes 667.6 million tons. Because it is a residual by-product of rice production, the annual world production of rice straw is around 650−975 million tons. Rice straw cannot be recycled in the soil owing to regular cropping practices and the availability of a limited sowing time for crops; therefore, straw can be an alternative biomass resource for effective valorization. Biomass is considered to be the most widely used source of renewable bioenergy because of its lower life-cycle greenhouse gas emissions compared with traditional fossil fuels [1]. Biomass has the potential to produce biofuels such as bioethanol, biogas, biodiesel, biosyngas, and biohydrogen on through the fermentation of various crop residues including wheat and corn [2]. For the production of bioenergy, rice straw has all of the characteristic properties owing to its surplus availability worldwide. For every kilogram of rice harvested, there is approximately 1−1.5 kg of rice straw. Every year, 50% of rice straw is burned in the field; the rest is used as fodder or left to decompose in landfills [3]. The practice of burning rice straw increases air pollution and affects public health by releasing harmful gases such as carbon monoxide, nitrogen and sulfur oxide, polycyclic aromatic hydrocarbons, and volatile organic compounds [4]. With an increase in the global population, there will be a rapid demand for biofuels in the future: approximately one-half of the total energy required for a developing country.

2. Rice production in Asian countries

Rice is a significant crop grown across the globe and an important staple food for half of the world's population. It accounts for over 20% of world's calorie intake. Globally, after corn and sugarcane, rice is the third most produced agricultural product [5]. According to the Food and Agriculture Organization, the world's rice production was 514.07 million metric tons in 2020−21. Over 90% of the world's rice was produced and consumed in Asian countries such as China, India, Indonesia, Bangladesh, Vietnam, and Japan. The annual growth rate for rice consumption in Asian regions has increased tremendously over more than half a century (1960−2022). China is one of the largest producers of rice, with approximately 148.3 million metric tons, followed by India, which produces 122.27 million metric tons. Among all developed and developing countries, India and China grow about 54% of the total rice demand worldwide. In addition to Asian countries, Brazil is the biggest producer of rice, cultivating about 12 million tons of rice in nearly 2 million hectares of land [6]. With increases in rice production, there has been an enormous increase in the residues of rice while harvesting. When the dry weight ratio of straw to grain is considered, it is 1.5 [7]. Therefore, 1120 million metric tons of rice straw is produced globally every year. After the harvest of rice, bulk residues (rice husks and rice straw) account for 1270 million metric tons. Because of this enormous quantity, it can be abundantly used as a renewable resource to produce biofuel, contributing to green evolution. However, the use of these crop residues is not easy because of its characteristics. Several pretreatment processes are essential for effective valorization.

3. Composition of rice by-products: rice straw and rice husks

Crop residues such as rice straw and rice husks are waste products that serve as a potential biomass feedstock for sustainable bioenergy production. There are two types of agricultural crop residue: field and process. Field residue consists of all of the plant parts, such as leaves, stalks, and stems of paddy, which are leftover materials after the harvest of any crop. Process residues are usable resources produced after the processing of crops, including husks, seeds, roots, bagasse, and molasses, which can be used as animal fodder or fertilizer to improving the soil quality [8]. The most common crop residues for biofuel production are rice straw, rice husks, corn, wheat straw, and bagasse (Table 4.1) because of their extensive availability [9]. The cells have two major components (1) cell walls consists of cellulose, hemicellulose, lignin (including monolignols such as p-coumaryl, coniferyl, and sinapyl alcohols) [10], silica (SiO_2), and cutin; whereas (2) cell contents include soluble carbohydrates (pentose and hexose) [11], fats, proteins, organic acids (uronic acid), and soluble ash. Rice straw consists of three components; cellulose (32%–47%), hemicellulose (19%–27%), and lignin (5%–24%) [12,13].

Rice straw and rice hull are lignocellulosic biomass that are in much demand in biofuel production depending on the composition. These two are not different in composition, but they differ in sulfur, nitrogen, and ash content. In terms of dry weight, the ash content of rice straw is less than that of rice hull by about 3%, which indicates that the combustible proportion of rice straw is higher than that of rice hull [15]. Similarly, the nitrogen and sulfur content of rice straw is more than that of rice hull. Thus, during combustion, a high amount of nitrogen oxide and sulfur oxide (NO_x and SO_x) is released. As a result, rice straw produces more NH_3, hydrocarbons (HCNs), and H_2S during pyrolysis. The heating value of rice straw is slightly less than that of rice hull because of the high amount of carbon and hydrogen and the lower nitrogen, oxygen, and sulfur content of rice hull.

4. Chemical components of different types of lignocellulosic biomass

Rice straw usually contains a high hemicellulose and cellulose content (i.e., approximately 9.3% more than rice husk), whereas rice hull constitutes 5.4% more lignin than rice straw. The weight losses of hemicellulose and cellulose pyrolysis occurs at 220–315 and 315–400°C, respectively, whereas that of lignin happens at a wider range of temperature, around 160–900°C [16]. This shows that the thermal reactivity of rice straw is much higher than that of rice hull, for which it is mainly preferred in thermal processes. In agricultural residue, wood, and herbaceous crops, the content of cellulose, hemicellulose, and lignin is 15%–30% [17]. Rice straw is a heterogeneous complex composed of cellulose and hemicellulose densely packed by layers of lignin. Cellulose is a linear chain of carbohydrate polymer prepared from repeating molecules of β-D-glucopyranose covalently linked through a glycosidic bond between the equatorial OH-group of C4 and C1 carbon atoms [18]. Hemicellulose is a mixture of various monosaccharides such as glucose, mannose, galactose, xylose, arabinose, galacturonic acid, and glucuronic acid residue polymerized to form a lower–molecular weight compound; whereas lignin is an amorphous polymer forming a mosaic structure consisting of phenylpropane and its precursor aromatic alcohols, such as coniferyl, sinapyl, and p-coumaryl alcohol [19]. Lignin serves as a rigid support to protect against microorganisms. It provides an internal transport system for water and nutrients.

TABLE 4.1 Different components of rice straw.

S. No.	Components of rice straw	Quantity (in wt%)
01	Cellulose	43
02	Hemicellulose	25
03	Lignin	12
04	Ash and silica	34–44
05	Glucose	19–22
06	Arabinose	2–4
07	Mannose	1.8–2
08	Galactose	0.4
09	Nonstructural compounds	16–20

Adopted from N. Vivek, L.M. Nair, B. Mohan, S.C. Nair, R. Sindhu, A. Pandey, N. Shurpali, P. Binod, Bio-butanol production from rice straw e recent trends, possibilities, and challenges, Bioresour. Technol. Rep. 7 (2019).

5. Potential of rice residue in biorefineries

Rice straw is the most abundant lignocellulosic residue used to produce bioethanol. Because of its surplus availability, its demand in biofuel production is increasing daily. In Asia, the major field residue is produced in a huge amount, approximately 667.6 million metric tons. This amount of rice straw can produce 205 billion liters of bioethanol approximately every year [20]. Through this, greenhouse gas emissions can be further reduced, decreasing air pollution. For rice straw, the alkali content is higher than that of rice husk, and is composed of Na_2O, MgO, CaO, and K_2O as the major constituents. Silica is present at approximately 74.67% in rice straw. The technology of converting these rice residues into bioethanol can be developed into two platforms. One is known as the sugar platform and other one is the synthesis gas (syngas) platform. For the sugar platform, only complex carbohydrates (cellulose and hemicellulose) are converted into fermentable sugars such as glucose, galactose, mannose, xylose, and arabinose. It is then fermented again to form ethanol with the help of enzymes or acids.

Various types of pretreatment processes use rice straw to produce bioenergy and biofuels (Fig. 4.1). Pretreatment processes are mandatory to increase the yield from biomass by enzymatic hydrolysis, resulting in greater production of biofuel [21]. During hydrolysis, several compounds are formed by the degradation of sugar and lignin, which termed hydrolysates. These hydrolysates are accompanied by some toxic compounds that need detoxification to increase the efficiency of fermentation of hemicellulose hydrolysates Several pretreatments such as biological, thermochemical, chemical, and physical processes are employed for the effective valorization of rice straw and associated waste (Table 4.2).

5.1 Physical pretreatment

Pretreatment process reduce the crystal nature of cellulose and the size of the biocomponents. Some commonly used physical pretreatment processes to degrade lignocellulosic residues are grinding and milling, microwave steaming, temperature, and pressure. Grinding and milling are the first steps of physical treatment of rice straw processing, which reduce the particle size to 0.2 mm [22,23]. Superfine grinding of steam-exploded biomass has been proved to be better than ground residue when hydrolyzed [24], although the energy required for the process has to be considered for large-scale processing. For grinding rice straw, wet disk milling proved to be better than ball milling in terms of glucose recovery and energy savings [25]. Advances in this field need to be implemented, such as enzymatic saccharification.

Microwave irradiation is also widely used as an efficient pretreatment approach owing to high heating efficiency and easy operation. This process potentially alters the structure of cellulose [26] to degrade lignin and hemicelluloses in lignocellulosic materials, including increasing the enzymatic susceptibility of lignocellulosic biomaterial [27]. When rice straw is treated by microwave irradiation, it exhibits the same hydrolysis rate and reducing sugar yield as those of raw straw. The cellulosic fraction of the lignocellulosic materials can be further degraded by this specific method of treatment to produce fragile fibers of low molecular weight (oligosaccharides and cellobiose) [28].

5.2 Chemical pretreatment

The most promising chemicals that can be used for chemical pretreatment of rice straw are alkali dilute acids and ammonia. Without chemical treatment, only enzymes are incapable of converting lignocellulose into fermentable sugars.

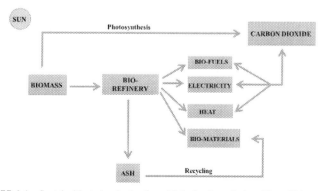

FIGURE 4.1 Sustainable technologies from biofuels via agricultural-based biomaterials.

TABLE 4.2 Different pretreatment techniques used for biomass conversion.

Type	Pretreatment approaches	Characteristic features
Physical	Mechanical extraction	Composed of crushing or grinding
		Produces solid fuel
	Briquetting	Also known as pelletizing
		Produces solid fuel
Chemical	Hydrolysis	Hydrolysis of carbohydrates (hemicellulose and cellulose); produces ethanol
	Solvent extraction	Primary and secondary metabolism
Biological	Anaerobic digestion	Produces methane and biogas
	Fermentation	Produces ethanol and butanol
Thermochemical	Torrefaction	200–300°C; produces bio-oil, biochar
	Hydrothermal liquefaction	200–374°C; produces bio-oil
	Pyrolysis	Fast pyrolysis (450–700°C)
		Slow pyrolysis (350–700°C)
		Produces biochar, bio-oil, and gas.
	Gasification	Atmospheric pressure and >700°C
		Produces syngas and fuel gas

5.2.1 Acid pretreatment

Acid pretreatment at ambient temperature is essential for the proper digestion of lignocellulosic compound to form a bioproduct. Usually, for this process, mineral acids are used such as HCl and H_2SO_4. Acid pretreatment solubilizes hemicellulose by making the complex carbohydrate more accessible to enzymes. Similarly, dilute acids of concentration at 0.25%–1% affects mainly hemicellulose with little impact on lignin degradation. These dilute acids work at an ambient temperature of 120–1800°C for 5–10 min. Usually, glycosidic bonds are broken down into monomeric forms using hydronium ions that are present in the acids, giving rise to open pores for further enzymatic reactions. This acid pretreatment process is followed by dilute acid pretreatment, which is a single-step process of hydrolysis of treated biomass by enzyme cellulase.

5.2.2 Alkali pretreatment

Alkali pretreatment is an efficient method of delignification by breaking the ester bonds between cellulose, hemicellulose, and lignin and avoiding the fragmentation of hemicellulose [29]. Alkali pretreatment causes less polymerization and crystallinity, resulting in a sharp increase in saccharification yield, owing to which less inhibitors are produced. This process involves alkaline solutions such as NaOH and KOH to break down lignin networks. Alkali pretreatment is performed at higher concentrations of 0.25%–1% for a relatively long time at higher temperatures of 120–1400°C. The main effect of sodium hydroxide on rice straw is that it breaks ester bonds cross-linking lignin and xylan through delignification, increasing the porosity of biomass [30]. When chopped rice straw was subjected to alkali pretreatment with 2% NaOH at 85°C for 1 h, it decreased lignin by approximately 36%. Although it has many advantages, several drawbacks are associated with this process; for example, alkali catalysts are an expensive and time-consuming process.

5.2.3 Pretreatment with oxidizing agents

Few oxidizing compounds such as hydrogen peroxide or peracetic acid are used to treat the lignocellulose biomass of rice straw. After pretreatment with such oxidizing agents, lignin and hemicellulose are removed, which makes cellulose more accessible to other enzymatic actions. During this oxidative pretreatment, several other processes occur, including the displacement of side chains, electrophilic substitution reactions, and oxidative cleavage of aromatic nuclei to decrease the crystalline nature of lignin [31]. Reports describe employing peracetic acid for the pretreatment of rice straw [32] (Toyama and Ogawa, 1975). The composition of treated rice straw with peracetic acid resulted in a slight loss of hemicellulose and cellulose, indicating that peracetic acid treatment caused little or no breakdown of the crystalline configuration of cellulose in rice straw.

5.2.4 Ammonia pretreatment

Ammonia is a highly selective, nonpolluting, and noncorrosive chemical that acts as an effective reagent for lignocellulosic biomass. It has high volatility; owing to this, its recovery is easy for further use. Because of its high selectivity of reaction with lignin, it causes cleavage of C—O—C bonds in lignin as well as with ether and ester bonds in the lignin—carbohydrate complex (Kim and Lee, 2009). A flow-through process called ammonia recycle percolation (ARP) was developed for pretreatment, in which ammonia is pumped through a bed of biomass maintained at 170°C. With this process, up to 85% delignification occurs [33]. Compared with other alkalis such as sodium hydroxide and ammonia, ammonia is highly selected for lignin removal because it has a significant swelling effect on lignocellulose. Another method of pretreatment (ammonia fiber expansion [AFEX]) uses anhydrous ammonia instead of aqueous ammonia. Similar to the ARP process, ammonia used in the AFEX process can be recovered and recycled owing to its high volatility. AFEX is reported to be an effective pretreatment process for rice straw that resulted in 3% sugar loss during pretreatment.

5.3 Biological pretreatment

The pretreatment approach using biological resources is an important, safe, and environmentally friendly process by which lignin can be removed from lignocellulosic biomass. It is the most advantageous treatment because it involves less energy and chemicals. This treatment uses microorganisms such as *Phanerochaete chrysosporium*, *Ceriporiopsis subvermispora*, *Phlebia subserialis*, *Pleurotus ostreatus*, *Streptomyces griseorubens JSD-1*, *Serpula lacrymans*, *Coniophora puteana*, *Meruliporia incrassate*, *Pleurotus eryngii*, *Irpex lacteus*, *Aspergillus niger*, and *Aspergillus awamori*. The most promising microorganism that can be used for biological pretreatment is white rot fungi, which belongs to the class Basidiomycetes [34]. In this approach of delignification, microorganisms act on mixtures of hydrolysates and toxic compounds to increase porosity by structurally loosening cells. Moreover, when a mixture of these five fungi (*A. niger*, *Aspergillus awamori*, *Trichoderma reesei*, *Phenerochaete chrysosporium*, and *Pleurotus sajor-caju*) was used for pretreatment and *Saccharomyces cerevisiae* (NCIM 3095) for fermentation, *A. niger* and *A. awamori* yielded a high amount of ethanol [35].

5.4 Thermochemical pretreatment

Thermochemical conversion techniques are preferred because the practical and sustainable energy solution prevents rising global warming and air pollution caused by insufficient energy resources and deteriorated atmospheric greenhouse effects. These factors have encouraged research and advances in various pathways of thermochemical processes for biomass conversion to bioproducts. Fundamentally, thermochemical conversion technology involves four pretreatment processes: pyrolysis, torrefaction, gasification, and hydrothermal liquefaction.

5.4.1 Pyrolysis

In pyrolysis, bulk and large molecules are broken down into various types of small molecules. The product of pyrolysis is categorized into three types: (1) liquid, which consists of tars and heavier HCNs and water; (2) solids, mostly composed of char; and (3) gases such as carbon dioxide, carbon monoxide, water, ethylene, benzene, ethane, and ethene [36]. Normally, the pyrolysis of biomass occurs at 350—700°C. It requires a pre-drying stage to carry out combustion properly. Pyrolysis can be of two types: slow and fast. In slow pyrolysis, alkaline earth metals are used as catalysts for biochar formation and gas yield, increasing the production of lighter hydrocarbons and reducing liquid bioproduct formation. In case of fast pyrolysis, the time for whole process to take place is less than 1 s, and it occurs at a high temperature, approximately 450—700°C. During pyrolysis, alkaline earth metals serve as a vital catalyst for bio-oil generation with more phenols and aromatic compounds.

5.4.2 Torrefaction

As a thermal process, torrefaction occurs under inert environmental conditions essential for improving the energy quality of solid biomass. It normally occurs at 200—300°C and takes 15—60 min. By this pretreatment process (torrefaction), lignocellulosic compounds are converted into solid biomass such as biochar and bio-oil. Several studies were conducted on the catalytic thermochemical conversion of biomasses, and positive results were reported. For example, for torrefaction, using alkali and alkaline earth metals as catalysts can increase the yield and quality of biochar. These alkaline earth metals help produce high heating value and high energy density bioproducts with improved thermal degradation properties.

5.4.3 Hydrothermal liquefaction

This pretreatment process is particularly used to convert biomass into a liquid form of bioproduct such as bio-oil. Unlike torrefaction, hydrothermal liquefaction works at a wider range of temperature (i.e., 250–374°C) and within a pressure range of 4–22 MPa to produce bio-oil or crude oil. It does not require pre-drying like pyrolysis. Through liquefaction, a high quality of bio-oil can be obtained for biofuel production. Here, homogeneous catalysts are used to produce an upgraded quality of bio-oil. These catalysts decrease char and tar formation during combustion, in which K_2CO_3 is a highly effective catalyst and NaOH is the least effective catalyst.

5.4.4 Gasification

Gasification is the process of converting biomass into synthesis gas, also known as syngas. It produces carbon monoxide (CO) and hydrogen (H_2) at higher temperatures, greater than 700°C. Catalyst-driven gasification implies the significantly high yield of gas with a valuable composition. Normally, alkaline earth metals increase the reaction rate in gasification. Catalysts help reduce tar formation during combustion, converting it into gas fuel more actively. Nickel is the most effective catalyst for tar reduction during the course of gasification.

6. Conversion of rice straw into biofuels

To decrease the exploitation of traditional sources of energy and meet global energy demands, the development of advanced technologies is essential to generate energy from renewable sources. Thus, renewable energy has grabbed the attention of numerous researchers [37]. Climatic changes such as global warming are risks to the environment caused by the burning or extensive exploitation of fossil fuels. In return, this has caused an energy crisis in the future. This problem has provoked researchers to look into developing an alternative source of energy such as biofuels to tackle global energy security [38]. A diverse category of feedstock may be exploited to produce biofuels from lignocellulosic biomass such as rice straw, wheat straw, rice hulls, and agricultural and forest residue.

Three generations of biofuels are classified based on the feedstock source used to produce them [39]: (1) first-generation of biofuels are derived from food crops such as corn, sugarcane, and rapeseed, which are primarily used to produce bioethanol and biobutanol through the fermentation of starch and sugars; (2) second-generation of biofuels are derived from nonfood crops such as lignocellulosic biomass; and (3) third-generation biofuels are derived from algal biomass such as bioethanol from microalgae and seaweed, biodiesel from microalgal species, and biohydrogen from algal microorganisms [40], owing to presence of a high amount of lipids and their better ability to fix carbon dioxide. Various other categories of feedstock are commonly exploited to produce biofuels, including rice straw, wheat straw, rice hulls [41], and agricultural and forest residue [42]. The most commonly used feedstocks are algae [43], *Jatropha* [10], palm, and sugarcane [44]. Among these, the postharvest residues of rice (i.e., rice straw) are more advantageous owing to their renewable nature and bulk presence because of the consecutive cropping seasons of rice [45]. Detailed applications of rice straw–based products are highlighted in Fig. 4.2.

6.1 Bioethanol

Bioethanol is a liquid biofuel produced from different biomass feedstocks. It is also known as ethyl alcohol (CH_3–CH_2–OH or EtOH) or grain alcohol. It is an oxygenated fuel that contains approximately 35% oxygen, which reduces particulate and NOx emissions from combustion [20,46]. Cellulose and hemicellulose, which typically comprise two-thirds of dry lignocellulosic biomass, are polysaccharides that can be hydrolyzed to sugars and eventually fermented to form bioethanol [47]. They have a higher-octane number, broader flammability limits, higher flame speeds, and higher heats of vaporization than gasoline. There are four stages in the production of lignocellulosic-based bioethanol: pretreatment, hydrolysis, fermentation, and product recovery (Fig. 4.3).

In addition, a number of approaches are used to improve ethanol production efficiency, including simultaneous saccharification and fermentation with an optimized enzyme cocktail and the xylose-fermenting fungus *Mucor circinelloides* [48]. Sarabana et al. [49] described a consolidated bioprocess in which rice straw was first pretreated with alkaline hypochlorite to facilitate cellulase production by *T. reesei* and then fermented with a culture of *S. cerevisiae* and *Aspergillus oryzae*.

FIGURE 4.2 Potential applications derived from rice straw.

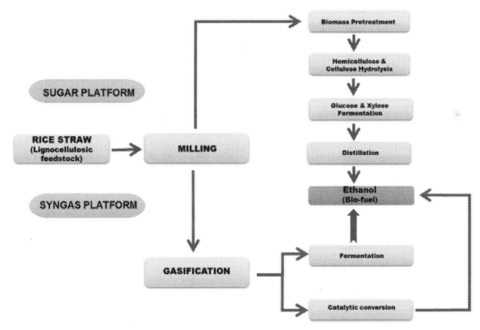

FIGURE 4.3 Basic technology for producing bioethanol from rice straw.

6.2 Biogas

Biogas is produced by the anaerobic digestion of organic materials that are decomposed by an assortment of microbes under oxygen-free conditions. Ideally, biogas can be generated by the anaerobic microbial degradation of a combination of treated rice straw and animal waste products. In the initial stage, straw treatment is a critical step in biogas production, and care is taken to optimize its pretreatment because of its economic value. At the beginning of the process, hydrolysis of extracellular enzymes occurs, produced by hydrolytic microbes, decomposing complex organic polymers to simple soluble monomers. Proteins, lipids, and carbohydrates are hydrolyzed to form amino acids and peptides, long-chain fatty acids, and sugars, respectively. These small molecules are then converted by fermentative bacteria known as acidogens into a mixture of volatile short-chain fatty acids and other products such as CO_2, H_2, and acetic acid. These components then serve as the direct substrates for methanogenesis, which is the last step of the anaerobic digestion process for CH_4 production.

6.3 Bio-oil

Bio-oil are a dark brown, free-flowing liquid produced by pyrolysis in which the biomass feedstock is heated in the absence of air [50]. Biomass serves as a unique resource for the sustainable production of bioderived chemicals and fuels to replace fossil fuel products. Although lignin is a major component of lignocellulosic materials, its complex cross-linking polymeric network makes it recalcitrant to current chemical technologies. Side-by-side catalysts are being developed that are beneficial for its depolymerization [51]. Naturally, bio-oils are a multicomponent mixture composed of cellulose, hemicellulose, and lignin. Fast pyrolysis for bio-oil production is attracting major interest because the oil can be stored, transported, and used for energy or chemicals, or as an energy carrier.

6.4 Other value-based products

Lignocellulosic biomass is used in various applications such as building materials, insulation boards, human food and animal feed, cosmetics, medicines, biopolymers, and fine chemicals [52]. Ideally, rice straw and rice hulls are considered source precursors for the production of value-added products. As biorefineries refers to conversion of biomass feedstocks into a multiple number of valuable chemicals, it also emphasize on generation of energy with minimal waste and emissions [53]. Also, biorefineries should be designed so that there is coordination between the production of both biofuels and high-value biobased products [54].

7. Challenges and future perspectives

A number of technological issues limit the feasibility of lignocellulosic biorefineries. Most current research focuses on evaluating critical factors such as the temperature, catalyst ratio, synthetic catalyst, gasification agent, and residence time, which have a vital role in the sustainable production of bioenergy and biofuels from agro-based residues such as rice straw. In addition, a number of potentially toxic elements and compounds produced during pretreatment, specifically heavy metals, metalloids, and polycyclic aromatic hydrocarbons, pose difficulties in the postprocessing of biomass [55]. The presence of a high oxygen, water, and acid content in biomass products poses a challenge to different pretreatment strategies. Also, crucial research areas such as the1 impact on the economy during the production of biofuels, 2improved efficacy, and3 less or no production of harmful gases or pollutants, or the proper processing of pollutants, need to be addressed for the sustainable generation of biofuels [56].

8. Conclusion

The global availability of rice straw in bulk amounts has encouraged researchers to make extensive efforts toward developing novel bioconversion technologies for producing biofuels and bioenergy, keeping in mind the adverse impact on the environment (air pollution caused by burning of rice residue) and human health (respiratory illness caused by increases in alarming levels of air pollution). The major challenge to utilize the rice straw for production of bioenergy, includes the adoption and application of technologies with generation of zero-waste. Considering the evolution of second-generation biofuels and their necessity, rice straw appears to be a promising and potential biomaterial for producing bioethanol owing to its abundance and attractive composition. The use of lignocellulosic biomass for bioethanol production makes the technology cost-effective and environmentally sustainable. The bioconversion of rice straw into fermentable sugars employing hydrolyzing enzymes is considered the most attractive alternative because of environmental concerns. Although there are several problems in developing an economically feasible technology because of its complex nature, high lignin, and ash content, several works are in progress to address the challenges and limitations of the technological approach.

Acknowledgments

The authors are thankful for the assistance and support extended by the head, School of Biotechnology, Kalinga Institute of Industrial Technology, Deemed-to-be-University, Odisha, for extending the infrastructure facilities.

References

[1] D. Tilman, R. Socolow, A. Foley, J. Hill, E. Larson, L. Lynd, S. Pacala, R. Williams, Beneficial biofuels and the food energy, and environment trilemma, Science 325 (5938) (2009) 270–271.

[2] L.K. Singh, G. Chaudhary, Advances in biofeedstocks and biofuels, in: Biofeedstocks and Their Processing, Ume 1, Wiley Scrivener Publishing, 2017.

[3] Q. Wang, Z. Wang, M. Kumar Awasthi, Y. Jiang, R. Li, X. Ren, J. Zhao, F. Shen, M. Wang, Z. Zhang, Evaluation of medical stone amendment for the reduction of nitrogen loss and bioavailability of heavy metals during pig manure composting, Bioresour. Technol. 220 (2016).

[4] S.I. Mussatto, I.C. Roberto, Optimal experimental condition for hemicellulosic hydrolyzate treatment with activated charcoal for xylitol production, Biotechnol. Prog. 20 (2004).

[5] P.-A. Per, in: P. Pinstrup-Andersen (Ed.), Food Price Policy in an Era of Market Instability: A Political Economy Analysis', Oxford University Press, 2015.

[6] Conab, Follow-Up of Brazil Crops 2015/2016-Grains-Eleventh Survey, 2016.

[7] R. Lal, World crop residues production and implications of its use as a biofuel, Environ. Int. 31 (2005).

[8] P. Shah Maulin, Microbial Bioremediation & Biodegradation, Springer, 2020.

[9] D. Sarkar, A. Bandyopadhyay, Adsorptive mass transport of dye on rice husk ash, J. Water Resour. Protect. 2 (2010).

[10] J.H. Grabber, How do lignin composition, structure, and cross-linking affect degradability? A review of cell wall model studies, Crop Sci. 45 (2005).

[11] H.V. Scheller, P. Ulvskov, Hemicelluloses, Annu. Rev. Plant Biol. 61 (2010).

[12] G. Garrote, H. Dominguez, J.C. Parajo, Autohydrolysis of corncob: study of non-isothermal operation for xylooligosaccharide production, J. Food Eng. 52 (2002).

[13] B.I. Maiorella, Ethanol industrial chemicals, Biochem. Fuels (1983).

[14] N. Vivek, L.M. Nair, B. Mohan, S.C. Nair, R. Sindhu, A. Pandey, N. Shurpali, P. Binod, Bio-butanol production from rice straw e recent trends, possibilities, and challenges, Bioresour. Technol. Rep. 7 (2019).

[15] Y.P. Zou, T.K. Yang, Rice Husk, Rice Husk Ash and Their Applications, Rice Bran and Rice Bran Oil. Elsevier, Amsterdam, 2019.

[16] H. Yang, R. Yan, H. Chen, D. Ho Lee, C. Zheng, Characteristics of hemicellulose, cellulose and lignin pyrolysis, Fuel 86 (12–13) (2007) 1781–1788.

[17] A.J. Ragauskas, C.K. Williams, B.H. Davison, G. Britovsek, J. Cairney, C.A. Eckert, W.J. Frederick, JR., J.P. hallett, D.J. Leak, C.L. Liotta, J.R. Mielenz, R. Murphy, R. Templer, T. Tschaplinski, The Path Forward for Biofuels and Biomaterials, 2006.

[18] D. Klemm Prof Dr, B. Heublein Dr, H.-P. Fink Dr habil, A. Bohn Dr, Cellulose: Fascinating Bio-Polymer and Sustainable Raw Materials, 2005.

[19] A.U. Buranov, G. Mazza, Lignin in straw of herbaceous crops, Ind. Crop. Prod. 28 (3) (2008) 237–259.

[20] M. Balat, H. Balat, C. Oz, Progress in bioethanol processing, Prog. Energy Combust. Sci. 34 (5) (2008) 551–573.

[21] T.L. Bezerra, A.J. Ragauskas, A review of sugarcane bagasse for second generation bio-ethanol and bio-power production, Bio-fuels, Bio-prod., Bio-refin. 10 (2016).

[22] N. Akhtar, K. Gupta, D. Goyal, A. Goyal, Recent advances in pretreatment technologies for efficient hydrolysis of lignocellulosic biomass, Environ. Prog. Sustain. Energy 35 (2016).

[23] G.L. Cao, S. Ya Chun, Z. Liang, S. Jin Zhu, C. Hua, Z. Jun Zheng, Biobutanol production from lignocellulosic biomass: prospective and challenges, J. Biorem. Biodegrad. 7 (4) (2016) 363.

[24] S. Jin, H. Chen, Superfine grinding of steam-exploded rice straw and its enzymatic hydrolysis, Biochem. Eng. J. 30 (2006).

[25] A. Hideno, H. Inoue, K. Tsukahara, S. Fujimoto, T. Minowa, S. Inoue, T. Endo, S. Sawayama, Wet disk milling pretreatment without sulfuric acid for enzymatic hydrolysis of rice straw, Bioresour. Technol. 100 (2009).

[26] C. Xiong, X. He, Z. Zhang, Microwave assisted extraction or saponification combined with microwave assisted decomposition applied in pre-treatment of soil or mussel sompus for the determination of polychlorinated biphenryus, Anal. Chem. Acta 413 (2002) 49–56.

[27] J. Azuma, F. Tanaka, T. Koshijima, Enhancement of enzymic susceptibility of lignocellulosic wastes by microwave irradiation, Japan 62 (4) (1984).

[28] M. Kumakura, I. Kaetsu, Effect of radiation pretreatment of bagasse on enzymatic and acid hydrolysis, Biomass 3 (3) (1983) 199–208.

[29] M. Gaspar, G. Kalman, K. Reczey, Corn fibre as a raw material for hemicellulose and ethanol production, Process Biochem. 42 (2007).

[30] H. Tarkov, W.C. Feist, A mechanism for improving the digestibility of lignocellulosic materials with dilute alkali and liquid ammonia, Adv. Chem. 95 (1969).

[31] Hon, Shiraishi, Wood and Cellulosic Chemistry, second ed., 2001 (Revised, and Expanded).

[32] M. Taniguchi, M. Tanaka, R. Matsuno, T. Kamikubo, Evaluation of chemical pretreatment for enzymatic solubilization of rice straw, Eur. J. Appl. Microbiol. Biotechnol. 14 (1982).

[33] C.M. Drapcho, N.P. Nhuan, T.H. Walker, Biofuels Engineering Process Technology, Mc Graw Hill Companies, Inc, 2008.

[34] M. Taniguchi, H. Suzuki, D. Watanabe, K. Sakai, K. Hoshino, T. Tanaka, Evaluation of pretreatment with Pleurotus ostreatus for enzymatic hydrolysis of rice Straw, J. Biosci. Bioeng. 100 (2005).

[35] S.J. Patel, R. Onkarappa, K.S. Shobha, Study of ethanol production from fungal pretreated wheat and rice straw, Internet J. Microbiol. 4 (2007).

[36] P. Basu, Chapter 5: Pyrolysis, in: B. G, P. B.T. (Eds.), Basu Pyrolysis and Torrefaction, second ed., 2013.

[37] B. Ghobadian, H. Rahimi, A.M. Nikbakht, G. Najafi, T.F. Yusaf, Diesel engine performance and exhaust emission analysis using waste cooking biodiesel fuel with an artificial neural network, Renew. Energy 34 (2009).

[38] M.H. Hassan, M.A. Kalam, An overview of bio-fuel as a renewable energy source: development and challenges, Procedia Eng. 56 (2013).

[39] R. Sindhu, P. Binod, A. Pandey, S. Ankaram, Y. Duan, M.K. Awasthi, in: S. Kumar, R. Kumar, A.B.T.-C.D. Pandey (Eds.), Chapter 5 - Biofuel Production From Biomass: Toward Sustainable Development, Elsevier, 2019.
[40] A. Sharma, S.K. Arya, Hydrogen from algal biomass: a review of production process, Biotechnol. Rep. 1 (2017).
[41] C.W. Murphy, A. Kendall, Life cycle analysis of biochemical cellulosic ethanol under multiple scenarios, GCB Bioenergy 7 (2015).
[42] Y. Hadar, Sources for lignocellulosic raw materials for the production of ethanol, Lignocell. Convers.: Enzymatic Microbial Tools Bio-ethanol Prod. (2013) 21−38.
[43] USEPA (United States Environmental Protection Agency), Inventory of U.S. Greenhouse Gas Emissions and Sinks 1990e2009, USEPA, Washington, DC, USA, 2011.
[44] M.S.U. Rehman, M.A. Umer, N. Rashid, I. Kim, J.I. Han, Sono-assisted sulfuric acid process for economical recovery of fermentable sugars and mesoporous pure silica from rice straw, Ind. Crop. Prod. 49 (2013).
[45] P.K. Das, B.P. Das, P. Dash, Chapter 13: Potentials of post-harvest rice crop residues as a source of bio-fuel, in: Refining Biomass Residues for Sustainable Energy and Bioproducts, 2020.
[46] P. Shah Maulin, Removal of Refractory Pollutants from Wastewater Treatment Plants, CRC Press, 2021.
[47] C.N. Hamelinck, G. van Hooijdonk, P.C.F. André, Ethanol from lignocellulosic biomass: techno-economic performance in short-, middle- and long-term, Biomass Bioenergy 28 (Issue 4) (2005) 384−410.
[48] M. Takano, K. Hoshino, Bioethanol production from rice straw by simultaneous saccharification and fermentation with statistical optimized cellulase cocktail and fermenting fungus, Bioresour. Bioprocess. 5 (2018) 16.
[49] S.S.H. Sarabana, A.I. Ramadan, A.M. Eldin, Consolidated bioprocess for ethanol production from rice straw and corn cobs by using fungi, Middle East J. Appl. Sci. 8 (2018).
[50] G.W. Huber, S. Iborra, A. Corma, Synthesis of transportation fuels from biomass, Chem., Catalys., Eng.; Chem. Rev. 106 (9) (2006) 4044−4098.
[51] A. Pineda, A. Lee, Heterogeneously catalyzed lignin depolymerization, Appl. Petrochem. 6 (2016) 243−256.
[52] N. Reddy, Y. Yang, Bio-fibers from agricultural byproducts for industrial applications, Trends Biotechnol. 23 (1) (2005) 22−27.
[53] A. Demirbas, Biofuels sources, biofuel policy, biofuel economy and global biofuel projections, Energy Convers. Manag. 49 (2009).
[54] A.J. Ragauskas, G.T. Beckham, M.J. Biddy, R. Chandra, F. Chen, M.F. Davis, B.H. Davison, R.A. Dixon, P. Gilna, M. Keller, P. Langan, A.K. Naskar, J.N. Saddler, T.J. Tschaplinski, G.A. Tuskan, C.E. Wyman, Lignin valorization: improving lignin processing in the biorefinery, Science 344 (2014).
[55] X. Tan, Y. Liu, G. Zeng, X. Wang, X. Hu, Y. Gu, Z. Yang, Application of biochar for the removal of pollutants from aqueous solutions, Chemosphere 125 (2015) 70−85.
[56] M. Ali, R. Sultana, S. Tahir, I.A. Watson, M. Saleem, Prospects of microalgal biodiesel production in Pakistan a review, Renew. Sustain. Energy Rev. 80 (2017).

Chapter 5

Biofuel production from algal biomass: Technologies and opportunities addressing the global energy crisis

Shibam Dey, Ishanee Mazumder, Chandrashish Roy, Omar Aweiss Ali, Subrat Kumar and Ritesh Pattnaik

School of Biotechnology, Kalinga Institute of Industrial Technology (Deemed-to-be University), Bhubaneswar, Odisha, India

1. Introduction

Fossil fuels are the primary worldwide source of energy in various sectors ranging from industries, transportation, businesses, households, and research. According to a report by Forbes, 84% of the world's energy consumption is a result of the combustion of fossil fuels [1]. By 2030, international agencies estimate that there will be a 53% rise in energy consumption. However, fossil fuels are nonrenewable resources of energy, and their extensive use owing to the surge in population leads to several problems such as climate change, a spike in crude oil prices, resource depletion, and issues with energy security [2,3]. The exhaustive use of fossil fuels has also posed a huge threat to the environment by causing global warming, leading to the destruction of flora and fauna ecosystems and high emissions of pollutants and greenhouse gases. These drawbacks have led researchers to find an alternative to fossil fuels. Biofuel is a viable alternative derived directly or indirectly from renewable biological materials called biomass. Biomass includes crops, agricultural and municipal solid waste, by-products from forestry, and microorganisms. Biofuels obtained from biomass include biodiesel, bioethanol, biogas, biohydrogen, and bioelectricity. As shown in Fig. 5.1, biomass-based biofuels can be broadly divided into three categories: first, second, and third generations.

First-generation biofuels are derived using starch fermentation of edible crops such as wheat, barley, maize, sugarcane, corn, potato, and beetroot, or derived chemically by using oil crops such as rapeseed, palm, soybean, coconut, and vegetable and animal oil [4]. They can be blended with petroleum-based fuels and show great efficiency in combustion engines. However, first-generation biofuels as an alternative to fossil fuels have not been successful owing to the use of food-based crops as feedstock, which affects global food security. Furthermore, to create sufficient amounts of biomass for both food and biofuel production, huge agricultural regions and yields are required. Thus, consequent harvesting may result in ecosystem degradation, water shortages, and air pollution [5].

Second-generation biofuels are primarily produced from inedible lignocellulosic feedstock, which includes food crop residue, municipal solid waste, and manure. Some examples of lignocellulosic biomass involved in biofuel production are *Sterculia foetida*, *Jatropha*, cassava, switchgrass, and *Miscanthus*. Second-generation biofuels have greater advantages over first-generation biofuels owing to their higher yield. Moreover, there is no competition with food production because the biomass that is used is inedible and thus reduces the land requirement. However, second-generation biofuels have major drawbacks because of inefficient technologies for commercial-scale biofuel conversion and difficulty during pretreatment [6].

To overcome the drawbacks of both first- and second-generation biofuels, third-generation biofuels began to be studied. Third-generation biofuel is derived from macroalgal and microalgal biomass, which is considered a viable alternative to fossil fuels. It can thrive under a variety of conditions with minimum requirements and can be harvested throughout the year. Macroalgae typically includes red and brown algae such as *Ulva*, whereas microalgae are mainly blue-green algae such as *dinoflagellates* and *Bacillariophyta*. Algae-based biofuels have the potential to substitute for petroleum-based fuels without affecting the food supply.

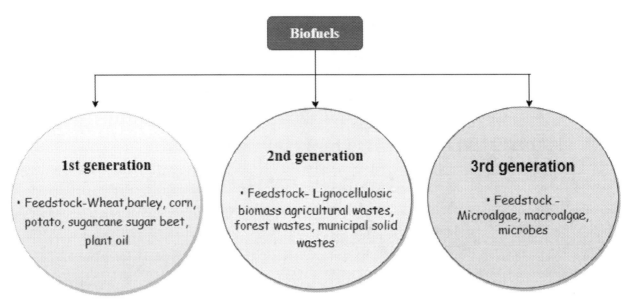

FIGURE 5.1 The different generations of biofuels.

2. Algae: a potential source of biofuels

Algae are a type of polyphyletic organism that can create biomass from light, water, and CO_2 [2,7]. Algae-based biofuel production has a number of advantages, including a rapid growth rate, highly efficient CO_2 reduction, lower water requirements than agricultural crops, efficient land use, and more economical farming [8]. Compared with other typical feedstock crops such as sugarcane bagasse and corn, algae potentially generate more energy per acre of land while emitting less NO_x than fossil fuels. All of these factors make algae-based biofuel a better energy resource. Algal biofuel cultivation does not require fertile soil because it can be cultivated anywhere with adequate sunlight and water. As a result, it may efficiently use unproductive or abandoned lands that were previously employed for intensive farming. It can also indirectly reduce biodiversity loss for native species, may be grown in wastewater, conserving freshwater resources, and does not require pesticides or fertilizers for production [9]. Algae is categorized as microalgae and macroalgae. All groups have varied levels of ash (18–55%), carbohydrates (25–60%), proteins (5–47%), and lipids (5%), which vary by species and are significantly affected by biotic and abiotic environment growth factors, including temperature and light.

2.1 Microalgae as a source of biofuel

Microalgae are unicellular photosynthetic microorganisms. Microalgae-derived biofuels have dominated research because of their relatively high lipid content and rapid growth rate. They have the potential to replace fossil fuels with a high oil yield of 136,900 L ha^{-1} per year. According to Correa and Beyer [9], microalgal cultivation methods are less likely to cause eutrophication. Despite all of these benefits, there are some drawbacks of microalgae-based biofuel production. Because of the limited light penetration in microalgal cultures and the tiny size of algal cells, harvesting algal biomasses is relatively expensive. Drying extracted algal biomass is also an energy-intensive operation owing to its high-water content. The Bacillariophyceae (diatoms) family is the most diverse and widespread. It is more suited to large-scale manufacturing because of the quick doubling time and ease of growth. Because of their high photosynthetic efficiency (PE) and capacity to synthesize lipids, microalgae are researched as a biodiesel feedstock.

2.2 Macroalgae as a source of biofuel

Macroalgae are a diverse and nonphylogenetic multicellular aquatic eukaryote. They can be differentiated into three major categories based on their photosynthetic pigmentation patterns: red (*Rhodophyta*), brown (*Phaeophyta*), and green (*Chlorophyta*) algae [10]. Several brown algae forms have been exploited in industry and energy production as an alternative to fossil fuels, whereas green algae is being researched for biodiesel generation. Algae can be grown in nearly any type of water, even wastewater [11]. Because macroalgae lacks lignin, it is easier to digest by microorganisms in the biorefinery process and to transform into a biofuel than land-based plants. Unlike microalgae, macroalgae is a multicellular plant with plant-like features, which makes harvesting easier [12]. It is composed mostly of carbohydrates, making it a strong option for biofuel generation such as biogas, bioethanol, and bio-oils. Macroalgae do not have a lot of lipids in them; thus, they are mostly valued for the natural sugars and other carbohydrates they contain. These materials can be fermented to create biogas or alcohol-based fuels.

3. Microalgal-based biofuel: technological perspectives

The cultivation techniques of algae are based on three major growth conditions: photoautotrophic, heterotrophic, and mixotrophic. Light is the sole source of energy in photoautotrophic growth, which is transformed to chemical energy through photosynthetic processes. This is the most frequent method of microalgae cultivation. The finest feature of the model is its ability to develop or generate fatty acids using carbon dioxide as a carbon source [13]. Compared with other growth methods, the phototrophic approach poses the least chance of contamination. To emulate growth, heterotrophic synthesis requires organic carbon sources as a carbon and energy source. Because light is not required for heterotrophic development, it has an advantage over phototrophic growth. When the density of the culture increases, the major issue with phototrophic is light penetration [14]. Therefore, heterotrophic development solves one of the most pressing issues. Compared with phototrophic growth, heterotrophic growth is more cost-effective. This strategy is also the most practical and promising way to boost productivity. When algae grow heterotrophically, they produce more oil and have better efficiency, but the danger of contamination is considerably higher than when they grow phototrophically.

Depending on the quantity of organic molecules and available light intensity, organisms in mixotrophic culture can flourish autotrophically or heterotrophically. Mixotrophic growth uses both organic and inorganic carbon, and photosynthesis necessitates the use of light. As a result, microalgae may survive in both environments. Organic molecules and carbon dioxide are used as carbon sources by microalgae, and the emitted carbon dioxide is also absorbed by photosynthesis. According to Park et al. [15], a mixotrophic culture increased biomass and lipid productivities. Moreover, the mixotrophic growth of several microalgae strains resulted in three to 10 times greater biomass output than phototrophic growth conditions.

4. Technologies for microalgal cultivation

Depending on the design parameters, microalgae cultivation techniques are classified as open or closed systems. Microalgae are grown in open systems such as ponds, lagoons, deep channels, and shallow circulation units. In closed systems, microalgae are grown in transparent containers and exposed to sunlight or artificial radiation to stimulate photosynthesis [16].

4.1 Open ponds

Microalgae cultivation is usually done in open ponds. The ponds are kept shallow to allow enough sunlight to reach the algae medium for effective photosynthesis. These ponds have paddle wheels that run 24 h/day to avoid settling and provide excellent gas circulation and liquid mixing. They are exposed to the environment, allowing liquid evaporation and process temperature regulation. Open-air systems are favorable because they are less expensive than photobioreactor (PBR) systems in terms of construction and operation, and they have a low energy requirement for mixing. However, these systems face a number of challenges, including the need for larger areas; the risk for contamination from external sources such as animals, other microalgae, bacteria, or algal grazers [17,18]; and growth control, which affects operational conditions such as temperature and water evaporation [19].

4.2 Photobioreactors

A PBR is a closed system that uses a recycling reactor to cultivate microalgae. PBRs are divided into three categories by Mussatto et al. [18]: tubular, flat, and bubble column. Open systems are less controllable than PBRs. PBRs offer better volumetric production than open-air ponds, according to Jorquera et al. (2010), because they are more effective at gathering solar energy and need less space. Furthermore, the PBR system allows for regulated culture conditions and protects microalgae from contamination. However, because of higher levels of energy consumption and reduced sunlight penetration owing to algal attachment, they are more expensive in terms of capital and operating expenses.

4.3 Two-stage hybrid system

The two-stage hybrid cultivation system incorporates the various growth stages of both open-pond and PBR systems, allowing it to benefit from the advantages of both systems. Beal et al. [20] described hybrid systems in which big PBRs continually deliver microalgae to open ponds. Furthermore, Zhou et al. [21] established an efficient heterophotoautotrophic two-stage microalgae cultivation system for low-cost wastewater treatment and the generation of sustainable biofuels from microalgal biomass.

4.4 Lipid induction technique

According to Chiaramonti et al. [22], unique growing approaches such as nitrogen and phosphorus restriction may boost oil output and effectiveness, depending on the microalgal strain, to manufacture diesel-like biofuels based on microalgal lipids. *Chlorella vulgaris* and *Nannochloropsis* sp. are two microalgae strains reported to do this. According to Miao et al. [23], *Chlorella protothecoides* produced under heterotrophic growth conditions promotes lipid buildup and yields 3.4 times more than autotrophic cultivation owing to the addition of glucose and the decrease in nitrogen.

5. Technologies for macroalgal cultivation

Macroalgae may be grown in a variety of ways, both onshore and offshore. Kelp growth, raft culture, and floating cultivation are examples of offshore cultivation. Nutrients available from saltwater are used to cultivate macroalgae in lagoons. Other macroalgae production methods include fixed off-bottom, long lines, and rock-based farming. Another form of cultivation is transplantation, in which species seedlings are allowed to develop indoors, and then cultivated in tanks and eventually placed into the sea using ropes.

Seawater has been extensively employed for culture in onshore cultivation techniques. It offers the advantages of a prohibitive degree of control over safety and a high product output. Onshore production is more adaptable to a larger range of macroalgae and more sustainable than offshore cultivation because marine animals are not harmed. Furthermore, with this method of culture, mixing is a possible component that promotes greater algal development.

6. Technologies for algal harvesting

The biomass is harvested after the algal cultivation stage to separate the biomass from the bioreactor effluent. Harvesting of algal products accounts for 20%–30% of the entire cost of biomass production [24]. As a result, developing a cost-effective harvesting system for the bulk production of biofuels from algae is critical. Filtration, centrifugation, sedimentation, flocculation, and floatation are some of the most used methods for harvesting algae.

6.1 Filtration

This approach can be used to filter out larger algae species, such as *Spirulina platensis*, but it is normally done on a small scale [25]. However, there are certain difficulties, because processing takes a long time and scaling up is difficult owing to frequent clogs.

6.2 Sedimentation

The first step in separating algae from water is sedimentation. Algae are allowed to settle and densify when agitation is finished. This approach is both cost-effective and energy-efficient. However, the method takes time and is not effective for small microalgal species [25].

6.3 Flocculation

Flocculation is causes algae to aggregate and form colloids by adding something to a combination of water and algae. Alum and ferric chloride are examples of chemical flocculants. When CO_2 is introduced to an algal environment, it causes the algae to flocculate on their own. This method is highly efficient and has a high volumetric capacity.

6.4 Flotation

This approach is based on material density differences. The algae will collect with the froth of bubbles at the top of algal systems, and there must be a mechanism to collect or remove the froth and algae off the top to separate it from the water. It is an expensive technology.

7. Physicochemical parameters for biofuel production from algae

The quantity of several algal metabolites such as lipid, carbohydrate, and protein present in algae is determined by the nutrient availability in algae and environmental factors such as temperature, light, pH, salinity, and CO_2 concentration. Macronutrients such as phosphorous, nitrogen, potassium, magnesium, and sulfur are essential for algal growth. Approximately 2.5—30 parts per million (ppm) of micronutrients and 2.5—4.5 ppm of trace elements such as cobalt, copper, boron, molybdenum, and zinc are essential for the growth of algae.

7.1 Effect of temperature

The impact of temperature on algal growth and biochemical composition is significant. Like many other microorganisms, the growth rate of algae is proportional to the increase in temperature. The cell growth rate increases by increasing the temperature until it reaches the optimal temperature. During the stationary phase at the ideal temperature, the algal growth rate decreases. The temperature needed for algal culture, based on the strain, area, and season, is 15—40°C. The composition of fatty acids in algae is affected by temperature. The total lipid content drops when the temperature rises, whereas the neutral lipid content rises. According to Santos et al. (2009), increasing the temperature from 20 to 25°C enhanced the carbohydrate content of *Nannochloropsis oculata* and *C. vulgaris* as a result of amylase activity. An increased temperature causes starch breakdown owing to the activity of amylase.

7.2 Effect of light

To produce energy in algal cells via photosynthesis, light is necessary. The amount of light energy that can be converted to chemical energy is referred to as PE, and it accounts for around 42.3% of total energy (photosynthetic active radiation). Cells use the stalled energy to generate storage molecules via the Kelvin cycle [26]. Photoinhibition occurs when the light intensity exceeds the optimal level, reducing biomass productivity and the CO_2 fixation rate. According to an experiment, the exponential growth rate and polysaccharide synthesis of *Porphyridium cruentum* were enhanced in the presence of blue light with a wavelength of 400—500 nm. The availability of light to the algal system is governed by mitigation within the growth chamber and internal self-darkening of cells; hence PE, depends not only on the irradiance level but also on the reactor layout [27].

7.3 Effect of pH

Another key component that influences algal development is the pH, which regulates the solubility and accessibility of CO_2 and other necessary nutrients in the growing medium. As a result, pH should be checked often throughout the culture phase. Because incorporating inorganic carbon by algae elevates the pH of the media during development, algal growth is best under neutral pH conditions. However, pH 7.0—9.0 is ideal for growing conditions. The elasticity of the cell wall is enhanced at a basic pH, which inhibits rupture and autospore release, extending the cell cycle.

7.4 Effect of salinity

The biomass and metabolic components of algae are also influenced by salinity in terms of sodium chloride (NaCl) concentration. High salinity aids in restoring turgor pressure and controls ion transport, activating stress proteins that help store osmoprotective solutes [28]. Depending on which metabolic pathway is activated by the stress, changing salinity influences the metabolite composition variably in the same species.

7.5 Effect of nutrients

Nutrients are thought to be the most significant aspects of achieving maximum biomass output on a pilot or industrial scale. Macronutrients including carbon, nitrogen, and phosphate, as well as vitamins such as B6 and B12 and trace elements such as iron salt, nickel, cobalt, manganese, zinc, and selenium are important nutrients for algal development. Phosphate aids in DNA and RNA backbone synthesis and adenosine triphosphate synthesis, whereas nitrogen is an essential source for protein and nucleic acids. Apart from CO_2, sodium bicarbonate and carbonates are prominent inorganic carbon sources; organic carbon sources include glucose, acetate, and glycerol [29]. According to Yang et al. [30], elevated CO_2 led to a rise in the specific growth rate of *Chlamydomonas reinhardtii*, *Chlorella pyrenoidosa*, and *Scenedesmus obliquus*. Algal development is hindered below a threshold CO_2 concentration, which is governed not only by CO_2 affinity and supply but also by the algal cell size and the presence of carbonic anhydrase.

Nitrogen and phosphorus, which account for 7–20% of dry cell weight, are critical macronutrients for algal development. Nutritional availability is governed by the N:P ratio. For example, with an N:P ratio of 15–30, *Chlorella* sp. showed the fastest growth rate [31]. The protein, nucleic acid, and chlorophyll makeup of algae is influenced by the nitrogen concentration. As a result, during nitrogen deficiency, algae's biomass, lipid, and starch content, as well as variations in the enzyme composition, are all greatly altered. In microalgae, phosphorus influences metabolic processes and energy transmission, accounting about 1% of dry cell weight. It frequently inhibits the development of algal cells. Phosphorus deficiency increases metabolite accumulation while decreasing chlorophyll concentration.

8. Genetically engineered algae for biofuel production

The analysis and manipulation of algal genomes in the endeavor to generate customized algal biofuel production strains have been made easier because of breakthroughs in the creation of genetic tools and in silico prediction capacity. Several strategies for raising microalgal biofuel productivity are being investigated, ranging from improving light use to modifying carbon flow routes for enhanced biomass accumulation and modulating lipid formation. Algal nuclear transformation is usually performed, similar to chloroplast transformation, with an emphasis on organelle transformation. Electroporation, biolistic particle bombardment, glass bead-assisted agitation, and the use of *Agrobacterium tumefaciens* [32] are the most prevalent transformation techniques for algal transformation. The use of selection markers such as antibiotic resistance genes including bleomycin [33], hygromycin, paromomycin, and streptomycin, as well as biochemical or fluorescent markers such as luciferase, β-glucuronidase, β-galactosidase, and green fluorescent protein [34], has substantially aided transformant separation.

8.1 Lipid augmentation by genetic alterations

Fatty acid synthesis relies heavily on algal chloroplasts. There are two forms of triacylglycerol (TAG) synthesis in algal systems: acetyl CoA-dependent and acetyl CoA-independent. Acetyl CoA is a key precursor in the de novo fatty acid synthesis pathway (Fig. 5.2). TAG may be increased by overexpressing genes or knocking down enzymes. The overexpression of thioesterase results in a rise in Acyl carrier protein concentration, which leads to less feedback inhibition and a better fatty acid profile. It is also possible to eliminate enzymes involved in the fatty acid β-oxidation pathway. The conversion of acetyl CoA to malonyl CoA occurs when Acetyl CoA Carboxylase (ACC) expression is increased, indicating that ACC genes regulate lipid production in non–lipid storage cells [35]. Pyruvate causes acetyl CoA to develop, which is then catalyzed by pyruvate dehydrogenase kinase. As a result of the downregulation of pyruvate dehydrogenase complex, the formation of acetyl CoA is reduced. Diacylglycerol acyltransferase is used as a target for increasing TAG in green algae. By restricting carbon flow for starch synthesis, starchless mutants of *C. reinhardtii* might be used for lipid improvement. Under nitrogen deficiency, the mutants accumulate fat rather than starch, without compromising the physiologic, biochemical, or phytochemical activities of the algae.

8.2 Genetic engineering to steer hydrogen production

The only microorganisms known to undertake photosynthesis and generate biohydrogen are blue-green algae and green algae. Algae produce biohydrogen by converting light energy into chemical energy while splitting water into molecular oxygen and hydrogen [36], or by serving as a dark fermentation substrate [37]. The major enzymes responsible for biological hydrogen generation are hydrogenase and nitrogenase; however, oxygen inhibits both enzymes' activity and destroys the hydrogen that is generated. Genetic engineering offers effective mechanisms to overcome the inhibitions. Possible strategies that might be used are modifying the algal genome or manipulating the enzymes involved (Fig. 5.3).

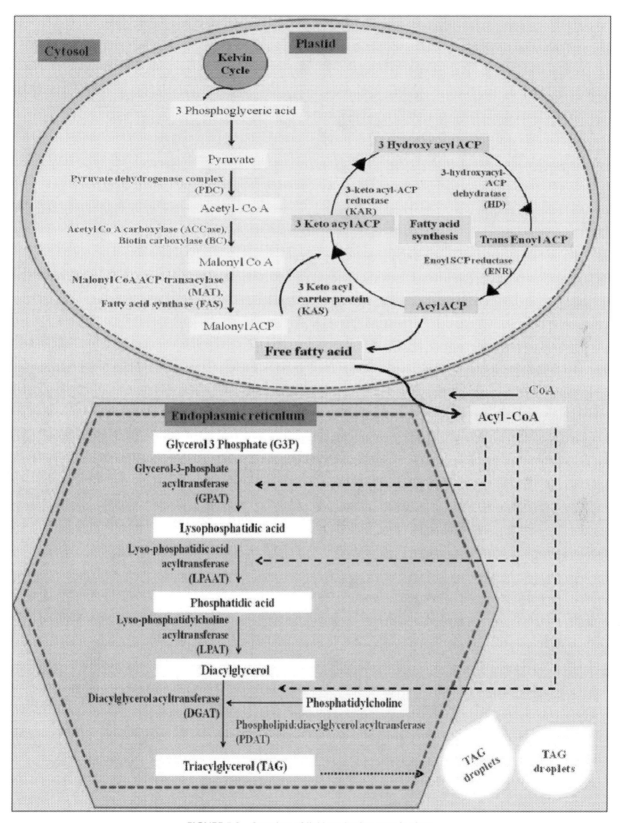

FIGURE 5.2 Overview of lipid production route in algae.

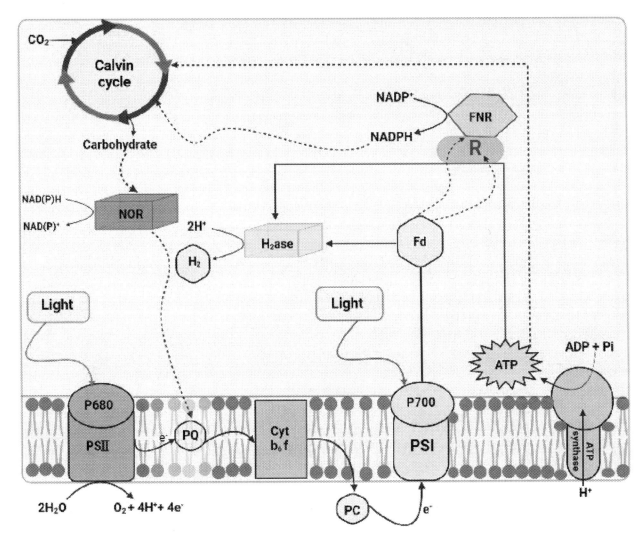

FIGURE 5.3 Direct and indirect biophotolysis by microalgae for hydrogen production. *ADP*, adenosine diphosphate; *ATP*, adenosine triphosphate; *Cyt*, cytochrome; *FNR*, flavin reductase; *NOR*, nitrous oxide reductase; *PSI*, photo system I.

9. Bioconversion technologies of algal biomass to biofuel

Any solid, liquid, or gaseous fuel derived from biological materials is referred to as a biofuel. Algal biofuels may be produced via a variety of biochemical and thermochemical techniques, as well as chemical processes and direct combustion. As shown in Fig. 5.4, anaerobic digestion, fermentation, gasification, transesterification, liquefaction, and pyrolysis are some procedures used to produce biomethane, bioethanol, biogas, biodiesel, biohydrogen, and bioelectricity. Gasification, pyrolysis, and hydrothermal liquefaction are some thermochemical processes that may be used to transform algal biomass into biofuels. Not only lipids or carbohydrates, but the whole algal biomass could be converted into biofuels using these techniques.

The use of microorganisms and/or enzymes to disintegrate algae into fuels usually involves biochemical conversion. Compared with thermochemical conversions, biochemical conversions are usually slower and use less energy. Anaerobic digestion, fermentation, and photobiological H_2 generation represent various biochemical conversion processes that typically necessitate pretreatment of the biomass, notably for anaerobic digestion and fermentation. The use of algae is predominantly for producing the following products.

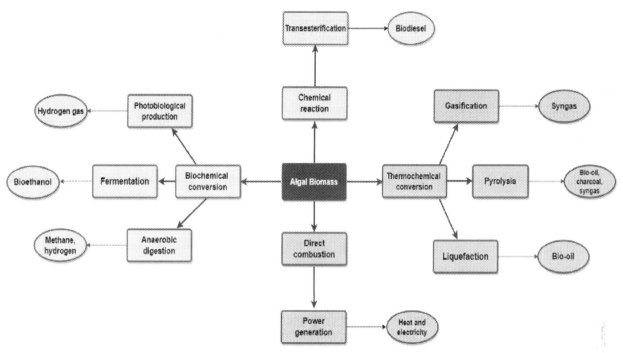

FIGURE 5.4 Biofuel conversion processes from algal biomass.

9.1 Biodiesel

Biodiesel, a low-emission fuel made from renewable biomass and waste lipids, is a substitute for petroleum-based diesel. A monoalcoholic transesterification process of triglycerides with a monoalcohol (most often methanol or ethanol) is used to produce biodiesel, which is catalyzed by alkali, acids, or enzymes. The amount of lipids in the algal biomass determines the quality and production of biodiesel generated, as well as its ability to replace petroleum-based fuel. Compared with fossil fuels, biodiesel has superior ignition qualities and emits 78% fewer pollutants and carbon dioxide [38]. Some microalgal species can store lipids to the extent that they make up a large amount of their biomass (30–50% dry weight basis); this makes them an attractive source of lipids for biodiesel generation. Because of the lack of triglycerides in macroalgae, it is not considered for biodiesel production. Macroalgae biodiesel has received little attention, and yields are significantly lower than those of microalgae [12]. However, compared with petroleum-based diesel, biodiesel has poorer oxidative stability and cold-weather performance difficulties, which make it unsuitable for use as a jet fuel. The increasing cost of algae-based biodiesel manufacturing is still a barrier to wide-scale commercialization.

9.2 Bioethanol

Algae are generally high in lipids and carbohydrates. Algal cell walls are made of cellulose, mannans, xylans, and sulfated glycans, which can be broken down chemically or enzymatically to create simple sugars and then are transformed to bioethanol under anaerobic circumstances using bacteria or yeast [39]. Pure ethanol is a good alternative to gasoline because it has around 66% of its energy of gasoline by volume but a better octane number [40]. Ethanol, with higher octane number, can burn at greater compression ratios with a faster combustion period, which is beneficial for the engine. Ethanol has a much higher heat of vaporization than gasoline, giving it a higher volumetric efficiency and power output [41]. Ethanol burns cleaner, emitting fewer carbon monoxide, hydrocarbons, and nitrogen oxides, and is theoretically compatible with compression-ignition engines. Algal bioethanol production involves numerous major mechanisms, including algal selection and cultivation, pretreatment operations, liquefaction, saccharification, fermentation, and distillation purification [42]. Brown algae has sparked a lot of interest and is explored on the industrial scale.

9.3 Biogas

Biogas, which consists of mostly methane and carbon dioxide, is produced primarily by anaerobic digestion and as a by-product of anaerobic fermentation. Because macroalgae have a lower lignocellulosic content, which makes biodegradation easier, they are more technically feasible than microalgae. Biogas generation from algae has more drawbacks than advantages. Drawbacks associated with using algae for biogas production include the generation of substances or compounds that are harmful to methanogens owing to the high level of ammonia, alkaline metals present in algae that can impede anaerobic digestion, and an undesirable C:N ratio in the algal biomass of some species [43]. The ideal C:N (20−30) ratio is required to have a greater methane output. When the C:N ratio is less than 20, the imbalanced ratio causes the bioreactor to produce more ammonia, which slows the pace of methane generation [44].

9.4 Biohydrogen production

Because of its high energy density, biohydrogen is a viable and long-term fuel source. For hydrogen, 2.2 lb produces the same amount of energy as 6.2 lb of gasoline [45]. Hydrogen is a valuable fuel with several uses in fuel cells, coal liquefaction, and heavy oil upgrading (e.g., bitumen). Heat treatment of *Laminaria japonica* at 170°C resulted in a hydrogen output of 109.6 mL g^{-1} chemical oxygen demand [46]. Hydrogen may be created biologically in a number of ways, including the steam reformation of bio-oils, dark and photofermentation of organic materials, and water photolysis stimulated by particular microalgal species.

9.5 Bioelectricity

The global use of coal is predicted to rise by nearly 56% by 2035 [47]. Algal biomass, which is regarded as the most promising source for cofiring, is an alternative to coal. Cofiring with biomass is a cost-effective method for generating power. Cultivating algae using CO_2 generated from power plants reduces CO_2 emissions significantly while generating extra electricity. By substituting algae biomass for coal in power plants, greenhouse gas emissions can be reduced. However, eutrophication is considerable as a result of the use of fertilizer during algal production. As a result, for the process to be more environmentally friendly, harmony between algae production and algae cofiring must be achieved. Also, synthesis gas (syngas) such as CO, CH_4, CO_2, and H_2, is regarded as a high-potential fuel source for generating heat, power, and transportation fuel for industrial use employing fuel cells, which can be produced via direct combustion, supercritical water gasification, hydrothermal gasification, and other conversion methods. Syngas has been used as a raw material in microbial fuel cells to generate power.

9.6 Biomass production and phycoremediation

A wide range of trace nutrients is needed to sustain algal growth, including carbon, nitrogen, and phosphorus. Fertilizer supplies are insufficient to fulfill the needs of industrial algae biorefineries, which are expanding rapidly. For this, alternative sources of essential nutrients must be used. Almost all nutrients needed for algal development may be found in wastewater. A further benefit of wastewater is the presence of local oleaginous microalgae, which seem to be well-adapted to current demanding conditions and have an important role in the stream's capacity to self-clean. As an example of how to use oleaginous biomass to reduce the nitrogen load in wastewater, a generic strategy is illustrated in Fig. 5.5.

The cultivation of microalgae in wastewater for the production of biomass is a promising prospect that is both economically and ecologically advantageous. The use of wastewater and saline water in conjunction with maximal recycling is the most promising option for ensuring the most efficient use of water resources. When it comes to nutrients, the current supply of nitrogen- and phosphorus-based fertilizers will not be able to meet the increased demand resulting from the widespread production of microalgae on a large scale. The production of synthetic fertilizers is an energy-intensive process that results in the emission of a significant amount of greenhouse gases into the environment. In conjunction with the use of saltwater and wastewater, the maximum amount of nutrient recycling may be possible, which may reduce the need for synthetic fertilizers as an external input. Furthermore, it was shown that when the nitrogen supply in the culture medium is restricted, microalgae tend to collect a bigger quantity of lipids in their biomass as a result. Consequently, incorporating such demanding circumstances into microalgae production not only enhances lipid content, it reduces the need for synthetic nitrogenous fertilizer, which is beneficial to the environment.

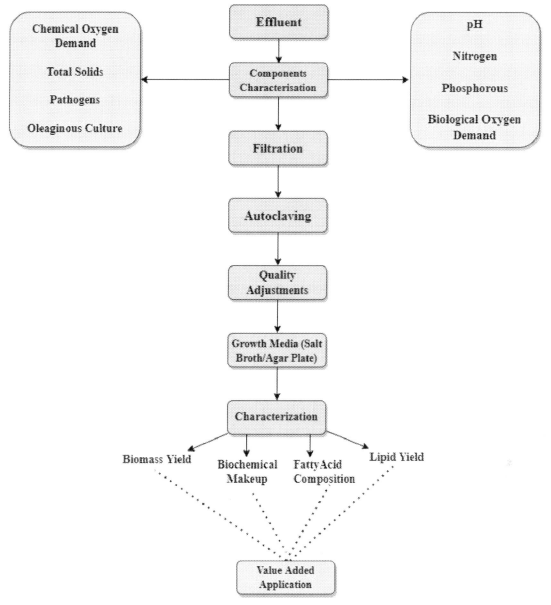

FIGURE 5.5 Schematic diagram of algae-based biomass production and wastewater treatment.

9.7 Value-added products

Algal oil is often valorized through the production of biofuels (biodiesel, renewable diesel, etc.) and other value-added products (polyunsaturated fatty acids, nutraceuticals, etc.). Carbohydrates, proteins, amino acids, pigments, and nutrients are abundant in deoiled biomass. The nutritional content (N, P, and K) of the biomass is recirculated into the culture media in a circular biorefinery. The deoiled biomass can be processed using a variety of procedures that are biomolecule-specific (carbohydrate saccharification and fermentation) or generic in nature (anaerobic digestion, gasification, pyrolysis, and so on). Fig. 5.6 depicts a simplified flowchart of the various processing routes for the production of different value-added products.

Using a biorefinery system, the combined production of biodiesel, biogas, and bioethanol from a given algal biomass increases the economy, energy, and sustainability of bioenergy production [48]. Transesterification produces glycerol, which has been demonstrated to be a suitable growth substrate for growing oleaginous microalgae. Glycerol is also an

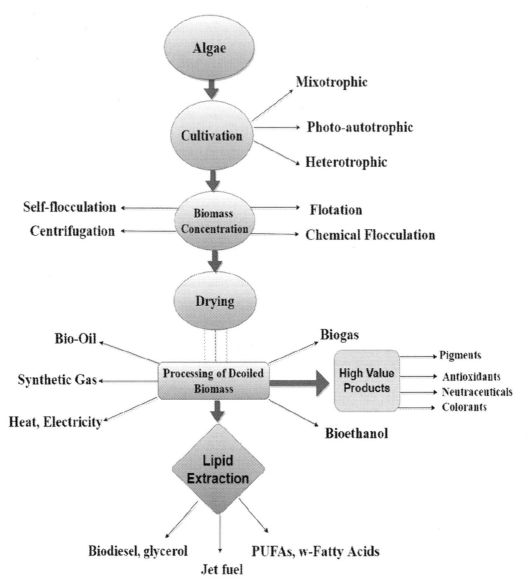

FIGURE 5.6 Products derived from an integrated algal biorefinery. *PUFAs*, polyunsaturated fatty acids.

important feed for industries and pharmaceutical firms, and it may be improved for use as a fuel additive. The main concept underlying the biorefinery model is the possibility of more product flexibility. To comprehend the technoeconomic feasibility of alternative products and production pathways better, it is critical to consider future fluctuations in the demand and cost of bioproducts.

10. Conclusion and future perspectives

Algae-based biofuel is considered a sustainable solution for the circular economy. The use of biofuel does not degrade the environment, and because algal biofuel is renewable, it enables a closed loop system. Cultivation of algae requires no arable land, so algae-based biofuel offers a feasible alternative to the circular economy because of flexibility regarding nonarable land cultivation and chemical-based medium conservation using wastewater resources. However, a considerable obstacle needs to be addressed for commercial viability of algal biofuel generation and sustainable technologies for efficient production toward the growing demand for energy.

Acknowledgments

The authors are thankful for the assistance and support extended by the head of the School of Biotechnology, Kalinga Institute of Industrial Technology, Deemed-to-be-University, Odisha, for extending the infrastructure facilities.

References

[1] R.P. Shukla, C. Rapier, M. Glassman, F. Liu, D.L. Kelly, H. Ben-Yoav, An integrated electrochemical microsystem for real-time treatment monitoring of clozapine in microliter volume samples from schizophrenia patients, Electrochem. Commun. 120 (2020) 106850.

[2] V. Kumar, I.S. Thakur, M.P. Shah, Bioremediation approaches for treatment of pulp and paper industry wastewater: recent advances and challenges, Microbial Bioremediat. Biodegrad. (2020) 1–48.

[3] Z. Tian-Yuan, W. Yin-Hu, Z. Lin-Lan, W. Xiao-Xiong, H. Hong-Ying, Screening heterotrophic microalgal strains by using the Biolog method for biofuel production from organic wastewater, Algal Res. 6 (2014) 175–179.

[4] J.T.T. Chye, L.Y. Jun, L.S. Yon, S. Pan, M.K. Danquah, Biofuel production from algal biomass, in: Bioenergy and Biofuels, CRC Press, 2018, pp. 87–118.

[5] L. Brennan, P. Owende, Biofuels from microalgae—a review of technologies for production, processing, and extractions of biofuels and co-products, Renew. Sustain. Energy Rev. 14 (2) (2010) 557–577.

[6] M.Y. Noraini, H.C. Ong, M.J. Badrul, W.T. Chong, A review on potential enzymatic reaction for biofuel production from algae, Renew. Sustain. Energy Rev. 39 (2014) 24–34.

[7] S. Banerjee, S. Banerjee, A.K. Ghosh, D. Das, Maneuvering the genetic and metabolic pathway for improving biofuel production in algae: present status and future prospective, Renew. Sustain. Energy Rev. 133 (2020) 110155.

[8] A. Demirbas, Use of algae as biofuel sources, Energy Convers. Manag. 51 (12) (2010) 2738–2749.

[9] D.F. Correa, H.L. Beyer, H.P. Possingham, S.R. Thomas-Hall, P.M. Schenk, Biodiversity impacts of bioenergy production: microalgae vs. first generation biofuels, Renew. Sustain. Energy Rev. 74 (2017) 1131–1146.

[10] S. Kraan, Mass-cultivation of carbohydrate rich macroalgae, a possible solution for sustainable biofuel production, Mitig. Adapt. Strategies Glob. Change 18 (1) (2013) 27–46.

[11] A.P. Peter, K.S. Khoo, K.W. Chew, T.C. Ling, S.H. Ho, J.S. Chang, P.L. Show, Microalgae for biofuels, wastewater treatment and environmental monitoring, Environ. Chem. Lett. 19 (4) (2021) 2891–2904.

[12] R. Maceiras, M. Rodrõ, A. Cancela, S. Urréjola, A. Sánchez, Macroalgae: raw material for biodiesel production, Appl. Energy 88 (10) (2011) 3318–3323.

[13] C.Y. Chen, K.L. Yeh, R. Aisyah, D.J. Lee, J.S. Chang, Cultivation, photobioreactor design and harvesting of microalgae for biodiesel production: a critical review, Bioresour. Technol. 102 (1) (2011) 71–81.

[14] J. Doucha, K. Lívanský, Production of high-density Chlorella culture grown in fermenters, J. Appl. Phycol. 24 (1) (2012) 35–43.

[15] K.C. Park, C. Whitney, J.C. McNichol, K.E. Dickinson, S. MacQuarrie, B.P. Skrupski, P.J. McGinn, Mixotrophic and photoautotrophic cultivation of 14 microalgae isolates from Saskatchewan, Canada: potential applications for wastewater remediation for biofuel production, J. Appl. Phycol. 24 (3) (2012) 339–348.

[16] S.A. Razzak, M.M. Hossain, R.A. Lucky, A.S. Bassi, H. De Lasa, Integrated CO_2 capture, wastewater treatment and biofuel production by microalgae culturing—a review, Renew. Sustain. Energy Rev. 27 (2013) 622–653.

[17] Y. Chisti, Biodiesel from microalgae, Biotechnol. Adv. 25 (3) (2007) 294–306.

[18] S.I. Mussatto, G. Dragone, P.M. Guimarães, J.P.A. Silva, L.M. Carneiro, I.C. Roberto, J.A. Teixeira, Technological trends, global market, and challenges of bio-ethanol production, Biotechnol. Adv. 28 (6) (2010) 817–830.

[19] R.R. Narala, S. Garg, K.K. Sharma, S.R. Thomas-Hall, M. Deme, Y. Li, P.M. Schenk, Comparison of microalgae cultivation in photobioreactor, open raceway pond, and a two-stage hybrid system, Front. Energy Res. 4 (2016) 29.

[20] C.M. Beal, L.N. Gerber, D.L. Sills, M.E. Huntley, S.C. Machesky, M.J. Walsh, C.H. Greene, Algal biofuel production for fuels and feed in a 100-ha facility: a comprehensive techno-economic analysis and life cycle assessment, Algal Res. 10 (2015) 266–279.

[21] M. Zhou, J. Wang, Biofuel cells for self-powered electrochemical biosensing and logic biosensing: a review, Electroanalysis 24 (2) (2012) 197–209.

[22] D. Chiaramonti, M. Prussi, M. Buffi, A.M. Rizzo, L. Pari, Review and experimental study on pyrolysis and hydrothermal liquefaction of microalgae for biofuel production, Appl. Energy 185 (2017) 963–972.

[23] X. Miao, Q. Wu, C. Yang, Fast pyrolysis of microalgae to produce renewable fuels, J. Anal. Appl. Pyrol. 71 (2) (2004) 855–863.

[24] Y. Li, M. Horsman, N. Wu, C.Q. Lan, N. Dubois-Calero, Biofuels from microalgae, Biotechnol. Prog. 24 (4) (2008) 815–820.

[25] L. Zhu, Y.K. Nugroho, S.R. Shakeel, Z. Li, B. Martinkauppi, E. Hiltunen, Using microalgae to produce liquid transportation biodiesel: what is next? Renew. Sustain. Energy Rev. 78 (2017) 391–400.

[26] T. Li, Y. Zheng, L. Yu, S. Chen, Mixotrophic cultivation of a Chlorella sorokiniana strain for enhanced biomass and lipid production, Biomass Bioenergy 66 (2014) 204–213.

[27] P. Chiranjeevi, S.V. Mohan, Critical parametric influence on microalgae cultivation towards maximizing biomass growth with simultaneous lipid productivity, Renew. Energy 98 (2016) 64–71.

[28] R. Radakovits, R.E. Jinkerson, A. Darzins, M.C. Posewitz, Genetic engineering of algae for enhanced biofuel production, Eukaryot. Cell 9 (4) (2010) 486–501.

[29] A. Juneja, R.M. Ceballos, G.S. Murthy, Effects of environmental factors and nutrient availability on the biochemical composition of algae for biofuels production: a review, Energies 6 (9) (2013) 4607–4638.

[30] L. Yang, J. Chen, S. Qin, M. Zeng, Y. Jiang, L. Hu, J. Wang, Growth and lipid accumulation by different nutrients in the microalga *Chlamydomonas reinhardtii*, Biotechnol. Biofuels 11 (1) (2018) 1–12.

[31] X. Xiong, K.M. Iris, L. Cao, D.C. Tsang, S. Zhang, Y.S. Ok, A review of biochar-based catalysts for chemical synthesis, biofuel production, and pollution control, Bioresour. Technol. 246 (2017) 254–270.

[32] K. Ullah, M. Ahmad, V.K. Sharma, P. Lu, A. Harvey, M. Zafar, S. Sultana, Assessing the potential of algal biomass opportunities for bioenergy industry: a review, Fuel 143 (2015) 414–423.

[33] M. Fuhrmann, W. Oertel, P. Hegemann, A synthetic gene coding for the green fluorescent protein (GFP) is a versatile reporter in *Chlamydomonas reinhardtii*, Plant J. 19 (3) (1999) 353–361.

[34] R. Kumar, G.R. Kumar, N. Chandrashekar, Microwave assisted alkali-catalyzed transesterification of *Pongamia pinnata* seed oil for biodiesel production, Bioresour. Technol. 102 (11) (2011) 6617–6620.

[35] J. Marín-Navarro, A.L. Manuell, J. Wu, S. P Mayfield, Chloroplast translation regulation, Photosynth. Res. 94 (2) (2007) 359–374.

[36] A.S. Fedorov, S. Kosourov, M.L. Ghirardi, M. Seibert, Continuous hydrogen photoproduction by *Chlamydomonas reinhardtii*, Appl. Biochem. Biotechnol. 121 (1) (2005) 403–412.

[37] A. Melis, H.C. Chen, Chloroplast sulfate transport in green algae–genes, proteins and effects, Photosynth. Res. 86 (3) (2005) 299–307.

[38] I.M. Atadashi, M.K. Aroua, A.A. Aziz, High quality biodiesel and its diesel engine application: a review, Renew. Sustain. Energy Rev. 14 (7) (2010) 1999–2008.

[39] A.R. Sirajunnisa, D. Surendhiran, Algae–A quintessential and positive resource of bioethanol production: a comprehensive review, Renew. Sustain. Energy Rev. 66 (2016) 248–267.

[40] P.S. Nigam, A. Singh, Production of liquid biofuels from renewable resources, Prog. Energy Combust. Sci. 37 (1) (2011) 52–68.

[41] H. Zabed, J.N. Sahu, A. Suely, A.N. Boyce, G. Faruq, Bioethanol production from renewable sources: current perspectives and technological progress, Renew. Sustain. Energy Rev. 71 (2017) 475–501.

[42] L.M. Chng, D.J. Chan, K.T. Lee, Sustainable production of bioethanol using lipid-extracted biomass from Scenedesmus dimorphus, J. Clean. Prod. 130 (2016) 68–73.

[43] M. Debowski, M. Zieliński, A. Grala, M. Dudek, Algae biomass as an alternative substrate in biogas production technologies, Renew. Sustain. Energy Rev. 27 (2013) 596–604.

[44] H.W. Yen, D.E. Brune, Anaerobic co-digestion of algal sludge and waste paper to produce methane, Bioresour. Technol. 98 (1) (2007) 130–134.

[45] S. Nagarajan, N.C. Skillen, J.T. Irvine, L.A. Lawton, P.K. Robertson, Cellulose II as bioethanol feedstock and its advantages over native cellulose, Renew. Sustain. Energy Rev. 77 (2017) 182–192.

[46] K.W. Jung, D.H. Kim, H.S. Shin, Fermentative hydrogen production from Laminaria japonica and optimization of thermal pretreatment conditions, Bioresour. Technol. 102 (3) (2011) 2745–2750.

[47] M. Kucukvar, O. Tatari, A comprehensive life cycle analysis of cofiring algae in a coal power plant as a solution for achieving sustainable energy, Energy 36 (11) (2011) 6352–6357.

[48] W. Lu, Z. Wang, X. Wang, Z. Yuan, Cultivation of *Chlorella* sp. using raw dairy wastewater for nutrient removal and biodiesel production: characteristics comparison of indoor bench-scale and outdoor pilot-scale cultures, Bioresour. Technol. 192 (2015) 382–388.

Chapter 6

Bioenergy production from algae: biomass sources and applications

Mostafa M. El-Sheekh[1], Ghadir Aly El-Chaghaby[2] and Sayed Rashad[2]

[1]Department of Botany, Faculty of Science, Tanta University, Tanta, Egypt; [2]BioAnalysis Lab., Regional Center for Food and Feed, Agricultural Research Center, Giza, Egypt

1. Introduction

Over decades, global energy consumption has risen, particularly in developing countries. Energy consumption in developed countries is substantial, especially since the use of petroleum-derived fuels increases emissions. Currently, the majority of energy used in daily activities is nonrenewable, that is, it is derived from fossil fuels [1]. Fossil fuels such as carbon, natural gas, and petroleum are commonly used in the transportation and energy sectors to meet the high energy demands of ongoing industrialization, economic growth, and rising populations. Many greenhouse gases released during the processing of fossil fuels have been discovered to trap heat and hasten global warming [2]. Bioenergy is a type of renewable energy that comes from biological sources and is burned directly or converted into a liquid or gas [3]. To overcome the issues encountered in the first and second generation biofuels, a suitable feedstock for biofuel production must not be competitive with food products [4]. Unlike terrestrial photosynthetic plants, algae are a good source of biofuel because they are not affected by rising food prices or the loss of biodiversity. The rapid growth rate of algae and their ability to survive in harsh conditions contribute to their high availability [5]. This chapter discusses the potential use of algae biomass for biodiesel, biofuel, biohydrogen, biomethane, biogas, and other bioenergy products. The chapter also highlights the effect of cultivation parameters and their optimization on algal biomass composition.

2. Algae cultivation conditions

A distinguishing characteristic of algae is their ability to grow on any land, in any environment, and freshwater is not required for cultivation. Seawater or sewage, on the other hand, can be utilized. Algae production systems are not predefined in shape or design, and they may be adjusted as an open or closed-loop cycle. Because algae production is not seasonal, all varieties of algae may be cultivated and collected all year [6,7].

Algae growing systems are classified as open or closed. The culture medium in an open cultivation configuration is freely exposed to the ambient air and direct sunlight, whereas in a closed system, the culture is under a roof (indoor) and is shielded from direct sunlight [8]. One of the main advantages of an open cultivation method is that it requires very little capital and running expenditures and a decreased energy required for culture mixing. On the other hand, open systems need a wide area to scale up and are vulnerable to contamination and poor weather conditions [9].

Whether growing algae in open ponds or closed photobioreactors, environmental factors must be carefully considered. The photosynthesis of algae is influenced by temperature, light, pH, and nutrients and development rate and the activity of their cellular metabolism and composition [10]. Light is the key abiotic characteristic impacting cell metabolism, since it is the force that propels photosynthesis. Photosynthetic organisms can accommodate changes in light intensity and spectrum. Although photosynthetic organisms have developed several ways to defend themselves from photodamage, too much light may limit microalgal production and have detrimental impacts on the quality of biofuels. The biomass composition of carbohydrates, lipids, and proteins is modulated by light quality, which significantly influences cell metabolism [11–13].

TABLE 6.1 Some typical algal cultivation conditions.

Condition	Typical range	Rational
Temperature	10–40°C	Growth dynamics are impacted below this range, whereas algal cells die above this range.
Light intensity	2500–5000 lux	Algae may perform photosynthesis only in this photosynthetically active radiation range (400–700 nm).
The pH of the culture media	6–10	Acidic pH inhibits nutrition absorption capacity and interferes with cellular functions, whereas alkaline pH decreases CO_2 affinity and slows cell cycle completion.

One of the most important environmental factors affecting algal growth rate, cell size, biochemical composition, and nutritional requirements is temperature. The temperature of microalgae cultures rises as a result of heat absorption from the light source used. As a result, for a large-scale outdoor culture, sunlight irradiance and temperature must be taken into account [14,15].

Algae require CO_2 as a carbon source. CO_2 absorption from the air into water is minimal, which impacts biomass production. In intense algal culture systems, the lack of or inadequate CO_2 will restrict and decrease development and algal output. CO_2 fixation in photosynthesis relies on the CO_2 concentration given during algal development. For effective algal development, the delivery of nutrients and other growth requirements to algae culture is crucial and highly critical. Organic carbon (C), nitrogen (N), phosphorus (P), and other nutrients, such as heavy metals, are the most frequent nutrients [16]. Some typical algae cultivation conditions are depicted in Table 6.1 [2,17,18].

3. Macro and microalgae biomass production

Algae are the world's fastest-growing organisms. Algae represent a significant source of biomass [19]. Microalgae are photosynthetic organisms that are microscopic in size (frequently found in marine or fresh water); they use carbon fixation to generate a variety of organic compounds such as lipids, carbohydrates, proteins, and vitamins [20]. They have great photosynthetic efficiency and can endure high carbon dioxide concentrations, resulting in around 50% percent of the carbon retained in microalgal cells [21]. Microalgae's simple structure allows for rapid cell development through one or two days or less with better carbon dioxide fixation efficiencies than terrestrial plants. Microalgae are one of the most efficient biological systems for converting solar energy into chemical energy, which can then be captured and converted into biomass. Microalgae species are over 50,000; among these, certain distinguishing characteristics, such as high lipid/sugar accumulation, and the ability to produce some high-value nutrition bioproducts, or a rapid growth rate [22].

Macroalgae are multicellular, large-sized algae visible to the naked eye [23]. Oceans and seas represent more than 70% of the Earth's surface, allowing for the sustainable development of macroalgae biomass feedstock. This is due to the fact that macroalgae growing does not interfere with agricultural land usage and does not necessitate the use of freshwater. Furthermore, macroalgae develop at a far faster pace than land-based crops. Over 10,000 macroalgae species have been documented worldwide, with just a dozen species being produced economically, while the remainder is harvested from the wild [24].

During algae cultivation, algae absorb carbon dioxide through photosynthesis and produce algae biomass (Fig. 6.1).

4. Algal biomass chemical composition

During their culture process, algae absorb carbon dioxide through photosynthesis to produce algal biomass [25]. Many algal biofuels and bioproducts processes rely on the composition of algal biomass [26,27]. Biomass chemical characterization improves identification and reduces overall uncertainty around individual process steps. The biochemical components of algal biomass, on the other hand, vary significantly depending on the strain and nutrient status of the algal culture medium. Also, the harvest timing significantly impacts overall reported fuel yields and downstream processing characteristics [28]. Algal cells change their metabolic composition as they grow and respond to environmental and physiological stimuli, which have been known for decades, and we use these exact physiological changes to demonstrate the challenges of measurement accuracy [29]. Criteria for selecting strains: because different algae strains have vastly

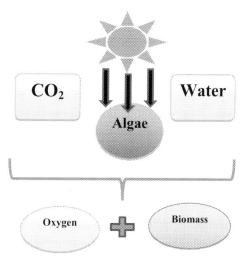

FIGURE 6.1 Algae biomass production.

different cellular contents and growth behavior, choosing the right one for the job is critical. The concentration of high-value side products in the cell, oil productivity, growth rate, optimal growth conditions, and scale-up potential are all factors to consider [14]. Following the selection of a strain, its growth parameters are adjusted to produce the desired product, such as biofuel or fine chemicals [30]. Algae while actively developing convert the sun's energy into a wide range of metabolites known as bioactive chemicals or biochemicals (Fig. 6.2). Algal biomass contains a high concentration of nutrients, particularly nitrogen and phosphorus. Proteins (6%–52%), lipids (7%–23%), and carbohydrates (5%–23%) make up the composition of algal biomass, which varies by species [31].

5. Macro and microalgae for bioenergy production

The first-generation biofuels are made from edible feedstocks, and as a result of using these resources to generate energy, food prices have increased [32]. Second-generation biofuel feedstocks are made from waste and dedicated lignocellulosic feedstocks like lignin, which are resistant to degradation but not economically viable at a large scale [33]. Micro- and

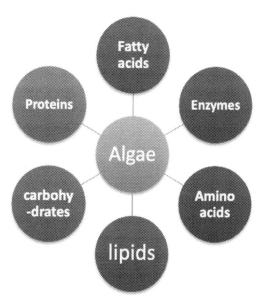

FIGURE 6.2 Algae biochemical compounds.

macroalgae are the third-generation biofuels feedstocks, which have more advantages than the previously mentioned two feedstocks [34].

Either micro or macroalgae are one of the most productive raw materials for biofuel production when measured in terms of productivity per unit area. Various products can be made from algal biomass, including bioenergy, chemicals, food, cosmetics, fertilizer, wastewater treatment agents, and CO_2 sequestration agents [35]. Algal biomass can be used as a raw material for biofuel production (bio-oil) via pyrolysis or biogas and bioethanol production via fermentation [36]. Macro and microalgae must meet certain criteria for bioenergy production, e.g., they must be highly productive, easily harvestable, able to withstand open ocean water currents, and produced at a cost that is comparable to or less than that of other available sources [37,38].

Macroalgae are a promising feedstock for bioenergy production because of their high biomass yield and superior production compared to various terrestrial crops [39]. Cultivating macroalgae at sea, which requires no arable land or fertilizer, could be a solution to the energy crisis. Macroalgae are primarily used to produce food and extract hydrocolloids, but they can also be used to make ethanol. Sugars are abundant in macroalgal biomass (at least 50%), which can be used to make ethanol fuel [40]. Macroalgae-derived biofuel is a brilliant replacement for currently used fossil fuels [35]. As a result, the cultivation and engineering of macroalgae have piqued the world's interest due to their potential as a replacement for conventional fossil fuels that are rapidly depleting [41].

Microalgae are being considered as a potential feedstock for next-generation biofuel production because they have the potential to be 10 to 20 times more productive than any other biofuel crop, and their large-scale cultivation does not require arable land or biodiverse landscapes [9]. Microalgae can be used to produce a variety of renewable biofuels. Methane is produced by anaerobic digestion of algal biomass, biodiesel is made from microalgal oil, and biohydrogen is created by a photobiological mechanism [42]. The large variety of algae species available gives a diverse set of starting strains for fuel generation. Table 6.2 depicts some alga species used for bioenergy production.

TABLE 6.2 Algae species for bioenergy production.

Algae	Product	Reference
Chlorella vulgaris	Biodiesel	[43]
Chlorella vulgaris	Biodiesel & biogas	[44]
Chlorella sorokiniana	Biodiesel	[45]
Chlorella vulgaris	Biodiesel	[46]
Chlorella vulgaris	Biodiesel	[47]
Chlorella pyrenoidosa	Biodiesel	[48]
Chlorella S4	Bioethanol	[49]
Desmodesmus sp.	Biodiesel	[50]
Desmodesmus sp.	Biofuel	[51]
Tetraselmis indica	Biodiesel	[52]
Laminaria sp.	Biogas	[53]
Botryococcus sp.	Biofuel	[54]
Tetradesmus obliquus	Bio-oil, and biogas	[55]
Chlamydomonas reinhardtii UTEX 2243 and *Chlorella sorokiniana* UTEX 2714	Biohydrogen	[56]
Microalgae consortium	Biodiesel	[57]
Desmodesmus sp., *D. Chlorella* sp., *C. vulgaris, brasiliensis, Botryococcus braunii,* and *B. terribilis*	Biodiesel	[55]
Spirulina	Biodiesel	[58]
Spirulina	Biodiesel	[59]

TABLE 6.2 Algae species for bioenergy production.—cont'd

Algae	Product	Reference
Chlorella sp.	Biohydrogen and methane	[60]
Chlamydomonas reinhardtii	Biohydrogen	[61]
Chlamydomonas reinhardtii	Biohydrogen	[62]
Chlamydomonas mutant	Biohydrogen	[63]
Spirulina sp. and Chlorella fusca	Biofuel	[64]
Chlamydomonas debaryana	Bio-oil	[65]
Botryococcus braunii	Biofuel	[66]
Chlorella sp.	Biofuel	[67]
Dunaliella salina	Biofuel	[68]
Nannochloropsis salina	Biomethane	[69]
Ulva lactuca	Bio-oil	[70]
Gracilaria edulis and Ulva lactuca	Bioethanol and bio-oil	[71]
Ulva rigida and Ulva intestinalis	Bioethanol	[72]
Gracilaria manilaensis	Biomethane	[73]
Kirchneriella lunaris	Biodiesel	[74]
Microcystis aeruginosa	Bioethanol	[75]
Caulerpa prolifera & Corallina elongata	Biogas	[76]
Cladophora sp.	Biodiesel	[77]
Gracilariopsis persica	Biomethane	[73]
Ascophyllum nodosum, Laminaria digitata, Laminaria hyperborea, Saccharina latissima and Saccorhiza polychides	Biomethane	[78]
Saccharina latissima	Biofuel	[79]
Phaeodactylum tricornutum	Biogas	[80]
Sargassum sp., Ulva fasciata, Turbinaria ornata, Gelidium sp. and Thalassia sp.	Biodiesel	[81]
Ellisolandia elongata and Caulerpa prolifera	Biogas	[76]
Saccorhiza polychides, Ascophyllum nodosum, L. digitata, Saccharina latissima, and Laminaria hyperborean	Biomethane	[78]
Ulva sp., Cladophorasp., and Spirogyra sp.	Biogas	[82]
Padina sp. Gelidium genus., Fucus, Saccharina, Ascophyllum, and Laminaria	Biogas	[83]

6. Methods for algal biomass conversion to bioenergy products

Algal biomass can be treated after harvesting and processed in a variety of ways (Fig. 6.3) to produce a diversity of end-use bioenergy products. Transesterification is the most common and widely utilized process for producing biodiesel, and it has proven viable for a variety of feedstock. The reaction of a triglyceride (lipid) with alcohol in the presence of a suitable catalyst, such as an acidic, alkaline, or enzyme-based mechanism, is known as transesterification; it produces fatty acid methyl esters with glycerol [84,85]. It is a reversible reaction with three steps, where long-chain triglycerides are first converted to diglycerides, then diglycerides to monoglycerides, and finally monoglycerides to esters [86]. Biodiesel is typically made from algal lipids by transesterification [85].

Fermentation of together glucose- and nonglucose-based sugars is required in the commercial synthesis of bioethanol from algae [87]. This metabolic reaction uses enzymes to convert an organic substrate into alcohol, and it is a biochemical process that converts the cellulose sugar or starch found in algae into bioethanol. Alginate, mannitol, and laminarin are

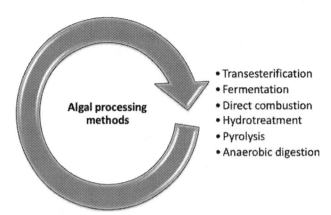

FIGURE 6.3 Algae processing for bioenergy production.

sugar components in algae that are converted to ethanol by yeast activity. Because algae have a high carbohydrate content, it may be used to make bioethanol even after it has been hydrolyzed [84].

Pyrolysis is the process of converting biomass into solid fuel (biochar), liquid fuel, and gaseous fuel products in the absence of oxygen. Pyrolysis can be gradual, quick, or instantaneous. It usually occurs at a rate of $300-700°C\ s^{-1}$. Pyrolysis is the most favored conversion technique due to the high ash content of algae, but the resulting oil still has concerns with acidity, viscosity, and stability [88]. According to some authors, the pyrolysis process looks to be a great choice for energy conversion since it allows for the use of a variety of organic matter sources and is not restricted by lipid content, as is the case with biodiesel manufacturing procedures. The pyrolysis process is based on the breakdown of organic molecules found in total biomass in the absence of oxygen and air pressure in a controlled environment, resulting in three phases: liquid (bio-oil), gas, and solid (char). Because biomass is pyrolyzed into different gases, biomass pyrolysis is considered a regenerative process. One of the gases produced is carbon dioxide, which is absorbed by the algae for development, making the process self-sustaining and contributing little to the greenhouse effect [89].

Anaerobic digestion is another process for converting algal biomass into bioenergy products. It is a process in which bacteria transform organic substrates into methane, carbon dioxide, and bacterial biomass through a series of events. For biomass with high water content, such as macroalgae, anaerobic digestion is the method of choice; it can tolerate biomass with high moisture content without the energy costs of dewatering and drying [90]. Current liquid biofuel production methods from algae produce roughly 60%–70% residual biomass as a byproduct. Anaerobic digestion not only produces biogas, but it can also restore important nutrients from algal biomass that has been lipid stripped. The biogas produced by the anaerobic digestion process of algal biomass can be used to generate onsite electrical power or thermal heat [91].

7. Benefits and limitations of using algae for bioenergy production purposes

Algal biomass has a number of advantages in the manufacture of biofuels: (a) ability to grow all year, resulting in higher algal oil productivity than conventional oil seed crops; (b) higher tolerance to high carbon dioxide content; (c) algae cultivation consumes very little water; (d) algal cultivation does not require herbicides or pesticides; (e) algal species have a very high growth potential in comparison to others; (f) apart from providing any additional nutrients, other sources of nutrient-rich wastewater, such as nitrogen and phosphorus can be used for algae production.; (g) the ability to thrive in tough environments such as saline, brackish, and coastal waters, which do not influence traditional agriculture [92,93].

According to Ref. [94], aquatic algae, which are classed as microalgae or macroalgae based on their shape and size, may provide viable biofuel production alternatives for the following reasons. Firstly, because of their quick biomass doubling time (as little as 3.5 h) and high productivities [up to 26,300 t dry weight (DW)/km²/yr], algae biofuel yields are 10–100 times greater than those from land-based crops in the same growing area. Second, because algae are aquatic organisms, they do not require arable land for widespread production. Third, algal biomass may be easily hydrolyzed for future hydrogen fermentation because it has little or no lignin.

On the other hand, the limitations of using algae as a feedstock for bioenergy production include low cell density in culture, inefficient and energy-intensive harvesting methods, and insufficient downstream processes that are currently impeding large-scale production of algal biofuels [95]. According to Behera et al. [93], algal biomass as a feedstock has a number of drawbacks, including a higher growing cost than conventional crops. Similarly, collecting algae requires a lot of energy, which accounts for about 20%—30% of the entire cost of manufacturing.

8. Economics of bioenergy production from algal biomass

Microalgae-based third-generation biofuel is a feasible global energy insecurity and climate change solution. Notwithstanding the fact that global algal biomass output is currently at 38 million liters per year, commercialization faces substantial financial obstacles. Cost-cutting measures, particularly for microalgae cultivation, have been largely ignored in recent studies [96].

The ability to use CO_2 gas from industrial plants such as ethanol, sugar, and other coal-burning plants with up to 90% conversion efficiency and the ability to use nutrients [97] from wastewater for microalgae growth while reducing environmental pollution are two of the most appealing features of using microalgal oil for biodiesel and biogas production [98].

In terms of industrial production and commercialization, biodiesel made from microalgae has proven to be a viable alternative to traditional feedstock. Microalgae have a high rate of growth and carbon sequestration, and can be grown without the use of arable soil in fresh and/or marine water [99]. They studied the technical and economic viability of producing biodiesel from *Phaeodactylum tricornutum* utilizing a scaled-up algal biomass production scenario, taking into account local realty prices and available technologies. Bianco—the economic analysis of microalgae biodiesel production—is based on capital investment, operating costs, and earnings or sales estimates, according to Vieira et al. [99]. It assumes that the two byproducts of the process (residual microalgae biomass and glycerol) are also marketed in addition to biodiesel. This method is used to assess the impact of the economic valorization of byproducts on the final cost of biodiesel production. The values used in the economic study were converted using the exchange rates on 2018—01-19, with € 1 equaling CLP 724 and US$ 1.22. The manufacture of biofuels from microalgal biomass has attracted interest due to numerous essential features of microalgae, including high oil content, ability to withstand harsh environments, and high photosynthetic efficiency [100]. They studied the economics of a microalgal biofuel production system that was conceptually connected with a real sugar plant, such that the industry's waste effluent and byproducts could be used as cheap supplies of nutrients and carbon dioxide.

Despite efforts to commercialize algal biodiesel production, the economic viability of growing algal biomass for low-value products such as biofuels remains a challenge [101]. Currently, the cost of producing microalgal biomass is significantly higher than that of other energy crops [102]. Genetic engineering for strain enhancement, biorefinery techniques for effective material and energy consumption, and the use of inexpensive nutrient and CO_2 sources are among the focused study topics for sustainable generation of biofuel from microalgae [98, 102]. Microalgal culture for biofuel production must be combined with wastewater (WW) treatment for low-cost nutrient and pollutant reduction [97]. Biofuel can be made from a range of different feedstocks, as well as algae. Algae is a promising feedstock for the production of liquid and gaseous biofuels that do not directly compete with food production. However, algae biofuels' long-term viability and economic viability have been questioned, owing to high carbon and fertilizer input needs and high growing and production costs. Algal genetics, carbon storage metabolism, photosynthesis, and physiology are all areas of study for algae, all of which may be improved and modified at a biological level, have the probability to make breakthroughs in algal biofuel practicality. Advances in genomic technology is being driven by technologies that allow for genetic and metabolic engineering, as well as the development of high-throughput tools for screening wild strains for acceptable biofuel traits [103—106].

9. Conclusion and future prospects

Due to the advantages they contain, commercial growth of marine microalgae is favored in nations physically endowed with saline coastal nonarable areas and harsh desert weather. Nonetheless, producing biodiesel from microalgae is not commercially viable. The energy gained from biodiesel was only 27% of the energy used in the biodiesel manufacturing process. However, the research should focus on reducing the cost of producing biodiesel from microalgae worldwide. Also, cooperation between the countries and research groups in all countries should be maximized to solve this problem and make biofuel from algae reliable. According to a recent economic analysis, coproducts are required to alance the cost of fuel production in order for algae biofuels to be cost-competitive with petroleum fuels. The coproduct suite must grow in tandem with fuel production while simultaneously optimizing the value of nonfuel precursors. The future use of marine microalgae or hypersaline lakes is important for safe freshwater for drinking and irrigation. The cultivation of microalgae in the treated wastewater is also a good trend of economic and environmental impacts.

References

[1] K. Ullah, M. Ahmad, V.K. Sofia, Sharma, P. Lu, A. Harvey, M. Zafar, S. Sultana, C.N. Anyanwu, Algal biomass as a global source of transport fuels: overview and development perspectives, Prog. Nat. Sci. Mater. Int. 24 (4) (2014) 329–339, https://doi.org/10.1016/j.pnsc.2014.06.008.

[2] S.R. Chia, H.C. Ong, K.W. Chew, P.L. Show, S.M. Phang, T.C. Ling, D. Nagarajan, D.J. Lee, J.S. Chang, Sustainable approaches for algae utilisation in bioenergy production, Renew. Energy 129 (2018) 838–852, https://doi.org/10.1016/j.renene.2017.04.001.

[3] S.A. Afolalu, O.O. Yusuf, A.A. Abioye, E. Moses, M.E. Emetere, S.O. Ongbali, O.D. Samuel, Biofuel; A sustainable renewable source of energy- A review, IOP Conf. Ser. Earth Environ. Sci. 665 (1) (2021), https://doi.org/10.1088/1755-1315/665/1/012040.

[4] S.J. Tudge, A. Purvis, A. De Palma, The impacts of biofuel crops on local biodiversity: a global synthesis, in: Biodiversity and Conservation, vol 30, Springer Netherlands, 2021, pp. 2863–2883, https://doi.org/10.1007/s10531-021-02232-5, 11.

[5] G. Gao, J.G. Burgess, M. Wu, S. Wang, K. Gao, Using macroalgae as biofuel: current opportunities and challenges, Bot. Mar. 63 (4) (2020) 355–370, https://doi.org/10.1515/bot-2019-0065.

[6] P. Shah Maulin, Microbial Bioremediation & Biodegradation, Springer, 2020.

[7] E.A. El Shenawy, M. Elkelawy, H.A.E. Bastawissi, M. Taha, H. Panchal, K.k. Sadasivuni, N. Thakar, Effect of cultivation parameters and heat management on the algae species growth conditions and biomass production in a continuous feedstock photobioreactor, in: Renewable Energy, vol 148, Pergamon, 2020, pp. 807–815, https://doi.org/10.1016/J.RENENE.2019.10.166. April.

[8] Y. Panahi, A.Y. Khosroushahi, A. Sahebkar, H.R. Heidari, Impact of cultivation condition and media content on *chlorella vulgaris* composition, in: Advanced Pharmaceutical Bulletin, vol 9, Tabriz University of Medical Sciences, 2019, p. 182, https://doi.org/10.15171/APB.2019.022, 2.

[9] R.R. Narala, S. Garg, K.K. Sharma, S.R. Thomas-Hall, M. Deme, Y. Li, P.M. Schenk, Comparison of microalgae cultivation in photobioreactor, open raceway pond, and a two-stage hybrid system, in: Frontiers in Energy Research, vol 4, Frontiers Media S.A, 2016, p. 29, https://doi.org/10.3389/FENRG.2016.00029/BIBTEX. Aug.

[10] A. Juneja, R.M. Ceballos, G.S. Murthy, Effects of environmental factors and nutrient availability on the biochemical composition of algae for biofuels production: a review, Energies 6 (9) (2013) 4607–4638, https://doi.org/10.3390/en6094607.

[11] A. Sánchez-Bayo, V. Morales, R. Rodríguez, G. Vicente, L.F. Bautista, Cultivation of microalgae and cyanobacteria: effect of operating conditions on growth and biomass composition, Molecules 25 (2020) 2834, https://doi.org/10.3390/MOLECULES25122834.

[12] P. Shah Maulin, Removal of Refractory Pollutants from Wastewater Treatment Plants, CRC Press, 2021.

[13] P. Shah Maulin, Removal of Emerging Contaminants through Microbial Processes, Springer, 2021.

[14] K.H. Chowdury, N. Nahar, U.K. Deb, K.H. Chowdury, N. Nahar, U.K. Deb, The growth factors involved in microalgae cultivation for biofuel production: a review, Comput. Water Energy Environ. Eng. 9 (4) (2020) 185–215, https://doi.org/10.4236/CWEEE.2020.94012.

[15] M.M. El-Sheekh, A. Abomohra, M. Abd El-Azim, R. Abou-Shanab, Effect of temperature on growth and fatty acids profile of the biodiesel promising microalga *Scenedesmus acutus*, Biotechnol. Agron. Soc. Environ. 21 (4) (2017) 233–239.

[16] S. Najiha Badar, M. Mohammad, Z. Emdadi, Z. Yaakob, Algae and their growth requirements for bioenergy: a review, Biofuels 12 (3) (2021) 307–325, https://doi.org/10.1080/17597269.2018.1472978.

[17] S. Anto, S.S. Mukherjee, R. Muthappa, T. Mathimani, G. Deviram, S.S. Kumar, T.N. Verma, A. Pugazhendhi, Algae as green energy reserve: technological outlook on biofuel production, Chemosphere 242 (2020) 125079, https://doi.org/10.1016/J.CHEMOSPHERE.2019.125079.

[18] FAO, Manual on the production and use of live food for aquaculture, in: P. Lavens, P. Sorgeloos (Eds.), FAO Technical Paper, Food and Agriculture Organization of the United Nations, 1996.

[19] K. Ullah, M. Ahmad, V.K. Sofia, Sharma, P. Lu, A. Harvey, M. Zafar, S. Sultana, Assessing the potential of algal biomass opportunities for bioenergy industry: a review, in: Fuel, vol 143, Elsevier, 2015, pp. 414–423, https://doi.org/10.1016/J.FUEL.2014.10.064. March.

[20] G. Muhammad, M.A. Alam, W. Xiong, Y. Lv, J.L. Xu, Microalgae biomass production: an overview of dynamic operational methods, in: Microalgae Biotechnology for Food, Health and High Value Products, January, Springer, Singapore, 2020, pp. 415–432, https://doi.org/10.1007/978-981-15-0169-2_13.

[21] Y. Chisti, Biodiesel from microalgae, in: Biotechnology Advances, vol 25, Biotechnol Adv, 2007, pp. 294–306, https://doi.org/10.1016/J.BIOTECHADV.2007.02.001, 3.

[22] M. Morales, L. Sánchez, S. Revah, The impact of environmental factors on carbon dioxide fixation by microalgae, in: FEMS Microbiology Letters, vol 365, Oxford Academic, 2018, p. 262, https://doi.org/10.1093/FEMSLE/FNX262, 3.

[23] M.I. Khan, J.H. Shin, J.D. Kim, The promising future of microalgae: current status, challenges, and optimization of a sustainable and renewable industry for biofuels, feed, and other products, in: Microbial Cell Factories, BioMed Central Ltd, 2018, https://doi.org/10.1186/s12934-018-0879-x.

[24] I.S. Tan, M.K. Lam, H.C.Y. Foo, S. Lim, K.T. Lee, Advances of macroalgae biomass for the third generation of bioethanol production, in: Chinese Journal of Chemical Engineering, vol 28, Elsevier, 2020, pp. 502–517, https://doi.org/10.1016/J.CJCHE.2019.05.012, 2.

[25] R. Prasad, S.K. Gupta, N. Shabnam, C.Y.B. Oliveira, A.K. Nema, F.A. Ansari, F. Bux, Role of microalgae in global CO_2 sequestration: physiological mechanism, recent development, challenges, and future prospective, Sustainability 13 (23) (2021), https://doi.org/10.3390/su132313061.

[26] M. Ismail, G. Ismail, M.M. El-Sheekh, Potential assessment of some micro and macroalgal species for bioethanol and biodiesel production, Energy Source A Recovery Util. Environ. Eff. (2020) 1–17, https://doi.org/10.1080/15567036.2020.1758853 (in press).

[27] J.S. Kruger, M. Wiatrowski, R.E. Davis, T. Dong, E.P. Knoshaug, N.J. Nagle, L.M.L. Laurens, P.T. Pienkos, Enabling production of algal biofuels by techno-economic optimization of Co-product suites, Front. Chem. Eng. 3 (2022), https://doi.org/10.3389/fceng.2021.803513.

[28] T. Mutanda, D. Naidoo, J.K. Bwapwa, A. Anandraj, Biotechnological Applications of microalgal oleaginous compounds: current trends on microalgal bioprocessing of products, Front. Energy Res. 8 (October) (2020), https://doi.org/10.3389/fenrg.2020.598803.

[29] S. Dickinson, M. Mientus, D. Frey, A. Amini-Hajibashi, S. Ozturk, F. Shaikh, D. Sengupta, M.M. El-Halwagi, A review of biodiesel production from microalgae, Clean Technol. Environ. Policy 19 (3) (2017) 637–668, https://doi.org/10.1007/s10098-016-1309-6.

[30] P.M. Schenk, S.R. Thomas-Hall, E. Stephens, U.C. Marx, J.H. Mussgnug, C. Posten, O. Kruse, B. Hankamer, Second generation biofuels: high-efficiency microalgae for biodiesel production, BioEnergy Res. 1 (1) (2008) 20–43, https://doi.org/10.1007/s12155-008-9008-8.

[31] B. Viswanathan, Biochemical routes for energy conversion, in: Energy Sources, Elsevier, 2017, pp. 357–368, https://doi.org/10.1016/B978-0-444-56353-8.00015-0.

[32] M.A.H. Khan, S. Bonifacio, J. Clowes, A. Foulds, R. Holland, J.C. Matthews, C.J. Percival, D.E. Shallcross, Investigation of biofuel as a potential renewable energy source, Atmosphere 12 (10) (2021), https://doi.org/10.3390/atmos12101289.

[33] R. Ahorsu, F. Medina, M. Constantí, Significance and challenges of biomass as a suitable feedstock for bioenergy and biochemical production: a review, Energies 11 (12) (2018), https://doi.org/10.3390/en11123366.

[34] M.E. Montingelli, S. Tedesco, A.G. Olabi, Biogas production from algal biomass: a review, in: Renewable and Sustainable Energy Reviews, vol 43, Elsevier, 2015, pp. 961–972, https://doi.org/10.1016/j.rser.2014.11.052.

[35] R. Rajkumar, Z. Yaakob, M.S. Takriff, Potential of the micro and macro algae for biofuel production: a brief review, Bioresources 9 (1) (2014), https://doi.org/10.15376/biores.9.1.1606-1633.

[36] S.O. Ebhodaghe, O.E. Imanah, H. Ndibe, Biofuels from microalgae biomass: a review of conversion processes and procedures, Arab. J. Chem. 15 (2) (2022) 103591, https://doi.org/10.1016/j.arabjc.2021.103591.

[37] N.K. Aliya, C.M. Jijeesh, K.C. Jisha, Algae: source of biofuel and phytoremediation. Bioenergy Crops, CRC Press, 2022, pp. 112–135.

[38] R. Araújo, F. Vázquez Calderón, J. Sánchez López, I.C. Azevedo, A. Bruhn, S. Fluch, M. Garcia Tasende, F. Ghaderiardakani, T. Ilmjärv, M. Laurans, M. Mac Monagail, S. Mangini, C. Peteiro, C. Rebours, T. Stefansson, J. Ullmann, Current status of the algae production industry in europe: an emerging sector of the blue bioeconomy, Front. Mar. Sci. 7 (2021), https://doi.org/10.3389/fmars.2020.626389.

[39] K.T.X. Tong, I.S. Tan, H.C.Y. Foo, M.K. Lam, S. Lim, K.T. Lee, Advancement of biorefinery-derived platform chemicals from macroalgae: a perspective for bioethanol and lactic acid, Biomass Convers. Biorefin. (2022), https://doi.org/10.1007/s13399-022-02561-7.

[40] T.V. Ramachandra, D. Hebbale, Bioethanol from macroalgae: prospects and challenges, Renew. Sustain. Energy Rev. 117 (2020) 109479, https://doi.org/10.1016/j.rser.2019.109479.

[41] S. Elhenawy, M. Khraisheh, F. AlMomani, M. Al-Ghouti, M.K. Hassan, From waste to watts: updates on key applications of microbial fuel cells in wastewater treatment and energy production, Sustainability 14 (2) (2022), https://doi.org/10.3390/su14020955.

[42] K. Wang, K.S. Khoo, K.W. Chew, A. Selvarajoo, W.-H. Chen, J.-S. Chang, P.L. Show, Microalgae: the future supply house of biohydrogen and biogas, Front. Energy Res. 9 (2021), https://doi.org/10.3389/fenrg.2021.660399.

[43] R. Chu, D. Hu, L. Zhu, S. Li, Z. Yin, Y. Yu, Recycling spent water from microalgae harvesting by fungal pellets to re-cultivate *Chlorella vulgaris* under different nutrient loads for biodiesel production, Bioresour. Technol. 344 (2022) 126227, https://doi.org/10.1016/j.biortech.2021.126227.

[44] M. Sakarika, M. Kornaros, *Chlorella vulgaris* as a green biofuel factory: comparison between biodiesel, biogas and combustible biomass production, Bioresour. Technol. 273 (2019) 237–243, https://doi.org/10.1016/j.biortech.2018.11.017.

[45] N. Hamidian, H. Zamani, Biomass production and nutritional properties of *Chlorella sorokiniana* grown on dairy wastewater, J. Water Process Eng. 47 (2022) 102760, https://doi.org/10.1016/j.jwpe.2022.102760.

[46] D. Xie, X. Ji, Y. Zhou, J. Dai, Y. He, H. Sun, Z. Guo, Y. Yang, X. Zheng, B. Chen, *Chlorella vulgaris* cultivation in pilot-scale to treat real swine wastewater and mitigate carbon dioxide for sustainable biodiesel production by direct enzymatic transesterification, Bioresour. Technol. 349 (2022) 126886, https://doi.org/10.1016/j.biortech.2022.126886.

[47] T. Fazal, M.S.U. Rehman, F. Javed, M. Akhtar, A. Mushtaq, A. Hafeez, A. Alaud Din, J. Iqbal, N. Rashid, F. Rehman, Integrating bioremediation of textile wastewater with biodiesel production using microalgae (Chlorella vulgaris). Chemosphere, 281 (2021), Article 130758, https://doi.org/10.1016/j.chemosphere.2021.130758.

[48] A. Jacob, B. Ashok, K.M. Usman, Production of *chlorella pyrenoidosa* biodiesel by heterotrophic pathway to improve CI engine output characteristics using statistical approaches, Process Saf. Environ. Protect. 160 (2022) 478–490, https://doi.org/10.1016/j.psep.2022.02.040.

[49] O. Jafari, M. Seyed, M. Azin, N. Moazami, Application of a statistical design to evaluate bioethanol production from chlorella S4 biomass after acid - thermal pretreatment, Renew. Energy 182 (2022) 60–68, https://doi.org/10.1016/j.renene.2021.10.019.

[50] E. Vimali, A. Senthil Kumar, N. Sakthi Vignesh, B. Ashokkumar, A. Dhakshinamoorthy, A. Udayan, M. Arumugam, A. Pugazhendhi, P. Varalakshmi, Enhancement of lipid accumulation in microalga desmodesmus sp. VV2: response surface methodology and artificial neural network modeling for biodiesel production, Chemosphere 293 (2022) 133477, https://doi.org/10.1016/j.chemosphere.2021.133477.

[51] Z. Huang, J. Zhang, M. Pan, Y. Hao, R. Hu, W. Xiao, G. Li, T. Lyu, Valorisation of microalgae residues after lipid extraction: pyrolysis characteristics for biofuel production, Biochem. Eng. J. 179 (2022) 108330, https://doi.org/10.1016/j.bej.2021.108330.

[52] N. Amit Kumar, S. Verma, J. Park, A.K. Jaiswal, U.K. Ghosh, R. Gautam, Utilization of nano-sized waste lime sludge particles in harvesting marine microalgae for biodiesel feedstock production, Nanotechnol. Environ. Eng. 7 (1) (2022) 99–107, https://doi.org/10.1007/s41204-021-00195-0.

[53] M.E. Montingelli, K.Y. Benyounis, B. Quilty, J. Stokes, A.G. Olabi, Optimisation of biogas production from the macroalgae *Laminaria* sp. at different periods of harvesting in Ireland, in: Applied Energy, vol 177, Elsevier Ltd, 2016, pp. 671–682, https://doi.org/10.1016/j.apenergy.2016.05.150. September.

[54] R. Sivaramakrishnan, S. Suresh, S. Kanwal, G. Ramadoss, B. Ramprakash, A. Incharoensakdi, Microalgal biorefinery concepts' developments for biofuel and bioproducts: current perspective and bottlenecks, Int. J. Mol. Sci. 23 (5) (2022), https://doi.org/10.3390/ijms23052623.

[55] G.F. Ferreira, L.F. Ríos Pinto, P.O. Carvalho, M.B. Coelho, M.N. Eberlin, R. Maciel Filho, L.V. Fregolente, Biomass and lipid characterization of microalgae genera *Botryococcus, Chlorella*, and *Desmodesmus* aiming high-value fatty acid production, Biomass Convers. Biorefin. 11 (5) (2021) 1675–1689, https://doi.org/10.1007/s13399-019-00566-3.

[56] J.- H. Hwang, W.H. Lee, Continuous photosynthetic biohydrogen production from acetate-rich wastewater: influence of light intensity, Int. J. Hydrog. Energy 46 (42) (2021) 21812–21821, https://doi.org/10.1016/j.ijhydene.2021.04.052.

[57] R. Chandra, S. Pradhan, A. Patel, U.K. Ghosh, An approach for dairy wastewaterremediation using mixture of microalgae and biodiesel production for sustainabletransportation. J. Environ. Manag. 297 (2021), 113210, https://doi.org/10.1016/j.jenvman.2021.113210.

[58] S. Ge, K. Brindhadevi, C. Xia, A. Salah Khalifa, A. Elfasakhany, Y. Unaprom, H. Van Doan, Enhancement of the combustion, performance and emission characteristics of *Spirulina* microalgae biodiesel blends using nanoparticles, Fuel 308 (2022) 121822, https://doi.org/10.1016/j.fuel.2021.121822.

[59] M. Haghighi, L.B. Zare, M. Ghiasi, Biodiesel production from *spirulina* algae oil over [Cu(H2PDC)(H2O)2] complex using transesterification reaction: experimental study and DFT approach, Chem. Eng. J. 430 (2022) 132777, https://doi.org/10.1016/j.cej.2021.132777.

[60] N. Toinoi, A. Reungsang, A. Salakkam, Co-digestion of sugarcane bagasse, microalgal biomass and cow dung for biohydrogen and methane production, Asia-Pacific J. Sci. Technol. 27 (02) (2022).

[61] S.J. King, A. Jerkovic, L.J. Brown, K. Petroll, R.D. Willows, Synthetic biology for improved hydrogen production in *chlamydomonas Reinhardtii*, Microb. Biotechnol. (2022), https://doi.org/10.1111/1751-7915.14024.

[62] M. Jesus Torres, D. González-Ballester, A. Gómez-Osuna, A. Galván, E. Fernandez, A. Dubini, Chlamydomonas-Methylobacterium oryzae cooperation leads to increased biomass, nitrogen removal and hydrogen production, Bioresour. Technol. 352 (2022) 127088, https://doi.org/10.1016/j.biortech.2022.127088.

[63] P. Liu, D. Ye, M. Chen, J. Zhang, X. Huang, Chlamydomonas mutant Hpm91 lacking PGR5 is a scalable and valuable strain for algal hydrogen (H2) production, bioRxiv (2022).

[64] P.Q.M. Bezerra, L. Moraes, T.N.M. Silva, L.G. Cardoso, J.I. Druzian, M.G. Morais, I.L. Nunes, J.A.V. Costa, Innovative application of brackish groundwater without the addition of nutrients in the cultivation of *Spirulina* and *Chlorella* for carbohydrate and lipid production, Bioresour. Technol. 345 (2022) 126543, https://doi.org/10.1016/j.biortech.2021.126543.

[65] E. Ansah, L. Wang, B. Zhang, A. Shahbazi, Catalytic pyrolysis of raw and hydrothermally carbonized Chlamydomonas debaryana microalgae for denitrogenation and production of aromatic hydrocarbons, Fuel 228 (2018) 234–242, https://doi.org/10.1016/j.fuel.2018.04.163.

[66] M.B. Tasić, L.F.R. Pinto, B.C. Klein, V.B. Veljković, R.M. Filho, Botryococcus braunii for biodiesel production, in: Renewable and Sustainable Energy Reviews 64, Elsevier Ltd, 2016, https://doi.org/10.1016/j.rser.2016.06.009.

[67] M. Mofijur, A.S.M. Rahman, L.N. Nguyen, T.M.I. Mahlia, L.D. Nghiem, Selection of microalgae strains for sustainable production of aviation biofuel, Bioresour. Technol. 345 (2022) 126408, https://doi.org/10.1016/j.biortech.2021.126408.

[68] R.K. Goswami, K. Agrawal, P. Verma, Microalgae dunaliella as biofuel feedstock and β-carotene production: an influential step towards environmental sustainability, Energy Convers. Manag. 13 (2022) 100154, https://doi.org/10.1016/j.ecmx.2021.100154.

[69] J. Ma, L. Li, Q. Zhao, L. Yu, C. Frear, Biomethane production from whole and extracted algae biomass: long-term performance evaluation and microbial community dynamics, Renew. Energy 170 (2021) 38–48, https://doi.org/10.1016/j.renene.2021.01.113.

[70] A. Amrullah, O. Farobie, A. Bayu, N. Syaftika, E. Hartulistiyoso, N.R. Moheimani, S. Karnjanakom, Y. Matsumura, Slow pyrolysis of *Ulva lactuca* (chlorophyta) for sustainable production of bio-oil and biochar, Sustainability 14 (6) (2022), https://doi.org/10.3390/su14063233.

[71] N. Sharmiladevi, A. Swetha, K.P. Gopinath, Processing of gracilaria edulis and ulva lactuca for bioethanol and bio-oil production: an integrated approach via fermentation and hydrothermal liquefaction, Biomass Convers. Biorefin. (2021) 1–9, https://doi.org/10.1007/s13399-021-01925-9.

[72] K. Ruangrit, S. Chaipoot, R. Phongphisutthinant, W. Kamopas, I. Jeerapan, J. Pekkoh, S. Srinuanpan, Environmental-friendly pretreatment and process optimization of macroalgal biomass for effective ethanol production as an alternative fuel using Saccharomyces cerevisiae, Biocatal. Agric. Biotechnol. 31 (2021) 101919, https://doi.org/10.1016/j.bcab.2021.101919.

[73] M.J. Hessami, S.M. Phang, J. Sohrabipoor, F.F. Zafar, S. Aslanzadeh, The bio-methane potential of whole plant and solid residues of two species of red seaweeds: *Gracilaria manilaensis* and *Gracilariopsis persica*, Algal Res. 42 (2019) 101581, https://doi.org/10.1016/j.algal.2019.101581.

[74] I.A. Nascimento, S.S.I. Marques, I.T.D. Cabanelas, S.A. Pereira, J.I. Druzian, C.O. de Souza, D.V. Vich, G.C. de Carvalho, M.A. Nascimento, Screening microalgae strains for biodiesel production: lipid productivity and estimation of fuel quality based on fatty acids profiles as selective criteria, 1, in: Bioenergy Research, vol 6Springer Science and Business Media, LLC, 2013, https://doi.org/10.1007/s12155-012-9222-2, 1–13.

[75] M.I. Khan, M.G. Lee, J.H. Shin, J.D. Kim, Pretreatment optimization of the biomass of *Microcystis aeruginosa* for efficient bioethanol production, Amb. Express 7 (1) (2017) 1–9, https://doi.org/10.1186/s13568-016-0320-y.

[76] E.A. Ouahid, R. Mohamed, O. meryam, H.S. Ali Yahya, L. Latrach, F. Soufiane, C. Abdelhafid, Evaluation of the energetic valorization of the lagoon and mediterranean algae (*caulerpa prolifera & corallina elongata*) by anaerobic digestion, in: Scientific African, vol 5, Elsevier B.V, 2019, https://doi.org/10.1016/j.sciaf.2019.e00111.

[77] K. Mureed, S. Kanwal, A. Hussain, S. Noureen, S. Hussain, S. Ahmad, M. Ahmad, R. Waqas, Biodiesel production from algae grown on food industry wastewater, in: Environmental Monitoring and Assessment, vol 190, Springer International Publishing, 2018, pp. 1–11, https://doi.org/10.1007/s10661-018-6641-3, 5.

[78] M.R. Tabassum, A. Xia, J.D. Murphy, Biomethane production from various segments of Brown seaweed, Energy Convers. Manag. 174 (2018) 855–862, https://doi.org/10.1016/j.enconman.2018.08.084.

[79] R. Lin, C. Deng, L. Ding, A. Bose, J.D. Murphy, Improving gaseous biofuel production from seaweed Saccharina latissima: the effect of hydrothermal pretreatment on energy efficiency, Energy Convers. Manag. 196 (2019) 1385–1394, https://doi.org/10.1016/j.enconman.2019.06.044.

[80] M.C. Ruiz-Domínguez, F. Salinas, E. Medina, B. Rincón, M.Á. Martín, M.C. Gutiérrez, P. Cerezal-Mezquita, Supercritical fluid extraction of fucoxanthin from the diatom Phaeodactylum tricornutum and biogas production through anaerobic digestion, Mar. Drugs 20 (2) (2022), https://doi.org/10.3390/md20020127.

[81] M.N. Fernando, R. Kapilan, Small scale biodiesel production from *Sargassum* sp. and optimization of conditions for yield enhancement, J. Sci. 11 (1) (2020) 23, https://doi.org/10.4038/jsc.v11i1.25.

[82] G. Gao, A.S. Clare, C. Rose, G.S. Caldwell, Ulva rigida in the future ocean: potential for carbon capture, bioremediation and biomethane production, GCB Bioenergy 10 (1) (2018) 39−51, https://doi.org/10.1111/gcbb.12465.

[83] J. Hinks, S. Edwards, P.J. Sallis, G.S. Caldwell, The steady state anaerobic digestion of *Laminaria hyperborea* − effect of hydraulic residence on biogas production and bacterial community composition, Bioresour. Technol. 143 (2013) 221−230, https://doi.org/10.1016/j.biortech.2013.05.124.

[84] O.M. Adeniyi, U. Azimov, A. Burluka, Algae biofuel: current status and future Applications, in: Renewable and Sustainable Energy Reviews, vol 90, Pergamon, 2018, pp. 316−335, https://doi.org/10.1016/J.RSER.2018.03.067.

[85] V.C. Akubude, K.N. Nwaigwe, E. Dintwa, Production of biodiesel from microalgae via nanocatalyzed transesterification process: a review, in: Materials Science for Energy Technologies, vol 2, Elsevier, 2019, pp. 216−225, https://doi.org/10.1016/J.MSET.2018.12.006, 2.

[86] A.K. Koech, A. Kumar, Z.O. Siagi, In situ transesterification of *spirulina* microalgae to produce biodiesel using microwave irradiation, J. Energy (2020) 1−10, https://doi.org/10.1155/2020/8816296, vol 2020 (December). Hindawi Limited.

[87] Q.A. Abdallah, B. Nixon, J.R. Fortwendel, The enzymatic conversion of major algal and cyanobacterial carbohydrates to bioethanol, in: Frontiers in Energy Research, vol 4, Frontiers Media S.A, 2016, p. 36, https://doi.org/10.3389/FENRG.2016.00036/BIBTEX.

[88] M.G. Saad, N.S. Dosoky, M.S. Zoromba, H.M. Shafik, Algal biofuels: current status and key challenges, Energies 12 (10) (2019), https://doi.org/10.3390/en12101920.

[89] F. Vargas e Silva, L.O. Monteggia, Pyrolysis of algal biomass obtained from high-rate algae ponds applied to wastewater treatment, in: Frontiers in Energy Research, vol 3, Frontiers Media S.A, 2015, p. 31, https://doi.org/10.3389/FENRG.2015.00031/BIBTEX. Jun.

[90] J.J. Milledge, B.V. Nielsen, S. Maneein, P.J. Harvey, A brief review of anaerobic digestion of algae for BioEnergy, Energies 12 (6) (2019) 1−22, https://doi.org/10.3390/en12061166.

[91] A.J. Ward, D.M. Lewis, F.B. Green, Anaerobic digestion of algae biomass: a review, in: Algal Research, vol. 5, Elsevier, 2014, pp. 204−214, https://doi.org/10.1016/J.ALGAL.2014.02.001, 1.

[92] M.M. El-Sheekh, Biodiesel from microalgae: advantages and future prospective Egyptian, J. Botany 61 (3) (2021) 669−671.

[93] S. Behera, R. Singh, R. Arora, N.K. Sharma, M. Shukla, S. Kumar, Scope of algae as third generation biofuels, Front. Bioeng. Biotechnol. 2 (2015) 90, https://doi.org/10.3389/fbioe.2014.00090.

[94] A. Xia, J. Cheng, W. Song, H. Su, L. Ding, R. Lin, H. Lu, J. Liu, J. Zhou, K. Cen, Fermentative hydrogen production using algal biomass as feedstock, in: Renewable and Sustainable Energy Reviews, vol 51, Pergamon, 2015, pp. 209−230, https://doi.org/10.1016/J.RSER.2015.05.076. November.

[95] H.D. Siegler, Process intensification for sustainable algal fuels production, in: Handbook of Algal Biofuels, Elsevier, 2022, https://doi.org/10.1016/B978-0-12-823764-9.00011-X.

[96] N. Rafa, S.F. Ahmed, I.A. Badruddin, M. Mofijur, S. Kamanngar, Strategies to produce cost-effective third-generation biofuel from microalgae, Front. Energy Res. 9 (2021) 749968.

[97] S. Broberg, V. Andersson, R. Hackl, Integrated Algae Cultivation for Biofuels Production in Industrial Clusters. Report Number: Arbetsnotat Nr 47, 2011.

[98] T.J. Lundquist, I.C. Woertz, N.W.T. Quinn, J.R. Benemann, A Realistic Technology and Engineering Assessment of Algae Biofuel Production, Vol October, Energy, California, 2010.

[99] M. Branco-Vieiraab, T.M. Mataa, A.A. Martinsa, M.A.V. Freitas, N.S. Caetanoa, Economic analysis of microalgae biodiesel production in a small-scale facility, Energy Rep 6 (8) (2020) 325−332, https://doi.org/10.1016/j.egyr.2020.11.156.

[100] D.T. Zewdie, A.Y. Abubeker, Techno-economic analysis of microalgal biofuel production coupled with sugarcane processing factories, S. Afr. J. Chem. Eng. 40 (2022) 70−79.

[101] S.R. Lyon, H. Ahmadzadeh, M.A. Murry, Algae-Based Wastewater Treatment for Biofuel Production: Processes, Species, and Extraction Methods. Biomass and Biofuels from Microalgae, Springer, Cham, 2015, pp. 95−115.

[102] L.M.L. Laurens, State of Technology Review − Algae Bioenergy, IEA Bioenergy, 2017, https://doi.org/10.1017/S0959270900002288.

[103] A. Abomohra, M. Wagner, M. El-Sheekh, D. Hanelt, Lipid and total fatty acid productivity in photoautotrophic fresh water microalgae: screening studies towards biodiesel production, J. Appl. Phycol. 25 (4) (2013) 931−936.

[104] A. Abomohra, M.M. El-Sheekh, D. Hanelt, Screening of marine microalgae isolated from the hypersaline Bardawil lagoon for biodiesel feedstock, Renew. Energy 101 (2017) 1266−1272.

[105] T. Driver, A. Bajhaiya, J.K. Pittman, Potential of bioenergy production from microalgae, Curr. Sustain. Renew. Energy Rep. 1 (2014) 94−103.

[106] M.M. El-Sheekh, A. Abomohra, H. El-Adel, M. Battah, S. Mohammed, Screening of different species of *Scenedesmus* isolated from Egyptian freshwater habitats for biodiesel production, Renew. Energy 129 (2018) 114−120.

Chapter 7

Microbial bioprospecting of biodiesel industry-derived crude glycerol waste conversion into value-added products

Hiren K. Patel[1,3], Nidhi P. Patel[1] and Maulin P. Shah[2]

[1]School of Science, P. P. Savani University, Surat, Gujarat, India; [2]Industrial Waste Water Research Lab, Environ Technology Limited, Ankleshwar, Gujarat, India; [3]School of Agriculture, P. P. Savani University, Surat, Gujarat, India

1. Introduction

Fossil energy use is the backbone of modern life, from automobiles to light bulbs. However, many individuals are concerned about refinery efficiency, global warming, air pollution, and fossil fuel resource depletion [1]. Biodiesel is an alternative fuel for use in place of diesel fuel and is quickly gaining popularity. US biodiesel consumption is growing, as annual biodiesel output was 2.83 times higher in 2012 (969 million gallons) than in 2010 (343 million gallons) [2]. Furthermore, according to the US Energy Development and Rules 2007, photosynthetically diesel generation was predicted to hit 36 billion gallons by 2022 under alternative energy standard RFS2 [3].

In a world where petroleum has become scarce and expensive, biofuels and chemicals produced from renewable resource feedstocks are needed to meet energy demand. One of the most serious problems with biofuels is their high production costs, which can be reduced if waste from biofuel manufacturing operations can be turned into lucrative coproducts [4]. Biofuel has been mandatory in many countries due to its ability to reduce net greenhouse gas (GHG) emissions [5]. The most frequent way of obtaining it is the transesterification of lipids and vegetable oils in the presence of an acid catalyst by methylation of fats by alcohol (typically methanol). A biofuel is a fatty acid methyl ester. Oilseed, rapeseed, soybean, and palm oils are the main substrates of sunflowers. Biodiesel is made worldwide, with geographical variations. Biofuel is the most important energy source in Brazil, for example, where it represented 80% of the generation of oil from soybeans in 2010 [6].

The amount of waste produced by biodiesel production has increased considerably in recent years. Although Europe remains the world's largest biodiesel producer, Brazil has witnessed the fastest growth in recent years, compared with the USA and Europe, with production increasing from 736 m^3 in 2005 to 2,400,000 m^3 in 2010. Omelets and crude glycerol, the two most prevalent residues, are growing in concert with the biodiesel industry. Pies are often used as animal feed or fertilizers, and are manufactured by crushing palms, seeds, and other plant products for oil exploration. This adds value to the biofuel manufacturing chain. The principal contaminants in crude glycerol, which is produced by the transesterification of fats and vegetable oils (triglycerides) to biodiesel, are methanol, salts, soaps, and water. The amount and existence of each impurity vary substantially from one company to the next due to various factors, such as cooking oil and reaction circumstances.

Sustained, long-term social and economic growth requires energy, and fossil fuel extraction accounts for most power consumption [7]. The fossil fuel reserves of our planet are limited. The depletion of environmentally damaging fossil fuels has compelled researchers worldwide to search for realistic renewable alternative energy sources to meet the energy requirements of all inhabitants, manufacturing industries, and transport industries while they reduce carbon emissions and address other environmental issues. When fossil fuel exploitation reaches its peak, supply will enter a final decline phase that triggers regional instability and compromises global energy security [8].

Biofuels can become a good replacement for harmful fossil energy by tackling the difficulties arising from their use, ensuring energy security by reducing the world's reliance on fossil fuels, meeting rising energy demand, and minimizing serious climate change [7]. Microbes can potentially produce biofuels and their typical functions, such as wastewater treatment, antibiotics, vaccinations, probiotics, and foodstuff fermentation. As a result, microbiological biofuels have piqued scientific interest as a preferable alternative energy source that benefits the environment. Microbes can produce bioethanol, biobutanol, biodiesel, biohydrogen, biogas, and other biofuels. Microbes have built-in pathways for converting a wide variety of substrates into biofuels. In the past, microbes played a crucial role in biomass saccharification and fermentation. Ligninolytic enzymes' potential role in digesting cellulosic biomass has piqued people's interest. Much work has been done to develop microbial fuels with a good dividend and sociocultural stability that can be used as a low-cost energy source.

Biodiesel is an alternative to diesel fuel that could be used in internal combustion engines (ICEs). It is made from plant oil (soybean oil, cottonseed oil, canola oil, corn oil), recovered cooking lubes or oils (yellow lubricant), animal fats (beef tallow, pork lard), or various mixtures of these. Plant-based cooking oils are the most common; however, animal fats can also be used. Previously used cooking oils are both biodegradable and renewable [9].

2. Biodiesel

Rudolf Diesel (Rudolf Christian Karl Diesel, a German mechanical engineer, March 18, 1858–September 29, 1913), the inventor of the diesel engine, tested vegetable oil for the first time in 1897, before the introduction of petroleum-based diesel fuel. Until 2001, biodiesel was used in the United States as an energy source only in small amounts. Since then, US biodiesel production and use have increased dramatically, thanks in part to government incentives and regulations to produce, promote, and use biodiesel.

The discovery of the transesterification of vegetable oils in the mid-19th century gave rise to biodiesel. It took another half-century before the world realized its significance as an energy source. During the 20th century, the push for global security among states was spurred by World War II and other regional upheavals. This chapter addresses the concept of a biodiesel fuel utilizing vegetable oils and animal fats. People used biofuel and pure vegetable oil during World War I, and energy crises prompted the widespread growth of these fuels near the end of the 20th century [10].

The feedstock, the final product, and the according to are all renewable in biodiesel. As a result, the term "carbon neutral" is frequently used to describe it. Biodiesel is safer in a crash or spill because of its renewability and high flash point. With today's technology, biodiesel may be manufactured from any plant- or animal-derived oil. However, some oils produce greater benefits from biofuel conversion than others, as previously stated [11]. Various vegetable oils and other lipid sources have been used as feedstocks worldwide for years. Finally, a suitable starting material for transesterification and subsequent biodiesel generation is required. One of these is triglycerides, which comprise three long strings of fatty acids joined by glycerin molecules as the beginning material. Fatty acids come in a variety of lengths and degrees of unsaturation.

Biofuel progression, including the use of biofuels, can be divided into four periods:

1. Straight vegetable oils (SVOs) were employed as a lamp fuel from antiquity through the mid-19th century.
2. SVOs were employed in ICEs during the 1930 and 1940s and chemically modified to generate biodiesel.
3. SVOs were employed as a petroleum substitute during the 1970s oil shortages.
4. Alternative fuels are sought to address global energy demand, agricultural production needs, and environmental concerns.

2.1 Chemistry of biodiesel transesterification of oils

Patrick and Duffy, two Irish chemists, reported the transesterification of oils in 1853.

Transesterification is a major industrial reaction used to make biodiesel and a variety of household goods like soaps and detergents. This method is used to make all biodiesel on the planet [12].

Triglycerides are molecules with three fatty acids connected to the hydroxyl groups and a glycerol molecule with a significant overall impact.

The chemical composition of the fatty acids connected to the glycerol determines the fat's general features. Because vegetable and nut oils are primarily triacyl glycerol or triglycerides, they are frequently used as biodiesel precursors. Fig. 7.1 illustrates triglyceride transesterification with methanol.

2.2 Chemistry of biodiesel transesterification of oils

Plant and seed oils have been around since 1500 BCE, according to historical records. Oils and fats were used for heat and electricity in the past, but the Ancient Egyptians also used scented oils for beauty treatments, religious activities, and

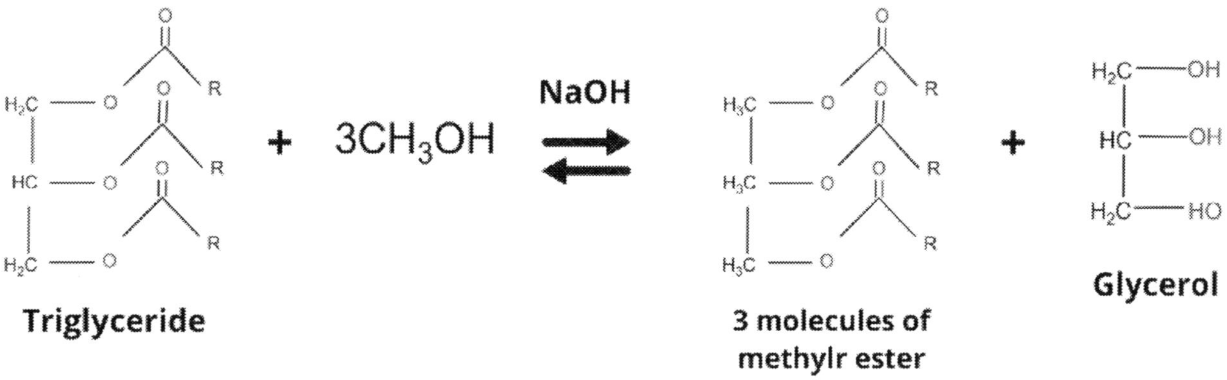

FIGURE 7.1 Transesterification of triglycerides with methanol.

healing. Furthermore, these oils had long been used in the food industry. Recent archaeological study in Galilee, Israel, has revealed that olive oil was harvested for a source of food as soon as 6000 BCE [13,14].

Olive trees have been cultivated in the eastern Mediterranean since 3500 BCE. The ancient Greeks used olive oil for nourishment, religious ceremonies, oil lamp fuel, and medicinal treatments. It was one of the region's main exports and is still a popular item in modern-day Italy and Greece. Plant and natural oils were used similarly by other ancient civilizations [15].

Castor oil was widely used in ancient Egypt, as evidenced by the Ebers Papyrus. The ancient book details the use of numerous plant components and derived oils for headaches, breathing, digestion, cosmetic procedures, and hair development. The oil was used for ceremonial purposes and as a source of fuel. The Smith Papyrus mentions almond oil combinations for skin and antiaging therapies [16,17].

As synthetic pharmaceutical chemistry improved in the 19th and early 20th centuries, medicinal herb oils fell out of favor [18]. Natural, plant, and essential oils have sparked fresh interest in their use as homeopathic and Eastern medicines in recent years. For example, a recent study compared the efficacy of lemongrass, pine, and clove oil with that of DEET. These oils were discovered to be up to 98% as effective as a common insecticide [19]. Plant oils are used for food, aesthetic, and therapeutic purposes in modern times, just as they were in ancient times. They are still used in developing nations as heating and lamp oil and as an energy source for emergency generators.

2.3 The transition to a fuel: 19th-century internal combustion engine use and fuel choices

Biodiesel was created and is now widely used thanks to ICEs. Even before the diesel engine was developed, multiple types of engines existed. Attempts at making an ICE were made throughout the 17th century. Lyle Cummins was a historian specializing in US history and recounted the history of these endeavors, during which the name of his book was revealed. In 1893, German engineer Rudolf Diesel invented the diesel engine. The concept and execution of appropriate temperatures are discussed in this study. The first biofuel was manufactured, according to historical accounts. Oil is used to power Diesel's 10-iron-cylinder truck. At the bottom is a flywheel. Peanut oil was used to power it for a long time. On August 10, 1893, in Augsburg, Germany, Diesel strove to discover alternative fuels for the first time. These engines are more efficient than typical gasoline engines [20]. "While the usage of vegetable oils for combustion fuels may appear insignificant now, such oils may become as vital as gasoline and coal-tar products in the future," said the author in a speech he gave a year before his death in 1912.

He filed for a patent for his design in the following years. Steam engines were ubiquitous at the time of Diesel's creation, despite their low efficiency of approximately 10%. Diesel's idea was a game changer at the time. In his ICE concept, fuel combustion and piston movement proceeded through an isothermal response. Despite being revolutionary within the sector, Diesel's first engine models were heavy and difficult to move, making them unsuitable for motors or trains.

2.4 Problems brought about by biodiesel

People who use biodiesel fuel have experienced a slew of issues. According to a thorough analysis, most of these issues are caused by low-quality biodiesel fuel and are nearly identical to the issues produced by low-quality petroleum diesel. However, a few issues (particularly cold-weather issues) are attributed to the inherent features of biodiesel fuel rather than low fuel quality. Fortunately, most of these issues may be prevented or reduced. The following are some of the most common biodiesel engine difficulties and their possible causes and treatments. This chapter is not intended to be a comprehensive repair manual; rather, it provides an overview of some performance difficulties associated with biodiesel fuel [21].

Problem: The fuel spray pattern is affected by deposits on injectors. Misfiring or a difficult start are the most common symptoms. This is most likely due to cold-weather operation with partially solidified fuel or fuel that has not been fully converted from oil to biodiesel. Vegetable oil tends to collect on injectors, specifically when the engine runs at part load.

Solution: Have the injectors cleaned by a competent mechanic—the exact nature of injectors makes them difficult to clean without specific expertise and equipment. Low-temperature flow-improving additions could enhance fuel performance in cold weather and help prevent future problems. Check to see if the fuel is devoid of pollutants and has been completely converted from biodiesel [22].

Problem: Injector pump deposits (varnish and gums) affect performance. Hard starting, diminished power, and misfiring are the most typical symptoms. It can be caused by either incompletely processed biodiesel fuel or biofuel that has completely oxidized.

Solution: A trained mechanic should clean the injector pump. Due to the exact nature of the pump's components, this work is not viable for the home mechanic, as it was with the injectors [23].

Problems: In cold conditions, the engine sometimes refuses to start or operates for only a few moments after starting. The filter is most likely clogged with hardened biodiesel particles.

Solution: You may wait for spring to arrive, or you could try heating the fuel filter with a 12 V jacket heater. Biodiesel can benefit from the "deicer" chemicals used in petroleum diesel. If you reside in a cold environment, you might consider adding a cold-weather additive to your fuel or installing a "preheater" to heat the fuel tank and filter. During the cooler months of the year, you may need to presoak your fuel by combining biodiesel with kerosene or cold-weather petroleum diesel [24].

3. Application of biodiesel

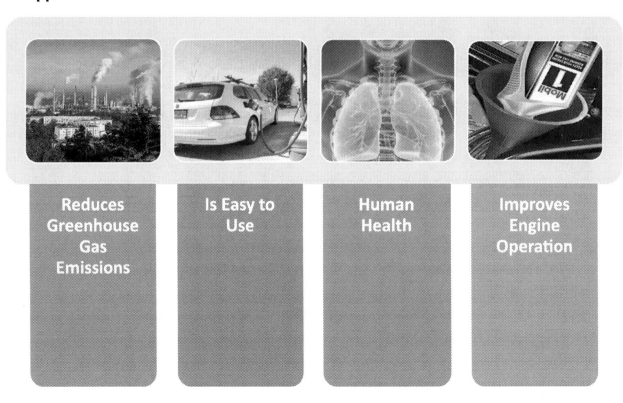

3.1 Biodiesel reduces greenhouse gas emissions

When biodiesel substitutes petroleum, it drastically reduces the life cycle of GHG emissions. According to a life cycle analysis conducted by the Argonne National Laboratory, GHG emissions from B100 appear to be 74% lower than those from petroleum diesel. The California Air Resources Board (CARB) recently released similar findings in a life cycle review of biodiesel from several sources [25].

As they grow, oilseed plants utilize CO_2 from the air to produce their stems, roots, leaves, and seeds. Oil is taken from oilseeds and converted into biodiesel. When biodiesel is burned, CO_2 and other pollutants are emitted and returned to the atmosphere. On balance, the CO_2 generated does not increase to the net level of carbon dioxide in the atmosphere because it is recycled by the next oilseed crop because it grows. Because fossil fuels and chemicals are used in agriculture and the biodiesel production process, a small percentage of the carbon exhaled comes from fossil fuels.

3.2 Biodiesel reduces tailpipe emissions

The pollution control catalysts and filters used in current diesel engines have been shown to be entirely compatible with biodiesel, resulting in substantial decreases in nitrogen oxide (NO_x) and particulate matter (PM) pollutants (sometimes called NTDEs). In the case of biodiesel, the effects are feedstock agnostic. The current biodiesel requirement is being examined to see if it adequately protects NTDE pollution control catalysts and filters.

NO_x emissions have been a concern with elderly diesel engines and biodiesel. Some of these environmental problems have been alleviated by replacing older engines with modern engines. According to CARB, the use of NTDEs will eradicate any fuel-related NO_x emissions [26,27]. NTDEs are automobiles that have received CARB certification after the 2010 model year. California introduced new rules for the use of biofuels in late 2015. A NO_x-reducing ingredient must be used in blends that exceed the concentration levels listed.

3.3 Biodiesel and human health

The effect of biodiesel and its mixes on human health is a hot topic of research. Diesel engine emissions of particulates and hydrocarbons may be harmful and/or carcinogenic. On this subject, there is a substantial amount of material available [28,29]. The Mining Safety and Security Administration of the US Dept. of Labor issued guidelines in 2011 that prohibit exposure of workers to diesel PM in deep mines. According to the Mine Safety and Health Administration, shifting from gasoline diesel fuels to higher biodiesel and its blend levels (B50–B100) considerably reduced PM emissions from underground diesel cars and substantially decreased worker exposure. In regions where humans are exposed to elevated amounts of diesel exhaust, even tiny amounts of biodiesel reduce PM pollution and give considerable medical and compliance benefits.

3.4 Biodiesel improves engine operation

Biodiesel improves the lubricity of gasoline and raises the cetane number even at low concentrations. Diesel vehicles rely on the friction modifier of the fuel to keep moving parts, such as petrol pumps and nozzles, from wearing out prematurely. The ASTM D975 diesel fuel specification was updated to add a lubricity requirement to account for ultralow sulfur in diesel's reduced natural lubricity (a maximum wear scar diameter on the high-frequency reciprocating rig [HFRR] test of 520 microns). At mix percentages as low as 1%, biodiesel can provide appropriate lubricity to diesel fuels with low natural lubricity.

3.5 Biodiesel is easy to use

Furthermore, simplicity is a major advantage of biodiesel. For B20 and lower mixes, no new technology or equipment improvements are necessary. B20 can be stored in petroleum diesel containers and pumped with diesel fuel pumps. Despite a few unique handling and use precautions, most B20 consumers should not have a problematic experience with the product.

4. Production of biodiesel

Biofuel essentially consists of mono-alkyl esters generated from animal fat or vegetable oil. When biodiesel is used as a fuel, the quantity of carbon emitted from the combustion chamber is equivalent to the amount of carbon absorbed by an animal or plant during its entire existence. As a result, emissions from green biofuel combustion will be minimal.

Diesel is currently the most dominant fuel in the world, with uses in transportation, agriculture, power generation, and some industrial applications. An alternative must be discovered to provide long-term energy security. Vegetable oil can be used as a diesel replacement as well as gasoline in diesel engines [30]. By 2015, biofuels will account for 27% of overall gasoline demand, reducing CO_2 emissions by almost 2.1 Gt per year [31].

4.1 Biodiesel sources

Industrial wastes (such as fly ash, coke, blast - furnace slag and steel slags, lime mud, and so on) are plentiful all over the world. As a result, those waste items can provide catalysis opportunities.

Alternative diesel fuels, including vegetable oil and fats, are extracted from renewable, renewable sources. Soybean, sunflower, palm, rapeseed, canola, cottonseed, and Jatropha are the oils used most often in biodiesel production. Waste vegetable oils and quasi-crude vegetable oils are considered prospective low-cost biodiesel sources because culinary vegetable oils are more expensive than diesel fuel. In India, using such palatable oil to make biodiesel is likewise not practical due to a large gap in the supply and availability of such oils. Because of natural property variations, animal fats have not been investigated to the very same extent as vegetable oils, despite their frequent mention. Because animal fats include more saturated fatty acids, they solidify at ambient temperature. Biodiesel comes from a variety of sources, including vegetable oils, nonedible oils, animal fats, and other biomass.

4.1.1 Waste cooking oil

Because of its low price and extensive availability, waste cooking oil (WCO) is widely used to make biodiesel [32,33]. WCO consists of TGs, or glycerides, which can be made up of animal fat or vegetable oil. WCO is classified as yellow and brown grease by Font de Mora et al. [34] and is made up of palm, canola, corn, sunflower, and other oils used in food preparation. WCO can also be categorized by the source from which it is gathered, such as eateries, fast food establishments, and private residences. WCO can be converted into biodiesel using alkali- and acid-based catalysts [35,36].

Researchers have recently concentrated on optimizing biodiesel production from WCO through influencing factors and new catalysts. Hamze et al. [37] investigated the effects of catalyst loading, column temperature, and catalyst concentration-to-oil molar ratio on the yield of biodiesel created from, for example, WCO. At a 1.4% wt. catalyst loading, 65°C reaction temperature, and 7.5:1 catalyst-to-oil molar ratio, they produced a maximum biodiesel yield of 99.38% wt. under optimal conditions, with catalyst concentration as the most crucial parameter. In a study by Gurunathan and Ravi, biodiesel from WCO was investigated using a heterogeneous catalyst, such as copper-doped zinc oxide nanocomposite [38]. Under ideal conditions, the highest biodiesel yield was 97.71% (w/w), with a nanocatalyst concentration of 12% (w/w), an oil-to-methanol ratio of 1:8 (v:v), a reaction temperature of 55°C, and a reaction period of 50 min.

4.1.2 Algal oil

Micro- and macroalgae are cultivated in both inorganic and organic environments since they need light, CO_2, and other inorganic elements to develop [39]. Microalgae growth in open and closed systems in wastewater treatment has improved algal biomass output [40]. Farming, harvesting, lipid extraction, and transesterification are the phases in the transesterification of microalgal species for biodiesel production [41]. Pretreatment is among the most important phases in accelerating the lipid extraction process and increasing the lipid yield from algae. Enzymatic hydrolysis, ultrasonication, increased homogenization, microwaves, bead beating, and chemical techniques such as microwaves are among the pretreatment methods. The biodiesel production pathway is illustrated in Fig. 7.2.

4.1.3 Vegetable oil

In many countries, biodiesel is manufactured from a wide variety of food and quasi-vegetable oils. The United States exports edible oil from soybeans for biodiesel synthesis, although European countries use mustard as a raw ingredient. Palm oil or coconut oil are used in tropical regions like Malaysia, whereas nonedible vegetable oils like Jatropha, Simarouba, and Karanja are often used in India. Papaya seed and stone fruit kernel oils are the most widely used non-vegetable oils in Australian biofuel generation [42]. Jatropha, a nonedible vegetable oil, is viewed as a commercially viable alternative to edible oil for making biodiesel because of its physicochemical qualities [43].

Karanja is a promising potential source of nonedible vegetable oil for biodiesel production. These trees thrive near a road, canals, and the margins of agricultural areas.

Another nonedible source for producing biodiesel is macaw oil (*Acrocomia aculeata*), which is abundant and native to tropical America, particularly Brazil, and has high manufacturing potential (4–6 ton oil/ha) [44].

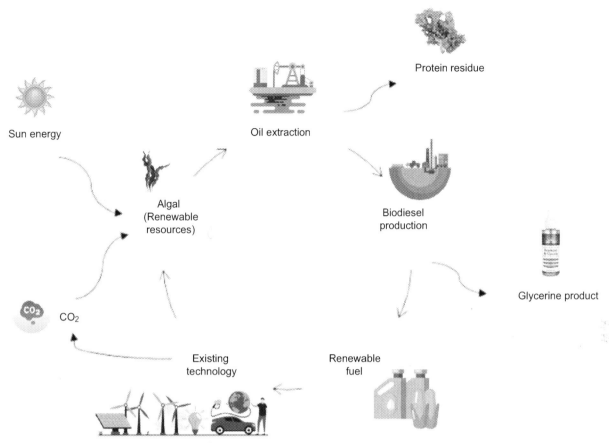
FIGURE 7.2 Algal biodiesel production pathway.

5. Crude glycerol waste from biodiesel as a by-product

The enormous growth of the transportation fuels sector, along with increased environmental concerns and limited crude oil supplies, has shifted attention to renewable energy. In recent years, biodiesel has attracted considerable attention as a viable alternative and ecological fuel, and its production capacity has increased dramatically.

5.1 Value-added opportunities for crude glycerol

Crude glycerol produced by biodiesel conversion increased from 200,000 tonnes in 2004 to 1.224 million tonnes in 2008 [45]. As a result, scientists must find new applications for refined and crude glycerol. Several papers have previously reported on directly using biodiesel production from biodiesel manufacturing. The following section reviews the use of feedstock in further detail.

5.1.1 Animal feedstuff

Since the 1970s, glycerol has been used as an animal feed additive. Glycerol's availability has also restricted its use in feeds [46]. The idea of using crude glycerol from biodiesel in feeds has been investigated in light of the recent increase in maize prices and excess crude glycerol. Glycerol is an excellent energy source with a high absorption rate. In the liver of animals, the enzyme glycerol kinases convert glycerol to glucose to generate energy. Crude glycerol specimens from multiple biodiesel producers have been tested as energy sources.

Although crude glycerol can be given to livestock feed, too much glycerin in the diet can disrupt an animal's normal physiological metabolism. Some works have been published on liquid product feed amounts and crude glycerol efficiency in animal feeds. During the growing stage, pigs' daily gains were influenced by their glycerol consumption. This effect was not present during the finishing stage. Supplementing lamb finishing diets with up to 15% dry matter crude glycerol increased feedlot efficiency, particularly during the first 14 days, with no discernable effect on carcass quality [47].

When crude glycerol was introduced to cattle finishing diets at a rate of 8% or less cell dry matter, weight and fat efficiency increased [48]. In addition to the discoveries mentioned above, a trademark [49] detailed methods to use or add feedstock into animal nutrition, as well as providing ideas.

Overall, using crude glycerol as an animal nutrition component offers great promise for replacing maize in animal diets, and it is receiving much attention. However, one must be cautious, as crude glycerol from biofuel may contain potentially harmful contaminants.

5.1.2 Feedstocks for chemicals

The most promising method for biological glycerol conversion is the anaerobic fermentative synthesis of 1,3-propanediol (1,3-PDO). In fed-batch cultures of *Klebsiella pneumoniae*, Mu et al. [50] found that crude glycerol could be used directly to produce 1,3-PDO. For crude glycerol produced by alkali- (51.3 g L^{-1}) and lipase-catalysis (53 g L^{-1}) and methanolysis of soybean oil, the variations between the final 1,3-PDO concentrations were modest. This meant that the crude glycerol composition had little impact on biological conversion, and thus, a low fermenting cost might be predicted.

Clostridium butyricum could also be used to convert crude glycerol to 1,3-PDO. On a synthetic medium, *C. butyricum* VPI 3266 reconstructed 1,3-PDO from crude glycerol. Commercial glycerol and unrefined glycerol have only minor differences.

Citric acid has been the focus of a few research studies on biodiesel production for citric acid synthesis. Synthesis from *Yarrowia lipolytica* ACA-DC 50109 using bioethanol produced both citric acid equivalent to that generated from a sugar-based complete medium and single-cell oil in addition to the citric acid.

5.1.3 Chemicals produced through conventional catalytic conversions

Oxygen-containing chemicals Crude glycerol might be used to produce the produced chemical (2,2-dimethyl-1,3-dioxolan-4-yl)methyl acetate, which can be used as a biodiesel component. It has the potential to boost biodiesel viscosity while meeting the flash point and oxidation stability standards for diesel and biodiesel fuels established by American and European Standards (ASTM D6751 and EN 14214, respectively). Other biodiesel additives could compete with this new chemical [51].

Hydrogen or synthetic gas Crude glycerol has indeed proved viable as a hydrogen or syngas generation alternative. Gasification was the most widely used method. Thermogravimetric and FTIR spectroscopic studies found that crude glycerol thermal breakdown is generally divided into four stages, with CO_2, H_2, CH_4, and CO being the most prominent gas products. Gasification with in situ CO2 removal proved efficient and effective.

The value-added potential of crude glycerol following biodiesel production is investigated in this study, especially as feed for animals and as a renewable chemical source. Nonruminant animals and pets, laying hens, and broilers have all benefited from these research discoveries. However, some concerns must be addressed before this biofuel chemical is widely used in animal feeds. First, crude glycerol's chemical structure changes significantly based on the biodiesel manufacturing processes and feedstocks used. Because different researchers used varied content of oil glycerol in their experiments, animal producers must exercise caution when selecting whether or not to use crude glycerol in their animal feed regimens. Pollutants in crude glycerol may affect how glycerol is transformed into other products significantly. In various biological conversion processes, pollutants in crude glycerol inhibit cell and fungal development, resulting in lower production rates and yields (compared with pure or commercialized glycerol under the same culture conditions). In classic catalytic conversions, impurities poison the catalysts, increasing char formation and lowering product yield.

6. Bioconversion of crude glycerol waste into value-added products

For biodiesel to remain sustainable, the crude glycerol produced during its production must be fully utilized. Large quantities of glycerol can be hazardous to the environment. Consequently, biotechnological processes must be applied to convert this crude glycerol into value-added products, allowing biodiesel producers to generate more money. Crude glycerol is a raw material in a range of industrial products, including biopolymers, polyunsaturated fatty acids, ethanol, hydrogen, and n-butanol. As a result, we have included several glycerol bioconversion technologies in this analysis of value-added industrial products.

Global energy consumption will rise because of population growth, industrialization, and humanity's desire for a better quality of life. However, given the limited nature of fossil energy resources and the environmental and climate concerns associated with their use, the development of renewable energy has become crucial [52,53]. Biodiesel is a biofuel that helps reduce GHG emissions, such as carbon dioxide. As a result, it may be useful in mitigating climate change and ensuring energy security. The esterification of vegetable oils or animal fats and long-chain alcohols produces biodiesel, which is a fatty acid methyl or ethyl ester.

During the transesterification process, 1 mol of triglyceride yields 3 mol of biodiesel (ester) and 1 mol of glycerol. On this premise, each batch of biofuel yields about 10% wt. glycerol. During the transesterification process, 1 mol of triglyceride yields 3 mol of biodiesel (ester) and 1 mol of glycerol. On this premise, each batch of biofuel yields about 10% wt. glycerol. Crude glycerol is glycerol that has been manufactured with various contaminants. The most common contaminants in crude glycerol are methanol, soap, free fatty acids, salt (inorganic salt residues via catalysts), unreacted mono-, di-, and triglycerol, and water [54].

Biodiesel, among the most widely used renewable energy sources, is produced in large quantities worldwide, yet produces a variety of residues and by-products that are economically and environmentally problematic. This report combines data on the current state of biofuel industry rejects and by-product transformation into biorefining-ready products. The investigation looked at glycerol, biofuel-washing municipal wastewater, and solid residues. Technologies are outlined to make it easier for researchers to obtain this information, and the most important experimental results and factors are presented.

One technique for valorizing crude glycerol involves conversion into microbial lipids via bioconversion. The development of microorganisms used in such conversions is hampered by contaminants in crude glycerol, such as methanol, salts, and soap. The research conducted for this thesis tackled these challenges and produced practical solutions to these issues.

Crude glycerol generated from biodiesel is typically used as animal feed or consumed as a low-energy fuel [1]. Despite the variety of contaminants present in crude glycerol, it is a useful energy source. According to Thompson (2006), the ranges of protein, lipids, and carbohydrates in crude glycerol are 0.06%−0.44%, 1%−13%, and 75%−83%, respectively. Incorporating a particular quantity of crude glycerol into the diets of ruminants and nonruminants has proved advantageous. Energy and nutrient values in crude glycerol samples vary from 1600 to 1650 kcal/lb, with metabolizing energy values ranging from 1500 to 1580 kcal/lb.

As a result, adding up to 6.7% crude glycerol to broiler diets has been demonstrated to improve feed conversion ratios [55]. Similarly, feeding cattle up to 15% crude glycerol enhanced their weight and feed efficiency significantly [1]. Certain contaminants found in crude glycerol, on the other hand, can have a deleterious impact on these animals. The prevalence of potassium impurities in crude glycerol, for example, has been linked to moist litter issues in broilers. Animals are toxic when higher quantities of methanol are present. As a result, decreased impurity crude glycerol can only be used as an animal feedstock, and it should be thoroughly examined before use in animal feeds. Furthermore, due to the numerous purification procedures required, producing crude glycerol with minimal levels of impurities is difficult and costly.

The energy generated by burning crude glycerol can be used to drive biodiesel synthesis or other founder systems. Glycerol does have a high auto-ignition threshold, is highly viscous, and has a poor heating value [56]. As a result, burning crude glycerol is technically challenging, inefficient in terms of energy, and commercially unviable. Furthermore, incomplete glycerol combustion results in extremely dangerous aldehydes (such as aromatic hydrocarbons), putting people and the environment in danger [57].

Due to the obstacles outlined in the preceding paragraphs, several researchers have looked at turning crude glycerol toward various value-added chemicals and products. The two methodologies for converting crude glycerol into useable chemicals or products are conventional chemical methods and bioconversion routes.

Many researchers have investigated chemical techniques for turning crude glycerol into various value-added compounds [58]. This method has produced hydrogen, 1,2-propanediol, acrolein, dihydroxyacetone, glyceric acid (GA), triacetin, monoglycerides, and other typical chemicals [59].

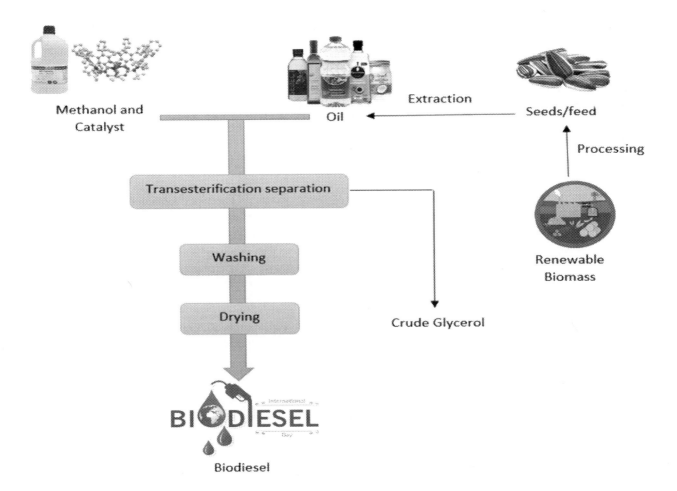

Another major component that can lead to the commercialization of technology is the usability of the final products obtained via such conversion.

The following are the most important utilities:

1. **1,3-PDO** is an organic compound produced by bacteria during anaerobic or microaerobic fermentation. It is used in adhesives and solvents [60,61]. Polytrimethylene terephthalates, adhesives, resins, films, laminates, and solvents all use 1,3-PDO ($C_3H_8O_2$) as a building block. It is also used as a wood paint and antifreeze [62]. The 1,3-PDO bioproduction technique was developed by Dupont Tate and Lyle Bio Products Co. in 2004, and a manufacturing unit with an annual capacity of 35 million pounds opened in London in 2019 [63]. The reductive glycerol route consists primarily of two cascade reactions: (1) glycerol to 3-hydroxypropionaldehyde (3-HPA; $C_3H_6O_2$) by glycerol hydrolase with coenzyme B12 and (2) 3-HPA to 1,3-PDO by propane diol oxidoreductase. Each aldehyde (3-HPA) conversion to one molecule of 1,3-PDO requires one molecule of NADH. The following equations demonstrate the stoichiometry of glycerol conversion to 1,3-PDO [62].

The platform chemical 1,3-PDO is used widely in adhesives, fiberboard, and food additives. Sustainable and nontoxic NADH regeneration is critical for the efficient synthesis of 1,3-PDO. For 1,3-PDO production using glycerol as an electron donor, ZVI (Zero-Valent Iron) can provide a reducing equivalent. *Klebsiella pneumoniae* has been examined as an electroactive strain capable of exchanging respiratory electrons via the cell membrane because it has a 1,3-PDO conversion route from glycerol. *K. pneumoniae* L17, an electroactive strain, and ZVI as an electron donor were used to create 1,3-PDO in this investigation.

With *K. pneumoniae* L17 as a reducing agent, the conversion of glycerol to 1,3-PDO was investigated. For the production of 1,3-PDO with glycerol as an electron donor, ZVI can give a reducing equivalent. As an exoelectrogen, *K. pneumoniae* has been widely investigated and has a 1,3-PDO synthesis route from glycerol. Compared with the control, using ZVI under anaerobic circumstances elevated 1,3-PDO levels 2.1-fold (13 mM). We looked at the best pH and ZVI levels. If more than 500 mM glycerol was supplied, 1,3-PDO synthesis was hindered by excessive lactate formation. The introduction of ferric ions produced identical amounts of 1,3-PDO as the control. The introduction of ferrous ions, on the other hand, resulted in approximately 4 mM more 1,3-PDO synthesis than ZVI. These findings indicate that adding affordable powdered iron, such as ZVI, to glycerol fermentation can boost platform chemical output.

2. **Lactic acid (2-hydroxypropanoic acid)** is made through the microbial conversion of glycerol, which is done through a fermentation process involving multiple bacteria. Lactic acid, as well as its salts and esters, could be used in the culinary, cosmetics, pharmaceutical, and agricultural industries [64].

Bioconversion has promise in producing biobased building blocks like 3-hydroxypropionic acid (3-HP). 3-HP recovery can be accomplished by reactive extraction, which can then be integrated into the bioconversion process. To our knowledge, no experimental data have been reported on the reacting extraction of 3-HP. The goal of this research was to investigate 3-HP extraction in n-decanol using tri-n-octylamine and Aliquat 336 as extractants. Lactic acid, which is its positional isomer, was also compared. Furthermore, the separation of 3-HP from bioconversion broths, both model and real, was investigated.

Mixed extractants have performed well in a variety of experimental settings, indicating a synergistic impact. The development of a compound between TOA, Aliquot 336, and carboxylic acid is one extraction possibility. When 3-HP was recovered from a genuine bioconversion broth, some cell-derived chemicals were suspected of adsorption to the soluble interface. This phenomenon, combined with the possibility of interaction between 3-HP and chemicals that can interact with the extraction solvent, results in yield (Y %) and specificity (3-HP) restrictions in the extraction. At pH = 3.2, for example, Y % and 3-HP were 89% and 35.9% for the model biological conversion broth, respectively, and 62% and 15.7% for the real broth.

Experimental research on tri-n-octylamine and Aliquat 336 reactive extraction in n-decanol revealed effective circumstances for recovering 3-HP. Surprisingly, the combined extractants had high distribution coefficients throughout a wide range of experimental settings, including various organic phase compositions, pH levels, and aqueous phase acid concentrations. This synergism's remarkably positive performance under low pH circumstances is a notable feature. However, the optimal pH level for 3-HP bioconversion should be considered when incorporating it into an integrated extractive process.

For the first time, suitable experimental values for the reacting separation of 3-HP were identified. More research will be done to better understand the mechanisms associated with an integrated extractive bioconversion system under real-world situations.

3. **Hydrogen** is made from crude glycerol via microbial fermentation. It is an energy carrier used in fuel cells [60,65]. Hydrogen is a possible environmentally beneficial fuel because it is the only waste fuel that creates water. The substrate for bacterial fermentation to produce fuel is crude glycerol. Because of the large range of substrates that may be used to ferment hydrogen, the energy consumption of bioenergy for hydrogen can be integrated with waste materials at the same time. Since hydrogen fuel only creates water as a by-product, it dramatically reduces CO_2, NO_x, PM, and other pollutants and is often followed by fossil fuels [66], with microbial fermentation-generated hydrogen as an acceptable option. Biohydrogen was created in an anaerobic reactor system at 30°C and an initial pH of 7.0 by increased concentrations of a crude glycerol founder with sanitary sewage by anaerobic consortium bacteria. The ingestion of 63.9% crude glycerol resulted in a higher H_2 production of 35.82 mmol L^{-1} [67]. Continuous hydrogen production (CHP) and hydrogen yield have both been improved by microbial inactivation (HY). Microbes are trapped using PVA-alginate. Immobilized microorganisms had the largest CHP and HY, at 64 mL/100 mL and 0.52 mol H_2/mol of glycerol, compared with suspended microorganisms, which had 9 mL/100 mL and 0.29 mol H_2/mol of glycerol [68].

4. **Trehalose** is a reducing sugar employed in therapeutic items as a stabilizer [65]. Trehalose is a disaccharide found in many species and comprises two glucose units joined by $\alpha -1, 1$ glycosidic link. Trehalose can defend the organism from cold, exhaustion, hypertonicity, ethanol cytotoxicity, and oxidation, in addition to the support it provides as a carbon and energy source [69]. Trehalose is widely used in the culinary, pharmaceutical, and cosmetic industries because of its unique properties [70]. Trehalose production has traditionally been divided into three stages [71]. Many bacteria have been found to have the trehalose synthase (TreS) (EC 5.4.99.16) pathway. Many TreS alleles have now been cloned and recombinantly expressed in *Escherichia coli*. Many TreS genes have been cloned and heterologously produced in *Escherichia coli* [72]. The properties of enzymes have been extensively researched. Maltose esterification to trehalose catalyzed by TreS49 is usually followed by maltose hydrolyzed to glucose. The ratio of isomerization to hydrolysis is proportional to the reaction conditions. The production of trehalose dropped while the yield of glucose increased at high temperatures. TreS has an optimum reaction temperature of around 35°C.

TreS is required for trehalose production because it catalyzes the reversible interconversion of 15 maltose to trehalose. As a result, understanding TreS's catalytic mechanism is critical in enzyme tuning and industrial applications.

Thermobaculum terrenum TreS (TtTreS) is a thermostable enzyme from *T. terrenum*. We used computer calculations and enzyme assays to investigate the makeup of the TtTreS catalytic site. The findings pointed to two double-displacement mechanisms comparable to glycoside hydrolase family enzymes. However, our findings imply that after the −1,4 glycosidic link is broken, glucose rotation is a critical element influencing the reaction orientation and conversion rate. In glucose rotation, the N246 molecule plays a crucial function. In addition, we developed a saturation mutation model for nonconserved amino acids in the nutrient gateway domain. Finally, compared with the wild-type enzyme, four TtTreS mutants (K136T, Y137D, K138N, and D139S) contributed to increased trehalose yield.

5. **Glyceric acid**, also known as 2,3-dihydroxypropionic acid, is an organic acid. Bioconversion of glycerol produces it. It is a monomer with several functions [1].

   Gluconobacter sp. produces GA as a minor by-product of DHA synthesis from glycerol. It also improves GA levels by applying varying volumes of CaCl2, identifying trace components contained in waste glycerol, and employing a microbe obtained from rotten apples. As a result, they claimed that GA might be mass-produced from waste glycerol. The starting pH of 6.0 promoted the growth of bacteria during fermentation, but as the conversion progressed, the pH dropped to 2.5, which tended to increase the formation of GA. The greatest precipitate of GA calcium salt solution was also found at various concentration ranges of 30%−70%, 50%, and 60%.

   Hydroxyacetic acid is another name for GA. Glycolic acid is used as a dyeing agent in the garment industry, a tanning agent in the food industry, and a flavoring agent, preservative, and skincare agent in pharmaceutical manufacturing. It is also employed in adhesives and plastics. Glycolic acid is commonly used to improve may include characteristics and produce gloss in emulsion polymers, solvents, and additives. The next section discusses the use of activated charcoal as a pretreatment to employ *Gluconobacter* sp. NBRC3259 to generate GA from crude glycerol. From 174 g L^{-1} of glycerol, 49.5 g L^{-1} of GA and 28.2 g L^{-1} of DHA were produced [73].

6. **Citric acid** is produced by fermentation. It is the weakest of all the organic acids. It is a tricarboxylic acid intermediate commonly used as a food preservative [1].

   Citric acid is a flavoring addition and preservative agent used in the food sector, with an estimated yearly output of over 800,000 tonnes and a predicted annual growth rate of 5% [74]. Citric acid generation from *Aspergillus niger* (fungus) submerged fermentation exploiting sugar as a medium is a well-developed technology. Glycerol, on the other hand, does not promote citric acid synthesis in *A. Niger* [75]. *Y. lipolytica* was studied for citric acid synthesis using glycerol as a substrate as an alternative to *A. niger* [76]. *Y. lipolytica* NCIM 3589 has been found capable of producing a terminal citric acid content of 77.4 g L^{-1} under ideal circumstances, which is similar to the fixed concentration achieved by *A. niger*. With an initial glycerol concentration of 200 g L^{-1}, *Y. lipolytica* Wratislavia AWG7 generated an even higher citric acid content of 139 g L^{-1} after 120 h.

   Citric acid is employed in ice creams as an emulsifier, in pharmaceuticals as a purifier, in cosmetics, and so on. Citric acid is a common acidulate, buffering agent, emulsifier, flavoring, preservative, and sequestrate in many industries, including food, drinks, pharmaceuticals, nutraceuticals, and cosmetics. Household detergents and dishwashing cleaners are the first prominent new uses, incorporating it as a joint producer with zeolites, notably in concentrated fluid detergents. Citric acid acts as a builder, strongly reducing Ca^{2+} and Mg^{2+} ions in water, but unlike phosphate builders, it does not promote the eutrophication of aquatic systems. *Y. lipolytica* SKY7 was used to create citric acid of 18.70, 12.0, and 8.30 g L^{-1} using glycerol supplies from three biodiesel industries: ROTHSAY, BIOLIQ, and BIOCARDEL [77]. Using crude glycerol as a substrate, *Y. lipolytica* strains Gut1 and Gut2 produced 42.5 ± 2.4 g L^{-1} isocitric acids [78]. Citric acid is a highly sought-after microbial product that is used in a variety of applications due to its low toxicity, primarily in the food and pharmaceutical industries [79]. Fungal feedstocks (with sucrose as well as molasses) had also previously been used to produce citric acid with *A. niger* [76]. Alternative fermentation procedures for the synthesis of citric acid, on the other hand, are desirable, given the rising demand for this value-added product [79]. In studies, a yeast strain of *Y. lipolytica* and numerous *Candida* species were recommended as an alternate fermentation technique that produces high citric acid yields. These citric acid fermentations were generated using batch culture fermentation with glucose, ethanol, plant oil, paraffin, and sucrose, and they were investigated using wild-type, mutant, and recombinant yeast stains. If any of these species and fermentations are successful, they may replace *Wlamentous* fungal species in conventional techniques.

7. **Docosahexaenoic acid (DHA)** is used in a variety of ways, including as a tanning beds agent in the cosmetic sector, pharmaceutical feedstock, and key component in the production of fine chemicals and novel biodegradable plastics [80]. DHA is a kind of omega-3 fatty acid. Several works have been published on using crude glycerol to generate DHA-rich microalgae by fermenting the alga *Schizochytrium limacinum*. The best range of crude glycerol concentrations for sustaining algal growth and DHA generation is 75−100 g L^{-1}. Heating and ammonia acetate concentration had a

substantial impact on algal DHA output. Heat and ammonium acetate were found to be best at 19.2°C and 1.0 g L^{-1}, respectively. Under optimum cultivation conditions, the greatest DHA yield was 4.91 g L^{-1} [81]. There were no significant differences in macro algal compositions when varied sources of crude glycerol were used. The algae produced exhibited a similar DHA content and nutritional profile as commercial algal biomass. This suggested that crude glycerol-derived algae might be used in omega-3-fortified meals or feeds [82]. Furthermore, DHA-containing microalgae have been created as omega-3 fatty acid substitutes for fish oil [83].

Through fungal fermenting with the fungus Pythium must provide, crude glycerol has been used to make fungal biomass that functioned as EPA-fortified meals or feeds. The EPA production and activity would exceed 90 and 14.9 mg L^{-1} per day, respectively, when grown in media containing 30 g L^{-1} crude glycerol and 1.0 g L^{-1} fungal extract. When compared to microalgae, the resultant EPA content was low. Future research must focus on improving culture conditions and creating high-cell-density cultivation methods [83]. Continuous culture was recently found to be an efficient method for examining the growth of bacterial and characteristics of microalgae on crude glycerol [84].

7. Microbial bioprospecting

Experimental strains of microorganisms have been used to identify a range of currently used industrially viable chemicals. Metagenomics, which employs a civilization method to collect the combined genomes of ecological microbial populations, provides resources for studying microbial compounds produced by a huge pool of bacteria that are known to exist in the environmental but are resistant to experimental culturing. Metagenomics approaches have been used to obtain new, highly specialized, tailor-made microbial metabolites for natural and industrial sustainability. Microbial metagenomics has a wide range of applications, including population ecology, pharmaceutical development, and environmental sustainability. As a result, the activity, efficacy, consistency, as well as specificity of real-time gene expression studies are presently being examined. Efforts combining researchers from several domains, such as microbial genetics, genomics, bioinformatics, and synthetic biology, will likely be critical to these unknown envoys' commendable promise for a sustainable society. This chapter explores contemporary discoveries in microbial metagenomics related to the development of industrially important compounds.

Microbes' crude and unfiltered extracts contain a variety of new metabolites in varied configurations. Bioprospecting is the effort expended to evaluate these active metabolites of biological origin [85]. Metagenomics' main goal is to find the genes responsible for generating the novel metabolite. The probability of finding novel metabolites depends mostly on the number of strains separated, the diversity among them, and their unique metabolite manufacturing method. Because of the complexity of the metagenomics sample, advanced and sensitive screening approaches are required to provide fast and reliable results for identifying genes encoding additional compounds from the pool of metabolomics library that is generated. As a result, rigorous evaluation of all elements is done before researching novel metabolites [86].

8. Application of microbial prospecting

Due to the depletion of petroleum supplies and growing concern about climate change, renewable energy sources have risen to the fore in the scientific community [96,97]. Biodiesel production from vegetable or animal fats through transesterification is a hot topic. Biodiesel is a biodegradable and environmentally friendly fuel that may be used without modification in diesel engines. Biodiesel has become one of the world's fastest-growing businesses, with worldwide biodiesel output expected to exceed 37 billion gallons in 2016 [87]. After a transesterification process, glycerol is a founder of biodiesel production, accounting for approximately 10% of the volume of the finished mixture. As global biodiesel production increases, so will global glycerol production, providing a new problem: Glycerol waste disposal [56,61,88]. Recycling crude glycerol is not only expensive but can also be wasteful and cause environmental issues [98]. As a result, it is critical to find better ways to use glycerol.

Pure glycerol is frequently used as a basic resource in the grooming, cosmetics, pharmaceuticals, and food sectors, and it can be turned into a range of value-added products. Since 2013, biodiesel has been the primary source of glycerol, accounting for about 1400 kt of propylene glycol production [89]. By 2020, the global glycerol industry was expected to expand to USD 2.52 billion. The glycerol used to manufacture biodiesel is called "crude glycerol" because it contains impurities such as methanol, fatty acid methyl esters, and salts left behind by the transesterification operation. Because of these contaminants, the purification price of crude glycerol is relatively high, making its use in these traditional industries impracticable given the low market rate [90,91]. Despite the modest demand for raw glycerol in these traditional sectors, it can be a reduced raw large and diverse group of applications in the rapidly growing renewable energy sector. Integration of waste glycerin into renewable power processes enhances those processes and helps reduce production costs. In addition, it is often environmentally favorable.

Several outstanding review studies in the literature address the developing problem of excess crude glycerol, each with a distinct focus. Papers [56,88], for example, discuss global glycerol production, the glycerol market, and its use in direct combustion. Ayoub and Abdullah [90] described the current state of glycerol manufacturing and its impact on the worldwide market, focusing on regional and demand-specific applications. Gupta and Kumar [10] advocated crude glycerol as a possible energy source and summarized the research done on this subject before 2012 but did not provide technical details. Stelmachowski et al. [92] concentrated on photocatalytic glycerol-to-hydrogen conversion techniques. For syngas production, Đurišić-Mladenović et al. [93] used a habitat agenda of glycerol with other biomass. Recently, using crude glycerol in polymeric was studied [94]. In the literature, the possibility of converting crude glycerol to high-valuation products has been thoroughly investigated [95].

9. Conclusion

Many solutions for the development of value-added goods have been investigated using the microbiological fermentation process. A progression can be seen from research that established wild-type strains capable of metabolizing glycerol to studies that manipulated these cultures to increase product output (of chemicals such as 1,3-PDO, ethanol, and citric acid). Concurrently, the metabolic engineering of nonnatural manufacturers of these chemicals, such as *E. coli*, has been examined. These discoveries have sparked research on the use of crude glycerol in the microbiological manufacture of high-value items, which led to the invention of crude glycerol engineering stains.

Glycerol used to be regarded as a precious commodity. Glycerol produces more lowering counterparts to be oxidized than other fuel sources, resulting in higher outputs of decreased molecules such as 1,3-PD, butanol, and ethanol. Crude glycerol, on the other hand, does not truly compete with other feedstocks, especially sugars, and should therefore be considered a supplement. This remarkable discovery will benefit both the biodiesel industry and the growth of biodistilleries and the entire biobased ecosystem.

10. Future prospects

The conversion of glycerin into methanol is a relatively novel process. Both the glycerin-to-ethanol and glycerin-to-methanol synthesis processes can be used. The market for biofuels is rapidly evolving. Rather than perceiving glycerin as a valuable resource, biofuel makers have come to regard it as a waste product. With their very own manufacturing processes, this waste product is a beneficial resource. This could have an impact on the market for other products for industries completely reliant on glycerin production, with the biofuels business as a supply source. Compared with traditional methods, the biotransformation/bioconversion of crude glycerol is known as catalysis. Has any domain received less attention? More emphasis is needed on a long-term strategy for biomass conversion in biodiesel production to polyurethane. These materials include foams, biopolyols, polyglycerols, and pigments.

More study is needed on crude glycerol-based bioconversions to enhance lactic acid, n-butanol, and PUFA yields. Only a few articles in the literature deal with crude glycerol preparation for improved use of MECs for hydrogen generation. Such additional study includes developing a collective elimination method for various contaminant types (such as acetic acid and butyric acid) that can be industrialized for near-term use in crude glycerol to hydrogen generation. More research is recommended on continuous microbial-producing hydrogen employing a coculture system. Additionally, genetically engineered microbial strains may boost crude glycerol's hydrogen production capacity.

References

[1] F. Yang, M.A. Hanna, R. Sun, Value-added uses for crude glycerol–a byproduct of biodiesel production, Biotechnol. Biofuels 5 (1) (2012) 1–10.

[2] L. Yu, S. Liang, R. Chen, K.K. Lai, Predicting monthly biofuel production using a hybrid ensemble forecasting methodology, Int. J. Forecast. 38 (1) (2022) 3–20.

[3] B.R. Moser, Biodiesel production, properties, and feedstocks, in: Biofuels, Springer, New York, NY, 2011, pp. 285–347.

[4] Y.H.P. Zhang, What is vital (and not vital) to advance economically-competitive biofuels production, Process Biochem. 46 (11) (2011) 2091–2110.

[5] D. O'Connor, Biodiesel GHG emissions, past, present, and future, Rep. IEA Bioenergy Task 39 (2011).

[6] M.C. Wildner, S.V. de Oliveira, O.M. Junior, As ações das cooperativas agrícolas frente à cadeia produtiva do Biodiesel no Rio Grande do Sul: evidências em direção ao desenvolvimento regional, Extensão Rural 25 (2) (2018) 92–111.

[7] A.P. Saravanan, T. Mathimani, G. Deviram, K. Rajendran, A. Pugazhendhi, Biofuel policy in India: a review of policy barriers in sustainable marketing of biofuel, J. Clean. Prod. 193 (2018) 734−747.
[8] F. Sarmiento, G. Espina, F. Boehmwald, R. Peralta, J.M. Blamey, Bioprospection of extremozymes for conversion of lignocellulosic feedstocks to bioethanol and other biochemicals, in: Extremophilic Enzymatic Processing of Lignocellulosic Feedstocks to Bioenergy, Springer, Cham, 2017, pp. 271−297.
[9] T.L. Alleman, R.L. McCormick, E.D. Christensen, G. Fioroni, K. Moriarty, J. Yanowitz, Biodiesel Handling and Use Guide (No. NREL/BK-5400-66521; DOE/GO-102016-4875), National Renewable Energy Lab.(NREL), Golden, CO (United States), 2016.
[10] N. Balasubramanian, K.F. Steward, Biodiesel: history of plant based oil usage and modern innovations, Substantia 3 (2) (2019) 57−71.
[11] N.N.A.N. Yusuf, S.K. Kamarudin, Z. Yaakub, Overview on the current trends in biodiesel production, Energy Convers. Manag. 52 (7) (2011) 2741−2751.
[12] D. Huang, H. Zhou, L. Lin, Biodiesel: an alternative to conventional fuel, Energy Proc. 16 (2012) 1874−1885.
[13] D. Namdar, A. Amrani, N. Getzov, I. Milevski, Olive oil storage during the fifth and sixth millennia BC at Ein Zippori, Northern Israel, Isr. J. Plant Sci. 62 (1−2) (2015) 65−74.
[14] D. Roccisano, J. Kumaratilake, A. Saniotis, M. Henneberg, Dietary Fats and Oils: Some Evolutionary and Historical Perspectives Concerning Edible Lipids for Human Consumption, 2016.
[15] P. Vossen, Olive oil: history, production, and characteristics of the world's classic oils, Hortscience 42 (5) (2007) 1093−1100.
[16] A. Hartmann, Back to the roots−dermatology in ancient Egyptian medicine, JDDG J. Deutschen Dermatol. Gesellschaft 14 (4) (2016) 389−396.
[17] B.B. Petrovska, Historical review of medicinal plants' usage, Phcog. Rev. 6 (11) (2012) 1.
[18] A.W. Jones, Early drug discovery and the rise of pharmaceutical chemistry, Drug Test. Anal. 3 (6) (2011) 337−344.
[19] M.F. Maia, S.J. Moore, Plant-based insect repellents: a review of their efficacy, development and testing, Malar. J. 10 (1) (2011) 1−15.
[20] G. Knothe, History of vegetable oil-based diesel fuels, in: The Biodiesel Handbook, AOCS Press, 2010, pp. 5−19.
[21] A.K. Agarwal, J. Bijwe, L.M. Das, Wear assessment in a biodiesel fueled compression ignition engine, J. Eng. Gas Turbines Power 125 (3) (2003) 820−826.
[22] P.V. Bhale, N.V. Deshpande, S.B. Thombre, Improving the low temperature properties of biodiesel fuel, Renew. Energy 34 (3) (2009) 794−800.
[23] G. Cambray, Helping Biodiesel become unstuck, Sci. Afr. (2007).
[24] M. Cetinkaya, Y. Ulusoy, Y. Tekìn, F. Karaosmanoğlu, Engine and winter road test performances of used cooking oil originated biodiesel, Energy Convers. Manag. 46 (7−8) (2005) 1279−1291.
[25] H. Rous, California Air Resources Board, 2014.
[26] M.P. Lammert, R.L. McCormick, P. Sindler, A. Williams, Effect of B20 and low aromatic diesel on transit bus NO_x emissions over driving cycles with a range of kinetic intensity, SAE Int. J. Fuels Lubr. 5 (3) (2012) 1345−1359.
[27] L.E.V. Regulations, T. Procedures, California Air Resources Board, 2015 (Sacramento, CA).
[28] A.A. Shvedova, N. Yanamala, A.R. Murray, E.R. Kisin, T. Khaliullin, M.K. Hatfield, S.H. Gavett, Oxidative stress, inflammatory biomarkers, and toxicity in mouse lung and liver after inhalation exposure to 100% biodiesel or petroleum diesel emissions, J. Toxicol. Environ. Health, Part A 76 (15) (2013) 907−921.
[29] S. Steiner, J. Czerwinski, P. Comte, O. Popovicheva, E. Kireeva, L. Müller, B. Rothen-Rutishauser, Comparison of the toxicity of diesel exhaust produced by bio-and fossil diesel combustion in human lung cells in vitro, Atmos. Environ. 81 (2013) 380−388.
[30] N.N. Mustafi, R.R. Raine, S. Verhelst, Combustion and emissions characteristics of a dual fuel engine operated on alternative gaseous fuels, Fuel 109 (2013) 669−678.
[31] N.T. Thanh, H.N. Murthy, K.Y. Paek, Optimization of ginseng cell culture in airlift bioreactors and developing the large-scale production system, Ind. Crop. Prod. 60 (2014) 343−348.
[32] M. Helmi, K. Tahvildari, A. Hemmati, A. Safekordi, Phosphomolybdic acid/graphene oxide as novel green catalyst using for biodiesel production from waste cooking oil via electrolysis method: optimization using with response surface methodology (RSM), Fuel 287 (2021) 119528.
[33] H.M. Khan, C.H. Ali, T. Iqbal, S. Yasin, M. Sulaiman, H. Mahmood, B. Mu, Current scenario and potential of biodiesel production from waste cooking oil in Pakistan: an overview, Chin. J. Chem. Eng. 27 (10) (2019) 2238−2250.
[34] F. Meisami, H. Ajam, M. Tabasizadeh, Thermo-economic analysis of diesel engine fueled with blended levels of waste cooking oil biodiesel in diesel fuel, Biofuels 9 (4) (2018) 503−512.
[35] S.E. Mahesh, A. Ramanathan, K.M.S. Begum, A. Narayanan, Biodiesel production from waste cooking oil using KBr impregnated CaO as catalyst, Energy Convers. Manag. 91 (2015) 442−450.
[36] Z. Ullah, M.A. Bustam, Z. Man, Biodiesel production from waste cooking oil by acidic ionic liquid as a catalyst, Renew. Energy 77 (2015) 521−526.
[37] H. Hamze, M. Akia, F. Yazdani, Optimization of biodiesel production from the waste cooking oil using response surface methodology, Process Saf. Environ. Protect. 94 (2015) 1−10.
[38] B. Gurunathan, A. Ravi, Biodiesel production from waste cooking oil using copper doped zinc oxide nanocomposite as heterogeneous catalyst, Bioresour. Technol. 188 (2015) 124−127.
[39] Y. Zhao, J. Wang, H. Zhang, C. Yan, Y. Zhang, Effects of various LED light wavelengths and intensities on microalgae-based simultaneous biogas upgrading and digestate nutrient reduction process, Bioresour. Technol. 136 (2013) 461−468.
[40] E.S. Salama, M.B. Kurade, R.A. Abou-Shanab, M.M. El-Dalatony, I.S. Yang, B. Min, B.H. Jeon, Recent progress in microalgal biomass production coupled with wastewater treatment for biofuel generation, Renew. Sustain. Energy Rev. 79 (2017) 1189−1211.

[41] J. Milano, H.C. Ong, H.H. Masjuki, W.T. Chong, M.K. Lam, P.K. Loh, V. Vellayan, Microalgae biofuels as an alternative to fossil fuel for power generation, Renew. Sustain. Energy Rev. 58 (2016) 180–197.

[42] M. Anwar, M.G. Rasul, N. Ashwath, M.N. Nabi, The potential of utilising papaya seed oil and stone fruit kernel oil as non-edible feedstock for biodiesel production in Australia—a review, Energy Rep. 5 (2019) 280–297.

[43] S. Thapa, N. Indrawan, P.R. Bhoi, An overview on fuel properties and prospects of Jatropha biodiesel as fuel for engines, Environ. Technol. Innovat. 9 (2018) 210–219.

[44] A. Da Silva César, F. de Azedias Almeida, R.P. de Souza, G.C. Silva, A.E. Atabani, The prospects of using Acrocomia aculeata (macaúba) a non-edible biodiesel feedstock in Brazil, Renew. Sustain. Energy Rev. 49 (2015) 1213–1220.

[45] W. Thurmond, Global biodiesel market trends, outlook and opportunities, in: Emerging Markets Online Global Energy and Biofuels Intelligence, 2008.

[46] B.J. Kerr, W.A. Dozier III, K. Bregendahl, Nutritional value of crude glycerin for nonruminants, in: Proceedings of the 23rd Annual Carolina Swine Nutrition Conference. Raleigh, NC, November 2007, pp. 6–18.

[47] P.J. Gunn, M.K. Neary, R.P. Lemenager, S.L. Lake, Effects of crude glycerin on performance and carcass characteristics of finishing wether lambs, J. Anim. Sci. 88 (5) (2010) 1771–1776.

[48] G.L. Parsons, M.K. Shelor, J.S. Drouillard, Performance and carcass traits of finishing heifers fed crude glycerin, J. Anim. Sci. 87 (2) (2009) 653–657.

[49] M. Cecava, P. Doane, D. Holzgraefe, N. Pyatt, U.S. Patent Application No. 12/107,997, 2008.

[50] Y. Mu, H. Teng, D.J. Zhang, W. Wang, Z.L. Xiu, Microbial production of 1, 3-propanediol by *Klebsiella pneumoniae* using crude glycerol from biodiesel preparations, Biotechnol. Lett. 28 (21) (2006) 1755–1759.

[51] E. Garcia, M. Laca, E. Pérez, A. Garrido, J. Peinado, New class of acetal derived from glycerin as a biodiesel fuel component, Energy Fuel. 22 (6) (2008) 4274–4280.

[52] R. Inglesi-Lotz, The impact of renewable energy consumption to economic growth: a panel data application, Energy Econ. 53 (2016) 58–63.

[53] N.L. Panwar, S.C. Kaushik, S. Kothari, Role of renewable energy sources in environmental protection: a review, Renew. Sustain. Energy Rev. 15 (3) (2011) 1513–1524.

[54] M. Pagliaro, M. Rossi, Glycerol: properties and production, Fut. Glycerol 2 (2010) 1–28.

[55] L. McLea, M.E.E. Ball, D. Kilpatrick, C. Elliott, The effect of glycerol inclusion on broiler performance and nutrient digestibility, Br. Poultry Sci. 52 (3) (2011) 368–375.

[56] C.A. Quispe, C.J. Coronado, J.A. Carvalho Jr., Glycerol: production, consumption, prices, characterization and new trends in combustion, Renew. Sustain. Energy Rev. 27 (2013) 475–493.

[57] J. Hassannia, Crude Glycerol Combustion System-Introduction and Commercialization Plan, 2011.

[58] O. Valerio, T. Horvath, C. Pond, M. Misra, A. Mohanty, Improved utilization of crude glycerol from biodiesel industries: synthesis and characterization of sustainable biobased polyesters, Ind. Crop. Prod. 78 (2015) 141–147.

[59] P. San Kong, M.K. Aroua, W.M.A.W. Daud, Conversion of crude and pure glycerol into derivatives: a feasibility evaluation, Renew. Sustain. Energy Rev. 63 (2016) 533–555.

[60] M. Anitha, S.K. Kamarudin, N.T. Kofli, The potential of glycerol as a value-added commodity, Chem. Eng. J. 295 (2016) 119–130.

[61] X. Luo, X. Ge, S. Cui, Y. Li, Value-added processing of crude glycerol into chemicals and polymers, Bioresour. Technol. 215 (2016) 144–154.

[62] C. Kim, Y.E. Song, J. Baek, H.S. Im, J.R. Kim, Zero-valent iron driven bioconversion of glycerol to 1, 3-propanediol using *Klebsiella pneumoniae* L17, Process Biochem. 106 (2021) 158–162.

[63] J.H. Lee, S. Lama, J.R. Kim, S.H. Park, Production of 1, 3-propanediol from glucose by recombinant *Escherichia coli* BL21 (DE3), Biotechnol. Bioproc. Eng. 23 (2) (2018) 250–258.

[64] A.A. Hong, K.K. Cheng, F. Peng, S. Zhou, Y. Sun, C.M. Liu, D.H. Liu, Strain isolation and optimization of process parameters for bioconversion of glycerol to lactic acid, J. Chem. Technol. Biotechnol. 84 (10) (2009) 1576–1581.

[65] V.K. Garlapati, U. Shankar, A. Budhiraja, Bioconversion technologies of crude glycerol to value added industrial products, Biotechnol. Rep. 9 (2016) 9–14.

[66] P. Prasertsan, C. Leamdum, S. Chantong, C. Mamimin, P. Kongjan, O. Sompong, Enhanced biogas production by co-digestion of crude glycerol and ethanol with palm oil mill effluent and microbial community analysis, Biomass Bioenergy 148 (2021) 106037.

[67] C.V. Rodrigues, K.O. Santana, M.G. Nespeca, A.V. Rodrigues, L.O. Pires, S.I. Maintinguer, Energy valorization of crude glycerol and sanitary sewage in hydrogen generation by biological processes, Int. J. Hydrogen Energy 45 (21) (2020) 11943–11953.

[68] Y. Chen, Y. Yin, J. Wang, Comparison of fermentative hydrogen production from glycerol using immobilized and suspended mixed cultures, Int. J. Hydrogen Energy 46 (13) (2021) 8986–8994.

[69] N. Al-Bader, G. Vanier, H. Liu, F.N. Gravelat, M. Urb, C.M.Q. Hoareau, D.C. Sheppard, Role of trehalose biosynthesis in Aspergillus fumigatus development, stress response, and virulence, Infect. Immun. 78 (7) (2010) 3007–3018.

[70] S. Ohtake, Y.J. Wang, Trehalose: current use and future applications, J. Pharmaceut. Sci. 100 (6) (2011) 2020–2053.

[71] M.L. Nuccio, J. Wu, R. Mowers, H.P. Zhou, M. Meghji, L.F. Primavesi, L.M. Lagrimini, Expression of trehalose-6-phosphate phosphatase in maize ears improves yield in well-watered and drought conditions, Nat. Biotechnol. 33 (8) (2015) 862–869.

[72] J. Liang, R. Huang, Y. Huang, X. Wang, L. Du, Y. Wei, Cloning, expression, properties, and functional amino acid residues of new trehalose synthase from *Thermomonospora curvata* DSM 43183, J. Mol. Catal. B Enzym. 90 (2013) 26–32.

[73] H. Habe, Y. Shimada, T. Fukuoka, D. Kitamoto, M. Itagaki, K. Watanabe, K. Sakaki, Production of glyceric acid by *Gluconobacter* sp. NBRC3259 using raw glycerol, Biosci. Biotechnol. Biochem. (2009), 0907091558-0907091558.

[74] R.W. Nicol, K. Marchand, W.D. Lubitz, Bioconversion of crude glycerol by fungi, Appl. Microbiol. Biotechnol. 93 (5) (2012) 1865−1875.

[75] P.F. Amaral, M.A.Z. Coelho, I.M. Marrucho, J.A. Coutinho, Biosurfactants from yeasts: characteristics, production and application, Biosurfactants (2010) 236−249.

[76] X. Fan, R. Burton, Y. Zhou, Glycerol (byproduct of biodiesel production) as a source for fuels and chemicals mini review, Open Fuel Energy Sci. J. 3 (1) (2010).

[77] L.R. Kumar, S.K. Yellapu, S. Yan, R.D. Tyagi, P. Drogui, Elucidating the effect of impurities present in different crude glycerol sources on lipid and citric acid production by Yarrowia lipolytica SKY7, J. Chem. Technol. Biotechnol. 96 (1) (2021) 227−240.

[78] D.A. Rzechonek, A. Dobrowolski, W. Rymowicz, A.M. Mirończuk, Aseptic production of citric and isocitric acid from crude glycerol by genetically modified *Yarrowia lipolytica*, Bioresour. Technol. 271 (2019) 340−344.

[79] A. Rywińska, W. Rymowicz, High-yield production of citric acid by *Yarrowia lipolytica* on glycerol in repeated-batch bioreactors, J. Ind. Microbiol. Biotechnol. 37 (5) (2010) 431−435.

[80] P.K. Dikshit, S.K. Padhi, V.S. Moholkar, Process optimization and analysis of product inhibition kinetics of crude glycerol fermentation for 1, 3-dihydroxyacetone production, Bioresour. Technol. 244 (2017) 362−370.

[81] Z. Chi, D. Pyle, Z. Wen, C. Frear, S. Chen, A laboratory study of producing docosahexaenoic acid from biodiesel-waste glycerol by microalgal fermentation, Process Biochem. 42 (11) (2007) 1537−1545.

[82] D.J. Pyle, R.A. Garcia, Z. Wen, Producing docosahexaenoic acid (DHA)-rich algae from biodiesel-derived crude glycerol: effects of impurities on DHA production and algal biomass composition, J. Agric. Food Chem. 56 (11) (2008) 3933−3939.

[83] S.K. Athalye, R.A. Garcia, Z. Wen, Use of biodiesel-derived crude glycerol for producing eicosapentaenoic acid (EPA) by the fungus Pythium irregulare, J. Agric. Food Chem. 57 (7) (2009) 2739−2744.

[84] S. Ethier, K. Woisard, D. Vaughan, Z. Wen, Continuous culture of the microalgae Schizochytrium limacinum on biodiesel-derived crude glycerol for producing docosahexaenoic acid, Bioresour. Technol. 102 (1) (2011) 88−93.

[85] M. Lahlou, The Success of Natural Products in Drug Discovery, 2013.

[86] S. Sharma, J. Vakhlu, Metagenomics as advanced screening methods for novel microbial metabolites, in: Microbial Biotechnology, CRC Press, 2018, pp. 58−77.

[87] M.K. Lam, K.T. Lee, A.R. Mohamed, Homogeneous, heterogeneous and enzymatic catalysis for transesterification of high free fatty acid oil (waste cooking oil) to biodiesel: a review, Biotechnol. Adv. 28 (4) (2010) 500−518.

[88] C.R. Coronado, J.A. Carvalho Jr., C.A. Quispe, C.R. Sotomonte, Ecological efficiency in glycerol combustion, Appl. Therm. Eng. 63 (1) (2014) 97−104.

[89] J. McNutt, J. Yang, Utilization of the residual glycerol from biodiesel production for renewable energy generation, Renew. Sustain. Energy Rev. 71 (2017) 63−76.

[90] M. Ayoub, A.Z. Abdullah, Critical review on the current scenario and significance of crude glycerol resulting from biodiesel industry towards more sustainable renewable energy industry, Renew. Sustain. Energy Rev. 16 (5) (2012) 2671−2686.

[91] A.B. Leoneti, V. Aragão-Leoneti, S.V.W.B. De Oliveira, Glycerol as a by-product of biodiesel production in Brazil: alternatives for the use of unrefined glycerol, Renew. Energy 45 (2012) 138−145.

[92] M. Stelmachowski, M. Marchwicka, E. Grabowska, M. Diak, The photocatalytic conversion of (biodiesel derived) glycerol to hydrogen-A short review and preliminary experimental results part 1: a review, J. Adv. Oxid. Technol. 17 (2) (2014) 167−178.

[93] N. Đurišić-Mladenović, B.D. Škrbić, A. Zabaniotou, Chemometric interpretation of different biomass gasification processes based on the syngas quality: assessment of crude glycerol co-gasification with lignocellulosic biomass, Renew. Sustain. Energy Rev. 59 (2016) 649−661.

[94] A. Hejna, P. Kosmela, K. Formela, Ł. Piszczyk, J.T. Haponiuk, Potential applications of crude glycerol in polymer technology−current state and perspectives, Renew. Sustain. Energy Rev. 66 (2016) 449−475.

[95] A. Talebian-Kiakalaieh, N.A.S. Amin, H. Hezaveh, Glycerol for renewable acrolein production by catalytic dehydration, Renew. Sustain. Energy Rev. 40 (2014) 28−59.

[96] M. Shah (Ed.), Microbial Bioremediation & Biodegradation, Springer, Singapore, 2020.

[97] M. Shah (Ed.), Removal of Refractory Pollutants from Wastewater Treatment Plants, CRC Press, USA, 2021.

[98] M. Shah (Ed.), Removal of Emerging Contaminants through Microbial Processes, Springer, Singapore, 2021.

Chapter 8

Microalgae-based biofuel synthesis

Mohamed Gomaa[1], Mustafa A. Fawzy[1,2] and Mostafa M. El-Sheekh[3]

[1]*Botany and Microbiology Department, Faculty of Science, Assiut University, Assiut, Egypt;* [2]*Biology Department, Faculty of Science, Taif University, Taif, Saudi Arabia;* [3]*Department of Botany, Faculty of Science, Tanta University, Tanta, Egypt*

List of abbreviations

CN Cetane number
FAs Fatty acids
HHV Higher heating value
HTL Hydrothermal liquefaction
IV Iodine value
L-CO$_2$ Liquid carbon dioxide
MUFAs Monounsaturated fatty acids
OS Oxidation stability
PUFAs Polyunsaturated fatty acids
SAFs Saturated fatty acids
Sc–CO$_2$ Supercritical carbon dioxide
SV Saponification value
UFAs Unsaturated fatty acids

1. Introduction

The search for renewable, environmentally benign, sustainable, and promising energy resources is an urgent demand due to the rapid increase in the world population and industrialization, as well as the rapid consumption of energy resources and associated environmental problems. It was estimated that the consumed energy will increase to 865 EJ by 2040 compared to 550 EJ in 2020 [1].

Biodiesel, a mixture of fatty acid esters produced by reacting oils with alcohol, represents a safe, carbon-neutral, renewable, and biodegradable energy resource. Biodiesel obtained by transesterifying oil from edible plants such as agricultural crops is referred to as first generation biodiesel, while that produced from nonedible plants and agricultural wastes is second generation biodiesel. Recently, the utilization of microalgae as a source of biodiesel is known as third generation [2].

Microalgae have several advantageous properties as promising biofuel feedstock which include: (1) high growth rate and biomass production and low freshwater consumption compared to terrestrial plants; (2) under certain conditions, several microalgae can bioaccumulate higher contents of lipids than oily plants; (3) Microalgae cultivation does not require fertilizers, herbicides, pesticides, or fertile land; (4) the effective biofixation of CO_2 by microalgae can help to reduce the greenhouse gases; (5) the coupling of biodiesel production and bioremediation of wastewaters is an environmentally sustainable process; (6) Beside lipids, the microalgal cells can coproduce several value-added compounds such as pigments, proteins, carbohydrates; (7) the biomass can be also used to produce other biofuels such as bioethanol, biohydrogen, bio-oil and biogas; (8) microalgae-based biodiesel has similar properties to the petrodiesel such as density, viscosity, flash point, heating value, and cold flow properties; (9) bio-oil synthesis by pyrolysis is more preferred from algae biomass than lignocellulosic biomass [2–4]. However, the utilization of microalgae for commercial biodiesel synthesis is still in its infancy. This is mainly related to the production costs, which are generally about two-fold higher than the petrodiesel [5]. Microalgal cultures are also susceptible to crashes at high temperatures or by contamination, especially in open pond systems.

Moreover, the synthesis of biodiesel by microalgae involves different sequential upstream and downstream processes. The downstream processes include the harvesting of the cultivated biomass, extraction, and conversion of lipids into biodiesel, which are still contributing about 60% of the total costs of biodiesel synthesis [6]. To ensure cost-effective biodiesel synthesis from microalgae, a few strategies should be followed such as (1) increasing the capturing of energy and CO_2 from the atmosphere; (2) increasing biomass and lipid productivities; (3) decreasing the energy consumed during harvesting, extraction, synthesis, and purification of biodiesel; (4) efficient sequential extraction of coproducts and (5) increasing the shelf-life of biodiesel. The cultivation of microalgae can be performed under different conditions viz., autotrophy, heterotrophy, and mixotrophy. Autotrophy relies only on the utilization of inorganic carbon sources, while both heterotrophic and mixotrophic conditions require organic carbon sources [7]. Despite the high cost of heterotrophic and mixotrophic modes, they can promote biomass and lipid productivities, which are fundamental parameters for efficient biodiesel production. Microalgae harvesting is a fundamental problem that contributes about 20%–30% of the total costs. Furthermore, lipid extraction is another key element that determines the efficiency of biodiesel synthesis and can affect the coproduct recovery.

The present chapter discusses the state-of-the-art in biodiesel synthesis from microalgae. The distinctiveness of this chapter is in its coverage of the most effective upstream and downstream methods for microalgae. The factors that affect biodiesel synthesis, cost, and properties are also discussed.

2. Upstream processes for biodiesel synthesis

2.1 Strain selection

Several factors can control lipid biosynthesis and accumulation by microalgae. The growth conditions can be controlled and optimized to not only manipulate the metabolic processes to increase lipid yield, but also to reduce the production cost [8]. The manipulation of growth conditions for wild strains is generally more preferred than mutagenesis and transgenic stains due to the instability of the mutant strains as well as its impact on environmental security. The properties and the production cost of microalgal biodiesel depend mainly on the producer strain. Several microalgal classes have been investigated for biodiesel production such as Cyanophyceae (e.g., *Arthrospira*), Chlorophyceae (e.g., *Chlamydomonas*, *Scenedesmus, Tetradesmus, Neochloris,* and *Dunaliella*), Trebouxiophyceae (e.g., *Botryococcus* and *Chlorella*), Chlorodendrophyceae (e.g., *Tetraselmis*), Bacillariophyceae (e.g., *Nitzschia*), Mediophyceae (e.g., *Chaetoceros*), Eustigmatophyceae (e.g., *Nannochloropsis*), Euglenophyceae (e.g., *Euglena*), Dinophyceae (e.g., *Alexandrium*), Pavlovophyceae (e.g., *Pavlova*), Xanthophyceae (e.g., *Tribonema*), Prymnesiophyceae (e.g., *Emiliana*) and Porphyridiophyceae (e.g., *Porphyridium*). However, the screening of microalgae for biodiesel synthesis should be performed by considering the natural habitat based on multiple fundamental criteria such as lipid productivity, fatty acid profile, robustness, and susceptibility to contamination.

2.1.1 Lipid productivity

Lipid productivity is the most important criterion for selecting microalgal species for commercial biodiesel production. Lipid content is generally related to the weight of cellular lipids to the dry weight of the biomass, while the lipid productivity is the product of lipid content and biomass productivity, which is usually expressed as mg lipid L^{-1} day^{-1} using the following equations [9,10]:

$$\text{Biomass productivity (mg L}^{-1} \text{ day}^{-1}) = (B_2 - B_1)/(t_2 - t_1)$$

$$\text{Lipid productivity} = \text{biomass productivity} \times \text{lipid content (\%)}/100$$

where B_1 and B_2 are the algal biomass (mg L^{-1}) at times 1 (t_1) and 2 (t_2), respectively.

These equations confirm that high cellular lipid contents are not sufficient for high biodiesel productivity unless it is accompanied by high biomass productivity. Accordingly, the simultaneous promotion of lipid contents and growth rates is a fundamental process for commercial biodiesel production. However, the processes for lipid accumulation and biomass production are inversely related due to the fact of high energy requirements for lipid biosynthesis [9]. In general, high cellular oil contents can decrease the cost during the extraction process, while a high growth rate can increase the yield of biodiesel per culture volume. Furthermore, microalgae with rapid growth require less space for cultivation owing to high cell density per area and are usually less prone to contamination by other microalgal species when cultivated in open ponds [11].

Several microalgae generally can bioaccumulate lipids under nutrient-depleted conditions such as nitrogen starvation. However, the algal biomass productivities under these conditions are relatively lower compared to nutrient replete cultures, leading to either the same or lower lipid productivity [9].

2.1.2 Cell wall composition

The cell wall constituents varied markedly between microalgal groups, and sometimes it represents a major barrier to the extraction of cellular metabolites and commercial utilization of microalgae. Several microalgal genera such as *Scenedesmus, Tetradesmus, Chlorella,* and *Nannochloropsis* contain algae, and cellulose in the cell wall, where the former is generally nonhydrolyzable [12]. Therefore, several mechanical, thermal, and electromagnetic-based techniques can increase the effectiveness of solvent-based recovery of lipid from algaenan-containing algae [13]. Consequently, the method applied for the disruption of the cell walls affects the final cost of biodiesel production.

On the other hand, some microalgae are lacking rigid cell walls such as *Dunaliella*. In this context, Yoo and coauthors indicated that the lipid recovery from wall-less mutant *Chlamydomonas reinhardtii* was markedly higher than the wild strain [14]. Accordingly, the extraction of lipids from wall-less algae could be more efficient compared with those with rigid cell walls.

Besides cell wall composition, the cell size, and shape are directly linked with the cultivation conditions as well as the effectiveness of cell disruption for lipid extraction [13]. Microalgae with smaller cell size are generally difficult to harvest, despite their advantageous high growth rates. The small size and the high content of liquid media can increase the production costs by 20%−30% [15].

2.1.3 Fatty acid profiling

Microalgal lipids may be categorized into two main classes as polar and nonpolar lipids. Polar lipids such as glycerophospholipids (glycerol-based phospholipids) and glycolipids (lipids connected to carbohydrates by a glycosidic bond) act as structural compounds in the cell membranes. Nonpolar (neutral) lipids include free fatty acids (FAs) and acylglycerols (esters formed from glycerol and FAs as mono-, di-, and triglycerides). The nonpolar lipids help in energy storage within the microalgal cells. The main component of these lipid classes is fatty acids which may be saturated (SFAs, without double bonds), monounsaturated (MUFAs, with one double bond), and polyunsaturated (PUFAs, with at least two double bonds). Furthermore, these FAs are categorized based on the length of the carbon chain into the short-chain (<6 carbon), medium-chain (6−12 carbon), long-chain (14−20 carbon), and very-long-chain (>20 carbon). In general, not all lipids produced by microalgae are characterized by adequate compatibility as a biodiesel feedstock, since they may have undesirable properties. Triacylglycerols are the most preferred lipids for biodiesel synthesis since they have a low unsaturation degree and can be easily converted into biodiesel [15]. Thus, the triacylglycerol determination rather than total lipid quantification is the clearest indication of the amount of suitable lipids for the transesterification reaction. The presence of a high concentration of free fatty acids is generally unfavorable since they form soaps upon reacting with alkali during the transesterification reaction and increase the downstream processing. However, microalgal oils with high content of free fatty acids can be hydrolyzed using acid or subjected to enzymatic hydrolysis.

Although neutral lipids in the form of TAGs are the main source of biodiesel, the total biodiesel yield may be markedly enhanced by utilizing all lipids, including polar lipids. It was hypothesized that the release of fatty acid moieties from polar lipids such as glycolipids and phospholipids and their conversion into biodiesel can decrease the undesirable impurities and increase yields as well as may contribute to the desirable fuel characteristics [16]. Thus, the determination of polar lipid content of microalgal oil is also an important process in terms of its impacts on the quality of the fatty acid profile.

The quality of biodiesel is related to the proportion of SFAs to MUFAs [17]. Microalgal lipids with high MUFA levels can effectively promote the combustion performance of the biodiesel, while higher SFAs and PUFAs can lead to instability of the biodiesel [18]. SFAs are generally characterized by high melting points, and thus can form precipitates when the biodiesel is cooled, especially when the levels of palmitic acid and stearic acid are high [19]. PUFAs are highly susceptible to auto-oxidation and decrease the longevity of biodiesel during storage but promote its cold-flow properties. Good-quality biodiesel should contain a 5:4:1 mass ratio of C16:1 (palmitoleic acid), C18:1 (oleic acid), and C14:0 (myristic acid) [15,19]. In addition, biodiesel with high C16 and C18 contents produces cetane numbers (47 minimum), kinematic viscosities (1.9−6.0 mm^2 s^{-1} at 40°C), and oxidative stabilities (3 minimum), which are in a good agreement within ASTM D6751 standard for biodiesel [20].

Previous studies evidenced that the fatty acid profile showed clear dissimilarities between microalgal classes. In general, SFAs and MUFAs are predominant in most algae examined. Specifically, the main FAs are C16:0 (palmitic acid) and C16:1 (palmitoleic acid) in the Bacillariophyceae, C16:0, and C18:1 (oleic acid) in the Chlorophyceae, C16:0, and

C18:1 in the Euglenophyceae, C16:0, C16:1 and C18:1 in the Chrysophyceae, C16:0 and C20:1 (eicosenoic acid) in the Cryptophyceae, C16:0, and C18:1 in the Eustigmatophyceae, C16:0, and C18:1 in the Prasinophyceae, C16:0 in the Dinophyceae, C16:0, C16:1, and C18:1 in the Prymnesiophyceae, C16:0 in the Rhodophyceae, C14:0 (myristic acid), C16:0 and C16:1 in the Xanthophyceae, and C16:0, C16:1 and C18:1 in cyanobacteria [21]. Chlorophyceae are generally recognized for their high levels of neutral lipids, thus may provide a good resource for lipid production and biodiesel synthesis. In general, C16:1 and C18:1 FAs are most favorable for biodiesel production. Microalgae in the Bacillariophyceae and Mediophyceae contain a high proportion of C14:0 and MUFAs. Thus, biodiesel produced from Mediophyceae is characterized by higher cold-flow characteristics than other classes [22].

Generally, the fatty acid profile of microalgae depends on a plethora of factors such as environmental factors relating to the growth conditions, physiological factors relating to the algal strain, and lipid extraction method and its conditions. All these factors should be taken into consideration when selecting an algal strain for biodiesel synthesis. However, it is hard to predict the variations in lipid profiles between species and isolates. Accordingly, blending lipid from different microalgae species has emerged as a suitable strategy to fulfill all the biodiesel requirements and produce biodiesel with standard quality [17].

2.1.4 Comparison between freshwater and marine microalgae

Despite the availability of seawater and low risk of contamination, the utilization of marine microalgae for biodiesel production still suffers from several drawbacks compared to freshwater microalgae. One disadvantage is linked to the fatty acid composition; marine microalgae have been shown to produce high levels of phospholipids than neutral lipids, which is unsuitable for biodiesel synthesis [3]. The site for the construction of production plants of marine microalgae is limited to the coastal areas to sufficiently utilize natural seawater in the cultivation. The salinity level of the marine culture should be regularly monitored and adjusted to avoid any possible inhibition in growth owing to salinity variation.

2.2 Algae cultivation and its effects on biodiesel production

Microalgae can possess several nutritional modes including photoautotrophic, heterotrophy, photoheterotrophy, and mixotrophy depending on the available sources of carbon and energy (Table 8.1). The selection of the cultivation mode is a critical process to achieving cost-effective biodiesel with high productivity.

2.2.1 Photoautotrophy

Photoautotrophic is the most common nutritional mode in algae since they can build their organic compounds by biofixing inorganic carbon (CO_2 and bicarbonate) by utilizing light energy through photosynthesis. The primary advantage of this nutritional mode is the biofixation of CO_2, which can help in mitigating climate change with simultaneous production of value-added products such as lipids from the produced algal biomass. In general, this mode is more suitable for open-pond cultivation systems without the risk of microbial contamination by heterotrophic fungi and bacteria since the cultures are free of exogenous organic compounds. Accordingly, photoautotrophism is generally considered as a cost and energy-saving mode of cultivation. However, to minimize the costs, outdoor cultures must be exposed to sunlight, thus finding a suitable site is fundamental to optimizing the process. Besides, the spatiotemporal variations in the sunlight irradiation can markedly affect the photoautotrophic production of algal biomass as well as lipid synthesis. Additionally, the cultivation of microalgae in a CO_2-rich environment can promote algal biomass productivity to a certain degree but can increase the costs of cultivation. The actual yield of photoautotrophic cultivation is far from theoretical yields owing to photon wastage (which can reach up to 80%). As the algal cell density increases, a shading effect aggravates owing to the exponential decrease in light penetration with increasing cell density [23]. Consequently, the light limitation can adversely affect both biomass and lipid productivity of the photoautotrophic cultures.

2.2.2 Heterotrophy

On the contrary to autotrophy, microalgae under heterotrophic mode can utilize organic carbons from the culture medium as a source of carbon and energy under dark conditions. Thus, light is no longer a limiting factor for growth, which permits an increase in algal cell density without the risk of photoinhibition. Subsequently, the biomass productivity of microalgae under heterotrophism is comparatively higher than that under photoautotrophism. Furthermore, cultivations under heterotrophism in the absence of light can be performed in simple photobioreactors, without the need for a high surface-to-volume ratio [23]. Nonetheless, the main disadvantages of heterotrophy are linked to the generation of CO_2 through the metabolism of the organic compounds, as well as the high risk of microbial contamination by heterotrophic bacteria and

TABLE 8.1 Types of nutritional modes found in algae.

Nutritional mode	Energy source	Carbon source	Advantages	Disadvantages
Photoautotrophy	Light	Inorganic carbon	Low risk of microbial contamination CO_2-biofixation Low cost Suitable for open pond systems	Low biomass and lipid productivity Light and CO_2-dependency
Heterotrophy	Oxidation of organic compounds	Organic compounds	High biomass and lipid productivity No risk of photoinhibition Simple reactor design	The high cost (cost can be reduced by natural carbon wastes) Generation of CO_2 High risk of microbial contamination Light-induced products are reduced
Mixotrophy	Light and oxidation of organic compounds	Inorganic and inorganic carbon	CO_2- biofixation High biomass and lipid productivity Reduction of light-dependency Light-induced products are produced	High cost (cost can be reduced by natural carbon wastes) High risk of microbial contamination

fungi. Accordingly, heterotrophy requires closed bioreactors and maintaining the axenic conditions of the microalgal cultures through sterilizing all the equipment and tools used during the production process. Furthermore, few species of microalgae can be effectively cultivated under heterotrophic conditions with high biomass productivity and product yields. According to a recent study, the cost of heterotrophic cultivation is generally similar to or close to the autotrophic conditions [24]. Furthermore, the cost of organic carbon could be further decreased by utilizing cost-effective natural carbon sources such as molasses, food waste, agricultural wastes, etc. For instance, the biodiesel production costs were halved when glucose was replaced by molasses in the heterotrophic conditions [24]. Moreover, heterotrophic cultivation of microalgae can produce high-quality biodiesel compared to autotrophic conditions. For instance, heterotrophic *Chlorella zofingiensis* tended to accumulate lipids with high triacylglycerol content along with an increase in MUFAs such as oleic acid compared to autotrophic cells [25].

2.2.3 Mixotrophy

Mixotrophic microalgae can simultaneously utilize both inorganic and organic carbon sources, and the source of energy comes from light absorption and oxidation of organic substrates. Thus, both photoautotrophic and heterotrophic metabolism is existing during mixotrophy without an apparent switch between them. Accordingly, mixotrophy combines the benefits of both photoheterotrophy and mixotrophy [7]. The utilization of CO_2 through photosynthesis helps to mitigate the CO_2 emissions. Therefore, mixotrophic cultivation has promising feasibility in environmental remediation for the simultaneous treatment of flue gases and organic wastewaters. Mixotrophic cultivation involves a gas exchange between cellular organelles; O_2 produced from photosynthesis in chloroplasts can be used in mitochondria for oxidative phosphorylation which in turn can generate CO_2 for photosynthesis. Hence, mixotrophy provides enough energy to the algal cells to form light and oxidation of organic substrates, which permits high algal biomass productivity and hyperaccumulation of certain algal metabolites such as pigments, lipids, proteins, and carbohydrates [9]. Generally, light is no longer a mandatory energy source during mixotrophic growth, which eliminates the photoinhibition of photoautotrophic cultivation. In general, the algal biomass productivities of several microalgae under mixotrophic conditions exceed those under photoautotrophic and heterotrophic conditions for the same species [23]. Besides, mixotrophic conditions have been shown to produce biodiesel with superior properties compared to autotrophic conditions [7,9]. Nonetheless, mixotrophic cultivation of microalgae has serious drawbacks relating to the cost of organic carbon and the susceptibility of the culture to contamination by heterotrophic microorganisms as well as the use of sterilized equipment, gases, and tools in the cultivation process. However, the use of cheap carbon wastes and the high algal biomass productivity can offset the added costs and make the large-scale synthesis of biodiesel more economical.

3. Downstream processes for biodiesel synthesis

3.1 Microalgae harvesting techniques

The development of suitable microalgae harvesting technology is critical for increasing the economic viability of biofuel generation. Efficient harvesting technology is the most difficult obstacle that requires to be surmounted to produce microalgae-based biofuel. This is attributed to the properties of microalgae cells including low cell density, small size, negative charge, and low biomass concentration in the culture systems. As a result, selecting the best harvesting method is reliant on both the features of the microalgae cells and the properties of the intended product.

Several techniques for harvesting microalgae biomass have been developed, including flotation, flocculation, centrifugation, and filtration. The cultivated biomass is primarily concentrated to about 2%−7% of total suspended solids through floatation or flocculation. Further dewatering of the biomass to obtain an algal cake with 15%−25% total suspended solids is usually performed by filtration or centrifugation [1]. These approaches can be used separately or in conjunction with other processes for increasing harvesting economics and efficiency. A suitable harvesting approach should be high in yield while being cheap in operating costs, be independent of microalgae species, and permit the recycling of cultivation media.

3.1.1 Flotation

Flotation is an anti-gravity-based separation technique in which the microalgal cells with low density float upwards and are concentrated at the surface of the medium, thereby it is referred to as inverted sedimentation. The flotation process is usually assisted by air bubbles that bind to the algal cells and convey them to the surface [1]. The efficiency of the flotation method is determined by the bubble size. By reducing bubble size, the efficiency of attachment and collision between the algal cells and the bubbles may be enhanced, resulting in greater harvesting effectiveness. Flotation is classified into dissolved and dispersed air flotation, electrolytic floatation, and dispersed ozone-based flotation.

In dissolved air floatation, the bubbles are formed when air dissolves in water under extremely high pressure, and the size of the produced bubbles ranged from 10 to 100 μm. While in the dispersed air flotation, the produced bubbles are with a size ranging between 700 and 1500 μm. Surfactants can be used during the flotation process to increase the size of microalgal particles, and the formation of aggregates, hence enhancing the flotation process efficiency. The cationic surfactants make the surface of bubbles positively charged and enhance the electrostatic interactions with algal cells as well as adsorbs to the cell surface making the cells hydrophobic, which promotes the interaction with bubbles [26]. Although dissolved air flotation is a successful flotation process, it is an energy-intensive method due to the large amount of energy required by the air compression process. Comparatively, the dispersed air floatation has various benefits since the separation units create air bubbles with surfactants without the need for an air compressor; as a result, it requires fewer extra materials and low energy demand compared to the dissolved air flotation process. In foam flotation, the operational energy consumption is reduced to $0.015\,kW\,h\,m^{-3}$ compared to $7.6\,kW\,h\,m^{-3}$ in case of the dissolved air flotation [26]. Furthermore, studies indicated that microalgae harvested by foam flotation using the surfactant cetyl trimethylammonium bromide had more lipid yield in relation to the one harvested by centrifugation. Besides, the profile of the FAs contained an increased level of SFAs and MUFAs, which indicated a high biodiesel quality [26]. These results were attributed to the adsorption of the surfactant on the algal cells and its effects on the lipid extraction through solubilizing the phospholipid bilayer membrane, thus increasing the lipid yield.

Electrolytic flotation works on the phenomenon known as electrolysis in which hydrogen bubbles are formed at the cathode during the electrolysis of water. The negatively charged microalgal cells interact with the positively charged hydrogen bubbles to form flocs and levitate to the surface. However, this process is more effective in harvesting marine microalgae compared to freshwater microalgae due to the high conductivity of seawater. Besides, it requires high energy consumption.

On the other hand, in the dispersed ozone floatation, the air is replaced by ozone. This method combines the harvesting of microalgal cells and their lysis by the oxidative damage caused by ozone. Accordingly, the use of ozone is a promising process that promotes lipid yield during extraction and increases the proportion of SFAs in the harvested feedstock [27].

3.1.2 Flocculation

Flocculation represents a promising and efficient method for harvesting small-sized cells. In this process, the cationic flocculants react with the negatively charged microalgal cells, resulting in the formation of aggregates defined as flocs that may be readily separated by sedimentation or any other harvesting approach.

However, the disadvantages of this technology may include economic or technical problems of flocculant toxicity, high energy requirement, or the inability to scale up [28]. Autoflocculation, bioflocculation, chemical flocculation, and electrochemical flocculation are the most often utilized methods of flocculation.

In autoflocculation, microalgal cells naturally aggregate in large flocs that separate by sedimentation. This process can occur in a variety of microalgae under unfavorable conditions such as culture aging, deficiency of nutrients, or pH-induced flocculation. The acidic and alkaline culture conditions have been stated to diminish the strength of negative charge on the cell surface, facilitating autoaggregation of microalgae cells. Accordingly, autoflocculation is an environmentally friendly process that does not require energy consumption or chemical addition. To date, few species of microalgae have been reported to be able to self-flocculate, however, they contain low lipid contents and are not recommended for biodiesel production. Therefore, the search for oleaginous microalgae able to self-flocculate can markedly reduce the cost of biodiesel synthesis. One study isolated a novel terrestrial strain of *Heveochlorella* sp. and indicated its superior properties as a biodiesel feedstock including autoflocculation with an efficiency of 85.16% after 2 h settling, high lipid content (42.63%), and good lipid profile [29].

The flocculation of microalgae with the help of other microorganisms such as fungi, bacteria, or algae and their exopolymers is known as bioflocculation. This method is an ecofriendly process that does not require energy consumption when compared to other methods and it has high efficiency for harvesting microalgae. Bioflocculation does not interfere with the subsequent downstream processes during biodiesel production and permits effective recycling of the spent medium. Bioflocculation using fungi can be induced by a cocultivation system which induces a synthetic symbiotic association between the targeted microalga and the cocultivated fungus. However, the key parameters of the coculture system are biomass and lipid productivity, which are altered in relation to the single system. Furthermore, a carbon source is mandatory for fungal growth, thus the cultivation may be performed under mixotrophic or heterotrophic conditions. Moreover, the coculture system has superior lipid yield and biomass production compared to monoculture. For instance, the co-cultivation of the oleaginous filamentous fungus *Aspergillus awamori* with *Chlorella minutissima* MCC 27 and *Chlorella minutissima* UTEX 2219 exhibited a 2.6–3.9-fold increase of biomass and 3.4–5.1-fold increase of total lipid yields compared to the single systems in the presence of glycerol as a carbon source [30]. Furthermore, C16:0 and C18:1 were the main FAs of the coculture oils, thereby the coculture system is a promising strategy for biodiesel production. The filamentous fungus helps in the entrapment of more than 90% of the algal cells in pellets, leading to an easier and economically feasible harvesting process. The mechanism of fungal-induced flocculation of microalgae is dependent on their positively charged cell walls as well as their filamentous structure and ability to form pellets. Similarly, bacteria could assist in bioflocculation, as a result of bacterial exopolysaccharides which attach to microalgae and form flocs. For instance, the poly-γ-glutamic acid broth produced by *Bacillus licheniformis* was revealed to be a 96% efficient harvester of *Desmodesmus* sp. [31].

Generally, the use of microalgae in the bioflocculation under a coculture system is more preferred than the use of fungi or bacteria. This is mainly attributed to the presence of long-chain FAs in the fungal cells and unsaturated fatty acids (UFAs) in the bacterial cells [32], which can decrease the quality of the produced biodiesel. Besides, bacteria and fungi have different growth requirements than microalgae. However, during alga-alga cocultivation, one of them must have an auto-flocculating ability or at least the coculture had enhanced flocculation, while both must have high biomass productivity and lipid contents. The flocculation efficiency of two poorly flocculating microalgae, *Desmodesmus* sp. ZFY and *Monoraphidium* sp. QLY-1, was enhanced to 85.33% after 4 h settling, which was better than that in monoculture (57.98% and 32.45%) [33]. This result was attributed to the enhancement of the exopolysaccharide content in the coculture. Furthermore, the quality of the biodiesel obtained from the coculture system is rich in C16 and C18 FAs, and its properties are agreed with international standards.

On the other hand, chemical flocculation includes adjusting the pH of the algal suspension or adding flocculants such as ferric sulfate, ferric chloride, aluminum sulfate, or organic polymers to the liquid media [34]. This method primarily encourages the aggregation of algal tiny cells into big aggregates, which will then settle via sedimentation. The harvesting efficiency of this process is influenced by the pH, flocculent dose, and algal surface charge [35].

The addition of inorganic flocculants can have adverse effects on the algal biomass characteristics, as well as the processing of algal biomass (extraction of microalgal lipids). Several studies have revealed the efficiency of flocculation using organic polymers compared to inorganic salts in microalgae harvesting. Among organic polymers, chitosan has emerged as a promising flocculant, which can nearly harvest 100% of the microalgal cells [35]. The use of chitosan has been also shown to allow the reuse of the spent medium with little growth inhibition. Kumar et al. [36] evaluated the FAs composition of the harvested *Scenedesmus* sp. using various flocculants (i.e., the cationic polymer and chitosan). The results indicated that the lipid yield and proportions of some FAs were promoted by flocculants in comparison to the use of alum. Low PUFAs content was in the case of alum-based harvesting, which may be related to oxidative cleavage of double bonds of PUFAs, and their conversion into SFAs. Furthermore, the use of cationic polymer exhibited no remarkable effects on the composition of FAs compared to centrifugation.

The harvesting of microalgae by transferring an electrical current via electrodes into an aqueous medium containing negatively charged algal cells is known as electrochemical flocculation. The electrodes may be sacrificial (made of metal ions that dissolve in the medium during electroflocculation) or nonsacrificial (made of nonreactive anodes and cathodes) [37]. The sacrificial electrodes are not preferred since the electrode is depleted and requires replacement as well as metallic contamination of the harvested algae. The development of electroflocculation at a large-scale production of biodiesel depends on the efficiency of the harvesting process and the energy requirement.

3.1.3 Centrifugation

It is another technique of microalgae harvesting from culture media by application of gravitational force. The major advantage of this technique is that it is rapid, effective, and simple to apply to any microalgal species at a high concentration rate. Moreover, it does not require the addition of flocculants, which may contaminate the harvested biomass. The harvesting efficiency is greatly dependent on the cell size as well as the density of the culture medium components [1]. Most laboratory centrifugation studies were performed at 500–1000 g force and revealed that around 80%–90% of microalgae can be separated within 2–5 min. In addition, large-scale centrifuges run at 5000–10,000 g and have a harvesting efficiency of 95% under optimum operating circumstances [1].

Although this method seems to be a viable option for high-value products, it is prohibitively expensive for low-value products generated in an integrated process, such as the production of biodiesel from microalgae. The fundamental cons of centrifugation may include high energy need, which is not cheap when it comes to producing biofuels. The energy consumption during centrifugation of microalgae at a feed rate of 1 L min^{-1} was reported to be 8 kW h m^{-3}. However, this value can be minimized by 10-folds by increasing the feed rate to 18 L min^{-1}, but this will make the harvesting process less efficient. Accordingly, the cost of the centrifugation-based harvesting at low efficiency and high flow rate of oleaginous algae was estimated to be \$0.864 L^{-1} oil [38]. In addition, the enormous centrifugal force employed in this method may induce cellular damage, rendering it unsuitable for some applications because sensitive nutrients may be lost. However, this effect is depending on the microalgal species employed, since some are more susceptible to such influences than others. Thereby, it is fundamental to optimize the conditions for centrifugation to minimize the expenses associated with the harvesting process.

3.1.4 Filtration

Filtration is defined as the separation of solid particles from a liquid by filtering them through a filter or membrane. Filtration has been successfully employed for harvesting microalgae with low cell density and small size like *Nannochloropsis* sp., *Chlorella* sp., and *Dunaliella* sp. [39,40]. It also has a high harvesting efficiency and low energy consumption due to the lack of chemicals' usage. However, the fundamental cons of filtration are clogging, and cake formation which is mainly problematic when harvesting small-sized microalgal cells. As a result, this strategy necessitates large economic costs. In order to enhance the performance of the filtration process and prevent membrane clogging, the filtration technique can be combined with additional approaches such as electromembrane usage, rotating disk, vibration, or aeration [40]. A study utilized submerged microfiltration for harvesting *Chlorella vulgaris* and *Phaeodactylum tricornutum* and reported energy consumption of 0.27 and 0.25 kW hm^{-3}, corresponding to 0.64 and 0.98 kW h kg^{-1}, respectively [39]. This value is generally more economical compared to the centrifugation technique, which typically consumes about 8 kW hm^{-3}. It was also estimated that when submerged microfiltration is combined with centrifugation to obtain a final concentration of 22% w/v, the consumed energy for the dewatering of *C. vulgaris* and *P. tricornutum* was 0.84 and 0.91 kW hm^{-3}, respectively [39]. An ultrafiltration membrane experiment combined with an air-assisted backwashing process was recently utilized to effectively harvest the microalga *Scenedesmus acuminatus*. The algal culture was concentrated from 0.5 to 136 g L^{-1} with a biomass harvesting efficiency of 93% and the consumed energy was estimated to be 0.59 kW h kg^{-1} dry biomass [41]. Moreover, microfiltration has been reported to have nonsignificant effects on both lipid yield and FAs composition compared to centrifugation-based harvesting [42].

3.2 Technologies for effective lipid extraction from algae

Lipid extraction of the harvested biomass is a critical step before the synthesis of biodiesel. The selected method of extraction can not only affect the yield of lipids but also determines the purity and the quality of the synthesized biodiesel. Furthermore, the lipid extraction efficiency contributes to a high impact on the final costs of the produced biodiesel.

After cell harvesting, extraction of lipids could be performed using either wet or dried biomass. The use of dry algal biomass can increase lipid yields but drying is generally unfavorable at large-scale production. Thus, to promote the Energy Returned on Energy Invested (EROEI), it is fundamental to develop a reliable extraction process based on wet biomass utilization with low cost and energy demands [16].

3.2.1 Cell disruption techniques

The hard cell wall structures present in several microalgae are a major barrier against the effective extraction of cellular lipids for biodiesel synthesis. Therefore, a pretreatment step to disrupt the cellular structure is a fundamental step especially when the wet biomass is used. Cell disruption processes can generally be grouped into mechanical and nonmechanical methods. Mechanical methods include bead milling, high-pressure homogenization, ultrasonication, microwave, and autoclave, while nonmechanical methods include the use of acids or bases, enzymes, and osmotic shock. The selection of the disruption method depends mainly on the cell size, cell wall structure, costs, and safety, and their effectiveness for releasing lipids without destructing other valuable components in the algal biomass. Mechanical-based disruption methods are usually energy-intensive, but they are generally distinguished to avoid chemical pollution and degradation of cellular products. The use of microwave irradiation was effective in disrupting the microalgal cells more effectively compared to other methods. Besides, microwave treatment is a promising fast process that usually contributes <1% of energy expenses during biofuel production [20]. Furthermore, the mechanical treatment is usually assisted with solvents to maximize lipid extraction efficiencies.

3.2.2 Conventional extraction methods

The conventional methods for extracting lipids from microalgae involve the use of a single or mixture of organic solvents at ambient temperature or high temperature in a Soxhlet extraction. The most used solvent mixtures are chloroform and methanol in a ratio of 1:2 (widely known as the Bligh and Dyer method) or 2:1 (commonly known as the Folch method) [43]. Hexane with or without a cosolvent is widely used as a less toxic alternative to the chloroform-methanol methods in the Soxhlet extraction. The main cons of the conventional extraction methods are linked to the toxicity of the organic solvents and environmental consequences as well as nonselectivity and high cost of the pure solvents. Therefore, several green solvents have emerged as promising alternatives for lipid extraction such as supercritical and liquid CO_2, and ionic liquids.

3.2.3 Simultaneous distillation and extraction process

In this process, the conventional extraction solvents are replaced by terpenes such as d-limonene, α-pinene, and p-cymene using Soxhlet coupled with solvent removal using Clevenger distillation. This method is effective in extracting lipids from wet biomass. Ref. [44] reported that p-cymene and pinene extracted similar oil yields compared to the conventional methods (Bligh and Dyer method and Soxhlet extraction using n-hexane) from *Nannochloropsis oculata* and *Dunaliella salina*, respectively. The same authors also confirmed that the energy required to extract lipids from 1 g biomass using SDEP is 2.15 kWh compared to 8.84 kWh in the case of the conventional Soxhlet extraction. Furthermore, this method does not alter the fatty acid composition and the contents of SFAs, MUFAs, and PUFAs in the extracted lipid, leading to high-quality biodiesel synthesis. Besides, the effectiveness and the low cost of SDEP, the process is ecofriendly since the aforementioned terpenes can be extracted from natural resources such as citrus peels and pine trees.

3.2.4 Supercritical fluid technology

Extraction of microalgal oils using supercritical fluid carbon dioxide (Sc-CO_2) is an environmentally benign and promising alternative process to the conventional excretion methods. The basic concept of this technology is achieving the supercritical phase of the substance at a temperature and pressure beyond its critical point, where the gaseous and the liquid phases can coexist. Sc-CO_2 extraction is usually performed at moderate temperatures (50–80°C) and pressure (20–30 MPa) [20]. The main pros of Sc-CO_2 include its good penetrability and extractability, high selectivity for extracting neutral lipids, short extraction period, and ease of separation and reuse. The Sc-CO_2 has less solvent consumption compared to the conventional methods. Furthermore, Sc-CO_2 can increase the yield of each FA along with increasing the MUFAs and decreasing SFAs contents within the extracted oil compared to solvent extraction methods [45]. The use of Sc-CO_2 generally eliminates the risk of FA degradation or oxidation during the extraction process. However, to extract polar lipids and increase the final yield, a cosolvent such as methanol, ethanol, or hexane is often used. The main cons of Sc-CO_2 are mainly linked to the high operation and equipment costs, which can be counterbalanced by the dispensing of the solvent recovery unit and the Sc-CO$_2$ can be easily separated by simple depressurizing and recycled with no greenhouse effect and avoiding the contamination of the residual biomass. However, the effectiveness of the Sc-CO$_2$ is generally low if the wet biomass is used compared to the dried one [45].

3.2.5 Liquid CO$_2$ technology

Liquid CO$_2$ (L-CO$_2$) is an innovative lipid extraction technology that has the same pros as Sc-CO$_2$ but can be effectively applied at low temperature (25°C) and pressure (15 MPa), which reduces the extraction costs [20]. The presence of water in the algal biomass does not affect the efficiency of the process, thus lipid extraction could be performed using wet microalgae biomass, thereby reducing the costs of drying. Furthermore, L-CO$_2$ has the same polarity as Sc-CO$_2$, therefore it has comparable selectivity to neutral lipids [20]. However, the use of L-CO$_2$ generally requires a two-stage extraction process to maximize the extraction efficiency. In the first stage, the algal biomass was disrupted using physical, chemical, or biological methods. Then L-CO$_2$ is pumped into the algal biomass in the presence of cosolvent such as methanol. Viner et al. compared the influence of different cell disruption techniques such as microwave, ultrasonication, grinding in liquid N$_2$, osmotic shock, and freeze drying on the efficiency of L-CO$_2$/methanol extraction of lipids from the algal slurry [20]. The results indicated that microwave pretreatment is an effective process to maximize the extraction of lipids using L-CO$_2$.

3.2.6 Ionic liquids

Ionic liquids (ILs) are salts in a liquid state consisting of asymmetric organic cations and inorganic or organic anions with a melting point below 100°C. ILs are characterized by fascinating properties such as low volatility, thermostability, tunability, and nonflammability [46]. The cationic moiety of the ILs is usually in the form of nitrogen-containing compounds such as imidazolium or pyridinium containing various alkyl substituents or in the form of alkyl-substituted ammonium, phosphonium, or sulfonium, while the anionic moiety varies from small molecules such as chloride, bromide, nitrate, or methanoate to amides, carboxylic acids, fluorinated structures, and esters [47]. ILs can dissolve the recalcitrant cellulose fibers within the cell wall of several microalgae; thereby, they have been utilized as green solvents for the cell disruption and extraction of both wet and dry microalgae with or without cosolvent. Ref. [47] screened several ILs for the disruption and extraction of lipids at room temperature from *Chlorella vulgaris* wet biomass. The study indicated that ILs containing small anions such as chloride, bromide, nitrate, or methanoate connected to short-chain alkyl cations are the most effective types in disrupting cell wall and extracting microalgal lipids. Lipids from *N. oculata* biomass containing 71.7% wt. water content were effectively extracted using Tetrakis (hydroxymethyl) phosphonium chloride with similar FAME yield and composition to the conventional method [46]. 1-Ethyl-3-methylimidazolium ethylsulfate has been also reported to effectively extract lipids from wet biomass (0%−82% wt. water) in the presence of methanol as a cosolvent in a short period (75 min) [47]. Furthermore, 1-ethyl-3-methylimmidazolium methyl sulfate as an extractive solvent in the presence of methanol and in combination with microwave treatment has been utilized for a single-step direct transesterification of *Nannochloropsis* sp. wet biomass (80% wt. water content) to synthesize high-quality biodiesel [48].

3.3 Techniques for biofuel production from algal biomass

Recently, microalgae have received a lot of attention as a potential resource for biofuel. Microalgae are regarded to be living cell factories for biofuel generation. The most widely used processes for converting microalgal biomass into biofuel are transesterification, pyrolysis, and hydrothermal liquefaction.

3.3.1 Two-step catalytic transesterification

Algal oil cannot be directly utilized as a biofuel due to its high viscosity, thereby it is converted into biodiesel by the transesterification process, which includes a reaction between TAGs and alcohol in the presence of acid or base to generate fatty acid esters and glycerol. Methanol is the most commonly used alcohol for industrial applications due to its low cost, although other alcohols such as propanol and ethanol are also usually utilized.

Transesterification is frequently catalyzed by homogenous catalysts such as alkalis, acids, or enzymes. Because of its simplicity, the base-catalyzed transesterification method is the most often utilized. Furthermore, it may be conducted under low pressure and temperature and produces high conversion (98%) in a short period of time. The most frequent alkalis used as catalysts in this process are NaOH and KOH. However, the use of alkali is limited if more than 0.5% wt. of the oil is FFAs, to prevent the saponification reaction with alkali [6].

On the other hand, acid-catalyzed techniques are appropriate for the transformation of feedstock rich in free FAs, but this process is less common, owing to the fact that acids are caustic, and the process is 4000 times slower than the alkaline transesterification, as a result, large volumes of alcohol and catalyst are required [6]. Several acids, including sulfuric acid and sulfonic acid, are usually utilized as catalysts in this process.

Lipase enzyme is another catalyst that may be employed in the transesterification process. It provides a high-purity product that is separated easily from glycerol. However, the enzyme cost remains quite expensive, posing a hurdle to its commercial application.

Transesterification of microalgal oil could also be performed using heterogenous catalysts such as CaO and MgO, which have the advantages of both acids and can simultaneously esterify and transesterify oils. Besides, heterogeneous catalysts are more environmentally benign, noncorrosive, and easy to separate and reuse [49].

3.3.2 In situ catalytic transesterification

In situ catalytic transesterification to produce biodiesel from biomass involves simultaneous extraction and transesterification in a single step. Accordingly, the direct transesterification of the biomass feedstock can markedly reduce the time and costs associated with the separate extraction and transesterification of algal oil. Ghosh et al. reported an effective lipid conversion (95%) of *Chlorella* sp. using in situ transesterification reaction at 65°C, 7 h, and 90 min using 4 M acid catalyst and 1:5 ratio of dried biomass/methanol [50]. The concentration of catalyst and the reaction period of the direct transesterification can be reduced by using microwave power or ultrasound radiation [6]. However, the presence of moisture in the biomass is a limiting factor during the in situ transesterification reaction. Besides, the need for high heating energy in the solvent extraction and recovery process is still a drawback for large-scale biodiesel synthesis.

3.3.3 In situ noncatalytic transesterification

In noncatalytic transesterification, supercritical alcohols are utilized for the simultaneous extraction and transesterification of algal oil. The penetration, solubility, and extractability of the supercritical alcohols are generally high, which increases the reaction rates and reduces time. In this case, the presence of moisture in the wet biomass can play a similar role as alcohol and promote product separation and reduce the conversion time [6]. Felix and coauthors optimized the supercritical methanol transesterification of wet *Chlorella vulgaris* (80% moisture) and indicated that the conditions that maximize the FAME yield to 74.6% with minimum energy consumption were 220°C in 2 h with an alga/methanol ratio of 1:8 (w/v) [51]. Even with the high efficiency and no waste generation, the use of supercritical alcohol in the transesterification of microalgae is still characterized by the high cost associated with high energy demands and a large amount of solvent.

3.3.4 Pyrolysis

Pyrolysis is a thermal treatment process that transforms biomass into gas, solid biochar, and liquid (bio-oil) fuels, at high-temperature conditions (350–700°C and above) in the absence of oxygen [52]. The efficiency of the pyrolytic product is highly affected by temperature, residence duration, and heating rate, in addition to moisture content, pressure, and type of biomass. The product of solid biochar is improved under the conditions of slow heating rate, low temperature, and long residence period. However, the short residence duration and high temperature improve the efficiency of gaseous fuels. In addition, the bio-oil product is increased at a short reaction period, high heating rates, and moderate temperatures [52].

Pyrolysis is classified into three categories slow, fast, and flash pyrolysis. The slow pyrolysis technique was carried out at 400°C, with very low heating rates (5–50°C min^{-1}) and reaction durations of many days. In contrast, fast pyrolysis uses a temperature range of 300–700°C and a short contact period (0.5–10 s) with a faster heating rate (10–200°C) to achieve higher products of bio-oil than slow pyrolysis. The Flash pyrolysis technique has attracted attention for producing bio-oil from algal biomass by employing short residence durations and extremely high temperatures (100–10,000°C per second) to minimize repolymerization of degenerated products, and around 75% bio-oil is obtained by this process [53].

Various publications on algae pyrolysis have been published as an effective method for biofuel production. According to Sotoudehniakarani et al., pyrolysis of *Chlorella vulgaris* generated about 47.7% wt. of bio-oil at 550°C mainly from lipids with a low number of oxygenated compounds and high calorific values, leading to desirable biofuel [54].

The major advantage of pyrolysis is that it is cost-effective, helps to reduce pollution, and the bio-oil is simple to store and transport. While the produced bio-oil has the disadvantages of being very viscous and lacking thermal stability, it also simulates diesel oil due to the presence of oxygenated molecules. However, bio-oil must be further upgraded to be utilized as a fuel replacement [53].

3.3.5 Hydrothermal liquefaction

The hydrothermal liquefaction (HTL) involves direct conversion of the wet algal biomass into liquid fuel at temperatures range of 250–450°C and pressures of 100–300 bar, with a contact time of 10–100 min and the presence of water and catalysts. It is noteworthy that, all nutrients of algae such as carbohydrates and proteins and not only lipids are converted into bio-oil by this process under the aforementioned conditions [55]. The HTL process is considered the most appropriate for the processing of outdoor cultivated algae immediately after harvesting. As a result, various microalgal biomasses may be transformed into bio-oils through this process.

The major benefit of utilizing the HTL technology for large-scale generation of biofuel is that the liquid phase formed in this method often has a greater concentration of essential nutrients. Therefore, the liquid solution might be reused for cultivations of microalgae. Furthermore, the direct transformation of wet algal biomass to bio-oil provides a more effective separation for the yield than the other alternatives [56]. Because of its higher physicochemical characteristics, HTL is considered an efficient technique with lower energy consumption. In addition, the produced bio-oil via hydrothermal liquefaction is generally 30%–50% dry wt. and contains 10%–11% wt. oxygen with a low higher heating value (32–35 MJ kg^{-1}) [57]. However, the higher concentration of oxygen results in a very acidic and viscous bio-oil with a low heating value. The quality of produced biocrude varies greatly based on the composition of feedstock and operating conditions.

The biofuel generated using this process has a high content of lower hydrocarbons with lower viscosity. As a result, the engine will emit SO_x and NO_x if it is run with this biofuel [57].

4. Properties of microalgal biodiesel

The transformation of algal biomass to biofuel is not ultimately the result, as the appropriateness of the produced biodiesel as an alternative to traditional diesel fuel must be also evaluated [58]. This is tested by assessing and comparing the physicochemical characteristics of biodiesel to international standards including EN 14214, ASTM D6751, and IS 15607 (Table 8.2). The goal of evaluating these criteria is to ensure that the biodiesel generated can be utilized as a cost-effective and ecofriendly alternative fuel.

The majority of the biodiesel generated from microalgal biomass was found to meet biodiesel criteria [7]. Furthermore, the biodiesel quality depends on the properties of fatty acid alkyl esters such as carbon chain length, chain branching, and unsaturation degree.

Iodine value, cetane number, saponification value, oxidation stability, kinematic viscosity, density, higher heating value, and cloud flow are the most important parameters to consider when evaluating the quality of biodiesel.

4.1 Iodine value

The iodine value (IV) is defined as the amount of iodine in grams that combines with 100 g of biodiesel. It depends on the percentage of UFAs and the number of double bonds found in the oil. The more double bonds in the FAs, the higher IV.

This parameter is related to the oxidation stability of the biodiesel, which is described by the value of induction time, which is the period before the beginning of biofuel oxidation. The longer the induction time, the better the oxidation resistance of the fuel. Therefore, the presence of long-chain FAs with a little number of double bonds improves biofuel quality and oxidation stability [9].

The IV of biodiesel was recorded in most microalgae as a low value. According to Ref. [60], the iodine value of *Scenedesmus incrassatulus* cultivated under optimized conditions was low (43.45 g I_2 100 g^{-1} fat), and found to match the European standard ($\leq$120 g I_2 100 g^{-1} fat). In another study, the mixotrophic cultivation of *Tetradesmus obliquus* using chitin hydrolysate exhibited a reduction in IV to 15.06 g I_2 100 g^{-1} fat compared to autotrophic conditions (48.45 g I_2 100 g^{-1} fat) [9]. In general, microalgae biodiesel has a higher UFAs level than traditional plant-derived biodiesel, although its IV is equivalent to vegetable oils. As a result, using microalgae as a feedstock for the production of biodiesel is not limited by this parameter.

4.2 Cetane number

The cetane number (CN) of biodiesel is an important parameter for evaluating its ability to combust. It is associated with the ignition delay time, which is the time between biodiesel injection and ignition. The higher the CN, the shorter the ignition delay period, which implies better combustion, reduced gas emissions, and simpler engine-startup [7].

The CN of biodiesel is greatly influenced by the structure of FAs. In general, it increases as carbon chain length increases and the degree of unsaturation decreases [9]. The oils of microalgal species with a high level of SFAs and MUFAs produce high-quality biodiesel [10]. Generally, microalgal biodiesel has a higher CN than petroleum-based diesel, which improves the engine performance and assists in the reduction of ignition delay time and entire biofuel combustion [61]. The specifications for minimum CN of biodiesel are 47 in ASTM D6751 and 51 in EN 14214 and IS 15607 (Table 8.2). Generally, higher CN is more favorable.

4.3 Saponification value

The saponification value (SV) of biodiesel is the amount of KOH in milligrams needed to saponify 1 g of oil sample. It is also a measure of the average molecular masses of all the FAs in the biodiesel. In general, SFAs with a long carbon chain have a lower saponification value than short-chain SFAs because they contain a lower number of carboxylic groups. However, the saponification value of UFAs increases as the degree of unsaturation increases [9].

4.4 Oxidation stability

Oxidation stability (OS) is described as the most fundamental and crucial parameter of biodiesel, which determines its stability throughout prolonged storage time. Biodiesel may often be kept for up to 6 months, and if it must be kept for an extended period, antioxidants can be used to increase its stability. The high content of UFAs in biodiesel fuel influences the oxidation stability, leading to rapid oxidation as a consequence of the combination of double bonds in the carbon chains of UFAs with oxygen to create free radicals, driving polymerization and further oxidation. Accordingly, a lower degree of unsaturation in carbon chains is beneficial for obtaining better OS values [59]. Furthermore, the oxidation of biofuel

TABLE 8.2 Properties of biodiesel as per EN 14214, ASTM D6751, and IS 15607 [59].

Biodiesel properties	EN 14214	ASTM D6751	IS 15607
Iodine value (g I_2 100 g^{-1} fat)	≤120	–	
Cetane number	≥51	≥47	≥51
Acid number (mg KOH g^{-1} fat)	≤0.8	≤0.5	≤0.5
Oxidation stability @ 110°C (h)	≥6	≥3	≥6
Kinematic viscosity @ 40°C ($mm^2 s^{-1}$)	3.5–5.0	1.9–6.0	2.5–6.0
Density @ 15°C (g cm^{-3})	0.86–0.9	–	0.86–0.9
Flashpoint (°C)	≥130	≥101	≥120
Copper strip corrosion @ 50°C	Class 1 max.	No. 3	Class 1 max.
Carbon residue (% mass)	≤0.3% mol mol^{-1}	≤0.05% mass	≤0.05% mass
Sulfated ash content (% mass)	≤0.02	≤0.02	≤0.02
Total sulfur	≤10 ppm	≤0.05% mass	50 mg kg^{-1}
Free glycerin (% mass)	≤0.02	≤0.02	≤0.02
Total glycerin (% mass)	≤0.25	≤0.24	≤0.25
Monoglycerides (% mass)	–	≤0.8	–
Diglycerides (% mass)	–	≤0.2	–
Triglycerides (% mass)	–	≤0.2	–
Methanol (wt.%)	–	≤0.2	≤0.2
Ester content (wt.%)	–	≤96.5	≤96.5
Linolenic acid methyl ester (% mass)	–	≤12	–
PUFAs (≥4 double bonds) (% mol mol^{-1})	≤1	–	–

ASTM, American Society for Testing and Materials; *EN*, European Committee for Standardization; *IS*, Indian standard.

produces byproducts that block filters or create deposits on injectors and fuel pumps, affecting the quality of biofuel and engine operation. Biodiesel fuel with high oxidation stability is required, particularly for those exposed to severe working conditions for extended periods at high temperatures. The oxidation stability of microalgal biofuel is often lower than that of conventional diesel, which limits its long-term storage.

4.5 Kinematic viscosity

The kinematic viscosity of engine oil is a measure of its resistance to flow. It is one of the most essential properties of biofuel, which impacts the performance of the fuel injection system. Viscosity is highly reliant on temperature, reducing at higher temperatures and allowing it to flow more easily. In general, increased viscosity at low temperatures results in lower fuel atomization and less precise engine performance [9]. Furthermore, the viscosity of oil increases with increasing carbon chain length and the degree of unsaturation, and it is also impacted by the double bond orientation, with the *cis*-orientation having a lower viscosity than the *trans*-orientation configuration [61]. The viscosity of microalgal biodiesel is greater than that of petroleum diesel, but it stands within the American, European, and Indian requirements [9].

4.6 Density

Density can give important information regarding biodiesel composition and performance properties, such as cost, smoke propensity, and efficiency of combustion. Since the engine's injection measure biofuel by volume, the density of the biodiesel determines how much fuel is injected and affects the energy content and proportion of air fuel inside the combustion tank.

The density is determined by the carbon chain length, the shorter the chain, the higher the density. The degree of unsaturation also influences density, which increases as the number of double bond increases [7].

The density of microalgal biodiesel is often somewhat greater than the conventional diesel, although they are within the range of international standards.

4.7 Higher heating value

The higher heating value (HHV) represents the thermal energy generated when 1 g of biodiesel is entirely burned at its initial temperature to form CO_2 and H_2O. The higher heating value of the biodiesel is correlated to its thermal conductivity and vapor heat capacity. These have an impact on the distribution of temperature and the proportion of air fuel, which decreases burning operation. It gradually decreases with increasing the degree of saturation owing to strong intramolecular interaction [55].

The limit of higher heating value for the international standards is not specified, but in general, the heating value for microalgal biodiesel ranged from 39 to 41 $MJ\,kg^{-1}$, which is approximately 10% less than petroleum diesel ($\sim 46\,MJ\,kg^{-1}$) [9].

4.8 Cold flow properties

The cold flow characteristics describe the flow pattern of biodiesel in cold weather. It is generally evaluated through the pour point, cloud point, and cold filter plugging point.

The pour point (PP) of oil is defined as the lowest temperature at which it solidifies and loses its flow characteristics. The cloud point (CP) is the temperature at which a liquid mixture turns cloudy owing to the production of solid crystals. Another important property of low-temperature biodiesel utilization is the cold filter plugging point (CFPP). It represents the lowest temperature at which a specific amount of biofuel flows through a wire mesh filter in a definite time.

The formation of crystals and solids at low temperatures results in the blockage of the fuel filter, which reduces the engine performance. SFAs have much greater PP, CP, and CFPP than UFAs. As a result, the SFAs alkyl esters crystallize more than unsaturated esters at higher temperatures. In addition, the quantity of SFAs influences the cold flow characteristics of biofuel [61]. As a result of the presence of SFAs esters, biodiesel exhibits poor cold flow characteristics as compared to conventional diesel in cold weather.

5. Future perspectives: challenges and opportunities

Microalgae are promising renewable and sustainable resource for the synthesis of biodiesel. However, there is still a long way for the commercial production of microalgae-based biofuels. Under the current technology, the production costs of biofuel are generally at least two times higher than petroleum-based diesel. Mixotrophic and heterotrophic cultivation of microalgae seems to be a promising alternative to autotrophic conditions in terms of high biomass and lipid productivities. Future research should focus on finding low-cost carbon sources and the sustainability of contamination should be controlled for the large-scale production of biomass using mixotrophic or heterotrophic mechanisms. The coupling of bioremediation of organically rich wastewater with biodiesel production should be improved and optimized to maximize biomass and lipid yield. The search for novel oleaginous microalgae that have the ability to self-flocculate can reduce the costs associated with the harvesting process. More efficient, feasible, and environmentally friendly techniques for extracting lipids from the wet biomass should be further explored. Energy consumption during the downstream processing of the biomass should be prioritized. Besides lipids, the utilization of microalgal products such as pigments, proteins, and carbohydrates under a biorefinery concept can markedly increase the market value of the microalgal biomass. Hence, great attention should be paid to the effective sequential extraction and processing of these products from microalgae.

6. Conclusion

The depletion of conventional fossil fuels makes the search for renewable and sustainable sources an obligate demand. Particularly, microalgae have been identified as promising microorganisms that can produce high-quality biodiesel with an adequate yield. The journey of biodiesel synthesis using the biomass of microalgae involves several upstream and downstream processes. Strain selection and cultivation conditions have direct effects on the proceeding downstream processes. Harvesting of biomass, extraction of lipid, and synthesis of biofuel are the main downstream processes that have contributed to the main portion of the production costs.

References

[1] M. Roy, K. Mohanty, A comprehensive review on microalgal harvesting strategies: current status and future prospects, Algal Res. 44 (2019) 101683.
[2] A.E. Atabani, M.M. El-Sheekh, G. Kumar, S. Shobana, Edible and nonedible biodiesel feedstocks, in: Clean Energy for Sustainable Development, Elsevier, 2017, pp. 507−556.
[3] I. Rawat, R. Ranjith Kumar, T. Mutanda, F. Bux, Biodiesel from microalgae: a critical evaluation from laboratory to large scale production, Appl. Energy 103 (2013) 444−467.
[4] M.M. Ismail, G.A. Ismail, M.M. El-Sheekh, Potential assessment of some micro- and macroalgal species for bioethanol and biodiesel production, Energy Sources, Part A Recov. Util. Environ. Eff. (2020) 1−17.
[5] A.T. Hoang, R. Sirohi, A. Pandey, S. Nižetić, S.S. Lam, W.-H. Chen, R. Luque, S. Thomas, M. Arōcō, V.V. Pham, Biofuel production from microalgae: challenges and chances, Phytochem. Rev. (2022), https://doi.org/10.1007/s11101-022-09819-y.
[6] J. Kim, G. Yoo, H. Lee, J. Lim, K. Kim, C.W. Kim, M.S. Park, J.-W. Yang, Methods of downstream processing for the production of biodiesel from microalgae, Biotechnol. Adv. 31 (2013) 862−876.
[7] M.A. Fawzy, M. Gomaa, Pretreated fucoidan and alginate from a brown seaweed as a substantial carbon source for promoting biomass, lipid, biochemical constituents and biodiesel quality of *Dunaliella salina*, Renew. Energy 157 (2020) 246−255.
[8] R.A.I. Abou-Shanab, M.M. El-Dalatony, M.M. EL-Sheekh, M.-K. Ji, E.-S. Salama, A.N. Kabra, B.-H. Jeon, Cultivation of a new microalga, *Micractinium reisseri*, in municipal wastewater for nutrient removal, biomass, lipid, and fatty acid production, Biotechnol. Bioproc. Eng. 19 (2014) 510−518.
[9] M. Gomaa, M.M.A. Ali, Enhancement of microalgal biomass, lipid production and biodiesel characteristics by mixotrophic cultivation using enzymatically hydrolyzed chitin waste, Biomass Bioenergy 154 (2021) 106251.
[10] M.M. El-Sheekh, H.M. Eladel, A.E.-F. Abomohra, M.G. Battah, S.A. Mohamed, Optimization of biomass and fatty acid productivity of *Desmodesmus intermedius* as a promising microalga for biodiesel production, Energy Sources, Part A Recov. Util. Environ. Eff. (2019) 1−14.
[11] M.J. Griffiths, S.T.L. Harrison, Lipid productivity as a key characteristic for choosing algal species for biodiesel production, J. Appl. Phycol. 21 (2009) 493−507.
[12] A. Karim, M.A. Islam, Z.B. Khalid, C.K.M. Faizal, M.M.R. Khan, A. Yousuf, Microalgal cell disruption and lipid extraction techniques for potential biofuel production, in: Microalgae Cultivation for Biofuels Production, Elsevier, 2020, pp. 129−147.
[13] M. Alhattab, A. Kermanshahi-Pour, M.S.-L. Brooks, Microalgae disruption techniques for product recovery: influence of cell wall composition, J. Appl. Phycol. 31 (2019) 61−88.
[14] G. Yoo, W.-K. Park, C.W. Kim, Y.-E. Choi, J.-W. Yang, Direct lipid extraction from wet *Chlamydomonas reinhardtii* biomass using osmotic shock, Bioresour. Technol. 123 (2012) 717−722.

[15] E. Nwokoagbara, A.K. Olaleye, M. Wang, Biodiesel from microalgae: the use of multi-criteria decision analysis for strain selection, Fuel 159 (2015) 241–249.
[16] T. Dong, E.P. Knoshaug, P.T. Pienkos, L.M.L. Laurens, Lipid recovery from wet oleaginous microbial biomass for biofuel production: a critical review, Appl. Energy 177 (2016) 879–895.
[17] A.F. Talebi, S.K. Mohtashami, M. Tabatabaei, M. Tohidfar, A. Bagheri, M. Zeinalabedini, H. Hadavand Mirzaei, M. Mirzajanzadeh, S. Malekzadeh Shafaroudi, S. Bakhtiari, Fatty acids profiling: a selective criterion for screening microalgae strains for biodiesel production, Algal Res. 2 (2013) 258–267.
[18] G. Li, J. Zhang, H. Li, R. Hu, X. Yao, Y. Liu, Y. Zhou, T. Lyu, Towards high-quality biodiesel production from microalgae using original and anaerobically-digested livestock wastewater, Chemosphere 273 (2021) 128578.
[19] I.A. Nascimento, S.S.I. Marques, I.T.D. Cabanelas, S.A. Pereira, J.I. Druzian, C.O. de Souza, D.V. Vich, G.C. de Carvalho, M.A. Nascimento, Screening microalgae strains for biodiesel production: lipid productivity and estimation of fuel quality based on fatty acids profiles as selective criteria, BioEnergy Res. 6 (2013) 1–13.
[20] K.J. Viner, P. Champagne, P.G. Jessop, Comparison of cell disruption techniques prior to lipid extraction from *Scenedesmus* sp. slurries for biodiesel production using liquid CO_2, Green Chem. 20 (2018) 4330–4338.
[21] Q. Hu, M. Sommerfeld, E. Jarvis, M. Ghirardi, M. Posewitz, M. Seibert, A. Darzins, Microalgal triacylglycerols as feedstocks for biofuel production: perspectives and advances, Plant J. 54 (2008) 621–639.
[22] G.R. Stansell, V.M. Gray, S.D. Sym, Microalgal fatty acid composition: implications for biodiesel quality, J. Appl. Phycol. 24 (2012) 791–801.
[23] A.P. Abreu, R.C. Morais, J.A. Teixeira, J. Nunes, A comparison between microalgal autotrophic growth and metabolite accumulation with heterotrophic, mixotrophic and photoheterotrophic cultivation modes, Renew. Sustain. Energy Rev. 159 (2022) 112247.
[24] J. Ruiz, R.H. Wijffels, M. Dominguez, M.J. Barbosa, Heterotrophic vs autotrophic production of microalgae: bringing some light into the everlasting cost controversy, Algal Res. 64 (2022) 102698.
[25] J. Liu, J. Huang, Z. Sun, Y. Zhong, Y. Jiang, F. Chen, Differential lipid and fatty acid profiles of photoautotrophic and heterotrophic *Chlorella zofingiensis*: assessment of algal oils for biodiesel production, Bioresour. Technol. 102 (2011) 106–110.
[26] T. Coward, J.G.M. Lee, G.S. Caldwell, Harvesting microalgae by CTAB-aided foam flotation increases lipid recovery and improves fatty acid methyl ester characteristics, Biomass Bioenergy 67 (2014) 354–362.
[27] M.T. Valeriano González, I. Monje-Ramírez, M.T. Orta Ledesma, J. Gracia Fadrique, S.B. Velásquez-Orta, Harvesting microalgae using ozoflotation releases surfactant proteins, facilitates biomass recovery and lipid extraction, Biomass Bioenergy 95 (2016) 109–115.
[28] Z. Yin, L. Zhu, S. Li, T. Hu, R. Chu, F. Mo, D. Hu, C. Liu, B. Li, A comprehensive review on cultivation and harvesting of microalgae for biodiesel production: environmental pollution control and future directions, Bioresour. Technol. 301 (2020) 122804.
[29] N. Cui, Y. Feng, J. Xiao, W. Ding, Y. Zhao, X. Yu, J.-W. Xu, T. Li, P. Zhao, Isolation and identification of a novel strain of *Heveochlorella* sp. and presentation of its capacity as biodiesel feedstock, Algal Res. 51 (2020) 102029.
[30] A. Dash, R. Banerjee, Enhanced biodiesel production through phyco-myco co-cultivation of *Chlorella minutissima* and *Aspergillus awamori*: an integrated approach, Bioresour. Technol. 238 (2017) 502–509.
[31] T. Ndikubwimana, X. Zeng, N. He, Z. Xiao, Y. Xie, J.-S. Chang, L. Lin, Y. Lu, Microalgae biomass harvesting by bioflocculation-interpretation by classical DLVO theory, Biochem. Eng. J. 101 (2015) 160–167.
[32] M.M. El-Sheekh, N.G. Allam, S.A. Shabana, M.M. Azab, Efficiency of lipid accumulating Actinomycetes isolated from soil for biodiesel production: comparative study with microalgae, Energy Sources, Part A Recov. Util. Environ. Eff. 39 (2017) 883–892.
[33] F. Zhao, J. Xiao, W. Ding, N. Cui, X. Yu, J.-W. Xu, T. Li, P. Zhao, An effective method for harvesting of microalga: coculture-induced self-flocculation, J. Taiwan Inst. Chem. Eng. 100 (2019) 117–126.
[34] A.E.-F. Abomohra, M. El-Sheekh, D. Hanelt, Pilot cultivation of the chlorophyte microalga *Scenedesmus obliquus* as a promising feedstock for biofuel, Biomass Bioenergy 64 (2014) 237–244.
[35] M. Nayak, N. Rashid, W.I. Suh, B. Lee, Y.K. Chang, Performance evaluation of different cationic flocculants through pH modulation for efficient harvesting of *Chlorella* sp. HS2 and their impact on water reusability, Renew. Energy 136 (2019) 819–827.
[36] S. Kumar Gupta, N.M. Kumar, A. Guldhe, F. Ahmad Ansari, I. Rawat, M. Nasr, F. Bux, Wastewater to biofuels: comprehensive evaluation of various flocculants on biochemical composition and yield of microalgae, Ecol. Eng. 117 (2018) 62–68.
[37] B. Matter, J. Seo, K. Lee, Oh, Flocculation harvesting techniques for microalgae: a review, Appl. Sci. 9 (2019) 3069.
[38] A.J. Dassey, C.S. Theegala, Harvesting economics and strategies using centrifugation for cost effective separation of microalgae cells for biodiesel applications, Bioresour. Technol. 128 (2013) 241–245.
[39] M.R. Bilad, D. Vandamme, I. Foubert, K. Muylaert, I.F.J. Vankelecom, Harvesting microalgal biomass using submerged microfiltration membranes, Bioresour. Technol. 111 (2012) 343–352.
[40] M. Rajvanshi, R. Sayre, Recent advances in algal biomass production, Biotechnol. Appl. Biomass (2021), https://doi.org/10.5772/intechopen.94218.
[41] L. Wang, B. Pan, Y. Gao, C. Li, J. Ye, L. Yang, Y. Chen, Q. Hu, X. Zhang, Efficient membrane microalgal harvesting: pilot-scale performance and techno-economic analysis, J. Clean. Prod. 218 (2019) 83–95.
[42] A.L. Ahmad, N.H.M. Yasin, C.J.C. Derek, J.K. Lim, Comparison of harvesting methods for microalgae *Chlorella* sp. and its potential use as a biodiesel feedstock, Environ. Technol. 35 (2014) 2244–2253.
[43] A.E.-F. Abomohra, W. Jin, M. El-Sheekh, Enhancement of lipid extraction for improved biodiesel recovery from the biodiesel promising microalga *Scenedesmus obliquus*, Energy Convers. Manag. 108 (2016) 23–29.

[44] C. Dejoye Tanzi, M. Abert Vian, F. Chemat, New procedure for extraction of algal lipids from wet biomass: a green clean and scalable process, Bioresour. Technol. 134 (2013) 271−275.
[45] Y. Li, F. Ghasemi Naghdi, S. Garg, T. Adarme-Vega, K.J. Thurecht, W. Ghafor, S. Tannock, P.M. Schenk, A comparative study: the impact of different lipid extraction methods on current microalgal lipid research, Microb. Cell Factories 13 (2014) 14.
[46] M. Olkiewicz, M.P. Caporgno, J. Font, J. Legrand, O. Lepine, N.V. Plechkova, J. Pruvost, K.R. Seddon, C. Bengoa, A novel recovery process for lipids from microalgæ for biodiesel production using a hydrated phosphonium ionic liquid, Green Chem. 17 (2015) 2813−2824.
[47] V.C.A. Orr, N.V. Plechkova, K.R. Seddon, L. Rehmann, Disruption and wet extraction of the microalgae *Chlorella vulgaris* using room-temperature ionic liquids, ACS Sustain. Chem. Eng. 4 (2016) 591−600.
[48] S. Wahidin, A. Idris, N.M. Yusof, N.H.H. Kamis, S.R.M. Shaleh, Optimization of the ionic liquid-microwave assisted one-step biodiesel production process from wet microalgal biomass, Energy Convers. Manag. 171 (2018) 1397−1404.
[49] M.O. Faruque, S.A. Razzak, M.M. Hossain, Application of heterogeneous catalysts for biodiesel production from microalgal oil—a review, Catalysts 10 (2020) 1025.
[50] S. Ghosh, S. Banerjee, D. Das, Process intensification of biodiesel production from *Chlorella* sp. MJ 11/11 by single step transesterification, Algal Res. 27 (2017) 12−20.
[51] C. Felix, A. Ubando, C. Madrazo, et al., Non-catalytic in-situ (trans) esterification of lipids in wet microalgae *Chlorella vulgaris* under subcritical conditions for the synthesis of fatty acid methyl esters, Appl. Energy 248 (2019) 526−537.
[52] G. Perkins, T. Bhaskar, M. Konarova, Process development status of fast pyrolysis technologies for the manufacture of renewable transport fuels from biomass, Renew. Sustain. Energy Rev. 90 (2018) 292−315.
[53] M. Saber, B. Nakhshiniev, K. Yoshikawa, A review of production and upgrading of algal bio-oil, Renew. Sustain. Energy Rev. 58 (2016) 918−930.
[54] F. Sotoudehniakarani, A. Alayat, A.G. McDonald, Characterization and comparison of pyrolysis products from fast pyrolysis of commercial *Chlorella vulgaris* and cultivated microalgae, J. Anal. Appl. Pyrol. 139 (2019) 258−273.
[55] F. Hossain, J. Kosinkova, R. Brown, Z. Ristovski, B. Hankamer, E. Stephens, T. Rainey, Experimental investigations of physical and chemical properties for microalgae HTL bio-crude using a large batch reactor, Energies 10 (2017) 467.
[56] Y. Guo, T. Yeh, W. Song, D. Xu, S. Wang, A review of bio-oil production from hydrothermal liquefaction of algae, Renew. Sustain. Energy Rev. 48 (2015) 776−790.
[57] R. Ganesan, S. Manigandan, M.S. Samuel, R. Shanmuganathan, K. Brindhadevi, N.T. Lan Chi, P.A. Duc, A. Pugazhendhi, A review on prospective production of biofuel from microalgae, Biotechnol. Rep. 27 (2020) e00509.
[58] M. El-Sheekh, A.E.-F. Abomohra, H. Eladel, M. Battah, S. Mohammed, Screening of different species of *Scenedesmus* isolated from Egyptian freshwater habitats for biodiesel production, Renew. Energy 129 (2018) 114−120.
[59] S. Deshmukh, R. Kumar, K. Bala, Microalgae biodiesel: a review on oil extraction, fatty acid composition, properties and effect on engine performance and emissions, Fuel Process. Technol. 191 (2019) 232−247.
[60] M.A. Fawzy, A.M. El-Otify, M.S. Adam, S.S.A. Moustafa, The impact of abiotic factors on the growth and lipid accumulation of some green microalgae for sustainable biodiesel production, Environ. Sci. Pollut. Res. 28 (2021) 42547−42561.
[61] G. Knothe, Production and properties of biodiesel from algal oils, in: Algae for Biofuels and Energy, Springer Netherlands, Dordrecht, 2013, pp. 207−221.

Chapter 9

A plausible scenario for the third generation of biofuels from microalgae

Rohit Saxena[1], Rosa M. Rodríguez-Jasso[1], Mónica L. Chávez-González[1], Cristóbal N. Aguilar[1], Cristina González-Fernández[2], Guillermo Quijano[3] and Héctor A. Ruiz[1]

[1]Biorefinery Group, Food Research Department, School of Chemistry, Autonomous University of Coahuila, Saltillo, Coahuila, Mexico; [2]Biotechnological Processes Unit, IMDEA Energy, Madrid, Spain; [3]Laboratory for Research on Advanced Processes for Water Treatment, Instituto de Ingeniería, Unidad Académica Juriquilla, Universidad Nacional Autónoma de México, Querétaro, Mexico

1. Introduction

In recent years, many countries have experienced challenges in population growth and industrial inflation that have hastened the depletion of fossil fuels, raising worries about global warming and motivating the quest for alternative, renewable, and sustainable energy sources. Biologically produced fuels have been identified as potential alternative energy sources to mitigate greenhouse gas emissions under the biorefinery concept. Biorefineries are based on the emerging interest in resource depletion (fossil-based) and associated environmental issues in the framework of bioeconomy and sustainability position analogous to the fossil-based refineries [1]. The International Energy Agency defined biorefining as the sustainable biomass processing into a spectrum of biobased products (food, feed, chemicals, materials) and bioenergy sources (biofuels, power, or heat). A biorefinery can maximize the value associated with biomass feedstocks from economic, social, or environmental perspectives [2]. Biofuels are being promoted as one of the most promising routes to lower CO_2 emissions and reduce the dependency on fossil fuels. Biofuel production from renewable sources is widely considered one of the most sustainable alternatives to petroleum-sourced fuels and a viable means for environmental and economic sustainability. The United States of America is at the forefront of the biofuel sector, intending to replace 20% of transportation fossil fuels with biofuel by 2022 [3].

Biofuels are usually grouped into different categories known as first-generation (sugarcane, sugar beet, corn), second-generation (agriculture and other organic residues), third-generation (algal feedstocks), and fourth-generation biofuels, depending on the feedstocks and conversion technology used under biorefinery production [3]. Biofuels of third and fourth generations are based on low-input autotrophic microbial feedstock, especially microalgae and cyanobacteria such as *Spirulina, Chlorella, Dunaliella, Botryococcus,* or *Haematococcus*. Innovative advancement of these technologies might allow for the generation of sustainable biofuel volumes while also fixing CO_2 via photosynthesis without competing for arable land. Indeed, both at the raw material level (fixing ambient CO_2) and at the process technology level (low CO_2 output), the production of this form of biofuel is predicted to be carbon negative, reducing CO_2 emissions and aiding climate change mitigation [4].

Several studies highlighted the potential and suitability of *Spirulina platensis* for the extraction of biofuels (such as bioethanol, biodiesel, and biohydrogen), proteins, and specific bioactive chemicals like phycocyanin, among other components largely [5]. *Spirulina platensis* is a microscopic photosynthetic microorganism capable of converting solar radiation, nutrients, and carbon dioxide into biochemical energy such as lipids, protein, and carbohydrates in a short period, satisfying their growing demands through two different processes called photosynthesis and chemosynthesis. Carbohydrate and lipid contents of microalgae are being processed into biofuels and other high value-added co-products such as phycocyanin, β-carotene, astaxanthin, and algal extracts for use in cosmetics [6].

Bioethanol is one of the most widely used alternative to fossil fuels, with the United States, Brazil, and other countries producing the most significant amounts to the date [7]. Bioethanol is a biofuel produced by biologic processes that include sugar and starch as key substrates [8]. Microalgae biomass is a suitable source of sugars for the production of third-generation bioethanol, as it overcomes the limitations of previous generations' feedstock, such as edible crops and lignocellulosic materials. To break down carbohydrate-rich feedstock and produce bioethanol, enzymatic hydrolysis (saccharification) and fermentation processes are required. During the saccharification process, enzymes work on complex polymers such as cellulose and hemicellulose to obtain simple sugars, which are further converted into bioethanol [9]. Sugars obtained from the saccharification process (i.e., glucose) can be then fermented by *Saccharomyces cerevisiae* into ethanol [8]. Biodiesel is a renewable, biodegradable, and environmentally friendly alternative to fossil fuels. Microalgae are primarily used for increased neutral lipid content (up to 60%—70%) and subsequent transesterification to make biodiesel [1]. Hydrogen (H_2) is another biofuel and energy carrier with large potential for replacing fossil fuels [10]. Hydrogen has a wide range of uses and has a significant potential for lowering fossil fuel usage in the power, chemical, and transportation sectors. The photo-fermentation of biohydrogen by microalgae is an interesting niche since this process can produce hydrogen gas from the most abundant resources, light and water [11]. Here, this chapter has attempted to summarize key breakthroughs in third-generation biorefineries employing microalgae as a feedstock for cultivation, harvesting, pretreatment techniques, enzyme saccharification, and fermentation of algal biomass to create biofuels, chemicals, and other high-value-added byproducts in this study.

2. Third-generation biorefinery

Third-generation biorefineries employ microbial cell factories to produce bioproducts using CO_2 from the atmosphere and renewable energy such as light, inorganic compounds from waste streams, and electricity supplied by sustainable sources such as solar cells and wind power. Compared to first-generation and second-generation biorefineries, third-generation biorefineries significantly reduce feedstock processing costs and offer far fewer security hazards to food and water systems and are thus gaining traction [12]. The fuels that may be obtained from algal biomass are known as third-generation biofuels. Microalgae, macroalgae, and cyanobacteria are among the microorganisms classified as algae. Bioethanol production from algae usually is dependent on different elements such as technology and the marine environment. The third generation has been developed from prospective feedstock that may be directly turned into energy. The lipid content of the microorganisms determines the biofuel amount generated from algae [13].

Microalgal biomass has an enormous capacity to create diverse metabolites appropriate for bioenergy, food additives, pigments, bioplastics, polymers, and chemicals. The algae-based biorefinery combines the large production of algal biomass for energy with various industrially relevant chemicals to ensure the microalgal industries' long-term viability. Microalgae biomass can be divided into three significant contents: lipid, carbohydrate, and protein, which can be converted into different biofuels and high-value-added bioproducts through biomass conversion processes like pretreatment (hydrothermal, physiochemical, physical, etc.), enzyme saccharification, a chemical reaction (transesterification), and biochemical conversion (fermentation, anaerobic digestion) [6]. Biorefineries allow for the full exploitation of algal biomass. In (Fig. 9.1), algae-based refineries are depicted.

3. Technologic applications of microalgae

3.1 Spirulina platensis

Spirulina platensis is the world's most extensive commercially produced microalgae and is a spiral-shaped filamentous cyanobacterium with different uses. Protein (greater than 50%), vitamins, vital amino acids, and minerals are all in *S. platensis* biomass. *S. platensis* cells may accumulate polysaccharides to over 70% of total dry weight under nitrogen-limited growth circumstances [6]. Hexoses make up most of these polysaccharides, making them a suitable feedstock for bioethanol synthesis. Surprisingly, *S. platensis* biomass has not been explored as a substrate for synthesizing microbial lipids [14]. Microalgae convert most CO_2 to fermentable sugars and lipids using solar energy. They may be tailored to acceptable climates and temperatures for mass-scale production in nonarable places, which reduces competition for agricultural land acquisition. Because of their quick growth rate and high protein, carbohydrate, and fat content, they have a lot of potential as a biofuel source. Moreover, the absence of lignin in third-generation feedstock reduces expenses for the pretreatment process required for optimum extraction of sugars from these feeds for biofuel production [6]. *S. platensis* is suggested as a sustainable and eco-friendly microalga useful for bioremediation, nitrification, and carbon dioxide fixation. It can perform autotrophic and heterotrophic metabolisms, and the resulting biomass constitutes a feedstock containing carbohydrates, lipids, and proteins (Table 9.1) [17]. The microalgal cell wall predominantly comprises cellulose, but it can also include pectin and sulfated polysaccharides. Intracellular starch is observed in the plastids, with a content ranging from

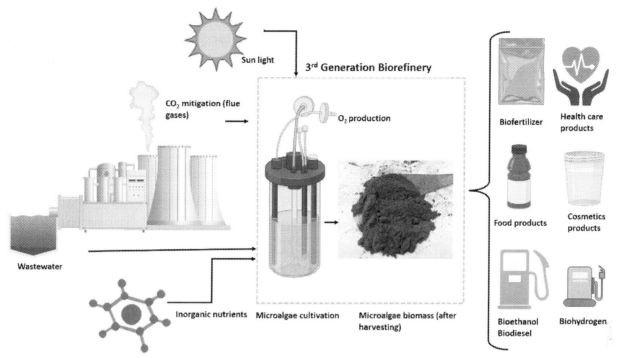

FIGURE 9.1 Third-generation biorefinery and its by-products.

TABLE 9.1 Biochemical composition of different microalgae.

Microalgae species	Protein %	Lipids %	Carbohydrates %	Ash %	Moisture %	References
A. platensis F&M-C256	63.9 ± 1.0	10.7 ± 0.56	12.8 ± 0.21	6.1 ± 0.10	7.9 ± 0.20	[15]
N. sphaeroides F&M-C117	50.8 ± 1.45	15.1 ± 1.19	14.5 ± 0.53	4.0 ± 0.25	7.8 ± 0.28	[15]
C. sorokiniana F&M-M49	51.3 ± 0.48	22.7 ± 2.0	15.5 ± 0.08	5.4 ± 0.11	8.5 ± 0.24	[15]
C. sorokiniana IAM C-212	39.9 ± 0.94	27.9 ± 1.30	10.7 ± 0.90	9.4 ± 0.37	7.5 ± 0.30	[15]
C. vulgaris Allma	56.8 ± 2.70	16.9 ± 2.83	5.9 ± 0.25	9.3 ± 1.47	4.9 ± 0.17	[15]
T. suecica F&M-M33(S)	18.3 ± 0.10	22.4 ± 1.15	36.8 ± 1.46	14.8 ± 0.47	6.1 ± 0.26	[15]
T. suecica F&M-M33(NR)	40.2 ± 0.51	28.5 ± 1.16	10.2 ± 0.20	15.7 ± 0.20	7.2 ± 0.14	[15]
P. purpureum F&M-M46	34.2 ± 0.10	13.1 ± 1.12	17.0 ± 1.72	22.0 ± 0.88	10.0 ± 0.39	[15]
P. tricornutum F&M-M40	38.8 ± 0.11	20.5 ± 0.54	11.0 ± 0.70	14.8 ± 0.12	8.0 ± 0.23	[15]
Tisochrysis lutea F&M-M36	42.9 ± 0.42	27.9 ± 3.25	8.6 ± 0.89	11.5 ± 0.27	6.3 ± 0.26	[15]
N. oceanica F&M-M24	43.1 ± 0.10	28.2 ± 2.04	14.3 ± 0.19	12.9 ± 0.84	7.2 ± 0.21	[15]
Scenedesmus almeriensis	12.93 ± 0.69	2.05 ± 0.12	4.51 ± 0.41	57.61 ± 2.20	8.89 ± 0.32	[16]
Dunaliella salina	10.03 ± 0.57	3.49 ± 0.10	25.31 ± 1.55	48.74 ± 2.50	6.63 ± 0.25	[16]
Nannochloropsis sp.	26.67 ± 0.10	15.30 ± 0.24	32.05 ± 0.70	8.31 ± 0.42	1.90 ± 0.05	[16]
Haematococcus pluvialis (green phase)	32.59 ± 1.20	3.24 ± 0.11	0.13 ± 0.01	29.49 ± 0.22	5.03 ± 0.12	[16]

20% to 50%. *Spirulina* has high nutritional values due to its content of proteins, essential amino acids, minerals, essential fatty acids, vitamins, and liposoluble antioxidants (vitamin E and carotenoids). Furthermore, *Spirulina* contains many functional bioactive ingredients with antioxidant and antiinflammatory activities, including phenolic phytochemicals and the phycobiliprotein C-phycocyanin (Table 9.2) [5].

3.2 Cultivation system

There are two methods for cultivating microalgae: open and closed systems (photobioreactors). Microalgae cultivation and extract products can be carried out in various growing methods (Table 9.3).

3.2.1 Open system

The open pond system is the conventional method for microalgae cultivation, like natural processes. In the 1960s, Ostwald was the first to describe an open raceway, which entails two outlets joined by bends. One or two paddle wheels are turned continuously to recycle the growth media and avoid precipitation of microalgae [33]. It is one of the oldest and most basic methods for large-scale microalgae cultivation (Fig. 9.2A). Because of its lower construction, maintenance, and operation costs, an open pond is widely used. Other advantages of using an open pond system include simple operation and maintenance, low energy demand, and ease of scale-up. Natural water bodies, such as lakes and ponds, and artificial water bodies, such as circular and raceway ponds, are found in some open ponds. However, despite the enormous culture, microalgae production from natural water has a lower cell density, necessitating a highly effective harvesting process. Furthermore, pollution, precipitation runoff, which influences salinity and changes pH, erosion of banks, which results in leaks, and increased water turbidity, limiting the productivity of microalgae in open ponds, impact the growing condition of microalgae [34].

3.2.2 Closed-system photobioreactor

Transparent materials such as borosilicate glass, polyethylene (PE), polyvinyl chloride (PVC), and polymethyl methacrylate are commonly used to make photobioreactors (PBRs). However, for PBR construction, borosilicate glass is the most utilized material. The algae species, culture site, ideal conditions, and economic considerations influence PBR design [35].

TABLE 9.2 Bioactive compounds from microalgae and their potential application.

Microalgae	Bioactive compounds	Activities and uses	References
Tetraselmis sp. Porphyridium cruentum Porphyridium purpueenum Chlorella	Polysaccharide	Antiinflammatory, immunomodulating, and antiviral properties as well as used as joint lubricants	[18,19]
Nannochloropsis gaditana, Scendesmus obliquus	EPA/DHA	Human consumption	[19]
Porphyra yezoensis	Pepsin	ACE inhibitory, antimutagenic, blood sugar reducing, calcium precipitation inhibition, lowering cholesterol, antioxidant, and improved hepatic function	[20,21]
Haematococcus pluvialis	Astaxanthin	Antioxidant Sunscreen protection	[19,22]
Dunaliella salina	β-carotenes	Antioxidant	[19,22,23]
Chlorella Skeletonema Porphyridium Nostoc flegelliforme	β-1,3-glucan	Free-radical collector Immune system booster Antiinflammatory	[19,22–24]
Porphyridium and Rhodella reticulata	Sulfated polysaccharides	Antioxidant	[19,25]

TABLE 9.3 Different types of microalgae cultivation systems in another medium and their extraction product.

Microalgae species	Medium	Growth condition	PBR	Growth OD	Extract product	References
Spirulina platensis	Beet vinasse in culture medium	Temp 30°C 150 rpm Fluorescent lamp	Tubular PBR (vertical airlift)	750 nm	Protein	[26]
Arthrospira maxima LJGR1	Z medium	Blue LED light intensity 100 µmols. s^{-2}m^{-1}	Open pond reactor	620/280 nm	Phycocyanin	[27]
Tribonema sp	BG 11	Temp 25°C Light 2500 lux	CSTR	—	Protein and carbohydrates	[28]
Spirogyra	Bold's basal medium	Temp 23°C Ext. high-pressure sodium vapor lamp intensity 100 µmols s^{-2}m^{-1}	Flat panel airlift PBR	—	Biomass	[29]
Scenedesmus dimorphus	BG 11	Temp 27 ± 2°C Light intensity 17 µmols s^{-2}m^{-1}	Airlift PBR	678 nm	CO_2/NO_2	[30]
Chlorococcum humicola TISTR 8641	BG 11	Temp 28–30°C LED light intensity 3500 lux 200 rpm	STR and airlift PBR	480 nm	Carotenoids and biomass	[31]
Dunaliella tertiolecta	Seawater-enriched medium	Temp 21°C ± 2°C pH 7.6 Humidity 15% ± 7% LED 3200 lux	Tubular PBR	—	Biomass	[32]

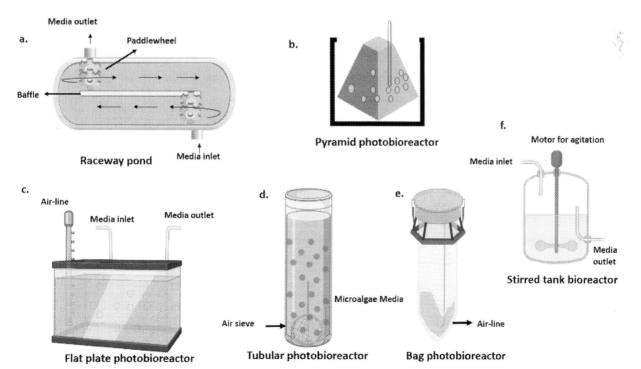

FIGURE 9.2 Microalgae cultivation systems: (A) raceway pond, (B) pyramid photobioreactor, (C) flat-plate photobioreactor, (D) tubular photobioreactor, (E) bag photobioreactor, (F) stirred-tank bioreactor.

3.2.2.1 Stirred-tank bioreactors

The stirred-tank bioreactor (STR) was first considered an appropriate reactor for photoautotrophic algaculture due to its construction as a conventional laboratory and industrial reactor, and *Selenastrum capricornutum*, *Laminaria saccharina*, *Euglena gracilis*, *Chlorella* sp., and *Synechococcus* sp. have all been effectively cultured in this type of reactor [35]. The impeller is mechanically moved by an electric motor, allowing adequate agitation of the culture (Fig. 9.2F). The CO_2 supply can be delivered through a sparger at the tank's bottom. The impellers spread the supplied gas evenly. As a result, the STR allows for excellent heat and mass transport, light dispersion, and homogenization mixing. To prevent mixing-ensured vertexing, baffles can be optimally placed along the interior wall of the tank. The culture medium fills just 70%–80% of the volume of STRs. Foam and any other liquid or gaseous forms awaiting discharge through the exhaust can be accommodated in the headspace. Foam generation can also be reduced in this design by using foam breakers [35]. Despite its mixing and mass transfer efficiency, STRs are energy-demanding systems, decreasing the cost-effectiveness of the algaculture.

3.2.2.2 Flat-plate photobioreactor

A flat-plate photobioreactor (FP PBR) is a general design with a rectangular box look and, as a result, a broad lighting area that ensures excellent light collection and optimal illumination of the grown cells. FP PBRs are 0.1–0.15-m spaced, parallel, flat transparent panels grouped into a rectangular channel (Fig. 9.2C) [35].

3.2.2.3 Bag photobioreactor

Due to high film extrusion temperatures, bag PBRs are especially attractive for commercial-scale production due to their low cost and functional sterility at startup. These bags can be fitted with aeration systems to improve yields (Fig. 9.2E) [36]. Large bag PBR systems are commonly used globally to cultivate microalgae for feed in the aquaculture industry. These PBRs have also been employed in wastewater treatment [37]. The most significant downside of bag PBRs is the difficulty in sterilizing the bags. Most bag PBRs are made of PE and may not tolerate autoclaving. Sterilization with ultraviolet light or gas is possible, but these methods come at a high price. Disposable bags that must be replaced regularly are used to obviate the necessity for sterilizing between cultures. Like other PBRs, the design considerations for bag PBRs are size, materials, method of aeration, and frame structure. Mainly, a bag PBR is constructed with PE and systems often hung vertically, and the growth medium is pumped through the top side of the PBR and flows down to the bottom of the PBR. In bag PBRs, like tubular PBRs, gas exchange is an important characteristic that occurs in a unit distinct from the photosynthetically active portion of the bag. Bag PBR systems provide the benefits of versatility, simplicity, and cost-effectiveness. While bag PBRs appear to be a potential alternative for large-scale microalgae production, we feel that more modification is required to make these systems industrially viable. A recent study investigated several PBR types: bag, flat plate, and bubble column for their efficacy at cultivating the marine benthic diatom *Cylindrotheca clostridium* [37].

3.2.2.4 Tubular photobioreactor

A tubular PBR is shaped like a tube and is commonly made of glass, PVC, or plastic, and it is the most prevalent type of PBR nowadays. This form of PBR is appealing because of its vast size and minimal surface area, making it ideal for use in an external application. The tube diameter spans from 10 to 60 mm, while its length extends from 10 to 100 m. To improve light penetration in a reactor, the diameter of the tube is maintained to be small. The recommended liquid velocities for mass culture are 0.2–0.5 m s^{-1} [38]. Mixing in a tubular PBR is usually work done by a sparger that creates bubbles. A tubular PBR comprises a column with an air sparger disk located at its bottom (Fig. 9.2D). At the top of the PBR, the freeboard regime functions for gas/liquid separation. Mixing is achieved by the turbulence created by air bubbles moving upward. Internal-loop airlift PBR typically comprises a transparent column, an internal column, and an air sparger. O_2- and CO_2-enriched air are introduced inside the internal column at the bottom [36]. To avoid excessive oxygen concentrations in the PBR system from accumulating, a degasser unit is attached to the tubes. Depending on the tube orientation, it is called horizontal, vertical, or inclined tubular PBR. Some disadvantages include dissolved oxygen accumulation, excessive power consumption, high temperature, high pH, CO_2, and O_2 gradients, expensive capital, operational expenses, and photo limits. Tubular photobioreactors include airlift and bubble columns. Under natural sunlight, this sort of PBR performs well [38].

3.2.2.5 Pyramid photobioreactor

These are new PBR designs series that are completely computerized and automated. They follow a pyramidal shape and have been successfully used to culture *Spirulina*, yielding a biomass concentration of 1.45 g L^{-1} d^{-1}. This yield is four times more than achieved in an open system. The pyramid design demands a relatively small area of land to set up. A place of 60 m^2 is enough to construct a pyramid PBR for a culture 100 t in mass. However, this design is still at the experimental stage and requires more time until concrete conclusions can be reached (Fig. 9.2B) [35].

3.2.2.6 Hybrid photobioreactor

Hybrid photobioreactors are devices that feature two or more photobioreactor configurations and are used to exploit the advantages of each chosen structure and minimize or even eliminate the disadvantages of each geometry. Researchers have looked at novel designs beyond the classic strategies discussed above to improve PBR operation and increase algal biomass. Hybrid systems aim to enhance the PBR's surface area, reduce dead zones, improve mixing, and deliver consistent illumination. The airlift-driven raceway reactor was one of the first hybrid designs. The system's producing space is a racetrack. An airlift tube with a riser and downcomer is utilized instead of a paddlewheel/pump to deliver the liquid level differential between the riser and downcomer. The top of the airlift system extends above the liquid in the racetrack, allowing flow in the raceway to be caused by the difference in liquid level (head). Pumping energy is conserved using this arrangement (Fig. 9.3) [37].

3.3 Harvesting

The industrial application of microalgae necessitates the efficient and low-cost harvesting of biomass from the culture medium. The strategies used to harvest microalgae biomass from the diluted algae growth medium include gravity sedimentation, flocculation, centrifugation, filtration, flotation, and electrocoagulation. The cost and energy consumption of microalgae harvesting are very high since the biomass density in the culture medium is generally low. Most microalgae cells carry a negative charge that keeps the stability of cells in a dispersed state. It was reported that the cost of the harvesting process even accounts for 30% of the total cost of microalgae biomass production [33,38]. As a result, it is possible to lower the cost of producing microalgae biomass [33].

The features of microalgae and the value of biomass products determine which harvesting procedures are used. The harvesting procedure is often separated into two steps: bulk harvesting and thickening. Separating biomass from the culture medium is the goal of bulk harvesting procedures, including flocculation, flotation, and gravity sedimentation. Thickening uses more energy than bulk harvesting since it involves filtering and centrifugation. The following is a list of microalgae harvesting technologies [33].

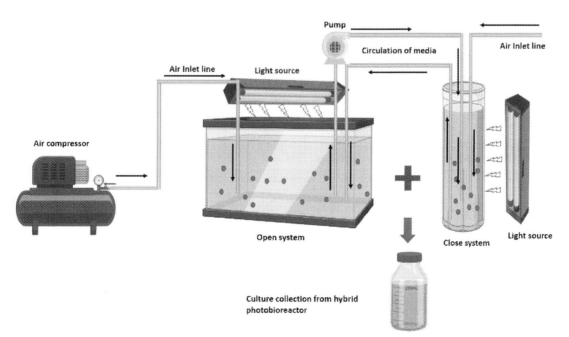

FIGURE 9.3 Hybrid photobioreactor for cultivation of microalgae.

3.3.1 Sedimentation

Sedimentation is a classic and straightforward harvesting process that involves gravity separation of microalgae cells. This low-cost method can be utilized to lower internal energy requirements during the earliest stage of biomass recovery. Using less than 0.1 kWh, conventional sedimentation systems may obtain a final solid concentration in the aqueous phase of between 1% and 3% [39]. Gravity is ideal for separating large particles like *Spirulina*. On the other hand, sedimentation might be extraordinarily sluggish if there are slight density or particle size changes [40,41]. Smaller particles, as a result, necessitate the inclusion of flocculants, which can boost recovery efficiency by up to 90% [42].

3.3.2 Flocculation

Flocculation is the process of solute aggregation that occurs when solutes in the algal growing media bind together. Flocculation is a high-cost method for harvesting microalgae; nonetheless, it has been regarded as one of the low-cost ways of harvesting microalgae [38]. Flocculation is the aggregation of suspended particles into bigger flocs that can be naturally separated by sedimentation or flotation. The electrostatic charges on the surface of the particles serve as the action mechanism. Negative accounts on cellular walls naturally cause individual cells to repel one another. As a result, the creation of flocs necessitates the inclusion of chemicals known as flocculants, which provide a strong enough attraction force and connect the suspended cells. Organic, inorganic, or biologic chemicals and altering pH can cause flocculation [43]. Chemical flocculants may have a detrimental impact on the quality of the intended final products, even though flocculation is an active process with biomass recoveries near 100%. The flocculant dose ranges between 501 and 600 mg [44].

3.3.2.1 Auto-flocculation

It is a laboratory-based version of the natural flocculation process. The chemical flocculation of microalgae cells in the presence of calcium and magnesium ions at a high pH, generally over 9, is known as auto-flocculation. Auto-flocculation is a low-energy process that may be triggered by CO_2 consumption through photosynthesis or by using inexpensive, nontoxic chemicals like NaOH. Microalgae extract CO_2 from the growth media during photosynthesis, raising the pH and neutralizing their surface (with a negative charge), resulting in spontaneous flocculation [45,46]. Induced auto-flocculation has the potential to attain high efficiency (above 90%). With the addition of a 1 M NaOH solution, auto-flocculation may obtain a recovery efficiency of over 90%. This approach, however, is not employed in business since it is uncontrollable and alters cell composition [47].

3.3.2.2 Chemical flocculation

Microalgae are frequently harvested using this flocculation process. It is one of the cheapest ways; however, it is only used as a pretreatment since microalgae are tiny. Electrolytes and synthetic polymers are used to flocculate and coagulate the cells. Because of the +3 charge, aluminum and ferric cations, aluminum sulfate, and ferric chloride are frequently utilized for charge neutralization [38]. When these compounds react with the $CaCO_3$ in the medium or effluent, the result is generally hydroxides such as aluminum hydroxide and ferric hydroxide. These hydroxides separate the biomass from the liquid as they settle through the media. Chitosan, cellulose, surfactants, cationic polyacrylamides, and artificial fibers are examples of cationic polymers that have been shown to be effective. Removing contaminants from separated algae is a crucial downside of this separation technology, making it inefficient and costly for commercial application. However, it could be helpful in the lab [38].

3.3.3 Filtration

Filtration is a practical approach for recovering giant algal cells, and it is ideal for large-sized microalgae. Filtration is the process of passing an algae-containing medium through filters that collect the algae and enable the medium to flow through. The medium is passed through the microfilters repeatedly until it accumulates a thick alga paste. Filtration at a low cost is frequently employed to extract filamentous algae. Vacuum filtration is commonly used for big particle sizes; ultrafiltration is chosen [38,42]. Membrane filtration is dependable and straightforward; however, it is inconvenient at dilute concentrations. This approach works well in small systems, provided the culture is preconditioned to 3%–4% biomass massively [38]. Conventional filtration is optimal for collecting microalgae with particle sizes greater than 70 m, such as *Spirulina*. Microalgae like *Scenedesmus*, *Dunaliella*, and *Chlorella*, on the other hand, require micro- or ultrafiltration membranes. Higher pressures are necessary to drive the liquid through the membranes in these processes, increasing energy needs as the pore size shrinks [40].

3.3.4 Centrifugation

Centrifugation may be used to separate any type of microalgae, with a recovery rate of more than 95% and a solid concentration of 10%−20%. It separates microalgae from the medium using centrifugal and centripetal forces. This approach may be used to separate almost all species of microalgae. This strategy is one of the most effective. However, it is highly energy intensive (about 8 kWhm^3), resulting in high running expenses. Depending on the strain and operating circumstances, it may impair the structure of the cells due to an increased gravitational force. A massive volume of culture takes a long time to complete, and the algal cell might be damaged by significant gravitational and shear stresses. The centrifugal force causes comparatively dense materials to settle down faster than they would under normal gravity. In the industry, centrifugation is the favored harvesting method, remarkably, when the additional value of the products compensates for the high processing costs [48].

3.3.5 Magnetic separation

It is an efficient separation process, such as removing arsenic and metals from wastewater. The magnetic field causes intrinsic paramagnetic movement of magnetic particles linked to products, which allows for separation. Magnetic beads can be used to quickly separate nonmagnetic target cells from the media under highly favorable circumstances [49]. Magnetic particles stick to microalgal cells and then use an external magnetic field to remove them from the growth medium. They can be coated with silica or cationic polyelectrolyte to attach the magnetic core to the microalgae cells and force them to aggregate. The critical element impacting harvesting efficiency is the organic polymers utilized for coating [50].

3.3.6 Flotation

Flotation is a gravity separation process in which air or gas bubbles are attached to solid particles and then carried to the liquid surface. Flotation is more beneficial and valuable than sedimentation in removing microalgae. Flotation can capture particles with a diameter of less than 500 μm by colliding between a bubble and a particle and the subsequent adhesion of the bubble and the particle. Based on bubble sizes used in the flotation process, the applications can be divided into dissolved air flotation, dispersed flotation, and electrolytic flotation [48].

3.3.7 Fungi-assisted sedimentation

The filamentous fungus may produce giant pellets, which can be used to capture algae. Fungi exhibit a variety of morphologies due to growing circumstances, including distributed hyphae, denser spherical aggregates, and loose hyphal aggregates [33]. Only denser spherical aggregates, sometimes known as pelleted fungus, have been shown to be successful in algae harvesting in prior investigations. As a result, the creation of fungal pellets, influenced by electrostatic force, hydrophobicity, and the unique interactions of fungal cell wall compositions, is critical for algae harvesting. The primary mechanism of fungi-assisted sedimentation in algae harvesting is that algal cells are co-pelletized into big fungal pellets, which are far more convenient to collect via a sieve than suspended cells. Fungi-assisted sedimentation uses substantially less energy in the harvesting process than centrifugation and electric flocculation [33].

3.4 Challenges in cultivation and harvesting

Despite the apparent benefits of microalgae as a biofuel, there are still several production and harvesting obstacles. The overall challenges in microalgal biomass are economic recovery. Some of the problems associated with the large-scale microalgae production and harvesting are the following. Cost of the PBR is very high; 50% of the total capital cost is from the PBR. Scale-up problems are due to complex design considerations. Harvesting microalgae is challenging because microalgae are very dilute. Typically, microalgae have a solid concentration ranging from 0.5 to 4 g L^{-1}, and their size petitely ranges from 1 to 30 μm [33].

3.5 Bioethanol production

Bioethanol is the most frequently used biofuel on the planet, and it is made from the fermentation of carbohydrate-rich plants [8]. Microalgae and cyanobacteria have also been shown to produce a lot of bioethanol since their carbohydrate content (up to 50% dry weight) may be digested and fermented to make a lot of bioethanol. Bioethanol synthesis from microalgal biomass typically involves raw material saccharification (hydrolysis), sugar fermentation, and distillation. The hydrolysis process is essential for improving biomass saccharification and releasing monomers of sugars from algae so that the fermentation process can make ethanol. Hydrolysis of microalgal carbohydrates using alkaline solutions (e.g., NaOH) converts polysaccharides to monosaccharides by disrupting H-bonding in these complex sugars and increasing cellulose microfibril accessibility. For industries, *Saccharomyces cerevisiae* is the primary decomposer in the fermentation process for ethanol generation. It has a solid capacity to survive and operate up to 10% ethanol concentration before degrading and stopping the fermentation process. Saccharification and fermentation can be carried out concurrently by generating microbial strains of amylase enzymes to manufacture bioethanol in a single step [9].

3.5.1 Pretreatment

In pretreatment, the raw material must be subjected to a pretreatment with high temperature and pressure, which is an essential step to the recalcitrance of the lignocellulosic biomass. Pretreatment consists of the fractionation of the biomass into its main components (cellulose, hemicellulose, and lignin) to facilitate the subsequent enzymatic and microbial attack. Various pretreatment methods have been developed, including physical, physicochemical, chemical, and biologic processes.

3.5.1.1 Hydrothermal pretreatment

Hydrothermal pretreatment (HTP) enables the transformation of wet biomass into bio-oil. HTP is typically carried out at high temperatures (100–250°C) and pressures without using any chemicals or organic solvents, with reaction times ranging from a few minutes to hours. The use of water as a reaction medium and reactant was a crucial aspect of HTP when compared with traditional biomass pretreatment processes. As a result, HTP may be able to achieve high selectivity of desirable products and be more ecologically friendly. HTP has been used to improve microalgal biorefinery operations by boosting biogas production rate, bioethanol yield, and intracellular bioactive component extraction efficiency [51]. In this scenario, the macromolecular composition of microalgae significantly influences bio-oil production. A combination of microalgae (*Nannochloropsis* and *Spirulina*, respectively, with low and high protein content) directly affected energy recovery during HTP [52].

3.5.1.2 Pulse/electric field pretreatments

It is a simple approach in which high electric potentials of 100–300 kV cm^{-1} are applied to cell cultures for a brief period of time, causing cell wall penetration and effective extraction of essential components such as proteins, carbohydrates, or other specific chemicals from the cell. Electroporation is another name for pulse field pretreatment. The key benefit is that it may be utilized for mediums with low and high cell concentrations. When an electric field is applied, the negatively charged cell wall experiences a dipole moment in the electric field direction, and it breaks when the electric field reaches a threshold. Perforations are created for the extraction of intracellular chemical molecules [17].

3.5.1.3 Microwave pretreatment

This form of pretreatment is employed in organic chemistry to speed up the reactions and extraction of more essential molecules. The research was conducted on the pretreatment of lignocellulosic wastes to create second-generation bioethanol and high-value chemicals [53]. Microwaves, which are known to heat water molecules by rotating dipoles as a polar molecule tries to align in a magnetic field [17,53], eventually vaporize water molecules and, as a result, disrupt hydrogen bonds in cells by exerting pressure on the cell walls. They also improve biomass solubilization by polarizing macromolecules, encouraging cellular component hydrolysis and protein structural alterations [53]. This heating is caused by two mechanisms: (a) dipole rotation, in which the polar molecules try to align in the rapidly changing electromagnetic field produced by microwaves, and (b) ionic conduction, which is caused by the instant superheating of the ionic substance caused by the friction of the ionic molecules generated by the movement that produces the electric field [17].

3.5.1.4 Ultrasound pretreatment

Sound waves pass through a liquid media, establishing compression and rarefaction zones and, as a result, causing a cavitation phenomenon that causes bubbles to develop in the elastic medium [17,53]. This method has long been used in organic chemistry to speed up chemical processes and extract bioactive chemicals from various plant species. Because of the low pressure inside the bubbles, they burst violently, releasing a considerable quantity of energy. It produces pressure and temperature differences across the liquid medium, resulting in hot patches. The collapsing bubbles have enough force to breach the cell walls of the microalgae, resulting in micro-jets and the solubilization of the cellular contents [17]. Meanwhile, the gas that cannot be kept in the bubbles condenses, releasing a great deal of energy, generating a catastrophic collapse, shock waves, and extremely high temperatures and pressures. The cavitation process enhances heat and mass transmission, creating hot spots that might cause a chemical reaction in the medium to accelerate [53].

3.5.1.5 Mechanical pretreatment

These are the most basic ways for releasing cell internals on an industrial scale, including bead and ball milling, grinding, high-pressure homogenization, and cavitation [53]. Attrition pressures operate on the cell walls during bead and ball milling, disrupting them. In liquid medium high-pressure homogenization, the same idea of attrition is applied. The cell culture is compressed and forced through an aperture, creating a high-pressure gradient along with the fluid flow, inducing viscous shear on the microalgae cell walls and disrupting them. On the other hand, cavitation techniques employ throttle valves to maintain this pressure gradient, which causes cavitation and the formation of bubbles. When these bubbles burst, they release a lot of energy in the form of pressure, waves, and heat, which disrupts the cell wall [17]. Mechanical pretreatments for microalgal biomass used in biodiesel production have been widely documented to help in lipid extraction. Similarly, mechanical pretreatment, such as ultrasound and microwave, has shown encouraging outcomes in biogas generation in terms of methane output [53].

3.5.1.6 Freezing/thawing pretreatment

In this pretreatment, cell culture is chilled to subzero temperatures, allowing water to crystallize into ice inside the cell. This transition quickly destroys the cell walls because of the increased volume of the ice crystals. Different proposed statements explain the underlying mechanism of cell damage by the freezing process. These include (a) extracellular constituents, such as electrolytes and other solutes, which are present in higher concentrations and are reported to damage cells by forming ice when exposed to water removal through the dehydration process, (b) water flow through osmosis via cell membrane also causes cell damage, (c) cell damage caused by highly concentrated extracellular components shrinking cells, and (d) slow cooling rate damages the cell by forming large external ice, as well as (e) fast cooling that causes intracellular ice crystals, harming the cell [53]. The procedure, however, has some drawbacks, including a considerable energy input, a time-consuming and expensive process, high pump maintenance costs, and, most crucially, the cell wall is not destroyed but is weakened.

3.5.1.7 Enzymatic pretreatment

Enzymatic pretreatment is favored over other methods because it has the benefit of being very selective and practical while also avoiding the formation of inhibitory by-products [53]. The composition of the microalgal cell wall influences enzymatic pretreatment. Enzymes' catalytic activity is determined by the cell wall component, such as cellulose, pectin, hemicellulose, glycoprotein, and so on. On the other hand, the cell wall components fluctuate depending on the microalgal strain, the algae ambiance, which includes medium growth conditions, nutrient content, and the stage of algal development. For the disintegration of cell walls, single enzymes or multienzyme mixes are available, with the most common single enzymes belonging to carbohydrates, such as amylases, cellulases, pectinases, and hemicelluloses, and the multienzyme mix or enzyme cocktail containing various combinations of lysozyme, protease, laccase, and carbohydrates [53].

3.5.2 Enzyme saccharification

Enzymatic saccharification is one of the most important processes in obtaining key monomer sugars like glucose and mannose for further fermentation and bioethanol synthesis [17]. In microalgae, the amylases enzyme is used for starch hydrolysis. Endo-amylase hydrolyzes the α-1,4-glycosidic bond and produces dextrin. Glucoamylase hydrolyzes the nonreducing end of starch since it presents α-(1−4) hydrolyzing activity. This enzyme also degrades dextrin to glucose and other oligosaccharides as it hydrolyzes 1,6-glycosidic bonds [54]. Amylase enzymes can increase the yield of fermentable sugars; for example, Lee et al. [55] observed an increase in sugar production from residual *D. tertiolecta* biomass treated with amylase compared with treatments using acid hydrolysis and cellulases. Cheng et al. [56] cultivated *Chlorella variabilis* under nitrogen-limited conditions to increase its starch content. However, enzymatic hydrolysis is considered the most promising strategy because it has several advantages over chemical saccharification, including high efficiency, low energy consumption due to the requirement of mild operation conditions, and no environmental damage. After all, enzymes are biodegradable, with no toxic compounds formation and normal deterioration and corrosion problems for the equipment.

3.5.3 Fermentation

The fermentation of carbohydrates may create bioethanol in algal biomass via yeast activity in the fermentation biorefinery pathway [57]. Pretreatment of algal biomass, enzyme hydrolysis of the carbohydrates acquired from algae biomass into monosaccharides, and fermentation of the monosaccharides into bioethanol are the procedures required to liberate the starches in the algae biomass. The term "pretreatment" of algal biomass refers to the process of physically or physiochemically breaking the algae cell wall before sugar extraction. After that, enzyme hydrolysis is used to break down the extracted complex sugars from algal biomass into monosaccharides (such as glucose). The monosaccharides (such as glucose) ae then fermented to produce bioethanol. Yeast is utilized as a biologic agent to transform simple carbohydrates like glucose into bioethanol during the fermentation process. The development of bioethanol from algal biomass is constantly linked to the development of biodiesel [58].

3.6 Biodiesel

Microalgae have a strong capacity for biomass production with high lipid content. Some microalgae have been found to contain lipid content ranging from 50% to 80% of their dry cell weight [7]. Microalgal oil with a high triglyceride content is a promising feedstock for biodiesel generation. It is a third-generation carbon-neutral biofuel that can absorb as much CO_2 during algae growth as is created during fuel burning, making it a feasible and long-term solution to climate change. Many techniques, such as micro-emulsification, catalytic cracking, and transesterification, produce fatty acids and alkyl esters (biodiesel). Transesterification is the most popular process for making biodiesel from oil, although others are expensive and yield low-quality fuel. With the aid of catalysts such as acids, alkalis, and enzymes, the transesterification process transforms natural algal lipids (triglycerides) into low-molecular-weight fatty acid alkyl esters. Sodium hydroxides, potassium hydroxides, and sodium methoxide are common alkali catalysts. Lipase is the enzymatic catalyst of choice; however, a heterogeneous inorganic catalyst is also used for esterification. Glycerol is a by-product of the transesterification process that may be utilized as a carbon source for heterotrophic and mixotrophic algae culture, with the glycerol being recycled for biodiesel synthesis [59].

3.6.1 Transesterification

Transesterification is a typical chemical biorefinery route that converts algae lipids into biodiesel. Transesterification is a multiple-step reaction, including three reversible steps in series, where triglycerides are transformed into diglycerides. Diglycerides are converted to monoglycerides and then converted to esters (biodiesel) and glycerol (by-product) [60]. This traditional transesterification reaction was a two-step method that began with chemical solvent extraction of lipids from algal biomass and ended with reaction optimization research [61]. As a result, algae strains rich in lipids are selected. Acid catalysts (e.g., HCl and H_2SO_4), alkaline catalysts (e.g., NaOH and KOH), and other heterogeneous inorganic catalysts are all extensively employed for transesterification (e.g., ZrO_2 and TiO_2). To lower free fatty acid content, algal oil was first esterified using an acid catalyst (H_2SO_4). The alkali catalyst was used for the second transesterification step (KOH). According to the findings, the two-step procedure improved biodiesel output to 86.1%, which was comparable to second-generation biomass (e.g., 90.6% biodiesel from rapeseed oil). Noncatalytic transesterification of lipids into biodiesel utilizing supercritical fluids such as water, CO_2, or alcohol, in addition to the catalytic process, is another option. As a result, a catalytic transesterification approach for algal biodiesel generation is expected to be advantageous in an algae biorefinery [61].

3.7 Biohydrogen from microalgae

As a clean energy source that promotes energy and environmental sustainability, hydrogen can help mitigate global climate change and energy scarcity. It is deemed one of the most idyllic energy types. Microalgae, third-generation feedstock, has emerged as a possible sustainable biohydrogen production alternative to first- and second-generation biomass [11]. Microalgae convert solar energy into biohydrogen under anaerobic conditions as facultative aerobic organisms. Biohydrogen ($BioH_2$) production from microalgae has been described as a highly appealing strategy for achieving carbon neutrality and bioenergy sustainability by producing a benign clean energy carrier. Green microalgae and cyanobacteria have excellent photosynthetic machinery to produce photobiological hydrogen from the most plentiful natural resources: sunlight and water [10]. $BioH_2$ generation from photosynthetic organisms has sparked a lot of excitement in the fuel sector. Furthermore, microalgal-based $BioH_2$ synthesis aids in the mitigation of current global warming by absorbing greenhouse gases, but it also aids in the treatment of wastewater.

3.8 High value-added by-product (phycocyanin)

Phycocyanin (PC) is a natural blue photosynthetic pigment, plant-based phycobiliprotein protein found in the cyanobacterium like *S. platensis*, which exhibits high bio functions nutritional values, antiinflammation, antioxidation, antitumor, neuroprotective effect, and immunological enhancement [62]. Phycocyanin is one of the most stable proteins and boosts the immune system in various organisms. It can be used in an immunoassay to track target cells because it has fluorescent properties. C-phycocyanin is a water-soluble fluorescent protein extracted from microalgae with a 615–620 nm emission spectrum. The purity of phycocyanin is crucial for its application in the food, pharmaceutical, and medical industries; however, the purification processes needed to achieve high purity can impact its bioactivity. Besides, as a nontoxic and noncarcinogenic natural coloring agent, phycocyanin has many applications in the food and cosmetic industries. C-phycocyanin is widely used as a natural pigment in the food and cosmetic industries, costing US$5000–33,000 g^{-1}. C-phycocyanin is a substitute for use as a novel pharmaceutical functionally [62].

4. Future perspective

Microalgae biomass encompasses various biochemical components such as protein, lipid, and carbohydrates, making them a feasible feedstock for biofuel production and high value-added bioproducts. The output of biofuel from microalgae is affected by various factors, including irradiance, temperature, mixing, pH, dissolved gases, qualitative and quantitative features of biomass, and photosynthetic efficiency. However, the most significant barrier to commercializing microalgae biofuels is increased production costs. To improve the biofuel economy, other high-value-added bioproducts must be extracted from microalgae biomass. The culture system is built of transparent materials like acrylics, plastics, or glass, but it will be costly. So, to keep costs down, chosen acrylics as the material. It is also worth noting that microalgae do not grow on the tubes' walls, which may obstruct sunlight transmission. Artificial lighting throughout the night helps algae grow biomass all day, increasing overall output.

This study will look at how traditional methods for producing microalgae biofuels have evolved, such as chemical and biochemical conversion routes and the extraction of diverse bioproducts from microalgae biomass for various purposes. In addition, case studies for different growing techniques, culture mediums, harvesting techniques, biofuel, and bioenergy production routes are discussed, emphasizing life cycle assessment studies on microalgae biorefinery. The algal biorefinery, which includes various bioproducts synthesis, opens up new possibilities for the valorization of microalgae biomass. Furthermore, in the biofuel production process, traditional or well-established new technologies may make better investments and boost the process profitability from a commercial aspect.

5. Conclusion

Microalgae can produce a variety of novel bioproducts. They can be used as an indigenous biologic source to help bridge the gap between environments and changing climatic conditions by creating environmentally friendly energy products with various uses in the food, medicine, and cosmetics industries. Microalgae and biofuel production under the biorefinery strategy is expected to enhance the microalgae biofuels' overall cost-effectiveness significantly. One can improve technologic processes like designing advanced PBR, biomass harvesting, drying, and other downstream processing techniques like pretreatment. The biochemical process is the essential thing that may lead to the microalgae's enhanced cost-effectiveness in a biofuels strategy.

Acknowledgments

This research project was supported by the Mexican Science and Technology Council (CONACYT, Mexico) with the Infrastructure Project – FOP02-2021–04 (Ref. 317250). The author Rohit Saxena thanks the National Council for Science and Technology (CONACYT, Mexico) for his Ph.D. Fellowship support (grant number: 1013150).

References

[1] C. Debnath, T.K. Bandyopadhyay, B. Bhunia, U. Mishra, S. Narayanasamy, M. Muthuraj, Microalgae: sustainable resource of carbohydrates in third-generation biofuel production, Renew. Sustain. Energy Rev. 150 (2021) 111464, https://doi.org/10.1016/j.rser.2021.111464.

[2] E de Jong, A. Higson, P. Walsh, M. Wellisch, Task 42 biobased chemicals - value added products from biorefineries, A Rep. Prep. IEA Bioenergy-Task 36 (2011).

[3] H.A. Alalwan, A.H. Alminshid, H.A.S. Aljaafari, Promising evolution of biofuel generations. Subject review, Renew. Energy Focus 28 (2019) 127–139, https://doi.org/10.1016/j.ref.2018.12.006.

[4] E.G. Bautista, C. Laroche, Arthrospira platensis as a feasible feedstock for bioethanol production, Appl. Sci. 11 (2021), https://doi.org/10.3390/app11156756.

[5] Y. Li, Z. Zhang, M. Paciulli, A. Abbaspourrad, Extraction of phycocyanin—a natural blue colorant from dried spirulina biomass: influence of processing parameters and extraction techniques, J. Food Sci. (2020), https://doi.org/10.1111/1750-3841.14842.

[6] W. Wu, J.S. Chang, Integrated algal biorefineries from process systems engineering aspects: a review, Bioresour. Technol. 291 (2019) 121939, https://doi.org/10.1016/j.biortech.2019.121939.

[7] E. Koutra, P. Tsafrakidou, M. Sakarika, M. Kornaros, Microalgal biorefinery, Microalgae Cultiv. Biofuels Prod. (2020) 163–185, https://doi.org/10.1016/b978-0-12-817536-1.00011-4.

[8] S. Agarwal, A. Kumar, Historical development of biofuels, in: Biofuels: Greenhouse Gas Mitigation and Global Warming, Springer, 2018, pp. 17–45.

[9] S.P. Choi, M.T. Nguyen, S.J. Sim, Enzymatic pretreatment of Chlamydomonas reinhardtii biomass for ethanol production, Bioresour. Technol. 101 (2010) 5330–5336, https://doi.org/10.1016/j.biortech.2010.02.026.

[10] H. Singh, D. Das, Biohydrogen from Microalgae, Elsevier Inc, 2020.

[11] J. Wang, Y. Yin, Fermentative hydrogen production using pretreated microalgal biomass as feedstock, Microb. Cell Factor. 17 (2018) 1–16, https://doi.org/10.1186/s12934-018-0871-5.

[12] Z. Liu, K. Wang, Y. Chen, T. Tan, J. Nielsen, Third-generation biorefineries as the means to produce fuels and chemicals from CO_2, Nat. Catal. 3 (2020) 274–288, https://doi.org/10.1038/s41929-019-0421-5.

[13] P. Ganguly, R. Sarkhel, P. Das, The Second- and Third-Generation Biofuel Technologies: Comparative Perspectives, INC, 2021.

[14] R. Kamal, Q. Huang, Q. Li, Y. Chu, X. Yu, S. Limtong, S. Xue, Z.K. Zhao, Conversion of arthrospira platensis biomass into microbial lipids by the oleaginous yeast cryptococcus curvatus, ACS Sustain. Chem. Eng. 9 (2021) 11011–11021, https://doi.org/10.1021/acssuschemeng.1c02196.

[15] A. Niccolai, G. Chini Zittelli, L. Rodolfi, N. Biondi, M.R. Tredici, Microalgae of interest as food source: biochemical composition and digestibility, Algal Res. 42 (2019), https://doi.org/10.1016/j.algal.2019.101617.

[16] A. Molino, A. Iovine, P. Casella, S. Mehariya, S. Chianese, A. Cerbone, J. Rimauro, D. Musmarra, Microalgae characterization for consolidated and new application in human food, animal feed and nutraceuticals, Int. J. Environ. Res. Publ. Health 15 (2018), https://doi.org/10.3390/ijerph15112436.

[17] J. Velazquez-Lucio, R.M. Rodríguez-Jasso, L.M. Colla, A. Sáenz-Galindo, D.E. Cervantes-Cisneros, C.N. Aguilar, B.D. Fernandes, H.A. Ruiz, Microalgal biomass pretreatment for bioethanol production: a review, Biofuel Res. J. 5 (2018) 780–791, https://doi.org/10.18331/BRJ2018.5.1.5.

[18] MF. de J. Raposo, R.M.S.C. De Morais, B. de Morais, A.M. Miranda, Bioactivity and applications of sulphated polysaccharides from marine microalgae, Mar. Drugs 11 (2013) 233–252.

[19] M.L. Mourelle, C.P. Gómez, J.L. Legido, The potential use of marine microalgae and cyanobacteria in cosmetics and thalassotherapy, Cosmetics 4 (2017), https://doi.org/10.3390/cosmetics4040046.

[20] P.A. Harnedy, R.J. FitzGerald, Bioactive proteins, peptides, and amino acids from macroalgae 1, J. Phycol. 47 (2011) 218–232.

[21] N.L. Ma, S.S. Lam, R. Zaidah, The application of algae for cosmeceuticals in the omics age, Genom., Proteom. Metabol. Nutraceut. Funct. Food. Second Ed (2015) 476–488, https://doi.org/10.1002/9781118930458.ch37.

[22] I. Hamed, The evolution and versatility of microalgal biotechnology: a review, Compr. Rev. Food Sci. Food Saf. 15 (2016) 1104–1123.

[23] M. Koller, A. Muhr, G. Braunegg, Microalgae as versatile cellular factories for valued products, Algal Res. 6 (2014) 52–63.

[24] P. Spolaore, C. Joannis-Cassan, E. Duran, A. Isambert, Commercial applications of microalgae, J. Biosci. Bioeng. 101 (2006) 87–96.

[25] M.F. de Jesus Raposo, A.M.B. De Morais, R.M.S.C. De Morais, Marine polysaccharides from algae with potential biomedical applications, Mar. Drugs 13 (2015) 2967–3028.

[26] D. Hernández, B. Riaño, M. Coca, M.C. García-gonzález, Saccharification of carbohydrates in microalgal biomass by physical, chemical and enzymatic pre-treatments as a previous step for bioethanol production, Chem. Eng. J. 262 (2015) 939–945, https://doi.org/10.1016/j.cej.2014.10.049.

[27] D.A. García-López, E.J. Olguín, R.E. González-Portela, G. Sánchez-Galván, R. De Philippis, R.W. Lovitt, C.A. Llewellyn, C. Fuentes-Grünewald, R. Parra Saldívar, A novel two-phase bioprocess for the production of Arthrospira (Spirulina) maxima LJGR1 at pilot plant scale during different seasons and for phycocyanin induction under controlled conditions, Bioresour. Technol. 298 (2020) 122548, https://doi.org/10.1016/j.biortech.2019.122548.

[28] S. Huo, X. Chen, F. Zhu, W. Zhang, D. Chen, N. Jin, K. Cobb, Y. Cheng, L. Wang, R. Ruan, Magnetic field intervention on growth of the filamentous microalgae Tribonema sp. in starch wastewater for algal biomass production and nutrients removal: influence of ambient temperature and operational strategy, Bioresour. Technol. 303 (2020) 122884, https://doi.org/10.1016/j.biortech.2020.122884.

[29] V. Vogel, P. Bergmann, Culture of Spirogyra sp. in a flat-panel airlift photobioreactor, 3 Biotech 8 (2018), https://doi.org/10.1007/s13205-017-1026-9.

[30] C.A. Arroyo, J.L. Contreras, B. Zeifert, C.C. Ramírez, CO2 capture of the gas emission, using a catalytic converter and airlift bioreactors with the microalga Scenedesmus dimorphus, Appl. Sci. 9 (2019), https://doi.org/10.3390/app9163212.

[31] T. Wannachod, S. Wannasutthiwat, S. Powtongsook, K. Nootong, Photoautotrophic cultivating options of freshwater green microalgal Chlorococcum humicola for biomass and carotenoid production, Prep. Biochem. Biotechnol. 48 (2018) 335–342, https://doi.org/10.1080/10826068.2018.1446152.

[32] J. Rebolledo-Oyarce, J. Mejía-López, G. García, L. Rodríguez-Córdova, C. Sáez-Navarrete, Novel photobioreactor design for the culture of Dunaliella tertiolecta — impact of color in the growth of microalgae, Bioresour. Technol. 289 (2019) 121645, https://doi.org/10.1016/j.biortech.2019.121645.

[33] W. Zhou, Q. Lu, P. Han, J. Li, Microalgae Cultivation and Photobioreactor Design, Elsevier Inc, 2020.

[34] J.S. Tan, S.Y. Lee, K.W. Chew, M.K. Lam, J.W. Lim, S.H. Ho, P.L. Show, A review on microalgae cultivation and harvesting, and their biomass extraction processing using ionic liquids, Bioengineered 11 (2020) 116–129, https://doi.org/10.1080/21655979.2020.1711626.

[35] E.T. Sero, N. Siziba, T. Bunhu, R. Shoko, E. Jonathan, Biophotonics for improving algal photobioreactor performance: a review, Int. J. Energy Res. 1–22 (2020), https://doi.org/10.1002/er.5059.

[36] L. Qin, M.A. Alam, Z. Wang, Open pond culture systems and photobioreactors for microalgal biofuel production, Microalgae Biotechnol. Dev. Biofuel Wastewater Treat. (2019) 45–74, https://doi.org/10.1007/978-981-13-2264-8_3.

[37] T.J. Johnson, S. Katuwal, G.A. Anderson, L. Gu, R. Zhou, W.R. Gibbons, Photobioreactor cultivation strategies for microalgae and cyanobacteria, Biotechnol. Prog. 34 (2018) 811–827, https://doi.org/10.1002/btpr.2628.

[38] S. Katuwal, Designing and Development of a Photobioreactor for Optimizing the Growth of Micro Algae and Studying its Growth Parameters. Electron Theses Dissertion, 2017.

[39] M.L. Gerardo, S. Van Den Hende, H. Vervaeren, T. Coward, S.C. Skill, Harvesting of microalgae within a biorefinery approach: a review of the developments and case studies from pilot-plants, Algal Res. 11 (2015) 248–262.

[40] J.J. Milledge, S. Heaven, A review of the harvesting of micro-algae for biofuel production, Rev. Environ. Sci. Bio/Technol. 12 (2013) 165–178.

[41] P. Shah Maulin, Microbial Bioremediation & Biodegradation, Springer, 2020.

[42] N. Pragya, K.K. Pandey, P.K. Sahoo, A review on harvesting, oil extraction and biofuels production technologies from microalgae, Renew. Sustain. Energy Rev. 24 (2013) 159–171.

[43] C. Wan, M.A. Alam, X.-Q. Zhao, X.-Y. Zhang, S.-L. Guo, S.-H. Ho, J.-S. Chang, F.-W. Bai, Current progress and future prospect of microalgal biomass harvest using various flocculation technologies, Bioresour. Technol. 184 (2015) 251–257.

[44] J. Kim, B.-G. Ryu, Y.-J. Lee, J.-I. Han, W. Kim, J.-W. Yang, Continuous harvest of marine microalgae using electrolysis: effect of pulse waveform of polarity exchange, Bioproc. Biosyst. Eng. 37 (2014) 1249–1259.

[45] S. Salim, N.R. Kosterink, N.D.T. Wacka, M.H. Vermuë, R.H. Wijffels, Mechanism behind autoflocculation of unicellular green microalgae Ettlia texensis, J. Biotechnol. 174 (2014) 34–38.

[46] P. Shah Maulin, Removal of Refractory Pollutants from Wastewater Treatment Plants, CRC Press, 2021.

[47] A.I. Barros, A.L. Gonçalves, M. Simões, J.C.M. Pires, Harvesting techniques applied to microalgae: a review, Renew. Sustain. Energy Rev. 41 (2015) 1489–1500.

[48] C.Y. Chen, K.L. Yeh, R. Aisyah, D.J. Lee, J.S. Chang, Cultivation, photobioreactor design and harvesting of microalgae for biodiesel production: a critical review, Bioresour. Technol. 102 (2011) 71–81, https://doi.org/10.1016/j.biortech.2010.06.159.

[49] S.-K. Wang, A.R. Stiles, C. Guo, C.-Z. Liu, Harvesting microalgae by magnetic separation: a review, Algal Res. 9 (2015) 178–185.

[50] Y. Zhao, W. Liang, L. Liu, F. Li, Q. Fan, X. Sun, Harvesting Chlorella vulgaris by magnetic flocculation using Fe_3O_4 coating with polyaluminium chloride and polyacrylamide, Bioresour. Technol. 198 (2015) 789–796.

[51] H.A. Ruiz, Foreword: high pressure processing for the valorization of biomass, RSC Green Chem. vii–viii (2017), https://doi.org/10.1039/9781782626763-FP007.

[52] H.A. Ruiz, M.H. Thomsen, H.L. Trajano, Hydrothermal processing in biorefineries: production of bioethanol and high added-value compounds of second and third generation biomass, Hydrotherm. Process Biorefiner. Prod. Bioethanol. High Added-Value Compd. Second Third Gener. Biomass (2017) 1–511, https://doi.org/10.1007/978-3-319-56457-9.

[53] S. Anto, S.S. Mukherjee, R. Muthappa, T. Mathimani, G. Deviram, S.S. Kumar, T.N. Verma, A. Pugazhendhi, Algae as green energy reserve: technological outlook on biofuel production, Chemosphere 242 (2020), https://doi.org/10.1016/j.chemosphere.2019.125079.

[54] C. Simas-Rodrigues, H.D.M. Villela, A.P. Martins, L.G. Marques, P. Colepicolo, A.P. Tonon, Microalgae for economic applications: advantages and perspectives for bioethanol, J. Exp. Bot. 66 (2015) 4097–4108, https://doi.org/10.1093/jxb/erv130.

[55] O.K. Lee, A.L. Kim, D.H. Seong, C.G. Lee, Y.T. Jung, J.W. Lee, E.Y. Lee, Chemo-enzymatic saccharification and bioethanol fermentation of lipid-extracted residual biomass of the microalga, Dunaliella tertiolecta, Bioresour. Technol. 132 (2013) 197–201.

[56] C. Chen, X. Zhao, H. Yen, S. Ho, C. Cheng, Microalgae-based carbohydrates for biofuel production, Biochem. Eng. J. 78 (2013) 1–10, https://doi.org/10.1016/j.bej.2013.03.006.

[57] S.-H. Ho, S.-W. Huang, C.-Y. Chen, T. Hasunuma, A. Kondo, J.-S. Chang, Bioethanol production using carbohydrate-rich microalgae biomass as feedstock, Bioresour. Technol. 135 (2013) 191–198.

[58] R. Harun, J.W.S. Yip, S. Thiruvenkadam, W.A.W. Ghani, T. Cherrington, M.K. Danquah, Algal Biomass Conversion to Bioethanol – a Step-by-step Assessment, 2014, pp. 73–86, https://doi.org/10.1002/biot.201200353.

[59] P.D. Patil, H. Reddy, T. Muppaneni, S. Deng, Biodiesel fuel production from algal lipids using supercritical methyl acetate (glycerin-free) technology, Fuel 195 (2017) 201–207.

[60] T.M. Mata, A.A. Martins, N.S. Caetano, Microalgae for biodiesel production and other applications: a review, Renew. Sustain. Energy Rev. 14 (2010) 217–232, https://doi.org/10.1016/j.rser.2009.07.020.

[61] C.G. Khoo, Y.K. Dasan, M.K. Lam, K.T. Lee, Algae biorefinery: review on a broad spectrum of downstream processes and products, Bioresour. Technol. 292 (2019) 121964, https://doi.org/10.1016/j.biortech.2019.121964.

[62] E. Kilimtzidi, S. Cuellar Bermudez, G. Markou, K. Goiris, D. Vandamme, K. Muylaert, Enhanced phycocyanin and protein content of Arthrospira by applying neutral density and red light shading filters: a small-scale pilot experiment, J. Chem. Technol. Biotechnol. 94 (2019) 2047–2054, https://doi.org/10.1002/jctb.5991.

Chapter 10

Microalgae as a promising feedstock for biofuel production

Sanaa M.M. Shanab[1], Mostafa M. El-Sheekh[2] and Emad A. Shalaby[3]

[1]*Department of Botany and Microbiology, Faculty of Science, Cairo University, Cairo, Egypt;* [2]*Department of Botany, Faculty of Science, Tanta University, Tanta, Egypt;* [3]*Department of Biochemistry, Faculty of Agriculture, Cairo University, Cairo, Egypt*

1. Introduction

The worldwide continuous use of fossil fuels in all sectors of life and development led to their depleting supplies and, at the same time, the increase of carbon dioxide emission in the atmosphere causing climatic changes and global warming. Biologically produced fuels are potential renewable alternative energy sources that can mitigate greenhouse gas (GHG) emissions and reduce the world dependence on petroleum-derived fuels [1,2].

Even though many fossil fuels have their origins in ancient carbon fixation, they are not considered biofuels by the general opinion since they include carbon that has been "out" of the carbon cycle for a long time. Because of factors such as growing oil costs, the necessity for augmented energy security, worries about GHG emissions from fossil fuels, and government subsidies, biofuels are gaining favor among the public and scientists [3]. Biofuels made from renewable resources are regarded as one of the most environmentally friendly substitutes fossil fuels.

Plants grown for crop production made up the first generation of biofuel; these plants are no longer restricted by the need for arable land to cultivate crops for the production of both food and energy. To solve this problem, scientists were directed toward lignocellulosic plants and their residues (as the second-generation biofuel), which are grown far from arable lands.

Microalgal biomass is a reliable and sustainable source of biofuels such as bio-oil, biodiesel, bioethanol, and biogas. One of the side effects of incorporating microalgal technologies into the business is that these species have a high ability to collect CO_2 throughout the application and biomass generation stages, resulting in lower CO_2 emissions [4,5].

Algae, the primary producer in aquatic ecosystems, undergo photosynthesis using CO_2, inorganic nutrients in the water, and solar energy to metabolize complex organic compounds such as proteins, carbohydrates, and lipids. Different algal species show variable lipid content, which may reach more than 50% of dry biomass under certain stress culture conditions [6,7].

Microalgae are a third-generation biofuel feedstock (Fig. 10.1) that do not compete with crops for land and food. The feedstock for the first generation of biofuels is edible oil, such as palm, soybean, corn, wheat, moringa, and vegetable oil [8]. However, the challenge with algae-based biodiesel is strain selection. There are millions of algal strains on the planet, but just a few are now being examined by scientists.

According to Harrison et al. [9], the microalgal lipids can be transesterified to biodiesel and biomass fermented to alcohols to be applied in different engines as spark-ignited engines, compression ignition engines, and aircraft gas turbine engines.

In comparison to agro-based fuels, the production of biofuels from microalgae is gaining widespread acceptability due to its economic viability and environmental sustainability. Algae-derived biofuels are currently considered the third-generation biofuel feedstock as algae grow rapidly, producing renewable, eco-friendly, biodegradable, and sustainable biomass that is not in competition with food crops and can be cultured on abandoned or not agriculture lands [10–13], on wasteland, or artificial ponds with treated wastewaters [14,15].

The current review focuses on the methods of algae cultivation and the pretreatment process for algal biomass for biofuel production in addition to illustrating the economic impact of algal fuels.

FIGURE 10.1 Photoautotrophic algal production unit (open and closed systems).

2. Microalgal biomass production

In the aquatic ecosystem, algae grow under natural conditions, absorbing sunlight and assimilating carbon dioxide (CO_2) from the air and nutrients in the ecosystem to synthesize complex organic compounds such as lipids, proteins, carbohydrates, pigments, enzymes, growth regulators, and other secondary metabolites of great importance.

Algal growth can follow three distinct production mechanisms, photoautotrophic, heterotrophic, and mixotrophic.

2.1 Photoautotrophic algal production

For long-scale algal production, only the photoautotrophic method is technically and economically feasible using either open ponds or closed photobioreactors techniques (Fig. 10.1) to harvest a large number of algal biomasses.

Selections of the algal species and optimization of the growth conditions are of basic and great importance before starting either technique.

2.1.1 Open pond production system

It is the most cost-effective way of producing huge amounts of algal biomass. It may be applied in any region with fewer energy input requirements and easier maintenance and cleaning, so it does not compete for land with existing crops [16].

Lakes and lagoons are examples of natural water ponds, while raceway ponds are the most typical manmade system [17,75]. It is normally made of concrete, but compacted earth-lined ponds are also popular (where white plastic may be used). To stabilize algae growth and productivity, they are made of closed-loop, oval-shaped recirculation tubes with mixing and circulation. In front of the continuously operating paddlewheel, algal broth and nutrients are supplied and circulated through the loop. The majority of CO_2 is absorbed from the surface air [18].

The chosen algal strain, climatic conditions, and land and water costs all influence the technical viability of any system [76]. *Dunaliella salina* (adapted to high salinity), *Spirulina* sp. (highly alkaline medium), and *Chlorella* sp. (adapted to nutrient-rich medium) are the commonly cultivated algal strains in the open pond system. *D. salina* was cultured for β-carotene in halophilic water of Hutt Lagoon.

An open pond system needs or requires an extremely selective environment to avoid contamination by other algal species.

Open ponds may be considered less efficient in biomass productivity when compared with closed photobioreactors [7,17]. This may be due to several determining factors that are uncontrollable as temperature fluctuations, evaporation losses, CO_2 deficiencies, light limitation, and inefficient mixing.

2.1.2 Closed photobioreactor system

It is used to produce high-value products from microalgae to be used in the cosmetics and pharmaceutical industry, and the photobioreactor technique avoids the risk of pollution and contamination that occurs in open pond systems as reported by Ugwu et al. [16].

Closed photobioreactors are composed of many arrays of straight transparent tubes in glass or plastic materials that can be deposited horizontally or vertically or as a helix [19].

An airlift system allows the O_2 to go out and CO_2 to go into the system in addition to a mean for biomass harvest. The second section is the solar receiver with a large-surface illuminated platform for algal growth. Algal cultures are circulated with a mechanical pump or airlift system, agitation, and mixing that cause gas exchange (O_2 and CO_2) in the tubes [16].

Tubular photobioreactors are more convenient for outdoor and massive algal cultivation and have the benefit of natural sunlight.

Column photobioreactor is compact and easy to operate with low cost. For the vertical column, aeration comes from the bottom and illumination either through the transparent wall or internally [20].

2.1.3 Hybrid production systems

It consists of two-stage cultivation, where the first is through a photobioreactor with controlled conditions and minimum contamination, and the second is the open ponds. Here, environmental stresses naturally stimulate algal production by transferring algal culture from the photobioreactor to the open pond [21] [77].

2.2 Heterotrophic algal production

In this process, algal growth can be achieved using an original carbon source as glucose in either photobioreactors or fermenters. Higher algal cell densities are achieved, and the step-up and harvest costs are minimal [22]. There is reported large-scale biodiesel production by heterotrophically cultivated *Chlorella protothecoides*. Also, Miao and Wu [23] studied the same species and found that heterotrophically cultured cells produced lipid content four times higher (55%) than autotrophic cells (15%) under the same conditions. It was concluded that heterotrophic cultured cells could produce higher biomass production that accumulates greater lipid content in cells [24].

2.3 Mixotrophic algal production

Light is not an absolute limiting factor of algal growth in many algal species that can grow either autotrophically or heterotrophically. This means that they can undergo photosynthesis as well as absorb organic carbon substrate material (i.e., mixotrophic) [78] as in the case of the cyanobacterium *Spirulina platensis* and the green *Chlamydomonas reinhardtii*. Photosynthetic metabolism uses light for growth, while aerobic respiration uses an organic carbon source. The addition of glucose to the medium during light and dark phases influences algal growth.

The growth of mixotrophic algae is higher than those cultured in the open pond but lower than for heterotrophic ones.

Chojnacka and Noworyta [25] compared the growth of *Spirulina* sp. in photoautotrophic, heterotrophic, and mixotrophic cultures. They found that mixotrophic cultures enhanced growth rate over both heterotrophic and autotrophic algae, allowing the integration of both photosynthetic and heterotrophic components during dark respiration and decreasing the quantity of organic carbon substance employed during growth.

This means that mixotrophic algal cultivation can be an important part of microalgae to produce biofuel.

2.3.1 Wastewater used for microalgal cultivation

Wastewater is rich in nitrogen, phosphorus, and organic contaminants. The use of treated wastewater for microalgal growth allows for a faster production rate, reduces the nutrient levels in the wastewater, reduces harvesting costs, and improves lipid production. Many applications of wastewater have been reported in literature using different microalgal species [26,79].

Sawayama et al. [27] used the microalga *Botryochoccus braunii* to absorb nitrate and phosphate from the primary treated sewage to produce rich carbohydrate biomass. While Marting et al. [28] used urban wastewater for the growth of *Scenedesmus obliquus*, where elimination of phosphorus and ammonium reached 98% and 100% in 94 and 183 h. On the other hand, Hodaifa et al. [80] used concentrated or diluted industrial wastewater from olive oil extraction for the cultivation of *Scenedismus obliquus*, which recorded 67.4% and 35.5% elimination of BOD respectively.

Mostafa et al. [29] demonstrated the feasibility of a wastewater treatment method that combines nitrogen removal and algal lipid production for use as a biodiesel feedstock. This study also provided information on the culture of nine algal strains for lipid production on secondary treated residential wastewater, as well as the potential of microalgae lipids for biodiesel generation by transesterification. Overall, it can be stated that using microalgal culture nutrient media is a feasible and cost-effective way of producing sustainable biodiesel and glycerol when compared with traditional growing methods.

3. Harvesting of microalgal biomass

It includes flocculation, filtration, flotation, and sedimentation by centrifugation. The selection of the algal species and the most appropriate harvesting method are of great importance for the cost-efficient economic production of algal biomass [81].

The selection of the harvesting technique depends on the microalgal characteristics such as size, density, and the target product value.

The concentration of the biomass takes place by centrifugation, filtration, or ultrasonic aggregation, this step requires more energy.

Flocculation is the first step used for microalgal aggregation before the application of any other harvesting methods [30]. Multivalent metal salts such as ferric chloride, aluminum sulfate or ferric sulfate, or chitosan may be used as flocculants [31,82] to neutralize the negative charge on algal cells. Ultrasonic may be used to optimize the aggregation efficiency as reported by Bosma et al. [32].

Unlike flocculation and the need of adding chemicals, flotation methods are based on the application of micro air bubbles that allow algal cells to float at the surface of the culture media (especially when there are increased lipid contents in the algal cells).

The most common harvesting technique for algal biomass is centrifugal sedimentation that depends on both the density and algal radius (suitable for large algal cells) as well as on the centrifugation velocity [83], especially in the case of algal biomass cultured on wastewater.

Filtration of algal biomass is preferable for larger algal species (such as *Spirulina*) and not used in smaller-sized microalgae such as *Chlorella, Scenedesmus,* or *Dunalliella* where membrane microfiltration or ultrafiltration techniques are more convenient, as reported by Petrusevski et al. [84].

Electrophoresis can be used where an electric field directs microalgae to the external part of the solution. Electrolysis of the water produced H_2 that carries the microalgae to the surface. This method has a high cost.

4. Extraction and purification of microalgal biomass

Oils were easily extracted from freeze-dried biomass, while difficult extraction was known from wet biomass, but it is expensive as well as the spray drying method [33]. Higher temperatures (more than 60°C) decrease both the triacyl glycerate and the lipid yield.

Combined ultrasound and electromagnetic pulse cause rupturing of algal cell walls, and the addition of CO_2 induce a lowering of the pH and separation of oils from algal biomass [85]. Cell disruption by high-pressure homogenizer, autoclaving, or addition of NaOH, HCL, or the use of alkaline lysis to the cells can be performed.

5. Conversion techniques/methods of algal biomass to biofuels

There are two categories of conversion techniques utilizing algal biomass that depend on the type (as shown in Fig. 10.2), the quantity of biomass, and the desired form of conversion of energy to the thermochemical and biochemical.

5.1 Thermochemical conversion

This means the thermal degradation of different organic compounds in the algal biomass to biofuel products. This may be achieved by different steps/process such as direct gasification, combustion, liquefaction, and pyrolysis, as reported by Takahara and Sawayama [34].

5.1.1 Gasification

Hirana et al. [86] reported that oxidation of *Spirulina* sp. at a temperature of 850–1000°C produced methanol (1 gm biomass produced 0.65 gm methanol). Whereas Minowa and Sawayama [35] gasified *Chlorella vulgaris* (at 1000°C) in a nitrogen cycling system to obtain methane-rich fuel.

5.1.2 Liquefaction

In this process, algal biomass under high temperature (300–350°C) and high pressure (5–20 Mpa) in the presence of hydrogen as catalyst can be converted to bio-oil as mentioned by Goyal et al. [87].

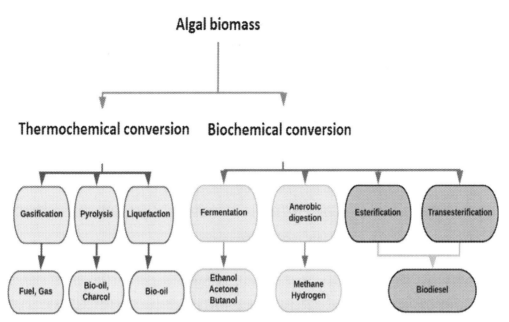

FIGURE 10.2 Biochemical and thermochemical conversion of algal biomass to biofuel.

Dote et al. [88] liquefied *Botryoccocus braunii* at 300°C, and *Dunaliella tertiolecta* was studied by Minowa et al. [36] to create liquid fuel.

5.1.3 Pyrolysis

In this process, algal biomass can be converted to bio-oil, syngas, and charcoal at a temperature range of 350−700°C in the absence of air. Flash pyrolysis at a moderate temperature of 500°C with short hot vapor exposure for 1 s, is the future technique to convert biomass to liquid fuels that can replace fossil fuels. Demirbas [37] and Miao and Wu [38] used the pyrolysis technique to convert *Chlorella protothecoides* and *Microcystis aeruginosa* to bio-oil. Results indicated that bio-oils from microalgae are of higher quality than those extracted from lignocellulosic materials [37,38].

5.1.4 Direct combustion

This process is concerned with the conversion of chemical energy in the biomass in presence of air to yield hot gases [87] in furnaces, boilers, or steam turbines at 800°C with a pretreatment process such as drying, chapping, or grinding.

5.2 Biochemical conversion

Biochemical conversion of algal biomass into biofuel includes anaerobic digestion, alcoholic fermentation, and photobiological hydrogen production.

5.2.1 Anaerobic digestion of algal biomass

It involves the breakdown of organic substances (wet algal biomass) to produce a gas (methane, CO_2, H_2S). Generally, it occurs in three stages: hydrolysis, fermentation, and methanogenesis. Hydrolysis includes the breakdown of complex compounds (carbohydrates) into soluble sugars. Fermentation of these sugars by fermentative bacteria will produce alcohols, acetic acid, volatile fatty acids, and a gas (H_2 and CO_2), which is the second step [39,40,89,90]. Methanogenesis is the generation of methane (CH_4) and CO_2 as reported by Cantrell et al. [91].

5.2.2 Alcoholic fermentation

Defatted biomass rich in sugars, starch, or cellulose, which is left after oil extraction, is converted to alcohol (ethanol) where hydrolysis of starch and cellulose to sugars is followed by the addition of water and yeast in fermenters where yeast breaks sugars to ethanol, as reported by Mckendry [89,90] and as shown in Fig. 10.3.

Alcohol must be purified or distilled to remove water, condensed to liquid form, and used in cars to replace petroleum. The solid left after this process may be used either for gasification or cattle feed [41].

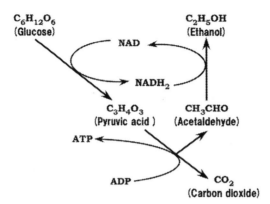

FIGURE 10.3 The chemical reaction for conversion of algal biomass to ethanol.

5.2.3 Photobiological hydrogen production

Photosynthetic H_2 production from water may be produced in two stages. In the first stage, algal photosynthesis occurs normally, while in the second stage, algae are grown in a sulfur-deficient condition that induces an anaerobic condition leading to hydrogen production [42].

6. Algal biofuel types

6.1 Biogas (biomethane production)

Anaerobic biochemical breakdown of all polymeric materials produces methane and CO_2. In the environment a diverse range of microorganisms coexist and interfere in the production of the end products as fermentative microbes. New digesters and advancements in the operation of several kinds of bioreactors are highlighted, as well as recent advances in the molecular biology of methanogens.

Methane fermentation is a multipurpose biotechnology able to convert virtually all forms of polymeric materials to methane and CO_2 under anaerobic circumstances. This is accomplished through the biochemical breakdown of polymers into methane and CO_2 in an environment where a diverse range of microorganisms coexist and produce reduced end products, including fermentative microbes (acidogens), hydrogen-producing, acetate-forming microbes (aceto gens), and methane-producing microbes (methanogens). At various stages of methane fermentation, anaerobes play a crucial role in generating a stable environment.

6.2 Bioethanol

Bioethanol is one of the most important renewable fuels due to its economic and environmental benefits. The increased use of bioethanol around the world as a renewable fuel may be due to (1) the depleting resources of fossil fuel and the emergence of biomass as a renewable energy. (2) Because of the use of fossil fuels, one of the most pressing concerns in this century is GHG emissions; biofuels may be a viable solution. (3) The price of petroleum is on the rise in the global market. (4) Petroleum reserves are limited, and some oil-importing countries have a stranglehold on them, putting the rest of the globe at risk. Recently El-Sheekh et al. [43] produced bioethanol from wastes such as wheat straw with the aid of fungi fermentation. They could produce and improve bioethanol by 3.6-fold after optimization conditions for commercial *Saccharomyces cerevisiae* on hydrolysate obtained from enzymatic saccharification of *Aspergillus niger* to 1% NaOH pretreated wheat straw. Furthermore, the produced bioethanol was blended with 10% and 20% in volume by diesel #1/wCO biodiesel commixture. Mixtures consisted of 50% diesel/50% biodiesel, 10% bioethanol/45% diesel/45% biodiesel, and 20% bioethanol/40% diesel/40% biodiesel that were tested as new fuel blends.

6.3 Biodiesel

Algal-based biofuels are gaining traction in the wake of expectations that crude oil prices may hit new highs. Transesterification of lipids to biodiesel is one of the sustainable, carbon-neutral fuel applications that use algal components [7].

Microalgae species with higher lipid content can be used to make sustainable energy products including jet fuel, biodiesel, and biogasoline [44,45]. Biodiesel is an environmentally friendly diesel fuel made from vegetable oils, animal

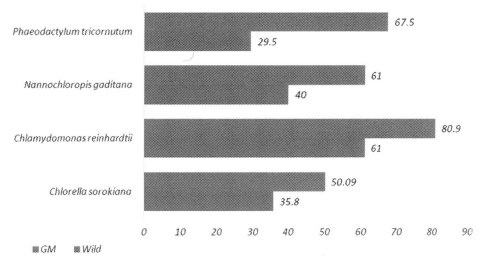

FIGURE 10.4 Comparison between some wild-type microalgae with genetically modified strains in bio-lipid production.

fats, or grease. It has a fatty acid alkyl esters chemical structure. Biodiesel emits far fewer hazardous pollutants into the atmosphere than fossil diesel. Furthermore, it burns cleaner and has less sulfur, resulting in lower emissions. It is more likely that since biodiesel is formed of renewable resources, it will compete with petroleum goods in the future [46,47]. To utilize biodiesel as a fuel, it must be combined with diesel fuel to make biodiesel-blended fuel (up to 20%). The "biodiesel" is made commercially by transesterification of triglycerides, the major components of l origin oils, in the presence of an alcohol (e.g., methanol) in addition to a catalyst (alkali or acid).

Lipids produced from oil crops (first generation) or nonfood lignocellulosic agricultural residues (second generation) or algal lipids (third generation) can be converted to biodiesel through a process called transesterification (as shown in Fig. 10.4), where a chemical reaction occurs between triglycerides and alcohol in the presence of a catalyst (acid or alkali) to produce the monoesters of the biodiesel, as reported by Sharma and Singh [92].

There are huge numbers of worldwide algal species, but only a few of them are currently investigated in researches. A few suitable microalgae strains for different biofuel generations and the fatty acid composition of their lipid content are illustrated in Tables 10.1 and 10.2.

TABLE 10.1 The main variation between the biofuel generations.

Biofuel generations		Main advantages and disadvantages	
First generation	Food sources	Require cultivable lands Negative impact on environment Negative effect on food security	
Second generation	Nonedible sources	No impact on food security No impact on biodiversity	
Third generation	Algal species	No impact on food security No impact on biodiversity Environmentally co-friendly Easily cultivated	
Fourth generation	Genetically modified algae	No impact on food security No impact on biodiversity Brackish and saline land can be used for cultivation	

TABLE 10.2 List of algae strains that are suitable for biofuel production.

Species	Lipid %	References
Scenedesmus bijuga	34.10–35.24	[48]
Chlorella saccharophila	41.71–44.73	[48]
Chlamydomonas reinhardtii	21–51	[49]
Chlorella sp.	19.28–28.79	[48]
Monoraphidium dybowskii	30.12–31.78	[50]
Chlorella vulgaris	24.28–31.04	[48]
Scenedesmus dimorphus	33.70–40.33	[48]
Chlorella pyrenoidosa	18.67–52.08	[48]

Algal biodiesel is considered a renewable, biodegradable, nontoxic alternative with similar physical and chemical characteristics to fossil fuel and comparable with the international standard EN!4214.

Algal biodiesel is considered more suitable for use in the aviation industry and reduces GHG emissions by up to 78% compared with those emitted by petroleum fuel.

All the world's scientists have directed their research to increase algal biomass production and especially the lipid contents through variable and optimized physical and chemical culture conditions, biomass collection, and transesterification of the produced lipids to obtain biodiesel of low cost and good quality that satisfies the international standards to replace the depleted fossil fuel in all sectors of life.

The quality and amount of lipids in the microalgae determine the quality of the biofuel produced [51]. Microalgae lipid quality and production are influenced by environmental factors such as growing time, nutrition availability, and illumination exposure [52]. These factors in the environment can be controlled. As a result, the microalgae strain chosen is the most important determinant of qualitative fatty acid profiles and lipid production rates. Indeed, previous research has revealed significant differences in the quantity and quality of lipids among microalgae strains. *Tetraselmis maculata*, for example, has a total lipid content of less than 4.5%, but *Schizochytrium* sp. has a total lipid content of more than 80%.

7. Genetically modified algae

To overcome the constraints of the first two generations, scientists and researchers focused on the production of biofuel by the third and fourth generations, using photosynthetic single-cell microalgae. Microalgae have been identified as a viable source of biofuel production, offering a better choice to meet current fuel demands due to their rapid biomass productivity, doubling time, high oil content, and ability to be cultivated on abandoned agricultural land [53].

Various experiments have been conducted to modify algal species to boost lipid and fatty acid content using genetic engineering methodologies such as single/multiple gene/key enzyme overexpression and deletion strategies [54]. Due to the quick availability of genome sequence information through a combination of proteomics, genomics, transcriptomics, and metabolomics, DNA sequencing is becoming more important in microalgae species.

Engineering the wild gene to construct intensified productivity and to personalize the final product is an important way out for algae biotechnologists, considering the algal potential in meeting fuel and food demands through exponentiation population. Algal biotechnology is recently dealing with engineering the wild genes to increase algal biomass productivity that covers both fuel and food demands. In this strategy, algal strains are selected and genetically modified for optimized growth and environmental stress tolerance producing larger algal biomasses with higher lipid content of promising quality for fuel production (Fig. 10.4). Biofuel produced from transgenic algae are more compatible with diesel engines than those originated from either lignocellulosic plants or crops [55]. An important strain of microalgae, as well as its genetic manipulation, is critical for increasing biofuel production.

Microalgae cultivation in wastewater (as one kind of abiotic stress that leads to gene expression and effect on chemical contents) allows necessary nutrient uptake for growth and biomass production and is considered an option for algal long-term biofuel production. To examine the proposed nutrition approach's sustainability and economic performance, a comparative life cycle assessment method and a techno-economic analysis are used. Our research is validated using two circumstances. Scenario one is based on a source-separated nutrient delivery technique in tertiary treatment with microalgae-integrated wastewater. Scenario two is based on a policy of non-separated point nutrient distribution and the use of microalgae in secondary wastewater treatment. The findings demonstrate that using a source-separated nutrition method can help reduce environmental consequences while also enhancing productivity [56].

The PROMETHEE-GAIA approach is developed and applied by Mofijur et al. [57] for selecting microalgae strains from which aviation fuel can be generated. A total of 19 criteria are used in the evaluation, with equal importance given to the following three: biomass production, lipid quality, and fatty acid composition, and biodiesel characteristics. The approach is used to evaluate 17 potential microalgae strains in this paper. The most suited strain for aviation fuel production is *Chlorella* sp. NT8a. The findings also reveal that unmodified biofuel from the best strain could fall short of all airplane fuel requirements. Microalgae-based fuel, in particular, failed to meet the density, heating value, and freezing point requirements of international jet fuel standards. These findings underscore the need for a comprehensive action plan that includes improvements in biofuel processing and modification.

Toxic materials generated by various businesses could be remedied in the future by the introduction of genetically engineered algae species. The issues surrounding the use of genetically modified algal species should be carefully considered. If a genetically modified algal species is transplanted to a new environment, it may not be able to adapt to the changing conditions. When natural type algae are mixed with genetically modified algae, new breeding can occur, which can be damaging to other creatures. Even if there are significant concerns regarding the use of genetically modified organisms, the current situation necessitates the total establishment of biofuel production employing modern technology in algae species. The use of genetic engineering technologies on algae is not restricted to the generation of biofuels [55].

Using genetically engineered algae species, the production of a variety of commercially relevant products can be boosted. As a result, the works related to algae cultivation will have market worth. The cost of bioremediation employing genetically modified algae species will be lower than that of traditional approaches. Because of the increased production of biofuels, there are more prospects for direct and indirect employment.

8. Algal fuel properties

The transesterification of algal oils and the characteristics of the produced biodiesel clearly remarked its efficiency used biofuel. Its properties mainly depend on the feedstock used as well as the conversion method employed.

To evaluate the biodiesel derived from algae, different aspects must be considered:

- The acid number indicates the corrosiveness of the oil.
- The iodine value indicates the degree of unsaturation of the biodiesel.
- The specific gravity and density indicate the energy efficiency of the fuel.
- The flashpoint expresses the lowest temperature at which the biodiesel vaporizes to form an ignitable mixture.
- The pour point indicates the low temperature at which the oil turns semisolid and loses its flow characteristics.
- Viscosity refers to the fluid's resistance to flow.
- The heating value indicates the released energy in the form of heat after the combustion of a compound.
- Cetane number refers to the ignition quality of the diesel engines that can operate efficiently.
- Variable fatty acid composition of the algal oil is subjected to transesterification process. Higher percentages of saturated fatty acid methyl esters refer to higher both oxidative stability and cetane number but at the same time have poor cold flow characteristics, as reported by Harrison et al. [9]. On the other hand, higher percentages of unsaturation in the fatty acid profile design better cold flow properties but lower oxidative stability and cetane number. So, the fatty acid composition (fatty acid methyl esters), the percentage of carbon chain length, and the degree of unsaturation have the most valuable effects on the fuel properties [58].

9. Economic importance of algae as biofuel sources

The depleting sources of fossil fuel together with the accumulation of GHG emissions in the environment, from its combustion, cause climate change and global warming. The search for renewable and sustainable energy from biomass starts with the seeded crop plants and then with the lignocellulosic plant residues, which do not require arable lands for cultivation and do not compete with the food crops.

Additional biomass-based fuels are necessary for liquid transesterification to biofuels used for transportation and to replace the depleted fossil fuels. Extreme treatment and production processes have been used for terrestrial feedstock. Initially, the outward benefits of such systems were favorable. Nonmarket benefits were intended to be the policy reinforcement (in the United States and Brazil, for example) [59]. However, the literature may exaggerate these benefits. The loss of significant carbon sinks due to ground clearance for crop cultivation, particularly in tropical places, results in a significant increase in GHG emissions.

Because of the influence of land clearance and conversion on food prices and supplies, the total social and economic gain of traditional biofuels is also unstable, resulting in the loss of nonmarket ecosystem services. These changes have far-reaching societal implications. Extra jobs and earnings from crop-based biofuel production, as well as increased access to gasoline, will compensate for higher food prices in impoverished areas in particular. Increased food costs frequently result in higher salaries for producers, especially those in low-income groups, of which there are many. However, because the benefits of feed crops cannot be distributed evenly throughout society, the distribution of revenue between net producers and net consumers of agricultural commodities remains an analytical problem that must be solved to determine the eventual impact on human well-being [60].

Algae, particularly microalgae, offer a new biofuel potential that does not appear to have the same level of negative development as other biofuels. Microalgae biodiesel, like most biomass-based biofuels, is not genuinely competitive with fossil fuels. However, the relative infancy of manufacturing and processing technologies may exacerbate this [61]. In addition to the possibility of advances that lower production costs, biomass could be used to produce other commodities, increasing financial profitability. However, only a few researches have been conducted on assigning viable organic fuel to determine the practicality of cultivating microalgae for biofuels. The produced algal biomass by advanced technologic techniques, in addition of biofuel production, can be used for many commodities and increasing financial profitability.

Microalgae production and transformation also have the disadvantage of being capital and resource intensive. Apart from the production and maintenance of artificial ecosystems, the facility needs a lot of electricity, water, and nutrients to generate enough biomass [62].

Microalgae have faster growth rates than terrestrial plants and can grow anywhere, in wastelands, in artificial ponds, raceways, or closed bioreactors using treated wastewater or saltwater and at the same time can benefit from solar energy in photosynthesis.

Biofuels from algal biomass are considered renewable and sustainable energy sources that can replace petroleum fuel. Algal biomass can be converted to different forms of energy using thermochemical transformation techniques generating gases, liquids, and solids fuels through liquefaction, gasification, and pyrolysis processes, as reported by Demirbas [63].

Energy conversion using thermochemical, chemical, and biochemical methods produces bio-oil, biodiesel, bioethanol, biomethane, and biohydrogen.

Intensive work and continual research are concerned with the optimal methods of algal culturing, biomass harvesting, oil extraction, and biomass conversion to different types of biofuels to optimize all the conditions affecting the large-size, low-cost, and good quality biofuels that can substitute fossil fuel in the newer future.

10. Applications of biomass

The algal cultivation followed by harvesting of the biomass, drying, and grinding of this biomass to powder can be utilized for many applications, including the drug industry, used as food supplements, pigments as natural coloring substances and antioxidants, anticancers, antivirals, biopolymers (bioplastic), biofuels (biohydrogen, bioethanol, biodiesel, bio-oil, biogas, biochar), and biofertilizers [64–74].

11. Conclusion

The world is entering a period of declining energy, and therefore there is a need for new and alternative energy sources. The algae are an auspicious substitute source for biofuel, including biodiesel, bioethanol, and biohydrogen. The researchers are looking for promising strains with high biomass, lipid, and fatty acids content. The cultivation conditions are the most significant factors that impact productivity and yield. In parallel with the screening for new algae strains, researchers are attempting to use genetic engineering techniques to improve the ability of algae to accumulate high content of lipids and fatty acids that are used for esterification to produce methyl esters (biodiesel). Another important factor that affects the efficiency and productivity of biofuel is the harvesting method. In conclusion, algae are a promising source of renewable energy that may compensate for the shortage of fossil fuel.

References

[1] P.J. Rajan, G.S. Anisha, N.K. Madhavan, A. Pandey, Micro and macroalgal biomass: a renewable source for bioethanol, Bioresour. Technol. 102 (2011) 186–193.

[2] S. Venkata Mohan, M. Prathima Devi, G. Mohamakrishna, N. Amarnath, M. Lenin Babu, P.N. Sarma, Potential of mixed microalgae to harness biodiesel from ecological water bodies with simultaneous treatment, Bioresour. Technol. 102 (2011) 1109–1117.

[3] E.A. Shalaby, Biofuel: Sources, Extraction, and Determination, 2012, https://doi.org/10.5772/51943.
[4] P. Shah Maulin, Microbial Bioremediation & Biodegradation, Springer, 2020.
[5] J.K. Yap, R. Sankaran, K.W. Chew, H.S. Halimatul Munawaroh, S.-H. Ho, J. Rajesh Banu, et al., Advancement of Green Technologies: A Comprehensive Review on the Potential Application of Microalgae Biomass Chemosphere 281 Article 130886, 2021.
[6] A. Abomohra, W. Jin, M.M. El-Sheekh, Enhancement of lipid extraction for improved biodiesel recovery from the biodiesel promising microalga *Scenedesmus obliquu*, Energy Convers. Manag. 108 (2016) 23–29.
[7] Y. Chisti, Biodiesel from microalgae, Biotechnol. Adv. 25 (2007) 294–306.
[8] T.M.I. Mahlia, Z.A.H.S. Syazmi, M. Mofijur, A.E.P. Abas, M.R. Bilad, H.C. Ong, et al., Patent landscape review on biodiesel production: technology updates, Renew. Sustain. Energy Rev. 118 (2020) 109526.
[9] B.B. Harrison, E.B. Mark, J.M. Anthony, Chemical and physical properties of algal methyl ester biodiesel containing varying levels of methyl eicosapentaenoic and methyl docosahexaenoic, Algal Res. 1 (2012) 57–69.
[10] G. Dragone, B. Femandes, A.A. Vicente, J.A. Teeiveira, Third-generation biofuels from microalgae, in: A. Mendez-Vilas (Ed.), Applied Microbiology and Microbial Biotechnology. Current Research, Technology and Education, Formatex Research Center, Spain, 2010.
[11] M.M. El-Sheekh, Biodiesel from microalgae: advantages and future prospective Egyptian, J. Bot. 61 (3) (2021) 669–671.
[12] M.M. El-Sheekh, A. El-Gamal, A.E. Bastawess, A. El-Bokhomy, Production and characterization of biodiesel from the unicellular green alga *Scenedesmus obliquus*, Energy Source Part A Recov. Util. Environ. Eff. 38 (8) (2017) 783–793.
[13] P. Shah Maulin, Removal of Refractory Pollutants from Wastewater Treatment Plants, CRC Press, 2021.
[14] R.A.I. Abou-Shanab, M.M. El-Dalatony, M.M. EL-Sheekh, M.-K. Jia, E. Salamaa, A.N. Kabraa, B.-H. Jeon, Cultivation of a new microalga *Micractinium reisseri* in municipal wastewater for nutrient removal, biomass, lipid and fatty acid production, Biotechnol. Bioproc. Eng. 19 (2014) 510–518.
[15] G. Vicente, B.L. Fernando, J.G. Francisco, R. Rosalia, M. Virhinia, A.R.F. Rosa, et al., Direct transformation of fungal biomass from submerged cultures into biodiesel, Energy Fuel 24 (2010) 3173–3178.
[16] C.U. Ugwu, H. Aoyagi, H. Uchiyama, Photobioreactors for mass cultivation of algae, Bioresour. Technol. 99 (2008) 4021–4028.
[17] M.M. El-Sheekh, S. Gheda, A.B. El-Sayed, A. Abo Shady, M. El-Sheikh, M. Schagerl, Outdoor cultivation of the green microalga *Chlorella vulgaris* under culture stress conditions as a feedstock for biofuel, Environ. Sci. Pollut. Control Ser. 26 (2019) 18520–18532.
[18] K.L. Terry, L.P. R Raymond, System design for the autotrophic production of microalgae, Enzym. Microb. Technol. 7 (1985) 474–487.
[19] A.P. Carvalho, L.A. Meireles, F.X. Malcata, Microalgal reactors: a review of enclosed system designs and performances, Biotechnol. Prog. 22 (2006) 1490–1506.
[20] I.S. Suh, C.G. Lee, Photobioreactor engineering design and performance, Biotechnol. Bioproc. Eng. 8 (2003) 313–321.
[21] M. Huntley, D. Redalje, CO_2 mitigation and renewable oil from photosynthetic microbes: a new appraisal, Mitig. Adapt. Strateg. Glob. Change 12 (2007) 573–608.
[22] K. Li, X.M. Li, et al., Natural promophenols from the marine red alga *Polysiphonia urceolata* (Rhodomelaceae): structural elucidation and DPPH radical-scavenging activity, Bioorg. Med. Chem. 15 (21) (2007) 6627–6631.
[23] X. Miao, Q. Wu, Biodiesel production from heterotrophic microalgal oil, Bioresour. Technol. 97 (2006) 841–846.
[24] M.M. EL-Sheekh, Lipid and fatty acids composition of photoautotrophically and heterotrophically grown *Chlamydomonas reihardtii*, Biol. Plant. 35 (3) (1993) 435–441.
[25] K. Chojnacka, A. Noworyta, Evaluation of *Spirulina* sp. growth in photoautotrophic, heterotrophic, and mixotrophic cultures, Enzym. Microb. Technol. 34 (2004) 461–465.
[26] W.M. Elakbawy, S.M.M. Shanab, E.A. Shalaby, Biological activities, and plant growth regulators producing from some microalgae biomass cultivated in different wastewater concentrations, Biomass Conv. Bioref. (2021), https://doi.org/10.1007/s13399-021-01610-x.
[27] S. Sawayama, S. Inoue, Y. Dote, S.-Y. Yokoyama, CO_2 fixation and oil production through microalga, Energy Convers. Manag. 36 (1995) 729–731.
[28] M.E. Martinez, S. Sanchez, J.M. Jimenez, F. El Yousfi, L. Munoz, Nitrogen and phosphorus removal from urban wastewater by the microalga *Scenedesmus obliquus*, Bioresour. Technol. 73 (2000) 263–272.
[29] S.S.M. Mostafa, E.A. Shalaby, G.I. Mahmoud, Cultivating microalgae in domestic wastewater for biodiesel production, Not. Sci. Biol. 4 (1) (2012) 56–65.
[30] A. Pandey, D.-J. Lee, Y. Chisti, C.R. Soccol (Eds.), Biofuels from Algae, Elsevier, B V, 2014, p. 338.
[31] E. Molina Grima, E.H. Belarbi, F.G. Acien Fernandez, A. Robles Medina, Y. Chisti, Recovery of microalgal biomass and metabolites: process options and economics, Biotechnol. Adv. 20 (2003) 491–515.
[32] R. Bosma, W.A. Van Spronsen, J. Tramper, R.H. Wijffels, Ultrasound is a new separation technique to harvest microalgae, J. Appl. Phycol. 15 (2003) 143–153.
[33] E. Molina Grima, J.A. Sanchez Perez, F. Garcia Camcho, J.M. Fernandez Sevilla, F.G. Acien Fernandez, Effect of the growth rate of on the eicosapentaenoic and docosahexaenoic acid content of *Isochrysis galbana* in chemostat culture, Appl. Microbiol. Biotechnol. 41 (1994) 23–27.
[34] K. Tsukahara, S. Sawayama, Liquid fuel production using microalgae, J. Jpn. Petr. Inst. 48 (2005) 251–259.
[35] T. Minowa, S. Sawayama, A novel microalgal system for energy production with nitrogen cycling, Fuel 78 (1999) 1213–1215.
[36] T. Minowa, S.Y. Yokoyama, Oil production from algal cells of *Dunaliella tertiolecta* by direct thermochemical liquefaction, Fuel 74 (12) (1995) 1735–1738.

[37] A. Demirbas, Oily product from mosses and algae via pyrolysis, Energy Source Part A: Recover. Util. Environ. Eff. 1556−7230 28 (10) (2006) 933−940.

[38] X. Miao, Q. Wu, et al., Fast pyrolysis of microalgae to produce remarkable fuels, J. Anal. Appl. Pyrol. 71 (2) (2004) 855−863.

[39] G.W. Abou El-Souod, E.M. Morsy, L. Hassan, M.M. El-Sheekh, Efficient saccharification of the microalga *Chlorella vulgaris* and its conversion into ethanol by fermentation, Iran. J. Sci. Technol. Trans. A-Sci. 46 (2021) 767−774.

[40] H. Gao, Y. Wang, Q. Yang, H. Peng, Y. Li, D. Zhan, H. Wei, H. Lu, M.M.A. Bakr, M.M. EI-Sheekh, Z. Qi, L. Peng, X. Lin, Combined steam explosion and optimized green-liquor pretreatments are effective for complete saccharification to maximize bioethanol production by reducing lignocellulose recalcitrance in one-year-old bamboo, Renew. Energy 175 (2021) 1069−1079.

[41] Y. Ueno, N. Kurano, S. Miyachi, Ethanol production by dark fermentation in the marine green alga Chlorococcum litoralee, J. Ferment. Bioeng. 12 (1998) 60057−60064.

[42] A. Melis, T. Happe, Hydrogen production green algae as a source of energy, Plant Physiol. 127 (2001) 740−748.

[43] M.M. El-Sheekh, M.Y. Bedaiwy, A. Aya, A.A. El-Nagar, M. ElKelawy, H.A. Bastawissi, Ethanol biofuel production and characteristics optimization from wheat straw hydrolysate: performance and emission study of DI-diesel engine fueled with diesel/biodiesel/ethanol blends, Renew. Energy 191 (2022) 591−607, https://doi.org/10.1016/j.renene.2022.04.076.

[44] A.R.K. Gollakota, N. Kishore, S. Gu, A review on hydrothermal liquefaction of biomass, Renew. Sustain. Energy Rev. 81 (2018) 1378−1392, https://doi.org/10.1016/j.rser.2017.05.178.

[45] M. Ismail, G. Ismail, M.M. El-Sheekh, Potential assessment of some micro and macroalgal species for bioethanol and biodiesel production, Energy Source Part A Recov. Util. Environ. Eff. (2020), https://doi.org/10.1080/15567036.2020.1758853 in press.

[46] A. Abomomra, M. Elasyed, S. Sakkimuthu, M.M. El-Sheekh, D. Hanelt, Potential of Fat, oil and grease (FOG) for biodiesel production: a critical review on the recent progress and future perspectives, Prog. Energy Combust. Sci. 81 (2020) 100868.

[47] M.A. Dube, A.Y. Tremblay, J. Liu, Biodiesel production using a membrane reactor, Bioresour. Technol. 98 (2007) 639−647.

[48] A. Liu, W. Chen, L. Zheng, L. Song, Identification of high-lipid producers for biodiesel production from forty-three green algal isolates in China, Prog. Nat. Sci. Mater. Int. 21 (4) (2011) 269−276, https://doi.org/10.1016/S1002-0071.

[49] S.T.L. Griffiths, M.J. Harrison, Lipid productivity as a key characteristic for choosing algal species for biodiesel production, J. Appl. Phycol. 21 (5) (2009) 493−507, https://doi.org/10.1007/s10811-008-9392-7.

[50] H. Yang, Q. He, C. Hu, Feasibility of biodiesel production and CO_2 emission reduction by *Monoraphidium dybowskii* LB50 under semi-continuous culture with open raceway ponds in the desert area, Biotechnol. Biofuels 11 (1) (2018) 1−14, https://doi.org/10.1186/s13068-018-1068-1.

[51] S. Ali, A. Paul Peter, K.W. Chew, H.S.H. Munawaroh, P.L. Show, Resource recovery from industrial effluents through the cultivation of microalgae: a review, Bioresour. Technol. 337 (2021) 125461, https://doi.org/10.1016/j.biortech.2021.125461.

[52] S.Y.A. Siddiki, M. Mofijur, P.S. Kumar, S.F. Ahmed, A. Inayat, F. Kusumo, I.A. Badruddin, T.M.Y. Khan, L.D. Nghiem, H.C. Ong, T.M.I. Mahlia, Microalgae biomass as a sustainable source for biofuel, biochemical and biobased value-added products: an integrated biorefinery concept, Fuel 307 (2022) 121782, https://doi.org/10.1016/j.fuel.2021.121782.

[53] M. Hannon, J. Gimpel, M. Tran, B. Rasala, S. Mayfield, Biofuels from algae: challenges and potential, Biofuels 1 (5) (2010) 763−784, https://doi.org/10.4155/bfs.10.44.

[54] S. Khan, P. Fu, Biotechnological perspectives on algae: a viable option for next generation biofuels, Curr. Opin. Biotechnol. 62 (2020) 146−152.

[55] B. Bharathiraja, J. Iyyappan, M. Gopinath, J. Jayamuthunagai, R. PraveenKumar, Transgenicism in algae: challenges in compatibility, global scenario and future prospects for next generation biofuel production, Renew. Sustain. Energy Rev. 154 (2022) 111829.

[56] P. Li, Y. Luo, X. Yuan, Life cycle and techno-economic assessment of source-separated wastewater-integrated microalgae biofuel production plant: a nutrient organization approach, Bioresour. Technol. 344 (2022) 126230.

[57] M. Mofijur, S.M. Ashrafur Rahman, L.N. Nguyen, T.M.I. Mahlia, L.D. Nghiem, Selection of microalgae strains for sustainable production of aviation biofuel, Bioresour. Technol. 345 (2022) 126408.

[58] M. Ramos, C.M. Fernandez, A. Casas, L. Rodriguez, A. Perez, Influence of fatty acid composition of raw materials on biodiesel properties, Bioresour. Technol. 100 (2009) 261−268.

[59] A. Gasparatos, P. Stromberg, K. Takeuchi, Sustainability impacts of first-generation biofuels, Anim. Front. 3 (2) (2013) 12−26.

[60] S. Msangi, Bioenergy and food security: synergies and trade-offs, Role Bioenergy Emerg. Bioecon. Resour. Technol. Sustain. Policy 355 (2018) 357−376.

[61] Z. Tu, L. Liu, W. Lin, Z. Xie, J. Luo, Potential of using sodium bicarbonate as an external carbon source to cultivate microalga innon-sterile condition, Bioresour. Technol. 266 (2018) 109−115.

[62] J. Fuhrman, H. McJeon, P. Patel, S.C. Doney, W.M. Shobe, A.F. Clarens, Food−energy-water implications of negative emissions technologies in a+ 1.5 C future, Nat. Clim. Change 10 (10) (2020) 920−927.

[63] A. Demirbas, Use of algae as biofuel sources, Energy Convers. Manag. 51 (2010) 2738−2749.

[64] M.L. Brown, K.G. Zeiler, Aquatic biomass and carbon dioxide trapping, Energy Convers. Manag. 34 (1993) 1005−1013.

[65] E.A. El-fayoumy, S.M. Shanab, E.A. Shalaby, Metabolomics and biological activities of *Chlorella vulgaris* grown under modified growth medium (BG11) composition, CMU J. Nat. Sci. 19 (2020) 91−123.

[66] E.A. El-Fayoumy, S.M. Shanab, H.S. Gaballa, M.A. Tantawy, E.A. Shalaby, Evaluation of antioxidant and anticancer activity of crude extract and different fractions of *Chlorella vulgaris* axenic culture grown under various concentrations of copper ions, BMC Complemen. Med. Ther. 21 (1) (2021) 1−16.

[67] M. Jau, S. Yew, P.S.Y. Toh, A.S.C. Chong, W. Chu, S. Phang, et al., Biosynthesis and mobilization of poly (3-hydroxybutyrate) [P(3HP)] by *Spirulina platensis*, Int. J. Biol. Macromol. 36 (2005) 144−151.

[68] S. Liang, L. Xueming, F. Chen, Z. Chen, Current microalgal health food R&D activities in China, Hydrobiol. (Sofia) 512 (2004) 45–48.
[69] A. Richmond, Handbook of Microalgal Mass Culture, CEC press, Boston, MA, USA, 1990.
[70] E.A. Shalaby, S.M.M. Shanab, V. Singh, Salt stress enhancement of antioxidant and antiviral efficiency of *Spirulina platensis*, J. Med. Plants Res. 4 (24) (2010) 2622–2632.
[71] E.A. Shalaby, S.M.M. Shanab, Comparison of DPPH and ABTS assays for determining antioxidant potential of water and methanol extracts of *Spirulina platensis*, Indian J. Geo-Mar. Sci. 42 (5) (2013) 556–564.
[72] S.M. Shanab, S.S. Mostafa, E.A. Shalaby, G.I. Mahmoud, Aqueous extracts of microalgae exhibit antioxidant and anticancer activities, Asian Pacific J. Trop. Med. 2 (2012) 608–615.
[73] L. Sharma, A.K. Singh, P, M. B, N. Mallick, Process optimization for poly-B-hydroxybutyrate production in a nitrogen-fixing cyanobacterium, *Nostoc muscorum* using response surface methodology, Bioresour. Technol. 98 (2007) 987–993.
[74] M. Shah (Ed.), Removal of Emerging Contaminants through Microbial Processes, Springer, 2021.
[75] C. Jimenez, B.R. Cossio, D. Labella, F.X. Niell, The feasibility of industrial production of Spirulina (Arthrospira) in southern Spain, Aquaculture 217 (2003) 179–190.
[76] M.A. Borowitzka, Algal biotechnology products and processes — matching science and economics, J. Appl. Phycol. 4 (1992) 267–279, https://doi.org/10.1007/BF02161212.
[77] L. Rodolfi, G.C. Zittelli, N. Bassi, G. Padovani, N. Biondi, G. Bonini, Microalgae for oil: Strain selection, induction of lipid synthesis and outdoor mass cultivation in a low cost photobioreactor, Biotechnol. Bioeng. 102 (1) (2008) 100–112.
[78] M.R. Andrade, J.A.V. Costa, Mixotrophic cultivation of microalga Spirulina platensis using molasses as organic substrate, Aquaculture 264 (2007) 130–134.
[79] T.J. Lundquist, Production of algae in conjunction with wastewater treatment. Paper presented at the National Renewable Energy Laboratory-Air Force Office of Scientific Research Joint Workshop on Algal Oil for Jet Fuel Production. February 19–21, 2008. Arlington, VA.
[80] G. Hodaifa, M.E. Martínez, S. Sánchez. Use of industrial wastewater from olive-oil extraction for biomass production of Scenedesmus obliquus. Bioresour. Technol. 99, (2008), 1111–1117.
[81] P.M. Schenk, S.R.T. Hall, E. Stephens, U.C. Marx, J.H. Mussgnug, C. Posten, O. Kruse, B. Hankamer, Second generation biofuels: high-efficiency microalgae for biodiesel production, Bioen. Res. 1 (2008) 20–43.
[82] R. Divakaran, V. Sivasankara Pillai, Flocculation of algae using chitosan, J. Appl. Phycol. 14 (2002) 419–422, https://doi.org/10.1023/A:1022137023257.
[83] Y. Nurdogan, W.J. Oswald, Tube settling of high-rate pond algae, Water Sci. Technol. 33 (7) (1996) 229–241.
[84] B. Petrusevski, A.N. van Breemen, Guy Alaerts, Optimisation of coagulation conditions for direct filtration of impounded surface water, J. Water SRT-Aquat. 44 (2) (1996) 93.
[85] A. Harel, Noritech Seaweed Biotechnologies Ltd, Algae World Conference 23 (2009) 44–53. Rotterdam, The Netherlands.
[86] R. Harun, M.K. Danquah, M. Forde-Gareth, Microalgal biomass as a fermentation feedstock for bioethanol production, J. Chem. Technol. Biotechnol. 85 (2010) 199–203, https://doi.org/10.1002/jctb.2287.
[87] H.B. Goyal, Diptendu Seal, R.C. Saxena, Biofuel from thermochemical conversion of renewable resources: a review, Renew. Sustain. Energy Rev. 12 (2) (2008) 504–517.
[88] Y. Dote, S. Sawayama, S. Inoue, T. Minowa, S.Y. Yokoyama, Recovery of liquid fuel from hydrocarbon-rich microalgae by thermochemical liquefaction, Fuel 73 (12) (1994) 1855–1857.
[89] P. McKendry, Energy production from biomass (Part 1): overview of biomass, Bioresour. Technol. 83 (1) (2002) 37–46.
[90] P. McKendry, Energy production from biomass (Part 2): conversion technologies, Bioresour. Technol. 83 (1) (2002) 47–54.
[91] K.B. Cantrell, T. Ducey, K.S. Ro, P.G. Hunt, Livestock waste-tobioenergy generation opportunities, Bioresour. Technol. 99 (2008) 7941–7953, https://doi.org/10.1016/j.biortech.2008.02.061.
[92] Y.C. Sharma, B. Singh, Development of biodiesel: current scenario, Renew. Sustain. Energ. Rev. 13 (2009) 1646–1651, https://doi.org/10.1016/j.biortech.2010.06.057.

Further reading

[1] M.A. Borowitzka, Commercial production of microalgae: ponds, tanks, tubes and fermenters, J. Biotechnol. 70 (1999) 313–321.
[2] L. Brennan, P. Owende, Biofuels from microalgae: a review of technologies for production, processing, and extraction of biofuels and co-products, Renew. Sustain. Energy Rev. 14 (2010) 557–577.
[3] X. Li, H. Xu, Q. Wu, Large scale biodiesel production from microalga *Chlorella protothecoides* through heterotrophic cultivation in bioreactors, Biotechnol. Bioeng. 98 (2007) 764–771.
[4] V. Patil, K.Q. Tran, H.R. Giselrod, Toward sustainable production of biofuels from microalgae, Int. J. Mol. Sci. 9 (2008) 1188–1195.

Chapter 11

Thermochemical conversion of biomass into valuable products and its modeling studies

R. Saravanathamizhan, V.T. Perarasu and K. Vetriselvan
Department of Chemical Engineering, A C Tech, Anna University, Chennai, Tamil Nadu, India

1. Introduction

The global energy requirement is steadily increasing due to growth of industrialization and population. Hence the conventional fossil fuels like coal, petroleum fuel, etc., are depleting very fast. Fossil fuels are the major fuel source that meets huge amount of energy needs. It is predicted that fossil fuels particularly crude oil, may be exhausted in another 200–300 years. Burning of fuel also leads to creation of poisonous gases which can lead to global warming. So it is essential to move toward an alternative source of energy. The alternative fuel should be focused on sustainable energy resources particularly on renewable energy and environmental friendliness. Among the renewable energy sources, biomass energy is playing important role nowadays. Utilization of biomass energy and biomass conversion to energy is the emerging research area and researchers are working to develop these technologies.

1.1 Biomass

Biomass is an alternate to renewable energy sources in th8e globe and is growing at a faster pace recently. Biomass energy is defined as the organic material that comes from plants including wood from different sources, agricultural and industrial residues, human and animal waste. The most commonly identified biomass resources are plant biomass from agricultural and food production. Biomass does not take more years to develop unlike fossil fuels.

Biomass is the oldest source of energy. Ancient day's people have burned wood to cook food and generate heat. Once the biomass is burnt it produces CO_2 and it is again taken up by the plants during photosynthesis and oxygen is released to atmosphere. Hence biomass energy falls under the renewable energy resource which can be used to generate energy when there is a demand. Biomass energy significantly reduces the greenhouse gas emissions, mitigating global warming effects. Some of the important sources of biomass are shown in Fig. 11.1.

1.2 Classification of biomass

Due to their different compositional characteristics and depending on purpose and scope, biomass can be grouped differently. The major classification of biomass is woody-based and non-woody-based biomass. In other words they are called as lignocelluloses and non-lignocelluloses biomass. Various sources of biomass is given below [1]

- Agricultural, forestry crops, and residues.
- Animal and human residues.
- Municipal solid wastes.
- Sea weeds and algae.
- Industrial residues.

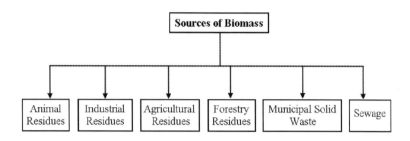

FIGURE 11.1 Important sources of biomass.

1.3 Characteristics of biomass

The biomass characteristics are evaluated by proximate analysis, ultimate analysis, and structural analysis. Proximate analysis such as fixed carbon, moisture, volatiles, and ash content are measured. In the ultimate analysis of biomass elemental compositions such as hydrogen, carbon, nitrogen, and oxygen are sulfur measured. The elemental composition of plant biomass varies from 42% to 47% of carbon, 40%–44% of oxygen, and 6% of hydrogen. The composition will vary depending upon the biomass source. Cellulose, hemicellulose, and lignin are measured as structural analysis. The biomass generated from cattle dung is rich in proteins and cereals contain high amount of starch. Biomass analysis and the estimation of their characteristics are essential for the design and operation of biomass conversion processing facilities [2,3]. The important physical properties such as particle size, density, moisture content, grindability, and flowability play an important role in the design and operation of the biomass reactor.

2. Biomass conversion techniques

Biomass energy conversion technology is growing rapidly to convert biomass in to valuable chemicals. Biomass resources need to be converted into useful products such as chemicals and fuels, whose application are ranging from domestic purpose to industrial purpose. The conversion technologies for biomass may be based on thermochemical, biochemical or physiochemical conversion processes. The details of classification of biomass conversion technologies and its methods are given in Fig. 11.2.

Variety of reactors and technologies have been used for conversion of biomass into value added products. The focus of the present chapter is to discuss the various thermal and catalytic conversion technologies using various types of reactors and the descriptions of several models developed for the same.

3. Thermochemical conversion

The energy and chemical products derived from biomass by the application of heat energy is called thermochemical conversion. It generally refers to the breakdown of organic matter present in the biomass by the application of heat energy into fuels such as in the form of solid, liquid, and gases.

Thermochemical conversion processes produce useful energy under the controlled temperature and oxygen (excess, controlled, and inert condition). These processes are more convenient and cost-effective than the direct thermal processes.

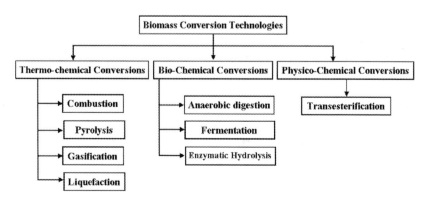

FIGURE 11.2 Biomass conversion technologies.

The conversion of biomass into energy produces producer gas, bio oil, and biochar. Thermochemical conversion processes include combustion, pyrolysis, gasification, and hydrothermal liquefaction process. The use of air and the products produced from biomass using these processes are depicted in Fig. 11.3.

3.1 Combustion

Combustion is an exothermic reaction process. While burning organic material, heat energy is liberated, and it can be utilized directly or to generate steam. In earlier days humans have used to generate heat energy through burning biomass and later it is used to generate power through steam. Wood is the most commonly used biomass and other biomass residues are straw, sawdust, bark residuals, shavings from sawmills, etc., whereas, energy crops such as switchgrass, poplar, and willow are used as a feedstock. Pelletized agricultural and wood residues biomass is used because they can be easy to handle. The best conventional and most commonly used technology is direct combustion for converting biomass to heat. Biomass fuel, during combustion, is burnt in excess air to generate heat.

$$\text{Biomass} + \text{Oxygen} \rightarrow \text{Carbon Dioxide} + \text{Water} + \text{Heat}$$

The biomass combustion products are carbon dioxide and heat energy. Minimization of emission gases and efficient burning are the environmentally acceptable biomass combustion systems.

3.1.1 Reactors for combustion

Biomass combustion process is an exothermic reaction and the product is heat energy which is utilized to produce electricity. Depending upon biomass property, different combustion reactor designs and firing parameters are important to get optimum efficiency.

Biomass combustion in a grate fire chamber is shown in Fig. 11.4. In the fired chamber, biomass is passed through the moving grate and combustion takes place before it reaches the end. Through the bottom of the grate, air is supplied and the

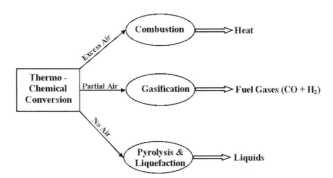

FIGURE 11.3 Methods and products of thermochemical conversion of biomass.

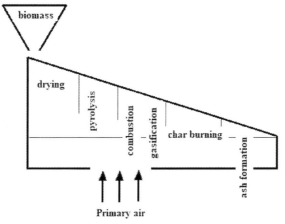

FIGURE 11.4 Grate fired chamber.

combustion occurs directly above the bed. Once the process is completed, the formed ash is removed to an ash pit at the end of the grate [4,5]. The reactors used for combustion process, that is, Fixed Bed (FB), Bubbling Fluidized Bed (BFB), and Circulating Fluidized Bed (CFB) combustion chambers are shown in Fig. 11.5.

In FB combustion chamber, the bed is fixed and air is passed from the bottom for combustion and the flue gas leaves through the top of the reactor after combustion. In fluidized bed combustion chamber, air is admitted below the bed of sand. There are two kinds of fluidized bed chambers, BFB and CFB. In BFB combustion, fluidization velocity is low and the particles of bed are kept under suspended condition. In CFB, the fluidization velocities are higher and the particles are carried along with air, out of the bed. Cyclone separators are generally provided at the top and bed particles are captured and reintroduced into the bed again. These beds (FB, BFB, and CFB) are capable of combusting wet biomass solids [6,7].

3.1.2 Significant studies on combustion

Combustion reaction is a complex process involving large number of reaction steps coupled with heat transfer, mass transfer, and fluid flow. Design and control of biomass combustion process require knowledge about biomass properties and influencing parameters. Jenkins et al. [8] summarized various properties of biomass such as moisture content, bulk density, thermal conductivity, heating value, composition of elements, specific gravity, and properties such as mechanical, acoustic, and electrical for a wide range.

Saidur et al. [9] has presented a detailed review relating to burning biomass in boilers. Their study includes various key parameters such as biomass composition, estimation of biomass heating value, and comparison with other fuel; co-firing of biomass with coal, economic, and social analysis and their impact, biomass transportation, densification and biomass associated problems for future. They concluded that biomass usage in boilers offers many benefits such as money saving, conservation of fossil fuel, and reduction in emission of hazardous gas.

Thermal conversion of biomass and its ashes' effects on equipment were presented by Nunes et al. [10]. They described ashing process causes metallurgical damage to boilers due to chloride gas. They have also analyzed various ways to prevent the problems caused by ash deposition in biomass-based fueled boilers [11]. Ash-related problems reduce energy conversion efficiencies by interfering with biomass combustion processes. Various additives can mitigate these ash-related issues and it can change the ash chemistry, decrease in concentration of problematic species, and raise melting temperatures of ashes.

Large volume of ash produced in the biomass combustion power generation is the problem in the world. The ash-related problems issues are slagging, corrosion and ash utilization, and the remedies such as co-firing, use of additives, and leaching have been studied intensively by Niu et al. [12].

Biomass is a renewable energy source that causes negative effect of NOx emission if the biomass is used in uncontrolled and massive way. It is an important and challenging task to control NOx emission. Mladenović et al. [13] presented and analyzed denitrification techniques such as pre-combustion, combustion, and post-combustion and also have shown the applicability of deNOx techniques for biomass combustion.

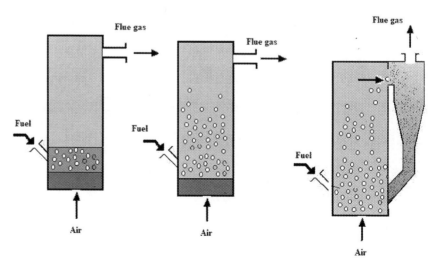

FIGURE 11.5 FB, BFB, and CFB combustion chambers.

3.2 Gasification

Gasification is a process of converting biomass into a gaseous product (producer gas) in the controlled oxygen environmental condition. This process involves a sequence of chemical and thermal reactions. This is achieved at high temperatures, with partial combustion, with a controlled steam or oxygen environment. The exhaust combustible gas was found to consist of carbon dioxide, carbon monoxide, hydrogen, methane, nitrogen, water along with contaminants like ash, tars and small char particles. The gas is purified and it is used in boilers, turbines and engines for power generation. Hence, gasification technology is considered as one of the methods to increase the energy production.

The basic gasification process involves devolatization, combustion, and reduction. Emission control is easy in gasification process compared to combustion process because the producer gas in gasification is at higher temperature and pressure than the exhaust gases produced in combustion. The gasification processes are carried out at temperatures ranging between 800 and 1100°C [14].

This technology is useful to reduce CO_2 emission, has the flexibility to use wide range of feedstock, produces energy, and generates wide range of fuels and chemicals. Thus, the focus of its function has taken an additional direction from creation of combined heat and power. The technical challenges involved in the biomass gasified product include cleaning the producer gas and its conversion to valuable products [15].

3.2.1 Reactors for gasification

Gasification is done in the gasifier and it is an endothermic reaction. The heat is supplied either directly or indirectly to the gasifier. They are four stages in gasification such as drying, pyrolysis, gasification, and combustion stages. The types of gasifier are fixed beds and fluidized beds and are discussed below.

3.2.1.1 Fixed bed

Based on the direction and entry of air flow, fixed beds are classified into updraft, down draft, or cross-draft.

In updraft gasifier, the biomass enters at the top and air enters from the bottom. At the gasifier top, the biomass gets dried with the contact of leaving hot gas. It then passes through pyrolysis zone in which the dried biomass is decomposed into volatiles, tar, and char. The volatile free biomass is then moving down toward gasifier zone and subsequently to the combustion zone which is the bottom most zone. Biomass oxidation takes place in the combustion zone and gases generated in this zone passes through the zone containing charcoal (which is produced by pyrolysis of biomass) and gets converted into producer gas and passes the drying zone. The reduction, pyrolysis, and drying zones effectively utilize the heat liberated in the combustion zone.

In downward type, both the biomass and air move downward in the lower section of the gasifier unit. As most of the tars are combusted in the gasifier, the product gas contains a low concentration of particulates and tars. The down draft gasifier is found to be ideal when clean gas is preferred.

In a cross-flow gasifier, the biomass is fed at the top of the unit and moves downward, whereas the air enters from the side of the gasifier. Product gas leaves from top of the unit at about the same level as the biomass feed location. A hot combustion and/or gasification zone forms around the air entrance and pyrolysis and drying zones get formed in the gasifier [16]. The typical reactor types are shown in Fig. 11.6.

3.2.1.2 Fluidized bed gasification

In this type, the biomass is in fluidized state which offers the advantage of uniform temperature distribution. The temperature uniformity is maintained using fine granular bed material into which air is circulated. There are two types of fluidized bed gasifiers: (a) bubbling fluidized bed gasifier, and (b) circulating fluidized bed gasifier [16]) are shown in Fig. 11.7.

In a bubbling fluidized bed type, the air is allowed through grate at the bottom. The bed material is kept above the grate through which the biomass is introduced. The temperature of the bed is maintained around 700−900°C by controlling the air/biomass ratio. The biomass is pyrolyzed in the hot bed producing gaseous compounds and tar [17].

Circulating fluidized bed gasifier operated continuous circulation of bed material between the reactor and separator. The ash is separated in the cyclone separator and bed material, char returns back to the reactor. These gasifiers are capable of handling large volume of biomass and can be operated at elevated pressures [18].

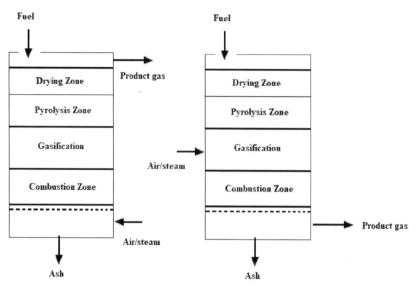

FIGURE 11.6 Updraft and down draft gasifiers.

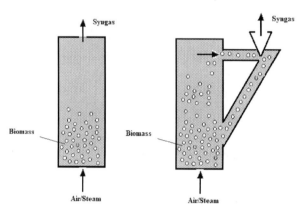

FIGURE 11.7 Bubbling and circulating fluidized bed gasifier.

3.2.2 Significant studies on gasification

An experimental study was conducted to produce hydrogen from biomass by Turn et al. [19] using a bench fluidized bed gasifier. They have conducted experiments to determine the effects of steam to biomass ratio, reactor temperature, and equivalence ratio on gas composition as well as yield.

Another remarkable technology is hydrothermal biomass gasification which converts the wet biomass (with natural water content) to hydrogen by a thermochemical reaction. The temperature needed is observed to be lower than for dry thermochemical reactions and no drying is necessary. Hydrothermal gasification of biomass is a potential method for the utilization of biomass having high water content. If biomass with 80% or more water content is heated up under pressure, it shows a high reactivity. Biomass without water and biomass with the natural water content (also called as green biomass) can be converted efficiently to gases. Burnable gases such as methane or hydrogen have been produced depending on the reaction condition. Recently, significant development was accomplished in various hydrothermal biomass gasification processes [20]. Water is used as reaction medium and it supports the fast degradation of the polymer.

Advanced concept for hydrothermal water gasification such as impregnation of nickel catalyst into lignocellulosic biomasses was carried out by Nanda et al. [21]. They gasified nickel-impregnated feedstocks in subcritical and supercritical water. Their study included the effect of residence time biomass-water ratio and temperature. They have concluded that their method showed high gas yields and hydrogen yields with greater carbon gasification efficiency when compared to non-catalytic gasification. They found that enhanced hydrogen production in supercritical water was due to the fact that nickel particles doped into the lignocellulosic matrix acted as nanocatalysts by offering active sites for enhancing.

An investigation to produce the biohydrogen and syngas by gasifying α-cellulose and other agriculture waste was conducted by Chang et al. [22]. They analyzed the syngas and biohydrogen—other products to maximize the yield. A kinetic model has also been proposed to determine the reaction order and activation energy.

Biomass gasification produces renewable hydrogen. Kirtay [23] highlighted various processes such as thermochemical and biological methods for the conversion of biomass into hydrogen gas.

Panwar et al. [24] reviewed gasification technology and scope of potential of byproducts like hydrogen and charcoal production and concluded that downdraft type biomass gasifier was found most appropriate for industrial applications and conversion of biomass through thermochemical conversion route helps to protect environment and ecology.

Ruiz et al. [25] carried out a review on current technology barriers for electricity generation through biomass gasification. Their study includes the selection of the gasifier types and various operating parameters. Though gasification is one of the promising technologies to convert biomass to gaseous fuel, the commercial utilization of biomass energy faces hurdles from a number of logistics and technological challenges.

Asadullah [26] pointed out various obstacles in all steps from the biomass collection to electricity generation and discussed the effects of parameters such as pretreatment and conversion of biomass to gas, cleaning and utilization of gas for power generation as well as in supply chain management, The degradation routes of biomasses such as cellulose and lignin at near and supercritical conditions was presented by Reddy et al. [27]. They have highlighted some homogenous and heterogeneous catalysts leading to water gas shift, methanation, and other sub-reactions during supercritical water gasification. The parametric impacts along with reactor configurations for maximum hydrogen production were discussed.

An assessment on the fundamentals such as feedstock types, the impact of different operating parameters, tar formation and cracking, and modeling approaches for biomass gasification were presented by Sikarwar et al. [28]. Other mechanisms for gasification, recent advances in gasification, and unique gasifiers along with new strategies were also discussed as a means to encourage this technology into alternative applications for greater flexibility and improved efficiency.

Ahmad et al. [29] highlighted the characteristics and performances of some types of gasifiers under different parameters that will have an effect on the yields of the end products as well as the composition of the gas. They have also discussed the various types of models used in the simulations and the optimization of the gasification conditions.

A review on the gasification technologies such as fixed bed reactors, fluidized bed reactors, entrained flow reactors, and biofuels obtained from syngas was reported by Molino et al. [30]. They have also reported that, according to the gasification technology used, the feedstock characteristics, and the operating parameters, the syngas composition varies.

Hydrogen production using various agents such as air, oxygen-enriched air, oxygen and steam were investigated by Shayan et al. [31] and compared from the perspectives of the first and second laws of thermodynamics. A parametric study was also further studied to evaluate the effects of parameters on the hydrogen concentration and calorific value of producer gas, energy and exergy efficiencies and exergy destruction rate. Their results indicated that the increased hydrogen production rates were connected respectively with oxygen, oxygen enriched air, steam, and air as the gasification agents. Also, it was concluded that for the gasification process the highest value of sensible energy efficiency was obtained for air gasification, whereas the highest exergy efficiency was noticed for steam gasification.

3.3 Pyrolysis

Thermal decomposition of biomass in the absence of oxygen is called pyrolysis. The products of pyrolysis operation are biochar, bio-oil, and gases including hydrogen, methane, carbon dioxide, and carbon monoxide. Depending on temperature and biomass residence time inside the pyrolysis operation we get different yield of products. At low temperature, less than 450°C, pyrolysis yields mainly biochar and at the high heating rate of 800°C, it yields mainly gases. At an intermediate temperature with elevated heating rate yields bio-oil. The process of pyrolysis gives a flexible and smart way of converting solid biomass into an easily stored and transported liquid.

3.3.1 Classification of pyrolysis process

Pyrolysis process is classified on the basis of operating conditions. The yield of biomass pyrolysis product mainly depends on chemical composition and the operating temperature [24]. The processes are classified as slow, fast, and flash pyrolysis.

- **Slow pyrolysis:** It is used for production of char. It involves slow heating of biomass over a longer period of time. The operating temperature of the slow pyrolysis operation is 290°C. Torrefaction is a slow pyrolysis process held from 230 to 300°C in the absence of oxygen.
- **Fast pyrolysis:** It mainly produces liquid biofuels such as bio-oil. In fast pyrolysis, biomass is thermally heated at elevated temperature 500–900°C in an inert condition. The bio-oil yield in the process is 60–75 wt%, biochar is 15–25 wt%, and 10–20 wt% of non-condensable gases.
- **Flash pyrolysis:** This operation involves heating of biomass at 450–600°C rapidly at very high heating rates in the absence of oxygen. The product depends on the conditions of pyrolysis. Temperatures of around 500°C with high heating rates and short vapor residence times (less than 1 s) maximizes liquid yield upto 80 wt%, whereas very rapid heating to temperatures around 700°C and vapor residence less than 1 s yield gas upto 80 wt%.

3.3.2 Reactors for pyrolysis

Pyrolysis reactors are very similar to gasifier. Fixed bed, circulating fluidized bed bubbling fluidized bed, rotating cone, vacuum pyrolysis reactor, microwave pyrolysis reactor, ablative reactor, auger reactor, plasma reactor, etc., are used for the pyrolysis operation. Among these fixed bed and fluidized bed gasifier are used widely. The process is carried out without air or in an inert atmospheric condition [32–36] Fluidized bed pyrolysis is shown in Fig. 11.8.

Pyrolysis operation is an endothermic process. Different types of pyrolysis processes give different types of products (bio-oil, biochar, and gases) based on their heating rate. The construction and operation of fluidized bed pyrolysis reactor is simple, and they offer good temperature control and temperature distribution to biomass particles. Frequently sand particles are used as bed material. Bubbling fluidized bed operation produces high quality of bio-oil with high oil yield. The char and vapor separation has been done in the cyclone separator. Selecting the appropriate size for the fluidized bed particle, the flow rate of gas and the residence time of vapor can be set to the desired level [37,38].

3.3.3 Significant studies on pyrolysis

There are substantial chances for production of conventional and unconventional fuels. There is always great importance to improve efficiency and reduce costs by following techniques such as the use of catalysts to improve either the yield or quality of gas, liquid fuels from biomass through pyrolysis and the integration of catalytic processes into the thermal conversion process. Bridgwater [39] described fast pyrolysis technologies with the use of catalysts in chemicals production. They also used catalytic processes to upgrade the pyrolysis products to higher quality and higher value fuels and chemicals. They found that removal or reinforcement of natural catalysts which influence the production of chemicals had a spectacular effect on product yield and composition. The vapors of pyrolysis were catalytically cracked over zeolites to give products which were further converted into gasoline and diesel.

Wood is the most preferred biomass source because of its consistency and comparability between tests. Mohan et al. [38] presented the recent progresses in the wood pyrolysis and reported the characteristics of the resulting bio-oils. The physical and chemical aspects of the resulting bio-oils using fast and slow wood/biomass pyrolysis with respect to wood structure and composition, rate of heating and residence time on the overall reaction rate, and the yield of the volatiles were discussed.

The pyrolysis of perennial shrub *E. rigida* was investigated by Pütün et al. [40] in a fixed-bed reactor with respect to catalyst ratio and atmosphere type. They found that catalyst ratio played much less responsibility when the pyrolysis was performed in inert compared with steam atmosphere. Implementation of pyrolysis in steam atmosphere caused noteworthy differences with respect to product characteristics. Use of steam instead of nitrogen produced less paraffins, also enriched

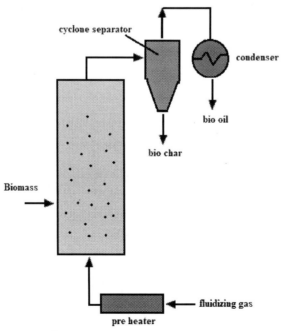

FIGURE 11.8 Fluidized bed pyrolysis reactor.

ketones, triterpenoid compounds, and carboxylic acids, while decreased formation of phenol. Their results further showed that the yield and composition of the oil were dependent on the ratio of catalyst and atmosphere of pyrolysis.

Fu et al. [41] studied the relationships between gas composition and char properties and pyrolysis temperature under high heating rates. The gases and chars produced during fast pyrolysis of maize stalk, cotton straw, rice straw, and rice husk at temperatures ranging from 600 to 1000°C were studied using FTIR spectroscopy, X-ray diffraction, nondispersive infrared technique, ultimate analysis, thermal conductivity detection method, helium density measurement, and N_2 adsorption method. They found that gas yield improved by more than 80% from 600 to 1000°C, while the char and liquid yield decreased. The CO and CH_4 content increased with temperature, while the CO_2 content decreased.

The effect of most important operating variables such as final pyrolysis temperature, residence times, inert gas sweeping, mineral matter, rate of biomass heating, moisture contents of biomass, and size of biomass particle on production of pyrolysis oil was studied by Akhtar and Saidina Amin [42]. Three constituents of biomass, cellulose, hemicellulose, and lignin, were pyrolyzed catalytically over SBA-15, Pt/SBA-15, AlSBA-15, and Pt/AlSBA-15 catalysts by Jeon et al. [43]. The catalytic performance was assessed by analyzing the composition of the products. SBA-15 was observed to have few acid sites, whereas AlSBA-15 had few acid sites that are advantageous for dehydration, cracking, decarbonylation, etc. The catalytic performances of AlSBA-15 and Pt/AlSBA-15 were superior to those of SBA-15 and Pt/SBA-15.

Catalytic fast pyrolysis of rice stalk over γ-Al_2O_3, CaO, and MCM-41 (mesoporous and macroporous catalysts) mixed with LOSA-1 (microporous catalysts) was conducted by Zhang et al. [44] in an internally interconnected fluidized bed. The maximum yield of aromatic + olefin (25.3%) was obtained with 10% γ-Al_2O_3/90% LOSA-1, which was increased by 39.8% compared to that obtained with pure LOSA-1. The selectivities of low-carbon components improved significantly with the addition of these catalysts.

Pyrolysis of the polymers constituting biomass for lignin conversion, cellulose, and hemicelluloses can be illustrated as the superposition of three main routes such as char formation, depolymerization and fragmentation, and secondary reactions. Their mechanisms and conversions were studied by Collard and Blin [45]. Some mechanisms were proposed to elucidate the formation of the main products and from the results they concluded that, the reactivity and energy content of the pyrolysis products are predictable.

Pyrolysis studies on conventional biomass such as corn cob, wheat straw, rice straw, and rice husk were carried out in fixed bed reactor at different temperatures by Biswas et al. [46]. The maximum bio-oil yields were 47.3, 36.7, 28.4, and 38.1 wt%, respectively. They observed that all bio-oil contents were composed of oxygenated hydrocarbons. The observed percentages of phenolic compounds in the corn cob bio-oil were higher than other bio-oils.

3.4 Liquefaction

Liquefaction of biomass is the conversion technique in which biomass is converted into liquid fuels using hot and pressurized water. The thermochemical conversion process such as combustion, gasification, and pyrolysis required dry biomass whereas hydrothermal liquefaction is suitable for handling wet biomass. Hydrothermal liquefaction operated at a temperature range from 250 to 350°C and pressure of 10–20 MPa.

In this condition the solid biopolymeric structures are broken down to mainly liquid components. In liquefaction process the biomass is fed into the reactor and the product is biocrude in the form of liquid. Biocrude production mainly depends on the lipid content of the biomass. In hydrothermal liquefaction process the feedstocks such as carbohydrates, protein, and starch can decompose using hot water at high pressure to yield biocrude.

3.4.1 Reactors for liquefaction

Batch and continuous types of reactors are used in hydrothermal liquefaction (HTL) process. Batch reactors are used in the laboratory and research purposes due to their simple construction and handling large variety of feedstocks. Wet-type solid biomass also is used in the HTL process and particle size does not play a vital role. In a continuous HTL reactor system, sample is directly sent to the reactor with desired temperature. The reaction time is maintained by the length of the reactor/volume of the reactor and flow rate of the feed stock to the reactor. The important difference between the liquefaction processes compared with thermochemical conversion process is that water or suitable solvent must be used as the reaction medium in the liquefaction process. Commonly water, ethanol, and menthol are used as a solvents; among these water is used widely because it is cheap, environmental friendly, and easy to separate [47].

3.4.2 Significant studies on liquefaction

Maldas and Shiraishi [48] studied biomass liquefaction process using alkalies and salts as the catalyst in the presence of phenol and H_2O. They reported that dissolution of biomass increased with increase in temperature. Fossil fuel energy are limited to overcome these two substitutions are practiced currently. The first one is 1st generation of biofuels which are derived from plant-based biomass converted to ethanol or biodiesel. The second one is 2nd generation biofuels via gasification of biomass with subsequent refining toward high-quality fuels. A third possible method is hydrothermal liquefaction, i.e., the direct conversion of biomass into liquid fuels without the gasification step. This one-step process is discussed and evaluated by Behrendt et al. [49] and various technical implementations were critically appraised by them.

Xu and Lad [50] studied water-soluble oil and heavy oil, by direct liquefaction without catalyst at temperatures of 280−380°C. They noticed that it produced heavy oils having caloric values of 30−35 MJ kg^{-1}, much higher than that of the crude wood sample.

Sun et al. [51] studied the hydrothermal liquefaction of paulownia with and without catalysts to investigate the catalyst's effects on the liquefaction process. The, Na_2CO_3 and Fe, were found to effectively increase the formation of heavy oil products. The experimental results observed that the maximum yield rate was 36.34% using Fe catalyst, and the minimum solid residue yield was obtained when Na_2CO_3 was used as catalyst.

Tekin and Karagöz [52] summarized and discussed non-catalytic and catalytic hydrothermal liquefaction of biomass. The author reported various experimental operating conditions such as temperature, biomass type, residence time, catalysts, and pressure have a significant effect on the product distributions. Guo et al. [53] provided a review and latest results on the hydrothermal liquefaction of microalgae. This is a promising process that converts different algal strains with high moisture content to high bio-oil yields with lower coke and lower energy consumption in comparison to other methods.

Liquefaction of biomass is widely investigated as a promising method to produce bio-oil or biocrude. Huang and Yuan [54] presented the recent research in the liquefaction of typical biomass from a new perspective and summarized. In the hydrothermal liquefaction operated in batch reactor but there is only limited information available on continuous flow for commercialization. Elliott et al. [55] reviewed hydrothermal liquefaction of biomass in continuous flow systems and developed a model for the process and mass and energy balances were determined with the recent results.

Hydrothermal liquefaction processes of three biomasses having different composition were conducted by de Caprariis et al. [47]. The effects of temperature and biomass composition on the bio-oil yield and quality were investigated. They reported that the bio-oil yield was a function of temperature and maximum bio-oil yield was highly dependent on the chemical composition of the biomass.

Hydrothermal liquefaction of biomass produced from domestic sewage treatment was studied by Couto et al. [56]. The operating parameters such as reaction time, temperature, and biomass/water ratio on the yield of bio-oil were studied and reported. Under the experimental condition solid residue was the largest byproduct due to the high ash content present in the biomass. The disadvantage of their process was the minimum nitrogen recovery in the bio-oil.

Durak [57] has carried out characterization of products obtained from hydrothermal liquefaction of biomass and compared to other thermochemical conversion methods. The author has observed that higher heating value of around 31.32 MJ kg^{-1} for the bio-oil which is higher heating value obtained by the pyrolysis and supercritical liquefaction method.

4. Modeling studies

Generally modeling is used to analyze the dynamic interactions between several components of a biological system, with the aim to understand the behavior of the system as a whole. As biomass technologies and the modeling are emerging fields in recent years, the following section describes various models proposed in the literature for conversion of biomass into value-added products [58].

4.1 Equilibrium models

Thermodynamic equilibrium models are used to predict the composition of product gas from gasifier. Equilibrium models are categorized in to stoichiometric models and nonstoichiometric models. In the biomass gasification, stoichiometric models used are based on equilibrium constants. In this model important chemical reactions are only considered and are used to estimate the end gas composition. This model results gives error in the prediction. This difficulty is overcome by the nonstoichiometric model which involves minimization of the Gibbs free energy. This model is more complex but it has the advantage of not needing chemical reactions [17,59].

Babu and Sheth [60] modeled char reactivity factor along the reduction zone of the biomass gasifier. The model predicted the steady temperature profile and composition of the reduction zone of the gasifier. The author observed that the char reactivity factor is the main parameter in the model and the value increased along the reduction bed length. The temperature and composition profiles in the reduction zone of the model are determined by solving the model equation using finite difference method. The model predictions are verified with the literature experimental data. Sharma [61] proposed an equilibrium model to predict the distribution of various gas species, reaction temperature, and unconverted char for a reduction reaction in downdraft gasifier. The author reported that the equilibrium model accurately predicts equilibrium constants and dry gas composition for reduction reactions.

Azzone et al. [62] prepared an equilibrium model for the thermochemical gasification of three agricultural residues such as corn stalks, sunflower stalks, and rapeseed straw. The model assumed that the transformations of biomass residues inside the gasifier can be described by a series of equilibrium reactions independent of each other. The process parameters such as pressure, operating temperature, etc., were analyzed and verified by literature data.

Chern et al. [63] developed an equilibrium model to predict exit temperature, gas composition, and char yield from the air blown downdraft gasification of wood. The model parameters such as char yield, gas composition, and gas exit temperature were simulated and the results were compared with experimental data.

Loha et al. [64] developed an equilibrium model to predict the gas composition in a fluidized bed gasifier. Steam was used as fluidizing agent for the gasification of rice husk. The developed equilibrium model has been extended to other biomasses to compare the performances. The effects of gasification temperature and steam/biomass were analyzed. The study reported that H_2 and CO production increased with increase in temperature but the CH_4 and CO_2 generation decreased. The author also observed that production of H_2, CO_2, and CH_4 increased with increase in steam/biomass ratio while CO decreased accordingly.

4.2 Kinetic models

A kinetic model is used to predict the exit gas composition, inside temperature distribution of the gasifier, gasifier performance, and gasifier configuration for a given operating condition. Gasification reaction kinetics and hydrodynamics of the gasifier are generally considered in the kinetic modeling. Kinetic modeling is more suitable at low operating temperatures compared to equilibrium model [65].

Kinetic modeling of biomass gasifier for the reduction zone was performed by Hameed et al. [66]. The model parameter such as temperature and initial compositions were simulated under isothermal and non-isothermal conditions for five different biomass materials. The authors reported that the maximum composition of syngas was obtained from peanut hull biomass under the isothermal condition and the final producer gas compositions were found to vary for all the five materials for the non-isothermal condition.

A kinetic model was developed for a circulating fluidized bed biomass gasifier by Miao et al. [67]. Hydrodynamics and biomass reaction kinetics were considered to predict the performance of the gasification process. The fluidized bed was divided into two sections, one at the bottom section of the gasifier called as dense phase region where most of the heterogeneous reactions occur. The top section is called dilute region where most of the homogeneous reactions occur. Each section was further divided into subunits in which mass and energy balance were applied. The model was able to predict the temperature and concentration profile along the length of the gasifier. Further it also predicted the exit gas composition and heating value, overall carbon conversion, and yield of produced gas. The developed model simulation results were compared with literature data and found good agreement with published data.

Sharma [61,68] developed a thermodynamic and kinetic modeling for the char reduction reaction in a downdraft gasifier. The model results were validated with literature data. The influence of bed length and reaction temperature in reduction zone has been studied and reported.

4.3 ASPEN Plus models

ASPEN PLUS is a process simulator used to model chemical processes that involve solid, liquid, and gaseous streams under defined condition by using mass and energy balance.

Modeling of biomass gasification using ASPEN Plus has grasped much attention in recent years.

Nikoo and Mahinpey [69] developed kinetics and hydrodynamics model for a bubbling fluidized bed gasifier using ASPEN Plus simulator for the pine saw dust biomass. The simulation results were compared with experimental results. Author observed that, hydrogen and carbon conversion increased with increase in temperature. Increase in steam to biomass ratio increased hydrogen and carbon monoxide production and decreased carbon dioxide and carbon conversion efficiency.

A process model was developed using ASPEN Plus for pressurized fluidized-bed gasifier by Hannula and Kurkela [70]. There were eight main blocks used to model the fluidized bed gasifier. They observed that the model appears to be appropriate for simulating gasification of pine sawdust, pine, and eucalyptus wood chips as well as forest residues, but were not suitable for pine bark or wheat straw.

Ramzan et al. [71] developed steady state ASPEN Plus model for gasification. A three-stage gasifier was modeled. The first stage is reduction of biomass moisture content and in the second stage biomass decomposition into elements. The third stage is biomass gasification reactions. It was then compared using experimental data and good agreement was observed.

Biomass gasification in a bubbling fluidized bed was modeled using ASPEN Plus simulator by Beheshti et al. [72] for hydrogen and syngas production. Aspen Plus simulates the process parameters and observed that the high temperature was more favorable for production of useful syngas and hydrogen yield. It was also noticed that equivalence ratio is the important factor and contributed to higher carbon conversion, tar reforming, and gas yield. The model was validated by experimental data and found to be in good agreement.

Steady state model was developed for biomass fired fluidized bed gasifier by Kaushal and Tyagi [73] using ASPEN Plus. Various ASPEN Plus unit blocks were combined to model the process within the modeling environment. The model parameters were studied to predict the gasifier performance.

4.4 CFD models

Computational Fluid Dynamics (CFD) model is used to study temperature profile of solid and gas phase of biomass combustion and gasification process. CFD models with the help of simulation were used to predict critical parameters required to control the thermochemical conversion of biomass process. These CFD techniques are likely to substitute empirical or semi-empirical models in large-scale biomass reactor design process in near term [74].

Jakobs et al. [75] developed a CFD model for a gasifier and the model equations were solved using finite volume method. Burner-nozzle is an important part in the gasifier for fuel conversion and syngas quality. In this work, atomizer quality for twin fluid nozzles as a function of gas velocity and reactor pressure was analyzed. The developed atomizers were used in the atmospheric entrained flow gasifier. Sauter Mean Diameter of the produced spray was found to be significantly influenced by gas velocity and reactor pressure. Increasing reactor pressure was found to increase the drop diameter and increasing gas velocity decreased the Sauter Mean Diameter. An influence of Sauter Mean Diameter on gasification process was observed from organic carbon, methane concentration, and temperature profiles at various positions.

A steady state three-dimensional CFD model to simulate biomass gasification was established by Liu et al. [76] in a circulating fluidized bed reactor. Hydrodynamics of the biomass gasifier is modeled using k-ε turbulence model coupled with kinetic theory of granular flow. The equation of continuity, motion, and energy were integrated with the kinetics of homogeneous and heterogeneous reactions to calculate mass and energy transfer in the gasifier. Velocity distributions, temperature, and concentration profile were studied using the CFD model and the simulation results were in good agreement with experimental data.

Nguyen et al. [77] studied a two-dimensional CFD model to predict the hydrodynamics of a dual fluidized bed gasifier in cold-mode. The circulation rate and hold up of solid holdup from CFD simulation were compared with a pilot-scale dual fluidized bed in the cold mode.

Askaripour [78] developed a two-dimensional CFD model for the coal gasification in a tapered fluidized bed gasifier to determine the effects of various parameters such as tapered angle, velocity, temperature, feed ratio, efficiency, product gas compositions. They found increasing tapered angle resulted in a decrease of the lower heating value, and higher heating value of the gas products, whereas the carbon conversion efficiency of gasification process increased. The cold gas efficiency of the gasifier increases as the tapered angle increases to 5 from 3 degrees, but further increase of the tapered angle from 5 to 11 degrees, it slightly varied. The results indicated that as the velocity of gasifying agent increases, lower and higher heating values of the product gas drop while carbon conversion efficiency of the gasifier enhances.

Kumar and Paul [79] performed a two-dimensional CFD model for a downdraft gasifier for rubber wood biomass. The model included four zones of the gasifier such as drying, pyrolysis, oxidation, and reduction. In CFD model, chemical kinetics was used and the results showed slight variation in the composition of synthesis gases. Hence the CFD model was revised and included frequency factor, activation energy, and water shift gas reaction. The results were compared with exit gas composition and were closer to the kinetic models' results. The validated CFD model was further used to investigate the effect of equivalence ratio. Different biomass feedstocks were tested for the gasification process and it was concluded that when compared to the other feedstocks, the rubber wood feedstock produced more synthesis gas. Their modeling approach showed a capable way to simulate the biomass gasification processes in downdraft gasifier.

4.5 ANN models

Artificial Neural Network (ANN) model is used to predict the gasification outputs in which a neural network is trained using different sets of experimental data simulating the human brain in terms of mathematical functions. Different layers in the ANN are called hidden layers and each hidden layer containing neurons is interconnected with other layers and exchange information among them. ANN is formed by an input layer, hidden layers, and an output layer. As a contemporary approach, ANNs are useful to obtain the solution of an extensive problem in science and engineering particularly. These characteristics make ANNs very attractive and useful, inspiring their use in the modeling of gasification processes. Therefore, biomass gasification can be conveniently simulated using the properly designed ANN [80].

Serrano et al. [80] developed an ANN model to calculate the exit gas composition and gas yield from bubbling fluidized bed biomass gasifier for different bed materials. Feed-forward and cascade-forward backpropagation network structures are used with one and two hidden layers for the modeling. The author observed that the feed-forward backpropagation network with two hidden layers predict product gas composition with a higher correlation coefficient with the experimental value.

ANN-based model was developed by Baruah et al. [81] to predict the exit gas composition in downdraft biomass gasifier of a fixed bed. The input parameters for the ANN model are C, H, O, ash, moisture content, and temperature. Exit gas composition are predicted for one hidden layer with five neurons for CH_4 and CO prediction, four neurons for CO_2, and three neurons for H_2 using a backpropagation algorithm. The author observed that the predicted values match satisfactorily with the experimental result.

Puig-Arnavat et al. [82] used an ANN model to study the gas composition and gas yield from the circulating and bubbling fluidized bed gasifier. Backpropagation neural network algorithm was used to calculate the exit gas composition. The results predicted using ANN model show high agreement with published experimental data.

George et al. [83] proposed an ANN model to predict the exit composition of gas in a bubbling fluidized bed gasifier. C, H, O content, moisture, ash, gasification temperature, and equivalence ratio are the input variables. The author observed single hidden layer having 15 neurons prediction showed good agreement compared with experimental results.

Li et al. [84] used ANN model to estimate the hydrogen gas composition from air-steam gasification of fluidized bed gasifier. Nine input variables such as steam/biomass ratio, ash content, moisture content, the amount of elements C, H, and O in the biomass, reaction temperature, biomass particle size, and equivalence ratio are used to estimate the product gas composition. One intermediate layer with two nodes predicted well the biomass composition.

5. Conclusion

Effective use of fossil fuels is the demand of the universe due to several reasons and most importantly due to fast depletion of fossil fuels. Hence more and more attention has been focused on alternate energy sources such as biomass energy. Biomass energy is abundant and available plenty in the world. Convert the biomass into useful form of energy is the growing technology in that aspects thermochemical conversion of biomass in to useful product is the growing and emerging field in the world. Hence, this chapter is focused on presenting the various thermochemical methods that are practiced globally for the conversion of biomass into valuable products.

Modeling is an important tool to study the mechanism of biomass conversions. The understanding of the behavior of reactors, design, operation and their optimization of parameters are time-consuming process compared to physical experimentation. For example, commissioning a biomass reactor at a location if the suggested feedstock is not available then the reactor has to be run on the different available feedstock and the best among them has to be found. Altering the feedstock in the biomass reactor becomes time consuming and expensive. Suppose a mathematical model of this system is available, then it is easy to identify the feedstock which gives the optimum output. In view to this, various models such as equilibrium, kinetic, ASPEN Plus, CFD, and ANN models and their widely differing complexities have been summarized and discussed in this chapter.

References

[1] A. Tursi, A review on biomass: importance, chemistry, classification, and conversion, Biofuel Res. J. 6 (2019) 962–979.
[2] J. Cai, Y. He, X. Yu, S.W. Banks, Y. Yang, X. Zhang, Y. Yu, R. Liu, A.V. Bridgwater, Review of physicochemical properties and analytical characterization of lignocellulosic biomass, Renew. Sustain. Energy Rev. 76 (2017) 309–322.
[3] P. Shah Maulin, Microbial Bioremediation & Biodegradation, Springer, 2020.
[4] P. Shah Maulin, Removal of Refractory Pollutants from Wastewater Treatment Plants, CRC Press, 2021.
[5] C. Yin, L.A. Rosendahl, S.K. Kær, Grate-firing of biomass for heat and power production, Prog. Energy Combust. Sci. 34 (2008) 725–754.
[6] A. Boriouchkine, V. Sharifi, J. Swithenbank, S.L. Jämsä-Jounela, A study on the dynamic combustion behavior of a biomass fuel bed, Fuel 135 (2014) 468–481.
[7] H. Thunman, B. Leckner, Co-current and counter-current fixed bed combustion of biofuel - a comparison, Fuel 82 (2003) 275–283.
[8] M. Jenkins, L. B., L. Bexter, R. T. Miles Jr., T. Miles R, Combustion properties of biomass flash, Fuel Process. Technol. 54 (1998) 17–46.
[9] R. Saidur, E.A. Abdelaziz, A. Demirbas, M.S. Hossain, S. Mekhilef, A review on biomass as a fuel for boilers, Renew. Sustain. Energy Rev. 15 (2011) 2262–2289.

[10] L.J.R. Nunes, J.C.O. Matias, J.P.S. Catalão, Biomass combustion systems: a review on the physical and chemical properties of the ashes, Renew. Sustain. Energy Rev. 53 (2016) 235−242.

[11] L. Wang, J.E. Hustad, Ø. Skreiberg, G. Skjevrak, M. Grønli, A critical review on additives to reduce ash related operation problems in biomass combustion applications, Energy Proc. 20 (2012) 20−29.

[12] Y. Niu, H. Tan, S. Hui, Ash-related issues during biomass combustion: alkali-induced slagging, silicate melt-induced slagging (ash fusion), agglomeration, corrosion, ash utilization, and related countermeasures, Prog. Energy Combust. Sci. 52 (2016) 1−61.

[13] M. Mladenović, M. Paprika, A. Marinković, Denitrification techniques for biomass combustion, Renew. Sustain. Energy Rev. 82 (2018) 3350−3364.

[14] D. Shen, R. Xiao, S. Gu, K. Luo, The pyrolytic behavior of cellulose in lignocellulosic biomass: a review, RSC Adv. 1 (2011) 1641−1660.

[15] A. Kumar, D.D. Jones, M.A. Hanna, Thermochemical biomass gasification: a review of the current status of the technology, Energies 2 (2009) 556−581.

[16] S.K. Sansaniwal, K. Pal, M.A. Rosen, S.K. Tyagi, Recent advances in the development of biomass gasification technology: a comprehensive review, Renew. Sustain. Energy Rev. 72 (2017) 363−384.

[17] T.K. Patra, P.N. Sheth, Biomass gasification models for downdraft gasifier: a state-of-the-art review, Renew. Sustain. Energy Rev. 50 (2015a) 583−593.

[18] P.N. Sheth, B.V. Babu, Experimental studies on producer gas generation from wood waste in a downdraft biomass gasifier, Bioresour. Technol. 100 (2009) 3127−3133.

[19] S. Turn, C. Kinoshita, Z. Zhang, D. Ishimura, J. Zhou, An experimental investigation of hydrogen production from biomass gasification, Int. J. Hydrog. Energy 23 (1998) 641−648.

[20] A. Kruse, Hydrothermal biomass gasification, J. Supercrit. Fluids 47 (2009) 391−399.

[21] S. Nanda, S.N. Reddy, A.K. Dalai, J.A. Kozinski, Subcritical and supercritical water gasification of lignocellulosic biomass impregnated with nickel nanocatalyst for hydrogen production, Int. J. Hydrog. Energy 41 (2016) 4907−4921.

[22] A.C.C. Chang, H.F. Chang, F.J. Lin, K.H. Lin, C.H. Chen, Biomass gasification for hydrogen production, Int. J. Hydrog. Energy 36 (2011) 14252−14260.

[23] E. Kirtay, Recent advances in production of hydrogen from biomass, Energy Convers. Manag. 52 (2011) 1778−1789.

[24] N.L. Panwar, R. Kothari, V.V. Tyagi, Thermo chemical conversion of biomass - eco friendly energy routes, Renew. Sustain. Energy Rev. 16 (2012) 1801−1816.

[25] J.A. Ruiz, M.C. Juárez, M.P. Morales, P. Muñoz, M.A. Mendívil, Biomass gasification for electricity generation: review of current technology barriers, Renew. Sustain. Energy Rev. 18 (2013) 174−183.

[26] M. Asadullah, Barriers of commercial power generation using biomass gasification gas: a review, Renew. Sustain. Energy Rev. 29 (2014) 201−215.

[27] S.N. Reddy, S. Nanda, A.K. Dalai, J.A. Kozinski, Supercritical water gasification of biomass for hydrogen production, Int. J. Hydrog. Energy 39 (2014) 6912−6926.

[28] V.S. Sikarwar, M. Zhao, P. Clough, J. Yao, X. Zhong, M.Z. Memon, N. Shah, E.J. Anthony, P.S. Fennell, An overview of advances in biomass gasification, Energy Environ. Sci. 9 (2016) 2939−2977.

[29] A.A. Ahmad, N.A. Zawawi, F.H. Kasim, A. Inayat, A. Khasri, Assessing the gasification performance of biomass: a review on biomass gasification process conditions, optimization and economic evaluation, Renew. Sustain. Energy Rev. 53 (2016) 1333−1347.

[30] A. Molino, V. Larocca, S. Chianese, D. Musmarra, Biofuels production by biomass gasification: a review, Energies 11 (2018) 1−31.

[31] E. Shayan, V. Zare, I. Mirzaee, Hydrogen production from biomass gasification; a theoretical comparison of using different gasification agents, Energy Convers. Manag. 159 (2018) 30−41.

[32] F. Campuzano, R.C. Brown, J.D. Martínez, Auger reactors for pyrolysis of biomass and wastes, Renew. Sustain. Energy Rev. 102 (2019) 372−409.

[33] S.S. Lam, H.A. Chase, A review on waste to energy processes using microwave pyrolysis, Energies 5 (2012) 4209−4232.

[34] W.M. Lewandowski, K. Januszewicz, W. Kosakowski, Efficiency and proportions of waste tyre pyrolysis products depending on the reactor type—a review, J. Anal. Appl. Pyrol. 140 (2019) 25−53.

[35] J. Ma, B. Su, G. Wen, Q. Yang, Q. Ren, Y. Yang, H. Xing, Pyrolysis of pulverized coal to acetylene in magnetically rotating hydrogen plasma reactor, Fuel Process. Technol. 167 (2017) 721−729.

[36] S. Sobek, S. Werle, Solar pyrolysis of waste biomass: Part 1 reactor design, Renew. Energy 143 (2019) 1939−1948.

[37] S.R.A. Kersten, X. Wang, W. Prins, W.P.M. Van Swaaij, Biomass pyrolysis in a fluidized bed reactor. Part 1: literature review and model simulations, Ind. Eng. Chem. Res. 44 (2005) 8773−8785.

[38] D. Mohan, C.U. Pittman, P.H. Steele, Pyrolysis of wood/biomass for bio-oil: a critical review, Energy Fuel. 20 (2006) 848−889.

[39] A.V. Bridgwater, Production of high grade fuels and chemicals from catalytic pyrolysis of biomass, Catal. Today 29 (1996) 285−295.

[40] E. Pütün, F. Ateş, A.E. Pütün, Catalytic pyrolysis of biomass in inert and steam atmospheres, Fuel 87 (2008) 815−824.

[41] P. Fu, W. Yi, X. Bai, Z. Li, S. Hu, J. Xiang, Effect of temperature on gas composition and char structural features of pyrolyzed agricultural residues, Bioresour. Technol. 102 (2011) 8211−8219.

[42] J. Akhtar, N. Saidina Amin, A review on operating parameters for optimum liquid oil yield in biomass pyrolysis, Renew. Sustain. Energy Rev. 16 (2012) 5101−5109.

[43] M.J. Jeon, J.K. Jeon, D.J. Suh, S.H. Park, Y.J. Sa, S.H. Joo, Y.K. Park, Catalytic pyrolysis of biomass components over mesoporous catalysts using Py-GC/MS, Catal. Today 204 (2013) 170−178.

[44] H. Zhang, R. Xiao, B. Jin, G. Xiao, R. Chen, Biomass catalytic pyrolysis to produce olefins and aromatics with a physically mixed catalyst, Bioresour. Technol. 140 (2013) 256−262.

[45] F.X. Collard, J. Blin, A review on pyrolysis of biomass constituents: mechanisms and composition of the products obtained from the conversion of cellulose, hemicelluloses and lignin, Renew. Sustain. Energy Rev. 38 (2014) 594–608.

[46] B. Biswas, N. Pandey, Y. Bisht, R. Singh, J. Kumar, T. Bhaskar, Pyrolysis of agricultural biomass residues: comparative study of corn cob, wheat straw, rice straw and rice husk, Bioresour. Technol. 237 (2017) 57–63.

[47] B. de Caprariis, P. De Filippis, A. Petrullo, M. Scarsella, Hydrothermal liquefaction of biomass: influence of temperature and biomass composition on the bio-oil production, Fuel 208 (2017) 618–625.

[48] D. Maldas, N. Shiraishi, Liquefaction of biomass in the presence of phenol and H_2O using alkalies and salts as the catalyst, Biomass Bioenergy 12 (1997) 273–279.

[49] F. Behrendt, Y. Neubauer, M. Oevermann, B. Wilmes, N. Zobel, Direct liquefaction of biomass, Chem. Eng. Technol. 31 (2008) 667–677.

[50] C. Xu, N. Lad, Production of heavy oils with high caloric values by direct liquefaction of woody biomass in sub/near-critical water, Energy Fuel. 22 (2008) 635–642.

[51] P. Sun, M. Heng, S. Sun, J. Chen, Direct liquefaction of paulownia in hot compressed water: influence of catalysts, Energy 35 (2010) 5421–5429.

[52] K. Tekin, S. Karagöz, Non-catalytic and catalytic hydrothermal liquefaction of biomass, Res. Chem. Intermed. 39 (2013) 485–498.

[53] Y. Guo, T. Yeh, W. Song, D. Xu, S. Wang, A review of bio-oil production from hydrothermal liquefaction of algae, Renew. Sustain. Energy Rev. 48 (2015) 776–790.

[54] H.J. Huang, X.Z. Yuan, Recent progress in the direct liquefaction of typical biomass, Prog. Energy Combust. Sci. 49 (2015) 59–80.

[55] D.C. Elliott, P. Biller, A.B. Ross, A.J. Schmidt, S.B. Jones, Hydrothermal liquefaction of biomass: developments from batch to continuous process, Bioresour. Technol. 178 (2015) 147–156.

[56] E.A. Couto, F. Pinto, F. Varela, A. Reis, P. Costa, M.L. Calijuri, Hydrothermal liquefaction of biomass produced from domestic sewage treatment in high-rate ponds, Renew. Energy 118 (2018) 644–653.

[57] H. Durak, Characterization of products obtained from hydrothermal liquefaction of biomass (*Anchusa azurea*) compared to other thermochemical conversion methods, Biomass Convers. Biorefin. 9 (2019) 459–470.

[58] F. Mafakheri, F. Nasiri, Modeling of biomass-to-energy supply chain operations: applications, challenges and research directions, Energy Pol. 67 (2014) 116–126.

[59] G. Schuster, G. Löffler, K. Weigl, H. Hofbauer, Biomass steam gasification—an extensive parametric modeling study, Bioresour. Technol. 77 (2001) 71–79.

[60] B.V. Babu, P.N. Sheth, Modeling and simulation of reduction zone of downdraft biomass gasifier: effect of char reactivity factor, Energy Convers. Manag. 47 (2006) 2602–2611.

[61] A.K. Sharma, Equilibrium modeling of global reduction reactions for a downdraft (biomass) gasifier, Energy Convers. Manag. 49 (2008a) 832–842.

[62] E. Azzone, M. Morini, M. Pinelli, Development of an equilibrium model for the simulation of thermochemical gasification and application to agricultural residues, Renew. Energy 46 (2012) 248–254.

[63] S.-M. Chern, W.P. Walawender, L.T. Fan, Equilibrium modeling of a downdraft gasifier I—overall gasifier, Chem. Eng. Commun. 108 (1991) 243–265.

[64] C. Loha, P.K. Chatterjee, H. Chattopadhyay, Performance of fluidized bed steam gasification of biomass—modeling and experiment, Energy Convers. Manag. 52 (2011) 1583–1588.

[65] D. Baruah, D.C. Baruah, Modeling of biomass gasification: a review, Renew. Sustain. Energy Rev. 39 (2014) 806–815.

[66] S. Hameed, N. Ramzan, Z.U. Rahman, M. Zafar, S. Riaz, Kinetic modeling of reduction zone in biomass gasification, Energy Convers. Manag. 78 (2014) 367–373.

[67] Q. Miao, J. Zhu, S. Barghi, C. Wu, X. Yin, Z. Zhou, Modeling biomass gasification in circulating fluidized beds, Renew. Energy 50 (2013) 655–661.

[68] A.K. Sharma, Equilibrium and kinetic modeling of char reduction reactions in a downdraft biomass gasifier: a comparison, Sol. Energy 82 (2008b) 918–928.

[69] M.B. Nikoo, N. Mahinpey, Simulation of biomass gasification in fluidized bed reactor using ASPEN PLUS, Biomass Bioenergy 32 (2008) 1245–1254.

[70] I. Hannula, E. Kurkela, A semi-empirical model for pressurised air-blown fluidised-bed gasification of biomass, Bioresour. Technol. 101 (2010) 4608–4615.

[71] N. Ramzan, A. Ashraf, S. Naveed, A. Malik, Simulation of hybrid biomass gasification using Aspen plus: a comparative performance analysis for food, municipal solid and poultry waste, Biomass Bioenergy 35 (2011) 3962–3969.

[72] S.M. Beheshti, H. Ghassemi, R. Shahsavan-Markadeh, Process simulation of biomass gasification in a bubbling fluidized bed reactor, Energy Convers. Manag. 94 (2015) 345–352.

[73] P. Kaushal, R. Tyagi, Advanced simulation of biomass gasification in a fluidized bed reactor using ASPEN PLUS, Renew. Energy 101 (2017) 629–636.

[74] R.I. Singh, A. Brink, M. Hupa, CFD modeling to study fluidized bed combustion and gasification, Appl. Therm. Eng. 52 (2013) 585–614.

[75] T. Jakobs, N. Djordjevic, S. Fleck, M. Mancini, R. Weber, T. Kolb, Gasification of high viscous slurry R&D on atomization and numerical simulation, Appl. Energy 93 (2012) 449–456.

[76] H. Liu, A. Elkamel, A. Lohi, M. Biglari, Computational fluid dynamics modeling of biomass gasification in circulating fluidized-bed reactor using the Eulerian-Eulerian approach, Ind. Eng. Chem. Res. 52 (2013) 18162–18174.

[77] T.D.B. Nguyen, M.W. Seo, Y. Il Lim, B.H. Song, S.D. Kim, CFD simulation with experiments in a dual circulating fluidized bed gasifier, Comput. Chem. Eng. 36 (2012) 48–56.

[78] H. Askaripour, CFD modeling of gasification process in tapered fluidized bed gasifier, Energy 191 (2020).
[79] U. Kumar, M.C. Paul, CFD modelling of biomass gasification with a volatile break-up approach, Chem. Eng. Sci. 195 (2019) 413–422.
[80] D. Serrano, I. Golpour, S. Sánchez-Delgado, Predicting the effect of bed materials in bubbling fluidized bed gasification using artificial neural networks (ANNs) modeling approach, Fuel 266 (2020).
[81] D. Baruah, D.C. Baruah, M.K. Hazarika, Artificial neural network based modeling of biomass gasification in fixed bed downdraft gasifiers, Biomass Bioenergy 98 (2017) 264–271.
[82] M. Puig-Arnavat, J.A. Hernández, J.C. Bruno, A. Coronas, Artificial neural network models for biomass gasification in fluidized bed gasifiers, Biomass Bioenergy 49 (2013) 279–289.
[83] J. George, P. Arun, C. Muraleedharan, Assessment of producer gas composition in air gasification of biomass using artificial neural network model, Int. J. Hydrog. Energy 43 (2018) 9558–9568.
[84] Y. Li, B. Yang, L. Yan, W. Gao, Neural network modeling of biomass gasification for hydrogen production, Energy Source, Part A Recover., Util. Environ. Eff. 41 (2019) 1336–1343.

Chapter 12

Rice straw for biofuel production

Pranjal P. Das, Ankush D. Sontakke and Mihir Kumar Purkait
Department of Chemical Engineering, Indian Institute of Technology Guwahati, Guwahati, Assam, India

1. Introduction

With the beginning of several challenges such as increased fuel consumption, global warming, fossil fuel exhaustion, and rising fuel prices, one must look for an alternative for effectively producing energy. Presently, biomass generates over 45 ± 10 EJ percent of the world's energy needs and is one of the renewable sources for biofuel production. Biomass contributes less in industrialized countries (<10%) than it does in developing ones, where it provides around 20%–30% of the total energy supply. Energy needs are expected to rise by 1.3% per year until 2040, making it urgent for all countries to strengthen their energy policy in order to ensure global energy security. Biofuels are one such viable contender for reducing the greenhouse gas (GHG) emissions, controlling biodiversity loss, improving food production, and bolstering the notion of sustainable development, which involves effective management of water and several other resources [1]. When it comes to the production of biofuels from biomass, various important considerations must be made, including environmental sensitivity and land-use exploitation. Biomass is a renewable bioenergy source, and its utilization for bioenergy production has addressed a variety of social issues and concerns. Compared to the conventional fossil fuels, the biomass possesses lower life cycle greenhouse gas (GHG) emissions with negligible conflict over food production such as (1) crop residues, (2) mixed cropping patterns, (3) perennial plant growth on degraded lands abandoned for agricultural use, (4) sustainable process of harvesting, and (5) municipal and industrial wastes [2]. Fossil fuel sources are depleting as a result of globalization, thereby posing a threat to the environment. As a result, it is imperative that more emphasis be placed on the sustainable use of renewable energy sources, especially biomass. Biofuels such as biodiesel, biohydrogen, bioethanol, biochar, biogas, and biosyngas can be effectively produced from biomass by fermenting sugar beet, corn, wheat, and agricultural residues. With the growing global population, the use of biofuels may undergo a rapid increase in the coming years, accounting for roughly 50% of total energy consumption in emerging countries by the year 2050. Furthermore, biofuels aid in the reduction of GHG emissions, resulting in a healthier environment. As a result, developing a renewable energy source is crucial to avert the current dire situations [3]. Rice (*Oryza sativa* L.) production has long been one of the most important agricultural activities in the world. It is the third most produced agricultural crop across the globe, behind sugarcane and corn. China and India together account for around 53% of global rice production. Overall, rice production has grown significantly in recent years, increasing from 660 million tons in 2007 to 746 million tons in 2014 and 760 million tons in 2017. Rice residues (rice straw, rice husk) that are left after harvesting, as well as the bulk amount, have considerably increased due to an increase in rice production. Such residues have been underutilized since the commencement of rice cultivation, thus, posing a threat to the environment and the economy. Cultivators often dispose of such residues using conventional procedures such as open-pit waste burning, which releases poisonous gases into the environment, causing air pollution and jeopardizing the public health owing to their carcinogenic nature [4]. A mixture of starch, proteins, cellulose, hemicellulose, lignin, inorganics, and extractives make up these rice residues. With such features and the ongoing energy and environmental challenges, researchers worldwide have shown significant interest in using lignocellulosic biomass to produce sustainable bioenergy. Rice straw possess enticing characteristics for the generation of bioenergy due to its abundant availability across the world. It is one of the most important crops, which produces a large amount of silica-enriched biomass. There is a generation of roughly 1.5 tons of rice straw for every ton of rice, till 2017. Almost half of rice straw is burned every year (an in-field method for utilizing rice straw), while the rest is utilized as fodder, or left to decompose in landfills. Burning emits hazardous gases such as volatile organic compounds (VOCs), carbon monoxide (CO), nitrous oxides (NOx), polycyclic aromatic hydrocarbons (PAHs), and suspended particulate matter which contributes significantly to global GHG emissions, thereby adversely affecting the environment and public health [5]. As a result, there is an urgent need to design or develop economic conversion techniques for effectively generating bioenergy from rice straw. The refractory nature of rice straw is one of the major challenges throughout its conversion into biofuel, which contributes to the sustainable generation of bioenergy. Due to its chemical composition, which includes cellulose (highly polymerized), hemicellulose (low molecular weight), and lignin (highly resistant to biodegradable approach); rice straw necessitates the use of a pretreatment process. However, rice contains a large amount of silica, which plays a crucial role in providing (1) polymeric characteristics to the cell walls, (2) mechanical strength, and (3) network remodelling of the cell wall, despite the fact that there is a scarcity of data regarding its effect on saccharification and fermentation processes. Thus, rice straw exploitation is comparably difficult, and improving the economics of such conversion techniques is also a critical concern [6].

This chapter provides the readers with an overview of several forms of agricultural residues, field management, composition, along with their negative effects on human health and environment. The lifecycle evaluation, and its corresponding SWOT analysis have been discussed elaborately for obtaining a better understanding of rice straw utilization for biofuel production. Different techniques for increasing the biofuel yield viz., anaerobic codigestion and pretreatment processes (physical, chemical, and biological) have also been covered in detail. The specific aim is to provide in-depth analysis on current research studies, biofuel globalization, possible biofuel feedstock, biofuel production methods, and applications of rice straw for generating biofuels, while adhering to the notion of a sustainable environment.

2. Crop residues

Currently, agricultural residues such as wheat straw, rice husk, and rice straw are considered to be promising biomass feedstocks for producing bioenergy and ensuring energy security without competing with the production of food for human consumption. Agricultural residues are the waste generated after processing and harvesting vegetative crops. There are two types of agricultural crop residues: (1) field residues (unusable materials of the crop after harvesting) such as seedpods, stalks, rice straw, leaves, and stems that are cultivated directly into the ground or after they have been burned. Proper management of such crop residues is required to improve irrigation efficiency and reduce soil erosion. (2) Process residues (useable materials after processing of crops) such as roots, bagasse, molasses, husks, and seeds. Such residues are utilized as fertilizers and animal feed to improve the soil texture. Cellulosic microfibrils embedded in a matrix of hemicellulose and lignin constitutes the cell wall of the plants [7]. The plant cells are made up of two crucial components: (1) cell wall consisting of lignin, cellulose and hemicellulose, silica and cutin; (2) inner chamber comprising of proteins, fats, organic acids, soluble ash, and carbohydrates [8]. The bulk of agricultural crop residues contain a chemical composition of lignin, polysaccharides, and proteins, accounting for 60%–80% of the dry mass. Rice husk, rice straw, maize straw, wheat straw, and sugarcane bagasse are the most prevalent crop residues for generating biofuels. Asia is the leading producer of rice and wheat straw, whereas America mostly produces maize straw and sugarcane bagasse. Different types of crop residues have different compositions in terms of dry matter weight percentage. Lignocellulosic materials (cellulose, hemicelluloses, and lignin) are not only the primary source of biofuel production but also the critical constituents of crop residues. Moreover, crop residues have varied heating values, ash content, and elemental composition. There are a variety of crop residues that are regarded as a viable candidate for use in numerous sectors, including farming, engineering, industry, and agriculture. On an average, the quantity of cellulosic fibers produced worldwide is around 1.3 gigatonne [9]. However, they have certain drawbacks, such as higher transportation costs compared to wood and the expenses related to the field collection of residues. There are several ways to benefit from agricultural crop residues including microorganism assisted soil nutrition, livestock, animal fodder, and waste burning (reduces pests and provides higher soil yield). Crop residues protect the ground from drying, thereby preserving the moisture content and increasing the shelf life of the soil. Other sources of crop residues, such as millet stalks (after being burnt), can be effectively used for culinary purposes in places where wood fuel is scarce. Furthermore, several parameters need to be considered during the generation of bioenergy (biodiesel, biogas, bioethanol, etc.) from agricultural crop residues/biomass/forest crop residues viz., (1) modifications in the land use, (2) wide range of yield gradients, (3) development of novel techniques, and (4) potential of food crops, lignocellulosic biomass, or forest residues to generate maximum biofuel yield [10].

3. Concept of rice straw

3.1 Composition of rice straw

Rice is one of the major agricultural crops, widely cultivated in Europe, Asia, Africa, and the USA, with a yearly production of around 730–1111 million tonnes. Rice straw is a byproduct obtained from rice cultivation and is the vegetative component of the rice plant. In the current scenario, rice straw has been extensively used for bioenergy generation viz., biohydrogen, biodiesel, biobutanol, and bioethanol for diverse applications. Lignin, cellulose and hemicellulose constitutes the lignocellulosic biomass. The composition of rice straw is shown in Table 12.1. Cellulose is a crucial component which comprises of monomeric D-glucose subunits connected via glycosidic linkages. Depending upon the types of structure, cellulose is further classified into amorphous (less organized) or crystalline (more ordered) form. The amorphous form is generally more suitable for the fermentation process [11]. On the other hand, hemicellulose is the carbohydrate component consisting of a complex structure. It is a mixture of various smaller polymeric components including uronic acids, hexose, and pentose. Hemicellulose connects the molecules of cellulose and lignin together, thereby providing tensile strength to the overall structure. Furthermore, lignin present in the rice straw plays an important role in protecting the mosaic structure of monolignols viz. sinapyl, coniferyl, and p-coumaryl alcohols [12].

3.2 Field management of rice straw

3.2.1 In-field rice straw management

Due to GHG emissions, air pollution, and deterioration in soil quality, the process of burning rice straw in an open field has a significant detrimental effect on the agriculture and environment. However, burning of rice straw produces around 98% less GHG emissions than incorporating the rice straw into flooded soil due to a reduction in methane emissions from the breakdown of the straw. This data does not take into consideration the emission of CO_2, but when included, it shows a decrease in the soil-carbon (SOC) sequestration of freshly incorporated rice straw. The reason may be attributed to an immediate reduction in the carbon content (90%) of rice straw during the CO_2 combustion process. The SOC sequestration plays a crucial role during the calculation of CO_2 emission [13]. For example, in China, a meta-assessment was conducted, and further, the effect of rice straw burning and its incorporation into the soil, consisting of sequestration, were compared. Switching from burning to incorporating rice straw into the soil has been shown to reduce the overall rice emissions in China by 34.21 Megaton of CO_2 equivalents per year (around 31%). This exhibits the high-end sequestration efficiency when the degraded soil is restored to its maximum SOC saturation capacity. This saturation allows for the reduction in CO_2 emissions, improvement in soil quality index, higher yields, and better drought resistance capacity. Despite the fact that burning of rice straw has an adverse effect on soil quality, SOC sequestration, and air quality, the farmers still continue to employ this practice for a variety of reasons such as easier tillage, less residual weed, and reduced costs. Keeping all of these in mind, development of sustainable alternatives and stringent government regulations should be urgently imposed [4,14].

TABLE 12.1 Composition of rice straw.

S. No.	Components found in rice straw	Quantity (wt%)
1.	Cellulose	43
2.	Hemicellulose	25
3.	Lignin	12
4.	Ash and silica	34–44
5.	Glucose	19–22
6.	Arabinose	2–4
7.	Mannose	1.8–2
8.	Galactose	0.4
9.	Nonstructural components	16–20

Reproduced with permission from A. Sharma, G. Singh, S.K. Arya, Biofuel from rice straw, J. Clean. Prod. 277 (2020) 124101. https://doi.org/10.1016/j.jclepro.2020.124101 @ Elsevier.

3.2.2 Off-field rice straw management

The methane emission from in-field addition of rice straw, along with nitrous oxide (77%) and methane (23%) emissions from manure management, can be effectively reduced by anaerobic composting of rice straw. Manure management accounts for almost 11% of global agricultural emissions. Thus, it contributes equally and significantly to GHG emissions. Aerobic composting leads to over 90% reduction in methane emission as compared to anaerobic composting. The emission of nitrous oxide occurs indirectly from manure, primarily due to ammonia volatilization, which gets converted to nitrous oxide followed by its release into the environment [15]. The use of SOC sequestration may increase due to its favourable conditions for reducing such emissions from composting. Similar to composting, the use of biochar is another approach for reducing the emission of methane, emerging from the addition of rice as an off-field alternative. Biochar has significant potential to reduce methane emissions than composting due to its ability to improve the SOC sequestration by 50% via converting rice straw into a stable and improved form of carbon. When burning of biochar and life-cycle assessment (LCA) of open-field were conducted, it was found that there have been around 50%–70% reduction in the overall carbon footprint along with the lowering of methane emissions during the rice cultivation process. Furthermore, it was observed that biochar reduces the GHG emissions by 40% for upland soils and 18% for paddy soils during the comparison of 29 meta-assessment studies [16]. Theoretically, it is the best alternative for reducing methane emissions and other adverse environmental effects during rice straw management. However, further data demonstrating the efficiency of biochar in reducing toxic emissions throughout the bulk volume of in-field management options are still needed.

4. Life cycle assessment (LCA) and SWOT analysis for biofuels production from rice straw

Crops are produced using a variety of cultivation techniques, crop management strategies, as well as harvest, and post-harvest activities. Rice cultivation may be classified into three distinct stages: (1) pre-planting, (2) plant growth and (3) post-cultivation [17]. LCA is a method for examining the energy balance and the environmental effects, starting from the gathering of raw materials until the finished product (ISO 14040:2006). This evaluation is useful for the simulation and computation of different environmental parameters viz., acidification of terrene land, ozone layer depletion, climate change, and eutrophication of marine and freshwater ecosystems. Furthermore, the LCA can be used for crop production and various agricultural systems including comparative analyses and evaluation of the best approach for different production techniques and activities based on specific environmental and economic factors viz., improving the manufacturing process, product development, commercialization, decision support, and strategic planning. The LCA framework includes (1) detailed description of the scope, purpose, and functional unit, (2) inventory review, (3) assessment of the environmental impact, and (4) interpretation of the generated data [4]. The transportation industry in India is largely dependent on nonrenewable liquid fuels, causing global climate concerns. Biofuels (especially those derived from food crops or agricultural residues) are considered as a viable alternative to the current challenges of transportation fuels viz,. gasoline and diesel. Parameters such as air pollution, GHG emissions, climate change and energy security contribute significantly toward the conversion of biofuels [18]. The framework and applications of LCA are shown in Fig. 12.1.

Low-cost cellulosic ethanol can be effectively generated at a high speed (within 24 h) from rice straw. Also, optimization of the process's byproducts can yield several commercial and valuable goods. The process starts with the bifurcation of lignocellulosic biomass into its three constituents, viz. lignin, cellulose, and hemicellulose. The process of ethanol generation from rice straw is classified into seven stages: (1) size depletion (3–4 cm to 300–500 mm), (2) pretreatment (removal of xylose from biomass), (3) hydrolysis of enzyme (polysaccharides into monosaccharides), (4) cofermentation (cell recycling), (5) distillation of ethanol (ethanol was separated from yeast cell), and (6) water treatment (obtaining purified water) [19]. The global warming potential (a quantified assessment of heat absorption by GHG) of ethanol is reported to be 2.81 kg^{-1} CO_2 eq. per liter, which indicates significant potential to reduce the inclusive effect by utilizing a sustainable approach to electricity generation, viz., hydro or solar. Another distinguishing aspect is the production of different value-added products such as silica, methanol, inorganics, and food-grade CO_2. In order to effectively exploit the agricultural residues for the generation of biofuels while adhering to the concept of sustainable development, it is essential to recognize and accurately understand the functional and structural changes that occur in a particular system at various stages, such as commercialization, policies, expenditure, and industries. The transformations during sustainable development are usually classified via a high degree of uncertainty, among other factors, thereby resulting in policy failures [20]. SWOT analysis is a collection of scientific methodologies that may be used to evaluate all conceivable aspects or approaches that either act as bottlenecks or favorable situations in order to prioritize the developmental plans in a particular industry. "Strengths,", "Weaknesses,", "Opportunities," and "Threats" constitute the SWOT analysis. The primary purpose

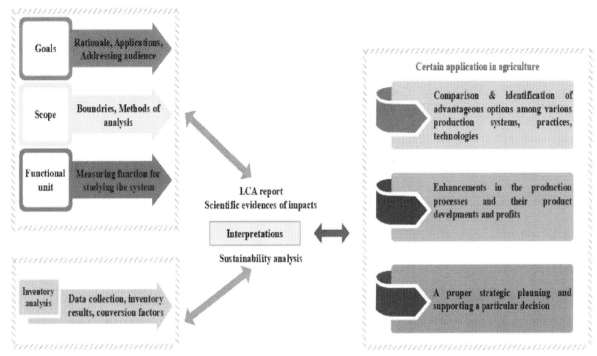

FIGURE 12.1 Life cycle framework and its applications. *Reproduced with permission from A. Sharma, G. Singh, S.K. Arya, Biofuel from rice straw, J. Clean. Prod. 277 (2020) 124101. https://doi.org/10.1016/j.jclepro.2020.124101 @ Elsevier.*

of an SWOT analysis is to simultaneously investigate the properties of external and internal systems to assist the operational resolutions effectively. SWOT analysis is divided into two categories: (1) internal parameters consisting of strengths and weaknesses, (2) external parameters consisting of opportunities and threats [21,22]. A SWOT analysis effectively aids in the development of strategic planning techniques to improve its efficiency and competitiveness during the management of agricultural residues for biofuels production. *Volvariella volvacea* is one of the variety of rice straw mushroom (RSM) that has been cultivated for over 20 years, however, its exploitation for mushroom production has been very limited. Fresh and processed forms of the mushroom are typically consumed through two channels: (1) export retails and (2) domestic retails. An SWOT analysis was conducted with the goal of examining better approaches for RSM value chains, which further aid in the exploration of opportunities to extend the subsector of value chains. This approach resulted in drawing the users' attention to the employment of rice straw from burning to producing value-added commercial goods [23].

5. Processes to improve the biofuel yield from rice straw

5.1 Pretreatment processes

5.1.1 Physical pretreatment

The digestibility of the lignocellulosic content in the biomass can be enhanced via physical pretreatment techniques like chipping, extrusion, ball or hammer milling, and irradiation. The particle size of RS plays a key role in the hydrolysis process. It was reported that the RS with a size >2 mm may confine the mass and heat transfer process during its hydrolysis within an anaerobic digester. Nevertheless, some reports stated higher production of methane from RS of size up to 10 mm [24]. Recently, Dai et al. (2019) reported a 50%–84% enhancement in the biomethane yield from RS of size 0.075–1.0 mm compared to 20 mm [25]. Zhu et al. (2010) stated that a reduction in particle size boosts up the enzymatic accessibility by improving the active surface area, porosity, and bulk density while reducing the polymerization degree and crystallinity of cellulose [26]. Consequently, a reduced size of the RS enhances the rate of cellulose degradation. In another study, Chen et al. (2014) have reported improvement in the surface area, porosity, bulk density of RS (~2.4 times of raw) and water-holding ability of RS via milling and extrusion pretreatment processes [27]. Further, the improvement in surface characteristics of RS resulted in the enhancement of the yield of biomethane by 29% and 72.20% for milling and extrusion pretreatment processes, respectively. The extrusion method owns advantages as compared to the milling process, as it comprises thermomechanical treatment of RS, which ultimately augments the biodegradation of RS. Dai et al. (2019) have also studied several RS characteristics, including dissolution ability, fractal dimension, and bioliquefaction, to explore the causes of the microbial response and enhanced biomethane yield in comminuted RS of diverse particle sizes in the AD process [25]. As per the investigation, particle size reduction boosted organic matter dissolution and bioliquefaction degree. It aided in bacterial community change and thriving specific microbial development, which promoted AD effectiveness. It was suggested that the rupturing of a siliceous mastoid layer of rice straw could increase methane output. However, smaller particle size can boost RS methane output, but extensive particle size reduction induces fast acidification, which debilitates RS-AD effectiveness. As a result, particle size must be optimized in combination with stable AD process performance. Moreover, failing to keep down lignin restricts the use of the physical pretreatment process for the RS-AD. The hydrothermal pretreatments are evolved as the most successful physical pretreatment techniques because of their ability to solubilize a wide range of biomass, from agriculture leftovers to algal biomass [28]. Several studies have reported improved biodigestibility of hydrothermally pretreated RS. Previously, Chandra et al. (2012) demonstrated that the hydrothermal pretreatment of RS followed by its neutralization with 5% NaOH could produce a higher methane quantity of 132.7 L kg^{-1}, which was ~122% higher as compared to untreated RS [29]. Similarly, Eskicioglu et al. (2017) investigated the influence of the hydrothermal pretreatment using CO_2 catalyst over five biomass resources, namely, maize stover, RS, Douglas fir bark, biomass sorghum and wheat straw for biohydrogen or biomethane production [30]. Aside from the wheat straw, other biomasses need rigorous pretreatment parameters, such as the pretreatment time of 30 min, the temperature of 150°C, onset CO_2 pressure of 50 bar, and the yield of methane obtained was 13 and 319 L kg^{-1} for the liquid and solid fractions, respectively. In addition to the hydrothermal pretreatment, the microwave pretreatment (temperatures ranging from 130 to 210°C, and exposure time ranging from 2 to 5 min) was investigated for boosting the yield of biomethane from RS. However, in light of the use of physical pretreatment processes arranged for RS-AD, a blend of thermal and mechanical pretreatment approaches might be the way to go forward [31].

5.1.2 Chemical pretreatment

The chemical pretreatment process employs inorganic and organic acids, which is a less energy-consuming approach to increase the disintegration of lignocellulosic biomass. The acids such as HCL, H_2SO_4 $C_4H_4O_4$, H_3PO_4, and bases like NaOH, KOH, and Ca $(OH)_2$ are frequently used for the pretreatment process. These chemicals were utilized for breaking down lignin-carbohydrate linkages in lignocellulosic biomass and saponification, which increases the possibility to enhance the yield of biomethane. Several reports have utilized numerous chemicals at varying doses to extensively hydrolyze the hemicellulose into monomers while rupturing the lignin content to a greater degree. NaOH is a frequently used alkali for rice straw pretreatment. Numerous investigations have demonstrated the significant applications of NaOH at various loading rates for pretreatment [32]. Previously, He et al. (2008) observed that the pretreatment process by solid-state NaOH of 6% loading had successfully increased RS biogas output [33]. When RS was pretreated with 6% NaOH, it yielded 27%–64% higher biogas as compared to the untreated RS at an organic loading rate (OLR) ranging from 35 to 80 g TS L^{-1}. However, the superior yield of biogas was recorded as 520 L kg^{-1} volatile solids (VS) for 50 g TS L^{-1} OLR. In a further study, they have explored the role of NaOH loading ranging from 4% to 10% (w/w) over the performance of AD on RS. It was observed that the lignocellulosic content was significantly solubilized by NaOH pretreatment at various loadings (cellulose: 14.2%–16.4%, hemicellulose: 35.2%–54.2%, lignin: 8%–44.5%). The results were remarkably similar to the prior investigation with a NaOH dosage of 6%. Shetty et al. (2017) examined an impact of pretreatment time and loading rate of NaOH with AD process conditions such as C/N ratio, pH, temperature, and HRT on biomethane generation from RS [34]. As a consequence, the RS pretreated with NaOH of 1% (w/v) loading rate and 180 min pretreatment time have yielded highest biomethane output of 278.3 L kg^{-1} VS, which was ~71% more than the untreated RS. The optimized process conditions include, pH 7, HRT of 15 days, temperature 37°C, and C/N ratio of 25. These optimized conditions were then tested for biomethane generation in a semi-continuous digester, which produced around 34% additional methane (303 L kg^{-1} VS) than the untreated rice straw. In conclusion, for any NaOH-based treatments, a lower NaOH concentration appears to produce significantly better outcomes in terms of methane output than a high load. It was suggested that, at larger NaOH loading, dissociated Na$^+$ ions advance the osmotic pressure, which may impede the methanogenic activity, consequently. In addition to NaOH, alternative alkalis like Na_2CO_3, NH_3, and Ca $(OH)_2$ were also investigated for enhancing RS biodegradability. Song et al. (2013a) improved several treatment parameters such as pretreatment period, chemical loading, and inoculum loading for pretreatment of RS using Ca $(OH)_2$ to improve the profitability of methane generation [35]. The optimized parameters resulted in an enhanced methane output of 225.3 L kg^{-1} VS for a pretreatment period of 5.98 d; 9.81% (w/w TS) loading of Ca $(OH)_2$; and inoculum concentration of 45.12%. Analogous findings were obtained with a methane production of 292 L kg^{-1} VS for rice straw pretreated using 0.5M

Na_2CO_3 at 110°C for 2 h^{-1} and 250.34 L kg^{-1} VS for RS pretreated by 4% NH_3. Overall, it was observed that the alkali compounds successfully broke down the glycosidic side chains and ester, decrystalize the cellulose, and removed the lignin. However, during pretreatment, alkali compounds are converted into irreversible salts, which limit their applicability in the pretreatment. As a result, a number of researchers explored the impact of organic and inorganic acids on rice straw and consequent biomethane production. For instance, Song et al. (2013b) discovered that 6.18 days of pretreatment with H_2O_2 (2.86%, w/w TS) could increase biomethane harvesting to 288 L kg^{-1} VS [36]. Amnuaycheewa et al. (2016) evaluated the influence of organic acids such as acetic acid ($C_2H_4O_2$), oxalic acid ($C_2H_2O_4$), and citric acid ($C_6H_8O_7$) on RS-AD to that of inorganic acid pretreatment [37]. They discovered that the citric acid pretreatment with 12.21% concentration at a temperature of 126.2°C and for 60 min pretreatment time had boosted the biogas output to 376.3 L kg^{-1} VS, which was much greater than the untreated RS. Mirmohamadsadeghi et al. (2014) investigated an impact of the organosolv pretreatment process over the RS in solid-state AD [38]. The pretreatment process was carried out at 150 and 180°C for 30 and 60 min, respectively, using sulfuric acid as a catalyst and 75% ethanol (organic solvent). The RS processed at 150°C for 60 min produced the higher biomethane production of 152.7 L kg^{-1} of carbohydrates (CH). Despite the fact that the pretreatment process with concentrated inorganic acid is highly productive compared to the dilute acid for hydrolyzing cellulose within the lignocellulosic biomass, it is hazardous and erosive, necessitating the use of specific nonmetal alloys in reactor design. As a result, pretreatment with dilute organic or inorganic acids is advised for an environmentally friendly process.

5.1.3 Biological pretreatment

The biological or microbial pretreatment process entails the decomposition of lignocellulosic biomass via fungi and bacteria such as brown, white, and soft rot fungi, either alone or in conjunction. The biological pretreatment system is renowned for its lower energy consumption as compared to the physical pretreatment processes and for being more environmentally friendly to than of chemical pretreatment processes. Under normal environmental conditions such as temperature, pressure, and pH, the bacteria secrete enzymes which degrade lignin, hemicellulose, and polyphenols, reducing the formation of inhibitory substances. This reduces the harmful impact of the AD process. White rot is potential fungi that had been employed for the RS pretreatment since it has a higher affinity toward lignin over cellulose [32]. Kainthola et al. (2019) explored RS pretreatment with three distinct fungal stains, *Pleurotus ostreatus* (PO), *Ganoderma lucidum* (GL), and *Phanerochaetechrysosporium* (PC), during a 4-week inoculation [39]. The sCOD, volatile fatty acids (VFAs), and biomethane potential of the pretreated RS were all examined. The maximum disintegration of RS was observed for the PC pretreated RS, which was ~36% larger than the untreated RS. The yield of methane was 339.31 L/kg VS for RS pretreated with PC. However, GL and PO pretreated RS had provided methane yields up to 269.99 and 295.91 L kg^{-1} VS, respectively. In addition to the fungal pretreatment process, Zhang et al. (2016) treated the RS using a microbial pretreatment process under anaerobic conditions with the help of rumen fluid at 39°C, 120 h^{-1} [40]. Rumen fluid pretreatment during the 24 h^{-1} was found to be an optimum pretreatment, presenting a 40% reduction in digestion time with the highest biomethane yield of 285.1 L kg^{-1} VS, which was 82.6% greater than untreated RS. Seesatat et al. (2021) used thermophilic lignocellulolytic bacteria obtained from the soil samples to degrade RS biologically [41]. The highest biomethane output of 177.93 L kg^{-1} TS^{-1} was obtained via deteriorated RS after 20 days; nevertheless, a degradation time of 15 days is proposed as the ideal interval for effective as well as economical generation of biomethane from the RS. Despite the superior benefits of microbial pretreatment, there are several drawbacks that must be addressed. The major obstacle which must be addressed is the augmentation of the incubation period of the pretreatment step. Table 12.2 represents the effects of different pretreatment processes on the production of biofuels from rice straw.

5.1.4 Combined pretreatment methods

The solitary pretreatment of RS-AD persists with numerous restrictions and intrinsic drawbacks, and it is nearly difficult to emphasize one over the other when it comes to its selection. To circumvent the constraints of a single pretreatment, combination pretreatment using two or more individual pretreatments has emerged as a promising technique for increasing biomass biomethane production. Recently, Sabeeh et al. (2020) studied the innovative pretreatment strategy using photocatalysis by employing titania (TiO_2) nanoparticles, as well as by the collective impact of photocatalytic pretreatment and NaOH on RS biomethane production [42]. It was observed that the photocatalytic pretreatment of rice straw with a 0.25 g L^{-1} concentration of TiO_2 resulted in a biomethane production of 397 L kg^{-1} VS, whereas the combination of 0.25 g L^{-1} TiO_2-1.5% w/v NaOH has provided a better yield of methane yield (645 L kg^{-1} VS). Such a hybrid system resulted in a 122% higher yield of methane as compared to untreated RS. Nevertheless, the effectiveness of the photocatalytic pretreatment process is still to be investigated. Zhang et al. (2015) recorded a methane output of 288 L/kg of RS

TABLE 12.2 Effects of various pretreatment processes on biofuel yield of rice straw.

Strategies	Optimum conditions	Optimum AD conditions	Methane yield of raw rice straw (L kg^{-1} VS)	Improved biomethane yield from treated rice straw (L kg^{-1} VS)	Improvement in biomethane yield compared to control (%)
Physical pretreatment process					
Milling pretreatment	Particle size: 0.075 mm^{-1}	T: 37°C; I/S: 0.5; C/N: 23.56; OLR: 44.2 gVS L^{-1}; HRT: 25 d	107.00	197.00	83.97
Hydrothermal pretreatment	Pretreatment at 200°C for 10 min; particle size: <1 mm^{-1}	T: 37°C; I/S: 1; C/N: 25; OLR: 42 gVS L^{-1}; HRT: 60 d	59.80	132.70	121.90
Extrusion pretreatment	Particle size: <2.5 mm^{-1}	T: 37°C; I/S: 0.4; C/N: 25; OLR: 50 kg m^{-3}; HRT: 60 d	135.00	227.30	72.20
Chemical pre-treatment process					
NaOH pretreatment	NaOH: 3% (w/w); T: 35°C; t: 120 h; particle size: <1 mm^{-1}	T: 37°C; I/S: 1; C/N: 25; OLR: 42 gVS L^{-1}; HRT: 60 d	69.80	74.10	23.90
NH$_3$ pretreatment	NH$_3$: 4%; moisture content: 70%; t: 7 d; T: 30°C; particle size: 5 mm^{-1}	T: 35°C; I/S: 15/65; C/N: NA; OLR: 65 gTS L^{-1}; HRT: 55 d	194.74	250.34	28.55
KOH pretreatment	Chemical loading: 10%, T: 50°C, t: 24 h^{-1}, particle size: <10 mesh	T: 35°C; I/S: 2; C/N: NA; OLR: 3.6 gVS L^{-1}; HRT: 23 d	240.00	323.00	34.60
Biological pretreatment process					
Fungal pretreatment	Moisture content: 75%; incubation time: 20 d; particle size: 2–3 cm^{-1}; fungus used: *Pleurotus ostreatus*	T: 37°C; I/S: 1; C/N: 31.7; OLR: 20% TS; HRT: 45 d	120.00	263.00	120.00
Rumen fluid pretreatment	T:39°C; incubation time: 24 h; particle size: 2–3 cm^{-1}	T: 35°C; I/S: 1; C/N: 25; OLR: NA; HRT: 30 d	156.13	285.10	82.60
Fungal pretreatment	Moisture content: 75%; incubation time: 20 d; particle size: 2–3 cm; fungus used: *Trichoderma reesei*	T: 35°C; I/S: 1; C/N: 31.7; OLR: 20% TS; HRT: 45 d	113.00	214.00	78.00
Combined pretreatment process					
Biological and chemical pretreatment	Pretreatment with CaO and liquid fraction of digestate; particle size: <20 mesh	T: 35°C; I/S: 1; C/N: NA; OLR: 50 gTS L^{-1}; HRT: 50 d	174.32	274.65	57.56
NaOH and microwave pretreatment	NaOH: 10% (w/w), soaking time: 24 h^{-1}, power: 720 W, T: 180°C; particle size: 3–5 cm^{-1}	T: 37°C; I/S: 1/1; C/N: NA; OLR: 7%–8% TS; HRT: 45 d	192.00	297.00	54.70
Grounded and thermal pretreatment	Combination of grinding: 10 mm; T: 110°C, ammonia treatment: 2%; particle size: 10 mm^{-1}	T: 35°C; I/S: NA; C/N: 25; OLR: 50–100 g L^{-1}; HRT: 24 d	197.60	244.80	23.88

Modified with permission from S. Kumar, T.C. D'Silva, R. Chandra, A. Malik, V.K. Vijay, A. Misra, Strategies for boosting biomethane production from rice straw: a systematic review, Bioresour. Technol. Rep. 15 (2021) 100813. https://doi.org/10.1016/j.biteb.2021.100813 @ Elsevier.

after RS was exposed to the cumulative effects of NaOH and extrusion pretreatment, which was 54% higher than the untreated RS [43]. Furthermore, Kaur and Phutela (2016) delignified RS using a combination of microwave and NaOH pretreatment techniques [44]. When RS was processed with 4% NaOH and microwaved for 30 min, it obtained a biogas output of 297 L kg^{-1} dry RS with the highest delignification (65%) by lowering silica content (88.7%). Owing to the accessibility of lignocellulose deteriorating microbes, anions, inorganic cations, and organic substances (amino acids, proteins, and sugars) that are else expelled in the environment, the reuse of these liquid fractions of digestate (LFD) has also been investigated with an amalgamation of other pretreatment processes. Recognizing the intrinsic properties of LFD, Guan et al. (2018) blended the LFD with calcium oxide (CaO) and ammonia solution (AS) to improve RS biodegradability [45]. CaO-LFD pretreated straw produced the greatest yield of methane (274.65 L kg^{-1} VS), which was nearly 58% greater than the control, along with 69.47% removal of VS. Additionally, Momayez et al. (2018) used biogas digestate discharge for the pretreatment of RS for 30–60 min at 130–190°C [46]. RS that has been pretreated was further utilized for simultaneous saccharification, enzymatic hydrolysis, fermentation, and dry and liquid AD. The RS was pretreated at 190°C for 60 min showed the greatest improvement in hydrolysis and ethanol yield. Correspondingly, methane yield was improved by 24% and 26% for RS digested by dry and liquid AD.

5.2 Anaerobic codigestion process

One of the major challenges with the single-digestion system of RS-AD is the larger C/N ratio. Adequate nutrients for microorganisms are not obtained through the hydrolysis of the substrate with a larger C/N ratio substrate, resulting in AD instability due to an excess buildup (VFAs). As a result, the idea of co-digestion arose and has been studied to optimize methane production utilizing various substrates, known as anaerobic co-digestion (AcoD). The nitrogen-enriched substrates like dairy waste, algal biomass, agricultural and food residue, sewage sludge, animal manure, aquatic plants, and others have been identified as promising co-substrates for RS-AD [47]. Recently, Shen et al. (2018) performed an AcoD investigation of pig manure (PM) and rice straw blended pretreated with formerly built cellulolytic microbial-consortium at 55°C for 30 h^{-1} to examine the synergic activity of co-digestion and biological pretreatment [48]. At an OLR of 2.5 kg^{-1} COD/(m^{-3} d), the optimal volumetric methane production rate (VMPR) and biomethane were 0.64 L L d^{-1}, and 0.45 L g^{-1} COD removed, respectively, which have been 62% and 37% higher in comparison to the untreated PM and RS blend under identical OLR conditions. For improved recovery of bioenergy, Abudi et al. (2016) codigested pretreated RS, thickened waste activated sludge (TWAS), and organic fraction of municipal solid wastes (OFMSW) [49]. TWAS was pretreated by thermal energy for 9 h^{-1} at 70°C and thermoalkaline process for 10 h^{-1} at 90°C with pH 11 in order to extract higher bioenergy. At the same time, RS was pretreated via 3% (w/w) H_2O_2 and 5% (w/w) NaOH distinctly for 5 days at 37°C. As a consequence, AcoD of thermoalkaline treated TWAS, OFMSW, and H_2O_2-treated RS at a ratio of 5:3:0.5 yielded a higher VS removal efficiency of ~77% and the highest biomethane output of 558.5 L kg^{-1} VS. Ye et al. (2013) blended RS, pig manure (PM), and kitchen waste (KW) at various mixing ratios with C/N ratio ranging from 17.20 to 47.00 and VS loading rate of 54 g L^{-1} [50]. A mixing ratio of 0.4:1:1.6 (RS: PM: KW) produced the maximum biomethane output of 383.9 L kg^{-1} VS with 55.76% removal efficiency of volatile solids. It was observed that the command on the AD process could be improved via a two-phase AD process. In this context, Chen et al. (2015) demonstrated AcoD of food waste (FW) and RS in a two-phase model [51]. The outcomes of their study revealed that the acidogenic fermentation via the butyric-acid route had boosted biodegradability by 95% and acidification by up to 36% for a biogas production of 535 L kg^{-1} VS. Recently, Ai et al. (2020) investigated the influence of solid digestate as of the first and second runs of RS-AD on mesophilic and thermophilic AD [52]. For mesophilic and thermophilic AD, codigestion of digestate and RS at a ratio of 1:1 resulted in enhanced production of methane (205 L kg^{-1} VS) with 151 L kg^{-1} of volatile solids, correspondingly. The continuous digestate recirculation offered higher hemicellulose and cellulose than the lignin, increasing the biogas production while decreasing the lag phase and improving the buffering capacity. Fig. 12.2 represents the schematic diagram of different stages involved in anaerobic digestion and energy balance analysis of rice straw.

5.3 Innovative process based on additives

The addition of supplements has arisen as a critical method for overcoming the inefficiency and instability of the AD process caused by washout of microflora, variety of substrates, nutritional shortage, inhibitory impact of diverse co-products, feedstock purity, and other factors. The supplemented additives like biological additives, micronutrients (Fe, Mo, Ni, W, Se, Co), macronutrients (N, P, S) and metal oxide nanoparticles (Al_2O_3, Fe_3O_4, SiO_2, TiO, ZnO) have been shown to increase biomethanation via offering microbial growth through electron transport, so sidestepping the harmful effects of inhibitory products delivering nutrients, maintaining optimal alkalinity and pH, eliminating H_2S, bioaugmentation, and

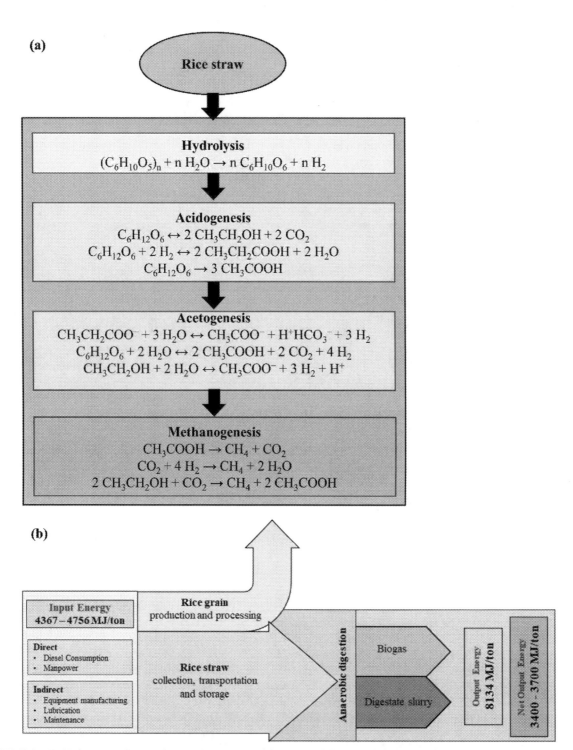

FIGURE 12.2 (A) Various stages involved in anaerobic digestion of rice straw and (B) energy balance analysis of rice straw-based biogas production. *Reproduced with permission from S. Kumar, T.C. D'Silva, R. Chandra, A. Malik, V.K. Vijay, A. Misra, Strategies for boosting biomethane production from rice straw: a systematic review, Bioresour. Technol. Rep. 15 (2021) 100813. https://doi.org/10.1016/j.biteb.2021.100813 @ Elsevier.*

CO_2 sequestration [32]. The innovative process based on cosubstrate and additives for improving the anaerobic digestion of rice straw is shown in Fig. 12.3. Mancini et al. (2018) investigated the collaborative effects of pretreatment of NaOH and trace element dosages of Ni, Co, and Se (100%, 200%, and 500% (µg g^{-1}TS)) [53]. The methane production of RS pretreated by 1.6% (w/w) of NaOH solution was 318 L kg^{-1} VS, which was 21.4% greater in comparison to the untreated

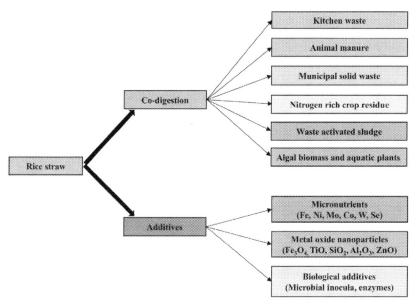

FIGURE 12.3 Innovative strategy based on cosubstrate and additives to improve the anaerobic digestion of rice straw. *Reproduced with permission from S. Kumar, T.C. D'Silva, R. Chandra, A. Malik, V.K. Vijay, A. Misra, Strategies for boosting biomethane production from rice straw: a systematic review, Bioresour. Technol. Rep. 15 (2021) 100813. https://doi.org/10.1016/j.biteb.2021.100813 @ Elsevier.*

RS. Although, the addition of trace elements was insignificant, increasing the biomethane output by just 3.8% and 3.5%, respectively, by a 100% dose of Ni and Co. The greatest methane output of 330 L kg^{-1} VS was found for NaOH pretreated RS by 100% Ni dosage (9 g g^{-1} TS), which was 26% higher as compared to the untreated. Recently, Tapadia-Maheshwari et al. (2019) studied the impact of several dietary and process factors over the function and structure of the microbial community to enhance methane production [54]. The highest biomethane yield of 274 L kg^{-1} VS was observed at the optimal conditions like a C/N ratio of 25, 1 mm^{-1} particle size, process temperature of 37°C, loading rate of 7.5%, pH of 7, supplementation of zinc (100 μM), and HRT of 21 days. To optimize the microbial population, Methanosarcina was chosen as a methanogenic bacterium, and the Bacteroides, Clostridium, and Ruminococcus were chosen as hydrolytic bacteria. The net profit was projected to be 12.48 and 5.62 lakhs by using the biogas yielded from 100 tons of rice straw in the form of bio-CNG and electricity, subsequently. Researchers have also focused on the use of nanoparticles as a supplement to enhance the RS-AD. Recently, Junaid et al. (2019) examined an impact of magnetite (Fe_3O_4) supplementation on AD in NaOH-treated and untreated rice straw [55]. The rice straw pretreated with 2% NaOH in 120 h^{-1} with a 120 ppm dose of Fe_3O_4 provided the highest yield of methane (585 L kg^{-1} VS), that was 129% greater in comparison to the untreated rice straw. Furthermore, the pretreated RS showed a positive net energy benefit of 3.76 MJ^{-1}.

6. Mathematical models and tools to scrutinize the sustainability of biofuels

It is essential to investigate the air pollutant emissions from biorefineries in order to reduce environmental pollution and promote human health, along with sustainable utilization of biomass. Numerous computational models and tools, such as chemical process design simulations, mathematical optimization, large-scale crop models, and life cycle assessment models have been developed owing to the advancements in the field of biofuel production and sustainability [56]. The very first step of analysis comprises of lab-scale tests and field trials, which include research on biomass cultivation techniques, factors needed for optimal biofuel production, and knowledge on reaction kinetics. The second step is to perform process-scale investigation, which includes crop-specific models such as Agro-IBIS, MISCANFOR, AUSCANE, ALMANAC, CANEGRO, MISCANMOD, Agro-BGC, EPIC, APSIM, WIMOVAC, 3 PG, and LPJmL, along with typical process design analysis. These models assist in increasing the biomass output, water requirements, carbon flow, nutrient cycling, and emission reduction, among other things [57]. The third step involves extending the analytical limits to consider energy and material flow as well as different emission levels across the entire integrated supply chain. This approach analyses the various effects related to the environment that are not considered in the typical process design boundaries. The LCA is a standard strategy for evaluating the environmental effect of a product or industrial process throughout its life cycle. The LCA is now widely used to scrutinize the environmental sustainability of biofuels. The final step of analysis examines the role of eco-friendly goods and services across the entire supply chain. Land-use is an essential aspect to consider during the production of biomass in a sustainable manner, particularly in terms of indirect effects on biodiversity and greenhouse gas emissions. As such, forest and agricultural residues can be considered a sustainable choice as there is no requirement of any additional land-use [58]. However, there are several uncertainties associated with the use of agricultural/forest residues such as (1) availability of natural agricultural lands to cultivate crops for generating biofuels, (2) total production from different agricultural activities, (3) use of animal-derived products, (4) uncertainty about the amount of accessible residues, (5) source of residues, (6) stringent criteria to use/regulate biomass for effective and sustainable bioenergy production, may have far-reaching consequences [59–61].

7. Conclusion

The widespread availability of rice straw in large quantities has prompted researchers to focus their efforts on developing novel bioconversion techniques for producing bioenergy and biofuels, while also considering the negative effects on the environment (air pollution from rice residue burning) and human health (respiratory disease from the alarming levels of air pollution). The property of rice straw, which comprises of cellulose, hemicelluloses, and lignin poses a substantial challenge for its use in producing bioenergy. This refractory structure allows biorefineries to utilize suitable pretreatment processes, thereby increasing the costs and further complicating the economic analyses. However, the high cost of biofuels is a substantial impediment to their commercial viability, which causes an economic imbalance in a country. Apart from refocusing the whole effort on the generation of bioenergy and value-added products from rice straw, it is highly essential to fine-tune the operating variables for optimum biofuel production to reduce the adverse economic effect on a nation. More research efforts should be carried out to address such challenges with the development of advanced and feasible techniques. Also, three major areas that all research and development sectors should focus during the production of biofuels are: (1) increased production efficiency, (2) less effect on the nation's economy, and (3) minimal or no production of toxic gases/contaminants, resulting in zero environmental pollution.

References

[1] D. Vamvuka, Bio-oil, solid and gaseous biofuels from biomass pyrolysis processes—an overview, Int. J. Energy Res. 35 (2011) 835–862, https://doi.org/10.1002/er.1804.

[2] D. Tilman, R. Socolow, J.A. Foley, J. Hill, E. Larson, L. Lynd, S. Pacala, J. Reilly, T. Searchinger, C. Somerville, R. Williams, Beneficial biofuels—the food, energy, and environment trilemma, Science (80-.) 325 (2009) 270–271, https://doi.org/10.1126/science.1177970.

[3] A. Demirbas, Biofuels sources, biofuel policy, biofuel economy and global biofuel projections, Energy Convers. Manag. 49 (2008) 2106–2116, https://doi.org/10.1016/j.enconman.2008.02.020.

[4] A. Sharma, G. Singh, S.K. Arya, Biofuel from rice straw, J. Clean. Prod. 277 (2020) 124101, https://doi.org/10.1016/j.jclepro.2020.124101.

[5] W. Wang, X. Wu, A. Chen, X. Xie, Y. Wang, C. Yin, Mitigating effects of ex situ application of rice straw on CH_4 and N_2O emissions from paddy-upland coexisting system, Sci. Rep. 6 (2016) 1–8, https://doi.org/10.1038/srep37402.

[6] Y.-F. Huang, S.-L. Lo, Utilization of rice hull and straw, in: J.B.T.-R.F.E. Bao (Ed.), Rice, fourth ed., AACC International Press, 2019, pp. 627–661, https://doi.org/10.1016/B978-0-12-811508-4.00019-8.

[7] M. Ali, M. Saleem, Z. Khan, I.A. Watson, The use of crop residues for biofuel production, in: D. Verma, E. Fortunati, S. Jain, X.B.T.-B. Zhang (Eds.), Biopolymer-Based Materials, and Bioenergy, Woodhead Publ. Ser. Compos. Sci. Eng., Woodhead Publishing, 2019, pp. 369–395, https://doi.org/10.1016/B978-0-08-102426-3.00016-3.

[8] A. Alemdar, M. Sain, Isolation and characterization of nanofibers from agricultural residues—wheat straw and soy hulls, Bioresour. Technol. 99 (2008) 1664–1671, https://doi.org/10.1016/j.biortech.2007.04.029.

[9] V. Smil, Crop residues: agriculture's largest harvest phytomass agricultural, Bioscience 49 (1999) 299–308.

[10] S. Yang, H. He, S. Lu, D. Chen, J. Zhu, Quantification of crop residue burning in the field and its influence on ambient air quality in Suqian, China, Atmos, Environ. Times 42 (2008) 1961–1969, https://doi.org/10.1016/j.atmosenv.2007.12.007.

[11] H.V. Scheller, P. Ulvskov, Hemicelluloses, Annu. Rev. Plant Biol. 61 (2010) 263–289, https://doi.org/10.1146/annurev-arplant-042809-112315.

[12] J.H. Grabber, How do lignin composition, structure, and cross-linking affect degradability? A review of cell wall model studies, Crop Sci. 45 (2005) 820–831, https://doi.org/10.2135/cropsci2004.0191.

[13] J. Chen, Y. Gong, S. Wang, B. Guan, J. Balkovic, F. Kraxner, To burn or retain crop residues on croplands? An integrated analysis of crop residue management in China, Sci. Total Environ. 662 (2019) 141–150, https://doi.org/10.1016/j.scitotenv.2019.01.150.

[14] F. Lu, X. Wang, B. Han, Z. Ouyang, X. Duan, H. Zheng, Net mitigation potential of straw return to Chinese cropland: estimation with a full greenhouse gas budget model, Ecol. Appl. 20 (2010) 634–647, https://doi.org/10.1890/08-2031.1.

[15] S.O. Petersen, M. Blanchard, D. Chadwick, A. Del Prado, N. Edouard, J. Mosquera, S.G. Sommer, Manure management for greenhouse gas mitigation, Animal 7 (2013) 266–282, https://doi.org/10.1017/S1751731113000736.

[16] J. Lehmann, J. Gaunt, M. Rondon, Bio-char sequestration in terrestrial ecosystems—a review, Mitig. Adapt. Stratg. Glob. Change 11 (2006) 403–427, https://doi.org/10.1007/s11027-005-9006-5.

[17] P.K. Das, B.P. Das, P. Dash, Potentials of postharvest rice crop residues as a source of biofuel, in: R.P. Kumar, E. Gnansounou, J.K. Raman, B. Baskar (Eds.), Refining Biomass Residues for Sustainable Energy and Bioproducts, Academic Press, 2020, pp. 275–301, https://doi.org/10.1016/B978-0-12-818996-2.00013-2.

[18] B.E. Dale, A new industry has been launched: the cellulosic biofuels ship (finally) sails, Biofuels Bioprod. Biorefin. 9 (2015) 1–3, https://doi.org/10.1002/bbb.1532.

[19] A. Sreekumar, Y. Shastri, P. Wadekar, M. Patil, A. Lali, Life cycle assessment of ethanol production in a rice-straw-based biorefinery in India, Clean Technol. Environ. Pol. 22 (2020) 409–422, https://doi.org/10.1007/s10098-019-01791-0.

[20] J. Markard, R. Raven, B. Truffer, Sustainability transitions: an emerging field of research and its prospects, Res. Pol. 41 (2012) 955–967, https://doi.org/10.1016/j.respol.2012.02.013.

[21] I. D'Adamo, P. Rosa, Current state of renewable energies performances in the European Union: a new reference framework, Energy Convers. Manag. 121 (2016) 84–92, https://doi.org/10.1016/j.enconman.2016.05.027.

[22] B. Wolfslehner, T. Baycheva-Merger, Evaluating the implementation of the Pan-European criteria and indicators for sustainable forest management—a SWOT analysis, Ecol. Indicat. 60 (2016) 1192–1199, https://doi.org/10.1016/j.ecolind.2015.09.009.

[23] M. Demont, T.T.T. Ngo, N. Van Hung, G.P. Duong, T.M. Dương, H.T. Nguyen, N.T. Hoang, M.C. Custodio, R. Quilloy, M. Gummert, Rice straw value chains and case study on straw mushroom in Vietnam's Mekong river delta, in: M. Gummert, N. Van Hung, P. Chivenge, B. Douthwaite (Eds.), Sustainable Rice Straw Management, Springer International Publishing, Cham, 2020, pp. 175–192, https://doi.org/10.1007/978-3-030-32373-8_11.

[24] S. Behera, R. Arora, N. Nandhagopal, S. Kumar, Importance of chemical pretreatment for bioconversion of lignocellulosic biomass, Renew. Sustain. Energy Rev. 36 (2014) 91–106, https://doi.org/10.1016/j.rser.2014.04.047.

[25] X. Dai, Y. Hua, L. Dai, C. Cai, Particle size reduction of rice straw enhances methane production under anaerobic digestion, Bioresour. Technol. 293 (2019) 122043, https://doi.org/10.1016/j.biortech.2019.122043.

[26] W. Zhu, J.Y. Zhu, R. Gleisner, X.J. Pan, On energy consumption for size-reduction and yields from subsequent enzymatic saccharification of pretreated lodgepole pine, Bioresour. Technol. 101 (2010) 2782–2792, https://doi.org/10.1016/j.biortech.2009.10.076.

[27] X. Chen, Y.L. Zhang, Y. Gu, Z. Liu, Z. Shen, H. Chu, X. Zhou, Enhancing methane production from rice straw by extrusion pretreatment, Appl. Energy 122 (2014) 34–41, https://doi.org/10.1016/j.apenergy.2014.01.076.

[28] F. Ahmad, E.L. Silva, M.B.A. Varesche, Hydrothermal processing of biomass for anaerobic digestion—a review, Renew. Sustain. Energy Rev. 98 (2018) 108–124, https://doi.org/10.1016/j.rser.2018.09.008.

[29] R. Chandra, H. Takeuchi, T. Hasegawa, Hydrothermal pretreatment of rice straw biomass: a potential and promising method for enhanced methane production, Appl. Energy 94 (2012) 129–140, https://doi.org/10.1016/j.apenergy.2012.01.027.

[30] C. Eskicioglu, F. Monlau, A. Barakat, I. Ferrer, P. Kaparaju, E. Trably, H. Carrère, Assessment of hydrothermal pretreatment of various lignocellulosic biomass with CO_2 catalyst for enhanced methane and hydrogen production, Water Res. 120 (2017) 32–42, https://doi.org/10.1016/j.watres.2017.04.068.

[31] J. Kainthola, M. Shariq, A.S. Kalamdhad, V.V. Goud, Enhanced methane potential of rice straw with microwave assisted pretreatment and its kinetic analysis, J. Environ. Manag. 232 (2019) 188–196, https://doi.org/10.1016/j.jenvman.2018.11.052.

[32] S. Kumar, T.C. D'Silva, R. Chandra, A. Malik, V.K. Vijay, A. Misra, Strategies for boosting biomethane production from rice straw: a systematic review, Bioresour. Technol. Rep. 15 (2021) 100813, https://doi.org/10.1016/j.biteb.2021.100813.

[33] Y. He, Y. Pang, Y. Liu, X. Li, K. Wang, Physicochemical characterization of rice straw pretreated with sodium hydroxide in the solid state for enhancing biogas production, Energy Fuel 22 (2008) 2775–2781, https://doi.org/10.1021/ef8000967.

[34] D.J. Shetty, P. Kshirsagar, S. Tapadia-Maheshwari, V. Lanjekar, S.K. Singh, P.K. Dhakephalkar, Alkali pretreatment at ambient temperature: a promising method to enhance biomethanation of rice straw, Bioresour. Technol. 226 (2017) 80–88, https://doi.org/10.1016/j.biortech.2016.12.003.

[35] Z. Song, G. Yang, X. Han, Y. Feng, G. Ren, Optimization of the alkaline pretreatment of rice straw for enhanced methane yield, BioMed Res. Int. 2013 (2013) 968692, https://doi.org/10.1155/2013/968692.

[36] Z. Song, G. Yang, Y. Feng, G. Ren, X. Han, Pretreatment of rice straw by hydrogen peroxide for enhanced methane yield, J. Integr. Agric. 12 (2013) 1258–1266, https://doi.org/10.1016/S2095-3119(13)60355-X.

[37] P. Amnuaycheewa, R. Hengaroonprasan, K. Rattanaporn, S. Kirdponpattara, K. Cheenkachorn, M. Sriariyanun, Enhancing enzymatic hydrolysis and biogas production from rice straw by pretreatment with organic acids, Ind. Crop. Prod. 87 (2016) 247–254, https://doi.org/10.1016/j.indcrop.2016.04.069.

[38] S. Mirmohamadsadeghi, K. Karimi, A. Zamani, H. Amiri, I.S. Horváth, Enhanced solid-state biogas production from lignocellulosic biomass by organosolv pretreatment, BioMed Res. Int. 2014 (2014) 350414, https://doi.org/10.1155/2014/350414.

[39] J. Kainthola, A.S. Kalamdhad, V. V Goud, R. Goel, Fungal pretreatment and associated kinetics of rice straw hydrolysis to accelerate methane yield from anaerobic digestion, Bioresour. Technol. 286 (2019) 121368, https://doi.org/10.1016/j.biortech.2019.121368.

[40] H. Zhang, P. Zhang, J. Ye, Y. Wu, W. Fang, X. Gou, G. Zeng, Improvement of methane production from rice straw with rumen fluid pretreatment: a feasibility study, Int. Biodeterior. Biodegrad. 113 (2016) 9–16, https://doi.org/10.1016/j.ibiod.2016.03.022.

[41] A. Seesatat, S. Rattanasuk, K. Bunnakit, P. Maneechot, P. Sriprapakhan, R. Artkla, Biological degradation of rice straw with thermophilic lignocellulolytic bacterial isolates and biogas production from total broth by rumen microorganisms, J. Environ. Chem. Eng. 9 (2021) 104499, https://doi.org/10.1016/j.jece.2020.104499.

[42] M. Sabeeh, Zeshan, R. Liaquat, A. Maryam, Effect of alkaline and alkaline-photocatalytic pretreatment on characteristics and biogas production of rice straw, Bioresour. Technol. 309 (2020) 123449, https://doi.org/10.1016/j.biortech.2020.123449.

[43] Y. Zhang, X. Chen, Y. Gu, X. Zhou, A physicochemical method for increasing methane production from rice straw: extrusion combined with alkali pretreatment, Appl. Energy 160 (2015) 39–48, https://doi.org/10.1016/j.apenergy.2015.09.011.

[44] K. Kaur, U.G. Phutela, Enhancement of paddy straw digestibility and biogas production by sodium hydroxide-microwave pretreatment, Renew. Energy 92 (2016) 178–184, https://doi.org/10.1016/j.renene.2016.01.083.

[45] R. Guan, X. Li, A.C. Wachemo, H. Yuan, Y. Liu, D. Zou, X. Zuo, J. Gu, Enhancing anaerobic digestion performance and degradation of lignocellulosic components of rice straw by combined biological and chemical pretreatment, Sci. Total Environ. 637–638 (2018) 9–17, https://doi.org/10.1016/j.scitotenv.2018.04.366.

[46] F. Momayez, K. Karimi, I.S. Horváth, Enhancing ethanol and methane production from rice straw by pretreatment with liquid waste from biogas plant, Energy Convers. Manag. 178 (2018) 290–298, https://doi.org/10.1016/j.enconman.2018.10.023.

[47] R. Singh, S. Kumar, A review on biomethane potential of paddy straw and diverse prospects to enhance its biodigestibility, J. Clean. Prod. 217 (2019) 295–307, https://doi.org/10.1016/j.jclepro.2019.01.207.

[48] F. Shen, H. Li, X. Wu, Y. Wang, Q. Zhang, Effect of organic loading rate on anaerobic co-digestion of rice straw and pig manure with or without biological pretreatment, Bioresour. Technol. 250 (2018) 155–162, https://doi.org/10.1016/j.biortech.2017.11.037.

[49] Z.N. Abudi, Z. Hu, N. Sun, B. Xiao, N. Rajaa, C. Liu, D. Guo, Batch anaerobic co-digestion of OFMSW (organic fraction of municipal solid waste), TWAS (thickened waste activated sludge) and RS (rice straw): influence of TWAS and RS pretreatment and mixing ratio, Energy 107 (2016) 131–140, https://doi.org/10.1016/j.energy.2016.03.141.

[50] J. Ye, D. Li, Y. Sun, G. Wang, Z. Yuan, F. Zhen, Y. Wang, Improved biogas production from rice straw by co-digestion with kitchen waste and pig manure, Waste Manag. 33 (2013) 2653–2658, https://doi.org/10.1016/j.wasman.2013.05.014.

[51] X. Chen, H. Yuan, D. Zou, Y. Liu, B. Zhu, A. Chufo, M. Jaffar, X. Li, Improving biomethane yield by controlling fermentation type of acidogenic phase in two-phase anaerobic co-digestion of food waste and rice straw, Chem. Eng. J. 273 (2015) 254–260, https://doi.org/10.1016/j.cej.2015.03.067.

[52] P. Ai, M. Chen, Y. Ran, K. Jin, J. Peng, A.E.-F. Abomohra, Digestate recirculation through co-digestion with rice straw: towards high biogas production and efficient waste recycling, J. Clean. Prod. 263 (2020) 121441, https://doi.org/10.1016/j.jclepro.2020.121441.

[53] G. Mancini, S. Papirio, G. Riccardelli, P.N.L. Lens, G. Esposito, Trace elements dosing and alkaline pretreatment in the anaerobic digestion of rice straw, Bioresour. Technol. 247 (2018) 897–903, https://doi.org/10.1016/j.biortech.2017.10.001.

[54] S. Tapadia-Maheshwari, S. Pore, A. Engineer, D. Shetty, S.S. Dagar, P.K. Dhakephalkar, Illustration of the microbial community selected by optimized process and nutritional parameters resulting in enhanced biomethanation of rice straw without thermo-chemical pretreatment, Bioresour. Technol. 289 (2019) 121639, https://doi.org/10.1016/j.biortech.2019.121639.

[55] M.J. Khalid, Zeshan, A. Waqas, I. Nawaz, Synergistic effect of alkaline pretreatment and magnetite nanoparticle application on biogas production from rice straw, Bioresour. Technol. 275 (2019) 288–296, https://doi.org/10.1016/j.biortech.2018.12.051.

[56] G.G. Zaimes, N. Vora, S.S. Chopra, A.E. Landis, V. Khanna, Design of sustainable biofuel processes and supply chains: challenges and opportunities, Processes 3 (2015) 634–663, https://doi.org/10.3390/pr3030634.

[57] S. Surendran Nair, S. Kang, X. Zhang, F.E. Miguez, R.C. Izaurralde, W.M. Post, M.C. Dietze, L.R. Lynd, S.D. Wullschleger, Bioenergy crop models: descriptions, data requirements, and future challenges, GCB Bioenergy 4 (2012) 620–633, https://doi.org/10.1111/j.1757-1707.2012.01166.x.

[58] G. Pourhashem, S. Spatari, A.A. Boateng, A.J. McAloon, C.A. Mullen, Life cycle environmental and economic tradeoffs of using fast pyrolysis products for power generation, Energy Fuel. 27 (2013) 2578–2587, https://doi.org/10.1021/ef3016206.

[59] R. Sindhu, P. Binod, A. Pandey, S. Ankaram, Y. Duan, M.K. Awasthi, Biofuel production from biomass: toward sustainable development, in: S. Kumar, R. Kumar, B. Pandey (Eds.), Current Developments in Biotechnology and Bioengineering, Elsevier, 2019, pp. 79–92, https://doi.org/10.1016/B978-0-444-64083-3.00005-1.

[60] M. Shah (Ed.), Removal of Refractory Pollutants from Wastewater Treatment Plants, CRC Press, USA, 2021.

[61] M. Shah (Ed.), Removal of Emerging Contaminants through Microbial Processes, Springer, Singapore, 2021.

Chapter 13

Biofuel production from algal biomass

Pranjal P. Das[1], Niladri S. Samanta[2], Simons Dhara[1] and Mihir Kumar Purkait[1]
[1]Department of Chemical Engineering, Indian Institute of Technology Guwahati, Guwahati, Assam, India; [2]Centre for the Environment, Indian Institute of Technology Guwahati, Guwahati, Assam, India

1. Introduction

Global energy consumption is rising due to industrial development and transformation, leading to excessive fossil fuel use. The generation of bioenergy from biomass has lately garnered interest owing to the abundance of biomass, scarcity of fossil fuels, and rising levels of greenhouse gases (GHG), mainly CO_2. Developing and using novel technology for biofuel generation and valuable products from biomass are hard challenges. The manufacture of biofuels from various biomasses is concerned with sustainability, affordability, and waste reduction. Biofuels will soon play a vital part in the vast fuel industry and ensure global energy security. Fossil fuels now account for roughly 70% of the global energy market, although global power consumption accounts for just 30% [1]. Although the significance of fuels is generally acknowledged, most technologies for generating power from renewable energy sources, like solar, hydroelectric, wind, geothermal, wave, and nuclear sources, are currently under development. Assuming the conditions mentioned earlier, we may consider biomass to produce energy like gaseous and liquid fuels and electricity. To lessen reliance on traditional fossil fuels, various possible feedstocks have been employed to produce biofuels. Plant biomass like maize, sugarcane, palm, and soybean are used to make first-generation biofuels, but it also causes environmental degradation, water shortages, and food versus fuel issues. Given the drawbacks of first-generation biofuels, the second- and third generation biofuels have emerged as viable alternatives, using agricultural, municipal, and plant wastes and microorganisms to make biofuels without harming the environment [2]. Bacteria, cyanobacteria, and microalgae are the microorganisms that may be used as cost-effective and sustainable feedstocks to manufacture biofuels and biomaterials, and they can potentially replace societal needs for fossil fuels. Most algae are photosynthetic creatures that live in water; however, a handful live on land [3].

Using sunlight and CO_2, they may generate biomass mainly consisting of lipids, carbohydrates, and proteins. It is calculated that 1.83 kg of CO_2 is needed to generate 1 kilogram of algal biomass. Algae and cyanobacteria have several advantages that make them possible contenders for biomaterials and biofuels production, including (1) oxygenic photosynthesis using water as an electron donor, (2) compared to oil seed crops, biomass production per acre is very high, (3) they are a nonfood feedstock, they can address food versus fuel conflicts, (4) there is no necessity for cultivable and fruitful farmland, (5) adaptation to various kinds of wastewater growth, and (6) developing a wide variety of items [4]. Several studies have demonstrated that algae and cyanobacteria can create a variety of fuels, including biodiesel, bioethanol, gasoline, and biohydrogen (H_2). Furthermore, complete algal biomass may be used in the Fischer-Tropsch process to produce syngas, hydrothermal gasification to yield methane (CH_4) and H_2, anaerobic digestion to produce CH_4, and cocombustion to generate power. As a result, algae may be used as a possible feedstock for the upcoming generation of sustainable biofuels and materials. Quite a few published studies and reviews detail the technology used in the biovalorization of algal biomass in this context. Still, the economic viability of the process remains a hurdle. Several hurdles, like environmental, financial, and biotechnological concerns, are required to be solved before this technique can be scaled up [5].

In light of the above, this chapter aims to detail the utilization of microalgae as a possible source of sustainable biofuels and valuable goods such as biodiesel, bioethanol, biochar, and biomethane. The many methods of biofuel production, as well as the different types of algae growth (open-air and closed systems), have been thoroughly discussed. In addition, the various technologies for converting algal feedstock to third-generation biofuels and value-added goods, as well as the potential and constraints associated with biofuel production, have been thoroughly reviewed and described.

2. Categories of biofuel generation

Biofuels are split into primary and secondary categories based on the way biomass is used. The primary biofuel is a feedstock that has not been treated in any way: animal fats, wood chips, forestry, and farmed crop wastes. This type of energy is often utilized in Third World nations for cooking, heating, and agricultural purposes. Traditional biomass is another name for primary biofuels. This type of fuel does not need any supply expenditures to be processed, and its application area is somewhat limited. In 2013, power derived from conventional biomass accounted for around 9% of global energy consumption. The most energy-intensive molecules (biodiesel, biomethanol, and biohydrogen) are extracted from biomass (the primary biofuel) to make secondary biofuel, which may be used to replace the omnipresent fossil fuel. At the same time, secondary biofuel may be classified as first, second, and third generation. The classification is according to the biomass used in fuel generation, procedures, and the chronological order in which the fuel first appeared on the global energy souk [6].

The first generation: Organic matter from food crops loaded with carbohydrates, sugars (sugar cane and sugar beet stems), and oils (rapeseeds, sunflower, soybeans seeds) is used to make first-generation bioalcohols and biodiesel. The benefit of this fuel's manufacture is the easy and relatively inexpensive processing technique, microbial fermentation. However, this biofuel has significant issues: (1) Low agricultural land efficiency: only a tiny portion of harvested biomass is usable for biofuel generation. (2) Land utilization dispute with the processed food industry [7].

The second generation: Second-generation biofuels are made from lignocellulosic biomass. In comparison to lipids and sugars, lignocellulose is abundant in plant biomass, the primary component of plant cell walls. Lignocellulose is made up of three ingredients: cellulose (40%–50%), hemicellulose (25–35%), and lignin (15%–20%). These components have some benefits for humans, but extracting particular components, particularly cellulose, is complex. Lignin is a polymer that is amorphous and composed of phenylpropane subunits. It gives cellulose rigidity and resistance to numerous hydrolytic enzymes. Dissolving lignin in alkaline-alcohol solutions may eliminate it. Lignin may be burnt to generate heat or energy, and it can also be chemically treated to yield valuable chemical compounds [8]. Hemicellulose is an amorphous polymer composed of various sugars, including D-mannose, D-glucose, D-xylose, D-galactose, and L-arabinose. It is conceivable to hydrolyze hemicellulose to produce sugar monomers, which may then be fermented to alcohols. Cellulose is a linear glucose polymer with some stiffness that makes hydrolysis difficult. Benefits of lignocellulose-enriched biomass include (1) land-use efficiency: relative to first-generation biofuel, most biomass is utilized for biofuel synthesis. (2) The effect on the food industry is reduced due to the use of nonedible crops. The complexity of the processes, on the other hand, is a drawback of this form of biomass [9].

The third generation: Algal biomass is linked to the third generation of biofuel. The utilization of algal feedstock for fuel synthesis is a pretty recent bioenergetics area. Based on the most recent data from many studies, algal biomass may collect a much higher quantity of lipids than oil plant biomass. This characteristic permits algae to be considered as a viable source of biodiesel. The primary distinction between plant and algal bioenergetics is the cultivation technique. Additional procedures are not required for plants other than agricultural development processes and a specific effort to create certain circumstances. But plant culture necessitates using a vital asset (fertile land) and produces a relatively poor output in terms of the volume of biomass generated to the amount of biofuel produced. Microalgae may develop in unfavorable environments for plant growth, such as salty soils and wastewater. As a result, the use of microalgae is being investigated as an intriguing alternative substrate for the manufacture of biofuels. On the other hand, algae need particular growing facilities known as bioreactors, the design of which is dependent on the variety of algae cultivated [10].

3. Algal biomass

3.1 Algal

Algae are a polyphyletic category of organisms, including prokaryotes (cyanobacteria) and eukaryotes (diatoms, green algae, brown algae, red algae). Algal metabolism varies across species. Several algae may alter metabolic pathways in response to changing environmental circumstances. This route is governed by carbon supply, energy source, and electron source. Mixotrophs are organisms that can take up carbon from both the atmosphere and organic molecules (autotrophy and heterotrophy). Algae are mixotrophic. Depending on the energy source, ATP synthesis might be phototrophic or chemotrophic. The organism sucks up the energy of the electromagnetic wave in the first situation. The organism gets energy by breaking high-energy molecules in the latter case [11]. Photoheterotrophs are heterotrophic species that can absorb the sun's energy, whereas chemoautotrophs are autotrophic organisms that get their energy via redox processes. Mixotrophy's algal ability might be beneficial in the production of third-generation biofuels. Microalgae may fix CO_2

from various sources, including the environment, soluble carbonates, and exhaust fumes. This characteristic distinguishes them favorably concerning biomass farming ease. Microalgae can transform light into chemical bond energy more effectively than complex plants [12]. Proteins, carbohydrates, and lipids are distributed differently among algae. *Spirulina maxima*, for example, contains 60%–71% protein by weight, *Porphyidium cruentum* has 40%–57% carbohydrates by weight, and *Scenedesmus dimorphus* contains almost 40% lipids by weight. Green algae species like *Botryococcus braunii* and *Chlorella protothecoides* have large quantities of glyceryl lipids and terpenoid hydrocarbons, which may be transformed into small hydrocarbons and used as significant crude oil. The biomass composition is also affected by the growth circumstances. Algal biomass may be an excellent raw material for synthesizing biodiesel or bioalcohol based on the presence of sugars and lipids. Microalgae may therefore be an excellent source of biofuel in certain circumstances [13].

3.2 Cultivation

The creation of affordable and effective microalgae culture methods is one of the most critical steps in the evolution of algal biomass from a goal of research to a competing product. The medium in which the organisms develop and reproduce is an essential component of such a system. It should include adequate origins of not just carbon, but also of several important metallic constituents: phosphorus and nitrogen in large numbers, and manganese, calcium, chlorine, iron, silicon, and magnesium in lower quantity, for effective biomass growth of the desired content. Nitrogen is usually obtained from nitrates; however urea and ammonium can also be used. Microalgae cultivation systems typically split into two categories namely open-air system and close system [9]. Table 13.1 represents the advantages and drawbacks of open-air and closed culture systems.

3.2.1 Open-air systems (ponds)

Open-air systems include the largest industrial algae growing systems. It has to do with the convenience and low cost of their design and maintenance; the length of their use and biomass increase efficiency are both greater than other kinds of biodigesters. The medium stirring process is critical in all open-air systems, including natural ponds. In a circular pond and raceway pond, for instance, the pivoting agitator and paddle wheel are essential components. Several kinds of open-air systems exist [14,15]:

TABLE 13.1 Benefits and limitations of open-air and closed culture systems.

Culture systems	Benefits	Limitations
Open-air system	Relatively economical; easy to clean up; easy maintenance; utilization of nonagricultural land; low energy inputs.	Little control of culture conditions; poor mixing, light and CO_2 utilization; difficult to grow algal cultures for long periods; poor productivity; limited to few strains; cultures are easily contaminated.
Tubular photobioreactors	Relatively cheap; large illumination surface area; suitable for outdoor cultures; good biomass productivities.	Gradients of pH, dissolved oxygen and CO_2 along the tubes; some degree of wall growth; requires large land space; photoinhibition.
Flat photobioreactors	Relatively cheap; easy to clean up; large illumination surface area; suitable for outdoor cultures; good biomass productivities; good light path; readily tempered; low oxygen build-up; shortest oxygen path.	Difficult scale-up; difficult temperature control; some degree of wall growth; hydrodynamic stress to some algal strains; low photosynthetic efficiency.
Column photobioreactors	Low energy consumption; readily tempered; high mass transfer; best exposure to light–dark cycles; low shear stress; easy to sterilize; reduced photoinhibition and photooxidation; high photosynthetic efficiency.	Small illumination surface area; shear stress to algal cultures; decrease of illumination surface area upon scale-up; expensive.

Modified with permission from R.A. Voloshin, M.V. Rodionova, S.K. Zharmukhamedov, T. Nejat Veziroglu, S.I. Allakhverdiev, Review: biofuel production from plant and algal biomass, Int. J. Hydrogen Energy 41 (2016) 17257–17273. https://doi.org/10.1016/j.ijhydene.2016.07.084. Elsevier.

(i) Artificial and natural ponds: They can only be utilized for microalgae production if the weather circumstances are ideal and the water has the appropriate ingredients in terms of pH and salinity. The cyanobacterium *Arthrospira*, for example, produces 40 tons of cyanobacterium each year in lakes around Lake Chad.
(ii) A raceway pond is a collection of loopback ponds, each of which is shaped like a racetrack. There is a paddle-wheel in the specific location that controls the flow of water.
(iii) Circular ponds are artificial circular shallow ponds with a central rotor mechanism for churning. They have a size limit: a pond with a large diameter no longer pays for itself due to energy costs for rotor operation.
(iv) A thin coating of medium falls on an inclined surface in incline ponds (cascade system). Gravity generates turbulence, which causes stirring.

The primary drawbacks of an open-air pattern are its susceptibility to time of day, season, and weather; heat control, which is complicated to apply; light; heterotrophic organism pollution, which reduces the lack of monocultural production chances and efficacy. Therefore, the biomass that is obtained is diverse [8].

3.2.2 Closed systems (photobioreactors)

Photobioreactors are closed devices that allow certain microalgae species to thrive under controlled circumstances. These circumstances frequently differ from those of the environment; and as such energy is required for its upkeep. There are three types of photobioreactors viz., tubular, flat, and column photobioreactors. In order to build a photobioreactor, the following points should be considered [16,17]:

(i) Light available to photosynthetic algae that is adequate for efficient photosynthesis and does not induce photoinhibition.
(ii) CO_2 levels that are optimal for synthesis of biomass do not result in CO_2-suppression impacts. Concentration of CO_2 varies between 2.3×10^{-2} M and 2.3×10^{-4} M.
(iii) Offering the required medium stirring. It boosts the amount of light and nutrients available to all cells. Without the motion, cells at the surface would consume additional light, while cells in the middle would become dark.

4. Techniques involved in transformation of third-generation biofuels from algal biomass

4.1 Extraction of oil or lipid from algal biomass

When compared to the cell walls of higher animals and plants, microalgal cell walls are comprised of high amount of lipids and fatty acids. Mechanical, enzymatic (biological), physical, and chemical approaches can be used to eliminate lipids or oils from algal biomass. At the commercial level, solvent extraction of lipids is neither environmentally friendly nor cost-effective, has negative health consequences, and alters the intrinsic quality of the finished goods [18]. Ryckebosch et al. (2012) used several solvent ratios to extract lipids from four several algal biomass and found that a chloroform:methanol ratio (v/v) of 1:1 offered a much higher yield [19]. Cryogenic and simple grinding with liquid nitrogen are two more well-known, efficient conventional lipids extraction procedures, but they are not cost-effective at the industrial level. Because it has been used to take out lipids from algal biomass with an ideal 0.5 mm bead size, bead milling is regarded the highly effective mechanical lipid extraction technique. Nonetheless, when it came to extracting lipids from *Chlorella vulgaris* biomass, this method proved inefficient. Mechanical pressing is used to extract oils from soybeans; however, owing to the low size of the microorganisms, this process is ineffective. At lesser scales, autoclaving and homogenization procedures have been proposed; but at larger scales, these processes are once again regarded uneconomical. Soxhlet is perhaps the most widely used process for extracting oil and lipids from food biomass and nutrients and chemicals from plant biomass. The procedure is impracticable due to the utilization of hazardous solvents in huge quantities, longer durations of extraction, and poorer lipid products. To date, various advanced physical processes have been employed to efficiently extract lipid from biomass, including supercritical fluid extraction (SFE), pressurized liquid extraction (PLE), the nanosecond pulse electric field (nsPEF), ultrasound-based extraction, and microwave-facilitated extraction [20,21]. SFE, wherein CO_2 is used as a solvent under critical circumstances, has been suggested as an environmentally friendly methodology for extracting fatty acid substances from biomass; similarly, PLE has been used to extract nonpolar and polar lipids from corn and oat at elevated pressure employing a variety of liquors. However, because of the enhanced energy needs to maintain the higher operating pressure and temperature, neither PLE nor SFE are advised for widespread use; also, this operation leads to reduction in quality of final product at higher temperatures. nsPEF was used to extract oil from

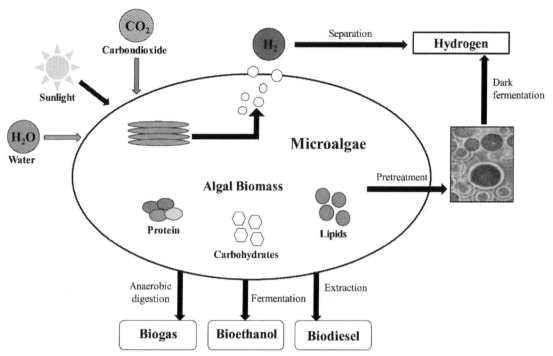

FIGURE 13.1 Pathways for conversion of algal biomass into biofuels and value-added products. *Reproduced with permission from M. Kumar, Y. Sun, R. Rathour, A. Pandey, I.S. Thakur, D.C.W. Tsang, Algae as potential feedstock for the production of biofuels and value-added products: opportunities and challenges, Sci. Total Environ. 716 (2020) 137116. https://doi.org/10.1016/j.scitotenv.2020.137116. Elsevier.*

Botryococcus braunii. The pulsed power technique might be utilized as an effective physical process in extraction of oil, according to experimental data, which showed that 50 pulses with energy consumption of 16.7 J mL^{-1} were sufficient for extraction of oil [22]. The pathways for the production of biofuels and other value-added products from algal biomass are shown in Fig. 13.1.

4.2 Lipids or oil transesterification

The end products of the transesterification of lipids or oils are biodiesel and glycerol, which are produced in the presence of a chemical catalyst, for instance, alkali and acid or a biological catalyst like alcohol and lipase. Various alcohols, namely butanol, methanol, propanol, ethanol, and amyl alcohol, are employed as cosolvents in the transesterification process, but ethanol and methanol are more favored at the commercial grade due to chemical and physical benefits as well as cost-effectiveness. The transesterification synthesis uses 3 M of alcohol for every 1 M of fatty acid, yielding 3 M of fatty acid methyl ester (FAME) and 1 M of glycerol as a derivative [23]. Because glycerol has a higher density than biodiesel, the fraction of glycerol can be removed from the batch of reaction at regular intervals to keep the reaction going. Once the reaction is completed, distillation and water washing are required to eliminate the catalytic, methanolic, and soap constituents from biodiesel; otherwise, they will cause engine choke and failure. The base catalyzed transesterification reaction is quicker than acid catalyzed transesterification, resulting in shorter processing times [24]. The transesterification reaction is classified into two categories on the basis of the number of stages required in the manufacture of biodiesel: in situ and extractive transesterification. Extractive transesterification is a multistep process that includes everything from cell drying through cell disruption, transesterification, lipid extraction, and ultimately downstream operations including refinement of biodiesel. This procedure is less appealing and useful due to the amount of stages needed. Furthermore, the feedstocks have a higher amount of water, which is particularly noticeable in the case of algal biomass, rendering the method unsustainable on an industrial scale. The extraction of lipids and transesterification were carried out in a one-step during in situ transesterification, resulting in a shorter processing time and less solvent consumption [25]. Fig. 13.2 represents the various types of lipids along with their extraction processes from algal biomass.

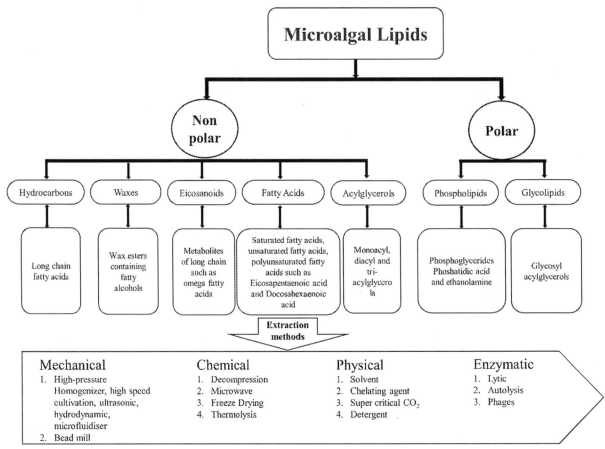

FIGURE 13.2 Various types of lipids and their extraction methods from algal biomass. *Reproduced with permission from M. Kumar, Y. Sun, R. Rathour, A. Pandey, I.S. Thakur, D.C.W. Tsang, Algae as potential feedstock for the production of biofuels and value-added products: opportunities and challenges, Sci. Total Environ. 716 (2020) 137116. https://doi.org/10.1016/j.scitotenv.2020.137116. Elsevier.*

4.3 Saccharification and pretreatment of algal biomass

Green algae such as *Chlorococcum* and *Spirogyra* have enormous quantities of carbohydrates in the form of polysaccharides in their cell walls. *C. vulgaris*, for example, has about 37%–55% cell dry weight (CDW) of carbohydrates, which can be used as a precursor material for bioethanol manufacture. Pretreatment and saccharification are required before microbial fermentation to produce biofuels from these feedstocks consisted high amount of saccharide. The simultaneous agitation procedure can create bioethanol by using possible microbial strains that can manufacture amylase enzyme and stimulate the saccharification method [26]. Nevertheless, very little article is accessible to emphasize manufacturing of bioethanol fermentation employing algal biomass as starting material, the procedures associated in the formation of bioethanol from microalgae are similar to those involved in the production of corn-based first-generation biofuel, as stated by Harun et al. (2014) [27]. Pretreatment is essential for removing undesirable materials such as lignin and permitting the extraction of polysaccharides from biomass, while enzymes act on such polysaccharides and transform them to monomers, as demonstrated by Behera et al. (2015) [28]. Biomass pretreatment with alkali and acid sources are generally accepted due to their low-energy consumption and high-efficiency for removing of undesirable contaminants and releasing sugar. Several different technologies, in addition to chemical pretreatments, have been used to make this process more environmentally friendly and sustainable, as reported by de Farias Silva et al. (2018) [29]. Other polymers found in algal biomass (alginate, mannitol, and fucoidan) require extra processing before being used in the synthesis of bioethanol. In another study, Yanagisawa et al. (2011) investigated the lignin-free concentration of polysaccharides in seaweeds (*Gelidium elegans, Alaria crassifolia, Ulva pertusa*), and found that these biomasses did not need to be pretreated before hydrolysis [30]. Both saccharification after pretreatment and direct saccharification (without pretreatment) have been researched using *Spirogyra* biomass in the manufacture of bioethanol. The first approach yielded 2% more (w/w) than the later, emphasizing the significance of *Spirogyra* in the manufacture of low-cost bioethanol. The pretreatment methods of algal biomass for the extraction of different compounds are shown in Table 13.2.

TABLE 13.2 Pretreatment methods of algal biomass for extraction of variety of compounds.

Pretreatment method	Operating conditions	Algal biomass	Extracted compounds
Acid hydrolysis	Acid hydrolysis H_2SO_4 1 M, 80–90°C, 120 min	Mix of microalgae (*Scenedesmus, Chlorella, Ankistrodesmus, Chlamydomonas*)	Carbohydrates
Hydrothermal water	1:13 (w/v), 147°C, 40 min	*Scenedesmus* sp.	Glucose
Enzymatic	Endogalactouronase 800 U/g, esterase 3600 U/g, protease 90 U/g, pH 6, 50°C, 24 h	*Scenedesmus obliquus*	Carbohydrates
High pressure homogenization	500–850 bar, 15 min	*Chlorococcum* sp.	Carbohydrates
Microwave	Acetone, 50 W, 56°C, 5 min	*Dunaliella tertiolecta*	Pigments
Alkaline-peroxide	H_2O_2 1%–7.5% (w/w), 50°C, 1 h	*Scenedesmus obliquus, Scenedesmus quadricauda, Nitzschia* sp., *Desmodesmus spinosus, Nitzschia palea*	Carbohydrates and byproducts
Freezing/thawing	3 cycles, 10 min freezing—80°C, 5 min thawing 37°C	*Synechocystis* sp.	Proteins
Pulsed electric field and solvents	1 cm electrode distance, 45 kV cm^{-1}, ethyl acetate/methanol/water	*Ankistrodesmus falcatus*	Lipids
Ionic liquid and solvent	Ionic liquid 1 h, ambient temperature, adding hexane mixture 30 s, 15 min	*Chlorella vulgaris*	Cell disruption and lipids

Modified with permission from M. Kumar, Y. Sun, R. Rathour, A. Pandey, I.S. Thakur, D.C.W. Tsang, Algae as potential feedstock for the production of biofuels and value-added products: opportunities and challenges, Sci. Total Environ. 716 (2020) 137116. https://doi.org/10.1016/j.scitotenv.2020.137116. Elsevier.

4.4 Microbial fermentation of algal biomass

Biofuels like biohydrogen and bioethanol are mostly produced by dark fermentation and yeast fermentation, severally. At the commercial grade, the generation of bioethanol utilizing yeast fermentation technology is widely known. Several aspects have been addressed to improve bioethanol production, including the screening of substrate selection, genetic manipulation, resilient strains, and modification, and minimal feedback restraint at increasing product concentrations. Following the pretreatment of algal biomass, the reclaimed sugar is transformed into bioethanol by microbial process. Simultaneous saccharification and fermentation (SSF) and separate hydrolysis and fermentation (SHF) are two distinguished fermentation techniques used in the manufacture of bioethanol [31]. The benefit of SHF is that it is conducted separately in two distinct reactors, with hydrolysis taking place in one reactor under each ideal state and the result of the hydrolysis being employed as the fermentation substrate in other reactor under each minimal condition. One of the key issues in the SHF process is the removal of wet biomass after hydrolysis, which restricts the output of bioethanol. The synthesis of biohydrogen from microalgae by dark fermentation with class of microorganisms or pure microbial strains is now attracting interest. However, due to the numerous and varied processes involved in dark fermentation, the biohydrogen generation potency is rather low [32]. Pretreatment is a key step in this process because it transforms polymers (carbohydrates) to monomers, making the monomers more accessible to bacteria. As a result, several chemical, biological, and physical pretreatments are commonly used in the dark fermentation method in order to transform monomers after depolymerization of carbohydrate polymers that result in generation of larger yields of biohydrogen from algal biomass [33].

4.5 Anaerobic digestion of microalgal biomass

The formation of biogas from microalgal biomass utilizing the anaerobic digestion method has proven to be a viable and long-term solution. The benefits of anaerobic digestion include the reduction of greenhouse gas emissions, the creation of biogas, and the production of organic manures. Acidogenesis, methanogenesis, hydrolysis, and acidogenesis are the four

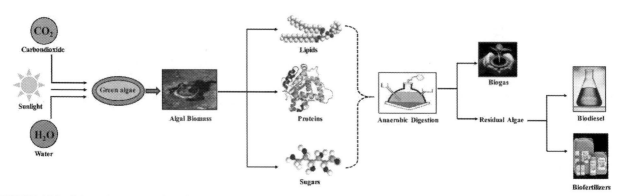

FIGURE 13.3 Schematic representation of anaerobic digestion process for biovalorization of algal biomass. *Reproduced with permission from M. Kumar, Y. Sun, R. Rathour, A. Pandey, I.S. Thakur, D.C.W. Tsang, Algae as potential feedstock for the production of biofuels and value-added products: opportunities and challenges, Sci. Total Environ. 716 (2020) 137116. https://doi.org/10.1016/j.scitotenv.2020.137116. Elsevier.*

key processes in anaerobic digestion, with methanogenesis being the rate limiting phase. The cost-effectiveness of biogas generation is closely linked to the formation of microalgal biomass and an effective digestion system. Depending on the cell makeup of the biomass, the anaerobic digestion yield is essential. Microalgae grown in low-nitrogen environments accumulate more fatty compounds behind the wall of cell, making the cell wall stiff and making anerobic reaction more difficult [34]. The C:N ratio is crucial in assessing the anaerobic process' viability. It was suggested that the C:N ratio of the microbial growth culture range from 20 to 30 for the anaerobic digestion method to increase methane output. Lipids or oils are employed as precursor components in the manufacturing of biodiesel from algal biomass, and following the lipids extraction, around 65% of the algal residual is deemed waste, which is a big difficulty for biofuel businesses. These microalgal leftovers, which are rich in carbohydrates and proteins, could be considered as source materials for anaerobic digestion operations to produce bio or natural gas [35]. The anaerobic digestion process for biovalorization of microalgal biomass is shown in Fig. 13.3.

5. Biofuels obtained from the biovalorization of algal biomass

5.1 Biomethane

The synthesis of biomethane (CH_4) from residual algal biomass by anaerobic digestion is commonly practiced since it creates a combination of gases wherein the CO_2 provides 30%–50% and CH_4 adds 50%–70%. Based on the algal species and research circumstances, the extraction efficiency of CH_4 from algal biomass ranges from 0.2 to 0.4 m^3 $CH_4.kg^{-1}$ or 0.024–0.6 L $CH_4.g^{-1}$ VS (volatile solid) [36]. Milledge et al. (2019) observed that the quantity of CH_4 differs even within the same species when producing biomethane from *Dunaliella* sp., with yields ranging from 0.063 to 0.323 L $CH_4.g^{-1}$ VS [37]. Biomethane synthesis is measured in biomethane potentials (BMPs), which are described as the maximal volume of CH_4 per gram of VS generated in batch lab experiments. The BMPs of *A. nodosum* and the cast brown seaweed *Saccharina latissimi* were found to be 166 L $CH_4.kg^{-1}$ and 342 L $CH_4.kg^{-1}$ of VS, respectively, as reported by Allen et al. (2015) [38]. Biomass loading rate, temperature, volume, time, algal cell wall composition, and bacterial strains, among other factors, all influence the biomethane yield produced from algae. The C:N ratio is another critical component that significantly impacts biomethane output. The C:N ratio of residual algae has been predicted to be low (6–9), and blending algal with greater carbonaceous biomass has been proposed for improved biomethane generation. This method should be combined with other biofuels and biomaterial manufacturing operations utilizing a biorefinery strategy to make this procedure affordable [39]. Bioethanol, biodiesel, biohydrogen, and biomethane, among others, are created from algal biomass, with the former two being the main outputs and the latter two side products in the biorefinery framework. With the evaluation of microalgae as prospective contenders for CO_2 sequestration and the generation of various biofuels, including CH_4, sustainability is expected to increase by combining the BECCS (bioenergy production with the Carbon Capture and Storage) approach.

5.2 Biodiesel

Triglycerides (TAGs) account for 95% of lipids in oleaginous microalgal cells, with mono and diglycerides and minor quantities of free fatty acids accounting for 5%. Oleic acid, palmitic acid, and stearic acid constitute the algae fatty acid content, which is comparable to the biodiesel standard. Oils or lipids are the primary basic materials used to make biodiesel. Many research papers have been published on the generation of fatty acids from the algal feedstock [40]. Enamala et al. (2018) found that several microalgal species yielded lipids in the range of 2.4%–62% of dry biomass [41]. According to Mitra et al. (2012), *Chlorella vulgaris* produces lipids in the range of 11.0%–43% [42]. *Scenedesmus* sp. ISTGA generated 20% lipids in dry biomass when cultivated with sodium bicarbonate (Na_2HCO_3). In additional investigations with Na_2HCO_3, *Chlorella*, *Haematococcus*, and *Scenedesmus* generated 14.0%–18.0% lipids. The method of transesterification is used to make biodiesel. The lipids and oils recovered from the algal feedstock are employed in a transesterification procedure where the alcohol to oil ratio is either 3:1 or 6:1 in alkali, acid, or biological lipase as a catalyst. The transesterification process generates methyl ester (biodiesel) as the final result and glycerol as a secondary item, which is isolated from biodiesel. Compared to saturated fatty acids, biodiesel made from microalgae has a larger ratio of unsaturated fatty acids, which is required for fuel engineering. Higher degrees of unsaturation improve cold flow characteristics, but they also elevate the likelihood of hydroperoxide formation, which causes acidification, polymerization, and the development of hydrophobic units and jellylike textures, clogging the filter [43]. The biodiesel made from algal biomass should meet American (ASTM-D-6751) and European Standard (EN-14214) requirements. Because algal biodiesel has a viscosity 10–20 times that of ordinary diesel, engine adjustment, or blending is an option. One kilogram of oil from algal biomass may provide 1 kg of biodiesel, a carbon-neutral biological output. Biodiesel generation from algal biomass is budget-friendly thanks to integrated biorefinery and biovalorization techniques, which produce numerous products from a single feedstock. After oil or lipid removal, the biomass residues may be utilized straight for combustion, power generation, and biomethane production through anaerobic digestion [44].

5.3 Bioethanol

Ethanol is a carbon-neutral fuel that is generally made from plant waste. Algal biomass may also be used to make bioethanol utilizing anaerobic fermentation and a variety of microorganisms, including yeast, bacteria, and fungus. *S. cerevisiae* and *Z. mobilis* are now being explored as possible microorganisms for bioethanol production through microbial fermentative methods. Various algal species contain a particular polymer called mannitol. This necessitates the presence of oxygen during bioethanol production from sugar, where *Zymobacter palmae* is used for fermentation. Agar is galactose and galactopyranose-based polymer found in some types of marine algae and utilized as a bioethanol feedstock [45]. Yanagisawa et al. (2011) employed *S. cerevisiae* IAM 4178 for fermentation to make 5.5% (wt/wt) ethanol from galactan and glucan isolated from agar weed [30]. The red algal biomass included galactose and glucose and galactose polymers depolymerized during fermentation. With mannitol and glucose sugar, brown seaweeds comprise roughly 14% extra polymeric carbohydrates as alginate. Horn et al. (2000) documented the generation of bioethanol from mannitol extracted from brown seaweed, with an ethanol yield of 0.38 g g^{-1} of mannitol [46]. Hon-Nami (2006) used *Chlamydomonas perigranulata* in an algal culture to produce bioethanol and butanediol at the same time [47]. Harun et al. (2010) exploited the remaining biomass after extracting lipids from *Chlorococcum* sp. to make 60% more bioethanol than the entire algal biomass [48]. This discovery reinforces the use of algal biomass in a biorefinery to manufacture bioethanol and biodiesel. Bioethanol manufacturing is a less energy-intensive method. Furthermore, the CO_2 generated may be employed to grow algal biomass during biofuel production, similar to the carbon cycle. Although commercialization is still a challenge, algal biomass may be regarded as a viable feedstock for the manufacture of bioethanol; nonetheless, technical developments may soon make this process cost-effective.

5.4 Biochar (BC)

BC is a carbon-containing substance made from biomass that has been thermally treated at a sensible heat with a restricted oxygen flow. From wet algal biomass, the technique of HTL produces algal BC at a suitable ambiance for a brief period. BC made from algae has less surface area and low carbon content than lignocellulosic BC, but it has a better cation tradeoff property. Because algal BC has a higher pH, it may be a superior material for reclaiming low pH soil. The greater nitrogen concentration and varied inorganic component configuration are also advantageous for soil strength. BC may also be used to remove organic and inorganic contaminants from wastewater because of certain particular functional groups. BC having specific functional groups can improve the remediation capability of harmful substances in soil and wastewater. Nonetheless, to produce BC and use it in wastewater remediation, the supply of cheap resources and the manufacturing process are still being developed [49]. Multiple researches have shown that algal BC has a higher overall output per unit of algae biomass than other feedstocks. BC production per unit dry mass of algal biomass varies between 8.1% and 62.4%. Under identical test settings, algal BC yields are lower than LCB, like pinewood, wheat straw, and plant waste. As the temperature for pyrolysis was raised from 300 to 750°C and the duration was extended from 10 to 60 min, the yield decreased [50]. Tag et al. (2016) observed a comparable pattern and hypothesized that the differences could be attributable to differences in the content and chemical structure of plant and algal biomasses, particularly in the cell wall structure [51]. Furthermore, algal biomass has a more significant ash concentration than plant biomass, affecting the output dispersal of pyrolysis operation. Salimi et al. (2019) recently investigated iron-catalyzed slow pyrolysis to produce olive-shaped magnetic BC with a 296.4 $m^2\,g^{-1}$ surface area and an enhanced carbon-based construction, making it suitable to use in Li-ion batteries as an electrode [52]. The ability to produce biochar and biofuels simultaneously utilizing biorefinery techniques makes this method appealing for future growth and investigation of algal biomass as a feedstock. However, few studies and papers detailing the restoration of algal biomass produced BC, and its use opens up new opportunities for scholars in this field.

6. Opportunities and challenges in the production of biofuels from algal biomass

Different potential approaches, for example, first-, second-, and third-generation biofuel generation techniques have been introduced in subsequent times to meet global energy needs with little environmental harm. Biogas, bioethanol, and biodiesel are primarily three forms of first-generation biofuels; all have well-established production technology and can be produced on a large scale. Because the biomasses needed to make first-generation biofuels include food crops and other important consumables, their manufacture raises the food versus fuel argument, which is a key roadblock in the generation of first-generation biofuel. Unlike first-generation biofuels, the manufacturing of second-generation biofuels is much more sustainable. Second-generation biofuels are often made from waste materials (waste oils, sludge) and lignocellulosic biomass, and they provide a feasible choice in terms of carbon neutrality to lower CO_2 emissions [53]. Technological difficulties, like the pretreatment for lignin removal must be solved before marketing second-generation biofuel. Festel et al. (2014) calculated and estimated the manufacturing budgets of first- and second-generation biofuels in contrast to current fossil fuel prices, considering various influencing factors like conversion costs, feedstocks, and crude oil prices [54]. As per the prototype, the most cost-effective is waste oil produced biodiesel (€Cent $55/L^{-1}$), followed by fossil fuel (€Cent $68/L^{-1}$), palm oil produced biodiesel (€Cent $81/L^{-1}$), and bioethanol from lignocellulosic biomass (€Cent $86/L^{-1}$) where the market price of crude oil is €100/barrel. However, this study suggested that various kinds of first- and second-generation biofuels can be produced more cost-effectively than fossil fuels. Still, it also raised serious doubts about whether an adequate feedstock supply could be available to meet rising energy requirements and trigger the switch to biofuels. The third-generation biofuel is getting a lot of recognition because algae are utilized as a source of fuel. Microalgae, unlike first-generation biofuel, do not cut down on the amount of food humans can eat. Global warming, GHG emissions, oil spills, wildfires, and other climatic problems are all caused by the continued use of traditional energy outlets. Microalgae are indeed a possible source of biofuel production that is good for the environment to fix these problems [55]. Apart from the many benefits of microalgae as a foundation for third-generation renewables and materials, relatively few cost-effective methods have developed, particularly in Third World nations. The increased manufacturing costs of lipids and other expenditures such as refining and capital investments are the principal impediment to the cheapness of third-generation biofuels production, resulting in a negative energy balance. From a technological and economic standpoint, the biorefinery or combined system, where complete algal biomass may be used to produce a variety of biomaterials and biofuels, has shown promising outcomes. Biofuels can be made from algal biomass after it has been stripped of their oils and sugars. This process can be done under different conditions, and various microbes can be used for each function, like fermentation and anaerobic digestion [56]. The anaerobic waste material left behind might be used as animal feed or manure. To be competitive with conventional fuels, third-generation biofuels must be 10 times cheaper than they are now. According to the United States Renewable Fuel Standards, 36 billion gallons of algal biofuel will be needed to meet bioenergy needs in 2022. Several elements must be considered to create this massive amount of bioenergy from algal biomass, including power consumption, water needs for algal growth and downstream operations, carbon, energy, and fertilizer supplies for algae culture. Third-generation biofuel technology is uneconomical because of the millions of tonnes of fertilizers such as nitrogen and phosphorus necessary for microalgae development. Furthermore, collecting the algal biomass and downstream operations add additional expenses to the process, making it difficult to sustain [57–59].

7. Conclusion

Algal biomass has promised to produce ecofriendly, efficient, and cheap biofuels and bioproducts. The method is more cost-effective when biorefinery systems are used, and activities like biodiesel, biohydrogen, and biofertilizer generation are integrated. The relevance of microalgal investigation has been proved by creating a variety of biofuels, including biohydrogen, biodiesel, biomethane, bio-oils, and biochar. Manufacturing is cost-effective because of procedures including anaerobic digestion, HTL, lipid extraction, gasification, transesterification, pyrolysis, and the integration of multiple operations. By reducing the usage of solvents, the microwave-aided lipid recovery process has been proved to be a better ecologically friendly approach. The anaerobic digestion process creates biomethane without requiring any extra external energy, and the process's secondary output is utilized as a biofertilizer to combat eutrophication in aquatic bodies. Bioethanol production by fermentation of algal biomass is less energy demanding than biodiesel, which is one more attractive technique. Although commercialization is still a challenge, algal biomass might be regarded as a viable source for constructing a wide variety of goods. However, in the coming years, technical developments may enable this procedure to be financially feasible in underdeveloped nations. The fact that so many different studies are being conducted in this field suggests that algal fuel is currently undergoing rapid development in all ways: increased growth rate, improved cultivation techniques, genetic modification of crops, and improvement of thermal and chemical technologies. The severity of the energy shortage: environmental concerns are linked to the usage of nonrenewable energy, while first- and second-generation biofuels have drawbacks. However, more effort must be made to improve algae culture and processing systems before microalgal biofuels may be commercialized. Nonetheless, we may anticipate that biofuels will be able to fulfill future energy needs.

References

[1] M. Kumar, Y. Sun, R. Rathour, A. Pandey, I.S. Thakur, D.C.W. Tsang, Algae as potential feedstock for the production of biofuels and value-added products: opportunities and challenges, Sci. Total Environ. 716 (2020) 137116, https://doi.org/10.1016/j.scitotenv.2020.137116.

[2] E.S. Shuba, D. Kifle, Microalgae to biofuels: 'Promising' alternative and renewable energy, review, Renew. Sustain. Energy Rev. 81 (2018) 743−755, https://doi.org/10.1016/j.rser.2017.08.042.

[3] M. Kumar, I.S. Thakur, Municipal secondary sludge as carbon source for production and characterization of biodiesel from oleaginous bacteria, Bioresour. Technol. Rep. 4 (2018) 106−113, https://doi.org/10.1016/j.biteb.2018.09.011.

[4] G. De Bhowmick, A.K. Sarmah, R. Sen, Zero-waste algal biorefinery for bioenergy and biochar: a green leap towards achieving energy and environmental sustainability, Sci. Total Environ. 650 (2019) 2467−2482, https://doi.org/10.1016/j.scitotenv.2018.10.002.

[5] Y.S. Shin, H. Il Choi, J.W. Choi, J.S. Lee, Y.J. Sung, S.J. Sim, Multilateral approach on enhancing economic viability of lipid production from microalgae: a review, Bioresour. Technol. 258 (2018) 335−344, https://doi.org/10.1016/j.biortech.2018.03.002.

[6] A. Demirbas, Biofuels sources, biofuel policy, biofuel economy and global biofuel projections, Energy Convers. Manag. 49 (2008) 2106−2116, https://doi.org/10.1016/j.enconman.2008.02.020.

[7] P. McKendry, Energy production from biomass (part 1): overview of biomass, Bioresour. Technol. 83 (2002) 37−46, https://doi.org/10.1016/S0960-8524(01)00118-3.

[8] R.A. Voloshin, M.V. Rodionova, S.K. Zharmukhamedov, T. Nejat Veziroglu, S.I. Allakhverdiev, Review: biofuel production from plant and algal biomass, Int. J. Hydrogen Energy 41 (2016) 17257−17273, https://doi.org/10.1016/j.ijhydene.2016.07.084.

[9] P.S. Nigam, A. Singh, Production of liquid biofuels from renewable resources, Prog. Energy Combust. Sci. 37 (2011) 52−68, https://doi.org/10.1016/j.pecs.2010.01.003.

[10] B. Wang, Y. Li, N. Wu, C.Q. Lan, CO2 bio-mitigation using microalgae, Appl. Microbiol. Biotechnol. 79 (2008) 707−718, https://doi.org/10.1007/s00253-008-1518-y.

[11] R. Slade, A. Bauen, Micro-algae cultivation for biofuels: cost, energy balance, environmental impacts and future prospects, Biomass Bioenergy 53 (2013) 29−38, https://doi.org/10.1016/j.biombioe.2012.12.019.

[12] O. Surriya, S.S. Saleem, K. Waqar, A. Gul Kazi, M. Öztürk, Bio-fuels: a blessing in disguise, in: M. Öztürk, M. Ashraf, A. Aksoy, M.S.A. Ahmad (Eds.), Phytoremediation for Green Energy, Springer Netherlands, Dordrecht, 2015, pp. 11−54, https://doi.org/10.1007/978-94-007-7887-0_2.

[13] N.H. Tran, J.R. Bartlett, G.S.K. Kannangara, A.S. Milev, H. Volk, M.A. Wilson, Catalytic upgrading of biorefinery oil from micro-algae, Fuel 89 (2010) 265−274, https://doi.org/10.1016/j.fuel.2009.08.015.

[14] F. Alam, A. Date, R. Rasjidin, S. Mobin, H. Moria, A. Baqui, Biofuel from algae-is it a viable alternative? Procedia Eng. 49 (2012) 221−227, https://doi.org/10.1016/j.proeng.2012.10.131.

[15] G. Abdulqader, L. Barsanti, M.R. Tredici, Harvest of arthrospira platensis from Lake Kossorom (Chad) and its household usage among the Kanembu, J. Appl. Phycol. 12 (2000) 493−498, https://doi.org/10.1023/a:1008177925799.

[16] A.P. Carvalho, L.A. Meireles, F.X. Malcata, Microalgal reactors: a review of enclosed system designs and performances, Biotechnol. Prog. 22 (2006) 1490−1506, https://doi.org/10.1021/bp060065r.

[17] L. Brennan, P. Owende, Biofuels from microalgae—a review of technologies for production, processing, and extractions of biofuels and co-products, Renew. Sustain. Energy Rev. 14 (2010) 557−577, https://doi.org/10.1016/j.rser.2009.10.009.

[18] R.V. Kapoore, T.O. Butler, J. Pandhal, S. Vaidyanathan, Microwave-assisted extraction for microalgae: from biofuels to biorefinery, Biology 7 (2018) 1−25, https://doi.org/10.3390/biology7010018.

[19] E. Ryckebosch, K. Muylaert, I. Foubert, Optimization of an analytical procedure for extraction of lipids from microalgae, J. Am. Oil Chem. Soc. 89 (2012) 189−198, https://doi.org/10.1007/s11746-011-1903-z.

[20] R.R. de Moura, B.J. Etges, E.O. dos Santos, T.G. Martins, F. Roselet, P.C. Abreu, E.G. Primel, M.G.M. D'Oca, Microwave-assisted extraction of lipids from wet microalgae paste: a quick and efficient method, Eur. J. Lipid Sci. Technol. 120 (2018) 1700419, https://doi.org/10.1002/ejlt.201700419.

[21] D.A. Nogueira, J.M. Da Silveira, É.M. Vidal, N.T. Ribeiro, C.A. Veiga Burkert, Cell disruption of chaetoceros calcitrans by microwave and ultrasound in lipid extraction, Int. J. Chem. Eng. (2018), https://doi.org/10.1155/2018/9508723.

[22] B. Hosseini, A. Guionet, H. Akiyama, H. Hosseini, Study of oil extraction from microalgae by pulsed power as a renewable source of green energy, IEEE Int. Pulsed Power Conf. 2017 (2018) 3518–3523, https://doi.org/10.1109/PPC.2017.8291277.

[23] S. Duraiarasan, M. Vijay, A comprehensive review on the potential and alternative biofuel, Res. J. Chem. Sci. 2 (2012) 71–82.

[24] M. Kumar, R. Morya, E. Gnansounou, C. Larroche, I.S. Thakur, Characterization of carbon dioxide concentrating chemolithotrophic bacterium Serratia sp. ISTD04 for production of biodiesel, Bioresour. Technol. 243 (2017) 893–897, https://doi.org/10.1016/j.biortech.2017.07.067.

[25] P.D. Patil, V.G. Gude, A. Mannarswamy, P. Cooke, N. Nirmalakhandan, P. Lammers, S. Deng, Comparison of direct transesterification of algal biomass under supercritical methanol and microwave irradiation conditions, Fuel 97 (2012) 822–831, https://doi.org/10.1016/j.fuel.2012.02.037.

[26] O.K. Agwa, I.G. Nwosu, G.O. Abu, Bioethanol production from *Chlorella vulgaris* biomass cultivated with plantain (*Musa paradisiaca*) peels extract, Adv. Biosci. Biotechnol. 08 (2017) 478–490, https://doi.org/10.4236/abb.2017.812035.

[27] R. Harun, J.W.S. Yip, S. Thiruvenkadam, W.A.W.A.K. Ghani, T. Cherrington, M.K. Danquah, Algal biomass conversion to bioethanol—a step-by-step assessment, Biotechnol. J. 9 (2014) 73–86, https://doi.org/10.1002/biot.201200353.

[28] S. Behera, R. Singh, R. Arora, N.K. Sharma, M. Shukla, S. Kumar, Scope of algae as third generation biofuels, Front. Bioeng. Biotechnol. 2 (2015) 1–13, https://doi.org/10.3389/fbioe.2014.00090.

[29] C.E. de Farias Silva, D. Meneghello, A.K. de Souza Abud, A. Bertucco, Pretreatment of microalgal biomass to improve the enzymatic hydrolysis of carbohydrates by ultrasonication: yield vs energy consumption, J. King Saud Univ. Sci. 32 (2020) 606–613, https://doi.org/10.1016/j.jksus.2018.09.007.

[30] M. Yanagisawa, K. Nakamura, O. Ariga, K. Nakasaki, Production of high concentrations of bioethanol from seaweeds that contain easily hydrolyzable polysaccharides, Process Biochem. 46 (2011) 2111–2116, https://doi.org/10.1016/j.procbio.2011.08.001.

[31] C. Xiros, E. Topakas, P. Christakopoulos, Hydrolysis and fermentation for cellulosic ethanol production, WIREs Energy Environ. 2 (2013) 633–654, https://doi.org/10.1002/wene.49.

[32] F. Offei, M. Mensah, A. Thygesen, F. Kemausuor, Seaweed bioethanol production: a process selection review on hydrolysis and fermentation, Fermentation 4 (2018) 1–18, https://doi.org/10.3390/fermentation4040099.

[33] J.J. Wang, Y.P. Han, J.Y. Chang, Z.Y. Chen, Light scattering of a Bessel beam by a nucleated biological cell: an eccentric sphere model, J. Quant. Spectrosc. Radiat. Transf. 206 (2018) 22–30, https://doi.org/10.1016/j.jqsrt.2017.10.025.

[34] V. Klassen, O. Blifernez-Klassen, D. Wibberg, A. Winkler, J. Kalinowski, C. Posten, O. Kruse, Highly efficient methane generation from untreated microalgae biomass, Biotechnol. Biofuels 10 (2017) 1–12, https://doi.org/10.1186/s13068-017-0871-4.

[35] B. Macura, S.L. Johannesdottir, M. Piniewski, N.R. Haddaway, E. Kvarnström, Effectiveness of ecotechnologies for recovery of nitrogen and phosphorus from anaerobic digestate and effectiveness of the recovery products as fertilisers: a systematic review protocol, Environ. Evid. 8 (2019) 1–9, https://doi.org/10.1186/s13750-019-0173-3.

[36] A. Rabii, S. Aldin, Y. Dahman, E. Elbeshbishy, A review on anaerobic co-digestion with a focus on the microbial populations and the effect of multi-stage digester configuration, Energies 12 (2019), https://doi.org/10.3390/en12061106.

[37] J.J. Milledge, B.V. Nielsen, S. Maneein, P.J. Harvey, A brief review of anaerobic digestion of algae for bioenergy, Energies 12 (2019) 1–22, https://doi.org/10.3390/en12061166.

[38] E. Allen, D.M. Wall, C. Herrmann, A. Xia, J.D. Murphy, What is the gross energy yield of third generation gaseous biofuel sourced from seaweed? Energy 81 (2015) 352–360, https://doi.org/10.1016/j.energy.2014.12.048.

[39] J.H. Mussgnug, V. Klassen, A. Schlüter, O. Kruse, Microalgae as substrates for fermentative biogas production in a combined biorefinery concept, J. Biotechnol. 150 (2010) 51–56, https://doi.org/10.1016/j.jbiotec.2010.07.030.

[40] R. Tripathi, J. Singh, I.S. Thakur, Characterization of microalga *Scenedesmus* sp. ISTGA1 for potential CO_2 sequestration and biodiesel production, Renew. Energy 74 (2015) 774–781, https://doi.org/10.1016/j.renene.2014.09.005.

[41] M.K. Enamala, S. Enamala, M. Chavali, J. Donepudi, R. Yadavalli, B. Kolapalli, T.V. Aradhyula, J. Velpuri, C. Kuppam, Production of biofuels from microalgae—a review on cultivation, harvesting, lipid extraction, and numerous applications of microalgae, Renew. Sustain. Energy Rev. 94 (2018) 49–68, https://doi.org/10.1016/j.rser.2018.05.012.

[42] D. Mitra, J.H. van Leeuwen, B. Lamsal, Heterotrophic/mixotrophic cultivation of oleaginous *Chlorella vulgaris* on industrial co-products, Algal Res. 1 (2012) 40–48, https://doi.org/10.1016/j.algal.2012.03.002.

[43] A. Demirbaş, Production of biodiesel from algae oils, energy sources, Part A recover, Util. Environ. Eff. 31 (2009) 163–168, https://doi.org/10.1080/15567030701521775.

[44] T.M. Mata, A.A. Martins, N.S. Caetano, Microalgae for biodiesel production and other applications: a review, Renew. Sustain. Energy Rev. 14 (2010) 217–232, https://doi.org/10.1016/j.rser.2009.07.020.

[45] E.T. Kostas, D.A. White, C. Du, D.J. Cook, Selection of yeast strains for bioethanol production from UK seaweeds, J. Appl. Phycol. 28 (2016) 1427–1441, https://doi.org/10.1007/s10811-015-0633-2.

[46] S.J. Horn, I.M. Aasen, K. Østgaard, Production of ethanol from mannitol by *Zymobacter palmae*, J. Ind. Microbiol. Biotechnol. 24 (2000) 51–57.

[47] K. Hon-Nami, A unique feature of hydrogen recovery in endogenous starch-to-alcohol fermentation of the marine microalga, *Chlamydomonas perigranulata*, Appl. Biochem. Biotechnol. 131 (2006) 808–828, https://doi.org/10.1385/ABAB:131:1:808.

[48] R. Harun, M. Singh, G.M. Forde, M.K. Danquah, Bioprocess engineering of microalgae to produce a variety of consumer products, Renew. Sustain. Energy Rev. 14 (2010) 1037−1047, https://doi.org/10.1016/j.rser.2009.11.004.

[49] M.I. Inyang, B. Gao, Y. Yao, Y. Xue, A. Zimmerman, A. Mosa, P. Pullammanappallil, Y.S. Ok, X. Cao, A review of biochar as a low-cost adsorbent for aqueous heavy metal removal, Crit. Rev. Environ. Sci. Technol. 46 (2016) 406−433, https://doi.org/10.1080/10643389.2015.1096880.

[50] I. Michalak, S. Baśladyńska, J. Mokrzycki, P. Rutkowski, Biochar from a freshwater macroalga as a potential biosorbent for wastewater treatment, Water (Switzerland) 11 (2019) 4−6, https://doi.org/10.3390/w11071390.

[51] A.T. Tag, G. Duman, S. Ucar, J. Yanik, Effects of feedstock type and pyrolysis temperature on potential applications of biochar, J. Anal. Appl. Pyrol. 120 (2016) 200−206, https://doi.org/10.1016/j.jaap.2016.05.006.

[52] P. Salimi, O. Norouzi, S.E.M. Pourhoseini, P. Bartocci, A. Tavasoli, F. Di Maria, S.M. Pirbazari, G. Bidini, F. Fantozzi, Magnetic biochar obtained through catalytic pyrolysis of macroalgae: a promising anode material for li-ion batteries, Renew. Energy 140 (2019) 704−714, https://doi.org/10.1016/j.renene.2019.03.077.

[53] T. Gomiero, Are biofuels an effective and viable energy strategy for industrialized societies? A reasoned overview of potentials and limits, Sustain. Times 7 (2015) 8491−8521, https://doi.org/10.3390/su7078491.

[54] G. Festel, M. Würmseher, C. Rammer, E. Boles, M. Bellof, Modelling production cost scenarios for biofuels and fossil fuels in Europe, J. Clean. Prod. 66 (2014) 242−253, https://doi.org/10.1016/j.jclepro.2013.10.038.

[55] B. Satari, K. Karimi, R. Kumar, Cellulose solvent-based pretreatment for enhanced second-generation biofuel production: a review, R. Soc. Chem. (2019), https://doi.org/10.1039/c8se00287h.

[56] S.N. Naik, V.V. Goud, P.K. Rout, A.K. Dalai, Production of first and second generation biofuels: a comprehensive review, Renew. Sustain. Energy Rev. 14 (2010) 578−597, https://doi.org/10.1016/j.rser.2009.10.003.

[57] W.H.L. Stafford, G.A. Lotter, G.P. von Maltitz, A.C. Brent, Biofuels technology development in Southern africa, Dev. South Afr. 36 (2019) 155−174, https://doi.org/10.1080/0376835X.2018.1481732.

[58] M. Shah (Ed.), Microbial Bioremediation & Biodegradation, Springer, Singapore, 2020.

[59] M. Shah (Ed.), Removal of Refractory Pollutants from Wastewater Treatment Plants, CRC Press, USA, 2021.

Chapter 14

Microalgae biofuels: a promising substitute and renewable energy

Indu Chauhan[1], Vivek Sharma[2], Pawan Rekha[3] and Lovjeet Singh[4]

[1]*Department of Biotechnology, Dr. B. R. Ambedkar National Institute of Technology Jalandhar, Jalandhar, Punjab, India;* [2]*Department of Chemistry, Banasthali Vidyapith, Newai, Rajasthan, India;* [3]*Department of Chemistry, Malaviya National Institute of Technology Jaipur, Jaipur, Rajasthan, India;* [4]*Department of Chemical Engineering, Malaviya National Institute of Technology Jaipur, Jaipur, Rajasthan, India*

1. Introduction

With the increased population and fast industrialization, global warming and demands for more energy as fuel are also increasing, which results in the depletion of fossil fuels [1]. Thus in the last few decades, production of renewable energy such as biofuels source has gained momentum, where they can be seen as an alternate and green energy to conventional fuels. Biofuel functions similar to nonrenewable fossil fuels but are derived from microbial, plant, or animal materials or wastes and generally cause less damage to the planet [2]. Depending upon sources and processing techniques, biofuels are classified into different generations, viz., first generation (produced from food crops), second generation (produced from nonfood crops), and third or advanced generation (produced from biomass) of biofuels [3]. However, the challenges in the production of biofuels are the increased food prices and large land requirement for cultivation, which limits biofuel production using crops. Therefore in last few decades, interest has been shifted toward biomass-based energy production such as use of microalgae. Microalgae owing to their unique potential can be used for production of a variety of biofuels such as biogas, biohydrogen, biodiesel, etc., and also can be used in phycoremediation of wastewater [4]. Use of various algal strains such as *Chlorella* and *Botryococcus braunii* have been reported so far for biodiesel, bioethanol, biohydrogen, biomethane, and bioelectricity production [5]. However selection of different strains of microalgae is exclusively dependent upon various factors such as oil concentration, production yield, and downstream processing [6] (Scheme 14.1).

2. Benefits of using microalgae for biofuel productions

For the production of renewable biofuels, microalgal biomass has been considered one of the potential alternative feedstocks. Microalgae owing to their advantages over other conventional biofuel feedstocks sources such as ability to grow in a harsh environment, high productivity, and high lipid contents are used as a suitable source for the production of different biofuels [7]. The concept of microalgae cultivation for biofuel manufacturing has become the spotlight nowadays due to several positive perspectives: (i) microalgae production do not interfere food chains, (ii) they have high carbohydrate, protein, and oil content, which are essential components of biofuel conversion, (iii) microalgae assimilate nutrients from polluted water and thus can easily be grown on wastewater, which in turn helps in waste water treatment, (iv) they have naturally sustainable growth in the presence of sunlight, (v) they can be cultivated in the waste and wet bare lands, and (vi) they help in decreasing CO_2 by utilizing it for photosynthesis respiration [8,9] (Table 14.1).

3. Cultivation and harvesting of microalgae for biofuel production

Microalgae grow rapidly and do not require fertile land. Furthermore, their cultivation can be done using effluents or domestic sewage water, which helps bioremediation of wastewater. Although it is easy to culture microalgae under laboratory conditions, their productivity in large-scale production is still difficult. For efficient microalgae culturing at a large

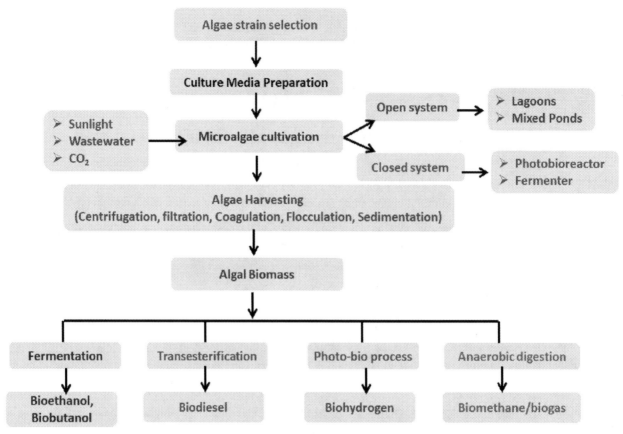

SCHEME 14.1 Microalgae for biofuel production.

TABLE 14.1 Carbohydrate–protein–lipid composition of various microalgae % dry weight [8].

Algae	Carbohydrate (%)	Protein (%)	Lipids (%)
Chlamydomonas reinhardtii	48	17	21
Chlorella sp.	56	22	19
Spirogyra sp.	20	55	16
Porphyridium cruentum	35	50	11
Spirulina platensis	60	12	8
Dunaliella salina	57	32	6
Bellerochea sp.	3	24	15
Chaetoceros sp.	2	18	18
Rhodomonas sp.	9	74	15
Scenedesmus sp.	18	56	12

scale, a system should possess an adequate light source, minimal contamination rate, low production cost, and high land efficiency [6]. To enhance the yield and growth of microalgal biomass, systems such as open ponds and photobioreactors (PBRs) are used.

3.1 Growth factors for microalgae cultivation

Light: Light is one of the most important factors that can affect microalgae culturing as microalgae grow photosynthetically. Artificial light, solar light, or both can be used as a light source to illuminate algal culture systems [10].

Naturally illuminated algal culture systems have large surface areas such as open ponds, horizontal/serpentine tubular airlifts, and flat plates. However in a closed-type system such as laboratory-scale tubular PBRs, cells are artificially exposed to light by illuminating with fluorescent lamps or other light distributors [11]. The density of cells also affects the light intensity requirement as higher cell densities need higher light intensities for their growth. However optimal light is required by every microalgal species. An increase in light intensity above a certain value reduces the biomass growth rate, which is known as photoinhibition [11].

Temperature: Temperature affects photorespiration and respiration more strongly than photosynthesis. Microalgae require an optimal value of temperature for their maximal growth. For microalgae cultures, an optimal temperature is between 20 and 24°C. Although, depending upon the species, cultured strain, and composition of the culture medium, temperature can still be varied. It has been observed that below 16°C, growth of microalgae slows down, whereas for a number of species, temperature of more than 35°C is fatal [11]. However due to the greenhouse effect, the temperature fluctuates between 10 and 45°C in temperate regions, which affects the growth of microalgae in outdoor photo-bioreactors. It has been observed that for each 10°C increase, photosynthesis, cell division and growth double, but microalgae growth rate sharply decreases on exceeding the optimal temperature [12].

Carbon dioxide (CO_2): Carbon is one of the major nutrient supplements required for the production of large quantities of algal biomass [10]. In addition to CO_2, nitrogen dioxide (NO_2) and various other pollutants from different sources are the nutrients for algae. However, according to PBR type and microalgal species, the amount of CO_2 required for microalgae growth varies. Usually for the maximum algae growth, low concentration of CO_2 is required, but some species can even survive with high concentration of CO_2 as well. *Chlorella* and *Euglena gracilis* are a couple of the high CO_2 tolerance species [13].

pH: The pH of cultures has important significance on the growth of microalgal biomass as it determines the solubility of minerals and CO_2 in the medium [14]. For microalgae culture, mostly commonly pH from 6 to 8 is preferred, but a tolerance level to the pH of the culture medium can also vary from species to species, which may affect their growth rate. Usually most microalgae tolerate wide pH intervals, but beyond pH 8, yield or growth of the culture is greatly reduced.

3.2 Bioreactors used for culturing

Cultivation of microalgae is done for the purposes of biomass production, wastewater treatment, pond filtration, and CO_2 fixation. However for the biofuel generation, the two most commonly used microalgae culturing systems are open pond and PBR.

3.2.1 Open ponds

For large-scale cultivation, open pond cultivation is one of the simplest and oldest ways. Natural water sources such as ponds and lakes and artificial water bodies such as raceway ponds are some of the types of open pond [15]. Naturally occurring ponds or lakes contain high pH, nitrate, and phosphate, which are required for the sufficient algal growth. Open raceway pond is a type of artificial open pond in which a culture vessel is made up of a concrete unit and compacted with the earth [14]. The culture system is mixed with the help of a paddle wheel or simple floater system, and biomass is harvested behind the paddle wheel after a complete circulation (Fig. 14.1A).

(a) (b)

FIGURE 14.1 (A) Open raceway pond and (B) closed tubular photobioreactor [14].

3.2.2 Photobioreactor

PBRs on the other hand are a closed culture system that prevents the direct penetration of light to the culture surface (Fig. 14.1B). Light in this system passes through transparent reactor walls to reach the culture system. This culture system does not allow direct exchange of gases and contaminants between the culture and atmosphere [10]. On the basis of structures, different types of PBRs are used for culturing. For production of microalgae, tubular (Fig. 14.1B), flat-plate reactors, and vertical column PBRs are most commonly used [11].

Although both open and closed photobioreactors are used for microalgae culturing, they have their own advantages and disadvantages (Table 14.2).

4. Harvesting of algal biomass

Harvesting of microalgae after culturing is done to dewater and concentrate the diluted algal suspension into a thick slurry that contains around 2%–7% (dry weight) algal suspension [17]. Harvesting methods that are most commonly used include physical, chemical, and biologic processes (Scheme 14.2). However, no single method is best for harvesting microalgae as each method has its own advantages and disadvantages [18].

4.1 Physical methods

Sedimentation: In this technique, cells are separated on the basis of their sizes, density, nutrient content, etc., under gravitational force. This method generates dilute slurry and should therefore be used before other dewatering methods such as centrifugation or filtration to preconcentrate the biomass [19,20].

Filtration: Filtration is used to separate solids from fluids by passing the samples through a permeable separator such as filter cloths or permeable membranes [19]. Microalgae in this technique are separated on the basis of their cell sizes. Also, by varying the pore sizes of a separator, cells are separated based on particle sizes or molecular weight [19]. Different techniques such as membrane filtration, vacuum filtration, etc., are employed for microalgae harvesting [19,20].

Centrifugation: It is the most commonly and widely used harvesting technology, which utilizes the centrifugal force for the separation of microalgae cells from a growth medium [20]. Although this harvesting method has close to 100% efficiency, it consumes more energy [21]. Different types of centrifuges such as a decanter centrifuge, disc bowl, nozzle type, solid ejecting, etc., are used for harvesting microalgae.

Flotation: This method works on the interaction of air bubbles with algae cells. Small bubbles of air are generated by mixing the supersaturated water with air into the culture. The small air bubbles attach with microalgae cells and bring them to the surface of water, leading to the formation of scum that can be later skimmed off [19,20]. However, because of the presence of negative charge on the surfaces of both air bubbles and microalgae, it is very difficult for them to interact. Therefore additives/surfactants are added in the culture medium to improve the interaction [20]. Based on the bubble sizes, different types of flotation systems are used such as dispersed air flotation, electro flotation, ozone flotation, etc., for microalgae harvesting [20].

TABLE 14.2 Advantages and disadvantages of open and PBR system [16].

S. No.	Bioreactor type	Advantages	Limitations
1.	Raceway pond	• Easy to construct • Relatively cheap • Easy maintenance • Good for mass cultivation	• High evaporative losses • High diffusion of CO_2 to the atmosphere • Poor light utilization • Requirement of large areas of land • Microbial and native algae contamination • Poor biomass productivity
2.	Tubular photobioreactor	• Relatively high biomass productivities • Less prone to contamination • Suitable for outdoor mass culture	• High initial investment costs • Toxic accumulation of O_2 fouling and overheating

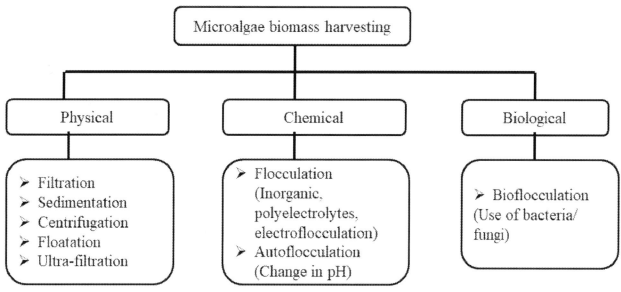

SCHEME 14.2 Methods used for microalgae biomass harvesting.

4.2 Chemical methods

Flocculation: Flocculation is a method in which additives/polymers are added to the microalgae culture, which causes algae to aggregate and form colloids [18]. Microalgal cells have negative charges on their surfaces, which makes them stable in dilute solution. Positively charged flocculants such as alum and ferric chloride are used to neutralize the negative charges present on the cell surface and thus destabilize them to form aggregates [18].

Autoflocculation (change in pH): In this method, pH of the solution is changed to cause the flocculation of the cells. Inorganic salt precipitates are used to increase the pH (alkaline) of the cell suspension, inducing flocculation of algae. Flocculation by increase in pH can also be induced naturally due to CO_2 depletion and by adding magnesium/calcium hydroxide [18,19].

4.3 Biologic method

Bioflocculation: It is a greener method of harvesting microalgae as no chemicals are required. In this technique, a mutual flocculation is induced by mixing two different species of microalgae [18]. Reports are available on using a mixture of microbes such as *Pseudomonas stutzeri* and *Bacillus cereus* for the harvesting of marine microalgae *Pleurochrysis carterae* [22]. Various microorganisms such as bacteria, algae, and fungi secrete extracellular biopolymers that act as a bioflocculant. Both freshwater and marine microalgae can be effectively harvested using poly (γ-glutamic acid) from *Bacillus subtilis* [23].

5. Biofuel production using microalgae

Microalgae consisting of higher amounts of lipid have faster growth rate and higher photosynthetic yield than their land plant counterpart and have emerged as a potential source for generation of biofuel.

5.1 Biohydrogen

Biohydrogen is also known as green hydrogen because no environmental pollutants are generated during its synthesis. Recently, use of microalgae as a substrate or via photolysis has been considered a promising approach for biohydrogen production [24].

Biophotolysis: Biophotolysis is the process where water (H_2O) dissociates into molecular hydrogen and oxygen in the presence of light in biologic systems. Hydrogen production in microalgae using biophotolytic process is classified into two types: direct and indirect. In direct biophotolysis, sunlight is captured by photosystem II protein complex present in thylakoid membrane of the cell, which utilizes this energy to split water into oxygen (O_2), protons (H^+), and electrons (e^-).

Then iron−sulfur proteins ferredoxin (Fd) and other intermediates mediate electron transfer to hydrogenase enzyme, which produces H_2 by catalyzing the proton−electron recombination reaction [24,25]. Whereas, in indirect biophotolysis, e^- generated from water splitting are used in the Calvin cycle carbon fixation and then subsequently stored in the form of endogenous reserve carbohydrates, which are later used to produce H_2. Nitrogen-fixing *cyanobacteria* (blue-green algae) usually show indirect biophotolytic production of hydrogen [25].

Fermentation: A fermentative metabolism has also been identified that can be used for the production of hydrogen under dark conditions. Here, the enzymatic activity of pyruvate:ferredoxin oxidoreductase in *Chlamydomonas reinhardtii* is responsible for the reduction of Fd and the passage of electrons toward hydrogenase enzyme [25].

5.2 Bioethanol

Bioethanol production can be classified into different generations depending upon process and raw materials used. The first generation of bioethanol is produced using starch- or saccharine-based raw material. Second and third generations of bioethanol are produced using lignocellulosic materials and microalgal biomass, respectively. Microalgae contain cellulose, starch, and glycogen, which are raw materials for production of ethanol [26]. For ethanol production, biomass or the cells are first pretreated to release their constituents, from which starch is then extracted using various physical methods such as ultrasonic, mechanical shear, etc., or by using enzymes to dissolve cell walls of biomass [27]. Over the last few years, bioethanol production as a fourth-generation biofuel has become an emerging technology where ethanol can be produced using genetically modified, i.e., *cyanobacteria*, without the breakdown of biomass. Herein, with the help of a process known as "photofermentation," ethanol can be directly produced using algae by utilizing sunlight, water, and nutrients [28].

5.3 Biodiesel

In the last few years, interest has gained momentum on using biodiesel as an alternative option to conventional fuel as it has properties very close to diesel. However, at present biodiesel costs are around double that of conventional diesel due to less availability of oil crops and high cost of raw sources [29]. Microalgae having higher oil content than other feedstock are now preferred as a green alternative fuel over the other sources for the generation of biofuels [29]. Both esterification and transesterification processes can be used for biodiesel production. However with these techniques, less yield of biodiesel is produced, and also the reaction takes a longer time [29]. Therefore, a two-stage process has been demonstrated by Rahman et al. for biodiesel production using algae *Spirulina maxima*, where in the first stage, free fatty acid content is reduced by algal esterifying algal oil with acid (H_2SO_4) catalyst for followed by transesterification using potassium hydroxide as alkaline catalyst to enhance biodiesel yield in the second stage (Fig. 14.2).

5.4 Biogas

Biogas is a fuel produced by the breakdown of organic waste by microorganisms in anaerobic digestion. The sources of biogas can be manure, food, and vegetable wastes. The composition of biogas consists of methane (50%−75%), carbon dioxide (25%−50%), nitrogen (0%−10%), hydrogen (0%−1%), and hydrogen sulfide (0%−3%) [30]. However depending upon raw sources and experimental conditions, the percentage of gases generated can vary. The use *C. reinhardtii* strain cc124 for the production of biogas per VDI 4630 guidelines of the Verein Deutscher Ingenieure (VDI, 2004) has been reported by Mussgnung et al. [31]. Similarly, Baltrėnas et al. have reported the production of biogas using *Cladophora glomerata*, *Chara fragilis*, and *Spirogyra neglecta* in batch fermentation technique (Fig. 14.3) [30].

6. Applications

6.1 Transport

Because of fossil fuel depletion and after the oil crises of 1973 and 1979, biofuels have started to be considered a substitute for transportation. On an energy content basis, just 2.7% of global road transport fuels are accounted for by biofuels. Although various organizations have started to produce biofuels for transportation purposes, still it is estimated that by 2030, only 7% of global transport fuels demand can be met by biofuels [32]. Among the various other biofuels, biodiesel has gained an interest to be used an alternative source of energy to replace petroleum-based fuel in transportation [33]. It has been reported that in both Malaysia and Colombia, palm oil biodiesel is finding a base of biodiesel policies for the transportation sector [33].

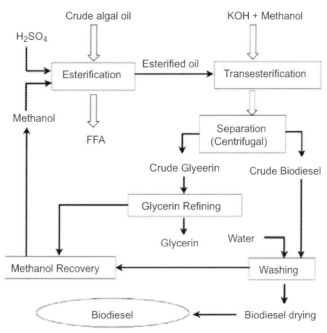

FIGURE 14.2 Steps involved in the synthesis of biodiesel using microalgae biomass [29].

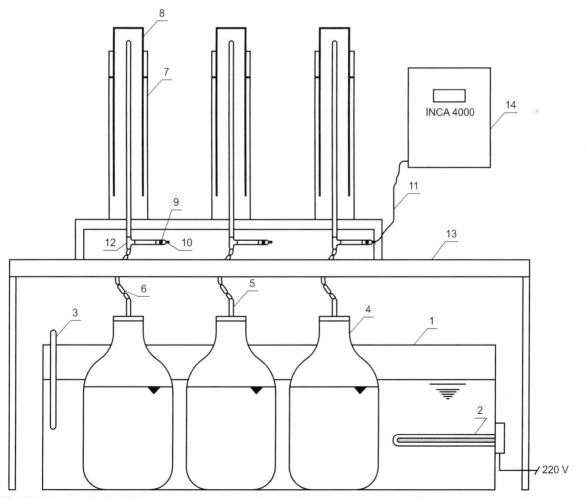

FIGURE 14.3 Scheme of the batch bioreactor stand: 1 — vessel with water, 2 — substrate heating device, 3 — temperature sensor, 4 — bioreactor, 5 — branch pipe for hose fixing, 6 — flexible hose, 7 — vessel with water, 8 — biogas accumulation vessel (PVC pipe), 9 — valve, 10 — branch pipe for gas discharge, 11 — elastic hose, 12 — T-socket, 13 — table, 14 — gas analyzer [30].

6.2 Aviation

The aviation sector determines the economy of a country. As aviation moves into the renewables era, hydrogen-powered aircraft might enter common use. Hydrogen used as a fuel in airplanes has several positive attributes such as zero emission of CO_2, global availability, safety, light weight, etc. In last few years, microalgae-based jet fuel has gained considerable attention to be used in the aviation sector [34]. Various approaches such as gasification with Fischer-Tropsch and sugar-to-jet have been used to produce bio-jet fuels [35]. *C. reinhardtii* is one the microalgal species that has been used to produce the biohydrogen [24].

6.3 Lubricants

Lubricants are used to reduce friction/wear in moving surfaces for the smooth operating of machines, while contributing to energy savings. Currently, crude oil–based lubrications are used worldwide [36]. However, production, usage, and improper disposal of these lubricants in the environment can have significant impacts. Therefore, nowadays interest is being shifted to use of vegetable oil–based green lubricants. Recently, use of *Chlorella* sp. and *Spirulina* microalgae due to their capacity to produce high-value lipids/oils for the production of lubricants has been demonstrated by Dziosa and Makawska and Xu et al., respectively [36–38].

6.4 Power/electricity generation

With the increase in population and improvement in lifestyle, demand of power consumption has also been increased. However, with the depletion of fossil fuel, it has become very difficult to meet this demand. Therefore, to solve the energy crisis and to prevent an environmental load, microbial fuel cells over the last few decades have gained interest to serve as an alternative sustainable energy source, delivering green energy and clean water. Microalgae species that have incorporated biofuel cells, i.e., the microalgae microbial fuel cells, convert solar energy into electricity and utilize CO_2 in a self-sustainable way [39]. Production of bioelectricity using *Chlorella pyrenoidosa* and Batik wastewater has been demonstrated by Polontalo et al. [40].

References

[1] F. Perera, Pollution from fossil-fuel combustion is the leading environmental threat to global pediatric health and equity: solutions exist, Int. J. Environ. Res. Publ. Health 15 (2018) 16.

[2] S. Mahapatra, D. Kumar, B. Singh, P.K. Sachan, Biofuels and their sources of production: a review on cleaner sustainable alternative against conventional fuel, in the framework of the food and energy nexus, Energy Nexus 4 (2021) 100036.

[3] A. Hayder, A. Alalwan, H. Alminshid Haydar, A.S. Aljaafari, Promising evolution of biofuel generations. Subject review, Renew. Energy Focus 28 (2019) 127.

[4] B. Koul, K. Sharma, M.P.S. Phycoremediation, A Sustainable Alternative in Wastewater Treatment (WWT) Regime, vol 25, Environmental Technology & Innovation, 2022, p. 102040.

[5] A. Tiwari, T. Kiran, Biofuels from microalgae, Adv. Biofuels Bioenergy. IntechOpen (2018), https://doi.org/10.5772/intechopen.73012.

[6] J.J. Milledge, B.V. Nielsen, S. Maneein, P.J. Harvey, A brief review of anaerobic digestion of algae for bioenergy, Energies 12 (2019) 1166.

[7] S.K. Gupta, F.A. Ansari, K. Bauddh, B. Singh, A.K. Nema, K.K. Pant, Harvesting of microalgae for biofuels: comprehensive performance evaluation of natural, inorganic, and synthetic flocculants, Green Technol. Environ. Sustain. (2017) 131.

[8] N. Hossain, T.M.I. Mahlia, R. Saidur, Latest development in microalgae-biofuel production with nano-additives, Hossain et al. Biotechnol. Biofuel. 12 (2019) 16.

[9] L.M.L. Laurens, M. Chen-Glasser, J.D. McMillan, A perspective on renewable bioenergy from photosynthetic algae as feedstock for biofuels and bioproducts, Algal Res. 24 (2017) 261.

[10] G.G. Satpati, R. Pal, Microalgae- biomass to biodiesel: a review, J. Algal Biomass Util. 9 (2018) 11.

[11] J.P. Bitog, I.-B. Lee, C.-G. Lee, K.-S. Kim, H.-S. Hwang, S.-W. Hong, I.-H. Seo, K.-S. Kwon, E. Mostafa, Application of computational fluid dynamics for modeling and designing photobioreactors for microalgae production: a review, Comput. Electron. Agric. 76 (2011) 131.

[12] M. Ras, J.-P. Steyer, O. Bernard, Temperature effect on microalgae: a crucial factor for outdoor production, Rev. Environ. Sci. Biotechnol. 12 (2013) 153.

[13] M. Fadhil, Salih. Microalgae tolerance to high concentrations of carbon dioxide: a Review, J. Environ. Protect. 2 (2011) 648.

[14] K.H. Chowdury, N. Nahar, U.K. Deb, The growth factors involved in microalgae cultivation for biofuel production: a review, Comput. Water Energy Environ. Eng. 9 (2020) 185.

[15] J.S. Tan, S.Y. Lee, K.W. Chew, M.K. Lam, J.W. Lim, S.-H. Ho, P.L. Show, A review on microalgae cultivation and harvesting, and their biomass extraction processing using ionic liquids, Bioengineered 11 (2020) 116.

[16] X. Zhang, J. Rong, H. Chen, C. He, Q. Wang, Current status and outlook in the application of microalgae in biodiesel production and environmental protection, Front. Energy Res. 2 (2014) 15.
[17] Z. Wen, J. Liu, F. Chen, Biofuel from microalgae, Comprehen. Biotechnol. 3 (2011) 127–133 (Second Edition).
[18] I. Branyikova, G. Prochazkova, T. Potocar, Z. Jezkova, T. Branyik, Harvesting of microalgae by flocculation, Fermentation 4 (2018) 12.
[19] K. Muylaert, D. Vandamme, I. Foubert, P.V. Brady, Harvesting of microalgae by means of 375 flocculation, in: Biomass and Biofuels from Microalgae, Springer International Publishing, 2015, pp. 251–273.
[20] A. Kumar, Current and future perspective of microalgae for simultaneous wastewater treatment and feedstock for biofuels production, Chem. Africa (2021), https://doi.org/10.1007/s42250-020-00221-9.
[21] Y.S.H. Najjar, A. Abu-Shamleh, Harvesting of microalgae by centrifugation for biodiesel production: a review, Algal Res. 51 (2020) 102046.
[22] A.K. Lee, D.M. Lewis, P.J. Ashman, Microbial flocculation, a potentially low-cost harvesting technique for marine microalgae for the production of biodiesel, J. Appl. Phycol. 21 (2009) 559.
[23] H.L. Zheng, Z. Gao, J.L. Yin, X.H. Tang, X. J Ji, H. Huang, Harvesting of microalgae by flocculation with poly (gamma-glutamic acid), Bioresour. Technol. 112 (2012) 212.
[24] A.R. Limongi, E. Viviano, M.D. Luca, R.P. Radice, G. Bianco, G. Martelli, Biohydrogen from microalgae: production and applications, Appl. Sci. 11 (2021) 14.
[25] A.H. Rather, A.K. Srivastav, A study on biohydrogen production based on biophotolysis from cyanobacteria, Ann. Romanian Soc. Cell Biol. 25 (2021) 12500.
[26] S.P. Cuellar-Bermudez, J.S. Garcia-Perez, B.E. Rittmann, R. Parra-Saldivar, Photosynthetic bioenergy utilizing CO_2: an approach on flue gases utilization for third generation biofuels, J. Clean. Prod. 98 (2015) 53.
[27] Z.M.A. Bundhoo, R. Mohee, Ultrasound-assisted biological conversion of biomass and waste materials to biofuels: a review, Ultrason. Sonochem. 40 (2018) 298.
[28] C.E. de Farias Silva, A. Bertucco, Bioethanol from microalgal biomass: a promising approach in biorefinery, Braz. Arch. Biol. Technol. 62 (2019) 14.
[29] M.A. Rahman, M.A. Aziz, R.A. Al-khulaidi, M. Islam, Biodiesel production from microalgae Spirulina maxima by two step process: optimization of process variable, J. Radiat. Res. Appl. Sci. 10 (2017) 140.
[30] P. Baltrėnas, A. Misevičius, Biogas production experimental research using algae, J. Environ. Health Sci. Eng. 13 (2015) 7.
[31] J.H. Mussgnug, V. Klassen, A. Schlüter, O. Kruse, Microalgae as substrates for fermentative biogas production in a combined biorefinery concept, J. Biotechnol. 150 (2010) 6.
[32] G.R. Timilsina, Biofuels in the long-run globalenergy supply mix for transportation, Philos. Trans. Royal Soc. A 372 (2014) 20120323, https://doi.org/10.1098/rsta.2012.0323.
[33] M.N.A.M. Yusoff1, N.W.M. Zulkifli1, N.L. Sukiman, O.H. Chyuan, M.H. Hassan, M.H. Hasnul, M.S.A. Zulkifli, M.M. Abbas, M.Z. Zakaria, Sustainability of palm biodiesel in transportation: a review on biofuel standard, policy and international collaboration between Malaysia and Colombia, BioEnergy Res. 14 (2021) 43.
[34] J. K Bwapwa, A. Anandraj, C. Trois, Possibilities for conversion of microalgae oil into aviation fuel: a review, Renew. Sustain. Energy Rev. 80 (2017) 1345.
[35] J.H.K. Lim, Y.Y. Gan, H.C. Ong, B.F. Lau, W.-H. Chen, C.T. Chong, T.C. Ling, J.J. Klemeš, Utilization of microalgae for bio-jet fuel production in the aviation sector: challenges and perspective, Renew. Sustain. Energy Rev. 149 (2021) 111396.
[36] L.I. Farfan-Cabrera, M. Franco-Morgado, A. González-Sánchez, J. Pérez-González, B.M. Marín-Santibáñez, Microalgae biomass as a new potential source of sustainable green lubricants, Molecules 27 (1205) (2022) 33.
[37] K. Dziosa, M. Makowska, A method for the preparation of lubricating oil from microalgae biomass, Tribologia 270 (2016) 33.
[38] Y. Xu, X. Hu, K. Yuan, G. Zhu, W. Wang, Friction and wear behaviors of catalytic methylesterified bio-oil, Tribol. Int. 71 (2014) 168.
[39] A. Kusmayadi, Y.K. Leong, H.-W. Yen, C.-Y. Huang, C.-D. Dong, J.-S. Chang, Microalgae-microbial fuel cell (mMFC): an integrated process for electricity generation, wastewater treatment, CO_2 sequestration and biomass production, Int. J. Energy Res. (2020) 12, https://doi.org/10.1002/er.5531.
[40] N.F. Polontalo, F. Joelyna, H. Hadiyanto, Production of Bioelectricity from microalgae microbial fuel cell (MMFC) using Chlorella pyrenoidosa and Batik Wastewater, IOP Conf. Ser. Mater. Sci. Eng. (2021) 1053, https://doi.org/10.1088/1757-899X/1053/1/012096.

Chapter 15

Biofuel production by catalysis

Vivek Sharma[1], Prashnasa Tiwari[1], Indu Chauhan[2] and Pawan Rekha[3]

[1]Department of Chemistry, Banasthali Vidyapith, Newai, Rajasthan, India; [2]Department of Biotechnology, Dr. B. R. Ambedkar National Institute of Technology Jalandhar, Jalandhar, Punjab, India; [3]Department of Chemistry, Malaviya National Institute of Technology Jaipur, Jaipur, Rajasthan, India

1. Introduction

In the recent years, fuels have been in high demand due to a variety of activities such as transportation (globally, transportation accounts for 25% of energy demand and nearly 62% of oil consumed) and power generation (electricity generation is the single largest use of fuel in the world, with more than 60% of power generated coming from fossil fuels), and furthermore, with prices of crude oil rising day by day and the world's growing energy demand is attached with an intensification in fossil fuel consumption. However, this source of energy is limited and is accompanied by pollution problems. As we know that fossil fuels are not renewable sources of energy and they will exhaust at some point, an accurate amount of the fuels reservoirs is not known. Therefore, in recent years, the scientific community has been very much interested to develop new and reliable alternative energy sources such as solar, wind, and biomass energy. Over the years, biomass has been the most ancient energy source, which have been used for light and heat generation. Although, the use of biomass in a traditional manner is tied with air pollution and low energy efficiency. However, the biomass-derived fuels, i.e., biofuels, can be the best alternative energy source for the modern era [1–3]. Hence, biofuels have attracted marked attention as these are renewable and sustainable sources of energy that can alleviate global warming by substituting fossil fuels [1].

According to the most recent statistics from the US Energy Information Administration for 2017, renewable energy sources provide 11% of overall energy demand and 17% of total power output in the United States [4]. As a result of rising transportation fuel consumption, Asia is the biggest biofuel consumer. According to the International Energy Agency, Brazil, China, Europe, and the United States consume roughly 90% of all biofuels produced worldwide, while Asian use is expected to increase in the early 2020s. Other countries around the world are adopting biofuels to diversify their energy assortment. Biodiesel is the best substitute for fossil diesel due to its biodegradability, nontoxic and renewable nature, and most importantly its similar properties to fossil diesel.

2. Different approaches for biofuel production

Biofuel is a liquid or gaseous fuel that is produced by using biomass through various conversion technologies. The pathways including fast and slow pyrolysis, fermentation, transesterification, and gasification for the conversion of biomass into biofuels are shown in Fig. 15.1 [5]. However, the use and production of biofuel has been a topic of discussion for many years as advanced techniques are not assimilated in easy steps.

In general, biochemical conversion and thermochemical conversion are the two primary methods for conversion of biomass into biofuels. Biochemical method mainly includes a fermentation process, whereas thermochemical conversion has two paths, i.e., catalytic and hydrothermal liquefaction. Catalytic liquefaction is comparable to hydrothermal liquefaction, but the catalytic conversion is a foremost path for producing biodiesel, bio-ethanol, bio-oil, bio-hydrogen, and biosyngas because to improve the quality of liquid products, a catalyst is utilized to minimize the residence time, functioning temperature, and pressure [6].

Broadly, homogeneous and heterogeneous are two types of catalysts used for biofuel production. The homogeneous catalysts used so far in biofuel production are KOH, H_2SO_4, and NaOH. The disadvantage of using a homogeneous catalyst is that it is difficult to recover product and catalyst. Moreover, the separation process includes a large number of

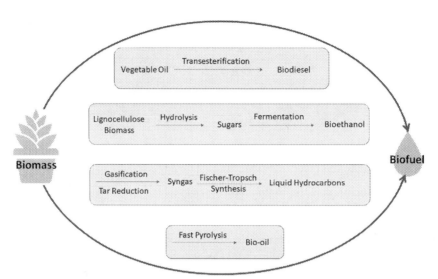

FIGURE 15.1 Primary paths for the conversion of biomass to biofuel.

steps. In comparison to homogeneous catalysts, heterogeneous catalysts provide easier and more effective separation of catalyst and product [5,7].

3. Catalysts used so far for biofuel production

There are many types of catalyst used so far for biofuel production such as homogeneous catalyst, heterogeneous catalyst, biocatalyst, and nanocatalyst. A nanocatalyst is a substance that has at least one nanoscale dimension either externally or internally with catalytic properties. Nanocatalyst have more active sites to react with reactants because they have a large surface area to volume ratio, which enhances the performance of the catalyst. Based on their physical state and ease of separation from the reaction mixture, nanocatalysts are classified in homogeneous and heterogeneous catalysts [8]. The aim of this chapter is to discuss recent advances in catalysis, specifically heterogeneous catalysts for biofuel production. The important heterogeneous catalysts for biofuel production are metal oxide nanocatalyst (e.g., ZnO), metal/alloy based nanoparticles (e.g., AuCu), and biomass-based catalyst, which are discussed herein.

3.1 Metal/alloy based nanocatalyst

Cu–Co, Au/Ag, and Mg/Al are a few examples of metal-based magnetic nanoparticles. The core metal particle may affect the lattice strain of the shell metal, which results in a shift of the shell metal's electronic band structure, so bimetallic nanoparticles have demonstrated considerable technologic potential in heterogeneous catalysis.

(A) Banerjee et al. have synthesized bimetallic gold silver core-shell nanoparticles (Au–Ag NPs) with particle size of 28 nm. These nanoparticles were investigated as catalysts in biodiesel production through transesterification of sunflower oil. The maximum yield of biodiesel was found to be 86.9% [9]. Transesterification reaction mainly consists of three reversible reactions: first, sunflower oil is converted into diglyceride, monoglyceride, and finally into glycerin and fatty acid methyl esters, as shown in Fig. 15.2. The core-shell structure of the nanoparticle was confirmed by TEM (transmission electron microscopy) and energy dispersive X-ray analysis. The confirmation for biofuel synthesis was done using Fourier-transform infrared spectroscopy. Fuel properties were determined by standard ASTM (American Society for Testing and Materials) protocols. Au–Ag NPs have shown very good catalytic activity under mild reaction conditions and moreover exhibited reusability for three cycles of reaction [9].

(B) Chen et al. synthesized Cu–Co bimetallic nanoparticles coated with carbon layers through direct heating of bimetallic oxide precursors incipiently deposited with polyethene glycol [10]. Polyethene glycol works as a carbon source for carbon layers and a reductant for metal species when applied to the precursors. The nanoparticles are protected against oxidation and deactivation by the deposition of carbon layers over them at the same time. The chemo-selective hydrogenolysis of 5-hydroxymethylfurfural to 2,5-dimethylfuran was achieved using this

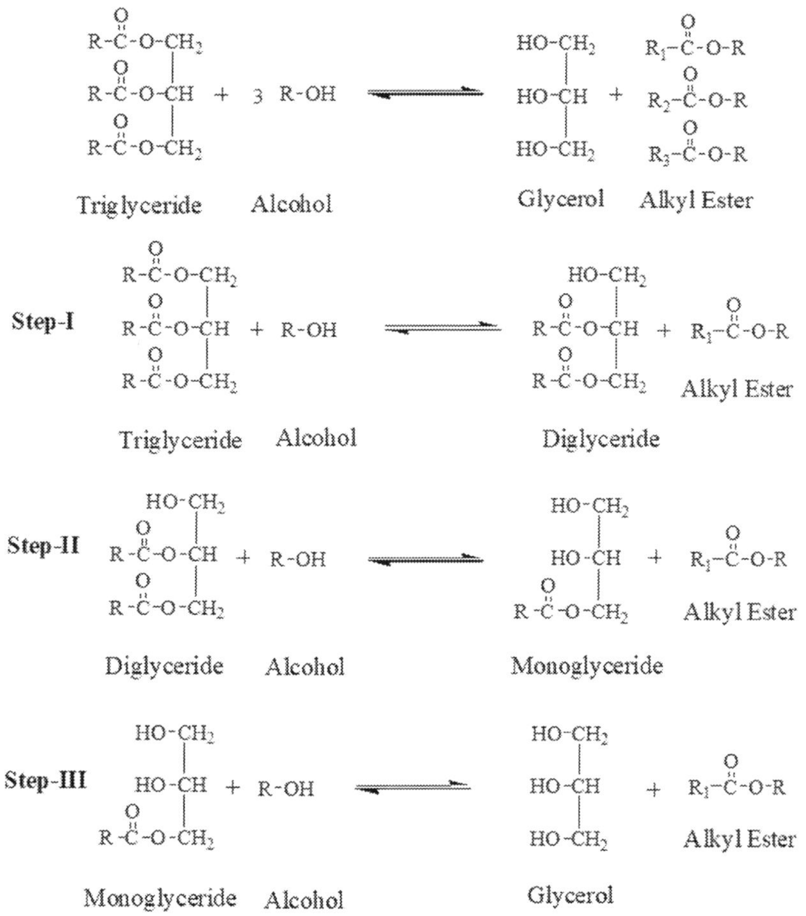

FIGURE 15.2 Chemical reactions involved in transesterification of glycerides.

nanocatalyst. Selective hydro-de-oxygenation of a platform molecule from cellulose yields 2,5-dimethylfuran biofuel. The yield of 2,5-dimethylfuran was 99.4%, which was higher than the results of supported noble metal catalysts. In six reaction cycles, the catalyst demonstrated good reusability and recoverability. TEM, X-ray diffraction (XRD), and X-ray photoelectron spectroscopy were used to characterize the bimetallic nanocatalyst [10].

(C) A co-precipitation approach was developed to make hydrotalcite-derived particles with a Mg/Al molar ratio of 3/1. The scanning electron microscopy (SEM) and atomic force microscopy (AFM) analysis indicates these particles of micro-size, even nano-size, were found based on particle size estimation using XRD data. Because of their high basicity, these nanoparticles are utilized as a catalyst in the production of biodiesel from jatropha oil. At optimal circumstances, a biodiesel production of 95.2% was attained. The catalyst is extremely reusable, and it may be reused up to eight times. By computation, these hydrotalcite-derived particles with Mg/Al had a catalytic dimension of 7.3 nm, but according to AFM analysis, they congregated to form a layered structure with a width of 0.941 mm and a thickness of 381 nm. Surface absorption of glycerol (by-product) and the collapse of the multilayer structure caused the catalyst to deactivate. The catalyst can be reused eight times after the glycerol has been removed from the surface [11].

(D) Hybrid nanocatalysts incorporating enzymes and metallic nanoparticles have also been utilized in biodiesel production. Xie et al. used 1-ethyl-3-(dimethylaminopropyl) carbodiimide as an activating agent to covalently immobilize magnetic Fe_3O_4 nanoparticles with lipase, and the bound lipase was used to catalyze the transesterification of vegetable oils with methanol to produce fatty acid methyl esters [12]. In comparison to free lipase, it was discovered that immobilized lipase was more resistant to temperature and pH inactivation. When 40% immobilized lipase was utilized, the conversion of soybean oil to methyl esters reached over 90% in the three-step transesterification. Furthermore, the lipase catalyst may be reused three times without losing its effectiveness [12].

3.2 Metal oxide nanoparticles

For the transesterification of oil, different metal oxides have been studied such as alkali earth metal oxides, transition metal oxides, mixed metal oxides, and supported metal oxides. These catalysts have emerged as potential heterogeneous catalysts [13]. The structure of metal oxide consists of positive metal ions (i.e., cations) that incorporate Lewis acid and negative oxygen ions (i.e., anions) that possess a Brønsted base [13].

In metal oxides, ZnO in particular is attractive because it is abundant. ZnO nanorods act as a catalyst for biofuel production from olive oil. By using a solution approach, ZnO nanorods were synthesized. The reaction rate of conversion achieved by ZnO nanorods is up to 94.8% at 150°C and the time taken for reaction is 8 h. The structure of ZnO nanorods' specific surface area and morphology were confirmed by gas adsorption analyzer and SEM, respectively [14]. $ZnAl_2O_4$ was successfully synthesized in supercritical water by using a flow-type apparatus and batch reactor with high specific surface area and used as catalyst for the conversion of glucose to H_2. For this catalytic conversion, glucose and H_2O_2 were used as model biomass and oxidant, respectively [15].

Some other metal oxides such as CaO and MgO are mixed with base mixed metal oxide to increase their catalytic activity. For the conversion of Chinese tallow seed oil into biofuel, Wen et al. in 2010 synthesized KF and CaO with the impregnation method. It has a porous structure with particle size of 30–100 nm. The conversion rate of this catalyst from Chinese tallow seed oil to biodiesel is 96.8% under the optimal conditions [16]. Reddy et al. prepared nano-crystalline calcium oxide for the conversion of soybean oil and poultry fat to biodiesel at room temperature via transesterification. On the other side under the same condition, laboratory-grade CaO shows no reaction with poultry fat and only 2% conversion in case of soybean oil [17].

Verziu et al. prepared the three different nanocrystalline MgO materials using simple, green, reproducible methods and involving nontoxic reagents [18]. These materials are very active, recyclable heterogeneous catalysts, and at low temperature, these can be used after transesterification of vegetable oil. These materials include the following methods of preparation:

- MgO (I) was prepared by Sol-gel method in which 4-methoxybenzyl alcohol was used as a template, followed by supercritical drying and calcination in air at 773 K.
- MgO (II) was made from a commercial MgO. The commercial MgO was boiled in water followed by drying at 393 K, and then dehydrated under vacuum at 773 K.
- To synthesize MgO (III), $Mg(OCH_3)_2$ was hydrolyzed in a methanol–toluene solution; then supercritical solvent was removed, resulting in the production of a $Mg(OH)_2$ aerogel that was dehydrated under vacuum at 773 K.

The structural and morphologic characterizations were carried out by the TDRIFT, TEM, and DR-UV-vis. This catalyst was used under different conditions such as autoclave, ultrasonication, and microwave. Out of these, microwave in comparison to autoclave or ultrasound condition provided selectivity to methyl ester and much higher conversion from sunflower and rapeseed vegetable oil to biofuels [18]. The synthesis of nano-NiO particles and their use as catalysts in biomass pyrolysis were investigated by Li et al. [19] To begin, precursors were made from an aqueous solution of nickel nitrate hexahydrate and urea, which were homogeneously precipitated. The precursor's formula was established to be $NiCO_3.2Ni(OH)_2.nH_2O$, and it could be transformed completely into NiO particles in an air environment at temperatures below 360°C. In biomass pyrolysis, nano-NiO particles outperformed micro-NiO particles in terms of catalytic efficiency. The results showed that the NiO catalyst may have a considerable impact on the pyrolysis process, with the primary decomposition reaction proceeding at a lower temperature and a significant reduction in the ultimate yield of char as well as the activation energy of biomass pyrolysis [19].

3.3 Biomass-based catalysts

Biomass is any simple organic material that comes from plants and animals, e.g., agricultural residue, bio-waste from industry, wood residue, etc. Any simple raw biomass can be converted into useful biochar. Recently biochar has been recognized as the green catalyst for the production of biofuels by altering its surface. Moreover, it can substitute metal-based catalysts and fossil fuel–driven carbon catalysts due to its ecofriendly nature.

The composition of plant biomass is mostly a carbon structure that can be converted into high-performance porous carbon material, which can be used as catalyst support. Catalysts based on biomass are nontoxic, biodegradable, ecofriendly, and have high specific surface area. These catalysts are used for the transesterification and esterification of vegetable oil to produce biodiesel. Heterogeneous catalysts derived from biomass are mainly divided into two types based on their nature: (i) heterogeneous base catalyst and (ii) heterogeneous acid catalyst [5,20]. There are three main pathways to transform biomass into a solid base catalyst for production of biodiesel, as shown in Fig. 15.3.

Various heterogeneous base catalysts have been synthesized from biomass. CaO catalysts have been synthesized from different biomass such as sea sand, eggshell, rice husk ash, and *Jatropha curcas* shell. Piker et al. synthesized the CaO catalyst from eggshell [21]. Eggshell-derived catalysts are used for the transesterification reaction of waste cooking oil and fresh soybean oil for biodiesel production, and recorded yield is 97% for both oils at room temperature and reaction time of 11 h. This catalyst exhibited five times reusability for waste cooking oil and 10 times for soybean oil. Moreover, this catalyst can be stored up to 3 months without any decrease in its catalytic activity and for 1 year with 10% decrease in yield [21].

Mucino et al. prepared a base catalyst from sea sand, where composition of that catalyst is mainly $CaCO_3$ [22]. The large particles of sea sand were eliminated by crushing and reduced to the size of <0.42 mm. In calcination at 800°C temperature, $CaCO_3$ completely converted to CaO. Sea sand–derived catalysts achieved 95.4% yield at 60°C and reaction time of 6 h [22].

Pandit et al. also utilize eggshells for the synthesis of highly active calcium oxide (CaO) nanocatalyst by easy synthesis methods using calcination–hydration–dehydration technique with average particle size of 75 nm and specific surface area of 16.4 $m^2\,g^{-1}$. This catalyst was able to achieve 86.41% biodiesel conversion at reaction time 3.6 h [23].

Waste obtuse horn shells were used for the synthesis of CaO catalyst for the transesterification of palm oil into biodiesel. The CaO catalyst is characterized by thermogravimetric analysis (TGA), BET surface area analysis, XRD, and SEM. TGA analysis showed weight loss of about 43%, which can be related to the decomposition of shell and conversion into CaO catalyst. Recorded yield of conversion of palm oil into biodiesel was 86.75% under reaction condition of 6 h. Recyclability of waste obtuse horn shells–derived catalyst was studied, and it showed long-lasting reusability in three cycles of reaction with conversion of more than 70% [24].

The catalytic activity of biomass-derived CaO catalyst can also be enhanced by doping with metal ions. It increases the basicity of catalyst and decreases the leaching of Ca^{2+} ion [25,26]. Boro et al. prepared CaO catalyst from *Turbonilla striatula* waste shell doped with barium [26]. Basicity of catalyst increased from 0.1 to 0.2375 mmol g^{-1} by doping of 1% $BaCl_2$ solution and transesterification reaction of waste cooking oil carried out at 60°C for 8 h and has shown the complete conversion of oil to biodiesel [25]. CaO catalyst was derived from *Crallus domesticus* shells synthesized by wet impregnation method and doped with the Mo and Zr. Mo–Zr/CS catalyst was used to synthesize biodiesel via transesterification reaction of waste cooking palm oil and reported 90.1% yield in 3 h. The derived catalyst exhibited excellent and long-lasting catalytic activity, used up to three cycles of reaction with more than 70% yield [26].

Chen et al. were able to generate a low-cost, high-efficiency catalyst by employing two biomass wastes, i.e., rice husk and eggshell, as raw material [27]. In this study, rice husk ash was employed as a catalytic support for the CaO generated by calcining eggshells. Rice husk ash was created by calcining dry rice husk for 4 h at 400, 600, and 800°C. In the

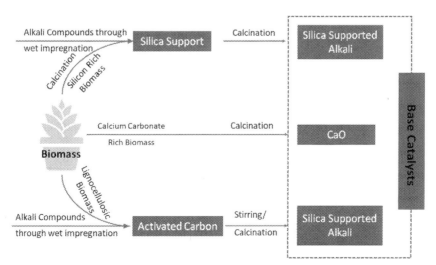

FIGURE 15.3 Synthesis paths to transform biomass into a solid base catalyst for biodiesel production.

meantime, dried eggshell was calcined for 4 h at 400°C to convert it to CaO. To make the basic catalyst, the CaO was wet impregnated with rice husk ash and calcined at 600, 800, and 1000°C. The catalyst had the highest basicity of 8.5 mmol g^{-1} (30% calcined egg loading, 800°C rice husk ash calcination temperature). The maximum biodiesel output was 91.5% when transesterification parameters were 7 wt% catalyst loading, 9:1 methanol to oil ratio, 65°C, and 4 h reaction time [27].

For heterogeneous acid catalysts, various lignocellulosic biomass, nonlignocellulosic biomass, and lignin-based biomass are transformed into catalysts via sulfonating method. Direct sulfonation by thermal treatment of carbon materials using concentrated H_2SO_4 is a very easy sulfonating pathway. It is the most common method to derive sulfonated solid acid catalyst from biomass and carbon materials. Carbonization and sulfonation of polycyclic aromatic hydrocarbons (naphthalene) were first introduced by Hara's group. Carbon materials with SO_3H functional group have been reported to act as a strong solid catalyst [28].

To synthesize solid heterogeneous acid catalyst from biomass for biodiesel production, first carbonization of biomass is performed to remove the volatile material and moisture, which results in a carbon ring structured material with randomly oriented carbon sheets. After that, using different types of sulfonating reagents such as concentrated sulfuric acid [29], fuming sulfuric acid [30], 4-benzenodiazonium sulfonate, and PTSA (*p*-toluenesulfonic acid) are used for the sulfonation [31]. Additionally, biomass-based carbon acid catalysts are mostly prepared by the direct sulfonation method, which is carried out by heating the mixture of catalyst support and sulfonating agent, where the most commonly used sulfonating agent is H_2SO_4 [20,29,32].

Ezebor et al. prepared a new catalyst from oil palm trunk and sugarcane bagasse via direct sulfonation method using concentrated sulfuric acid at 150°C for different times (2, 4, 6, 8, and 10 h). The total acid density of catalyst and yield of biodiesel increased with the increment of sulfonation time from 2 to 6 h. This catalyst showed 10.2%—12.5% loss in activity after six repeating cycles [29]. Zhou et al. prepared the catalyst from bamboo-derived carbon via sulfonation, and they examined this catalyst for 1—5 h at 105°C, but the catalyst with only 2-h sulfonation shows maximum conversion, which is 98.4% [33]. In situ partial carbonization and sulfonation of biomass (microalgae residue, bio glycerol, de-oiled seed cake) also produced the catalyst. In this, mixture of biomass and H_2SO_4 simultaneously were heated at 120°C hydrothermally. Catalyst derived from microalgae residue shows higher biodiesel yield due to efficient dissemination of catalyst in methanol and better approach of reactants to active sites in transesterification or esterification reaction. Besides microalgae residues—based catalysts, catalysts derived from de-oiled seed cake also showed higher biodiesel yield of 94%, which is higher than the commercially derived activated carbon catalysts with 75% biodiesel yield [34—36].

Liu et al. synthesized acid catalyst directly from biomass in one-step synthesis that includes both carbonization and sulfonation. This synthesis method is cost effective, straightforward, less time consuming, and can be performed at room temperature. The catalyst was prepared by mechanically mixing of H_2SO_4 and distiller grains at 20°C temperature for 1.4 h, and rate of catalytic conversion was attained 97%. The recyclability study showed a decrease in catalytic conversion up to 50.7% after the five cycles because the catalyst faces a leaching problem, as it is a catalyst synthesized through low-temperature sulfonation.

Apart from direct sulfonation by using concentrated H_2SO_4, researchers have examined and reported different pathways with other sulfonating agents to make the synthesis easy and to reduce production cost of catalysts. Fuming sulfuric acid is also reported for sulfonation of biomass-derived carbon material. A solid acid catalyst prepared by Liu et al. from carbonized corn straw by using fuming sulfuric acid (fuming sulfuric acid is a solution of different composition of sulfur trioxide dissolved in sulfuric acid and also referred to as oleum). This catalyst achieved 98% conversion in transesterification reaction of oleic acid and methanol at 333 K for 8 h [30]. Pure sulfur trioxide is also employed as a sulfonating agent in addition to fuming sulfuric acid. Katsner et al. prepared the catalyst by sulfonating wood chip carbon with sulfur trioxide [37]. This catalyst was able to achieve 97% biodiesel conversion [37]. PTSA is also used as sulfonating agent to produce catalysts from biomass. PTSA sulfonation of sawdust-derived activated carbon was studied by Wang et al. Esterification reaction of high acetic acid content bio-oil with alcohol using synthesized catalyst gave 86.6% conversion [31].

4. Comparative highlights of catalytic efficiency of developed catalysts for biofuel production

Various catalyst have been explored for the biofuel production so far [9—37]. The catalytic efficiency, their feedstock, along with their reaction time and reaction conditions of these catalysts have been summarized in Table 15.1.

TABLE 15.1 Comparative highlights of catalytic efficiency of various explored catalysts for biofuel production.

S. No.	Catalyst	Feedstock	Reaction time	Alcohol to oil ratio	Conversion	Refs.
1.	Au–Ag nanoparticles	Sunflower oil	2 h	5:1	86.90%	[9]
2.	Cu–Co bimetallic nanoparticles	5-Hydroxymethylfurfural	–	–	99.4%	[10]
3.	Mg/Al nanoparticles	Jatropha oil	1.5 h	5:1	95.2%	[11]
4.	Fe_3O_4 nanoparticles	Soybean oil	12 h	1.5:1	90%	[12]
5.	ZnO nanorods	Olive oil	8 h	50:1	94.8%	[14]
6.	KF-CaO	Chinese tallow seed oil	2.5 h	12:1	91%	[16]
7.	CaO	Soybean oil	24 h	1:33	99%	[17]
8.	MgO	Sunflower oil	40 min	1:4	99%	[18]
9.	CaO (egg shell)	Waste cooking oil (1.8% FFA)	11 h	6:1	97%	[21]
10.	CaO (sea sand)	Pretreated used cooking oil	6 h	12:1	95.4%	[22]
11.	CaO (obtuse horn shell)	Refined palm oil	6 h	12:1	86.75%	[24]
12.	Ba doped CaO (*Turbonilla striatula* waste shell)	Pretreated used cooking oil	11 h	6:1	100%	[25]
13.	Mo–Zr doped CaO (waste gallus domestic shell)	Waste cooking palm oil	3 h	15:1	90.1%	[26]
14.	Rice husk egg supported CaO	Refined palm oil	4 h	9:1	91.5%	[27]
15.	Oil palm trunk	Waste oil	4 h	1.17 mL min^{-1}	80.6%	[29]
16.	Sugarcane bagasse	Waste oil	4 h	1.17 mL min^{-1}	83.2%	[29]
17.	Wood chip-derived carbon	Palmitic acid/methanol	6 h	1:6	97%	[31]
18.	Bamboo	Oleic acid/ethanol	2 h	1:7	98.4%	[33]
19.	De-oiled seed cake	Oleic acid/methanol	2 h	1:12	94%	[33]
20.	Corn straw	Oleic acid/methanol	4 h	1:3	92%	[37]

5. Conclusion

In this chapter, the potential of biofuel as a possible substitute of fossil fuels is discussed. Subsequently, the different approaches for biofuel production and the role of catalysts in biofuel production have been summarized. Broadly two types of catalysts, i.e., homogeneous and heterogeneous catalysts, have been used for the production of biofuels. Among these catalysts, heterogeneous catalyst is best fitted due to ease of separation, recyclability, and high catalytic efficiency. In heterogeneous catalysts, metal oxide nanoparticles, alloy/metal-based nanocatalysts, and biomass-based catalysts have been used so far for biofuel production. But the biomass-based catalysts can be considered the most potential catalysts owing to use of bio-waste raw material in the preparation of such type of catalysts, and most significantly, the biomass is produced in a large quantity in a country that is mainly dependent on agricultural activities. Overall, the catalysts are essential for commercial biofuel production and have the potential to revolutionize biofuel production both qualitatively and quantitatively to fulfill future energy demands.

References

[1] L. Ou, S. Banerjee, H. Xu, H.C.U. Lee, M.S. Wigmosta, T.R. Hawkins, Utilizing high-purity carbon dioxide sources for algae cultivation and biofuel production in the United States: opportunities and challenges, J. Clean. Prod. 321 (2021) 128779.

[2] M. Akial, F. Yazdani, E. Motaee, D. Han, H. Arandiyan, A review on conversion of biomass to biofuel by nanocatalysts, Biofuel Res. J. 1 (2014) 16−25.
[3] A. Demirbas, Biofuels sources, biofuel policy, biofuel economy and global biofuel projections, Energy Convers. Manag. 49 (2018) 2106−2116.
[4] D. Gielen, F. Boshell, D. Saygin, M.D. Bazilian, N. Wagner, R. Gorini, The role of renewable energy in the global energy transformation, Energy Strategy Rev. 24 (2019) 38−50.
[5] K. Velusamy, J. Devanand, P.S. Kumar, K. Soundarajan, V. Sivasubramanian, J. Sindhu, D.-V.N. Vo, A review on nano-catalysts and biochar-based catalysts for biofuel production, Fuel 306 (2021) 121632.
[6] M. Verma, S. Godbout, S.K. Brar, O. Solomatnikova, S.P. Lemay, J.P. Larouche, Biofuels production from biomass by thermochemical conversion technologies, Int. J. Chem. Eng. 2012 (2012) 1−18.
[7] K. Saoud, Nanocatalyst for biofuel production: a review, Green Nanotechnol. Biofuel Prod. (2018) 39−62.
[8] S. Hashmi, S. Gohar, T. Mahmood, U. Nawaz, H. Farooqi, Biodiesel production by using CaO-Al2O3 Nano catalyst, Int. J. Eng. 2 (2016) 42−49.
[9] M. Banerjee, B. Dey, J. Talukdar, M.C. Kalita, Production of biodiesel from sunflower oil using highly catalytic bimetallic gold-silver core-shell nanoparticle, Energy 69 (2014) 695−699.
[10] B. Chen, F. Li, Z. Huang, G. Yuan, Carbon-coated Cu-Co bimetallic nanoparticles as selective and recyclable catalysts for production of biofuel 2,5-dimethylfuran, Appl. Catal., B 200 (2017) 192−199.
[11] X. Deng, Z. Fang, Y.H. Liu, C.L. Yu, Production of biodiesel from Jatropha oil catalyzed by nanosized solid basic catalyst, Energy 36 (2011) 777−784.
[12] W. Xie, N. Ma, Enzymatic transesterification of soybean oil by using immobilized lipase on magnetic nano-particles, Biomass Bioenergy 34 (2010) 890−896.
[13] A. Refaat, Biodiesel production using solid metal oxide catalysts, Int. J. Environ. Sci. Tech. 8 (2011) 203−221.
[14] M.M. Molina, ZnO Nanorods as Catalyts for Biodiesel Production from Olive Oil, ProQuest Dissertations Publishing, 2013.
[15] M.W. Levy, K. Sue, Synthesis of nanophased metal oxides in supercritical water: catalysts for biomass conversion, Int. J. Appl. Ceram. Technol. 3 (2006) 337−344.
[16] L. Wen, Y. Wang, D. Lu, S. Hu, H. Han, Preparation of KF/CaO nanocatalyst and its application in biodiesel production from Chinese tallow seed oil, Fuel 89 (2010) 2267−2271.
[17] C.R.V. Reddy, R. Oshel, J.G. Verkade, Room-temperature conversion of soybean oil and poultry fat to biodiesel catalyzed by nanocrystalline calcium oxides, Energy Fuels 20 (2006) 1310−1314.
[18] M. Verziu, B. Cojocaru, J. Hu, R. Richards, C. Ciuculescu, P. Filip, Sunflower and rapeseed oil transesterification to biodiesel over different nanocrystalline MgO catalysts, Green Chem. 10 (2008) 373−381.
[19] J. Li, R. Yan, B. Xiao, D.T. Liang, D.H. Lee, Preparation of nano-NiO particles and evaluation of their catalytic activity in pyrolyzing biomass components, Energy Fuels 22 (2008) 16−23.
[20] Z.E. Tang, S. Lim, Y.L. Pang, H.C. Ong, K.T. Lee, Synthesis of biomass as heterogeneous catalyst for application in biodiesel production: state of the art and fundamental review, Renew. Sustain. Energy Rev. 92 (2018) 235−253.
[21] B.T.,N.P. Piker, A. Gedanken, A green and low-cost room temperature biodiesel production method from waste oil using egg shells as catalyst, Fuel 182 (2016) 34−41.
[22] G.G. Mucino, R. Romero, A. Ramlrez, S.L. Martlnez, R. Baeza-Jimenez, R. Natividad, Biodiesel production from used cooking oil and sea sand as heterogeneous catalyst, Fuel 138 (2014) 143−148.
[23] P.R. Pandit, M.H. Fulekar, Egg shell waste as heterogeneous nanocatalyst for biodiesel production: optimized by response surface methodology, J. Environ. Manag. 198 (2017) 319−329.
[24] S.L. Lee, Y.C. Wong, Y.P. Tan, S.Y. Yew, Transesterification of palm oil to biodiesel by using waste obtuse horn shell-derived CaO catalyst, Energy Convers. Manag. 93 (2015) 282−288.
[25] J. Boro, L.J. Konwar, A.J. Thakur, D. Deka, Ba doped CaO derived from waste shells of T striatula (TS-CaO) as heterogeneous catalyst for biodiesel production, Fuel 129 (2014) 182−187.
[26] N. Mansir, S.H. Teo, U. Rashid, Y.H. Taufiq-Yap, Efficient waste *Gallus domesticus* shell derived calcium-based catalyst for biodiesel production, Fuel 211 (2018) 67−75.
[27] G.Y. Chen, R. Shan, J.F. Shi, B.B. Yan, Transesterification of palm oil to biodiesel using rice husk ash-based catalysts, Fuel Process. Technol. 133 (2015) 8−13.
[28] M. Hara, T. Yoshida, A. Takagaki, T. Takata, J.N. Kondo, S. Hayashi, A carbon material as a strong protonic acid, Angew. Chem. Int. Ed. 43 (2004) 2955−2958.
[29] F. Ezebor, M. Khairuddean, A.Z. Abdullah, P.L. Boey, Oil palm trunk and sugarcane bagasse derived heterogeneous acid catalysts for production of fatty acid methyl esters, Energy 70 (2014) 493−503.
[30] T. Liu, Z. Li, W. Li, C. Shi, Y. Wang, Preparation and characterization of biomass carbon-based solid acid catalyst for the esterification of oleic acid with methanol, Bioresour. Technol. 133 (2013) 618−621.
[31] C. Wang, Y. Hu, Q. Chen, C. Lv, S. Jia, Bio-oil upgrading by reactive distillation using p-toluene sulfonic acid catalyst loaded on biomass activated carbon, Biomass Bioenergy 56 (2013) 405−411.
[32] S.I. Akinfalabi, U. Rashid, R. Yunus, Y.H. Taufiq-Yap, Synthesis of biodiesel from palm fatty acid distillate using sulfonated palm seed cake catalyst, Renew. Energy 111 (2017) 611−619.

[33] Y. Zhou, S. Niu, J. Li, Activity of the carbon-based heterogeneous acid catalyst derived from bamboo in esterification of oleic acid with ethanol, Energy Convers. Manag. 114 (2016) 188−196.

[34] K. Neumann, K. Werth, A. Martin, A. Gorak, Biodiesel production from waste cooking oils through esterification: catalyst screening, chemical equilibrium and reaction kinetics, Chem. Eng. Res. Des. 107 (2016) 52−62.

[35] E.M. Santos, A.P. d C. Teixeira, F.G. da Silva, T.E. Cibaka, M.H. Araujo, W.X.C. Oliveira, F. Medeiros, A.N. Brasil, L.S. Oliveira, R.M. Lago, New heterogeneous catalyst for the esterification of fatty acid produced by surface aromatization/sulfonation of oilseed cake, Fuel 150 (2015) 408−414.

[36] X. Fu, D. Li, J. Chen, Y. Zhang, W. Huang, Y. Zhu, J. Yang, C. Zhang, A microalgae residue based carbon solid acid catalyst for biodiesel production, Bioresour. Technol. 146 (2013) 767−770.

[37] J.R. Kastner, J. Miller, D.P. Geller, J. Locklin, L.H. Keith, T. Johnson, Catalytic esterification of fatty acids using solid acid catalysts generated from biochar and activated carbon, Catal. Today 19 (2012) 122−132.

Chapter 16

Potential of lignin as biofuel substrate

Sagarjyoti Pathak and Hitesh S. Pawar

DBT-ICT Centre for Energy Biosciences, Institute of Chemical Technology, Mumbai, Maharashtra, India

1. Introduction

1.1 Current energy status globally

The global energy demand is increasing exponentially. The U.S. Energy Information Administration (EIA), based on International Energy Outlook 2019 (IEO2019), predicts a nearly 50% increase by 2050 [1]. The primary reason for the can be attributed to the economic development and continuous growth in the global population, with estimates from the UN Population Division showing it will reach close to 10 billion by the latter part of the mid-21st century [2]. According to the EIA, there will be an increase in global energy consumption by 30% in the industrial sector, 40% in transportation, 79% in generating electricity, and 40% in natural gas by 2050 [1]. Although renewables will make up 27% of the energy demand in the next 2 to 3 decades, liquid fuels will meet 28% of the global energy demand [3].

1.2 Climate change—a serious threat worldwide

The dawn of the industrial era enabled humankind to realize the potential of fossil fuels. The economic development that it brought, along with political support, led to its overexploitation in the 20th century [4]. This took a toll on the environment, leading to its considerable degradation. Not only did it result in severe depletion of the limited fossil fuel reserves, but their extensive combustion also led to a significant increase in the level of greenhouse gases (GHGs) emitted. These gases, which comprise carbon dioxide (CO_2), methane (CH_4), nitrous oxide (N_2O), and ozone (O_3), possess the ability to emit and absorb infrared radiation, increasing the surface temperature of the Earth by 20–34°C. Since their absence would have made the Earth's surface temperature reach only −19°C, their existence in the atmosphere is essential. Keeping a mean global temperature of 14°C ensures the survivability of life forms. However, their exorbitant concentration in the current atmosphere is a concern, as it has resulted in global warming and associated changes in the climate such as an intensification of the water cycle (leading to flooding and drought in many regions), changes in rainfall patterns (higher latitudes experiencing more rainfall and subsequently lowering the amount in large parts of the subtropics), altered monsoon rainfall in many regions, rises in sea level impacting coastal flooding and erosion, increased permafrost thawing, losses in biodiversity, impaired human health, etc. Reports show that among other GHGs, CO_2 contributes most to climate change. According to the Intergovernmental Panel on Climate Change (IPCC) 2021 report, it could take 20–30 years to stabilize global temperatures, even with substantial and sustained reductions in emissions of CO_2. Between 1850 and 1900, there has been approximately a 1.1°C warming of the planet caused by the emission of GHGs due to human activities. Moreover, at such a rate, the global temperature is expected to increase by 1.5°C in the next 20 years [5]. To put this into perspective, the last time levels were this high in the Earth's atmosphere was about a million years ago [6]. The trajectory that we are currently in projects to increase 4.1–4.8°C warming by the end of this century if no climate policies are put into action [7]. Several agreements have been put forward for the realization of the detrimental effects and the need to exercise control over further depletion. The first one of these was the United Nations Convention on Climate Change (UNFCC) in May 1992, followed by the Kyoto Protocol in December 1997 and the Paris Agreement in December 2015 which stated "*holding the increase in the global average temperature to well below 2°C above pre-industrial levels and pursuing efforts to limit the temperature increase to 1.5°C above pre-industrial levels*" [8]. The main motive behind all these agreements is to employ green technologies to mitigate the problems associated with climate change, uncertain supply, and environmental pollution to make the Earth more sustainable for current and future generations.

Since the generation of power plays an imperative role in rapid economic growth, urbanization, industrialization, and transport infrastructure, the utilization of alternative, renewable, clean, and carbon-neutral sources has become crucial. This would ensure energy security, sustainability, social and economic development, and provide a way to overcome the instability of markets and prices of fossil fuels. Therefore, using renewable sources by replacing fossil fuels has become a firm political target for both developed and developing economies. Water, wind, and solar are the eminent sources that could reduce the reliance on fossil fuels. Out of the total supply of primary energy, 81% comes from fossil fuels, nuclear energy accounts for 5%, and 14% is contributed by renewable sources (this includes 70% contribution from biomass) [9]. In the case of transportation, batteries cannot wholly substitute liquid fuels because some transportation methods such as aviation and heavy-duty vehicles require dense energy sources. Moreover, even though increasing the use of hybrid and electric cars could reduce fuel consumption, technologies such as hydrogen and fuel cells present particular challenges. One of these is the dependence on electricity. The other is the difficulty in producing hydrogen due to the heavy energy requirement, and security, making their use questionable. Therefore, fuels will likely be the preferred energy source in the transport sector. An ideal alternative is biomass, more specifically lignocellulosic biomass, which possesses the capacity to produce liquid biofuels. The reason for this is that it is the most abundant terrestrial biomass. Also, transportation biofuels usually produce energy that exceeds the amount required in its production and distribution. Having a positive energy balance, therefore, favors its use in the transport sector.

1.3 Potential biomass feedstocks for energy production

Biomass, primarily obtained from plants and animals is a renewable organic material. Some of the most common biomass feedstocks for the generation of energy are [10]:

Energy crops: Plants exclusively grown for their wood from marginal lands for energy generation like switchgrass, miscanthus, hybrid poplar, willow, algae, prosopis, leuceana, bamboo, etc.

Argo-industrial wastes comprising of wastes from paper mills, molasses from sugar refineries, pulp wastes from food processing units, textile fiber waste, etc.

Agricultural wastes including straws of cereals and pulses, stalks of fiber crops, seed coats of oil seed, crop wastes like sugarcane trash, rice husk, coconut shell, etc.

Municipal solid waste (MSW), which covers the biodegradable wastes such as food and kitchen waste, green waste, paper, inert waste like fabric, clothes, etc.

Forest wastes such as logs, chips, barks, leaves, forest industry-based product like sawdust.

1.4 Availability and abundance of biomass

Globally, the energy picture is changing rapidly in favor of renewable energy. The global energy report of IRENA, showing the energy roadmap by 2030 (REmap 2030) [11], summarizes from its findings that utilizing the realizable potential of all renewable energy technologies could contribute to 36% of the global energy mix by 2030. This would cause the renewable energy share to be doubled compared to 2010 levels. Biomass has, in this regard, an auspicious future. As biomass has potential in all sectors, it could account for 60% of total final renewable energy by 2030. In today's world most of the demand for biomass is in cooking and heating. Fifty-three exajoules (EJ), i.e., more than 60% of the total global biomass demand in 2010, was used in the residential and commercial buildings sectors. Out of this total, much if it was in traditional uses of biomass for cooking and heating. The manufacturing industry (15%), power and district heating (8%), and transport (9%) sectors accounted for about one-third of the total. Over time, biomass applications could also change. If the whole potential of biomass is implemented, its global demand could double to 108 EJ by 2030. For the production of power and generation of heat directly, about a third of this total would be used. In the transport sector for the generation of biofuels around 30% would be consumed. The manufacturing industry and building sectors would take up half of the remainder for heating applications. The main way to increase the share of biomass use in the manufacturing industry and power sectors will be in the generation of combined heat and power (CHP). As per REmap 2030, it has been estimated that the United States, China, India, Brazil, and Indonesia would altogether account for 56% of the total biomass demand. Supply of biomass globally is predicted to vary between 97 and 147 EJ per year. Agricultural residues and waste would have a contribution of approximately 40% (33–66 EJ) of the total supply. The rest of the potential suppliers of energy would include energy crops (33–39 EJ) and forest products and residues (24–43 EJ). Asia and Europe (including Russia) are the most important and largest suppliers (43–77 EJ). By 2030, to meet the growing demand globally, international trade of biomass will play a crucial role. The costs of biomass supply in the domestic sector are estimated to be between as low as USD 3 in the case of agricultural residues and as high as USD 17 per GJ for energy crops. Many challenges need to be overcome in the supply and demand of biomass, its international trade, and also substitution of its traditional uses to achieve such high growth rates. Moreover, with estimates of doubling in demand for bioenergy between 2010 and 2030, there is growing importance in considering the sustainability of biomass including environmental, economic, and societal aspects. In order to have a sustainable and affordable bioenergy system, the present national and international initiatives/partnerships, along with the energy and resource policies, need to be expanded to deal with the challenges present in the use and supply chain of biomass.

To sum up, lignin holds promise in being able to mitigate the environmental and sustainability issues with the existing petroleum-based products and refineries, and also issues with regard to increasing dependency on fossil fuels and economic viability. Out of the total $1.85-2.4 \times 10^{12}$ tons of renewable feedstock (lignocellulosic biomass [LCBM]), 20% is lignin which is also produced in surplus quantities from biorefineries and paper industries. It is one of the most abundant renewable sources of aromatics and other important organic compounds that can be directed into the production of platform chemicals, value-added products, and biofuel. At present, the pulp and paper mills generate 50–70°million tons of lignin. However, this number will certainly increase due to the mandatory production of 60 billion gallons of biofuel by the renewable fuel standard program (RFS). The necessary calculation amounts to the requirement for LCBM to be 0.75°billion tons, which would generate 0.225°billion tons of lignin-rich by-product. Therefore, this chapter emphasizes (1) the structural complexity of lignin at the molecular level, (2) established routes to extract bulk lignin, (3) characterization of the isolated lignin, and (4) the potential methods to convert lignin into biofuel precursors and their further upgradation.

2. Generations of biofuel

2.1 First-generation biofuels

Various food crops like wheat, corn, sugar, beet, etc. serve as the source of first-generation biofuels (or conventional biofuels). There are three main types of commercially available biofuels which fall under this category. These are biodiesel, bioethanol, and biogas, produced by different technologies. Transesterification of fats, residual oils, and vegetable oils produces biodiesel. Through some chemical changes and slight modifications, it can be substituted in diesel-powered engines. The chemical reaction of the transesterification process involves a reaction between triglycerides and alcohol (e.g., biomethanol), along with a catalyst or enzyme to generate biodiesel and glycerol. Bioethanol, substituted for gasoline, is produced by the fermentation of sugar or starch. When used as a feedstock for ethyl tert-butyl ether (ETBE), it mixes more easily with gasoline. Biogas, produced form organic materials by anaerobic digestion, is comprised of methane and carbon dioxide.

There are many issues which have prevented widespread adoption of first-generation biofuels even though they have several advantages. The first major concern is the food-versus-fuel competition that arises when food crops from agricultural lands are implemented in the generation of fuels. This would lead to an increase in food prices. The issue becomes more prominent in arid and insular regions where the prevailing environment does not make its production technically feasible. Fokaides et al. [12] justified that in arid and insular environments the price of energy crop seed when compared to the production of conventional fuels exceeds the determined feasible price. Hence it would not be possible to produce biofuels at a price considered competitive with that of fossil fuels. Further, it was concluded by GIS (Geographic Information System) analysis that under semiarid and subtropical environments, the promotion of energy crops is extremely limited and the contribution from biofuels in any case is insufficient to satisfy the expected need.

2.2 Second-generation biofuels

In order to overcome the challenge of first-generation biofuels, a major focus currently is to derive bioenergy from alternative nonfood sources. Agricultural and forest residues, industrial wastes, lignocellulosic biomass, and nonfood crop feedstocks fall into this category. Biofuels derived from such sources are known as second-generation biofuels (or advanced biofuels). Their production usually involves a pretreatment stage followed by physical, thermochemical, and biochemical technologies. Since the conversion step is highly facilitated by the pretreatment step, which prepares the biomass properties like size, moisture, density, etc., it is very important to carry out this process. Fig. 16.1 shows a general scheme for the utilization of lignocellulosic biomass, with a particular emphasis on lignin as the substrate to yield fuel precursors.

Briquetting, pelletizing, and fiber extraction are the most common physical conversion processes. The loose biomass is converted into high-density solid blocks by the briquetting method, while fine-particle raw material is compacted by the application of pressure to pellets by pelletization. Fibers from biomass residues have potential to be used as a burning fuel and therefore, in fiber extraction, fibers are extracted from it [13].

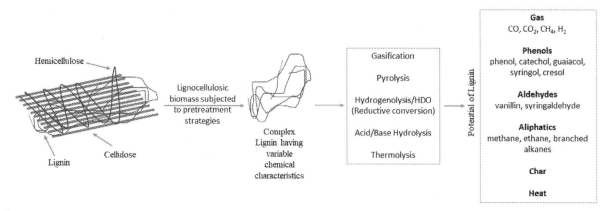

FIGURE 16.1 Overview of the general steps taken to isolate lignin from lignocellulosic biomass and its further conversion into compounds of economic importance and precursor molecules for biofuel.

In the production of second-generation biofuels, pyrolysis, direct combustion, liquefaction, and gasification are the main thermodynamic chemical processes available. Biomass is converted into solid, liquid, and gaseous fractions by restricting the supply of air in pyrolysis [14]. Moreover, on the basis of operating conditions, pyrolysis can be classified as slow or fast. Production of solid biofuels is seen to be favored in slow pyrolysis, whereas carrying out the process in a fast manner produces gaseous and liquid (bio-oil) biofuels [15]. The process of partial oxidation of biomass at high temperatures (e.g., 800–900°C) in the presence of a gasification medium like steam, air, or oxygen to convert it into synthetic gas or syngas (a combustible, gaseous fuel) is known as gasification. The syngas thus obtained is primarily used in the production of intermediate chemicals and fuels [13]. In liquefaction, a liquid product with various ranges of molecular weights is produced by repolymerizing the depolymerized or broken-down small, unstable, and reactive molecules of high lignin content biomass. It is necessary to aid this process through the use of solvent, syngas, and catalysts for the necessary conversion to a heavy fuel-oil product.

Burning of biomass in excess air is a direct combustion method to produce heat. This results in volatilization of vapors from biomass which are combustible and burn in flames. This method is employed to convert biofuel to heat; however there also are secondary-conversion technologies to achieve production of power.

Production of second-generation biofuels can also be carried out through common biochemical pathways yielding ethanol and other alcohols. The anaerobic process to convert the glucose (or carbohydrates) in organic wastes through fermentation takes place by a series of chemical reactions. It is essential that hydrolysis by enzymatic treatment and fermentation later, or simultaneous saccharification and fermentation (SSF) process, be done after a pretreatment step to increase the sugar yield [16].

2.3 Third- and fourth-generation biofuels

Fuels derived from algae fall under the category of third-generation biofuels. Due to their ability to convert solar energy into chemical energy they possess a special perspective in the production of biofuel. They also hold the abilities to not compete with food crops, pose negligible environmental impacts, and mitigate the challenges that the world is currently facing from GHGs [13]. Through the transesterification process the algae oil is converted to biofuel known as third-generation biofuel.

Despite the seeming potential of this generation of biofuel to eliminate land use and food conflict, it demands a lot of progress in research to improve its viability and the amount of energy produced [13].

Biofuels produced by technologies with the capability to convert optimized biomass feedstock into biofuels are known as fourth-generation biofuels. The primary feature of these fuels is the capture and sequestration of CO_2. Production of such biofuels proceeds by consumption of CO_2 that is higher than the amount produced during their use [17].

3. What is lignin chemically?

3.1 Role of lignin in the biological living world

When plants started life on the land, they had to adjust their physiology to adapt to specific conditions such as desiccation, UV radiation, and nutrient rarefaction. The most successful species to exploit terrestrial environments were seed plants such as the conifer phylum and angiosperm of tracheophytes. Xylem tissues with lignin helped this first plant group in the transport of water and nutrients from roots to leaves. Tracheids belonging to some characteristic cells of xylem helped in water transportation. Metabolism of the lignin precursor led to the formation of cuticles that protected plants from drought. Lignin within which were embedded cellulose microfibrils acted as a waterproofing agent. When subjected to negative pressures, the enhanced intermolecular hydrogen bonding between cellulose microfibrils provided greater stiffness to the cells and prevented collapsing of the plant. According to scientists, during the Silurian period, around 440 million years ago, lignin first appeared since xylem-like elements were found but the origin of lignin could be estimated to date back to 250 million years ago from the Mesozoic plant fossils. By using a single enzyme, chemical, or biological method it is not possible to decompose natural lignin. This is because of the various types of linkages and the non-regular macromolecular structure. The presence of such a structure provides protection to the plant against chemical, mechanical, and biological forces of nature. Lignin, along with providing structural support, also aids in the transport of moisture and nutrients. Compressive strength, hydrophobicity in woody biomass of cell walls of xylem, binding, encrusting, and water transport can be affected by varying the lignin content, subunit constituents, and localization.

3.2 Synthesis of lignin in the living system

Lignin isolation was described for the first time in 1956 [18]. By treating spruce wood with dioxane−water (96:4) for 6 weeks, lignin called milled wood lignin was extracted. The precursors of lignin, also called monolignols, like p-coumaryl, coniferyl, and sinapyl alcohols, were found. The chemical composition mainly consisted of the presence of an aromatic ring in hydroxycinnamic alcohols along with a side chain of three carbons denoted as α, β, and γ. The numbering is done by assigning the side chain attached to the aromatic ring as 1 and the phenol carbon as 4. The degree of substitution on the aromatic ring on positions 3 and 5 differs in monolignols. Hence, the lignin structure can be decomposed into guaiacyl units (G), derived from coniferyl alcohol, having one aryl-OCH_3 group, syringyl units (S), derived from sinapyl alcohol, having two aryl-OCH_3 groups, and p-hydroxyphenyl units (H), derived from p-coumaryl alcohol, having no OCH_3 groups. Adler, in 1977, proposed the complete lignin structure for the first time, describing it as a "highly branched biopolymer containing various methoxy, carboxylic, phenolic and aliphatic hydroxyl, and carbonyl functional groups" [19]. Moreover, the lignin content varies from one plant source to another and, on this basis, deriving lignin can be broadly classified as hardwood with a presence of 27−33 wt.% lignin containing 25%−50% G and 50%−70% S, softwood with a presence of 18−25 wt.% lignin consisting largely of G units (80%−90%), and grass having 17−24 wt.% lignin containing all three aromatic units: S (25%−50%), G (25%−50%), and H (10%−25%). While the lignin content, in general, is found to be lower in agricultural residues comprising of rice straw, bagasse from sugarcane, and wheat straw, its content is higher in forest residues.

The S/G ratio determines the structure of hardwood and in the case of softwood it is mainly composed of G units. G-type lignin is prevalent in the S2 layer of hardwood vessels. The presence of the latter component with a condensed structure enables the plant and consequently the cell walls to tolerate high compressive forces and thereby provides strength and enables the movement of fluids. A complex enzymatic pathway leads to the formation of lignin polymer. The deamination of phenylalanine molecules serves as the starting point to initiate biosynthesis of monolignols through the shikimate pathway. The formation of monolignols proceeds by the involvement of more than 10 enzymes. After their synthesis in the cytosol, they are moved to the cell walls where they are oxidized by oxidase enzymes like laccase and peroxidase forming phenoxy radicals. Several types of radicals are formed due to the resonance-stabilization of phenoxy radicals which then react making lignin through a polymerization process known as lignification. It is at the cell corner of the middle lamella that the process of polymerization starts and from there it progresses across the secondary wall toward the lumen. Lignification is not only affected by the biosynthesis of monolignol but also impacted by the time between polymerization and its deposition in the cell walls upon exportation. Although this aspect is important for the final structure of lignin it is not clearly understood. Further, with the discovery of the enzyme monolignol transferase (PMT) a debate regarding the export of trimers or oligomers by plant cells for lignification was ignited. In the first step of lignification, the C-β of one of the oxidized monolignols forms a dimer with the second oxidized monomer at either its C5 (in G- and H-units) or the C-β [20]. The length of the chain progressively increases as new monolignol radicals are added to the end. The formation of cross-linking points take place by the coupling of phenolic hydroxyl−C5, coupling of C5−C5, or by the generation of dibenzodioxin structures [21]. This coupling results in the formation of many covalent bonds such as the most common diphenyl methane β−1 carbon−carbon bonds in ether β−O−4, and diphenyl ether like pinoresinol β−β, phenylcoumaran β−5, dibenzodioxin, 4−O−5, and 5−5 [22]. C−β seems to be the most reactive site as it leads to a lot of β−β, β−O−4, and β−5 bonds. Moreover, it has been estimated that nearly 50% of the lignin bonds belong to the β−O−4 ether type. It is also the most fragile bond that is broken during the pulping process. Radicals of sinapyl alcohol are restricted to form covalent bonds in the β—, 4—O—, and the position at C1. This is because they are unable to couple at the C5 position. When compared to softwood lignin, hardwood lignin has significantly more β—β′, β—O—4′-bonds and fewer bonds at C5, but almost the same amount of β—1′ bonds. Due to these features, it is believed that with respect to softwood lignin, hardwood lignin is less branched and more linear. Such differences are the key behind what makes Kraft pulp easily made from hardwoods and not softwoods. The existence of sinapyl alcohols in hardwood lignin gives rise to a greater proportion of methoxy groups. The radical coupling process of lignin formation is described as dehydrogenative polymerization, which is considered to be an entirely random process. Through molecular dynamic simulations some authors have shown the existence of a certain level of order in lignin structure due to the β—O—4 guaiacyl bonds parallel to the microfibrils of cellulose and hemicellulose.

3.3 Lignin carbohydrate complex (LCC)

Several studies have shown the occurrence of covalent bonds between carbohydrates and lignin [23]. Lignin carbohydrate complex (LCC) refers to the intricate assembly of lignin, cellulose, and hemicellulose. The covalent bond is mainly found to exist between lignin and xylan and glucomannan hemicelluloses. Also, pectin can form an ester linkage with lignin but there is no such evidence to prove this. In the middle lamella region, it has been suggested that pectin surrounds lignin and there is formation of a globular pectin—lignin complex [24]. In regulating or controlling the shape of lignin in the middle lamella, pectin appears to play a role. What makes the LCC structure significantly more compact and complex is the hydrogen bond that lignin has with cellulose. From this complex structure and the tough and resilient covalent bonds to isolate lignin, it is necessary to first break such bonds. The study by Lawoko et al. showed that almost 90% of the lignin in a pulp of spruce Kraft was linked covalently to carbohydrates, with the glucomannan—lignin complex being the major one [25]. Salmén and Olsson proposed an arrangement wherein one type of lignin is surrounded by glucomannan and, on the other hand, xylan surrounds another type of lignin [26]. The three types of covalent linkages of lignin-carbohydrates that exist are benzyl ether, benzyl ester, and glycosidic bonds. Hydrolysis of the ester type of bonds occurs most readily in alkali media, while the etherified unit of benzyl ether is quite stable at alkaline conditions of pulping also. When considering the flexibility of fibrils, it is the covalent bonds between lignin and xylan that impart this property.

4. Isolation and extraction of lignin from lignocellulosic biomass

The pretreatment step is mainly concerned with the breakdown or weakening of the bonds between the lignin and LCC carbohydrates. Lignin in LCBM is tightly intercalated into the polysaccharide portion of the biomass. As has been already discussed in the preceding section about LCC, many covalent bonds between lignin and hemicellulose make it recalcitrant to pretreatment strategies. Therefore, extracting and isolating the total lignin in its native state is difficult. When the LCBM undergoes the pretreatment step, along with deconstruction of the three major components, lignin also gets partially depolymerized. This results in the generation of phenolic monomers and oligomers, along with lignin of smaller carbon chains. However, the major challenge is to isolate the lignin components at a sufficiently high yield and with minimum alteration in its chemical structure. Once depolymerized, the reactive intermediates participate in unwanted side reactions. They condense or repolymerize, forming new C—C bonds and different compounds with undesired chemical properties. Condensation also generates large quantities of solid residues with insufficient applications and adds to the difficulty of the process. These problems contribute to the challenges of acquiring well-defined compounds that would serve as promising candidates for fuel and aromatics for value-added products. However, as it is not feasible to obtain the large macromolecules as a whole, the pretreatment step is of utmost necessity. It degrades the large macromolecules into smaller fractions, facilitating their solubilization from the LCBM.

4.1 Pretreatment technologies to separate and depolymerize lignin

All the existing pretreatment technologies, in general, follow two principles to extract the lignin portion of LCBM. One is to extract lignin by digesting the cellulose and hemicellulose fractions and the other one is its extraction by fractionation methods to get rid of the other components. In the former strategy, hot water and dilute sulfuric acid aid in the digestion of the polysaccharide by enzymatic hydrolysis. The latter one generally uses alkali solutions with calcium, sodium, or potassium as the metal ion to target the removal of lignin from the biomass. The comparatively weak linkage and the one that is the most abundant, β—O—4 (aryl ether), is most easily broken down in almost all the pretreatment methods. While the stronger ones, β—β′, β—1, β—5, 4—O—5, and 5—5 pose resistance to the degradation process. Disruption of such linkages can be done by subjecting the LCBM to a combination of extraction technologies. Moreover, since this step is often associated with the formation of a wide range of undefined lignin intermediates, appropriate selection of the pretreatment method is essential. Variation in the obtained phenolic monomers, aromatics, and lignin structure is also observed when the feedstock is changed or different extraction methods and severity are used for the same biomass. Hence, the choice of optimum technique ensures its proper utilization in second-generation biorefineries.

The biorefinery approach is beneficial because, whatever might be the chosen route, it involves high energy consumption and costly chemicals. As much as 20% of the entire conversion is achieved through different approaches [27]. Through effective pretreatment, it is desired to (1) have maximum yield of each of the three major components to funnel the important compounds, thus generated, into different units of the biorefinery, (2) treat a wide variety of different lignocellulosic feedstocks, (3) have efficient recovery of lignin to carry out combustion, (4) reduce the formation of inhibitors, and (5) lower the operational costs and energy consumption.

4.1.1 Physical methods

4.1.1.1 Pretreatment by mechanical extrusion

This method serves as the most common method to disintegrate lignocellulosic biomass into smaller cross-sections and disrupt the crystalline and amorphous cellulose matrix. The shear force generated by the screw blades along with the high temperature in the barrel of the extruder aids in the process of disruption. Single-screw extruders and twin-screw extruders, among the many types of extruders, find much use in the pretreatment and enhance its hydrolysis in the presence of enzymes [28]. However, the significant cost associated with the energy requirement makes this pretreatment strategy unsuitable to be scaled up for industrial processes. Some advantages of this process are the enhanced monitoring of the controlled process, product formation with negligible degradation of sugar, and sugar recovery from a wide range of biomass [29]. Coupling extrusion with other methods like ammonia fiber explosion, bio-extrusion, chemical treatment with alkali or acid, vacuum extrusion, and steam explosion can further enhance sugar recovery. When used in combination it has been seen that the pretreatment cost is also reduced to an extent.

4.1.1.2 Pretreatment by milling

Through milling, it is desired to reduce the size of biomass and thereby increase the surface area. This is achieved through the processes of grinding, chipping, and/or milling. The crystalline structure of the cellulose in biomass is disrupted by applying shear forces. After the process of chipping, biomass size of 10–30 mm is obtained, which when subjected to grinding and milling to decrease the size of the particles to 0.2 mm [30]. However, Mosier et al. [31] observed that neither the yield nor the hydrolysis rate improved significantly on reducing the size below 0.4 mm. The rate of feeding of biomass, machinery parameters, amount of moisture in the biomass, size of the biomass initially, properties and efficiency of the machine, and time are some of the factors that are taken into account to achieve effective milling [32]. The type of biomass that is fed into the system also determines the milling time and the kind of milling used. Commonly used milling methods are two-roll milling, colloid milling, vibratory milling, and hammer milling [30]. In order to achieve better result in milling, newer and better techniques of milling that are different from the traditional approaches have been adopted, such as vibratory milling. Such recent techniques increase the degradability of chips of spruce and aspen. To enable lower energy consumption, wet disk milling is widely employed. This results in fiber generation and also improves cellulose hydrolysis considerably. The advantages of the milling process as compared to other physical pretreatment methods include low energy requirement, shorter processing time, and a lower water requirement.

4.1.1.3 Pretreatment by sonication

With this method, ultrasonic waves are made to pass through a fluid with lignocellulosic biomass. Microcavities are generated within the lignocellulosic biomass by these acoustic waves which results in pressure variation. The net effect is degradation of the biomass through the formation of shock waves that generate gas bubbles, which further collapse inside the microcavity. It is the shock waves that cause biomass degradation in a mechanical manner. Maximum cavitation and easy degradation are achieved within the biomass by keeping the ultrasound frequency between 10 and 100 kHz. Moreover Yachmenev et al. reported that the rate of hydrolysis by enzymes increases by almost 200% when ultrasonic waves are passed through a cellulosic suspension at 50°C [33]. The factors which affect the sonication mode of pretreatment are configuration and geometry of the reactor, frequency of the ultrasound, length of exposure to ultrasonic frequency, properties of biomass, kinetics, and the solvent in which the biomass is present and through which the ultrasound is passed [34]. Since the pretreatment can be carried out at a lower temperature and also a short time, it could be more desirable by reducing the cost [35].

4.1.2 Chemical methods

4.1.2.1 Pretreatment by the alkaline method

A reliable, comparatively easier method with high potential to cleave the alkyl–aryl linkages of lignin in lignocellulosic biomass is treatment by alkaline solutions such as ammonium hydroxide, lime, sulfite, sodium hydroxide, and calcium hydroxide [36]. Sodium hydroxide has been found to be a potent delignifying agent with the ability to provide this effect under a wide variety of conditions, and it is therefore preferred in most cases. The method progresses by increasing porosity and thereby the associated surface area. Another added advantage is that it does not cause elimination of reducing sugars and carbohydrates. Further, this treatment enables better action of hydrolysis by enzymes by making the biomass more digestible, which in turn results in maximizing the yield of fermentable sugars. As a matter of fact, alkalis are less corrosive solutions in comparison to acids like sulfuric acid employed in acidic pretreatment. The main disadvantage lies in the significant time requirement, which demands the operation be carried out for from several hours to up to 1 day [35].

4.1.2.2 Pretreatment using acids

Many studies targeting bioethanol production have reported the acidic mode of pretreatment that mainly targets degradation of hemicellulose [37]. As hemicellulose is converted into soluble sugars, the structure of the biomass also changes. Increasing the temperature improves the conversion process significantly. The mechanism of action allows maximum recovery of monomeric sugars by degrading the polysaccharide and lignin linkages. Phosphoric acid, sulfuric acid, and acetic acid are among the acids that are generally employed. Loading of the solid biomass, time of residence, temperature, and importantly the concentration of the acid are some of the factors that need to be considered. The reason for using this process lies in the assured conversion of amorphous cellulose and the degradation of the matrix of lignocellulose. However, the generation of inhibitors in the process adds to the recovery cost and energy demand. It has been seen that the use of dilute acid results in the formation of more glucose from cellulose, thereby requiring a lower amount of acid but an increased temperature. Hence, the dilute acid treatment is chosen more often than concentrated acid for hydrolysis.

4.1.2.3 Pretreatment using ionic liquids (ILs)

The way in which the anions and cations of ionic liquids depolymerize the cellulose and lignin of lignocellulosic biomass mean that it becomes strongly soluble in the solvent. However, the process exerts an undesirable effect on the activity of cellulase and therefore work is in progress to make the enzyme more tolerant. Temperature and biomass loading in this method are seen to affect the rate of hydrolysis. There exist numerous ionic liquids in this pretreatment strategy, some of them are pyrrolidinium-based $[(C_4N)Xn]^+$, pyridinium-based $[(C_5N)Xn]^+$, phosphonium-based $[PX_4]^+$, imidazolium-based $[(C_3N_2)Xn]^+$, sulfonium-based $[SX_3]^+$, and ammonium-based $[NX_4]^+$ ILs. The acidic ionic liquid category has been proved to have a strong capability in the pretreatment process. The extraction of lignin becomes sufficiently easy due to the breaking of ether linkages resulting in lignin depolymerization [38]. Since in this case mostly water is used and there is very limited use of chemicals it might seem to be advantageous, however, the difficulty in the recovery of ionic liquids making it nonrecyclable coupled with a greater energy requirement make the entire method costly.

4.1.2.4 Pretreatment using organic solvent

The ability of organic solvents to separate cellulose, hemicellulose, and lignin into different fractions allows easy recovery and reuse of solvents using this process. This process can be carried out even without the use of a catalyst. As per the literature, organic solvents make the extraction and conversion of biomass easier. Alcohol, dioxane, propionic acid, phenol, acetone, esters, amines, and formaldehyde are some of the prevalently used organic solvents. It is desired to use such solvents because of their generally low boiling points. Some of the preferred ones are alcohols like methanol and ethanol due to the convenience of their recovery. This is a widely used method that can be operated at a temperature between 100 and 250°C. The advantages of this method include the purity with which cellulose and hemicellulose can be extracted with negligible degradation. However, there is an economic concern regarding the production of pure lignin and related chemical derivates together with coproducts of high value [39].

4.1.3 Physico-chemical methods

4.1.3.1 Pretreatment by steam explosion

This process involves subjecting the biomass to high pressure (0.69–4.83 MPa) and high temperature (160–260°C) through the use of saturated steam in a reactor. The time for which the biomass is kept exposed to such a condition, or the residence time, can range between seconds and minutes [35]. Pretreatment by this method functions by causing autohydrolysis when the vapors condense on entering into the biomass. The presence of acetyl groups within the biomass then reacts with the condensed steam to form organic acids that facilitate the disintegration of hemicellulose and the cleavage of glycosidic bonds. Once the matrix of lignocellulosic biomass loses its integrity, the hemicellulose solubilizes. However, the net cost increases due to the need for high amounts of energy and water. The process also leads to generation of inhibitors which is another major disadvantage [35].

4.1.3.2 Pretreatment by hot water

Hot water or hydrothermal pretreatment makes use of water in the liquid state at a temperature of 140–240°C to depolymerize hemicellulose. This leads to solubilization of the formed products in the liquid state while leaving the cellulose intact in the solid state. Lignin undergoes depolymerization and polymerization simultaneously due to the temperature of the water. Since the temperature of glass transition of lignin in aqueous conditions is between 80 and 100°C [35], relocalization of lignin occurs, improving the digestibility of cellulose as this migration promotes access to cellulose microfibrils. The lack of a requirement for chemical reagents results in a drop in the overall process cost and makes it comply with the principles of green technology. It must be noted that such a process results in the formation of inhibitors when subjected to harsh conditions [40].

4.1.3.3 Explosion by supercritical CO_2

Supercritical CO_2, with its non-emissive property of organic vapors serves as a green solvent that is also non-flammable. Having 31°C as the critical temperature and 1071 psi as the critical pressure, it allows for easy separation and subsequent recovery after the pretreatment process. When it is charged into the biomass at high pressure through a small opening, it acts on the cellulose and hemicellulose explosively. This results in an increase in the surface area due to the disintegration of cellulose and hemicellulose [41]. Modulating some of the parameters like the solvent flow rate, temperature, pressure, and size with respect to the temperature bed are attributed to the enhanced yield. As no toxic substances are produced, and there is easy separation with no emission of greenhouse gases, this strategy is cost-effective and eco-friendly. This pretreatment method is able to attain greater efficiency when carried out along with milling and extrusion [42].

4.1.3.4 Pretreatment by ammonia fiber expansion (AFEX)

Expansion of lignocellulosic biomass with ammonia fiber is another pretreatment strategy that exploits treating the biomass with pressurized anhydrous or gaseous ammonia within a vessel. With a residence time of 30 min and a temperature of 90°C, to 1 kg of biomass, 1–2 kg of liquid ammonia is added [35]. Such application of ammonia at high pressure and temperature results in the removal of acetyl groups, decrystallization of cellulose and finally hemicellulose depolymerization in the biomass. This method provides a cost-effective strategy for pretreatment because of the recyclability of ammonia for many cycles, easy recovery of and low cost of ammonia, and no production of inhibitors, enabling easy downstream processing.

4.1.3.5 Pretreatment by microwave

Pretreatment can also be done by exposing biomass to electromagnetic microwave radiation that has chemical, physical, and biological effects on the biomass. Irradiation of microwaves leads to vibration of polar molecules and ionic movements, causing collisions and heating. However, it is not feasible to implement this in an industrial scale due to the associated cost even though it does not produce any toxic products. Moreover, the effectiveness of the method is largely determined by the properties of the lignocellulosic biomass such as the dielectric constant [43].

4.1.3.6 Pretreatment by wet oxidation

In the presence of oxygenated air and at an elevated temperature, higher than 120°C, the biomass is subjected to water or hydrogen peroxide for 30 min. Biomass with a significantly high lignin content finds importance in this pretreatment method. The reaction time, temperature, and oxygen pressure considerably influence the effectiveness of this method. When the temperature is increased beyond 170°C water starts to act as an acid and thereby catalyzes hydrolytic reactions. This leads to oxidation of the lignin and conversion of hemicellulose into pentose sugar monomers, whereas cellulose remains unaffected and remains almost in the native state. In order to reduce the formation of inhibitors like furfurals and corresponding aldehydes and to enhance the breaking down of hemicellulose, the temperature is decreased by adding chemical compounds like sodium carbonate and alkaline peroxide. The storage problem of oxygen due to its inflammable tendency along with the significant cost of hydrogen peroxide makes this process unsuitable to be scaled up to an industrial level.

4.1.4 Biological methods

Biomass pretreatment can also be targeted by using microorganisms that have the power to degrade the lignocellulosic components through their inbuilt potential of produce lignolytic and cellulolytic enzymes. These enzymes cause delignification of the biomass, leaving the rest of the polysaccharide for hydrolysis. Fungal species serve as a good source for this purpose. Their ability to synthesize a wide variety of different enzymes contributes significantly to the degradation process. Lignin can be removed by employing the properties of white rot fungi, whereas on the other hand, to degrade cellulose and hemicellulose, brown rot fungi have been found to have tremendous potential. They carry out this function through enzymes like endoglucanases and β-glucosidases. There is also another class of fungi, called soft rot fungi, which degrades cellulose through endo-1,4-β-glucanase, 1,4-β-glucosidase, and exo-1,4-β-glucanase enzymes. They also possess laccase enzyme which enables incomplete degradation of lignin.

Since a single microorganism is unable to degrade the complex biomass properly, a collection of different microbial species (microbial consortia) is used to enhance the process. The presence of a huge diversity in the microbial population helps to fulfill the task by interacting with each other and acting either cooperatively or competitively in a different manner [44]. Some of the popular microorganisms that the consortia are comprised of are *Aspergillus*, *Trichoderma*, *Penicillium*, and *Humicola* among the fungal community and *Cellulumonas* and *Cytophada* from the bacterial community [45]. A notable advantage of adopting this method is that they do not have any negative impact on the environment as there is no need for chemicals and also the energy required is low. Despite this advantage, the high incubation time that affects the cost makes the process not viable at an industrial level [35].

4.2 Commercially available lignin

Lignin, the main source of which is lignocellulosic biomass, is undoubtedly an abundant raw material with an amount of approximately 100 billion tons, that is replenished each year. Out of this total quantity, about 1.5–1.8 billion tons of lignin contributes to industrial lignin each year. It is well known that the pulp and paper industry, which separates 50–70 million tons of lignin, is the major contributor. However, only a small percentage of the black liquor (extracted lignin) is used as a fuel in industrial boilers to generate heat in the pulping process, the rest is considered as waste and so discarded. Scientists and specialists in the field claim that the potential for lignin is not fully utilized, with only US$300 million as the current level of business [46].

4.2.1 Sulfur-bearing process

Kraft lignin and lignosulfonate lignin are mainly synthesized in the pulp and paper industry, while extracting lignin from cellulose falls under the category of sulfur lignin.

4.2.1.1 Kraft lignin via the Kraft process

In 1879, the Kraft pulping technology first came into action and since then its widespread use in as much as 80% of the chemical processes of pulping has made it one of the major technologies used by pulping industries. The reason behind such widescale adoption lies in the easy separation of the chemicals required in the process. In the Kraft process, biomass is treated with two major chemicals, sodium hydroxide (NaOH) and sodium sulfide (Na_2S). The aqueous solution of these two compounds degrades and break downs lignin into smaller parts. This results in its dissolution, producing a black liquor or pulping liquor. Generally, the black liquor thus produced was combusted in the recovery step of Kraft pulp mills. Lignin being significantly energy dense therefore served as the fuel. However, the lignin content in black liquor is found to be much greater than the amount demanded as an energy source in the mills. The need to channel the excess lignin initiated its production at the commercial scale. Different strategies for lignin isolation from black liquor started to be developed, and MeadWestvaco has been marketing 20,000 tons of pine Kraft lignin with the name of Indulin lignin every year since the 1940s [47]. Also, in North Carolina (USA), Domtar Plymouth Mill in 2013 installed its own technology called the LignoBoost to produce up to 25,000 tons per year of Kraft lignin, also from pine species [48]. Sora Enso (Sunila Mill in Finland) improved the LignoBoost technology even further and since 2015 their plant started producing 50,000 tons per year of Kraft lignin. FPInnovations, in 2015, established the LignoForce technology with the potential to generate 30 tons per day of Kraft lignin [49]. They arranged the entire setup of the Hinton Pulp Mill at Alberta in Canada at the West Fraser Timber Company. The cost of lignin produced through such technologies has a commercial value in the range of between US$250–600/ton [50]. If we look into the chemical composition of Kraft lignin, the literature reports it to be composed of 1.0–2.3 wt.% sugar, 3.0–6.0 wt.% moisture, 1.0–3.0 wt.% sulfur, with a molecular weight less than 25,000 g mol^{-1} [51]. The ash content of Kraft lignin could be as high as 30% and it can be lowered to 0.5–3.0 wt.% by treating it with a solution of dilute sulfuric acid. Another characteristic property observed in the extracted Kraft lignin is the presence of phenolic hydroxy groups at a higher percentage along with biphenyl structures. The extracted fraction also contains condensed lignin which is determined by the process duration. When the process is continued for a longer time the condensed fraction increases [52].

4.2.1.2 Lignosulfonate lignin via sulfite pulping

For over a century, the operation of the sulfite process has found importance in the manufacturing of different grades of paper. Aqueous sulfur dioxide having a pH of 1–2 along with a sulfite base using metal ions of sodium, magnesium, calcium, and ammonium is needed to achieve sulfite pulping [53]. As the reaction progresses, a high amount of sulfur gets integrated into the lignin, which promotes the attachment of sulfonate groups on the aliphatic part of the lignin's phenylpropane (C9) unit to the benzylic carbon atom [53]. The lignin gets dissolved mainly by hydrolysis and the sulfonation reaction. As per reports from the literature it has been observed that the variation in the number of sulfonate substitutions per mole of C9 unit can range between 0.4 and 0.5 mol^{-1} [54]. Lignosulfonate has been estimated to be produced annually at one million tons [51]. It has significant importance in the industry. the main producers of it globally are DomsjoFabriker (Sweden), Tembec (Canada), La Rochette Venizel (France), Borregaard LignoTech (Norway), and Nippon (Japan) [51]. The market value of lignosulfonate is from US$180 to 500/ton, and it is composed of approximately 5.8 wt.% moisture, the ash content is 4.5–8.0 wt.%, and the sulfur content is 3.5–8.0%. Its molecular weight has been reported to be less than 15,000 $g\ mol^{-1}$ [55]. However, the use of the sulfite process nowadays is gradually becoming less due to the greater versatility of the Kraft process and also the low availability of lignosulfonate.

4.2.2 Sulfur-free process

A popular and more recent class of industrial lignin is the sulfur-free lignins. After fractionation through an extraction process they have been found to possess lower molecular weight and the structure remains similar to native lignin. The properties exhibited by them make a good source of low-molecular-weight phenol or aromatic compounds. Sulfur-free lignins are further divided into the organosolv lignin obtained by solvent pulping and soda lignins obtained by alkaline pulping.

4.2.2.1 Soda lignin via soda pulping

The soda-anthraquinone or the soda pulping process leads to the formation of soda lignin. Since this process does not involve the use of sulfur in any form, the soda lignin thus obtained is sulfur-free. This is a major advantage over the Kraft process [56]. A total of 13%–16% sodium hydroxide at a pH of 11–13 is a major chemical that needs to be added [57]. Yet another advantage over other methods of lignin extraction is that it has a comparatively lower negative impact on the environment [56]. The soda lignin may also include the presence of nitrogen and silica when lignocellulosic biomass like flax, straws, and bagasse are used as raw material [58]. As the existence of high concentration of silica from non-woody sources results in its precipitation along with lignin, the use of such sources for the extraction of lignin is difficult with this approach. Therefore, when using this method with nonconventional sources, the extracted lignin is sufficiently low in quality and purity [56].

In this industrial method the production capacity ranges approximately between 5000 and 10,000 tons per year [54] and it has a market value of US$200–300 per ton [59]. Among all the industries involved in the production of soda lignin one of the most notable is GreenValue SA, which is a Swiss company that stand apart from the rest in the annual production from wheat straw. It produces close to 10,000 tons per year [51]. The composition of the extracted lignin comprises of no sulfur, sugar from 1.5 to 3.0 wt.%, moisture having 2.5–5.0 wt.%, and 0.7–2.3 wt.% ash [56]. When non-wood sources are subjected to alkaline pulping, in some literature it has been said that the concentration of the phenolic hydroxy group increases and the soda lignin shows a low glass transition temperature. The amount of impurities in the form of ash and sugar also contributes to the thermal properties of this type of lignin [60]. The source of biomass and the mode of operation also influence the temperature-dependent behavior of soda lignin. Also, in the case of soda lignin obtained by treating wheat straw using this process, leads to a lower Tg point as compared to that of hemp. The molecular weight of soda lignin is less than 15,000 $g\ mol^{-1}$ [51,55] and in recent years it has drawn attention because it can be potentially converted into phenols of low molar mass and aromatic compounds [61]. In the case of Kraft and lignosulfonate type of lignin the presence of sulfur in a chemically bonded state does not allow for their easy removal. This does not present itself as a problem in this sulfur-free method, making the process less costly. However, one major disadvantage is the significant amount of nitrogen and ash in soda lignin [60].

4.2.2.2 Organosolv lignin via solvent pulping

The principle in this case employs treating the biomass with ethanol, acetone, methanol, or other mixtures of organic solvent at a temperature of 100–250°C to separate the lignin and hemicellulose [62]. As the linkages of β-ether get degraded it causes the solubilization of lignin. This in turn makes the extraction process much simpler and also enables better activity of acid catalysts like hydrochloric acid, sulfuric acid, and organic acids like formic acid or acetic acid. The deacetylation of hemicellulose also sometimes generates the much-needed acetic acid in the reaction mixture itself [63]. The solvent in this approach is recovered by carrying out distillation with water and the precipitation of organosolv lignin is achieved through the use of solvents. In the above-mentioned solvents, the associated benefit is that almost all the above-mentioned solvents can be recovered, reused, and recycled. Some of the most desired solvents in this process are those with lower molecular weight and boiling point such as alcohols (methanol and ethanol) [64]. The functioning here is exerted by the breaking down of bonds between carbohydrates and lignin that leaves lignin in its native form. On comparing between the organosolv and Kraft processes of lignin extraction, it has been reported that in both cases the extent of delignification and the number of β-ether linkages in the extracted lignin stands out as almost identical [63]. The lignin so extracted has been found to have a sugar weight percentage of around 1.0–3.0, an ash weight percentage of 1.7, moisture weight percentage of 7.5, and no sulfur [56]. The molecular weight is less than 5000 g mol^{-1} which is less than for the other types of lignin.

To include the associated drawbacks of this process, one of the most significant is that the cost involved in solvent recovery is quite high and they also pose a risk to the environment. These solvents are also unstable at higher temperatures [51]. Another disadvantage is that the parameters of the process need to be adjusted with the change in solvent as it makes the system more complex, such as, for example, the need to increase the pressure to attain optimization. Some reports even claim that the biological method of pretreatment mentioned earlier and enzymatic hydrolysis inhibit the organosolv process. As already mentioned, it is not possible to recover 100% of the solvent [64]. Although the purity and quality of organosolv lignin is more than the sulfite and Kraft lignin, the entire production strategy is challenging and therefore acts as a barrier to its wide-scale implementation.

Among the major companies worldwide those worth mentioning are Lignol in Canada, and Dedini, DECHEMA/Fraunhofer, and CIMv in France. It has been estimated that the annual production of organosolv lignin stands close to 3000 tons per year, generating a market value of US$300–520 per ton [51].

5. Elucidation and characterization of the extracted lignin

The complex phenylpropanoid matrix of LCBM with extreme diversity and versatility results in the formation of a large variety of different organic compounds after its pretreatment. Reaction of the reactive intermediates, pretreatment strategy, and fractionation severity, coupled with the observed variation in the case of different biomasses, also contributes significantly to the pool of unpredictability. Also, the age and the part of the plant that has been chosen as feedstock add to the variability in the lignin fraction. There are no definite means to elucidate the complete structural chemistry of lignin to date. However, with advancements in instrumentations and different analytical techniques, most of it can be determined. In recent years, tremendous effort has been exerted by governments to reduce reliance on nonrenewable resources and develop technologies that would promote growth in biorefineries. Tons of lignin generated from the pulp and paper industry and bioethanol factories will encourage the establishment of more biorefineries if utilized to the fullest potential. However, this can be truly envisioned if further research can reveal the complex lignin and the LCBM, which would allow harnessing the full advantage of each component. Therefore, the stepping stone to this is complete characterization of the purified and extracted lignin. Knowing the composition and structure helps decide on the most effective extraction method, design and plan the whole process with ideal conditions, and choose the most appropriate route to valorize lignin.

It is not possible to have lignin in its original form, i.e., lignin cannot be isolated intact from LCBM. In order to study the structure of lignin with a strong resemblance to native lignin, the lignin must be isolated by using the least harsh conditions. The pretreatment conditions and extraction severity should be chosen in such a way that there are minor alterations to its structure and that also without compromising on the yield of lignin. Isolation techniques for milled wood lignin (MWL) [65], cellulolytic enzyme lignin (CEL) [66], and enzymatic mild acidolysis lignin (EMAL) [67] allow the attainment of very close to the aforementioned criteria. Among all these lignin, the dioxane/water-treated ball-milled wood lignin results in Bjorkman lignin [65], representing native lignin most closely. However, the yield is significantly lower (i.e., <30%). In CEL, the ball-milled lignin is treated with cellulases to degrade the carbohydrates while leaving the lignin for extraction by solvent. The purity of lignin is a major concern in this case due to the associated carbohydrates as impurities. However, the yield is greater, at up to 50%. EMAL is obtained if, after hydrolyzing the lignin enzymatically, mild acid is added for further hydrolysis of the solid lignin residues. It has improved yield and purity. However, the techniques employed to attain these three types of lignin can be implemented only in the lab-scale, due to economic constraints when scaled up to the level of industrial biorefineries. Thermomechanical pulp lignin (TMPL) is also found to preserve the structure of native lignin because of the low requirement for chemicals in the process. Therefore, the isolated lignin has minimum modifications. In Table 16.1, the different methods to characterize lignin are described.

6. Routes of lignin conversion

The existing paths for lignin conversion are very similar to those of petroleum refining. The approaches that are widely taken are pyrolysis, gasification, transformation by oxidation or reduction, hydrolysis by treating with acid or alkali, and hydrocracking. In gasification of lignin, the whole lignocellulosic biomass is used for the generation of synthetic gas without any prior pretreatment. Further, Fischer−Tropsch synthesis and methanol synthesis can then be applied to promote the conversion of synthetic gas into chemicals or fuels. In contrast, the use of other strategies is compulsorily preceded by breaking down the lignin biopolymer of molecular weight in tens of thousands of Daltons in a targeted manner to obtain small fragments. Moreover, the processes like hydrolysis, hydrocracking, and pyrolysis are largely followed by another step that aims to achieve the predetermined and desired final products by either largely removing some of the groups or regenerating or choosing to keep certain selected functional groups. By using particular catalysts, one can also successfully convert it into products of high quality, and that too in a single-pot reaction. The catalysts are so chosen that through their mechanism of action, some of the targeted bonds are cleaved while leaving the functions and rest of the linkages and molecules unaffected.

6.1 Pyrolysis method to convert lignin

It is the decomposition process that is achieved by the application of thermal energy. To successfully carry out this process, the ideal temperature range is 450−650°C, and the process is able to proceed with the complete absence of oxygen. The vapors released in the pyrolysis process must be quickly quenched to convert 75% of the biomass into bio-oil, while the rest results in the formation of solid char and exists as noncondensable gas. A mixture of a number of different types of oxygenated organic compounds along with ketones, acids, aldehydes, monolignols, phenols like guaiacol, syringol, and catechol, and benzene and the derivatives formed by its alkylation are altogether present in bio-oil. Due to the high concentration of oxygen (in the form of oxygenated compounds), acidic nature, highly viscous nature, and low volatility of bio-oil, it is not possible to use it as a fuel directly. However, upgrading technology, called hydrodeoxygenation, exists to make it a fuel-ready substituent. In situ upgradation of bio-oil is achieved by adding suitable and appropriate catalysts while carrying out pyrolysis. The gases liberated in the process that cannot be condensed involve the presence of low-molecular-weight hydrocarbons like CH_4, C_2H_2, C_2H_4, C_3H_6, etc., and CO and CO_2. These are mainly generated through the reformation of alkyl chains and functional groups like COOH and C=O. A solid residue remains on completion of the process, known as bio-char. Its chemical nature is an aromatic polycyclic structure that is a product of rearrangement by intra- and intermolecular derivatives of lignin degradation intermediates that usually contain benzene rings. Bio-char finds importance as an adsorbent, biofertilizer, and as a support for catalysts.

The factors that determine the yield of bio-oil and its composition are temperature, size of the particles, catalyst, and the rate of heating. The production of biofuel is seen to improve at an enhanced heating rate $>100°C \, s^{-1}$, while the amount of char formation increases considerably with a drop in the rate of heating [81]. At a greater heating rate multiple chemical bonds are cleaved at the same time, releasing a number of volatile compounds by degrading lignin. This, in turn, leads to a decrease in the char quantity because of less formation of stable oligomers. These oligomers undergo rearrangement with one another to generate char. The particle size of the feed can have an indirect impact on the rate of heat transfer and it changes with a change in particle size. In multiple studies, it has been claimed that the concentration of volatile compounds decreases when large particles are used. A particle size above $1 \, mm^{-1}$ increases the amount of nonvolatiles. Another crucial parameter is the final temperature that affects the concentration of the products formed through pyrolysis in a distinct manner. Due to the heterogeneous nature, a wide range of temperature coverage is achieved when conversion is carried out by pyrolysis. The maximum degradation takes place from 360 to 400°C [82]. A point worth mentioning is that at a temperature greater than 500°C, bio-oil degrades significantly to form biogas. Introduction of catalysts such as NaCl, KCl, Na_2CO_3, K_2CO_3, and some Lewis acids, like $ZnCl_2$, has been found to improve the process and also give a better distribution of products.

TABLE 16.1 Lignin characterization through different instrumental methods.

Sl. no.	Method	Property elucidated	Features	References
1.	Gel permeation chromatography (GPC); vapor pressure osmometry (VPO); ultrafiltration; mass spectrometry; light scattering	Analysis of molecular weight (mainly)—weight average molecular weight (mw) and number average molecular weight (Mn); lignin heterogeneity through lignin polydispersity index	GPC—most advantageous wide range molecular weight detection, quantitative analysis of even mg-sized fraction, less processing time, endures both natural and synthetic polymers	[68,69]
2.	^{1}H NMR	Structural characterization; quantitative detection of carboxylic acids; CH_2, CH_3, $-OCH_3$, groups; phenolics; monolignol types; β—O—4 substructures; aliphatics	High signal to noise (S/N) ratio in a much shorter time duration of the experiment; signal overlapping is a major problem, due to short spectral range (∼δ 12–0 ppm) and structural complexity of lignin	[70,71]
3.	^{13}C NMR	Detailed lignin structure, functional groups, noncondensed and condensed aromatic groups, monolignol ratio, aldehydic groups, β—O—4 interunit bonds	Has a wide spectral range and lower signal overlap, therefore, better resolution; the problem is that it requires a higher experimental time, attributed to less natural isotopic quantity of nucleus of ^{13}C	[72,73]
4.	^{31}P NMR	Quantitative detection of groups: hydroxyl, p-hydroxyphenyls, carboxylic, aliphatic, phenolic hydroxyls	Relatively shorter experimental time, requirement of small sample size	[74]
5.	2D HSQC (heteronuclear single quantum coherence) NMR	Semiquantitative technique to investigate relative amounts of monolignol ratios, interunit linkages, and other structural features in native and genetically modified plants and also after pretreatment	Most frequently used method for lignin structural characterization, greater sensitivity of ^{13}C nuclei, no signal overlap	[75,76]
6.	FTIR spectroscopy	Detection of typical lignin functional groups: methoxyl, aromatic C—H, aliphatic C—H, carbonyl, carboxyl, hydroxyl; ratio of phenolic hydroxyl to aliphatic hydroxyl groups; characteristic aromatic framework absorbance at 1500–1560 cm^{-1}; G, S, and H units; characterization of variation in chemical structure of wood from same species but from different geographical locations	Most extensively used to determine functional groups based on chromophoric compounds; technique is extremely sensitive, noninvasive, nondestructive, rapid; versatile method to identify structural properties of lignin	[77]
7.	Raman spectroscopy	Sister technique of FTIR spectroscopy with bands in absorption spectrum very close to FTIR; determination of lignin chemical structure; structural changes due to dilute acid pretreatment via confocal Raman microscopy	In situ analysis on plant cell wall excludes sample preparation; wet samples having water can also be used for structural characterization	[78]
8.	UV spectroscopy	Semiquantitative analysis of lignin purity and degraded lignin via extinction coefficient (EC); components in isolated lignin and acid-soluble lignin; phenolic hydroxyl groups	Simple and accurate method based on Lambert—Beer's law; low EC value signifies higher abundance of nonlignin components in isolated lignin	[79,80]

6.2 Lignin conversion by acid catalysis

In the past, acid catalysts were used as a pretreatment strategy in biomass and wood pulping processes. They caused the cleavage of aryl-ether bonds, mainly the linkages of α- and β-aryl ether. Almost 70% of the linkages present in lignin are composed of α- and β-aryl ether, and therefore, this method of lignin depolymerization has become quite prevalent in recent years. After the completion of this process, monomeric, dimeric, or oligomeric derivatives of lignin are present in the reaction mixture. As the activation energy of the linkages in α- and β-aryl ether are comparatively lower than the C—C and biphenolic ether bonds, they are the first to be disrupted. While comparing α-aryl ether bonds and β-aryl ether bonds, it has been found that hydrolysis of the former takes place much more rapidly than in the latter. This is because the activation energy is from 80 to 118 kJ mol^{-1} for α-aryl ether bonds and 148—151 kJ mol^{-1} for β-aryl ether bonds.

In order to find the best-suited acids to be used in the process, a number of acids have been experimented with, such as acidic ionic liquids, organic acids, mineral acids, zeolites, and Lewis acids like $ZnCl_2$, $AlCl_3$, $FeCl_3$, $InCl_3$, and metal triflates have found importance in lignin depolymerization. The two properties that contributes to the catalytic activity of Lewis acid are (1) formation of Brønsted acid through the reaction between Lewis acid and water enhances the hydrolysis of lignin ether bonds and (2) the metal center of Lewis acid promotes the depolymerization of lignin by acting on the oxygen atoms in ether bonds. The "one-pot" ionic liquids have been preferred in recent years [83] due to their ability to solubilize and convert lignin through the general reaction mechanism of catalysis by Brønsted acid [84]. Different anions and cations can be used to improve the effectiveness of catalytic activity and solubility. Further, use of metal chlorides in ILs have also been found to have a positive impact on the lignin conversion.

The chemical composition and structure of the substrate, along with the solvent, also determine the acid hydrolysis of lignin. Solvents that provide a better effect on the yield of lignin monomers include phenols, alcohols, and alkylbenzenes. These have been found to even change the distribution of products. Some of the solvents possess the ability to stabilize the intermediates of lignin degradation by entrapping them in situ. The hydrolysis activity of aryl-ether bonds is also influenced by the substituents at the side chain of alkyl groups and the methoxy and hydroxyl groups on the aromatic rings. When a hydroxyl group of phenol is present in the substrate, the cleavage of the C—O bond in β—O—4 linkage is enhanced by almost two-fold [85]. The remarkable increase in the rate is because of the stabilization of important carbocation intermediates. In native lignin, a common substituent is the methoxy group which can make the rupture of β—O—4 linkage easier by preventing the formation of the stable phenyl-dihydrobenzofuran [85]. In the literature it has also been claimed that the rate of hydrolysis of α—O—4 increases by almost one-fold in magnitude when the hydroxyl groups of the α-carbon in lignin's alkyl side chain are methylated [86].

6.3 Lignin conversion through base catalysis

Here, the most apparent linkage of lignin, β—O—4, is targeted for cleavage by using base catalysts such as $Mg(OH)_2$, $Ca(OH)_2$, NaOH, and KOH. For this purpose, the breaking down of all such bonds in the biopolymer generates phenolic compounds like acids, aldehydes, and alcohols. Much like that observed in lignin degradation by acid hydrolysis, this process is also affected by the differences in the chemical structures of substrates. Hence, different reactivities are seen with the change in biomass feedstock. This property was particularly studied by Miller et al. who experimented with different models of lignin such as anisole, syringole, and phenol in an alkaline water solution [41]. Their study showed that syringol having a greater number of substitutions is less stable than anisole with a smaller number of substitutions. It was successfully shown that the ratio of NaOH to that of lignin decided the lignin conversion and its subsequent degradation. A molar ratio of 1.5—2 exhibited the greatest extent of decomposition of lignin [41]. Other factors on which the yield and selectivity of the products in this process are dependent include the base concentration, time, solvent, and temperature [84]. Due to the solvolysis effect from ether linkages, the process happens more quickly in alcohol or phenolic solvents as compared to that observed in water. When the reaction is carried out for a longer period of time at a greater thermal environment, the formation of lignin monomers was much higher [84]. Apart from the use of alkaline catalysts of homogeneous nature, layered double hydroxides (LDHs), an example of solid base catalysts among many of this type, have also been used in this process [87].

6.4 Lignin conversion through reductive catalysis

Lignin as we know is a polymer of biological origin. It contains an excess of oxygenated functional groups along with ether bonds. To make lignin suitable for use as a biofuel, the two important chemical reactions are depolymerization followed by deoxygenation. For the purpose of depolymerization, the above-introduced methods of pyrolysis and hydrolysis in the presence of acids or bases can be carried out, but the end products are compounds that are heavily oxygenated. Such compounds possess an internal tendency to undergo quick polymerization to generate chemically and thermally unstable degradation products [88]. Hence, for the productive utilization of these compounds, the oxygenated compounds need to be modified by some chemical reaction. The reductive chemical process in presence of H_2, optimized thermal conditions, and suitable catalysts, popularly known as the hydrodeoxygenation (HDO) method, enables the desired conversion of lignin depolymerized derivatives. Coupling the two reactions, i.e., depolymerization and deoxygenation, generates stable compounds with improved heating value ready for direct incorporation with the already-existing fossil fuels [89]. A number of different types of reactions are associated with lignin HDO, and these include isomerization, repolymerization, dehydration, hydrocracking, decarboxylation, hydrogenolysis, dimerization, hydrogenation, etc. A suitable HDO process with the ability to generate suitable biofuel candidates must proceed by targeting the selective breakdown of lignin into dimers or monomers while removing the oxygen at the same time, so that carbon loss is negligible.

As has been mentioned in the preceding paragraph, HDO is a catalytic process, the two categories of catalysts that have been adequately tested are those that are made from noble metals like Pt, Rh, Re, Ru, Pd, etc., and the earth-abundant metals like Cu, Mo, Zn, Co, Fe, Ni, etc. [90]. Bimetallic and bifunctional catalysts have shown improved activity of catalysts as well as allowed the adjustment of selectivity toward product formation. Currently there is a strong need to find and develop catalysts with better stability, low cost, improved activity, and selectivity with respect to formation of the desired fuel compounds. The catalyst should also be such that it consumes a lower amount of the costly H_2.

Bimetallic catalysts have proved to be a promising choice because of their better catalytic power as compared to the previously mentioned metals. When Zn and Pt/C were present together, the HDO conversion of β−O−4 lignin models was much more effectively converted to aromatic monomers than that observed when only Pd/C was present [91]. A number of NiM (where M is Ru, Rh, and Pd) type bimetallic catalysts were made by Yan et al. that catalyzed the selective hydrogenolysis of C−O bonds in lignin [92]. In their study, they reported that their discovery of catalyst Ni85Ru15, exhibited better activity at a comparatively lower temperature of 100°C and lower pressure of H_2, i.e., 1 bar, by causing hydrogenolysis of C−O bonds of β−O−4 type [92]. Metal-acid type bifunctional catalysts have also been proved to possess strong potential in the reductive conversion of lignin [90]. In a research study it was found that acidic zeolite (HY) in combination with metal catalyst (Ru/Al_2O_3) resulted in the formation of jet fuel grade hydrocarbons [45]. It was found that two such types of catalysts were the key behind such the reactions. However, there was corrosion and a problem in recycling of catalyst when homogeneous Brønsted acids or many compositions of several catalysts were tested. Hence, it was conclusively established that heterogeneous bifunctional catalysts such as Ni/HZSM-5, Pt/HY, Ru/HZSM-5, and Pd/HY had a positive impact on the efficiency of the catalyst. When integrated with acidic supports and active metals such catalysts also showed an overall better result. As acid zeolites possess relatively higher temperature stability and greater tolerance to high concentrations of acid, HY and HZSM-5 are selected as active metal supports.

One of the major drawbacks of this conversion strategy is its complexity. Along with the catalyst, several other factors such as the source of lignin, catalyst loading, and the solvent needs are crucial considerations. In the study performed by Xu et al. using birch wood, they found that about 50 wt.% of lignin monomers were obtained when Ni/C was used as a catalyst in the presence of common types of alcohols [93]. Omar et al. attempted to elaborate this finding by testing the same over larger substrates [91]. From their study, they concluded that yield is significantly dependent on substrates and catalyst loading [91]. Ethanol, ethylene glycol, and methanol are among the important alcohols that have a major impact on the reaction. From a more recently performed research it was found that selective breaking of some weak ether bonds like α−O−4 and β−O−4 took place in corn stover lignin extracted by the dilute alkali method. Here, the corn stover lignin was depolymerized first, followed by the selective cleavage under HY zeolite and Al_2O_3 in aqueous phase [45]. The high loading capacity of depolymerized lignin species served as the best precursor to obtain a range of hydrocarbons that can be used as jet fuel.

6.5 Oxidative conversion of lignin

The economic advantage that is offered by the use of oxygen- or peroxide-based chemicals has made this chemical conversion stand out from other valorization strategies. With the change in catalysts and use of different oxidants, the mechanism to depolymerize lignin also changes. Some of the most preferred oxidants are chlorine peroxyacids (also chlorine dioxide), hydrogen peroxide, and ozone. When oxygen is used as the oxidant, it leads to the formation of syringaldehyde/acid, p-hydrobenzaldehyde/acid, and vanillaldehyde/acid through the radical, nucleophilic, and electrophilic mechanism between the lignin derivatives as of the degradation products [94]. Meanwhile, in the case of hydrogen peroxide being used as the oxidant, similar reactive compounds are formed, but mainly through the radical and nucleophilic mechanisms. The major products thus formed are acids, quinines, and aldehydes. Due to the cost advantage and easy operability with oxygen, it has served as the most reliable candidate and there also exist several studies in the past that used this oxidant. Reactive oxygen species like $O_2^{\bullet}$, $HO^{\bullet}$, $HOO^{\bullet}$, and HOO^- have been found to be present along with O_2 in this chemical reaction. The three main paths of lignin conversion through this approach are condensation, opening of ring, and cleavage of side chain [94].

This method has also exploited the use of many types of catalysts which mainly fall into the following three categories: organic catalysts, metal inorganic catalysts, and metal organic catalysts. In this conversion process, metal oxide catalysts, polyoxometalates, metal catalysts, metal ion catalysts, and composite metal oxides are used as catalysts that majorly result in the formation of acids, ketones, and phenolic aldehydes as products [94]. Although metal catalysts are widely used in the reduction conversion of lignin via the HDO method, there are only some of them which can also cause oxidative conversion of lignin. This includes the transition metals which possess the ability to get easily oxidized at an elevated thermal range and in an oxygenated atmosphere. Some noble metals which are generally stable have also been included in the oxidative conversion. In the ionic state the transition metals which hold their electron in the 4d vacant orbital serve as metal ion catalysts [94]. A high yield of phenols and carboxylic acids is produced when metal ion catalysis is carried out in peroxygen or oxygen in an aqueous phase [95].

The low cost, easy modification, and high activity of metal oxide make it a very good choice for implementation in this method. Distinct active sites may be formed in nonstoichiometric compounds by the generation of anionic or cationic vacancy in the oxides, which allows for the catalytic reaction. In this case the reaction takes place by an electrophilic mechanism which is activated by the formation of metal–oxygen bonds in metal oxides. When two different metals are used as the catalyst, it is known as composite oxide which exhibits a strong catalytic effect in lignin oxidation due to the synergistic effect from both compounds [87].

7. Challenges in lignin conversion

All in all, even though a lot of progress has taken place with respect to lignin valorization, the extensive work carried out in the related and allied areas has not been able to successfully elucidate the complete native structure of lignin. Therefore, it has not been possible yet to unlock the full potential of lignin by following a well-defined economical protocol. The root of the problem lies at the initial stage of lignin extraction and characterization via the existing methodologies and the technical drawbacks of analytical methods. While carrying out lignin isolation by following such pathways, its structure changes significantly due to a number of modifications in the lignin matrix and thus elucidating the native lignin structure has always remained unsolved. To have an effective conversion, it is crucial that the lignin isolation techniques proceed at a low temperature, greater yield, and a small number of condensed lignin structures. Technical lignins mostly generated from the conventional pulp and paper mills render their further conversion challenging due to the highly condensed structures as impurities. Moreover, most of the biorefinery techniques aim to convert the carbohydrate part of the biomass, leading to altered lignin structure. Some of the methods to isolate lignin with negligible changes in the native structure are Bjorkman lignin where extraction is done by dioxane/water or ball-milled wood lignin [65]. However, these methods demand higher energy use with a lower yield of lignin, i.e., less than 30%. In the extraction of lignin by use of the ball-milled method, cellulases are first added to enable the separation of carbohydrates followed by extraction with solvents. Although the yield obtained here is quite high there is a severe compromise in the purity of isolated lignin. The reason for this is the presence of carbohydrates and their derivatives as impurities. Such contamination has a seriously negative impact on the hydrodeoxygenation method of lignin conversion and therefore to avoid such consequences it is absolutely vital to employ the most suitable pretreatment strategy. The choice also determines the effectiveness with which lignocellulosic biomass can be separated into cellulose, hemicellulose, and lignin fractions.

Selectively converting lignin also poses a major challenge due to the tough and hardy nature of the polymer in which all the components are integrated with one another in a thick and tight "glue." As has been already discussed in the previous section, lignin is derived by random coupling of the three monolignols, sinapyl, coumaryl, and coniferyl alcohol, and the biosynthesis take place through a radical mechanism that results in a strong, robust, and stable three-dimensional matrix, that is highly resilient to both chemical and biological degradation. The heterogeneous structure of the biopolymer also contributes to this effect. When lignin is to be isolated from a different source there is a need to select a different pretreatment strategy and even if the pretreatment strategy remains unchanged, the characteristics of the isolated lignin changes with a change in biomass, making it difficult to establish a defined standard. The C–O and C–C bonds present between the aromatic centers are associated with a wide bond dissociation energy that adds to the complications in selectively depolymerizing lignin. Even the several types of functional groups such as the hydroxyl group in phenols, some of the terminal groups in aldehyde, and the methoxyl groups add to the difficulty in the selective isolation of lignin. At ambient temperatures the lignin is sparingly soluble in most of the commonly available solvents. This is attributed to its amorphous nature and comparatively high molecular weight. Moreover, while degrading lignin, reactive low-molecular-weight intermediates are formed that possess a strong tendency to get involved in some of the side reactions and hence obtaining the compounds of interest at sufficiently high purity becomes challenging. Finally, even to this day, there exists neither any method to aptly characterize the structure of isolated lignin, nor is it possible to accurately find the theoretical conversion yield from technical lignin.

8. Conclusions and future outlook

Lignin holds immense potential to be used as a biofuel substrate, and the reasons for justifying this claim are:

1. Available in abundant quantities from agricultural and forest residues and from existing biorefineries and paper and pulp mills. New biorefineries can also be developed, keeping in mind the platform chemicals obtained from lignin. Not only based on fuel perspective but also high-value chemicals (toluene, phenols, xylene or BTX, benzene), and economically sustainable products (activated carbon, polymers, carbon fibers) can be formulated. Moreover, this approach is more profitable because most of the lignin, even now, is used to generate power in the paper industry. However, before this could be done, the water needs to be removed from the wet lignin, making the whole process have an energy efficiency that is less than that of coal.
2. Low O:C ratio compared to its competitors, hemicellulose and cellulose, i.e., it has a much lower percentage of oxygen ($\sim 36\%$ against $\sim 50\%$).
3. High energy content and an aromatic structure along with chain units of nine-carbon monomers make it ready to be used as a drop-in fuel with proper pretreatment and upgrading of the converted lignin.

Although the above-mentioned points may be in favor of lignin, one major challenge that remains is that this complex macromolecule possesses a long carbon chain ($\sim C_{800-900}$) which exceeds the useable limit as a fuel ($\sim C_{6-20}$). Therefore, lignin must be (1) first depolymerized to reduce its carbon length, (2) further hydrogenated to increase the H:C ratio, and (3) further deoxygenated, i.e., the O:C ratio needs to be made even lower to make it fuel-ready. Merely treating degraded lignin with acids or alkali is a futile way to convert it into the desired molecules. The main reason for this can be attributed to the reactivity of the lignin intermediates and monomers. They condense and repolymerize to give unpredictable substances. Therefore, the entire conversion must be carried out in a directional manner so that lignin can be used to produce the product of interest. With all the advancements and research that have taken place to date, the following approaches need to be adopted to transfer its energy into fuel:

1. Selective reduction of the wide range of C—O—C bonds that would cause its cleavage (i.e., generation of smaller carbon chains and subsequent lowering of the molecular weight) and an increase in the fuel-compatible hydrocarbon content. The reduction is carried out by hydrogenation. To increase the selectivity and not have a significant impact on the cost factor, an ideal catalyst or a combination of catalysts needs to be selected. Since the harsh conditions can be turned to mild ones, the choice of a catalyst plays a significant role in making the overall process economic. Hence, the hydrogenolysis or HDO method needs to be standardized. Upgrading of degraded lignin by use of zeolite is another effective route to follow.
2. Optimization of the existing catalysts, such as bimetallic or bifunctional catalysts, or the design of new catalysts with improved properties that would reduce the requirement for costly hydrogen. It must also be ensured that it performs hydrogenation up to a certain limit and that not all the hydrocarbons are fully saturated. This is because unsaturation is crucial in bio-oil containing aromatics, olefins, etc. Hydrogen wastage is also minimized.
3. Use of stabilizers or blockers (e.g., diols, boric acid, phenol, p-cresol, and 2-naphthol) to prevent reactive intermediates from forming excessive solid char.
4. Employing appropriate solvents, preferably organic solvents (alcohols, formic acid, etc.) that reduce char formation and solubilize lignin and its derived products much more easily.

In conclusion, despite the immense potential of lignin as a biofuel substrate, the high cost associated with the entire process, lack of adequate trials and testing to confirm its potential, and the challenge in obtaining lignin because of the lack of lignin biorefineries are the key reasons for the failure of the realization of lignin's potential in the truest sense. Moreover, there exist many loopholes in understanding how different chemical reactions affect the reactivity and structure of the isolated lignin. Although a wide variety of potential catalysts have been developed in recent years that improve the selectivity, efficient utilization of solvents, and altering the lignin structure as per the need, more research studies need to be conducted to employ it economically on an industrial scale.

List of abbreviations

AFEX Ammonia fiber expansion
CEL Cellulolytic enzyme lignin
EMAL Enzymatic mild acidolysis lignin
ETBE Ethyl tert-butyl ether
G Guaiacyl units
GHGs Greenhouse gases
GIS Geographic Information System
GPC Gel permeation chromatography
H *p*-Hydroxyphenyl units
HDO Hydrodeoxygenation
ILs Ionic liquids
IPCC Intergovernmental Panel on Climate Change
LCBM Lignocellulosic biomass
LCC Lignin carbohydrate complex
LDHs Layered double hydroxides
MSW Municipal solid waste
MWL Milled wood lignin
RFS Renewable fuel standard program
S Syringyl units
SSF Simultaneous saccharification and fermentation
TMPL Thermomechanical pulp lignin
UNFCC United Nations Convention on Climate Change
VPO Vapor pressure osmometry

References

[1] EIA Projects Nearly 50% Increase in World Energy Usage by 2050, Led by Growth in Asia—Today in Energy—U.S. Energy Information Administration (EIA). https://www.eia.gov/todayinenergy/detail.php?id=41433. (Accessed 12 January 2022).

[2] World Population Prospects—Population Division—United Nations. https://population.un.org/wpp/Download/Standard/CSV/. (Accessed 16 January 2022).

[3] Global Energy Demand to Grow 47% by 2050, with Oil Still Top Source: Us Eia | S&P Global Platts. https://www.spglobal.com/platts/en/market-insights/latest-news/oil/100621-global-energy-demand-to-grow-47-by-2050-with-oil-still-top-source-us-eia. (Accessed 16 January 2022).

[4] Y. Dahman, C. Dignan, A. Fiayaz, A. Chaudhry, An introduction to biofuels, foods, livestock, and the environment, in: Biomass, Biopolymer-Based Materials, and Bioenergy: Construction, Biomedical, and Other Industrial Applications, Elsevier, 2019, pp. 241–276, https://doi.org/10.1016/B978-0-08-102426-3.00013-8.

[5] Climate Change Widespread, Rapid, and Intensifying—Ipcc—Ipcc. https://www.ipcc.ch/2021/08/09/ar6-wg1-20210809-pr/. (Accessed 16 January 2022).

[6] H. Ritchie, M. Roser, CO_2 and Greenhouse Gas Emissions. https://ourworldindata.org/co2-and-other-greenhouse-gas-emissions. (Accessed 16 January 2022).

[7] H. Ritchie, M. Roser, Future Greenhouse Gas Emissions, OurWorldInData.org, 2020. https://ourworldindata.org/future-emissions. (Accessed 16 January 2022).

[8] Masson-Delmotte, et al., IPCC, 2018: Global Warming of 1.5°C. An IPCC Special Report on the Impacts of Global Warming of 1.5°C above Pre-industrial Levels and Related Global Greenhouse Gas Emission Pathways, in the Context of Strengthening the Global Response to the Threat of Climate Change, Sustainable Development, and Efforts to Eradicate Poverty, Intergovernmental Panel on Climate Change, 2018. https://www.ipcc.ch/site/assets/uploads/sites/2/2019/06/SR15_Full_Report_High_Res.pdf. (Accessed 16 January 2022).

[9] REN21. Renewables 2018, Global Status Report; Renewable Energy Policy Network for the 21st Century, 2019. https://www.ren21.net/wp-content/uploads/2019/08/Full-Report-2018.pdf. (Accessed 11 February 2022).

[10] Biomass Explained—U.S. Energy Information Administration (EIA). https://www.eia.gov/energyexplained/biomass/. (Accessed 22 February 2022).

[11] I. Renewable Energy Agency, IRENA's Global Renewable Energy Roadmap, REmap 2030: Summary of Findings, 2014. www.irena.org/remap. (Accessed 18 March 2022).

[12] P.A. Fokaides, L. Tofas, P. Polycarpou, A. Kylili, Sustainability aspects of energy crops in arid isolated island states: the case of Cyprus, Land Use Pol. 49 (2015) 264–272, https://doi.org/10.1016/J.LANDUSEPOL.2015.08.010.

[13] W.H. Liew, M.H. Hassim, D.K.S. Ng, Review of evolution, technology and sustainability assessments of biofuel production, J. Clean. Prod. 71 (2014) 11–29, https://doi.org/10.1016/J.JCLEPRO.2014.01.006.

[14] P. McKendry, Energy production from biomass (Part 2): conversion technologies, Bioresour. Technol. 83 (1) (2002) 47–54, https://doi.org/10.1016/S0960-8524(01)00119-5.
[15] P.A. Fokaides, P. Polycarpou, Renewable Energy: Economics, Emerging Technologies and Global Practices, Nova Science Publishers, 2013. https://novapublishers.com/shop/renewable-energy-economics-emerging-technologies-and-global-practices/. (Accessed 22 February 2022).
[16] J.M. Romero-García, L. Niño, C. Martínez-Patiño, C. Álvarez, E. Castro, M.J. Negro, Biorefinery based on olive biomass. State of the art and future trends, Bioresour. Technol. 159 (2014) 421–432, https://doi.org/10.1016/J.BIORTECH.2014.03.062.
[17] C. Baskar, S. Baskar, R.S. Dhillon, Biomass Conversion: The Interface of Biotechnology, Chemistry and Materials Science, in: Biomass Conversion: The Interface of Biotechnology, Chemistry and Materials Science, 2012, pp. 1–465, https://doi.org/10.1007/978-3-642-28418-2.
[18] A.B.-S. Papperstidn, Studies on finely divided wood. Part 1. Extraction of lignin with neutral solvents, in: ci.nii.ac.jp vol. 1, 1956, pp. 477–485. July 1956, https://ci.nii.ac.jp/naid/10022025137/. (Accessed 22 February 2022).
[19] E. Adler, Lignin chemistry—past, present and future, Wood Sci. Technol. 11 (3) (1977) 169–218, https://doi.org/10.1007/BF00365615.
[20] J. Ralph, et al., Lignins: natural polymers from oxidative coupling of 4-hydroxyphenyl-propanoids, Phytochem. Rev. 3 (1–2) (2004) 29–60, https://doi.org/10.1023/B:PHYT.0000047809.65444.A4.
[21] L. Zhang, G. Gellerstedt, J. Ralph, F. Lu, NMR studies on the occurrence of spirodienone structures in lignins, J. Wood Chem. Technol. 26 (1) (2006) 65–79, https://doi.org/10.1080/02773810600580271.
[22] C. Crestini, M. Crucianelli, M. Orlandi, R. Saladino, Oxidative strategies in lignin chemistry: a new environmental friendly approach for the functionalisation of lignin and lignocellulosic fibers, Catal. Today 156 (1–2) (2010) 8–22, https://doi.org/10.1016/J.CATTOD.2010.03.057.
[23] R. Deshpande, et al., The reactivity of lignin carbohydrate complex (LCC) during manufacture of dissolving sulfite pulp from softwood, Ind. Crop. Prod. 115 (2018) 315–322, https://doi.org/10.1016/J.INDCROP.2018.02.038.
[24] M. Lawoko, Unveiling the structure and ultrastructure of lignin carbohydrate complexes in softwoods, Int. J. Biol. Macromol. 62 (2013) 705–713, https://doi.org/10.1016/J.IJBIOMAC.2013.10.022.
[25] M. Lawoko, R. Berggren, F. Berthold, G. Henriksson, G. Gellerstedt, Changes in the lignin-carbohydrate complex in softwood kraft pulp during kraft and oxygen delignification, Holzforschung 58 (6) (2004) 603–610, https://doi.org/10.1515/HF.2004.114/MACHINEREADABLECITATION/RIS.
[26] L. Salmén, A.M. Olsson, Interaction between hemicelluloses, lignin and cellulose: structure-property relationships, J. Pulp Pap. Sci. 24 (3) (1998) 99–103.
[27] F.R. Amin, et al., Pretreatment methods of lignocellulosic biomass for anaerobic digestion, Amb. Express 7 (1) (2017), https://doi.org/10.1186/S13568-017-0375-4.
[28] J. Zheng, L. Rehmann, Extrusion pretreatment of lignocellulosic biomass: a review, Int. J. Mol. Sci. 15 (10) (2014) 18967–18984, https://doi.org/10.3390/IJMS151018967.
[29] A. Duque, P. Manzanares, M. Ballesteros, Extrusion as a pretreatment for lignocellulosic biomass: fundamentals and applications, Renew. Energy 114 (2017) 1427–1441, https://doi.org/10.1016/J.RENENE.2017.06.050.
[30] A. Kumar, Anushree, J. Kumar, T. Bhaskar, Utilization of lignin: a sustainable and eco-friendly approach, J. Energy Inst. 93 (1) (2020) 235–271, https://doi.org/10.1016/J.JOEI.2019.03.005.
[31] N. Mosier, et al., Features of promising technologies for pretreatment of lignocellulosic biomass, Bioresour. Technol. 96 (6) (2005) 673–686, https://doi.org/10.1016/J.BIORTECH.2004.06.025.
[32] M. Jędrzejczyk, E. Soszka, M. Czapnik, A.M. Ruppert, J. Grams, Physical and chemical pretreatment of lignocellulosic biomass, Second Third Generat. Feedstocks: Evol. Biofuel. (2019) 143–196, https://doi.org/10.1016/B978-0-12-815162-4.00006-9.
[33] V. Yachmenev, B. Condon, T. Klasson, A. Lambert, Acceleration of the enzymatic hydrolysis of corn stover and sugar cane bagasse celluloses by low intensity uniform ultrasound, J. Biobased Mater. Bioenergy 3 (1) (2009) 25–31, https://doi.org/10.1166/JBMB.2009.1002.
[34] M.J. Bussemaker, D. Zhang, Effect of ultrasound on lignocellulosic biomass as a pretreatment for biorefinery and biofuel applications, Ind. Eng. Chem. Res. 52 (10) (2013) 3563–3580, https://doi.org/10.1021/IE3022785/SUPPL_FILE/IE3022785_SI_001.PDF.
[35] D. Batista Meneses, G. Montes de Oca-Vásquez, J.R. Vega-Baudrit, M. Rojas-Álvarez, J. Corrales-Castillo, L.C. Murillo-Araya, Pretreatment methods of lignocellulosic wastes into value-added products: recent advances and possibilities, Biomass Convers. Biorefin. 12 (2) (2020) 547–564, https://doi.org/10.1007/S13399-020-00722-0.
[36] Y. Chen, M.A. Stevens, Y. Zhu, J. Holmes, H. Xu, Understanding of alkaline pretreatment parameters for corn stover enzymatic saccharification, Biotechnol. Biofuels 6 (1) (2013), https://doi.org/10.1186/1754-6834-6-8.
[37] J.C. Solarte-Toro, J.M. Romero-García, J.C. Martínez-Patiño, E. Ruiz-Ramos, E. Castro-Galiano, C.A. Cardona-Alzate, Acid pretreatment of lignocellulosic biomass for energy vectors production: a review focused on operational conditions and techno-economic assessment for bioethanol production, Renew. Sustain. Energy Rev. 107 (2019) 587–601, https://doi.org/10.1016/J.RSER.2019.02.024.
[38] A.M. Asim, et al., Acidic ionic liquids: promising and cost-effective solvents for processing of lignocellulosic biomass, J. Mol. Liq. 287 (2019) 110943, https://doi.org/10.1016/J.MOLLIQ.2019.110943.
[39] K. Zhang, Z. Pei, D. Wang, Organic solvent pretreatment of lignocellulosic biomass for biofuels and biochemicals: a review, Bioresour. Technol. 199 (2016) 21–33, https://doi.org/10.1016/J.BIORTECH.2015.08.102.
[40] H.A. Ruiz, et al., Engineering aspects of hydrothermal pretreatment: from batch to continuous operation, scale-up and pilot reactor under biorefinery concept, Bioresour. Technol. 299 (Mar. 2020), https://doi.org/10.1016/J.BIORTECH.2019.122685.
[41] J.E. Miller, L. Evans, J.E. Mudd, K.A. Brown, Batch Microreactor Studies of Lignin Depolymerization by Bases. 2. Aqueous Solvents, 2002, https://doi.org/10.2172/800964.

[42] A.W. Bhutto, et al., Insight into progress in pre-treatment of lignocellulosic biomass, Energy 122 (2017) 724–745, https://doi.org/10.1016/J.ENERGY.2017.01.005.

[43] S.S. Hassan, G.A. Williams, A.K. Jaiswal, Emerging technologies for the pretreatment of lignocellulosic biomass, Bioresour. Technol. 262 (Aug. 2018) 310–318, https://doi.org/10.1016/J.BIORTECH.2018.04.099.

[44] F. Liu, E. Monroe, R.W. Davis, Engineering microbial consortia for bioconversion of multisubstrate biomass streams to biofuels, Biofuels—Challeng. Opportun. (Nov. 2018), https://doi.org/10.5772/INTECHOPEN.80534.

[45] H. Wang, H. Wang, E. Kuhn, M.P. Tucker, B. Yang, Production of jet fuel-range hydrocarbons from hydrodeoxygenation of lignin over super Lewis acid combined with metal catalysts, ChemSusChem 11 (1) (2018) 285–291, https://doi.org/10.1002/CSSC.201701567.

[46] N.S.-F. & Sullivan, undefined, High-value opportunities for lignin: unlocking its potential, Greenmater.fr (2012). Accessed: February 22, 2022, https://www.greenmaterials.fr/wp-content/uploads/2013/01/High-value-Opportunities-for-Lignin-Unlocking-its-Potential-Market-Insights.pdf.

[47] F. Kong, S. Wang, J.T. Price, M.K.R. Konduri, P. Fatehi, Water soluble kraft lignin–acrylic acid copolymer: synthesis and characterization, Green Chem. 17 (8) (2015) 4355–4366, https://doi.org/10.1039/C5GC00228A.

[48] A.E. Rodrigues, P.C. de O.R. Pinto, M.F. Barreiro, C.A. Esteves da Costa, M.I. Ferreira da Mota, I. Fernandes, Chemical pulp mills as biorefineries, Integrat. Appro. Added-Value Produc. Lignocellulosic Biorefin. (2018) 1–51, https://doi.org/10.1007/978-3-319-99313-3_1.

[49] Z. Hu, X. Du, J. Liu, H.M. Chang, H. Jameel, Structural characterization of pine kraft lignin: BioChoice lignin vs Indulin AT, J. Wood Chem. Technol. 36 (6) (2016) 432–446, https://doi.org/10.1080/02773813.2016.1214732.

[50] C. Abbati De Assis, et al., Techno-economic assessment, scalability, and applications of aerosol lignin micro- and nanoparticles, ACS Sustain. Chem. Eng. 6 (9) (2018) 11853–11868, https://doi.org/10.1021/ACSSUSCHEMENG.8B02151/ASSET/IMAGES/ACSSUSCHEMENG.8B02151.SOCIAL.JPEG_V03.

[51] A. Tribot, et al., Wood-lignin: supply, extraction processes and use as bio-based material, Eur. Polym. J. 112 (2019) 228–240, https://doi.org/10.1016/J.EURPOLYMJ.2019.01.007.

[52] A. Eraghi Kazzaz, Z. Hosseinpour Feizi, P. Fatehi, Grafting strategies for hydroxy groups of lignin for producing materials, Green Chem. 21 (21) (2019) 5714–5752, https://doi.org/10.1039/C9GC02598G.

[53] B.M. Matsagar, S.A. Hossain, T. Islam, Y. Yamauchi, K.C.W. Wu, A novel method for the pentosan analysis present in jute biomass and its conversion into sugar monomers using acidic ionic liquid, JoVE 2018 (136) (2018), https://doi.org/10.3791/57613.

[54] J.H. Lora, Industrial commercial lignins: sources, properties and applications, Monom. Polym. Compos. Renew. Resour. (2008) 225–241, https://doi.org/10.1016/B978-0-08-045316-3.00010-7.

[55] Z. Strassberger, S. Tanase, G. Rothenberg, The pros and cons of lignin valorisation in an integrated biorefinery, RSC Adv. 4 (48) (2014) 25310–25318, https://doi.org/10.1039/C4RA04747H.

[56] A. Vishtal, A. Kraslawski, Challenges in industrial applications of technical lignins, Bioresources 6 (3) (2011) 3547–3568, https://doi.org/10.15376/biores.6.3.vishtal.

[57] S. Guadix-Montero, M. Sankar, Review on catalytic cleavage of C–C inter-unit linkages in lignin model compounds: towards lignin depolymerisation, Top. Catal. 61 (3–4) (2018) 183–198, https://doi.org/10.1007/S11244-018-0909-2/FIGURES/11.

[58] K. Wörmeyer, T. Ingram, B. Saake, G. Brunner, I. Smirnova, Comparison of different pretreatment methods for lignocellulosic materials. Part II: influence of pretreatment on the properties of rye straw lignin, Bioresour. Technol. 102 (5) (2011) 4157–4164, https://doi.org/10.1016/J.BIORTECH.2010.11.063.

[59] L.A. Zevallos Torres, et al., Lignin as a potential source of high-added value compounds: a review, J. Clean. Prod. 263 (2020) 121499, https://doi.org/10.1016/J.JCLEPRO.2020.121499.

[60] J.H. Lora, W.G. Glasser, Recent industrial applications of lignin: a sustainable alternative to nonrenewable materials, J. Polym. Environ. 10 (1) (2002) 39–48, https://doi.org/10.1023/A:1021070006895.

[61] S. Laurichesse, L. Avérous, Chemical modification of lignins: towards biobased polymers, Prog. Polym. Sci. 39 (7) (2014) 1266–1290, https://doi.org/10.1016/J.PROGPOLYMSCI.2013.11.004.

[62] X. Zhao, K. Cheng, D. Liu, Organosolv pretreatment of lignocellulosic biomass for enzymatic hydrolysis, Appl. Microbiol. Biotechnol. 82 (5) (2009) 815–827, https://doi.org/10.1007/S00253-009-1883-1.

[63] A. Shrotri, H. Kobayashi, A. Fukuoka, Catalytic conversion of structural carbohydrates and lignin to chemicals, Adv. Catal. 60 (2017) 59–123, https://doi.org/10.1016/BS.ACAT.2017.09.002.

[64] P. Alvira, E. Tomás-Pejó, M. Ballesteros, M.J. Negro, Pretreatment technologies for an efficient bioethanol production process based on enzymatic hydrolysis: a review, Bioresour. Technol. 101 (13) (2010) 4851–4861, https://doi.org/10.1016/J.BIORTECH.2009.11.093.

[65] A. Björkman, Lignin and lignin-carbohydrate complexes extraction from wood meal with neutral solvents, Ind. Eng. Chem. 49 (9) (1957) 1395–1398, https://doi.org/10.1021/IE50573A040/ASSET/IE50573A040.FP.PNG_V03.

[66] B. Jiang, T. Cao, F. Gu, W. Wu, Y. Jin, Comparison of the structural characteristics of cellulolytic enzyme lignin preparations isolated from wheat straw stem and leaf, ACS Sustain. Chem. Eng. 5 (1) (2017) 342–349, https://doi.org/10.1021/ACSSUSCHEMENG.6B01710/SUPPL_FILE/SC6B01710_SI_001.PDF.

[67] A. Tolbert, H. Akinosho, R. Khunsupat, A.K. Naskar, A.J. Ragauskas, Characterization and analysis of the molecular weight of lignin for biorefining studies, Biofuels Bioprod. Biorefin. 8 (6) (2014) 836–856, https://doi.org/10.1002/BBB.1500.

[68] A.v. Gidh, S.R. Decker, T.B. Vinzant, M.E. Himmel, C. Williford, Determination of lignin by size exclusion chromatography using multi angle laser light scattering, J. Chromatogr. A 1114 (1) (2006) 102–110, https://doi.org/10.1016/J.CHROMA.2006.02.044.

[69] H. Lange, F. Rulli, C. Crestini, Gel permeation chromatography in determining molecular weights of lignins: critical aspects revisited for improved utility in the development of novel materials, ACS Sustain. Chem. Eng. 4 (10) (2016) 5167–5180, https://doi.org/10.1021/ACSSUSCHEMENG.6B00929/ASSET/IMAGES/MEDIUM/SC-2016-00929H_0010.GIF.

[70] N. Jiang, Y. Pu, A.J. Ragauskas, Rapid determination of lignin content via direct dissolution and 1H NMR analysis of plant cell walls, ChemSusChem 3 (11) (2010) 1285–1289, https://doi.org/10.1002/CSSC.201000120.

[71] E. Tiainen, T. Drakenberg, T. Tamminen, K. Kataja, A. Hase, Determination of phenolic hydroxyl groups in lignin by combined use of 1H NMR and UV spectroscopy, Holzforschung 53 (5) (1999) 529–533, https://doi.org/10.1515/HF.1999.087/MACHINEREADABLECITATION/RIS.

[72] M.Y. Balakshin, E.A. Capanema, R.B. Santos, H.M. Chang, H. Jameel, Structural analysis of hardwood native lignins by quantitative 13C NMR spectroscopy, Holzforschung 70 (2) (2016) 95–108, https://doi.org/10.1515/HF-2014-0328/MACHINEREADABLECITATION/RIS.

[73] Y. Pu, F. Chen, A. Ziebell, B.H. Davison, A.J. Ragauskas, NMR characterization of C3H and HCT down-regulated alfalfa lignin, BioEnergy Res. 2 (4) (2009) 198–208, https://doi.org/10.1007/S12155-009-9056-8.

[74] Y. Pu, S. Cao, A.J. Ragauskas, Application of quantitative 31P NMR in biomass lignin and biofuel precursors characterization, Energy Environ. Sci. 4 (9) (2011) 3154–3166, https://doi.org/10.1039/C1EE01201K.

[75] A. Jensen, Y. Cabrera, C.W. Hsieh, J. Nielsen, J. Ralph, C. Felby, 2D NMR characterization of wheat straw residual lignin after dilute acid pretreatment with different severities, Holzforschung 71 (6) (2017) 461–469, https://doi.org/10.1515/HF-2016-0112/MACHINEREADABLECITATION/RIS.

[76] W. Chen, D.J. McClelland, A. Azarpira, J. Ralph, Z. Luo, G.W. Huber, Low temperature hydrogenation of pyrolytic lignin over Ru/TiO2: 2D HSQC and 13C NMR study of reactants and products, Green Chem. 18 (1) (2015) 271–281, https://doi.org/10.1039/C5GC02286J.

[77] L. Moghaddam, et al., Structural characteristics of bagasse furfural residue and its lignin component. An NMR, py-GC/MS, and FTIR study, ACS Sustain. Chem. Eng. 5 (6) (2017) 4846–4855, https://doi.org/10.1021/ACSSUSCHEMENG.7B00274/SUPPL_FILE/SC7B00274_SI_001.PDF.

[78] U.P. Agarwal, R.S. Reiner, A.K. Pandey, S.A. Ralph, K.C. Hirth, R.H. Atalla, Raman Spectra of Lignin Model Compounds, fs.usda.gov, 2005. https://www.fs.usda.gov/treesearch/pubs/20194. (Accessed 19 April 2022).

[79] F. Xu, R.C. Sun, M.Z. Zhai, J.X. Sun, J.X. Jiang, G.J. Zhao, Comparative study of three lignin fractions isolated from mild ball-milled *Tamarix austromogoliac* and *Caragana sepium*, J. Appl. Polym. Sci. 108 (2) (2008) 1158–1168, https://doi.org/10.1002/APP.27761.

[80] S.O. Prozil, D.v. Evtuguin, A.M.S. Silva, L.P.C. Lopes, Structural characterization of lignin from grape stalks (*Vitis vinifera* L.), J. Agric. Food Chem. 62 (24) (2014) 5420–5428, https://doi.org/10.1021/JF502267S/ASSET/IMAGES/MEDIUM/JF-2014-02267S_0008.GIF.

[81] F.X. Collard, J. Blin, A review on pyrolysis of biomass constituents: mechanisms and composition of the products obtained from the conversion of cellulose, hemicelluloses and lignin, Renew. Sustain. Energy Rev. 38 (2014) 594–608, https://doi.org/10.1016/J.RSER.2014.06.013.

[82] M. Carrier, et al., Quantitative insights into the fast pyrolysis of extracted cellulose, hemicelluloses, and lignin, ChemSusChem 10 (16) (2017) 3212–3224, https://doi.org/10.1002/CSSC.201700984.

[83] J. Dai, A.F. Patti, L. Longé, G. Garnier, K. Saito, Oxidized lignin depolymerization using formate ionic liquid as catalyst and solvent, ChemCatChem 9 (14) (2017) 2684–2690, https://doi.org/10.1002/CCTC.201700632.

[84] C. Li, X. Zhao, A. Wang, G.W. Huber, T. Zhang, Catalytic transformation of lignin for the production of chemicals and fuels, Chem. Rev. 115 (21) (2015) 11559–11624, doi: 10.1021/ACS.CHEMREV.5B00155/ASSET/IMAGES/ACS.CHEMREV.5B00155.SOCIAL.JPEG_V03.

[85] M.R. Sturgeon, et al., A Mechanistic investigation of acid-catalyzed cleavage of aryl-ether linkages: implications for lignin depolymerization in acidic environments, ACS Sustain. Chem. Eng. 2 (3) (2014) 472–485, https://doi.org/10.1021/SC400384W/SUPPL_FILE/SC400384W_SI_001.PDF.

[86] A.W. Pelzer, et al., Acidolysis of α-O-4 aryl-ether bonds in lignin model compounds: a modeling and experimental study, ACS Sustain. Chem. Eng. 3 (7) (2015) 1339–1347, https://doi.org/10.1021/ACSSUSCHEMENG.5B00070/SUPPL_FILE/SC5B00070_SI_001.PDF.

[87] M.R. Sturgeon, et al., Lignin depolymerisation by nickel supported layered-double hydroxide catalysts, Green Chem. 16 (2) (2014) 824–835, https://doi.org/10.1039/C3GC42138D.

[88] S. van den Bosch, et al., Reductive lignocellulose fractionation into soluble lignin-derived phenolic monomers and dimers and processable carbohydrate pulps, Energy Environ. Sci. 8 (6) (2015) 1748–1763, https://doi.org/10.1039/C5EE00204D.

[89] L.P. Xiao, et al., Catalytic hydrogenolysis of lignins into phenolic compounds over carbon nanotube supported molybdenum oxide, ACS Catal. 7 (11) (2017) 7535–7542, https://doi.org/10.1021/ACSCATAL.7B02563/SUPPL_FILE/CS7B02563_SI_001.PDF.

[90] P. Sirous-Rezaei, et al., Mild hydrodeoxygenation of phenolic lignin model compounds over a FeReOx/ZrO2 catalyst: zirconia and rhenium oxide as efficient dehydration promoters, Green Chem. 20 (7) (2018) 1472–1483, https://doi.org/10.1039/C7GC03823B.

[91] I. Klein, C. Marcum, H. Kenttämaa, M.M. Abu-Omar, Mechanistic investigation of the Zn/Pd/C catalyzed cleavage and hydrodeoxygenation of lignin, Green Chem. 18 (8) (2016) 2399–2405, https://doi.org/10.1039/C5GC01325A.

[92] J. Zhang, et al., A series of NiM (M = Ru, Rh, and Pd) bimetallic catalysts for effective lignin hydrogenolysis in water, ACS Catal. 4 (5) (2014) 1574–1583, https://doi.org/10.1021/CS401199F/SUPPL_FILE/CS401199F_SI_001.PDF.

[93] Q. Song, et al., Lignin depolymerization (LDP) in alcohol over nickel-based catalysts via a fragmentation–hydrogenolysis process, Energy Environ. Sci. 6 (3) (2013) 994–1007, https://doi.org/10.1039/C2EE23741E.

[94] R. Ma, Y. Xu, X. Zhang, Catalytic oxidation of biorefinery lignin to value-added chemicals to support sustainable biofuel production, ChemSusChem 8 (1) (2015) 24–51, https://doi.org/10.1002/CSSC.201402503.

[95] R. Gao, Y. Li, H. Kim, J.K. Mobley, J. Ralph, Selective oxidation of lignin model compounds, ChemSusChem 11 (13) (2018) 2045–2050, https://doi.org/10.1002/CSSC.201800598.

Chapter 17

Biohydrogen production from dark fermentation of lignocellulosic biomass

Prajwal P. Dongare and Hitesh S. Pawar
DBT-ICT Centre for Energy Biosciences, Institute of Chemical Technology, Mumbai, Maharashtra, India

1. Introduction

Fossil fuel energy consumption in India and the world is increasing annually, according to energy statistics of India, such as natural gas from 55 billion m^{-3} (in 2015–16) to 64.14 billion m^{-3} (in 2019–20), as shown in Table 17.1 and Fig. 17.1. For fossil fuels, prices increase with increases in consumption because they are a non-renewable resource and emit greenhouse gases (CO_2, NO_x, and SO_2), which result in several health-related problems. The increased demand for eco-friendly, sustainable, renewable, high-energy fuels led to an alternative fuel that is clean, carbon-neutral, renewable, and eco-friendly, and has a high-energy content [1]. This is bioenergy/biofuel, in which biohydrogen has a high energy content of about 122 MJ kg^{-1} [2]. It is a triple of other hydrocarbon fuels, carbon-neutral, biodegradable, and eco-friendly, and it does not produce greenhouse gases because the combustion of biohydrogen results in water. Biohydrogen simply means the production of hydrogen with the help of micro-organisms or fermentation.

In India, the most important topic is the burning of crop residues, which affects air pollution and results in health problems. Countries such as India, Bangladesh, Indonesia, and Myanmar generate 500, 72, 55, and 19 million tons per year of agricultural waste, respectively [3] (National Policy for Management of Crop Residues report). According to the Inter-governmental Panel on Climate Change, the highest contributing states are Uttar Pradesh, Punjab, and Haryana [4,5]. Therefore, renewable abundant resources such as second-generation biomass, that is, lignocellulosic biomass (LB) (agricultural residue, energy crops, agroindustrial, and forest trees), especially dark fermentation (DF), have significant potential for fulfilling the future energy supply and current problems of waste treatment at a higher rate. LB is a second-generation biomass because it does not interfere with the food chain. LB is a dense complex that includes cellulose, lignin, and hemicellulose (CHL), mainly with cellulose present at a higher content. CHL varies in composition in different types of biomass. For biohydrogen yield (BHY), these complex molecules need to be broken down for use by micro-organisms [1] because micro-organisms can use only simple molecules efficiently. So, how can we overcome such a problem? The

TABLE 17.1 Energy statistics of crude oil, coal, and natural gas during the year 2014–2020.

Sr. No	Year	Crude oil (metric million tons)	Coal (million tons)	Natural gas (billion cubic meters)
1	2014–15	223.24	822.13	51.30
2	2015–16	234.86	836.73	52.51
3	2016–17	245.36	847.22	55.70
4	2017–18	251.96	898.50	59.17
5	2018–19	257.20	968.36	60.79
6	2019–20	254.39[a]	942.63[a]	64.14

[a]Decreased value may be due to COVID-19.

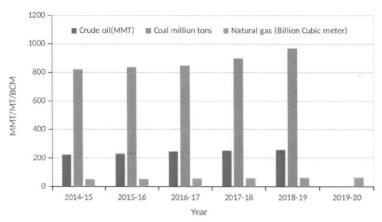

FIGURE 17.1 Energy statistics of India from 2014 to 2020. *BCM*, billion cubic meters; *MMT*, metric million tons.

answer is to employ different pre-treatment methods. Because of crystallinity and heterogeneity, the direct consumption of LB for biohydrogen production is challenging and requires pre-treatment methods In which lignin is difficult to break down. To find an answer, multiple pretreatments have been developed, including alkaline pre-treatment [1]. It was found that rice straw biomass has a low lignin content, nearly 13%, with the substrate showing yield with sewage sludge of 21 mL of H_2/g. Moreover, research found that switchgrass (lignin of 12%–18%) in which no chemical method was employed had an effective hydrogen yield (HY) in DF of 11.2 mmol of H_2 g^{-1} using the inoculum extreme thermophile *Caldicellulosiruptor saccharolyticus* DSM 8903 [6,7]. Also, rye, wheat straw, and so on, showed high comparative lignin values, which can be most effectively digested by alkaline pre-treatment. Starch is effective for reducing cost owing to its inexpensiveness and availability. There are different pre-treatment methods such as physical, chemical, biological, and combination/physicochemical. Reginatto et al. [6] elaborated on fermentative H_2 production without chemical or biological pre-treatment of LB, (2) with pre-treatment of LB, and (3) with hydrolysates of LB from a pre-treatment process step followed by hydrolysis enzymatically. There are few publications that use no pre-treatment and BHYs from untreated LB. Most micro-organisms require pre-treatment; therefore, this method is excluded on an industrial scale. In this method, scientists mostly estimated good results using thermophilic micro-organisms compared with mesophilic ones [1]. The yield ranges from 0.12 to 11.2 mmol of H_2 $gram^{-1}$-substrate, in which 11.2 mmol of H_2 $gram^{-1}$-substrate comes from switchgrass [8]. The second method uses pre-treatment, in which the major drawback is the formation of inhibitors, which negatively affects HY. Such drawbacks can be overcome using a resistant strain or techniques covered in the Inhibitor section. *Clostridium beijerinckii* strains are resistant to inhibitors more efficiently than other non-clostridial and clostridial bacteria [6]. Enzymatic hydrolysis is essential for monomer production because it is an environmental process. Generally, physical, chemical, biological, and physicochemical/combinational pre-treatments are available. The third pretreatment method followed by hydrolysis enzymatically is efficient method for saccharifying LB in DF. However, researchers need to determine the substrate type, the conditions for pre-treatment employed, and the hydrolysates that could inhibit fermentative biohydrogen production. Therefore, it is vital to develop strategies for pre-treatment to avoid or overcome inhibitor problems or effect detoxification or simultaneous saccharification and fermentation (SSF), simultaneous saccharification and cofermentation (SSCF), resistant strains, or enzymatic hydrolysis, to show effectiveness and cost-effectiveness.

HYs result in higher rates, and different types of organic waste can be enriched with carbohydrates as substrate, resulting in minimum cost for hydrogen production. It has a low biohydrogen (1.0–2.5 minimum moles per mole) yield, which can be increased by optimizing different pre-treatment methods [1]. Producing a co-product along with biohydrogen from biomass can make the whole process more economically viable compared with conventional means. Substrates for photofermentative bacteria are limited, but DF can use the wide range of utility of different substrates. DF biohydrogen production is mainly due to the hydrogenase enzyme, which has a robust inhibiting effect by oxygen. An outline of the process for understanding biohydrogen production is present in Fig. 17.2, in which LB, a complex of CHL, is directly used for H_2 production, a traditional method, H_2 production followed by fermentation, and a third pretreatment followed by enzymatic hydrolysis. Enzymatic hydrolysis employs various types of treatment: separate hydrolysis and fermentation, SSF, and consolidated bioprocessing (CBP) [9]. In SHF fermentation and the hydrolysis of enzymes are in different reactors, SSF means fermentation and hydrolysis as a co-culture in a reactor, and CBP means enzyme production, fermentation, and hydrolysis as a co-culture in a reactor, which is cost-effective [9]. After enzymatic hydrolysis and

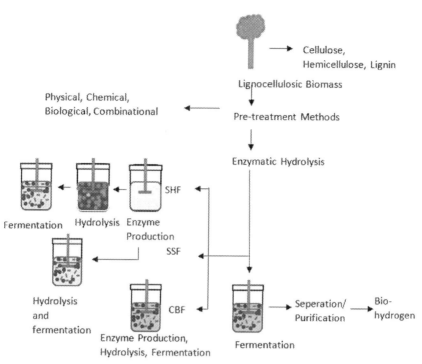

FIGURE 17.2 Overview of biohydrogen production. *CBF*, combined biohydrogen fermentaion; *SHF*, separate hydrolysis and fermentation; *SSF*, simultaneous saccharification and fermentation.

fermentation, a purification strategy is also important for separating gases. Membrane separation or pressure swing adsorption is used; an overview of these is covered in the Separation section.

In the batch operation of DF, a closed batch operation gives a yield of 441 Y(H_2/S) [10]. Biohydrogen produced through DF is even more expensive owing to the requirement for pre-treatment methods in LB. Adding molecules such as zeolite improves the HY by 10%−26% in the DF of carbon-5/carbon-6 and biomass of *Sargassum* sp. [11].

This chapter emphasizes different strategies, advances, and information regarding pre-treatment technologies, how to convert LB efficiently into biohydrogen, and strategies that increase the yield. Various methods have been developed to increase hydrogen production, such as sequential dark photofermentation, CBP, and novel cellulase enzyme development, the use of different materials, improving and optimizing cellulase systems, mixing cultures, and employing a genetic engineering approach.

1.1 Lignocellulosic biomass for dark fermentation

LB is a renewable resource and the most abundant one on earth. It can be obtained from different sources such as energy crops, agricultural residues, forest trees, and agroindustrial waste. Therefore, LB as a substrate can eliminate fossil fuel dependency in the near future.

1.2 Sources of lignocellulosic biomass

Salakkam et al. divided LB into two types: non-woody and woody. Woody biomass includes hardwoods and softwoods. Hardwoods are composed of beech, maple, and teak; softwoods include cedar, juniper, and pine. Non-woody second biomass includes agricultural residues. Energy crops of agroindustrial waste can be categorized into LB. For our convenience, we can categorize it into four types: agricultural waste, energy crops, agroindustrial waste, and forest trees. Energy crops do not affect the food market and are not involved in food application. The US Department of Energy estimated that bioenergy crops and crop residues will be prime sources of herbaceous LB in the future. In contrast, short-rotation forest residues and woody crops are key sources of woody biomass.

Bioenergy perennial crops grown on non-forested land will mitigate in land-use change [1]. Perennial crops are also helpful for decreasing leaching and water runoff and require low input. According to such benefits, switchgrass is a potential LB for biohydrogen production. Categories of biomass include [12]: (1) woody, agricultural, aquatic, and herbaceous crops, which come under energy crops [12]; (2) animal waste and crops, which come under agricultural residues; (3) logging residues, mill waste, shrub, and tree residues, which come under forestry waste; and (4) sewage sludge, industrial waste, and municipal solid waste, which comes under industrial or municipal waste.

1.3 Characteristics of lignocellulosic biomass

LB is complex in structure; it includes CHL, which is inaccessible to microorganisms, because of which different pre-treatment techniques have been developed. LB is complex owing to the complex nature of CHL. It is important to know about CIIL so that one can use appropriate enzymes and techniques.

Cellulose is the crucial constituent of LB. It is made of glucose monomers that are cross-linked with β-1,4-glycosidic bonds (β-1,4-glucan). In-plant cellulose includes a part with a crystalline (organized) and amorphous structure. Cellulose has a higher molecular weight than hemicellulose [13]. Glucose is the most important substrate for fermentation and HY compared with xylose. The packed chain β-1,4-glucan is complex with a long structure known as micro-fibrils that complex into lattices, which cause the significant problem of cellulose fibers inaccessible by cellulase. Glucose theoretically can produce 4 mol H_2 g^{-1} of the substrate when acetate is the product and 2 mol H_2 g^{-1} when butyrate is the product [6]. However, the yield was found to be low experimentally.

Hemicellulose is also the most crucial component in DF. Hemicellulose is a heteropolymer of hexoses (mannose, glucose, and galactose), pentoses (xylose and arabinose), and uronic acids (galacturonic acid and 4-O-methyl-glucuronic acid) [14]. Xylose is dominant among hemicellulose constituents. It shows a connection with cellulose and lignin, resulting in the formation of the network in a more rigid structure. Nimz et al. [13] demonstrated that LB solvability relies on pH, temperature, and moisture.

Because of the heteromeric nature of hemicellulose, it is found in a variety of types of biomasses [14]. In hardwoods, glucuronoxylan is major hemicellulose, in which xylose is linked in the backbone by β-1,4-glycosidic linkage. A major hemicellulose in softwoods is galactoglucomannan, which consists of glucose and a mannose backbone, with a side chain group of galactose and acetyl. For grasses, glucuronoarabinoxylan, in which xylose is the backbone, is most commonly found in hemicellulose. The lower component in wood is glucomannan [14]. Xylose is a major monomer component for biohydrogen production from hemicellulose. Theoretically, When xylose is a substrate and acetate is a product, HY is 3.33 H_2 g^{-1}, and when butyrate is a fermentation product, HY is 1.33 H_2 g^{-1}. Xylose has approximately 20%–30% of plant biomass [6]. Xylan extraction from hemicellulose can occur in an alkaline or acid environment, although glucomannan is generally extracted in an acidic, stronger alkaline for xylan extraction. Among CHL, hemicelluloses are thermally and chemically sensitive [13]. During pre-treatment of thermochemical hemicellulose, at first, side groups present in hemicellulose react; after that, the hemicellulose backbone reacts.

Lignin is only the non-sugar moiety of LB. It is the second most important one after cellulose. Lignin is abundant in the cell wall. It is a heteropolymer (i.e., amorphous composed of three phenylpropane parts of guaiacyl [G], p-hydroxyphenyl [H], and syringyl [S] and held by various linkages) [1]. Lignin makes the structure more complex. Lignin and hemicellulose are linked in such a way that cellulose is inaccessible owing to the linkage. Lignin results in additional strength and enhances wall hydrophobicity, which prevents the action of hydrolytic enzymes, all of which result in making lignin degradation tough. The purpose of lignin in the plant is to provide support, a wide range of resistance toward microbes, and impermeability, and to combat stress. Hemicellulose and lignin are similar because they usually begin as soluble in water at nearly 180°C under optimal situations [15]. Lignin solubility is based on the precursor, however [15]. Hemicellulose and lignin externally cover cellulose, owing to which there is low access to the enzyme (i.e., cellulose), which means cellulase cannot reach the active site and results in low bioproduct yield. After pre-treatment, it shows accessibility by the enzyme, as displayed in Fig. 17.3.

1.4 Compositional variation in cellulose, hemicellulose, and lignin content in lignocellulosic biomass

Pre-treatment methods and enzymes vary according to different CHL contents. If lignin is greater in concentration in any biomass, appropriate techniques are used to reduce or nullify this presence. Alkaline pre-treatment best reduces lignin content [1]. CHL content differs with biomass sources. From a literature review, it was found that corn stover contains 38, 21.5–24, and 11–19; napier grass 45.7, 33.7, and 20.6; rice straw 31, 22, and 13; switchgrass 31–45, 20–31, and 12–18; bagasse 38.2, 27.1, and 20.2; wheat straw 29–38, 21–30, and 16–24; eucalyptus 38.0–45.0, 12.0–13.0, and 25.0–37.0; giant reed straw 33, 18–19, and 24.5; barley straw 36–37, 24–25, and 25–26; rye straw 38, 36.9, and 17.6–25;

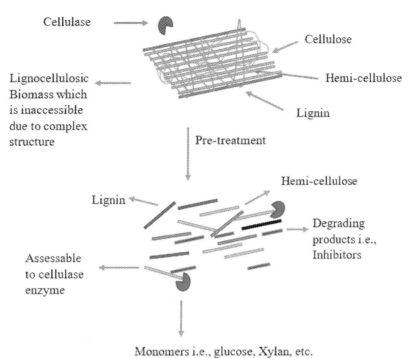

FIGURE 17.3 Pre-treatment overview.

sunflower stalk 31, 15.6, and 29.2; sugarcane bagasse 31−42, 25−41.6, and 20−24.8; grasses 25−40, 25−50, and 10−30; hardwoods 40−55, 24−40, and 18−25; corncobs 45, 35, and 15; softwoods 45−50, 25−35, and 25−35; banana waste 13.2, 14.8, and 14; rice husk 40.26, 12.54, and 25.40; cornstalk 34.45−27.55, and 21.81; corn bran residue 10.54, 40.57, and 1.06; waste sorghum leaves 28.56, 29.18, and 3.94; sweet sorghum stover 37.42, 20.43, and 16.28; oat straw 34.8−39, 26.7, and 9; pine tree wood 39.48, 22.10, and 37.11; pine tree wood pellets nearly 39, 22, and 37; wastepaper 60−70, 10−20, and 5−10; spent mushroom compost 31.2, 6.9, and 14.0; newspaper 40−55, 25−40, and 18−30; deoiled jatropha waste nearly 14, 24, and 30; cashew apple bagasse 20.56, 10.17, and 35.26; peanut shell 46.1, 5.6, and 27.8; sorted refuse 60, 20, 20; Cotton seed hairs 80−95, 5−20, and 0; rice bran 35, 25, and 17; sorghum stalk 27, 25, and 11; coconut fiber 36−43, 1.5−2.5, and 41−45; poplar 49.9, 17.4, and 18.1; corn fiber 14.28, 16.8, and 8.4; and pine wood nearly 46, 8, and 29, respectively [16−18]. A review of the literature showed that the CHL content varies with different biomass and in the same sources in different papers. This may be because of environmental and developmental parameters. Raw material rice straw and wheat straw are effective sources of bioethanol and gas production. A detailed review on wheat straw biomass is given by Tian et al. [19]. Cornstalk is a well-known LB substrate studied for biohydrogen productivity [6]. Switchgrass results in the highest BHY: 11.2 mmol biohydrogen per gram of substrate along with the inoculum extreme thermophile *C. saccharolyticus* [8].

2. Role of pretreatment

Because of the highly ordered structure of LB, enzymes and microorganisms during fermentation cannot be accessed, as shown in Fig. 17.3. To make them accessible, the biomass needs to be pretreated. Pretreatment such as degradation crystallinity and lignin breakdown makes cellulase enzymes accessible. To make fermentable sugar, it needs to be cut from the lignin barrier. Pretreatment help to increase the surface area so that more surface area will be interactive for enzymatic activity. Different pretreatment methods have been developed to overcome the lignocellulosic complex biomass problem and achieve a cost-effective strategy. Different strategies have different advantages and disadvantages. For example, the mechanical/physical method produces a low sugar yield and has a high equipment cost, and the crystallinity of cellulose can be reduced with no inhibitor production or use of chemicals. An effective pretreatment method (1) gives the maximum yield of fermentable sugar, (2) gives less or no loss of carbohydrate, (3) gives less or no inhibitor formation, (4) requires less energy as input, and (5) is environmentally as well as cost-effective [16]. In addition, the low cost of catalysts and the production of high-value lignin products are important. Appropriate pretreatment can help to reduce downstream processing costs.

3. Pretreatment methods for dark fermentation

LB is the most abundant on earth and is a readily available, low-cost source for biohydrogen (biofuel) production [1] owing to its constitution (CLH). Because LB is a more complex ordered structure, microorganisms cannot use such a biomass. How can we make it accessible? Different pretreatments help to access biomass as it breaks down and make it readily available for fermentation. However, in the case of chemical fermentation, it forms undesirable side products such as hydroxymethylfurfural (HMF), furans, and phenolics, which we will cover in the Inhibitor section. It is crucial to select such strategies to make the whole process more economical. Different novel pretreatment methods are developed by scientists to find cost-effective strategies. Pretreatment methods must fulfill certain requirements including (1) increasing the sugar content, (2) lowering the inhibitor content, (3) avoiding the loss of carbohydrates, and (4) finding a cost-effective strategy. The following section includes different methods and their advantages and disadvantages. Fig. 17.3 displays an understanding of pretreatment.

3.1 Physical pretreatment method for dark fermentation

Physical methods consist of milling, chilling, extrusion, microwave, and ultrasonic units, which increase the accessible surface area of the LB by breaking down the shape and size of LB. For reduction into small pieces, chipping and milling methods are generally used. Milling reduces the acceptable size of particles, which ultimately improves its enzymatic hydrolysis and causes no production of inhibitors. Such techniques require higher energy input, owing to which these methods were not used for industrial purposes. Microwave pretreatment cause the fragmentation and swelling of LB. Shahzadi et al. [20] stated that microwave-assisted pretreatment increases the breakdown of LB and increases efficiency with a catalyst such as acid and alkali. A mesh depletion size of particles below 40 mesh shows slight increase in the yield of biohydrogen. The drawback of microwave technology is that it allows low radiation penetration in bulk products. In the physical method, toxic biomass is negligible. BHY was estimated to be 1947 mL-H_2 L^{-1}-substrate from oil palm trunk (OPT) by microwave-assisted acid hydrolysate using *Thermoanaerobacterium thermosaccharolyticum* KKU19 [21].

3.2 Chemical pretreatment methods for dark fermentation

Chemical treatment includes acid, alkaline, ionic liquid (IL), organosolv, oxidative, and metal chlorides. Inhibitory compounds are produced in this treatment. Since 1819, sulfuric and hydrochloric acids have been used, but then various diluted acids and alkali were discovered, which are now used for LB pretreatment. Cao et al. [22] estimated that corn stover hydrolysate, obtained by the pretreatment of dilute sulfuric acid using *T. thermosaccharolyticum* W16, resulted in an HY of 2.24 mol of H_2 mol^{-1} of sugar. Anaerobic granular sludge and activated sludge give HY 627 and 822 mL of H_2 L^{-1} of the substrate by the diluted H_2SO_4 hydrolysate of corn stover under thermophilic conditions, resulting in 3.7 and 2.3 times greater than those obtained under mesophilic conditions [1].

Acid pretreatment disintegrates cellulose from the hemicellulose fraction and degrades lignin slightly and forms inhibitors (i.e., furan derivatives). These methods corrode equipment, a disadvantage of this method. Sometimes enzymatic hydrolysis is not required during fermentation, which is advantageous. Acid pretreatment at ambient temperature enhances anaerobic digestibility. Acid pretreatment by oxalic acid dihydrate results in a 42% reduction in sugar [23], and H_2SO_4 2% results in a 93.1% glucose yield [24]. Acid treatment gives a high sugar yield but can increase the equipment cost, inhibitor formation, and by-product generation [25].

Diluted acid pretreatment is more advantageous than acid pretreatment because it reduces inhibitor formation. Dilute acid hydrolysis is most commonly used for DF, but inhibitors should be neutralized before the DF of LB. Diluted acid pretreatment slightly reduces lignin and hydrolyzes both cellulose and hemicellulose [1]. The 90% glucose yield was obtained from diluted sulfuric acid [26]. The most commonly employed method for DF with a cost-effective strategy and to eliminating hazards is dilute acid treatment with enzymatic treatment.

Alkaline pretreatment method effectively splits ester linkages in the lignin; hemicellulose also increases the solubilization rate. Solvation and saphonication are the first reaction in alkaline pretreatment [27], resulting in better accessibility to the enzyme. At high alkali concentrations, dissolution, end group peeling, alkaline hydrolysis, and polysaccharide decomposition and degradation can occur [27]. The alkaline method causes redistribution, solubilization, lignin condensation, lignin removal, and alteration of the cellulose structure to become denser and thermodynamically more stable. In alkaline pretreatment, cold alkaline treatment [28] results in 59.1% lignin removal, 88% digestibility of xylan achieved by aqueous ammonia [29], and a 10% sodium hydroxide effect into a 73.8% glucose yield [30]. Alkaline pretreatment is mainly used for lignin removal from LB [28]. It includes sodium hydroxide, oxidative alkaline, potassium hydroxide, aqueous ammonia, and calcium hydroxide compounds. Among them, NaOH and sodium carbonate are used most. Xylan can be removed selectively with KOH [27]. Alkaline pretreatment is advantageous because it has a minimal cost, but it requires a long retention time and the production of inhibitors [1]. Drawbacks include a long time requirement and salt formation with this method [29]. Compared with acid treatment, it does not require expensive equipment, a higher sugar yield, and low inhibitor formation [25].

IL pretreatment is a technique composed of cations and anions. It is a salt at room temperature andit is considered a green solvent [31]. It forms hydrogen bonding with the cellulose hydroxyl group, resulting in easy precipitation. IL is a more advantageous method than other chemical pretreatment methods because of its melting point of less than 100°C, low toxicity, strong polarity, difficulty in being oxidized, low vapor pressure. It avoids environmental pollution, is a good solvent, is stable, dissolves cellulose [25], is easy to synthesize and recover, and is nonvolatile [31]. The drawback or disadvantage of this pretreatment is its toxicity to cellulase [31], and that it is more viscous in pretreatment processing [32]. In IL, treatment with 1-butyl-3-methylimidazolium methylsulfate ($MeSO_4$–H_2SO_4) resulted in 77% digestibility of cellulose [31] and 1-ethyl-3-methylimidazolium acetate (CH_3COO) resulted in 76% yield of glucose; IL was used more than 20 times [32], as shown in Table 17.2.

The use of oxidative compounds is called as oxidative pretreatment, in which hydrogen peroxide or peracetic acid are the key reagents. The major objective of oxidative pretreatment method is removal of lignin and hemicellulose. Reaction can occurred during the oxidative method are, electrophilic substitution, side chains displacement and oxidative cleavage. As a result of the oxidation of lignin, inhibitors form. To remove lignin, peracetic acid was estimated to be selective [33]. Hydrolysis increased from 6.8% when untreated to 98% in peracetic acid pretreatment. For better delignification, H_2O_2 was used, with a 1% H_2O_2 and biomass weight ratio of 0.25 at pH 11.5–11.6 [27].

The technique using organic solvent is called the organosolv process. The organosolv technique os used to break down lignin and hemicellulose. In this method, lignin dissolves inorganic solvent and cellulose is precipitated as a solid. This method is highly advantageous in terms of lignin removal, but it is a limited approach because of the cost and hazards. In organosolv pretreatment [19], 5%–10% formic acid has 40% yield of glucose, and glycerol Organosolv results in 65% lignin removal [19], as shown in Table 17.2. The organic solvent can break down lignin and hemicellulose internal bonds. Among methanol, acetone, glycerol, and ethylene organic solutions, glycerol is the most important one [19].

TABLE 17.2 Different approaches to dark fermentation of pretreatment methods.

Sr. no.	Pretreatment method	Results	References
1.	Physical method	Reduces particle size, increases surface area, inhibitor eliminated, more power required.	[20,21]
2.	Chemical method	High sugar yield, inhibitor produced, alkaline method lowers inhibitor, economical cost.	
2.1.	Acid	Oxalic acid dihydrate results in 42% reducing sugar; H_2SO_4 2% results in 93.1% glucose yield	[23,24]
2.2.	Dilute acid	Dilute sulfuric acid results in 90% glucose yield	[26]
2.3.	Alkaline	Cold alkaline treatment 59.1% lignin removal; aqueous ammonia effect results in 88% xylan digestibility and NaOH 10% effect in 73.8% glucose yield.	[28–30]
2.4.	Ionic liquid	1-Butyl-3-methylimidazolium methylsulfate coupled with H_2SO_4 treatment results in 77% cellulose digestibility	[31]
		CH_3COO results in 76% yield of glucose.	[32]
2.5.	Oxidative	Hydrolysis increased from 6.8%, which is untreated, to 98% in peracetic acid pretreatment.	[33]
2.6.	Organosolv	5%–10% formic acid results in 40% glucose yield, and glycerol organosolv results in 65% lignin removal.	[19]
3.	Biological method	Inhibitor problem eliminated; environmental process, optimum condition required	
		Combination of rot fungi results in 36.27% lignin reduction.	[34]
4.	Combinational	Combination of liquid hot water plus biological pretreatment of Populus tomentosa results in 92.33% hemicellulose removal, resulting in 2.66-fold enhancement in glucose yield compared with just liquid hot water pretreatment.	[35]

3.3 Biological pretreatment methods

The biological method means the use of microorganisms or enzymes. Different microorganisms have different levels of inhibitor tolerance capacity. In biological pretreatment methods, microorganisms and enzymes have greater BHYs from LB because the production of inhibitors is not involved. Before enzymatic hydrolysis, the biological method is important for pretreatment [36]. White and soft rot fungi degenerate lignin and hemicellulose from activities of laccase, lignin peroxidase, and manganese peroxidase. The activity of laccase is estimated for the hydrolysis of lignin [37], while the activity of brown rot fungi were proposed to degrade cellulose [36]. Tian et al. [19] Microorganism *Phanerochaete chrysosporium* has the ability to reduce 26.45% fermentable sugars and 28.93% lignin content. In white rot fungi, the hyphae white initially solubilizes the mucilaginous surface of straw materials; then the hyphae move inside and initiate fiber enzymes (i.e., hemicellulases, exoglucanases, endoglycanases) and degrade them, resulting in easily digestible biomass. Biological methods have advantages; for instance, they eliminate chemical use and reduce high-power requirements, which are required during physical pretreatment. The biological method is an effective strategy compared with other methods. However, a survey found that combined methods have better biohydrogen production, according to data available in the literature [35,38]. *Clostridium* species are most widely cited for biohydrogen production. *Clostridium* species hydrolyze LB owing to the extracellular cellulosome enzyme complex. Cellulosome enzyme complexes have a wide range of activities. The biological method is environmentally friendly and requires lower energy compared with other techniques. A long pretreatment or hydrolysis time is the principal disadvantage of the biological technique. Sweet sorghum stalks with coupled cultures of *Clostridium thermocellum* DSM7072 and DSM572 have a BHY of 5.1 mmol biohydrogen per gram of substrate [39]. SSF needs a long time. Thermophilic microorganisms can eliminate the problem of SSF [18]. Lime-pretreated OPT of SSF by *T. thermosaccharolyticum* KKU19 had a greater HY of 60.22 mL of H_2 per gram pretreated OPT. In comparison, fungal pretreatment of cornstalk had a BHY of 89.3 mL of H_2 per gram of substrate by SSF using *T. thermosaccharolyticum* W16 [1]. 4.08 mmol g^{-1} HY from 5.0 g L^{-1} Cassava residue by using *Clostridium lentocellum* Cel10 [40], and it degraded the biomass without needing chemical pretreatment.

Different rot fungi strain have different results. It was found that 30.6% lignin removal with 30 days' retention time of pretreatment of the *Phlebia brevispora* strain of rot fungi [19], *I. lacteus* strain of rot fungi in 21 days result in 54% sugar yield [19], a *Ceriporiopsis subvermispor* stain retention time of 10 weeks resulted in 60% pretreatment yield [19], a *Pycnoporus anguineus* strain with a retention time of 10 days resulted in 38.7% lignin removal [41], an *I. lacteus* strain with a retention time of 21 days resulted in 66% glucose yield [37], an *I. lacteus* strain with a retention time of 10–46 days resulted in a 4 cellulose/6 lignin ratio [42], a *Poria subvermispora* strain with a retention time of 21 days resulted in 69% glucose yield [37], and a *Trametes trogii* MT strain with a retention time of 20 days resulted in 21.9% lignin removal [19]. From these observations of rot fungi strains, 21 days is ideal for *I. lacteus*. Moreover, a combination of rot fungi results in 36.27% lignin reduction [34], as shown in Table 17.2. Zhang et al. [40] showed that a *C. lentocellum* Cel10 strain with cassava residue LB is a suitable substrate with an HY of 4.08 mmol g^{-1}, which is 81.6% conversion, compared with carboxymethylcellulose (CMC) as the substrate, which resulted in a 5.4 g L^{-1} H_2 with 5.0 g L^{-1} concentration, a 92% conversion. This strain was isolated from *Ailuropoda melanoleuca* waste. *A. melanoleuca* is a panda found in China and Tibet (Vocabulary.com). The high yield may be due to the use of the number of substrates (i.e., glucose, fructose, arabinose, xylose, cellobiose, sucrose, xylan, CMC, galactose, Avicel), but not mannose. DF had 42.80% improved efficiency in fermentative bacteria and *Alcaligenes*. *Citrobacter* had 36.08–141.42 mL (g TS) $^{-1}$ [43] and 2.58% efficiency in energy conversion, which shows the effective production of biohydrogen from corn stover [44]. Fig. 17.1 gives an overview of biohydrogen production. When Zhang et al. [40] used *C. lentocellum* Cel10 as inoculum with rice straw and cassava residue as substrate (with no chemical pretreatment), the result was 3.0 mmol of biohydrogen per gram of substrate and 4.08 mmol H_2 g^{-1} substrate HY, respectively. Substrate hexose resulted HY of 1.9–3.5 mol H_2 mol^{-1}-substrate with *T. neapolitana* from different feedstocks [45].

3.4 Combination of pretreatment methods

The combination method is advantageous compared with others in terms of yield. Physicochemical treatment in which liquid hot water (LHW), steam explosion, and subcritical water are potential methods for breaking down the compositional structure of the LB, but the production of degrading products lowers enzyme and microorganism activity. A combination of the process is best in the survey. The article [38] found that the combined chemical and biological treatment (i.e., 2% H_2SO_4 and *P. ostreatus*) resulted in higher lignin digestion than the single pretreatment step. Most literature use combined the techniques of pretreatment. The sodium hydroxide pretreated enzymatic hydrolysate of corn stover using the *T. thermosaccharolyticum* W16 strain resulted in HY 108.5 mmol H_2 L^{-1} of the substrate [1]. LHW plus biological pretreatment of *Populus tomentosa* ended up in 92.33% hemicellulose removal [35], which resulted in a 2.66-fold enhanced glucose yield compared with just LHW pretreatment, as shown in Table 17.2.

In steam pretreatment, in which LB is treated in a large vessel with steam under pressure and high-level temperature, after retention time-released steam and biomass have cooled, the main objectives of steam explosion are to make the cellulose available for enzymatic activity, solubilize the hemicellulose, and neglect the formation of inhibitors, among other physicochemical pretreatment methods. Saratale et al. [18] found a cost-effective way but had a problem with lignin breakdown and producing an inhibitory effect.

In LHW, instead of steam, LHW is used. The objectives of LHW are estimated to make cellulose available for enzymatic activity, solubilize hemicellulose, and neglect the inhibitor problem. In LHW, pH 4−7 eliminates degradation outcomes [46]. LHW pretreatment results in a xylan yield and decreases when the solid concentration increases. Weil et al. [46] found a two- to fivefold enhancement in enzymatic substrate hydrolysis subsequent to LHW treatment. The addition of an external acid to thermal and LHW pretreatment improves the yield, which helps the solvability of the hemicellulose, minimizes the temperature, and results in a better hydrolyzable substrate. Therefore, the addition of acid with thermal pretreatment achieves a better result. Like alkaline pretreatment, for thermal pretreatment in which lime pretreatment is common at temperatures of 100−150°C and a lime concentration of about 0.1 Ca(OH)$_2$ g per substrate, lime pretreatment results in an acetyl valve and a lignin valve, openings that provide better accessibility for the enzyme [27]. Therefore, comparing data in our observation combinational method is more effective than using a single method, as shown in Table 17.2.

4. Inhibitor formation and solution strategies to overcome inhibitor formation for dark fermentation

The presence of inhibitors is a major constriction of pretreatment methods during DF [6]. Inhibitors adversely affect BHY by interfering with the pathway. Xylose and pentose are the most important products of carbohydrate degradation resulting from hemicellulose and cellulose. In addition to these derivatives, furans such as furfural and HMF, and formic and acetic acid (organic acids) are produced; acetic acid and phenolic inhibitors are produce from lignin, as shown in Fig. 17.4. Thus, it is essential to reduce the formation of carbohydrate degradation and reduce inhibitors, which are toxic to fermentative microorganisms. The route of formation of such inhibitors is estimated in Fig. 17.4. At higher temperatures, LB degrades hemicellulose (i.e., pentoses), and to a lesser extent, hexoses, which inhibit fermentation. Furan derivative HMF and furfural are produced with a rise in pretreatment temperature, acidic concentration, and pretreatment retention time. The produced furan derivatives react or polymerize to produce sub-products. Lignin degrades to produce phenolics, vanillin, syringaldehyde [9], and, to some extent, acetic acid. An inhibitor or suppressor that might interrupt the pathway, the activity of the cell, or fluidity of cell membrane, might be toxic to microbial cells [22]; therefore, we may estimate that degree of pretreatment based on the raw material nature and microbe inoculated [22]. Inhibitors from the H$_2$SO$_4$ hydrolysis of corn stover resulted in furan (0.85 g L^{-1}), phenolic compounds (0.1 g L^{-1}), and (1.85 g L^{-1}) acetate [22]; Furfural (0.60 g L^{-1}) and HMF (0.25 g L^{-1}) were produced. Major phenols in the hydrolysate were syringaldehyde and vanillin at 0.06 g L^{-1} and 0.04 g L^{-1}, respectively [22]. These inhibitors severely affect fermentation.

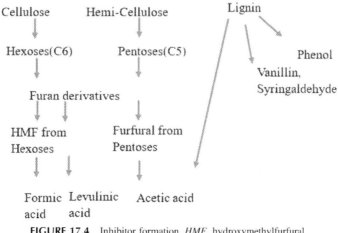

FIGURE 17.4 Inhibitor formation. *HMF*, hydroxymethylfurfural.

It was stated that 1.0 g L^{-1} phenol concentration suppressed *Clostridium butyricum* metabolism and did not show an effect when nearly 0.6 g L^{-1} phenol concentration was present [6].

How can one overcome such inhibitor criticality? There are different ways by which we can overcome such problems, such as by (1) using a resistant strain. From the literature review survey, we can estimate that different microorganisms have different tolerance levels. For example, *C. beijerinckii* strains have resistance to inhibitors more efficiently than do other nonclostridial and clostridial bacteria. Various strains of *Saccharomyces cerevisiae* are tolerant to inhibitors, which has been shown in a comparison between the performance of *S. cerevisiae* and *S. cerevisiae* ATCC 96581 in the fermentation of spent sulfite liquor [47]. (2) The detoxification method can be used [47]. Detoxification means removing specific inhibitors before fermentation. There are three types of detoxification: biological, chemical, and physical. Enzymatic detoxification by laccase or peroxidase is used before fermentation, but it increases the cost of the process. Therefore, integrating immobilization results in reuse, reduces cost, and improves the process economy owing to complete substrate degradation before fermentation; however, a paramount disadvantage of this method is the time requirement. Thermophilic bacteria overcome temperature problems in fermentation, because temperature changes are a major problem in fermentation. Treatment of the laccase enzyme assists in removing phenolic monomers and phenolic acids. (3) SSF, CBP, and SSCF methods may be used, as shown in Fig. 17.1 [25]. SSCF and SSF are reasonable choices using biohydrogen from LB. In such methods, the products of hydrolyzed are used immediately by fermentative microbes for the BHY (i.e., simultaneously), because of which the concentration of monosugars remains minimal in the medium and has less inhibition, resulting in increased efficiency and BHY [25]. Such complete synchronized substrate use results in an energetically effective and economical process at the industrial level. Cellulose hydrolysis and the fermentation of hydrogen in a single bioreactor are limited because optimal conditions are different and thermophilic microorganisms may eliminate this technical challenge of SSF for enzymatic saccharification and fermentation. SSF needs a long time. The problem of SSF can be overcome by thermophilic microorganisms [18]. Lime-pretreated OPT of SSF by *T. thermosaccharolyticum* KKU19 had a greater HY of 60.22 mL of H$_2$ per gram of pretreated OPT, whereas fungal pretreatment of cornstalk had a BHY of 89.3 mL of H$_2$ per gram of substrate by SSF using *T. thermosaccharolyticum* W16 [1]. Moreover, (4) in enzymatic treatment, cellulase and xylanase degrade biomass and peroxidase, manganese peroxidase, and laccase degrade lignin or microorganisms, to be used as biological pretreatment. This method is classified as an enzymatic detoxification method, as described earlier. Fungi microorganisms (white for lignin and soft rot for hemicellulose breakdown) [36] and bacteria that produce laccase and peroxidase are effective in reducing the inhibitor concentration, and the use of hemicellulase can help to remove inhibitors indirectly by removing lignin and hemicellulose. Quéméneur et al. [48] showed that enzymatic hydrolysis before fermentation resulted in 11.06−19.63 mL-H$_2$ g^{-1} of volatile solids (VS) and wheat straw without an enzymatic process of 5.18−10.52 mL-H$_2$ g^{-1}. Thus, enzymatic pretreatment is preferred. However, Talluri et al. [8] showed for 11.2 mmol of biohydrogen per gram of switchgrass substrate, the result was due to the substrate and inoculum used (i.e., *C. saccharolyticus*), which is thermophilic.

5. Strategies for dark fermentation

5.1 Pretreatment followed by enzyme hydrolysis

In enzyme hydrolysis, the application of enzymes for hydrolysis includes cellulase/xylanase, and so on, for cellulosic material degradation. Cellulase and xylanase are used to degrade LB, as shown in Fig. 17.3. The enzymatic hydrolysis method is an efficient and economically friendly process [16]. LB hydrolysis is used by cellulase enzymes, but characteristics such as porosity, fiber crystallinity, and the content of LB affect degradation [18]. The yield of biohydrogen was 1947 mL of H$_2$ per liter of substrate with OPT by microwave-assisted acid hydrolysate using *T. thermosaccharolyticum* KKU19 [21], and when enzymatic hydrolysate of lime-pretreated OPT was used, it was 2179 mL of H$_2$ per liter of a substrate using *T. thermosaccharolyticum* KKU19 [1]. Therefore, a two-stage method is more effective than a single one, but optimization is an essential parameter of mixed culture production enhancement. Enzymatic hydrolysis after pretreatment is the most efficient strategy for increasing the saccharification yield of LB. Hydrolysates could inhibit fermentative H$_2$ production. Different researchers used pretreatment following an enzymatic strategy that resulted in increased sugar concentration for biohydrogen production in hydrolysates [49]. Pan et al. [49] demonstrated that pretreatment followed by enzymatic hydrolysis resulted in an increase in total soluble sugars (i.e., 562.1 ± 6.9 mg per gram of total volatile solids [TVS]). High HY from the hydrolysate with the mixed condition of anaerobic culture resulted in 209.8 mL of H$_2$ per gram of TVS. One author showed that sunflower stalks with dilute acid had adverse effects on hydrogen production because of the presence of inhibitors. LB with added enzymes in DF improves the BHY in one-stage fermentation. Quéméneur et al. [48] evaluated the enzymatic treatment of biohydrogen production. The author used two experimental designs: the direct addition of enzymes (one system) and fermentation after enzymatic hydrolysis (a two-stage system). Quéméneur et al. [48] demonstrated that treatment with enzyme was advantageous compared with other methods regarding yield. Pretreated cornstalk had 4.74 mmol-H_2 per gram of substrate, whereas untreated cornstalk had 2.17 mmol of biohydrogen per gram of the substrate, and when hydrolysate was used, it had 5.93 mmol of H_2 per gram of the substrate [6]. Because of the cost of an enzyme, strategies were developed to enhance activity, including rational design and directed evolution by random mutagenesis (error-prone polymerase chain reaction, DNA shuffling, etc.). Such biotechnology and genetic engineering strategies enhance BHY by enhancing substrate specificity.

5.2 Natural strategy relevant to dark fermentation

Rumen digestion, or our digestion system, is similar to the fermentation system. After taking in food or feeding, our teeth crush and break the material, which may be correlated with physical treatment (i.e., milling to minimize the size). After that, salivation takes place by salivary amylase, which is the enzymatic process in humans. Digestion in the stomach owing to the high acid concentration is the same as acid treatment. The material further dissociates into small forms, which can be correlated with chemical treatment. After that, certain reactions take place and the absorption of useful material takes place in the body (i.e., product formation).

Digestion in the rumen starts by grinding and moistening the ingested material. Then, in the rumen there is interaction with various microorganisms in which both hydrolysis and fermentation into energy carrier molecules take place. Weimer et al. [50] described the efficient use of energy in cattle for digest LB (<2% of the consumption of energy in feed). No commercialized bioconversion system is close to achieving digestion performance in terms of the conversion rate and energy efficiency. Research suggests that conditions (pH, temperature, microbial community, etc.) of the rumen and its strategies (i.e., mechanisms of ruminant fermenters) should be mimicked to improve anaerobic digesters.

Weimer et al. [50] integrated a process consisting of DF (i.e., anaerobic digestion). The rumen is where LB is cut down biologically and chemically through microbial activity and fermentation. This process is analogous to anaerobic digestion (i.e., DF). The teeth are where LB breaks down mechanically via chewing, which is analogous to milling. After this, the mouth and rumen connection near the esophagus for biomass transportation is analogous to the pumping process, and these cycles continue until all is digested. Cotreatment is termed combining fermentation with the milling of partially digested biomass. These combinations reduce the energy and cost of the process. More effort has been taking place to understand the rumen system. According to our perspective, it is a model system for understanding DF to increase the efficiency of the anaerobic process.

5.3 Separation

In biohydrogen production, the major problem is to reduce biohydrogen with increased amounts of dissolved biohydrogen. How can one use such an operation so that it can effectively remove biohydrogen? Mass transfer is important in such operations. There are certain methods such as physical and chemical ones for separation. The pressure swing adsorption method is sometimes used; it is considered an adsorption method [51] in which pressure act as a driving force. As name suggests, the adsorption of gas takes place. At higher pressures, greater gas is adsorbed, and when pressure is released, gas is released. For example, zeolite adsorbent molecules are used to remove nitrogen [11] because they attracts nitrogen. Membrane separation techniques are mostly employed for gas separation. It is important to select an appropriate membrane module in such operations. Polydimethylsiloxane (PDMS) is widely used in biohydrogen separation. It is considered a glassy polymer. A publication [52] demonstrated that PDMS is based on cross-flow membrane separation. Some research found that other methods are also effective, such as the absorption method for biohydrogen, a two-stage chemical absorption method. In a short and good published paper by Nor Azira and Umi Aisah [53] the author used the concept of chemical adsorption. The author used methyldiethanolamine to absorb of CO_2, but it is well-known that it requires activation, and activation done by piperazine and subsequent biohydrogen was purified with the use of NaOH. Conventionally, continuously stirred tank reactors (CSTR) were used, but with increasing needs and a cost-effective strategy, certain bioreactors were preferred, such as the anaerobic membrane bioreactor, in which the integration of the bioreactor and membrane was allowed [54]. Membrane separation techniques have the problem of fouling, but an increase in temperature decreases fouling. Thus, we can use such a concept according to our understanding and predict that the use of thermophilic microbes will be a more effective strategy for researchers. According to our literature survey, under DF conditions, membrane fouling problems may not mostly occur owing to the concept that thermophilic conditions are available in thermophilic microorganisms. The load can be reduced by the appropriate use of the membrane separation module, but cost criteria also need to be remembered. The petrochemical industry mainly uses by-products that form during DF and biohydrogen.

6. Comparative study with dark fermentation

There is a difference in the terminology of hydrogen and biohydrogen, because hydrogen production can be done from metabolic routes such as electrochemical and thermochemical methods. Biohydrogen production means the production of hydrogen from the biological route (i.e., from microorganisms (fermentation), biophotolysis, and bioelectrical systems) [2]. However, the term is used in most cases. The hydrogenase enzyme has a crucial role in biohydrogen fermentation.

Biophotolysis involves oxygenic photosynthetic microbes such as cyanobacteria and green microalgae are used in which biohydrogen is produced with water and sunlight [2]. Biophotolysis is direct or indirect. There are two ways of fermentation: dark and photofermentation. Hydrogen production occurs naturally by organic matter decomposition. Three types of microorganisms have the role of biohydrogen: algae or Cyanobacteria, photosynthetic bacteria, and DF bacteria. The hydrogenase enzyme has reactive problems with oxygen, and the anaerobic/dark method has advantages, among others [18]. Photofermentation includes nonsulfur bacteria. Photofermentation has high substrate conversion efficiency, uses organic substrates, and requires less free energy but requires a high activation energy [18]. Theoretically, photofermentation produces 12 moles of hydrogen per gram of substrate, as shown in Eq. (17.1):

$$C_6H_{12}O_6 + 6H_2O + Light \rightarrow 6CO_2 + 12H_2 \tag{17.1}$$

Photofermentation routes include the presence of light and specific substrates. Compared with photofermentation, DF is more potent and promising. The crucial disadvantage of DF is that it has a low yield, which can be enhanced by applying sequential DF and photofermentation, described in yield enhancement strategies. The maximum substrate quantity in cofermentation was lower in the BHY of DF. Metabolically, biohydrogen can be produced from glucose and xylose in DF. When glucose (a product of cellulose) is used as a substrate, it follows the glycolysis pathway to produce pyruvate; ultimately, H_2, CO_2, and acetate are formed from pyruvate ferredoxin oxidoreductase and ferredoxin-dependent hydrogenase enzyme. When Xylose (a product of hemicellulose) is used as a substrate, xylose isomerase, xylulokinase, epimerase, isomerase, transketolases, and transaldolases. This results in pyruvate, which is converted to H_2, CO_2, and acetate [6]. The consumption of reducing power can produce butyric acid rather than acetate, which can reduce the BHY, as described subsequently [6]. Adding ferric oxide nanoparticles divert the reaction toward acetate over the ethanol metabolic pathway [9]. DF requires little energy, a greater rate, and wide range specificity, and is economical [18].

7. Dark fermentation advantages and stoichiometry

DF means biohydrogen production. H_2 is hazardous when reacting with an oxidizer such as oxygen. Because hydrogenase is oxygen-sensitive and separation is a significant problem, we cannot use direct biophotolysis. How can one resolve such a problem when the aim is to produce hydrogen from an economic perspective? The simple solution is DF. DF is regarded as an acidogenic process of fermentation under an anaerobic process in which anaerobic microorganisms decompose organic feedstocks and neglect such problems. An anaerobic process takes place in the absence of oxygen. The production of by-products and their use are advantageous. Therefore, DF is a potent option over other methods. The use of a broad sense of the substrate high production rate of biohydrogen over photofermentation is another advantage. In DF, acidogens convert monomers into organic acids. Acetogenic bacteria use these acids and convert them into acetate, carbon dioxide, and hydrogen. For BHY, it is necessary to cease the process of methanogenesis and accumulate biohydrogen in the previous steps to produce biohydrogen [2]. In DF, most cited anaerobic microorganisms are *Clostridium* species, *Enterobacter* spp. (anaerobic, facultative bacteria), and the bacterial combination of organic waste of (sludge, animal feces, etc.) [18]. *T. saccharolyticum* W16 with both glucose and xylose resulted in an HY of 2.37 moles of H_2 per gram of the substrate [1]. Theoretically, we can conclude which will be better for HY as substrate. When glucose is used as a substrate for fermentation and an acetate fermentation product, it is 4 moles of H_2/g, and when butyrate is a fermentation product, it is 2 mol H_2 g^{-1}. When coculture is used in DF, mixed acids are produced that lower HY to 2.5 H_2 g^{-1}, as estimated in Eqs. (17.2–17.4), respectively. Experimentally, the lower yield is estimated, ranging from nearly 2 to 2.4 mol mol^{-1} [18]:

$$C_6H_{12}O_6 + 2H_2O \rightarrow 2CH_3COO\text{-} + 2CO_2 + 4H_2 \tag{17.2}$$

$$C_6H_{12}O_6 + 2H_2O \rightarrow CH_3CH_2CH_2COO\text{-} + 2CO_2 + 2H_2 \tag{17.3}$$

$$4C_6H_{12}O_6 + 6H_2O \rightarrow 3CH_3CH_2CH_2COOH + 2CH_3COOH + 8CO_2 + 10\,H_2 \text{ (i.e., 10/4 } H_2\,g^{-1}) \qquad (17.4)$$

When xylose is used as a fermentation substrate and the fermentative product is acetate, it will give an HY of 3.33 of $H_2\,g^{-1}$, and when butyrate is the fermentation product, an HY of 1.66 of H_2/g of xylose will be given, as shown in Eq. (17.5) [6] and (6), respectively:

$$C_5H_{10}O_5 + 1.66H_2O \rightarrow 1.66CH_3COO\text{-} + 1.66CO_2 + 3.33H_2 \qquad (17.5)$$

$$C_5H_{10}O_5 + 1.66H_2O \rightarrow 1.66CH_3CH_2CH_2COO\text{-} + 1.66CO_2 + 1.66H_2 \qquad (17.6)$$

Therefore, from this information, we can estimate a directly related relationship between the product formed and the HY. However, the experiment shows that the practical yield is always lower than the estimated theoretically calculated yield, and according to the survey, it may be caused by using some microbes from the growth perspective. The major disadvantage of DF is the lower yield in DF. Research is ongoing to increase yield by optimization, develop cost-effective pretreatment methods, and develop new strategies. The performance of DF depends on factors that include the type of inoculum, feedstock source, temperature, pH, reactor type, and HRT.

8. Factor affecting dark fermentation

(a) Type of inoculum: In DF, two types of inoculum are used, pure or mixed (co-culture). Anaerobic *Clostridium* species and *Enterobacter* species facultative anaerobic are the pure cultures most widely used to produce biohydrogen. Single cultures are most important because they have a high production rate and HY. A pure culture requires highly ascetic and sterile conditions, resulting in an increased price. The problem can be overcome by a co-culture inoculum (i.e., a mixed culture). A mixed culture is easy to operate, but the occurrence of hydrogen consumers limits the use of mixed culture. Mixed culture problems can be eliminated by using specific chemicals and long-chain fatty acids [1]. Sources of mixed culture are anaerobic sludge, animal dung, municipal solid soil, waste, and compost.

(b) Feedstock: More than 20% total solids run on batch-type reactors, whereas feedstock includes a 2%−12% solids content that can be run on CSTRs [1]. Higher feedstock inhibits the production rate in mixed cultures. Different LBs can produce different levels of HY with varied pretreatment methods and conditions. A DF of high solids and fiber content feedstocks generates less water than conventional anaerobic digestion process (AD)'s low solids [55].

(c) Nitrogen and phosphorous effect: For the growth of microorganisms, nitrogen has a vital role, which is generally classified as inorganic and organic. An ammonium inorganic nitrogen concentration of $0.1-7.0\,g\,L^{-1}$ is widely used [56]. Organic nitrogen such as peptone, yeast extract, and steep corn liquor can be used. A high BHY was estimated when organic nitrogen was applied.

(d) Temperature factor: As described earlier, an optimal temperature is essential in the overall yield of biohydrogen. Different microorganisms require different temperatures. The metabolism of microorganisms generates temperatures that can affect microorganism development; in scale-up operations it is essential to maintain the optimal temperature. Different optimum temperatures were used for different strains. DF operates in a broad variety of ranges: thermophilic, mesophilic, or hyperthermophilic (i.e., 40−65°C, 25−40°C, or > 80°C, respectively). Thermophilic has a greater BHY compared with mesophilic conditions. The required temperature varies according to the inoculum and substrate types. With an increase in temperature from 60 to 77°C, H_2 production and growth and the rate of glucose consumption increased, but there was no significant variation from 77 to 85°C [45].

(e) Hydraulic retention time: The hydraulic retention time (HRT) for biohydrogen production in DF is short because of the course of the high rate of fermentation of thermophilic microorganisms. HRT means the total time taken in fermentation. Hydrogen-producing microorganisms under thermophilic conditions need a short retention time [1] compared with methanogens. A low retention time (i.e., a high organic loading rate) is favorable for biohydrogen production, which eliminates methanogens because it eliminates slow-growing bacteria. Hyperthermophilic bacteria (*T. neapolitana*) show that higher HRT for growth suspended biomass and low HRT immobilized biomass (i.e., 0.25−60 h) [57]. The HRT for the solid substrate is higher than for the liquid-type substrate because the time to a hydrolyzing substrate containing a high solid is longer. The literature survey found that DF production with thermophilic microorganisms has a short HRT.

(f) Hydrogen partial pressure: High hydrogen partial pressure (HPP) results in reversibly reducing and oxidizing ferredoxin because it affects hydrogenase activity. High HPP reduces the biohydrogen rate, which can be reduced by agitation and enlarging the headspace volume. The advantage of DF is the simple reactor configuration [2]. H_2 production and increase in yield were estimated by sparging of N_2 compared with the condition in which there was no sparging.

(g) pH: In biological pretreatment, pH has a significant role in carrying out the function. *Thermotoga neapolitana* had inhibited growth at 4.5 pH, whereas a pH shift from 4.0 to 5.5 increased H_2 [58]. In the laboratory when monitoring of pH, it was suggested to use tris(hydroxymethyl) aminomethane, piperazine-N,N'-bis(2-ethanesulfonic) acid and 4-(2-hydroxyethyl)-1-piperazineethanesulfonic acid, respectively [45], but at that large scale the economic aspects were neglected. However, the same effect can be achieved with more suitable products such as CO_2 [58]. The production of lignocellulosic enzymes is affected by the pH. Laccase enzymes show crucial activity in LB degradation owing to the breakdown of the lignocellulosic bridge. Laccase is greatly influenced by a pH change in the medium. A change in pH results in a change the three-dimensional structure of the enzyme. Fungi, which is white rot, grew well at pH 4.0–5.0 [7].

9. Cultivation techniques of bioreactor

It is crucial to know the cultivation techniques in DF. In the literature survey, most research was done in batch and continuous operation and little in feed-batch mode. Rittmann et al. [10] provided such information regarding this DF. A quantitative comparison is significant, which can be possible when the result presents a C-molar mass in the scientific community. Rittmann et al. [10] showed that substrate conversation capacity $Y(H_2/s)$ is maximal at 441 mol mol^{-1} in the closed batch cultivation technique. The batch fermentation technique is the most promise because of the closed system. Food waste generates 593 mL of biohydrogen per gram of carbohydrate and 39.14 mL of biohydrogen per gram food waste [59]. Nazlina et al. [59] described different effects on BHY from food waste. The ground wheat waste solution was 25.7 mL H_2 per grams per cells per hour. In research, anaerobic sludge as a microorganism with batch cultivation at pH 5.0 and 5.9 was 1.6 mol of H_2 mol^{-1} of glucose and 2.4 mol of H_2 mol^{-1} of glucose, respectively [60,61]. *C. butyricum* at 30°C and pH 5.6 and 6.5 was 2.0 mol of H_2 mol^{-1} of glucose and 1.9 mol of H_2 mol^{-1} of glucose [62]. In the chemostat culture strategy, when *C. saccharolyticus* DSM 8903 was used, the maximum yield of H_2/substrate was 4.0 mol mol^{-1} [45]. The maximum hydrogen evolution reaction (HER) of 77 mmol L^{-1} h was estimated for *Enterobacter cloacae* II BT-08 [63]. The maximum qH_2 of 35 mmol g^{-1} h is estimated for *Caldicellulosiruptor kristjanssonii* DSM 12,137 [64]. Anh et al. [65] showed that the biohydrogen rate of cells that were immobilized reached 1.84 ± 0.1 mol of H_2/mol of xylose and 5.64 ± 0.19 mmol H_2 L h^{-1}, respectively, which were higher than for free cells: 1.7–1.3-fold. Bioreactors such as fluidized bed reactors (FBRs), CSTRs, up-flow anaerobic sludge blanket (UASB), packed bed reactors (PBRs), anaerobic sequencing batch reactors, and membrane biological reactors have excellent prospects for the process of DF [45]. In CSTRs, sound mixing is achieved, which depends on the reactor's input power and geometry. It helps to generate a more significant mass transfer and shows biomass washout, which is minimal in the attached LB reactor (i.e., PBRs, UASBs, and FBRs). These contain bacteria by inert support, providing a greater concentration of cells and resulting in a high solids time of retention, more efficient mass transfer, and a high tolerance for shock loads [45]. A brief overview of thermophilic bacteria *T. neapolitana* and different bioreactor operations is available in Ref. [45]. The high-yield H_2/substrate was shown for a recombinant strain of ATCC 25,755, including 2.15 mol mol^{-1} [66]. Few papers are available on fed-batch fermentation, which might be because organic acid end products (i.e., alcohols) accumulate a massive presence, growth, and kinetics of H_2 inhibition. It can be applied only using broth exchange or cell separation systems [66].

10. Yield enhancement strategies

From this information, we know that biohydrogen has a high rate but low BHY. Therefore, efforts are being made in pretreatment, optimization, and genetic engineering approaches to increase BHY, including inhibitor suppression strategies and selecting techniques or strategies to minimize inhibitor problems and help to increase BHY, as discussed previously.

Sequential/coupling method: More focus is currently directed toward increasing the yield of DF. In the coupling method, sequential DF and photofermentation is used and sequential anaerobic bacteria in DF and purple nonsulfur bacteria in photofermentation are used. Because DF theoretically produces $4H_2$ g^1 of the substrate, the by-product, volatile fatty acid(s) from DF, is used for photofermentation by purple nonsulfur bacteria, which produces $10H_2$/g of the substrate. Then, the total of $14H_2$/g of the substrate can be produced, as described theoretically in Eqs. (17.7) and (17.8):

$$C_6H_{12}O_6 + 2H_2O \rightarrow 2CH_3COO- + 2CO_2 + 4H_2 \tag{17.7}$$

$$CH_3CH_2CH_2COOH + 6H_2O + Light \rightarrow 4CO_2 + 10H_2 \tag{17.8}$$

$$\text{Total } 14H_2$$

Cornstalk DF from cow dung by mixed culture resulted in an HY of 192.9 mL of H_2 per gram of TVS. When sequentially combined with photofermentation of *Rhodobacter sphaeroides* HY01, BHY was increased to around 401.5 mL of H_2 per gram of TVS [55]. The BHY obtained by pretreated corncob from DF with cocultures or mixed culture using dairy manure has an HY of 120.2 mL of H_2 per gram of substrate, and after sequentially combined photofermentation of the effluent it was 713.6 mL of $H_2\ g^{-1}$ [67].

Application of enzymatic treatment after physical or chemical pretreatment: Song et al. [43] described enzyme hydrolysis plus pretreatment of acid using *Alternanthera philoxeroides* to discharge reducing sugars. The discharged reducing sugar was 0.35 $g\ g^{-1}$ of *A. philoxeroides*, and when enzymolysis was applied, it was 0.56 $g\ g^{-1}$ of *A. philoxeroides*. Increasing the sugar content (i.e., glucose) increased the BHY. Moreover, bacterial optimization increased the yield of biohydrogen. Even after pretreatment, some lignin was present or bound. To overcome such problems, neutralization or laccase or enzymatic treatment is essential.

Combinational pretreatment method: Combining the different pretreatment methods along with biological pretreatment or different pretreatments of LB can enhance accessibility to obtain more reducing sugar and improve H_2 production. It was shown that 46 mL g^{-1} $VS_{initial}$ for only the base level of pretreatment and pretreatment with acid and ultrasonication pretreatment resulted in 118 mL g^{-1} $VS_{initial}$ [68]. Rice straw was pretreated with hyperthermophilic H_2 fermentation (*T. neapolitana*) even though that the combination of NH_3 and diluted H_2SO_4 increased the digestibility of rice straw and HY [69]. Menon et al. [25] described SSF, SSCF, and CBP. SSCF and SSF are feasible options for biohydrogen production from lignocellulosic materials. In such methods, the hydrolyzed products are used immediately by fermentative bacteria for biohydrogen production. Thus, the presence of a concentration of monosugar remains minimal in the medium and suppression is lessened, resulting in increased efficiency and BHY. All the monosacharides substrate results in an energetically effective and economical method which can be viable at the industrial scale. Cellulose hydrolysis and hydrogen fermentation are limitations in a single bioreactor because optimal conditions are different between them, and thermophilic microorganisms could eliminate these technical challenges of SSF for enzymatic SF [1]. CBP means the cooperation of the production and hydrolysis of enzymes in a single reactor fermentation (i.e., CBP), which has advantages from an economical perspective. It reduces the cost compared with other processes. CBP further improves in a pure culture fermentation Weimer et al. [50], suggested but there has been no experimental demonstration of its performance. Anaerobic pure culture fermentation of LB integration mf Mixed treatment with CBP improves the biogas yield, reduces particle size, and improves biomass solubilization as a result of the increased accessibility of plant cell walls from pure culture fermentations of *C. thermocellum*. DF with mixed cultures is the conventional form of CBP because all processes occur in one fermenter.

Mixed substrate: The use of pure substrate is not economical or feasible. Media with complementary cosubstrate characteristics may be helpful to achieve increased BHY and eliminate additional nitrogen use. The use of mixed substrate helps to increase biohydrogen production and maintain pH levels. Lopez-Hidalgo et al. [70] determined that when hydrolysates of wheat straw and cheese whey were used as cofermentation, the high production rate increased from 66.6 to 89.5 $H_2\ L^{-1}$ per hour [70]. Zagrodnik et al. [71] showed that *Clostridium* with starch as cosubstrate enhanced biohydrogen from 0.3 to 1.2 $H_2\ L^{-1}$ medium. Li et al. [72] and Pachapur et al. [73] found an increased yield when using an additional substrate. When the concentration of substrate is high, cofermentation results in reduced BHY in DF, which may be due to the enhanced concentration of metabolic products.

Engineering of pathway: Pathway engineering is the most promising method to improve desirable products in the host by altering the pathway of microorganisms. It requires knowledge about biosynthesis and degradation of the pathway. Introducing a gene that produce biohydrogen or overexpression, expression, or introduction of the gene in a microorganism where it cannot produce the desired product is helpful to increase biohydrogen production. When lactate can be a fermentation product, there is no H_2 production. By identifying the ability of *Caldicellulosiruptor bescii* to can grow on LB without treatment, Cha et al. [74] removed the LDH gene from *C. bescii*, which enhanced H_2 production by 21%−34% compared with a wild strain, but the result was minimal compared with the treated method. Therefore, more efforts are required to hydrolyze or increase microbial activity. Yang et al. reported that when the nfnAB gene of *Thermoanaerobacterium aotearoense* SCUT27 is deleted, it increases the yield of H_2 production by 41% [9]. Zhang et al. [44] reported that, when the formate lyase activator gene was expressed in *E. cloacae*, there was increased biohydrogen production, around 188%, for fermentation reaction time of 24h from cornstalk hydrolysate [9]. To suppress butyrate formation, it is important to change the flux of butyrate formation so that more yield will be achieved. Research might place more emphasis on this strategy in the near future.

Use of supporting or additional materials; Adding materials could increase biohydrogen production. Silva et al. [11] showed that zeolite improves BHY by 10%−26% from DF of carbon-5/carbon-6-sugars and *Sargassum* species biomass. The use of additional materials or nanomaterials may be helpful to increase H_2 production. Nanomaterials such as gold, iron, silver, titanium oxide, and copper increase HY, perhaps because they may act as cofactors or oxygen scavengers. Zhang et al. [44] stated that the use of nano-ferrihydrite resulted in 2.55 mol of H_2 mol^{-1} of glucose [9].

Cell-free and synthetic pathway biotransformation: The cell-free technique is emerging because cell systems do not easily optimize temperature, and so on. Biohydrogen production through a cell-free pathway system includes [75]: (1) the reduction of phosphorylation for glucose-1-phosphate by phosphorylase; (2) phosphoglucomutase to convert the molecule to glucose-6-phosphate; (3) 10 enzymes for production into 12 NADPH/G-6-P through pentose pyrophosphate (PPP); and (4) BHY by hydrogenase. It was found that 3.92 mmol H_2 h^{-1} L biohydrogen production reached the reactor. Cellobiose as a substrate at an HRT of 150 h for a total reaction, had a BHY of 11.2 mol mol^{-1} of anhydroglucose (cellobiose), with a theoretical yield of 93.1%. The analysis shows the yield is greater compared with the microbial system. Synthetic pathway biotransformation includes (1) pathway reconstruction, (2) the selection of the enzyme, (3) engineering of the enzyme, (4) production of the enzyme, and (5) engineering of the process [75].

Producing resistant microbes: Microbes have a critical role in effective production with a cost-efficient strategy, so the total industrial yield depends on how efficient the microbe is. Identification of the microbe is an important task. People are trying to develop strategies for the highly efficient throughput screening of microbes. Degradation of LB is important in biohydrogen production by DF. Different microbes have different levels of inhibition. Thus, identifying an efficient microbe that grows under inhibitor concentration is an interesting problem. Can we identify microbes for high-throughput screening and ultimately increase the yield? The answer is that we can develop a strategy by which microbe identification can be made easier: that is, by identifying microbes that grow in that inhibitory condition by making them resistant. Resistance can be achieved by growing such microbes in an inappropriate inhibitory environment. For example, *C. butyricum* grows at a certain level of phenol concentration. We can use that microbe, which is highly resistant in a high inhibitory concentration, we can go with error-prone polymerase chain reaction to increase the efficiency, or we can go to soil metagenomics.

Biosensor; Much effort is being made to identify microbes with high capacity for high-throughput microbe screening. From an understanding of an analysis of the data, we are given a perspective for research. Our strategy is as follows: (1) first, identify the product or the side by-product in the research. In our case, the side product acetate is produced with 4 mol of biohydrogen (theoretical). (2) Identify the microbe that can degrade the product of interest or a related side by-product. In our case, acetate is degraded by, for example, *Clostridium kluyeri*, to organic acids. (3) Then, treat that microbe (e.g., *C. kluyeri*) with that product (e.g., acetate) according to varying concentrations, starting with zero. This will activate the gene of interest, which will degrade the product, or we can use a model microbe to degrade our product. (4) After concentration treatment, we can compare the treated and untreated genetic material by 16srRNA sequencing to find what gene is converted. (5) If we know the gene directly by prior knowledge available in the literature, the previous steps may be eliminated. (6) Cloning with green fluorescent protein (GFP) and insert into the host after knowing the gene. Then, we can develop a construct that includes GFP gene and the gene that degrades the product and develop a recombinant DNA and insert it into the *Escherichia coli* cell (host). The *E. coli* cell will be helpful because it is a model microbe, and most of the information and tools are available. (7) After insertion and identification of the construct, we can use it as a biosensor so that when acetate is produced, biohydrogen is also produced, but because our biosensor has a gene that degrades acetate and GFP under the control of the promoter, when acetate is available in the environment, the gene will activate. Depending upon the acetate produced, it will produce fluorescent and an appropriate microtiter identified by fluorescent-activated cell sorting, and the microbe with greater efficiency can be identified. We were inspired by this concept from the research paper by Sana et al. [76].

11. Conclusion and future perspectives

Biohydrogen is the cleanest high-energy source. Upon combustion, only water vapor is released. Because of the minimal abundance of fossil fuels, it is essential to change to other sources that are renewable. LB is a great option for achieving such an opportunity to produce high-energy molecule biohydrogen because of its abundant ability and renewable property. India and the world are facing the significant problems of air pollution and waste treatment. During winter, the people of Delhi (the capital of India) face problems caused by the burning of waste of harvested crops. To resolve such a problem, DF through LB can be helpful. H_2 production not only mitigates emission problems, it offers environmental sustainability.

DF is an anaerobic method in which the cost of production is reduced owing to the lack of a requirement for oxygen and less handling. A high rate of production is advantageous. However, even the cost of biohydrogen is high, which can be reduced industrially by increasing the BHY. It is essential to consider the cost of a variety of pretreatment methods for LB. Pretreatment methods are not viable because they produce inhibitory by-products. Inhibitors can be reduced by appropriate biological pretreatment, detoxification strategies, and so on. Efforts are being made to make the whole process viable.

Alkaline pretreatment is most effective; it reduces the lignin content and produces sugar in high quantities compared with acid pretreatment. Novel methods are being developed for biohydrogen production, such as methods combining biology, chemistry, and physics. A combinational method will be more effective than a single one. Microbes cannot produce biohydrogen under such an ordered structure (i.e., LB); they require unordered structures such as monomers. The biomass yield can be increased and make the whole process viable by sequential photofermentation after DF. Pretreatment followed by enzymolysis is important to convert the complex to monomers. CBP will be most helpful, but needs to search for a microbe or genetically engineer one with a high rate and resistance to temperature and can operate in a consolidated bioprocess. A large portion of microbes is available in an environment with an enormous capacity. Isolating appropriate microbes and screening them will be helpful to achieve the task of higher H_2 production through genetic engineering. The strain with a higher capacity to use substrates such as glucose and hemicellulose derivatives will be helpful to achieve a high BHY task, which can be done by genetic engineering or by optimizing conditions or screening different microbes and identifying the capacity of each. The action of additional molecules can increase the yield by acting as a cofactor. It is necessary to characterize energy requirements for milling in cotreatment, techno-economic analysis of processes to assess industrial scale feasibility, and the development of efficient handling of high-fiber solid content feedstocks. The addition of different unit operations can increase the chances of the process, according to an economic perspective. The future depends on sustainability and economic and environmental values. There is an attempt to change fuel dependency to biohydrogen dependency. Extensive research is going into biohydrogen production. Fossil fuel dependency will be overcome more economically owing to biohydrogen production in coming years. A lot of efforts has been take into the present book chapter to make an overall understanding of research aspects and various reported strategies of biohydrogen production for both academia and industry researchers.

List of abbreviations

BHY Biohydrogen yield
CBP Consolidated bioprocessing
CHL Cellulose, hemicellulose, lignin
CSTR Continuously stirred tank reactors
DF Dark fermentation
FBR Fluidized bed reactor
HMF Hydroxymethylfurfural
HRT Hydrogen retention time
IL Ionic liquid
LB Lignocellulosic biomass
OPT Oil palm trunk
PBR Packed bed reactor
PDMS Polydimethylsiloxane
SSCF Simultaneous saccharification and co-fermentation
SSF Simultaneous saccharification and fermentation
TVS Total volatile solids
UASB Up-flow anaerobic sludge blanket reactor

References

[1] A. Salakkam, P. Plangklang, S. Sittijunda, M.B. Kongkeitkajorn, S. Lunprom, A. Reungsang, Bio-Hydrogen and Methane Production From Lignocellulosic Materials, 2019. www.intechopen.com.

[2] A.I. Osman, T.J. Deka, D.C. Baruah, D.W. Rooney, Critical challenges in biohydrogen production processes from the organic feedstocks, in: Biomass Conversion and Biorefinery, Springer, 2020, https://doi.org/10.1007/s13399-020-00965-x.

[3] K. Bhawan, N. Delhi, National Policy for Management of Crop Residues (NPMCR) Incorporation in Soil and Mulching Baling/Binder for Domestic/industrial as Fuel, Government of India Ministry of Agriculture Department of Agriculture & Cooperation (Natural Resource Management Division), 2014.

[4] S. Bhuvaneshwari, H. Hettiarachchi, J.N. Meegoda, Crop residue burning in India: policy challenges and potential solutions, Int. J. Environ. Res. Publ. Health 16 (5) (2019), https://doi.org/10.3390/ijerph16050832.

[5] P. Shah Maulin, Microbial Bioremediation & Biodegradation, Springer, 2020.

[6] V. Reginatto, R.V. Antônio, Fermentative hydrogen production from agroindustrial lignocellulosic substrates, Braz. J. Microbiol. 46 (2) (2015) 323−335, https://doi.org/10.1590/S1517-838246220140111. Sociedade Brasileira de Microbiologia.

[7] P. Shah Maulin, Removal of Refractory Pollutants From Wastewater Treatment Plants, CRC Press, 2021.

[8] S. Talluri, S.M. Raj, L.P. Christopher, Consolidated bioprocessing of untreated switchgrass to hydrogen by the extreme thermophile Caldicellulosiruptor saccharolyticus DSM 8903, Bioresour. Technol. 139 (2013) 272–279, https://doi.org/10.1016/J.BIORTECH.2013.04.005.

[9] S.K. Bhatia, S.S. Jagtap, A.A. Bedekar, R.K. Bhatia, K. Rajendran, A. Pugazhendhi, C.v. Rao, A.E. Atabani, G. Kumar, Y.H. Yang, Renewable biohydrogen production from lignocellulosic biomass using fermentation and integration of systems with other energy generation technologies, Sci. Total Environ. 765 (2021) 144429, https://doi.org/10.1016/J.SCITOTENV.2020.144429.

[10] S. Rittmann, C. Herwig, A comprehensive and quantitative review of dark fermentative biohydrogen production, in: Microbial Cell Factories, vol. 11, 2012, https://doi.org/10.1186/1475-2859-11-115.

[11] R.M. Silva, A.A. Abreu, A.F. Salvador, M.M. Alves, I.C. Neves, M.A. Pereira, Zeolite addition to improve biohydrogen production from dark fermentation of C5/C6-sugars and Sargassum sp. biomass, Sci. Rep. 11 (1) (2021), https://doi.org/10.1038/s41598-021-95615-1.

[12] M. Ni, D.Y.C. Leung, M.K.H. Leung, K. Sumathy, An overview of hydrogen production from biomass, Fuel Process. Technol. 87 (5) (2006) 461–472, https://doi.org/10.1016/j.fuproc.2005.11.003.

[13] H.H. Nimz, Wood-chemistry, ultrastructure, reactions, Holz als Roh- Werkst. 42 (8) (1984), https://doi.org/10.1007/bf02608943.

[14] F. Peng, P. Peng, F. Xu, R.C. Sun, Fractional purification and bioconversion of hemicelluloses, Biotechnol. Adv. 30 (4) (2012), https://doi.org/10.1016/j.biotechadv.2012.01.018.

[15] O. Bobleter, Hydrothermal degradation of polymers derived from plants, Progr. Polymer Sci. 19 (5) (1994), https://doi.org/10.1016/0079-6700(94)90033-7.

[16] Z. Anwar, M. Gulfraz, M. Irshad, Agro-Industrial Lignocellulosic Biomass a Key to Unlock the Future Bio-Energy: A Brief Review, 2014, https://doi.org/10.1016/j.jrras.2014.02.003.

[17] A.-D. Isamar, Thesis- Anaerobic Digestion of Lignocellulosic Biomass via Cotreatment: a Techno-Economic Analysis, 2019.

[18] G.D. Saratale, S.-D. Chen, Y.-C. Lo, R.G. Saratale, J.-S. Chang, Outlook of biohydrogen production from lignocellulosic feedstock using dark fermentation-a review, J. Sci. Ind. Res. 67 (2008).

[19] S.Q. Tian, R.Y. Zhao, Z.C. Chen, Review of the pretreatment and bioconversion of lignocellulosic biomass from wheat straw materials, Renew. Sustain. Energy Rev. 91 (2018) 483–489, https://doi.org/10.1016/j.rser.2018.03.113. Elsevier Ltd.

[20] T. Shahzadi, S. Mehmood, M. Irshad, Z. Anwar, A. Afroz, N. Zeeshan, U. Rashid, K. Sughra, Advances in lignocellulosic biotechnology: a brief review on lignocellulosic biomass and cellulases, Adv. Biosci. Biotechnol. 05 (03) (2014), https://doi.org/10.4236/abb.2014.53031.

[21] S. Khamtib, P. Plangklang, A. Reungsang, Optimization of fermentative hydrogen production from hydrolysate of microwave assisted sulfuric acid pretreated oil palm trunk by hot spring enriched culture, Int. J. Hydrog. Energy 36 (21) (2011), https://doi.org/10.1016/j.ijhydene.2011.05.117.

[22] G. Cao, N. Ren, A. Wang, D.J. Lee, W. Guo, B. Liu, Y. Feng, Q. Zhao, Acid hydrolysis of corn stover for biohydrogen production using Thermoanaerobacterium thermosaccharolyticum W16, Int. J. Hydrog. Energy 34 (17) (2009) 7182–7188, https://doi.org/10.1016/J.IJHYDENE.2009.07.009.

[23] L. Schneider, J. Haverinen, M. Jaakkola, U. Lassi, Solid acid-catalyzed depolymerization of barley straw driven by ball milling, Bioresour. Technol. 206 (2016), https://doi.org/10.1016/j.biortech.2016.01.095.

[24] W.H. Chen, B.L. Pen, C.T. Yu, W.S. Hwang, Pretreatment efficiency and structural characterization of rice straw by an integrated process of dilute-acid and steam explosion for bioethanol production, Bioresour. Technol. 102 (3) (2011), https://doi.org/10.1016/j.biortech.2010.11.052.

[25] V. Menon, M. Rao, Trends in bioconversion of lignocellulose: biofuels, platform chemicals & biorefinery concept, Prog. Energy Combust. Sci. 38 (4) (2012) 522–550, https://doi.org/10.1016/J.PECS.2012.02.002.

[26] J. Zhang, S. Shao, J. Bao, Long term storage of dilute acid pretreated corn stover feedstock and ethanol fermentability evaluation, Bioresour. Technol. 201 (2016) 355–359, https://doi.org/10.1016/J.BIORTECH.2015.11.024.

[27] A.T.W.M. Hendriks, G. Zeeman, Pretreatments to enhance the digestibility of lignocellulosic biomass, in: Bioresource Technology vol. 100, Elsevier Ltd, 2009, pp. 10–18, https://doi.org/10.1016/j.biortech.2008.05.027.

[28] J.C. García, M.J. Díaz, M.T. Garcia, M.J. Feria, D.M. Gómez, F. López, Search for optimum conditions of wheat straw hemicelluloses cold alkaline extraction process, Biochem. Eng. J. 71 (2013), https://doi.org/10.1016/j.bej.2012.12.008.

[29] X. Li, T.H. Kim, N.P. Nghiem, Bioethanol production from corn stover using aqueous ammonia pretreatment and two-phase simultaneous saccharification and fermentation (TPSSF), Bioresour. Technol. 101 (15) (2010) 5910–5916, https://doi.org/10.1016/J.BIORTECH.2010.03.015.

[30] M.C. Coimbra, A. Duque, F. Saéz, P. Manzanares, C.H. Garcia-Cruz, M. Ballesteros, Sugar production from wheat straw biomass by alkaline extrusion and enzymatic hydrolysis, Renew. Energy 86 (2016), https://doi.org/10.1016/j.renene.2015.09.026.

[31] N. Sathitsuksanoh, A. George, Y.-H.P. Zhang, New lignocellulose pretreatments using cellulose solvents: a review, J. Appl. Chem. Biotechnol. 88 (2) (2013) 169–180. https://www.academia.edu/2292174/New_lignocellulose_pretreatments_using_cellulose_solvents_a_review.

[32] A.M. da Costa Lopes, K.G. João, D.F. Rubik, E. Bogel-Łukasik, L.C. Duarte, J. Andreaus, R. Bogel-Łukasik, Pre-treatment of lignocellulosic biomass using ionic liquids: wheat straw fractionation, Bioresour. Technol. 142 (2013) 198–208, https://doi.org/10.1016/J.BIORTECH.2013.05.032.

[33] L.C. Teixeira, J.C. Linden, H.A. Schroeder, Alkaline and peracetic acid pretreatments of biomass for ethanol production, Appl. Biochem. Biotechnol. Part A Enzyme Eng. Biotechnol. (1999) 77–79, https://doi.org/10.1385/ABAB:77:1-3:19.

[34] M. Monrroy, I. Ortega, M. Ramírez, J. Baeza, J. Freer, Structural change in wood by brown rot fungi and effect on enzymatic hydrolysis, Enzym. Microb. Technol. 49 (5) (2011), https://doi.org/10.1016/j.enzmictec.2011.08.004.

[35] W. Wang, T. Yuan, K. Wang, B. Cui, Y. Dai, Combination of biological pretreatment with liquid hot water pretreatment to enhance enzymatic hydrolysis of Populus tomentosa, Bioresour. Technol. 107 (2012), https://doi.org/10.1016/j.biortech.2011.12.116.

[36] S. Sun, S. Sun, X. Cao, R. Sun, The role of pretreatment in improving the enzymatic hydrolysis of lignocellulosic materials, Bioresour. Technol. 199 (2016) 49–58, https://doi.org/10.1016/J.BIORTECH.2015.08.061.

[37] D. Salvachúa, A. Prieto, M. López-Abelairas, T. Lu-Chau, Á.T. Martínez, M.J. Martínez, Fungal pretreatment: an alternative in second-generation ethanol from wheat straw, Bioresour. Technol. 102 (16) (2011), https://doi.org/10.1016/j.biortech.2011.05.027.

[38] J. Yu, J. Zhang, J. He, Z. Liu, Z. Yu, Combinations of mild physical or chemical pretreatment with biological pretreatment for enzymatic hydrolysis of rice hull, Bioresour. Technol. 100 (2) (2009), https://doi.org/10.1016/j.biortech.2008.07.025.

[39] M.S. Islam, C. Zhang, K.Y. Sui, C. Guo, C.Z. Liu, Coproduction of hydrogen and volatile fatty acid via thermophilic fermentation of sweet sorghum stalk from co-culture of Clostridium thermocellum and Clostridium thermosaccharolyticum, Int. J. Hydrog. Energy 42 (2) (2017), https://doi.org/10.1016/j.ijhydene.2016.09.117.

[40] L. Zhang, Y. Li, X. Liu, N. Ren, J. Ding, Lignocellulosic hydrogen production using dark fermentation by Clostridium lentocellum strain Cel10 newly isolated from Ailuropoda melanoleuca excrement, RSC Adv. 9 (20) (2019) 11179–11185, https://doi.org/10.1039/C9RA01158G.

[41] N.J. Feng, H.M. Zhai, Y.Z. Lai, On the chemical aspects of the biodelignification of wheat straw with Pycnoporus sanguineus and its combined effects with the presence of *Candida tropicalis*, Ind. Crop. Prod. 91 (2016), https://doi.org/10.1016/j.indcrop.2016.07.035.

[42] S.Q. Tian, R.Y. Zhao, Z.C. Chen, Review of the pretreatment and bioconversion of lignocellulosic biomass from wheat straw materials, in: Renewable and Sustainable Energy Reviews 91, Elsevier Ltd., 2018, pp. 483–489, https://doi.org/10.1016/j.rser.2018.03.113.

[43] W. Song, L. Ding, M. Liu, J. Cheng, J. Zhou, Y.Y. Li, Improving biohydrogen production through dark fermentation of steam-heated acid pretreated Alternanthera philoxeroides by mutant Enterobacter aerogenes ZJU1, Sci. Total Environ. 716 (2020), https://doi.org/10.1016/j.scitotenv.2019.134695.

[44] T. Zhang, D. Jiang, H. Zhang, Y. Jing, N. Tahir, Y. Zhang, Q. Zhang, Comparative study on bio-hydrogen production from corn stover: photo-fermentation, dark-fermentation and dark-photo co-fermentation, Int. J. Hydrog. Energy 45 (6) (2020), https://doi.org/10.1016/j.ijhydene.2019.04.170.

[45] N. Pradhan, L. Dipasquale, G. D'Ippolito, A. Panico, P.N.L. Lens, G. Esposito, A. Fontana, Hydrogen production by the thermophilic bacterium Thermotoga neapolitana, Int. J. Mol. Sci. 16 (6) (2015) 12578–12600, https://doi.org/10.3390/ijms160612578. MDPI AG.

[46] J.R. Weil, A. Sarikaya, S.L. Rau, J. Goetz, C.M. Ladisch, M. Brewer, R. Hendrickson, M.R. Ladisch, Pretreatment of corn fiber by pressure cooking in water, Appl. Biochem. Biotechnol. Part A Enzyme Eng. Biotechnol. 73 (1) (1998), https://doi.org/10.1007/BF02788829.

[47] E. Palmqvist, B. Hahn-Hägerdal, Fermentation of lignocellulosic hydrolysates. I: inhibition and detoxification, Bioresour. Technol. 74 (1) (2000) 17–24, https://doi.org/10.1016/S0960-8524(99)00160-1.

[48] M. Quéméneur, M. Bittel, E. Trably, C. Dumas, L. Fourage, G. Ravot, J.P. Steyer, H. Carrère, Effect of enzyme addition on fermentative hydrogen production from wheat straw, Int. J. Hydrog. Energy 37 (14) (2012a) 10639–10647, https://doi.org/10.1016/J.IJHYDENE.2012.04.083.

[49] C.M. Pan, H.C. Ma, Y.T. Fan, H.W. Hou, Bioaugmented cellulosic hydrogen production from cornstalk by integrating dilute acid-enzyme hydrolysis and dark fermentation, Int. J. Hydrog. Energy 36 (8) (2011), https://doi.org/10.1016/j.ijhydene.2011.01.114.

[50] P.J. Weimer, J.B. Russell, R.E. Muck, Lessons from the cow: what the ruminant animal can teach us about consolidated bioprocessing of cellulosic biomass, Bioresour. Technol. 100 (21) (2009), https://doi.org/10.1016/j.biortech.2009.04.075.

[51] S. Manish, R. Banerjee, Comparison of biohydrogen production processes, Int. J. Hydrog. Energy 33 (1) (2008) 279–286, https://doi.org/10.1016/j.ijhydene.2007.07.026.

[52] N. Nemestóthy, K. Bélafi-Bakó, P. Bakonyi, Enhancement of dark fermentative H_2 production by gas separation membranes: a review, in: Bioresource Technology, vol. 302, Elsevier Ltd, 2020, https://doi.org/10.1016/j.biortech.2020.122828.

[53] A.M. Nor Azira, A. Umi Aisah, Purification of biohydrogen from fermentation gas mixture using two-stage chemical absorption, E3S Web of Conferences 90 (2019), https://doi.org/10.1051/e3sconf/20199001012.

[54] M. Aslam, R. Ahmad, M. Yasin, A.L. Khan, M.K. Shahid, S. Hossain, Z. Khan, F. Jamil, S. Rafiq, M.R. Bilad, J. Kim, G. Kumar, Anaerobic membrane bioreactors for biohydrogen production: recent developments, challenges and perspectives, Bioresour. Technol. 269 (2018), https://doi.org/10.1016/j.biortech.2018.08.050.

[55] H. Yang, B. Shi, H. Ma, L. Guo, Enhanced hydrogen production from cornstalk by dark- and photo-fermentation with diluted alkali-cellulase two-step hydrolysis, Int. J. Hydrog. Energy 40 (36) (2015) 12193–12200, https://doi.org/10.1016/J.IJHYDENE.2015.07.062.

[56] A. Bisaillon, J. Turcot, P.C. Hallenbeck, The effect of nutrient limitation on hydrogen production by batch cultures of *Escherichia coli*, Int. J. Hydrog. Energy 31 (11) (2006), https://doi.org/10.1016/j.ijhydene.2006.06.016.

[57] S.H. Kim, S.K. Han, H.S. Shin, Optimization of continuous hydrogen fermentation of food waste as a function of solids retention time independent of hydraulic retention time, Process Biochem. 43 (2) (2008), https://doi.org/10.1016/j.procbio.2007.11.007.

[58] H. Fang, H. Liu, Granulation of a hydrogen-producing acidogenicsludge. http://hub.hku.hk/handle/10722/110866, 2001.

[59] H.M.Y. Nazlina, A.R.N. Aini, F. Ismail, M.Z.M. Yusof, M.A. Hassan, Effect of different temperature, initial pH and substrate composition on biohydrogen production from food waste in batch fermentation, Asian J. Biotechnol. 1 (2) (2009), https://doi.org/10.3923/ajbkr.2009.42.50.

[60] B.F. Belokopytov, K.S. Laurinavichius, T.v. Laurinavichene, M.L. Ghirardi, M. Seibert, A.A. Tsygankov, Towards the integration of dark- and photo-fermentative waste treatment. 2. Optimization of starch-dependent fermentative hydrogen production, Int. J. Hydrog. Energy 34 (8) (2009), https://doi.org/10.1016/j.ijhydene.2009.02.042.

[61] A. Cakr, S. Ozmihci, F. Kargi, Comparison of bio-hydrogen production from hydrolyzed wheat starch by mesophilic and thermophilic dark fermentation, Int. J. Hydrog. Energy 35 (24) (2010) 13214–13218, https://doi.org/10.1016/J.IJHYDENE.2010.09.029.

[62] J. Masset, S. Hiligsmann, C. Hamilton, L. Beckers, F. Franck, P. Thonart, Effect of pH on glucose and starch fermentation in batch and sequenced-batch mode with a recently isolated strain of hydrogen-producing Clostridium butyricum CWBI1009, Int. J. Hydrog. Energy 35 (8) (2010), https://doi.org/10.1016/j.ijhydene.2010.01.061.

[63] N. Kumar, D. Das, Continuous Hydrogen Production by Immobilized *Enterobacter cloacae* IIT-BT 08 Using Lignocellulosic Materials as Solid Matrices, 2001. www.elsevier.com/locate/enzmictec.

[64] A.A. Zeidan, P. Rådström, E.W.J. van Niel, Stable coexistence of two Caldicellulosiruptor species in a de novo constructed hydrogen-producing co-culture, Microb. Cell Factories 9 (2010) 102, https://doi.org/10.1186/1475-2859-9-102.

[65] N. Anh, T. Viet Bui, T. Anh Ngo, H. Thi Viet Bui, Biohydrogen production using immobilized cells of hyperthermophilic eubacterium Thermotoga neapolitana on porous glass beads, J. Technol. Innovat. Renew. Energy 2 (3) (2013) 231–238, https://doi.org/10.6000/1929-6002.2013.02.03.4.

[66] X. Liu, Y. Zhu, S.T. Yang, Construction and characterization of ack deleted mutant of Clostridium tyrobutyricum for enhanced butyric acid and hydrogen production, Biotechnol. Prog. 22 (5) (2006) 1265–1275, https://doi.org/10.1021/BP060082G.

[67] H. Yang, L. Guo, F. Liu, Enhanced bio-hydrogen production from corncob by a two-step process: dark- and photo-fermentation, Bioresour. Technol. 101 (6) (2010), https://doi.org/10.1016/j.biortech.2009.10.078.

[68] E. Elbeshbishy, H. Hafez, B.R. Dhar, G. Nakhla, Single and combined effect of various pretreatment methods for biohydrogen production from food waste, Int. J. Hydrog. Energy 36 (17) (2011), https://doi.org/10.1016/j.ijhydene.2011.02.067.

[69] T.A.D. Nguyen, K.R. Kim, M.S. Kim, S.J. Sim, Thermophilic hydrogen fermentation from Korean rice straw by Thermotoga neapolitana, Int. J. Hydrog. Energy 35 (24) (2010), https://doi.org/10.1016/j.ijhydene.2009.11.112.

[70] A.M. Lopez-Hidalgo, Z.D. Alvarado-Cuevas, A. de Leon-Rodriguez, Biohydrogen production from mixtures of agro-industrial wastes: chemometric analysis, optimization and scaling up, Energy 159 (2018), https://doi.org/10.1016/j.energy.2018.06.124.

[71] R. Zagrodnik, K. Seifert, Direct fermentative hydrogen production from cellulose and starch with mesophilic bacterial consortia, Pol. J. Microbiol. 69 (1) (2020), https://doi.org/10.33073/pjm-2020-015.

[72] W. Li, C. Cheng, G. Cao, N. Ren, Enhanced biohydrogen production from sugarcane molasses by adding Ginkgo biloba leaves, Bioresour. Technol. 298 (2020), https://doi.org/10.1016/j.biortech.2019.122523.

[73] V.L. Pachapur, S.J. Sarma, S.K. Brar, Y. le Bihan, G. Buelna, M. Verma, Biohydrogen production by co-fermentation of crude glycerol and apple pomace hydrolysate using co-culture of Enterobacter aerogenes and clostridium butyricum, Bioresour. Technol. 193 (2015), https://doi.org/10.1016/j.biortech.2015.06.095.

[74] M. Cha, D. Chung, J.G. Elkins, A.M. Guss, J. Westpheling, Metabolic engineering of Caldicellulosiruptor bescii yields increased hydrogen production from lignocellulosic biomass, Biotechnol. Biofuel. 6 (1) (2013), https://doi.org/10.1186/1754-6834-6-85.

[75] W. Guo, J. Sheng, X. Feng, Mini-review: in vitro metabolic engineering for biomanufacturing of high-value products, in: Computational and Structural Biotechnology Journal vol 15, Elsevier B.V, 2017, pp. 161–167, https://doi.org/10.1016/j.csbj.2017.01.006.

[76] B. Sana, K.H.B. Chia, S.S. Raghavan, B. Ramalingam, N. Nagarajan, J. Seayad, F.J. Ghadessy, Development of a genetically programed vanillin-sensing bacterium for high-throughput screening of lignin-degrading enzyme libraries, Biotechnol. Biofuels 10 (1) (2017), https://doi.org/10.1186/s13068-017-0720-5.

Chapter 18

Theoretical insight into methanol sono-conversion for hydrogen production

Aissa Dehane, Leila Nemdili, Slimane Merouani and Atef Chibani

Laboratory of Environmental Process Engineering, Department of Chemical Engineering, Faculty of Process Engineering, University Salah Boubnider Constantine 3, El Khroub, Algeria

1. Introduction

Over the past several decades, the consumption of fossil fuels has continuously expanded due to the rapid rise in global population and the industrial revolution of today's society [1]. Nowadays, governments recognize the need of using sustainable energy sources to increase their energy reserves and reduce carbon emissions [2]. In order to satisfy present and future energy needs without increasing carbon emissions, hydrogen can be used as an energy vector. Hydrogen is abundant in the environment and has the largest energy capacity (i.e., 120.7 kJ g^{-1}) of any known fuel [3]. Indeed, gasification, partial oxidation, and steam reforming were the main ways that hydrogen (H_2) was made from fossil fuels [4–6]. Even while these techniques are incredibly effective, they are not environmentally friendly because they release CO_2 [7]. Alternative methods for producing hydrogen, such as water electrolysis, biological photosynthesis, and photocatalysis, have been developed as clean technologies [1,8,9].

Scientists have been attracted to employ ultrasound as a clean technique for hydrogen production because of the high-energy phenomena generated by power ultrasound propagation in water (water sonolysis) [10–20]. Furthermore, pure water sonolysis has been determined to provide lower hydrogen yields (<40 μM min^{-1}) [10]. To overcome this issue, a variety of chemicals can be added to enhance the sonolytic hydrogen producing process, either in the saturating gas or the sonicated water. For example, the inclusion of hydrocarbons including methane and ethane massively improves the generation of hydrogen [21]. Methane and ethane can pass through the bubble's hot gas phase, and their pyrolysis produces an excessive quantity of hydrogen [21]. Alcohols, on the other hand, perform the same role as CH_4 within the bubble and follow a similar pyrolytic process [22,23]. On the other hand, the concentration of these additives needs to be carefully monitored since excessive amounts of it can lower (or suppress) the generation of of hydrogen.

Recently, many investigations have been carried out on the sono- production of hydrogen in the presence of methanol [10,24–27]. Generally, the oxidation process was decreased when methanol is included in the sonicated solution, although hydrogen generation is considerably increased [22,28,29]. In order to (i) increase the yield of hydrogen and (ii) improve the conversion capacity of methanol inside the acoustic bubble, several researchers have been interested in those findings.

In this chapter, the sono-production of hydrogen by methanol sonolysis are discussed. The impact of the crucial medium and operational conditions (such as methanol concentration, argon saturation degree, ultrasound frequency, and acoustic power) on the hydrogen generation yield by methanol sono-pyrolysis within an acoustic bubble was highlighted. Before starting, an overview of the instruments and concepts of sonochemistry is presented.

2. The sonochemical process

Fig. 18.1 depicts the most physical and chemical effects of ultrasound in aqueous solution. Cavitation bubbles created by water sonolysis grow and collapse violently in a matter of microseconds, creating intense pressures and temperatures

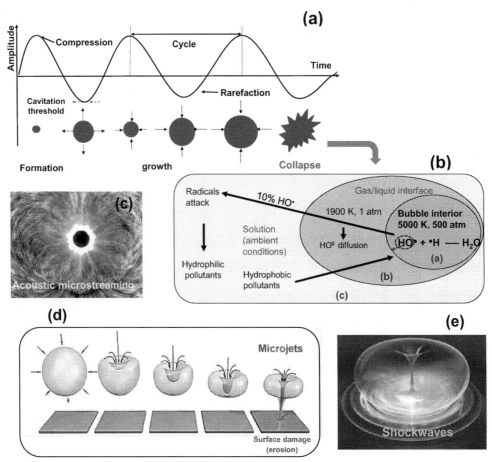

FIGURE 18.1 Acoustic cavitation (A) and its chemical (B) and physical consequences (C, D, and E). *From A. Dehane, S. Merouani, O. Hamdaoui, M. Ashokkumar, Sonochemical and sono-assisted reduction of carbon dioxide: a critical review, Chem. Eng. Process—Process Intensif. (2022) 109075. https://doi.org/10.1016/j.cep.2022.109075.*

(about ∼5200 K and ∼500 atm) inside the bubbles [30,31]. Due to the this phenomena, known as acoustic cavitation, ultrasound has several applications, ranging from cleaning [32], polymers synthesizing [33], pollutants destruction [34,35], and nanomaterials manufacturing [36,37] to biomedical uses [38–40] and food sciences [41,42]. The bubbles may be inertial (i.e., transient) and develop for one acoustic cycle (or less) before exploding during the compression phase of the wave, or they may be stable and bounce around their equilibrium (ambient) radius for multiple cycles [43].

In fact, inertial cavitation provides the most promising sonication performance [43]. The fast collapse of these bubbles is quite adiabatic, which leads to high energy chemistry in the interior of the bubble and its surrounding (Fig. 18.1A and B). Dissolved gases and evaporating H_2O molecules decompose (react) to form many reactive species including •OH, H•, HO_2• and O_3 [44,45]. Radicals can recombine, react with gaseous species (inside the bubble), or diffuse beyond the bubble [46]. Organic or inorganic volatile substances which have evaporated can also be pyrolyzed or incinerated inside the hot cavity. As a result, the sonochemical reaction zones are (i) the bubble gas phase (Fig. 18.1B), (ii) the bubble/solution interface, and (iii) the solution bulk, while the oxidation mechanisms are (i) the gas-phase pyrolytic oxidation for volatile compounds and (ii) reaction with radicals, i.e., HO•, at the bubble–solution interface and in the liquid bulk for nonvolatile compounds [47,48]. Only around 10% of radicals may reach the solution bulk, although the •OH concentration is greater at the liquid shell of the acoustic bubble [49–51].

The major sonication-related physical consequences are the acoustic microjet (Fig. 18.1D), the microstreaming field (Fig. 18.1C) and the shock wave damage (Fig. 18.1E) [47]. A microjet, which is a stream of quickly moving liquid oriented to the surface of the solid, is generated during implosion as a result of a self-reinforcing bubble deformation [52]. Localized erosion/pitting of the surface is caused by the high microjet velocity (i.e., ∼100 m s^{-1} [53]). Photographic confirmation of the ultrasonic pitting was obtained [52], and the solid surface area has improved due to the development of new fine surface material [52]. Shock waves are also created by the cavity collapse [54]. Shock waves remove particles that are only weakly attached to a solid surface and separate slightly agglutinated grains through existing fissures (increasing surface area) [54]. Sonoluminescence was also produced by the extreme bubble implosion.

For conducting sonochemical reactions, two types of ultrasonic reactors are most frequently used [55–58]: (i) probe type (also known as horn), in which the solution is directly irradiated by a submerged horn (Fig. 18.2A), and (ii) a standing wave–based type, in which the transducer is attached to the reactor's bottom surface. For the standing wave type, there are two configurations: the indirect sonication bath system (Fig. 18.2B), in which the beaker solution is immersed, and the direct sonication bath (Fig. 18.2C), within which the ultrasonic waves move directly through the solution. All kinds of ultrasonic reactors are constructed with unique cooling/heating systems in order to keep the sonicated liquid's temperature constant.

3. Some literature data

Experimental research demonstrates hydrogen's production (from water sonolysis) using diluted aqueous alcohol solutions. Water–methanol mixtures were sonolyzed at 1 MHz in an argon environment by Buettner et al. [59]. H_2 was produced in deionized water at a rate of 22 $\mu M\ min^{-1}$. When methanol is added to the solution, the H_2 rate rises to 100 $\mu M\ min^{-1}$ at 5% (v/v) of methanol and 150 $\mu M\ min^{-1}$ at 10% (v/v), but as methanol concentrations increase, the molar yield of H_2 gradually decreases until no H_2 was identified at around 80% (v/v) of methanol. Penconi et al. [60] found a 1.4-fold increase in the rate of H_2 production while utilizing 38 kHz ultrasound in water and water/ethanol (20% v/v) under argon saturation. Rassokhin et al. [22] reported that when methanol concentration increased from 0.001 to 0.50 M in the sonicated solutions, the generation rates of the sonolysis products (H_2, CH_4, and CO) increased (H_2: from 33.3 to 167.0 $M\ min^{-1}$), where the molar yield of these species is as follows: $H_2 > CO > CH_4$. Rassokhin et al. [22] have

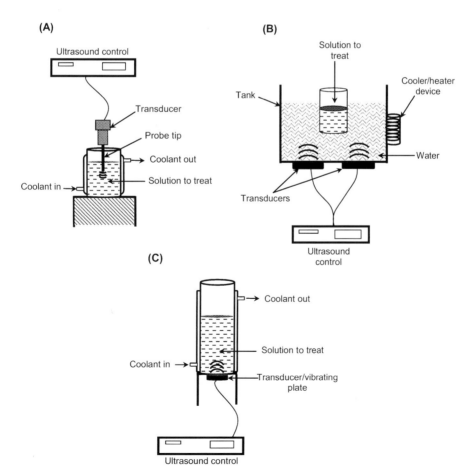

FIGURE 18.2 Schemes of the main types of lab-scale sonoreactors. (A) Probe system, (B) bath system (indirect sonication), and (C) standing wave reactor with direct sonication. *From S. Merouani, O. Hamdaoui, Sonochemical treatment of textile wastewater, in: M.P. Inamuddin, A. Asiri (Eds.), Water Pollution and Remediation: Photocatalysis, Springer-Nature Switzerland, 2021. https://doi.org/10.1007/978-3-030-54723-3_5.*

demonstrated that the optimal medium temperature for the maximum hydrogen production is retrieved depending on the mole fraction of methanol in the bulk solution ($[H_2]_{max} = 200$ μM min^{-1} at ~30°C and $x_{MeOH} = 0.018$), where the rise of methanol concentration reduces the temperature at which the maximal H_2 formation rate is reached.

Various numerical investigations have focused on sonochemical hydrogen production [10,12,24,26,27]. These researches have demonstrated promising results of clean and efficient hydrogen production. Recent reviews on sonochemical hydrogen production by Merouani et al. [10,61] are noteworthy. In a numerical investigation of the influence of the saturating gas (O_2 and Ar) on the sono-pyrolysis of CH_3OH, Kerboua et al. [17] found that solution containing 40% CH_3OH produced the highest molar yield, despite of the amount of argon used. Kerboua et al. [17] revealed that the highest molar output is achieved at a methanol concentration of 40% and an argon molar fraction of 40%. According to Kerboua et al. [26], a solution with 20% methanol (v/v) saturated with 70% concentration of argon provides the best results for H_2 production and CH_3OH conversion. As a result, 99.9% of the methanol is eliminated and roughly 60% of the methanol is converted to hydrogen. Dehane et al. [62] investigated the impact of methanol dosage (in the bulk phase) on the maximum sonochemical performance for hydrogen production, CH_3OH consumption, and the range of active bubbles using a comprehensive mathematical model for single bubble sono-pyrolysis of methanol. They revealed that the range of active bubble sizes for CH_3OH consumption is mostly unaffected by the dose of aqueous phase methanol (0%–100% (v/v)). Nevertheless, in the case of hydrogen production, the active bubble range progressively decrease for methanol concentrations greater than 20%. The greatest performance for CH_3OH conversion, hydrogen production and the width of active bubble sizes (for H_2 production and CH_3OH disintegration) is achieved at 80% argon (in the initial gas matrix) with a methanol concentration of 7%–20%. However, Dehane et al.'s research [62] is not sufficient because all findings were obtained for a single matrix of operational parameters (frequency: 355 kHz, In = 1 W cm^{-2}, and $T_{liq} = 20$°C). Theoretical researches [17,26,62] on methanol sono-decomposition indicate that the impacts of operational conditions, such as ultrasound frequency, acoustic intensity and bulk temperature have not yet been studied.

4. Methanol pyrolysis model

The model was developed by our research group and it is available with results in [62,63]. Only the model's principle outcomes will be discussed herein. Briefly, the principal mechanism of methanol degradation is pyrolysis within acoustic bubbles due to its high volatility [62]. The model is based on a set of ordinary differential equations that take into account chemical processes, thermal conduction both within and outside of a bubble, and nonequilibrium evaporation and condensation of water vapor and methanol at the bubble wall. Bubble interactions are not considered. The chemical pathways used to examine the interior bubble chemistry for the Ar–O_2–CH_3OH bubble are shown in Table 18.1. Applying relevant experimental data from the literature [62,63], the model has been validated. For instance, the model's predicted bubble temperatures for an argon–methanol aqueous solution at various methanol dosages were compared to those empirically reported by Rae et al. [64]. Especially for lower alcohol dosages, the model effectively reflects the data on methanol's effect on bubble temperature [63]. The model predicted 4570 K for 0 mM methanol compared to Rae et al. [64] data of 4600 K, indicating a divergence of less than 1% [63]. For 60 and 150 mM of methanol, unappreciable deviations of ~1.8 and 2.5% were also obtained. Nevertheless, with higher CH_3OH concentrations (> 200 mM), a comparatively higher deviation is observed. Furthermore, the model performed an excellent fit of reflecting the overall tendency of the bubble temperature variation with alcohol dosage. Comparing the results of H_2 production obtained theoretically using the model to those retrieved experimentally by Buettner et al. [59] at 1000 kHz [62,63] has provided additional evidence of the model's accuracy.

5. Optimal gas conditions

In order to clearly understand how methanol affects hydrogen sono-production, the optimum bubble composition (argon content) should be established. At 355 kHz and 1 w/cm^2, Dehane et al. [62,63] estimated the evolution of hydrogen production by a single bubble as well as the corresponding methanol conversion as a function of the molar quantity of argon (within the bubble) and methanol concentration in the bulk liquid [% (v/v)] (Fig. 18.3). For the majority of methanol concentrations, the optimum values are found at 80% and 60%–80% of argon mole fraction (within the bubble) for the generation of hydrogen (Fig. 18.3A) and the conversion of methanol (Fig. 18.3B), respectively. At 3% (v/v) of CH_3OH and 70% of argon mole fraction (Fig. 18.3B), which corresponds to 9.46×10^{-19} mol for the highest generation of hydrogen (Fig. 18.3A), the maximum conversion of methanol (1.38×10^{-17} mol) is recovered.

On the other hand, at 7% (v/v) of methanol and 80% argon (mole fraction), the maximum yield of hydrogen is achieved, providing a value of 1.13×10^{-18} mol (19.45% of increase compared to the previous case (Fig. 18.3A). In comparison to the

TABLE 18.1 Adopted reaction scheme inside a collapsing $Ar-O_2-CH_3OH$ bubble.

N°	Reaction	A	N	E_a	ΔH
1	$H_2O + M \rightarrow H^\bullet + {}^\bullet OH + M$	1.912×10^7	−1.83	28.35	508.82
2	$CH_3OH + O_2 \rightarrow CH_2OH + H_2O_2$	2.0×10^7	0.0	1.88×10^5	196.492
3	$CH_3OH + HO_2 \rightarrow CH_2OH^\bullet + H_2O_2$	8.0×10^7	0.0	8.11×10^4	−35.508
4	$CH_3OH + H^\bullet \rightarrow CH_2OH^\bullet + H_2$	1.35×10^{-3}	3.2	1.46×10^4	−34.057
5	${}^\bullet OH + {}^\bullet OH \rightarrow HO_2{}^\bullet + H^\bullet$	1.08×10^{05}	0.61	1.51×10^5	154.957
6	$HO_2{}^\bullet + {}^\bullet OH \rightarrow H_2O + O_2$	2.89×10^{07}	0.00	-2.08×10^{03}	−291.451
7	$HO_2{}^\bullet + O \rightarrow {}^\bullet OH + O_2$	2.00×10^{07}	0.00	0.00	−224.131
8	$HO_2{}^\bullet + H^\bullet \rightarrow H_2O + O$	3.10×10^{07}	0.00	7.20×10^{03}	−22.957
9	$HO_2{}^\bullet + H^\bullet \rightarrow H_2 + O_2$	1.66×10^{07}	0.00	3.44×10^{03}	−230.293
10	$HO_2{}^\bullet + H^\bullet \rightarrow 2\,{}^\bullet OH$	7.08×10^{07}	0.00	1.23×10^{03}	−155.613
11	$CH_2OH^\bullet + O_2 \rightarrow CH_2O + HO_2{}^\bullet$	5.00×10^{06}	0.00	0.00	−79.404
12	$CH_3OH + {}^\bullet OH \rightarrow CH_3O^\bullet + H_2O$	4.4×10^{00}	2.0	6.3×10^3	−54.215
13	$CH_3OH + {}^\bullet OH \rightarrow CH_2OH^\bullet + H_2O$	1.44×10^{00}	2.0	-3.51×10^3	−95.215
14	$CH_3O^\bullet + M \rightarrow CH_2OH^\bullet + M$	1.00×10^{08}	0.00	8.00×10^{04}	−38.000
15	$CH_3O^\bullet + M \rightarrow CH_2O + H^\bullet + M$	7.78×10^{07}	0.00	5.65×10^{04}	126.297
16	$CH_3O^\bullet + O_2 \rightarrow CH_2O + HO_2{}^\bullet$	4.28×10^{-19}	7.60	-1.48×10^{04}	−117.404
17	$CH_2O + HO_2{}^\bullet \rightarrow HCO^\bullet + H_2O_2$	4.11×10^{-02}	2.50	4.27×10^{04}	−48.976
18	$CH_2O + {}^\bullet OH \rightarrow HCO^\bullet + H_2O$	3.90×10^{04}	0.89	1.70×10^{03}	−128.159
19	$CH_2O + O \rightarrow HCO^\bullet + {}^\bullet OH$	3.50×10^{07}	0.00	1.47×10^{04}	−60.839
20	$CH_2O + H^\bullet \rightarrow HCO^\bullet + H_2$	5.74×10^{01}	1.90	1.15×10^{04}	−67.001
21	$HCO^\bullet + O_2 \rightarrow CO + HO_2{}^\bullet$	7.58×10^{06}	0.00	1.72×10^{03}	−140.530
22	$HCO^\bullet + M \rightarrow CO + H^\bullet + M$	1.86×10^{11}	−1.00	7.11×10^{04}	−65.171
23	$CO + {}^\bullet OH \rightarrow CO + H^\bullet + CO_2$	4.40×10^{00}	1.50	-3.10×10^{03}	−212.853
24	$2\,HO_2{}^\bullet \rightarrow H_2O_2 + O_2$	3.02×10^{06}	0.00	5.80×10^{03}	−212.268
25	$H_2O_2 + M \rightarrow 2\,{}^\bullet OH + M\ k_0\ k_\infty$	8.15×10^{17} 2.62×10^{13}	−1.92 −1.39	2.08×10^{05} 2.15×10^{05}	262.356
26	$H^\bullet + O_2 + M \rightarrow HO_2{}^\bullet + M\ k_0\ k_\infty$	5.75×10^{07} 4.65×10^{00}	−1.40 0.44	0.00 0.00	−205.701
27	$H_2 + O \rightarrow {}^\bullet OH + H^\bullet$	5.06×10^{-02}	2.67	2.63×10^{04}	6.162
28	$H^\bullet + O_2 \rightarrow {}^\bullet OH + O$	3.52×10^{10}	−0.70	7.14×10^{04}	68.518
29	$CH_2OH^\bullet + H^\bullet \rightarrow CH_2O + H_2$	3×10^7	0.0	0.0	−309.578
30	$2\,{}^\bullet OH \rightarrow H_2O_2$	1.08×10^5	0.61	1.51×10^5	154.957
31	$CH_2OH^\bullet + {}^\bullet OH \rightarrow CH_2O + H_2O$	2.4×10^7	0.0	0.0	−370.7
32	${}^\bullet CH_3 + HO_2{}^\bullet \rightarrow CH_4 + O_2$	3.61×10^6	0.0	0.0	−233.852

From A. Dehane, S. Merouani, O. Hamdaoui, Methanol sono-pyrolysis for hydrogen recovery: effect of methanol concentration under an argon atmosphere, Chem. Eng. J. 426 (2021) 130251. https://doi.org/10.1016/j.cej.2021.130251.

previous case (1.38×10^{-17} mol), this production (1.13×10^{-18} mol) results in a 1.3×10^{-17} mol methanol conversion, a reduction of 5.79% (Fig. 18.3B). Over the full range of methanol concentration (0%−100% (v/v)), it is indicated that the majority of maximal productions for the total yield of acoustic cavitation are produced at 80% (mole fraction) of argon inside the bubble [62]. Furthermore, the overall production of the bubble was slightly less than the maximum conversion of methanol. Figs. 18.3A and B show that the bubble performs at its maximum efficiency at 80% argon and 20% oxygen, with 7%

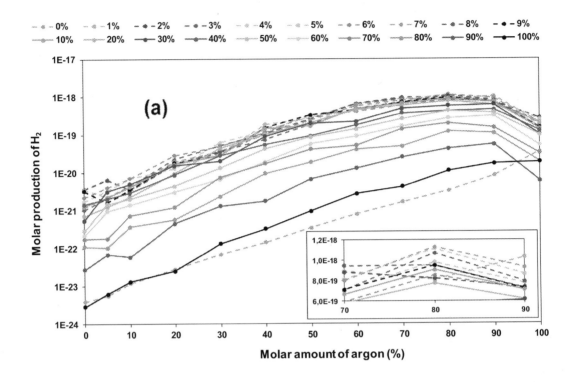

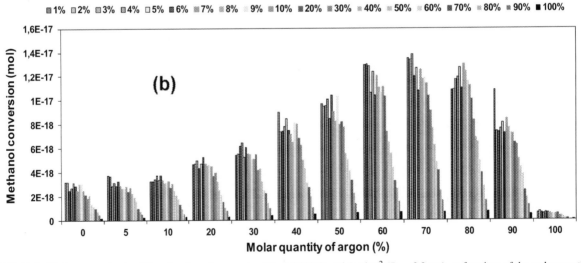

FIGURE 18.3 Hydrogen production (A) and methanol conversion (B), at 355 kHz and 1 w/cm² ($R_0 = 3.2$ μm), as functions of the molar quantity of argon (inside the bubble) and methanol concentration in the bulk liquid [% (v/v)]. *From A. Dehane, S. Merouani, O. Hamdaoui, Methanol sono-pyrolysis for hydrogen recovery: effect of methanol concentration under an argon atmosphere, Chem. Eng. J. 426 (2021) 130251. https://doi.org/10.1016/j.cej. 2021.130251.*

(v/v) methanol and 80% argon being the optimal ratios for maximum hydrogen generation (1.13×10^{-18} mol) and methanol conversion (1.3×10^{-17} mol). The greatest amount of hydrogen produced at this level, which is 1.13×10^{-18} mol, is raised by 32.7 times compared to the maximum value obtained in the absence of methanol (and 100% Ar), i.e., 3.46×10^{-20} mol.

Figs. 18.3A and B indicate that the concentration of methanol (in solution) may be increased until 20%, so at that point the yield of hydrogen rises by 22.4 (7.74×10^{-19} mol) in comparison to the situation in which there is no methanol (and 100% Ar). Methanol's conversion within the bubble is 22.3% (1.12×10^{-17} mol) and 13.84% (1.01×10^{-17} mol), respectively, at 10% and 20% (v/v) of methanol in the bulk solution, compared to the conversion rate that occurred with its maximum at 7% (1.3×10^{-17} mol).

6. Ultrasound frequency effect

Dehane et al. [63] have evaluated the yield of H_2 generation and CH_3OH conversion at 1 W cm^{-2} in the presence (20%, v/v) and absence of methanol in the liquid phase as functions of ultrasonic frequency (from 213 to 1000 kHz) (Fig. 18.4A and B). Initially, as expected, an increase in wave frequency has a negative impact on the molar generation of hydrogen, the peak temperature of the bubble and the conversion of methanol, either in the presence or absence of methanol (Fig. 18.4A and B). These phenomena have been widely discussed in several experimental and theoretical

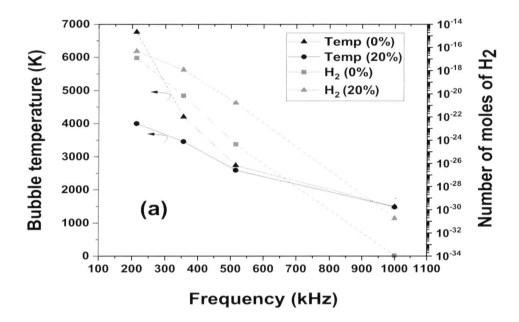

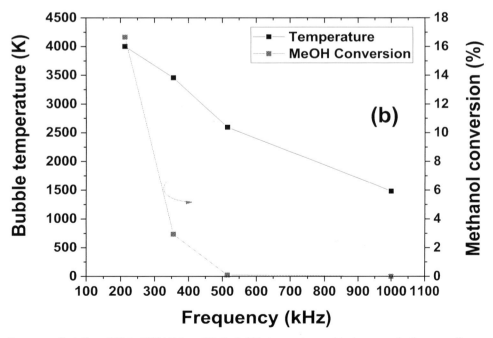

FIGURE 18.4 Frequency effect (from 213 to 1000 kHz) on (A) the bubble temperature and hydrogen production per collapse, and (B) methanol conversion (I_n = 1 W cm^{-2}, T_{liq} = 20°C). *From A. Dehane, S. Merouani, A. Chibani, O. Hamdaoui, Influence of processing conditions on hydrogen sonoproduction from methanol sono-conversion: a numerical investigation with a validated model, Chem. Eng. Process—Process Intensif. 179 (2022) 109080. https://doi.org/10.1016/j.cep.2022.109080.*

investigations [65–68]. This tendency was mainly attributed to the decrease of bubble collapse's intensity (and activity) proportionally with the increase in wave frequency [65,67,69]. In other words, a milder collapse occurs as the wave frequency rises because less water vapor and, if any methanol molecules remain present in solution, few of them enter the acoustic cavitation during the expansion phase. As a result, the bubble temperature at the end of its compression period will be lower, which implies that fewer molecules of hydrogen (and methanol conversion) will be produced by water vapor dissociation (and methanol combustion). Moreover, due to an increase in the bubble's heat capacity and the endothermic dissociation of methanol, the addition of methanol (20% in solution) drops the bubble's maximum temperature more (particularly at 213 and 355 kHz) [63,70]. The peak temperature of the bubble is slightly affected by the presence of methanol, as can be seen in Fig. 18.4A and B, for wave frequencies greater than or equal to 515 kHz. For example, at 515 and 1000 kHz, the maximal temperatures are 2593 K (2741 K in the absence of methanol) and 1481 K (1489 K in the absence of methanol), respectively. This may be attributed to a rise in ultrasonic frequency, where a reduced bubble lifetime results in less methanol being entrapped within the bubble at the moment of collapse. Because methanol increases the bubble's heat capacity and the number of endothermal reactions its existence will only have a minor impact on the intensity of bubble collapse (its maximum temperature) [65]. On the other hand, regardless of the applied frequency, the yield of H_2 in the presence of methanol is higher than that recovered in the absence of methanol. Experimentally, Rassokhin et al. [22] (724 kHz) and Buttner et al. [59] (1000 kHz) both revealed a similar trend. This behavior (increase in H_2 yielding) is mostly attributed to methyl alcohol's ability to scavenge hydroxyl radicals and its association with hydrogen atoms. As a result, the amount of $^\bullet OH$ radicals that react with hydrogen atoms is lowered proportionally as hydrogen generation is increased. When CH_3OH is present or absent, a close generation of H_2 is produced at 213 kHz. This is attributed to the significant drop in the bubble temperature (4000 K) in the methanol-containing system. Consequently, less H_2 is generated at 213 kHz. However, a sharp decline in hydrogen generation is shown for ultrasound frequencies higher than 355 kHz and in the presence of methanol: 1.16×10^{-18} mol at 355 kHz, 1.7×10^{-21} mol at 515 kHz, and 1.8×10^{-31} mol at 1000 kHz (Fig. 18 4A).

The maximal conversion of methanol (per collapse) is depicted in Fig. 18.4B as a function of its maximum quantity at the end of bubble growth. At first look, these conversions ($<16.7\%$) appear to be quite low in comparison to the recorded spectrum of bubble temperatures (4000–1481 K) throughout the whole frequency range (213–1000 kHz). Furthermore, it is evident from Fig. 18.4B that significant amounts of methanol are converted, particularly in the region of ultrasonic frequency from 213 (1.73×10^{-16} mol, 85.22%) to 355 kHz (7.22×10^{-18} mol, 71.2%), compared to the entire production of the bubble [from 2.03×10^{-16} to 1.52×10^{-22} mol]). The comparison of Fig. 18.4A and B indicates that the range of ultrasonic frequency between 213 and 355 kHz may provide the highest amount of methanol breakdown and hydrogen generation. This conclusion is supported, in particular, by the fact that the number of bubbles increases according to ultrasonic frequency increase [61,71,72], which may enhance the molar yield of H_2 and the conversion rate of methanol.

7. Acoustic intensity effect

Fig. 18.5B displays the molar yield of hydrogen generation and the bubble temperature as functions of acoustic intensity (0.7, 1, and 1.5 W cm^{-2}) at 355 kHz. At 0.7 W cm^{-2}, it is found that the bubble temperatures are almost the same in both cases, i.e., the presence and absence of CH_3OH, while the generation of hydrogen is higher when methanol is present in the bulk liquid. This obviously demonstrates that the molar generation of the acoustic bubble depends on both the maximum amount of water vapor and volatile species (methanol) present at the end of the rarefaction phase as well as the peak temperature of the bubble. In other words, even if the temperatures in the presence and absence of methanol were nearly the same, the favorable effects of methanol on the bubble (limiting $H^\bullet$ and $^\bullet OH$ recombination and promoting the production of H_2 molecules) exceed the temperature drop (resulting from the increase in its heat capacity and the combustion of methanol). A similar behavior was observed when volatile CCl_4 is present during the bubble implosion [73]. In the absence of methanol, an increase in acoustic intensity from 1 to 1.5 W/cm^2 results in an increase in the production of H_2 (and bubble temperature) from 6.92×10^{-21} mol (4205 K) to 7.36×10^{-18} mol (6646 K). Despite the rise in acoustic intensity from 1 (H_2: 1.16×10^{-18} mol, T_{max}: 3456 K) to 1.5 W cm^{-2} (H_2: 8.97×10^{-18} mol, T_{max}: 3801 K), it is clear that the presence of methanol amortizes the rise in the molar production of hydrogen (and the bubble's peak temperature). This is evidently due to the detrimental effects of evaporating methanol within the bubble at 1.5 W cm^{-2} on both the bubble temperature and the molar yield of H_2 (by raising the bubble heat capacity and the combustion of CH_3OH). This is due to the fact that increased acoustic intensity leads to higher maximum radius (R_{max}), which permits massive volumes of water vapor and methanol (more volatile than water) to penetrate into the bubble. As a result, thanks to the presence (within the bubble) of significant amounts of methanol and water vapor, a milder collapse is obtained. Similar results were obtained by Yasui et al. [74] for the maximum bubble temperature at 1 MHz and by Ferkous et al. [75] for the yield of $^\bullet OH$ radicals at

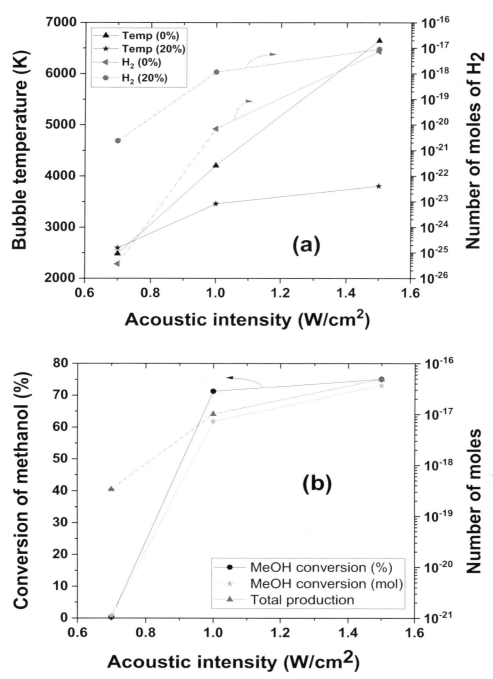

FIGURE 18.5 Acoustic intensity effect on (A) the bubble temperature and hydrogen production and (B) the quantity of converted methanol ($f = 355$ kHz, $T_{liq} = 293.15$ K). *From A. Dehane, S. Merouani, A. Chibani, O. Hamdaoui, Influence of processing conditions on hydrogen sonoproduction from methanol sono-conversion: a numerical investigation with a validated model, Chem. Eng. Process—Process Intensif. 179 (2022) 109080. https://doi.org/10.1016/j.cep.2022.109080.*

585, 860 and 1140 kHz. This finding shows that an optimum acoustic intensity of 1 W cm^{-2} is recommended for an efficient sono-generation of hydrogen.

According to Fig. 18.5B, the conversion of methanol is shown to be promoted between 0.7 W cm^{-2} (1.1×10^{-21} mol, 0.33% of the total bubble production) and 1 W cm^{-2} (7.22×10^{-18} mol, 71.24% of the total bubble yield). Nonetheless, increasing I_n to 1.5 W cm^{-2} seems inefficient for methanol decomposition (3.72×10^{-17} mol, or 75.09% of the total bubble production), owing to a decrease in the maximum bubble temperature at this ultrasonic intensity (1.5 W cm^{-2}) (3801 K, Fig. 18.5A). These outcomes are in line with the results of Alippi et al. [76] and Weissler et al. [77] for the yield of iodine in a sonicated solution of CCl_4 for electrical powers in the range of 200–450 W and from 0 to 600 W, respectively. It is clear from Figs. 18.4A and 18.5B that the acoustic intensity needs to be adequately adjusted for maximal H_2 production and CH_3OH conversion with respect to the supplied energy into the sonoreactor.

Acknowledgements

This study was supported by the Ministry of Higher Education and Scientific Research of Algeria (project No. A16N01UN250320220002) and the General Directorate of Scientific Research and Technological Development (GD-SRTD).

8. Conclusion

The acoustic production of hydrogen and methanol conversion can be controlled by the dissociation of methanol within bubbles. The optimal combination for the maximum H_2 generation and CH_3OH conversion is reached at 7% (v/v) of methanol and 80% of argon, which was used to achieve the highest sonochemical efficiency of bubbles.

On the other hand, the peak temperature of the bubble, the generation of hydrogen and the conversion of methanol all decrease proportionally with an increase in the frequency, whether methanol is present or not. The addition of methanol has a slight effect on the maximum temperature of the bubble for ultrasound frequencies equal to or higher than 515 kHz. Moreover, regardless of the employed frequency, the yield of H_2 in the presence of methanol is higher than that obtained when it is absent in the bulk liquid. In the range from 213 to 355 kHz, a maximal efficiency is observed for hydrogen production and methanol conversion. In addition, an accurate control of acoustic intensity is required for maximum H_2 generation and CH_3OH conversion in the sonicated solution.

References

[1] M. Kaddami, M. Mikou, F. Ezzahra Chakik, M. Kaddami, M. Mikou, Effect of operating parameters on hydrogen production by electrolysis of water, Int. J. Hydrogen Energy 42 (2017) 2–9, https://doi.org/10.1016/j.ijhydene.2017.07.015.

[2] G. Zini, P. Tartarini, Wind-hydrogen energy stand-alone system with carbon storage: modeling and simulation, Renew. Energy 35 (2010) 2461–2467, https://doi.org/10.1016/j.renene.2010.03.001.

[3] S. Merouani, O. Hamdaoui, Y. Rezgui, M. Guemini, Mechanism of the sonochemical production of hydrogen, Int. J. Hydrogen Energy 40 (2015) 4056–4064, https://doi.org/10.1016/j.ijhydene.2015.01.150.

[4] I. Dincer, C. Acar, Review and evaluation of hydrogen production methods for better sustainability, Int. J. Hydrogen Energy 40 (2014) 11094–11111, https://doi.org/10.1016/j.ijhydene.2014.12.035.

[5] A. Haryanto, S. Fernando, N. Murali, S. Adhikari, Current status of hydrogen production techniques by steam reforming of ethanol: a review, Energy Fuels 19 (2005) 2098–2106.

[6] I. Dincer, Green methods for hydrogen production, Int. J. Hydrogen Energy 37 (2011) 1954–1971, https://doi.org/10.1016/j.ijhydene.2011.03.173.

[7] I. Dincer, C. Acar, Innovation in hydrogen production, Int. J. Hydrogen Energy 42 (2017) 14843–14864, https://doi.org/10.1016/j.ijhydene.2017.04.107.

[8] M. Ni, M.K. Leung, D.Y.C. Leung, K. Sumathy, A review and recent developments in photocatalytic water-splitting using TiO_2 for hydrogen production, Renew. Sustain. Energy Rev. 11 (2007) 401–425, https://doi.org/10.1016/j.rser.2005.01.009.

[9] D. Das, T.N. Veziroglu, Advances in biological hydrogen production processes, Int. J. Hydrogen Energy 33 (2008) 6046–6057, https://doi.org/10.1016/j.ijhydene.2008.07.098.

[10] S. Merouani, O. Hamdaoui, The sonochemical approach for hydrogen production, Sustain. Green Chem. Process Their Allied Appl. Nanotechnol. Life Sci. (2020) 1–29, https://doi.org/10.1007/978-3-030-42284-4.

[11] H. Harada, Sonochemical reduction of carbon dioxide, Ultrason. Sonochem. 5 (1998) 73–77, https://doi.org/10.1016/S1350-4177(98)00015-7.

[12] H. Islam, O.S. Burheim, B.G. Pollet, Sonochemical and sonoelectrochemical production of hydrogen, Ultrason. Sonochem. 51 (2019) 533–555, https://doi.org/10.1016/j.ultsonch.2018.08.024.

[13] S.S. Rashwan, I. Dincer, A. Mohany, An investigation of ultrasonic based hydrogen production, Energy 205 (2020) 118006, https://doi.org/10.1016/j.energy.2020.118006.

[14] S.S. Rashwan, I. Dincer, A. Mohany, A unique study on the effect of dissolved gases and bubble temperatures on the ultrasonic hydrogen (sonohydrogen) production, Ulrason. Sonochem. 5 (2020) 20808–20819, https://doi.org/10.1016/j.ijhydene.2020.05.022.

[15] K. Kerboua, O. Hamdaoui, Energetic challenges and sonochemistry: a new alternative for hydrogen production? Curr. Opin. Green Sustain. Chem. 18 (2019) 84–89, https://doi.org/10.1016/j.cogsc.2019.03.005.

[16] K. Kerboua, O. Hamdaoui, Sonochemical production of hydrogen: enhancement by summed harmonics, Chem. Phys. (2018), https://doi.org/10.1016/j.chemphys.2018.11.019.

[17] K. Kerboua, O. Hamdaoui, Oxygen-argon acoustic cavitation bubble in a water-methanol mixture: effects of medium composition on sonochemical activity, Ultrason. Sonochem. 61 (2020) 104811, https://doi.org/10.1016/j.ultsonch.2019.104811.
[18] K. Kerboua, O. Hamdaoui, Numerical estimation of ultrasonic production of hydrogen: effect of ideal and real gas based models, Ultrason. Sonochem. 40 (2018), https://doi.org/10.1016/j.ultsonch.2017.07.005.
[19] S. Merouani, O. Hamdaoui, Y. Rezgui, M. Guemini, Computational engineering study of hydrogen production via ultrasonic cavitation in water, Int. J. Hydrogen Energy 41 (2015) 832−844, https://doi.org/10.1016/j.ijhydene.2015.11.058.
[20] S. Merouani, O. Hamdaoui, The size of active bubbles for the production of hydrogen in sonochemical reaction field, Ultrason. Sonochem. 32 (2016) 320−327, https://doi.org/10.1016/j.ultsonch.2016.03.026.
[21] E.J. Hart, C.H. Fischer, A. Henglein, Sonolysis of hydrocarbons in aqueous solution, Int. J. Radiat. Appl. Instrum. C Radiat. Phys. Chem. 36 (1990) 511−516, https://doi.org/10.1016/1359-0197(90)90198-Q.
[22] D.N. Rassokhin, G.V. Kovalev, L.T. Bugaenko, Temperature effect on the sonolysis of methanol/water mixtures, J. Am. Chem. Soc. 117 (1995) 344−347, https://doi.org/10.1021/ja00106a037.
[23] C.M. Krishna, Y. Lion, T. Kondo, P. Riesz, Thermal decomposition of methanol in the sonolysis of methanol-water mixtures. Spin-trapping evidence for isotope exchange reactions, J. Phys. Chem. 91 (1987) 5847−5850, https://doi.org/10.1021/j100307a007.
[24] M.H. Islam, J.J. Lamb, K.M. Lien, O.S. Burheim, J.-Y. Hihn, B.G. Pollet, Novel fuel production based on sonochemistry and sonoelectrochemistry, ECS Trans. 92 (2019) 1−16, https://doi.org/10.1149/09210.0001ecst.
[25] H. Islam, O.S. Burheim, B.G. Pollet, M.H. Islam, O.S. Burheim, B.G. Pollet, H. Islam, O.S. Burheim, B.G. Pollet, M.H. Islam, O.S. Burheim, B.G. Pollet, Sonochemical and sonoelectrochemical production of hydrogen, Ultrason. Sonochem. 51 (2018) 1−86, https://doi.org/10.1016/j.ultsonch.2018.08.024.
[26] K. Kerboua, O. Hamdaoui, S. Al-Zahrani, Sonochemical production of hydrogen: a numerical model applied to the recovery of aqueous methanol waste under oxygen-argon atmosphere, Environ. Prog. Sustain. Energy 40 (2021) e13511, https://doi.org/10.1002/ep.13511.
[27] K. Kerboua, S. Merouani, O. Hamdaoui, A. Alghyamah, H. Islam, H.E. Hansen, B.G. Pollet, How do dissolved gases affect the sonochemical process of hydrogen production? An overview of thermodynamic and mechanistic effects—on the "hot spot theory", Ultrason. Sonochem. 72 (2021) 105422, https://doi.org/10.1016/j.ultsonch.2020.105422.
[28] V.A. Henglein, R. Schulz, Der Einfluß organischer Verbindungen auf einige chemische Wirkungen des Ultraschalls, 1953.
[29] K. Okitsu, H. Nakamura, N. Takenaka, H. Bandow, Y. Maeda, Y. Nagata, Sonochemical reactions occurring in organic solvents: reaction kinetics and reaction site of radical trapping with 1, 1-diphenyl-2-picrylhydrazyl, Res. Chem. Intermed. 30 (2004) 763−774, https://doi.org/10.1163/1568567041856864.
[30] T.G. Leighton, The Acoustic Bubble, Academic press, London, UK, 1994.
[31] T.J. Mason, D.I. Peters, Practical Sonochemistry: Power Ultrasound Uses and Applications, second ed., Woodhead Publishing, Cambridg, UK, 2002 https://doi.org/10.1533/9781782420620.1.
[32] N.S. Mohd-Yusof, B. Babgi, M. Aksu, J. Madhavan, M. Ashokkumar, Physical and chemical effects of acoustic cavitation in selected ultrasonic cleaning applications, Ultrason. Sonochem. 29 (2016) 568−576, https://doi.org/10.1016/j.ultsonch.2015.06.013.
[33] B.M. Teo, F. Grieser, Applications of ultrasound to polymer synthesis, in: D. Chen, S.K. Sharma, A. Mudhoo (Eds.), Handbook on Applications of Ultrasound: Sonochemistry for Sustainability, second ed., Tylor& Francis grou, 2012, pp. 475−500.
[34] C. Pétrier, The use of power ultrasound for water treatment, in: J.A. Gallego-Juarez, K. Graff (Eds.), Power ultrasonics: applications of high-intensity ultrasound, Elsevier, 2015, pp. 939−963, https://doi.org/10.1016/B978-1-78242-028-6.00031-4.
[35] D. Chen, Applications of ultrasound in water and wastewater treatment, in: D. Chen, S.K. Sharma, A. Mudhoo (Eds.), Handbook on applications of ultrasound: sonochemistry for sustainability, second ed., Tylor& Francis group, 2012, pp. 373−405.
[36] K. Okitsu, M. Ashokkumar, F. Grieser, Sonochemical synthesis of gold nanoparticles: effects of ultrasound frequency, J. Phys. Chem. B 109 (2005) 20673−20675, https://doi.org/10.1021/jp0549374.
[37] J.H. Bang, K.S. Suslick, Applications of ultrasound to the synthesis of nanostructured materials, Adv. Mater. 22 (2010) 1039−1059, https://doi.org/10.1002/adma.200904093.
[38] D. Dalecki, Biological effects of microbubble-based ultrasound contrast agents, in: E. Quaia (Ed.), Contrast Media in Ultrasonography, Springer, Berlin, Heidelberg, 2005, pp. 77−85.
[39] D. Dalecki, Mechanical bioeffects of ultrasound, Annu. Rev. Biomed. Eng. 6 (2004) 229−248, https://doi.org/10.1146/annurev.bioeng.6.040803.140126.
[40] D. Dalecki, D.C. Hocking, Advancing ultrasound technologies for tissue engineering, in: M. Ashokkumar (Ed.), Handbook of Ultrasonics and Sonochemistry, Springer Science+Business Media, Singapore, 2015, pp. 1−26.
[41] F. Chemat, Zill-E-Huma, M.K. Khan, Applications of ultrasound in food technology: processing, preservation and extraction, Ultrason. Sonochem. 18 (2011) 813−835, https://doi.org/10.1016/j.ultsonch.2010.11.023.
[42] S. Kentish, H. Feng, Applications of power ultrasound in food processing, Annu. Rev. Food Sci. Technol. 5 (2014) 263−284, https://doi.org/10.1146/annurev-food-030212-182537.
[43] K. Yasui, Fundamentals of acoustic cavitation and sonochemistry, in: M.A. Pankaj (Ed.), Theoretical and Experimental Sonochemistry Involving Inorganic Systems, Springer Dordrecht, New York, 2011, pp. 1−29.
[44] K. Makino, M.M. Mossoba, P. Riesz, Chemical effects of ultrasound on aqueous solutions. Evidence for •OH an •H by spin trapping, J. Am. Chem. Soc. 104 (1982) 3537−3539, https://doi.org/10.1021/ja00376a064.

[45] E.J. Hart, A. Henglein, Sonochemistry of aqueous solutions: H_2-O_2 combustion in cavitation bubbles, J. Phys. Chem. 91 (1987) 3654–3656, https://doi.org/10.1021/j100297a038.
[46] A. Compounds, Y.G. Adewuyi, Sonochemistry: environmental science and engineering applications, Ind. Eng. Chem. Res. 40 (2001) 4681–4715, https://doi.org/10.1021/ie010096l.
[47] K.S. Suslick, Y. Didenko, M.M. Fang, T. Hyeon, K.J. Kolbeck, W.B. McNamara III, M.M. Mdleleni, M. Wong, Acoustic cavitation and its chemical consequences, Philos. Trans. R Soc. A 357 (1999) 335–353, https://doi.org/10.1098/rsta.1999.0330.
[48] C. Petrier, Y. Jiang, M.F. Lamy, Ultrasound and environment: sonochemical destruction of chloroaromatic derivatives, Environ. Sci. Technol. 32 (1998) 1316–1318, https://doi.org/10.1021/es970662x.
[49] A. Henglein, Chemical effects of continuous and pulsed ultrasound in aqueous solutions, Ultrason. Sonochem. 2 (1995) 115–121, https://doi.org/10.1016/1350-4177(95)00022-X.
[50] G. Mark, A. Tauber, H. Schuchmann, D. Schulz, A. Mues, C. Von Sonntag, OH-radical formation by ultrasound in aqueous solution—Part II: terephthalate and Fricke dosimetry and the influence of various conditions on the sonolytic yield, Ultrason. Sonochem. 5 (1998) 41–52, https://doi.org/10.1016/S1350-4177(98)00012-1.
[51] A. Tauber, G. Mark, H.-P. Schuchmann, C. von Sonntag, Sonolysis of tert-butyl alcohol in aqueous solution, J. Chem. Soc., Perkin Trans. 2 (2) (1999) 1129–1136, https://doi.org/10.1039/a901085h.
[52] M.F. Mousavi, S. Ghasemi, Sonochemistry: a suitable method for synthesis of nano-strured materials, in: F.M. Nowak (Ed.), Sonochemistry: Theory, Reactions, Syntheses, and Applications, Nova Science Publishers, New York, USA, 2010, pp. 1–62.
[53] K.S. Suslick, S.J. Doktycz, E.B. Flint, On the origin of sonoluminescence and sonochemistry, Ultrasonics 28 (1990) 280–290, https://doi.org/10.1016/0041-624X(90)90033-K.
[54] W. Lauterborn, C.D. Ohl, Cavitation bubble dynamics, Ultrason. Sonochem. 4 (1997) 65–75, https://doi.org/10.1016/S1350-4177(97)00009-6.
[55] K. Yasui, T. Tuziuti, Y. Iida, Dependence of the characteristics of bubbles on types of sonochemical reactors, Ultrason. Sonochem. 12 (2005) 43–51, https://doi.org/10.1016/j.ultsonch.2004.06.003.
[56] K. Yasui, Acoustic Cavitation and Bubble Dynamics, 2018, https://doi.org/10.1007/978-3-319-68237-2.
[57] F. Grieser, P.-K. Choi, N. Enomoto, H. Harada, K. Okitsu, K. Yasui, Sonochemistry and the Acoustic Bubble, 2015, https://doi.org/10.1016/B978-0-12-801530-8.01001-X.
[58] S. Asgharzadehahmadi, A. Aziz, A. Raman, R. Parthasarathy, Sonochemical reactors: review on features, advantages and limitations, Renew. Sustain. Energy Rev. 63 (2016) 302–314, https://doi.org/10.1016/j.rser.2016.05.030.
[59] J. Buettner, M. Gutierrez, a. Henglein, Sonolysis of water-methanol mixtures, J. Phys. Chem. 95 (1991) 1528–1530, https://doi.org/10.1021/j100157a004.
[60] M. Penconi, F. Rossi, F. Ortica, F. Elisei, P.L. Gentili, Hydrogen production from water by photolysis, sonolysis and sonophotolysis with solid solutions of rare earth, gallium and indium oxides as heterogeneous catalysts, Sustain. Times 7 (2015) 9310–9325, https://doi.org/10.3390/su7079310.
[61] K. Kerboua, O. Hamdaoui, Void fraction, number density of acoustic cavitation bubbles, and acoustic frequency: a numerical investigation, J. Acoust. Soc. Am. 146 (2019) 2240, https://doi.org/10.1121/1.5126865.
[62] A. Dehane, S. Merouani, O. Hamdaoui, Methanol sono-pyrolysis for hydrogen recovery: effect of methanol concentration under an argon atmosphere, Chem. Eng. J. 433, Part 2 (2022) 133272, https://doi.org/10.1016/j.cej.2021.133272.
[63] A. Dehane, S. Merouani, A. Chibani, O. Hamdaoui, Influence of processing conditions on hydrogen Sonoproduction from methanol sono-conversion: a numerical investigation with a validated model, Chem. Eng. Process—Process Intensif 179 (2022) 109080, https://doi.org/10.1016/j.cep.2022.109080.
[64] J. Rae, M. Ashokkumar, O. Eulaerts, C. Von Sonntag, J. Reisse, F. Grieser, Estimation of ultrasound induced cavitation bubble temperatures in aqueous solutions, Ultrason. Sonochem. 12 (2005) 325–329, https://doi.org/10.1016/j.ultsonch.2004.06.007.
[65] A. Dehane, S. Merouani, O. Hamdaoui, A. Alghyamah, A complete analysis of the effects of transfer phenomenons and reaction heats on sono-hydrogen production from reacting bubbles: impact of ambient bubble size, Int. J. Hydrogen Energy 46 (2021) 18767–18779, https://doi.org/10.1016/j.ijhydene.2021.03.069.
[66] P. Kanthale, M. Ashokkumar, F. Grieser, Sonoluminescence, sonochemistry (H_2O_2 yield) and bubble dynamics: frequency and power effects, Ultrason. Sonochem. 15 (2008) 143–150, https://doi.org/10.1016/j.ultsonch.2007.03.003.
[67] S. Merouani, O. Hamdaoui, Y. Rezgui, M. Guemini, Sensitivity of free radicals production in acoustically driven bubble to the ultrasonic frequency and nature of dissolved gases, Ultrason. Sonochem. 22 (2015) 41–50, https://doi.org/10.1016/j.ultsonch.2014.07.011.
[68] A. Brotchie, F. Grieser, M. Ashokkumar, Effect of power and frequency on bubble-size distributions in acoustic cavitation, Phys. Rev. Lett. 102 (2009) 1–4, https://doi.org/10.1103/PhysRevLett.102.084302.
[69] K. Yasui, T. Tuziuti, T. Kozuka, A. Towata, Y. Iida, Relationship between the Bubble Temperature and Main Oxidant Created inside an Air Bubble under Ultrasound, J. Chem. Phys. 127 (2007) 154502, https://doi.org/10.1063/1.2790420.
[70] S. Gireesan, A.B. Pandit, Modeling the effect of carbon-dioxide gas on cavitation, Ultrason. Sonochem. 34 (2017) 721–728, https://doi.org/10.1016/j.ultsonch.2016.07.005.
[71] S. Merouani, H. Ferkous, O. Hamdaoui, Y. Rezgui, M. Guemini, A method for predicting the number of active bubbles in sonochemical reactors, Ultrason. Sonochem. 22 (2015) 51–58, https://doi.org/10.1016/j.ultsonch.2014.07.015.
[72] B. Avvaru, A.B. Pandit, Oscillating bubble concentration and its size distribution using acoustic emission spectra, Ultrason. Sonochem. 16 (2009) 105–115, https://doi.org/10.1016/j.ultsonch.2008.07.003.

[73] A. Dehane, S. Merouani, O. Hamdaoui, Effect of carbon tetrachloride (CCl_4) sonochemistry on the size of active bubbles for the production of reactive oxygen and chlorine species in acoustic cavitation field, Chem. Eng. J. 426 (2021) 130251, https://doi.org/10.1016/j.cej.2021.130251.

[74] K. Yasui, T. Tuziuti, Y. Iida, H. Mitome, Theoretical study of the ambient-pressure dependence of sonochemical reactions, J. Chem. Phys. 119 (2003) 346, https://doi.org/10.1063/1.1576375.

[75] H. Ferkous, S. Merouani, O. Hamdaoui, Y. Rezgui, M. Guemini, Comprehensive experimental and numerical investigations of the effect of frequency and acoustic intensity on the sonolytic degradation of naphthol blue black in water, Ultrason. Sonochem. 26 (2015) 30–39, https://doi.org/10.1016/j.ultsonch.2015.02.004.

[76] A. Alippi, F. Cataldo, A. Galbato, Ultrasound cavitation in sonochemistry: decomposition of carbon tetrachloride in aqueous solutions of potassium iodide, Ultrasonics 30 (1992) 148–151.

[77] A. Weissler, W. cooper Herbert, S. Snyder, Chemical effect of ultrasonic waves: oxidation of potassium iodide solution by carbon tetrachloride, J. Am. Chem. Soc. 72 (1950) 1769–1775, https://doi.org/10.1021/ja01160a102.

[78] A. Dehane, S. Merouani, O. Hamdaoui, M. Ashokkumar, Sonochemical and sono-assisted reduction of carbon dioxide: a critical review, Chem. Eng. Process—Process Intensif. (2022) 109075, https://doi.org/10.1016/j.cep.2022.109075.

[79] S. Merouani, O. Hamdaoui, Sonochemical treatment of textile wastewater, in: M.P. Inamuddin, A. Asiri (Eds.), Water Pollution and Remediation: Photocatalysis, Springer-Nature Switzerland, 2021, https://doi.org/10.1007/978-3-030-54723-3_5.

Chapter 19

Upscaling of LaNi$_5$-based metal hydride reactor for solid-state hydrogen storage: numerical simulation of the absorption−desorption cyclic processes

Atef Chibani, Aissa Dehane, Leila Nemdili and Slimane Merouani

Laboratory of Environmental Process Engineering, Department of Chemical Engineering, Faculty of Process Engineering, University Salah Boubnider Constantine 3, El Khroub, Algeria

1. Introduction

Because of the fast rise of the world population and the industrial revolution of today's society, fossil fuel use has steadily increased over the last few decades [1]. Governments have grown aware of the need to use sustainable energies to enhance their energy reserves and minimize carbon emissions [2]. The use of hydrogen as an energy source can help in meeting current and future energy demands while avoiding additional carbon releases. Hydrogen is the most abundant element in the universe and has the highest mass energy density of any fuel (120.7 kJ g^{-1}) [3]. Although hydrogen generation techniques are already well developed, the primary problem that has to be solved to make hydrogen technology commercially viable is the safe storage of hydrogen in containers with the proper weight and volume. These technologies will have a significant impact on the future of hydrogen as a practical energy source.

Recently, there has been a lot of effort in developing high-efficiency hydrogen storage methods [4]. Solid-state hydrogen storage materials (metal hydride, MH) have received substantial attention in comparison to conventional hydrogen storage systems (i.e., cryogenic hydrogen and compressed hydrogen gas) due to their high volumetric storage density, security, and cheaper cost [5,6]. To provide a wide surface area for hydrogen storage, MHs have a porous structure [7]. While various storage materials, including TiMn$_2$, alanates, nitrogen hydride, metal-organic frameworks, and borohydrides, have been studied [8−13], LaNi$_5$ has proven to be the most useful due to its low density and high hydrogen storage capacity (1.4 wt%) [14−16]. Many published works have been conducted on using this material for hydrogen storage [5,17−29]. Even though various reactor designs have been treated, it is clear that there is a significant lack of research on hydrogen storage using large-scale MH structures, with the majority of studies focused on lab-scale to pilot-scale reactors. The design and simulation of hydrogen storage/release in industrial-scale MH reactors are essential for understanding the process behavior in such big structures.

Hydrogen absorption/desorption with MHs is an exothermic/endothermic processes. Additionally, an MH is in a porous form in order to provide a large surface area for the reaction. Consequently, the rates of hydrogen absorption and desorption are strongly controlled by heat and mass transfer inside the MH bed [30]. Thus, a theoretical study of MH vessels is strongly needed because it will enable a complete understanding of heat and mass transport phenomena in MH vessels as well as the development of computer-aided tools for design and optimization [5].

We recently published many interesting papers that simulated hydrogen storage performance in a large-scale, multipipe LaNi$_5$-based MH reactor using Ansys Fluent software [15,31−34]. The most important outcomes from these studies were discussed in this chapter, with special emphasis on mass and heat transfer during the absorption—desorption cyclic processes. The absorption—desorption processes were investigated in relation to the operating pressure and the thermal conductivity of the MH.

2. Previous works

Multiple studies have addressed the issue of heat and mass transfer in MH-based hydrogen storage. Studies have suggested various designs for hydrogen storage systems to facilitate effective heat transfer [14]. These designs are as diverse as the heat transfer enhancements themselves and include the addition of fins, cooling tubes, cooling coils, etc. [35,36]. Mellouli et al. [26,27,37] developed several mathematical models for detailed analysis of MH bed operations. Nam et al. [23] Mohammadshahi et al. [25] and Kyoung et al. [5] have developed three-dimensional hydrogen absorption/desorption models to precisely study the hydrogen absorption reaction and resultant heat and mass transport phenomena in $LaNi_5$-H_2 MH hydrogen storage vessels. Gambini et al. [24] reported a dynamic analysis model aimed at describing the hydrogen absorption and desorption of two MHs ($LaNI_5$-H_2 and MgH_2). They found that $LaNi_5$ alloy has faster kinetics than MgH_2 does, but the latter is capable of storing a much larger volume of hydrogen gas. Thanks to a two-dimensional axisymmetric model for hydrogen absorption and desorption, Busqué et al. [22,38] have studied the effect of metal properties and boundary conditions on charging and discharging performance. Their results show that the properties that most influence the charging performance are: absorption rate constant, activation energy, and thermal conductivity. Several other models describing the storage/destocking of hydrogen in MH reactors of different geometries are available in the literature [28,29,39,40].

Concerning the form of reactors, several geometric shapes were highlighted. Shafiee and McCay [14] have reviewed the different reactor configurations for MH hydrogen storage systems, and Srinivasa Murthy [30] has reviewed the importance of optimal thermal design of the storage device and effective thermal conductivity on sorption performance. The most simplistic one is the cylindrical MH bed reactor, as used by Busqué et al. [22,38]. Muthukumar et al. [41] have tested an MH storage device employing a conventional outer cooling jacket for transferring heat from/to the MH bed and found that the cooling fluid temperature is a strong influencing parameter for the absorption of hydrogen. Linder et al. [42] have used a capillary tube bundle heat exchanger. Mohan et al. [43] considered embedded filters and tubes in a cylindrical storage device and simulated the absorption process. Dhaou et al. [44] have experimentally investigated the effects of a finned spiral tube heat exchanger on the sorption performance of a cylindrical hydrogen storage device filled with 1 kg of $LaNi_5$ alloy. Anbarasu et al. [45] have performed experiments on two different designs of the storage device with 36 and 60 embedded cooling tubes using $LmNi_{4.91}Sn_{0.15}$ alloy. Bhouri et al. [46] have optimized the hydrogen absorption process in a multitubular hydrogen storage container equipped with longitudinal fins with sodium alanate as the hydriding alloy.

Globally, despite the variety of the proposed reactor configurations, it is surprisingly regarded that most of them are of lab scale to pilot scale only. However, there is scarce data in the literature concerning the large-scale MH reactors.

3. Physical description of the unit

The reactor is made up of a multipipe heat exchanger that is 1 m in height and 1 m in width. Fig. 19.1 depicts the geometric shape of the $LaNi_5$ porous-bed reactor. It is made up of 19 filling/discharging pipes of 10 cm diameter, which are arranged as shown in Fig. 19.1B. A rectangular region of 6020 cm has been omitted from each lateral side of the reactor in order to increase the exchange area and lower the cooling/heating energy by using HTF with a low heat transfer coefficient (h = 1000 W m^{-2} K^{-1}). The reactor's exterior frontal side was cooled/heated with water of h (heat transfer coefficient) equal to h = 3000 W m^{-2} K^{-1} [41,47]. During the charging phase of the reactor (absorption), the initial bed temperature as well as that of the HTF was 293.15 K. However, during the reverse operation (desorption), the initial block and HTF temperature were 333 K.

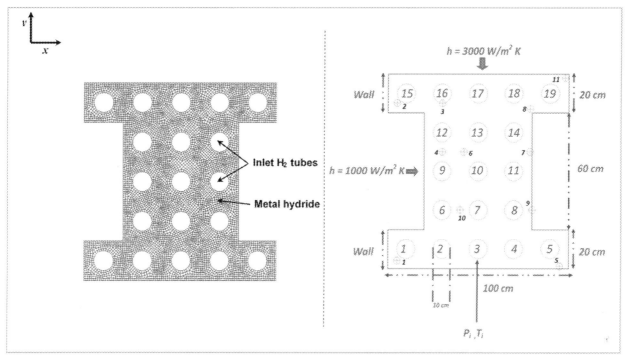

FIGURE 19.1 Reactor geometry (A) and geometries with boundary conditions (B).

4. Model

The primary objective of our research [15,32] was to address the issue of heat and mass transfer in the large-scale MH reactor shown in Fig. 19.1. The equations governing the model, the boundary conditions, the thermophysical properties of hydrogen and $LaNi_5$ metal, as well as the adopted assumptions made for simulating the absorption/desorption of hydrogen in the reactor shown in Fig. 19.1 are available in Refs. [15,32]. The reader can return to these references to get more details. The mathematical model and the resolution of thermal issues, as well as the processes of chemical reactions during hydrogen charging/discharging in the reactor, were solved using the energy model of the Ansys Fluent software.

The model was verified using hydrogen adsorption data of Busqué et al. [22] (cylindrical lab-scale $LaNi_5$ alloy reactor) and that of Muthukumar et al. [48] ($MmNi_{4.6}Fe_{0.4}$ bed) for the desorption cycle. The conidtions of the two studies with the validation results, avaiblabe in [15,32], are summarized in Fig. 19.2A and B. Our model suited the adsorption data perfectly (Fig. 19.2A). Besides, while the agreement during the desorption phase is less than ideal (Fig. 19.2B), the simulation findings typically match the experimental data, capturing the major empirical points during hydrogen desorption.

5. Analysis of the absorption process (hydrogen storage)

Fig. 19.3A and B shows the variation of the average bed temperature as a function of time during hydrogen storage process in the multipipe MH reactor for different conditions of thermal conductivity (λ) and loading pressure (P), whereas Fig. 19.3C and D display the corresponding time evolutions of the average hydrogenation degree of the metal for up to 3000 s, which is enough to attain the maximum storage capacity in most cases. In all cases, the external cooling process (HTF: 293.15 K) is considered from the beginning of the absorption process (at time t = 0).

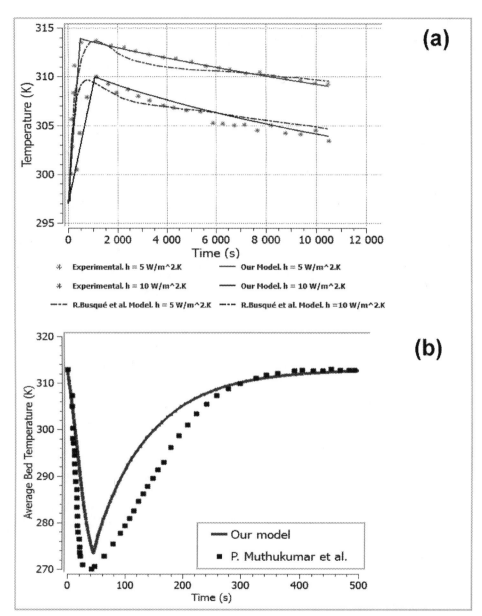

FIGURE 19.2 Model validation with: (A) Busqué's et al. [22] absorption data for natural and forced convective systems [conditions—MH: LaNi$_5$, P (H$_2$ pressure) = 0.65 atm, T$_f$ (cooling fluid temperature) = 297 K, t$_{amb}$ = 296, C$_a$ = 2 s^{-1}, E$_a$ = 27,200 J mol^{-1}, h = 5 and 10 W m^{-2} K^{-1}] and (B) Muthukumar's et al. [48] desorption data [conditions—MH: MmNi$_{4.6}$Fe$_{0.4}$, P$_d$ = 1 atm, T$_f$ = 297 K, t$_{amb}$ = 310 K, C$_d$ = 475 s^{-1}, E$_a$ = 25,000 J mol^{-1}, h = 1000 W m^{-2} K^{-1}]. *From A. Chibani, S. Merouani, C. Bougriou, L. Hamadi, Heat and mass transfer during the storage of hydrogen in LaNi5-based metal hydride: 2D simulation results for a large scale, multi-pipes fixed-bed reactor, Int. J. Heat Mass Trans. 147 (2020) 118939. https://doi.org/10.1016/j.ijheatmasstransfer.2019.118939, and A. Chibani, S. Merouani, A. Dehane, C. Bougriou, Hydrogen recovery from big-scale porous metal hydride (MH) reactor: impact of pressure and MH-thermophysical properties, Energy Storage Sav. 1 (2022). https://doi.org/10.1016/j.enss.2022.07.005.*

In Fig. 19.3A and C, the effect of the thermal conductivity (λ, 3.19–9 W m^{-1} K^{-1}) is shown for a charging pressure of 40 bar. For each λ, the temperature and concentration profiles follow three regimes: (1) the material temperature increased suddenly from the ambient temperature (293.15 K) up to 323 K within $\sim$50 s and then (2) the temperature continuously increases with lesser slope than the first phase until a maxima, T$_{max}$ = 322 K, at = 1000 s. The similar behavior was observed for hydrogen concentration (Fig. 19.3C). The metal hydrogenation degree at T$_{max}$ is at about 77% of its maximum capacity while a lower percentage of 51% was recorded at 50 s. Throughout the period of phases (1) and (2), the

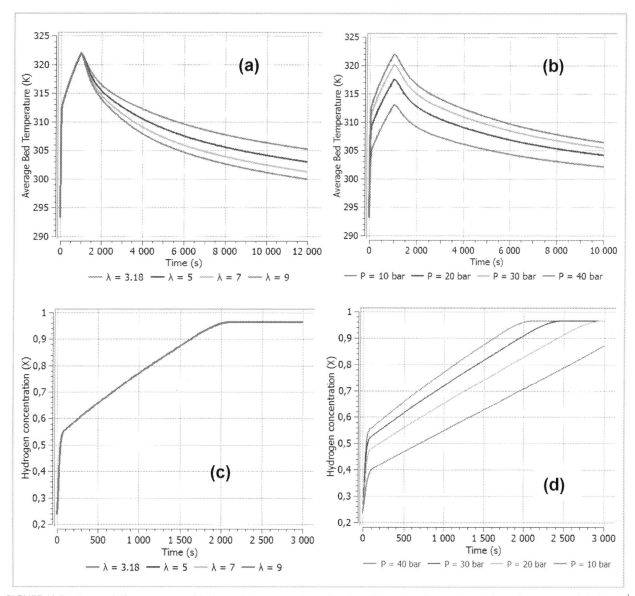

FIGURE 19.3 Average bed temperature and hydrogen bed concentration as function of time for different values of thermal conductivity λ [in W m^{-1} K^{-1}] and charging pressure P [in bar]. Simulation conditions: (A, C) P = 40 bar, C_a = 59.187 s^{-1}, and E_a = 21,179.6 J mol^{-1}, (B, D) λ = 3.18 W m^{-1} K^{-1}, C_a = 59.187 s^{-1}, and E_a = 21,179.6 J mol^{-1}. *From A. Chibani, S. Merouani, C. Bougriou, L. Hamadi, Heat and mass transfer during the storage of hydrogen in LaNi5-based metal hydride: 2D simulation results for a large scale, multi-pipes fixed-bed reactor, Int. J. Heat Mass Trans. 147 (2020) 118939. https://doi.org/10.1016/j.ijheatmasstransfer.2019.118939.*

thermal conductivity has no observed impact on both bed temperature and hydrogen concentration profiles. All these findings agreed qualitatively with the results of several researchers [19,28,38,45,49,50], obtained for lab-scale MH reactors of different configurations.

The absorption of hydrogen is an exothermic process that releases a significant amount of heat, particularly at the first stage of absorption when the material is empty on hydrogen. Therefore, the sudden increase in temperature (t < 50 s) is due to the reaction heat liberated during hydrogen absorption. The regime (3) started for t > 1000 s; the temperature drops exponentially with longer time under the action of the external convective cooling of the reactor. During this period, the hydrogenation fraction of the material continues to increase up reaching maxima of ∼96% at 2000 s and then taking a plateau, which indicates the stop of the absorption process. Very similar behavior was reported by Gkanas [49] for hydrogen storage in $MmNi_{4.6}Al_{0.4}$ hydride. Note that λ has no effect on hydrogen concentration in this period, which is in accordance with the results of Chibani et al. [19] for a triple concentric MH system. However, during phase (3), the higher the metal conductivity, the lower was the corresponding bed temperature, which is attributed to the fact that higher λ could assure higher rate of heat dissipation, i.e., through the porous medium, due to the lower conductive resistance at higher λ [19]. Therefore, at fixed conditions of coolants, a longer time is required for cooling the MH if lower thermal conductivity is used. Another important statement that can made from the effect of λ is that it is not necessary to achieve the maximum storage capacity at the maximum bed temperature (only 77% was stored at T_{max}), which is excellently consistent with the results of Afzal et Sharma [50] who conducted a noteworthy work on large-scale MH design–based hydrogen storage reactor.

In Fig. 19.3B and D, the effect of hydrogen charging pressure (10–40 bar) is shown. As seen, higher bed temperatures and H_2-bed concentrations were associated with higher charging pressures. The maximum bed temperature, achieved at t = 1000 s, increased from 313 K at 10 bar to 317 K at 20 bar, 320 K at 30 bar, and 322 K at 40 bar. At this instant, the metal hydrogenation degree increased from 55% at 10 bar to 65% at 20 bar, 71% at 30 bar, and 77% at 40 bar. Additionally, the time required for attaining the maximum storage capacity increased from 1000 s at 40 bar to 2400 s at 30 bar, 2900 s at 20 bar, and >3000 s for 10 bar. The same behavior has been reported for the effect of supplied pressure on hydrogen storage in different lab-scale systems [18,19,38,51]. In fact, it is well known that increasing gas pressure will increase the amount of hydrogen present in the porous media and this accelerated the adsorption rate of the gas on the active sites of the material, which resulted in higher heat flux released from the exothermic reaction process.

In Fig. 19.4, the distribution of hydrogen concentration in the MH is shown for different times during the vessel charging, when the supply pressure is 40 bar. Initially (a t = 30 s), the hydrogenated zones are located at the inside wall of the bloc; the H_2 concentrations at these zones are relatively higher than the interior of the bed reactor (Fig. 19.4A). As the time advanced (t = 100 s), the hydrogenated zone was extended to central areas delimited by charging pipes (Fig. 19.4B). However, at these zones, the maximum charging capacity of metal has not been attained yet. At 700 s, hydrogen concentration started to become homogenous in the MH (Fig. 19.4C). The maximum bed temperature at this instant is ~320 K. At t = 2000 s, the system is in the cooling phase (t > 1000 s), the MH attained its maximum hydrogenation capacity, and the mean metal temperature at this time is 317 K (Fig.19 4D).

6. Analysis of the desorption process (hydrogen discharging)

Fig. 19.5 shows the time evolutions of the overage bed temperature (for up to 20,000 s) and dehydrogenation kinetics (for 3000 s) for different discharging pressures and the thermal conductivity (λ). In all cases, the external heating process (HTF: 333 K) is considered from the start of the desorption process (at t = 0).

The impact of hydrogen discharging pressure (10–40 bar) is depicted in Fig. 19.5A and C. The rate of hydrogen desorption rose considerably as the discharging pressure was raised. This occurrence was related to a significant reduction in the average bed temperature at tremendous pressure, particularly at the beginning of the desorption process. At 100 s, 10 bar of pressure desorbed 30% of the adsorbed H_2, whereas applying 20, 30, and 40 bar resulted in desorbed percentages of 55%, 60%, and 65%, respectively. During the desorption process, the minimum temperature dropped from 300 K at 10 bar to 280 K at 20 bar, 255 K at 30 bar, and 240 K at 40 bar. Therefore, increasing the discharging pressure assures a greater flow of heat transfer from the reactor's exterior heated walls to the MH system. In reality, a higher discharge pressure means a lower pressure inside the porous MH bed, which promotes H_2 desorption (i.e., this event is precisely like augmenting the temperature of the external heating source).

The impact of the MH thermal conductivity (3.18–9 W m^{-1} K^{-1}) is demonstrated in Fig. 19.5B and D for a discharging pressure of 40 bar. Thermal conductivity had no influence on the dehydrogenation kinetics of the MH or the average MH temperature for up to 1500 s, which corresponds to the time required for full dehydrogenation of the MH (Fig. 19.5B). After that, thermal conductivity had a significant impact on heat transfer, with higher λ ensuring a greater flow of heat from the heated walls of the reactor to the MH bed. When higher thermal conductivity is selected, the MH bed quickly returns to its starting temperature. Chibani et al. [15,19] previously demonstrated that for a triple concentric and big MH system, λ has no influence on the adsorbed hydrogen concentration during reactor charging. However, when the reactor was filled, the higher the metal conductivity, the lower the associated bed temperature, which is ascribed to the fact that higher λ produces a faster rate of heat dissipation, i.e., via the porous channel, due to the lower conductive resistance in

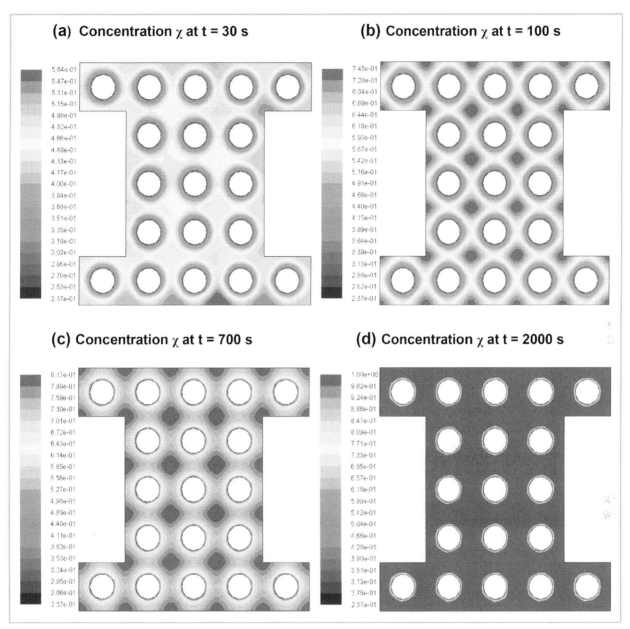

FIGURE 19.4 Distribution of hydrogen concentration in the storage vessel at different charging times: (A) 30 s, (B) 100 s, (C) 700 s, and (D) 2000 s (conditions: P = 40 bar, $\lambda = 3.18$ W m^{-1} K^{-1}, $C_a = 59.187$ s^{-1}, and $E_a = 21,179.6$ J mol^{-1}). *From A. Chibani, S. Merouani, C. Bougriou, L. Hamadi, Heat and mass transfer during the storage of hydrogen in LaNi5-based metal hydride: 2D simulation results for a large scale, multi-pipes fixed-bed reactor, Int. J. Heat Mass Trans. 147 (2020) 118939. https://doi.org/10.1016/j.ijheatmasstransfer.2019.118939.*

this case. As a result, at fixed heating/cooling conditions, the MH bed requires more time to go back to its initial temperature [6,24].

Fig. 19.6 depicts the gas velocity distribution during the dehydrogenation of the material hydride at various periods (1, 5, 20, and 1200 s). The discharging pressure is set to 40 bar, and the other simulation parameters are $\lambda = 3.18$ W m^{-1} K^{-1}, $C_d = 10$ s^{-1}, and $E_d = 16,420$ J mol^{-1}. It is seen that hydrogen rapidly discharges from the material bloc to the tubes. It is also evident that greater velocities are usually seen close around the tubes, and they decrease with further penetration in the vessel material. The highest hydrogen velocity rose with increasing time from 1

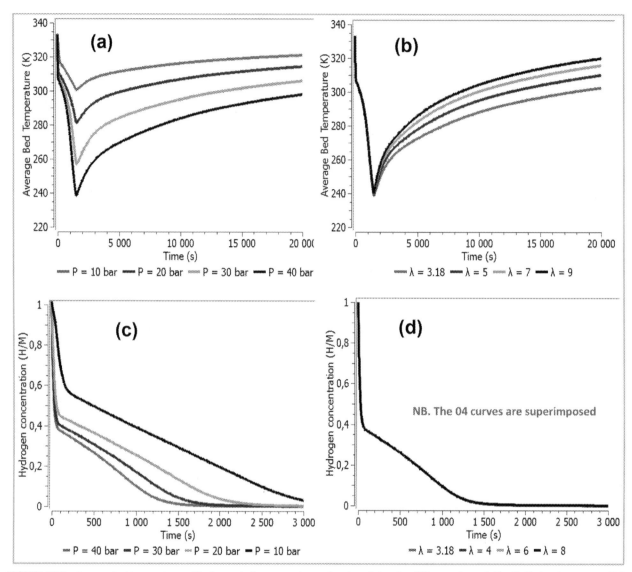

FIGURE 19.5 Average bed temperature and hydrogen bed concentration as function of time for different values of: (A, C) discharging pressure P [in bar] and (B, D) thermal conductivity λ [in W m^{-1} K^{-1}]. Simulation conditions: (A, C) P = 10–40 bar, C_d = 10 s^{-1}, E_d = 16,420 J mol^{-1}, and λ = 3.18 W m^{-1} K^{-1}, (B, D) P = 40 bar, λ = 3.18–8 W m^{-1} K^{-1}, C_d = 10 s^{-1}, and E_d = 16,420 J mol^{-1}. *From A. Chibani, S. Merouani, A. Dehane, C. Bougriou, Hydrogen recovery from big-scale porous metal hydride (MH) reactor: impact of pressure and MH-thermophysical properties, Energy Storage Sav. 1 (2022). https://doi.org/10.1016/j.enss.2022.07.005.*

to 20 s and subsequently dropped at 1200 s. Their values are as follows: 6.91 × 10^{-2} m s^{-1} at 1 s, 4.98 × 10^{-1} m s^{-1} at 5 s, 9.49 × 10^{-1} m s^{-1} at 20 s, and finally 1.25 × 10^{-1} m s^{-1} at 1200 s. The increase in gas velocity at the early stage of the desorption cycle (up to 20 s) was ascribed to the faster dehydrogenation process at this stage as the material is loaded with hydrogen, implying that a rapid disequilibrium of hydrogen from the material sites to pores occurs, an event that was previously evidenced (i.e., Fig. 19.4A) where a fast dehydrogenation rate is achieved (∼60% at 100 s). However, more than 90% of H$_2$ was released by 1200 s (mean dehydrogenation degree at 1200 s is 93%, as shown in Fig. 19.4A). As a result, the hydrogen velocity may be reduced since there are not enough hydrogen molecules in the porous medium.

Fig. 19.7 depicts the distribution of hydrogen content in the MH at various periods throughout the discharging phase when the destocking pressure is 40 bar. Initially (t = 1 s), the bloc is still charged with hydrogen (χ ∼ 97%−100%) (Fig. 19.7A); at this moment, the desorption rate is in its early stages. After 5 s the H$_2$ concentration in the bloc was

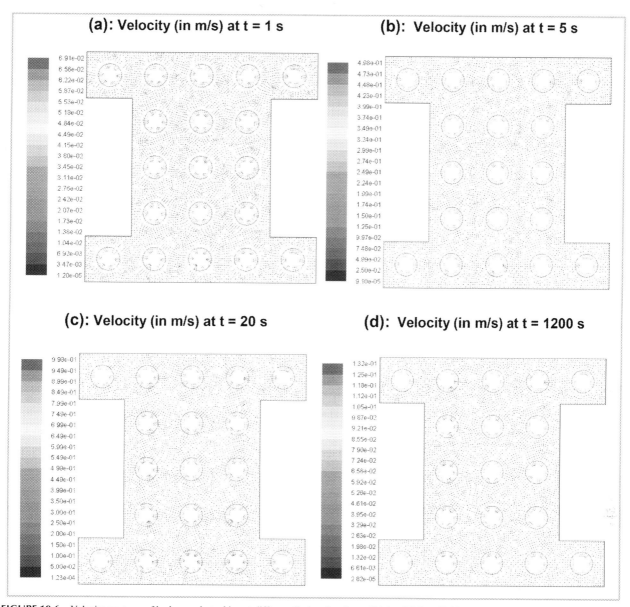

FIGURE 19.6 Velocity contours of hydrogen destocking at different discharging times: (A) 1 s, (B) 5 s, (C) 20 s and (D) 1200 s (simulation conditions: $\lambda = 3.18$ W m^{-1} K^{-1}, $C_d = 10$ s^{-1}, and $E_d = 16{,}420$ J mol^{-1}). *From A. Chibani, S. Merouani, A. Dehane, C. Bougriou, Hydrogen recovery from big-scale porous metal hydride (MH) reactor: impact of pressure and MH-thermophysical properties, Energy Storage Sav. 1 (2022) (in press). https://doi.org/10.1016/j.enss.2022.07.005.*

reduced to 72%–86%, and the majority of the H_2 adsorbed at the inside wall of the bloc was released (i.e., since this zone is very next to the heating wall). The majority of the remaining concentrations were located in the bloc's interior. Because of the importance of the desorption process at this time, the bed temperature was reduced to 320 K (Fig. 19.5A). As time goes up, the bloc interior dehydrogenated by 50% (temperature at 315 K) compared to the starting condition (t = 0 s), while a comparatively larger concentration of 60%–70% persisted surrounding the tubes (Fig. 19.7C). The MH bed was evacuated on hydrogen at t = 1200 s, and only a minor concentration was observed surrounding the tubes (Fig. 19.7D). The bed temperature is at its lowest point (235 K) at 1200 s (Fig. 19.5A). After this stage, there is almost no mass transfer, although the external fluid still heats the reactor.

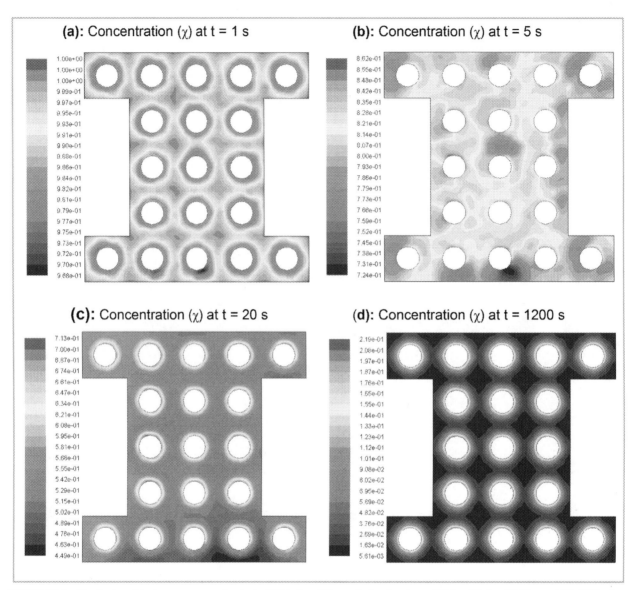

FIGURE 19.7 Distribution of hydrogen concentration in the MH bed at different discharging times: (A) 1 s, (B) 5 s, (C) 20 s, and (D) 1200 s (conditions: $\lambda = 3.18$ W m^{-1} K^{-1}, $C_d = 10$ s^{-1}, and $E_d = 16{,}420$ J mol^{-1}). *From A. Chibani, S. Merouani, A. Dehane, C. Bougriou, Hydrogen recovery from big-scale porous metal hydride (MH) reactor: impact of pressure and MH-thermophysical properties, Energy Storage Sav. 1 (2022). https://doi.org/10.1016/j.enss.2022.07.005.*

7. Conclusion and future perspectives

A two-dimensional model was used to simulate hydrogen charging/discharging from a large configuration LaNi$_5$-MH reactor. Mass and heat exchanges were predicted under various operating pressures and MH thermal conductivity. The (desorption) process produces a sudden increase (drop) in bed temperature during the early stage of the reaction. Increasing the applied pressure resulted in significant improvements in heat transfer, increasing the hydrogen absorption/desorption rate and allowing the reactor to charging/discharging in less time. Furthermore, increasing the metal thermal conductivity improved heat transfer but did not affect the absorption/desorption rate of hydrogen.

Finally, in this work, we solely looked at heat transfer inside the MH reactor, where the adsorption/desorption processes occur. The system's external temperature is considered to be constant. Besides, the use of thermal units based on phase change material (PCM) to manage the reaction heat resulted/used during the exothermal/endothermal hydrogen absorption/desorption is another technological solution requiring deep attention. With this tool, the cost of hydrogen storage unit could be reduced significantly. PCMs can replace the traditional HTF cooling/heating system, with the advantage of acting in dual roles, i.e., heat absorbers during hydrogen charging and heat suppliers during the discharging. These materials holding great latent heat can stock important heat and transfer it with solid−liquid phase change (melting/solidification).

Nomenclature

C reaction rate constant, s^{-1}
E activation energy, $J\ mol^{-1}$
h convection heat transfer coefficient, $W\ m^{-2}\ K^{-1}$
P hydrogen pressure, Pa
T temperature, K
t time, s
V velocity, $m\ s^{-1}$
wt maximum weight percentage of hydrogen into material, %

Greek symbols

a Absorption
d Desorption
Ref. Reference
χ Hydrogen concentration
λ Thermal conductivity, $W\ m^{-1}\ K^{-1}$

References

[1] F. Chakik, M. Kaddami, M. Mikou, Effect of operating parameters on hydrogen production by electrolysis of water, Int. J. Hydrog. Energy 42 (2017) 2−9, https://doi.org/10.1016/j.ijhydene.2017.07.015.

[2] G. Zini, P. Tartarini, Wind-hydrogen energy stand-alone system with carbon storage: modeling and simulation, Renew. Energy 35 (2010) 2461−2467, https://doi.org/10.1016/j.renene.2010.03.001.

[3] S. Merouani, O. Hamdaoui, Y. Rezgui, M. Guemini, Mechanism of the sonochemical production of hydrogen, Int. J. Hydrog. Energy 40 (2015) 4056−4064, https://doi.org/10.1016/j.ijhydene.2015.01.150.

[4] M. Pasini, C. Corgnale, B.A. Van Hassel, T. Motyka, S. Kumar, K.L. Simmons, Metal hydride material requirements for automotive hydrogen storage systems, Int. J. Hydrog. Energy 8 (2017) 1−11, https://doi.org/10.1016/j.ijhydene.2012.08.112.

[5] S. Kyoung, S. Ferekh, G. Gwak, A. Jo, H. Ju, Three-dimensional modeling and simulation of hydrogen desorption in metal hydride hydrogen storage vessels, Int J. Hydrog. Energy (2015) 14322−14330, https://doi.org/10.1016/j.ijhydene.2015.03.114.

[6] K.J. Kim, G. Lloyd, A. Razani, K.T. Feldman, Development of $LaNi_5/Cu/Sn$ metal hydride powder composites, Power Technol. 99 (1998) 40−45.

[7] D.V. Blinov, D.O. Dunikov, A.N. Kazakov, Measuring the gas permeability of a metal hydride bed of the $LaNi_5$ type, Alloy 54 (2016) 153−156, https://doi.org/10.1134/S0018151X1506005X.

[8] Z. Chen, X. Xiao, L. Chen, X. Fan, L. Liu, S. Li, H. Ge, Q. Wang, Development of Ti-Cr-Mn-Fe based alloys with high hydrogen desorption pressures for hybrid hydrogen storage vessel application, Int. J. Hydrog. Energy 38 (2013) 12803−12810, https://doi.org/10.1016/j.ijhydene.2013.07.073.

[9] H. Leng, T. Ichikawa, H. Fujii, Hydrogen storage properties of Li-Mg-N-H systems with different ratios of $LiH/Mg(NH_2)_2$, J. Phys. Chem. B 110 (2006) 12964−12968, https://doi.org/10.1021/jp061120h.

[10] F. Liang, J. Lin, Y. Wu, L. Wang, Enhanced electrochemical hydrogen storage performance of Ti[sbnd]V[sbnd]Ni composite employing $NaAlH_4$, Int. J. Hydrog. Energy 42 (2017) 14633−14640, https://doi.org/10.1016/j.ijhydene.2017.04.202.

[11] Y. Yang, Y. Liu, Y. Li, X. Zhang, M. Gao, H. Pan, Towards the endothermic dehydrogenation of nanoconfined magnesium borohydride ammoniate, J. Mater. Chem. A. 3 (2015) 11057−11065, https://doi.org/10.1039/c5ta00697j.

[12] B. Panella, M. Hirscher, H. Pütter, U. Müller, Hydrogen adsorption in metal-organic frameworks: Cu-MOFs and Zn-MOFs compared, Adv. Funct. Mater. 16 (2006) 520−524, https://doi.org/10.1002/adfm.200500561.

[13] D. Ji, H. Zhou, Y. Tong, J. Wang, M. Zhu, T. Chen, A. Yuan, Facile fabrication of MOF-derived octahedral CuO wrapped 3D graphene network as binder-free anode for high performance lithium-ion batteries, Chem. Eng. J. 313 (2017) 1623−1632, https://doi.org/10.1016/j.cej.2016.11.063.

[14] S. Shafiee, M.H. McCay, Different reactor and heat exchanger configurations for metal hydride hydrogen storage systems—a review, Int. J. Hydrog. Energy 41 (2016) 9462−9470, https://doi.org/10.1016/j.ijhydene.2016.03.133.

[15] A. Chibani, S. Merouani, C. Bougriou, L. Hamadi, Heat and mass transfer during the storage of hydrogen in $LaNi_5$-based metal hydride: 2D simulation results for a large scale, multi-pipes fixed-bed reactor, Int. J. Heat Mass Trans. 147 (2020) 118939, https://doi.org/10.1016/j.ijheatmasstransfer.2019.118939.

[16] A.N. Kazakov, I.A. Romanov, S. V Mitrokhin, E.A. Kiseleva, Experimental investigations of AB_5-type alloys for hydrogen separation from biological gas streams, Int. J. Hydrog. Energy 45 (2020) 4685−4692, https://doi.org/10.1016/j.ijhydene.2019.11.207.

[17] R. Busqué, R. Torres, J. Grau, V. Roda, Mathematical modeling, numerical simulation and experimental comparison of the desorption process in a metal hydride hydrogen storage system, Int. J. Hydrog. Energy 43 (2018) 16929–16940, https://doi.org/10.1016/j.ijhydene.2017.12.172.

[18] F. Askri, M. Ben Salah, A. Jemni, S. Ben Nasrallah, A new algorithm for solving transient heat and mass transfer in metal–hydrogen reactor, Int. J. Hydrog. Energy 34 (2009) 8315–8321, https://doi.org/10.1016/j.ijhydene.2009.07.072.

[19] A. Chibani, C. Bougriou, S. Merouani, Simulation of hydrogen absorption/desorption on metal hydride $LaNi_5$-H_2: mass and heat transfer, Appl. Therm. Eng. 142 (2018) 110–117, https://doi.org/10.1016/j.applthermaleng.2018.06.078.

[20] H. Choi, A.F. Mills, Heat and mass transfer in metal hydride beds for heat pump applications, Int. J. Heat Mass Trans. 33 (1990) 1281–1288.

[21] G. Cacciola, A. Freni, F. Cipitò, Finite element-based simulation of a metal hydride-based hydrogen storage tank, Int. J. Hydrog. Energy 34 (2009) 8574–8582, https://doi.org/10.1016/j.ijhydene.2009.07.118.

[22] R. Busqué, R. Torres, J. Grau, V. Roda, Effect of metal hydride properties in hydrogen absorption through 2D-axisymmetric modeling and experimental testing in storage canisters, Int. J. Hydrog. Energy 42 (2017) 19114–19125, https://doi.org/10.1016/j.ijhydene.2017.06.125.

[23] J. Nam, J. Ko, H. Ju, Three-dimensional modeling and simulation of hydrogen absorption in metal hydride hydrogen storage vessels, Appl. Energy 89 (2012) 164–175, https://doi.org/10.1016/j.apenergy.2011.06.015.

[24] M. Gambini, T. Stilo, M. Vellini, High temperature metal hydrides for energy systems Part B: comparison between high and low temperature metal hydride reservoirs, Int. J. Hydrog. Energy 42 (2017) 1–11, https://doi.org/10.1016/j.ijhydene.2017.03.227.

[25] S.S. Mohammadshahi, T. Gould, E.M.A. Gray, C.J. Webb, An improved model for metal-hydrogen storage tanks—Part 2: model results, Int. J. Hydrog. Energy 41 (2016) 3919–3927, https://doi.org/10.1016/j.ijhydene.2015.12.051.

[26] S. Mellouli, F. Askri, H. Dhaou, A. Jemni, S. Ben Nasrallah, Numerical simulation of heat and mass transfer in metal hydride hydrogen storage tanks for fuel cell vehicles, Int. J. Hydrog. Energy 35 (2010) 1693–1705, https://doi.org/10.1016/j.ijhydene.2009.12.052.

[27] S. Mellouli, F. Askri, H. Dhaou, A. Jemni, S. Ben Nasrallah, Numerical study of heat exchanger effects on charge/discharge times of metal–hydrogen storage vessel, Int. J. Hydrog. Energy 34 (2009) 3005–3017, https://doi.org/10.1016/j.ijhydene.2008.12.099.

[28] L. Tong, J. Xiao, T. Yang, B. Pierre, Complete and reduced models for metal hydride reactor with coiled-tube heat exchanger, Int. J. Hydrog. Energy 44 (2019) 15907–15916, https://doi.org/10.1016/j.ijhydene.2018.07.102.

[29] D.G. Oliva, M. Fuentes, E.M. Borzone, G.O. Meyer, P.A. Aguirre, Hydrogen storage on $LaNi_{5-x}Sn_x$. Experimental and phenomenological model-based analysis, Energy Convers. Manag. 173 (2018) 113–122, https://doi.org/10.1016/j.enconman.2018.07.041.

[30] S. Srinivasa Murthy, Heat and mass transfer in solid state hydrogen storage: a review, J. Heat Trans. 134 (2012) 031020, https://doi.org/10.1115/1.4005156.

[31] A. Chibani, S. Merouani, C. Bougriou, A. Dehane, Heat and mass transfer characteristics of charging in a metal hydride-phase change material reactor with nano oxide additives: The large scale-approach, Appl. Therm. Eng. 213 (2022) 118622, https://doi.org/10.1016/j.applthermaleng.2022.118622.

[32] A. Chibani, S. Merouani, A. Dehane, C. Bougriou, Hydrogen recovery from big-scale porous metal hydride (MH) reactor: impact of pressure and MH-thermophysical properties, Energy Storage Sav. 1 (2022) in press, https://doi.org/10.1016/j.enss.2022.07.005.

[33] A. Chibani, S. Merouani, C. Bougriou, The performance of hydrogen desorption from a metal hydride with heat supply by a phase change material incorporated in porous media (metal foam): heat and mass transfer assessment, J. Energy Storage 51 (2022) 104449, https://doi.org/10.1016/j.est.2022.104449.

[34] A. Chibani, S. Merouani, N. Gherraf, I. Ferhoune, Y. Benguerba, Numerical investigation of heat and mass transfer during hydrogen desorption in a large-scale metal hydride reactor coupled to a phase change material with nano-oxide additives, Int. J. Hydrog. Energy 47 (2022) 14611–14627, https://doi.org/10.1016/j.ijhydene.2022.02.171.

[35] P. Muthukumar, A. Kumar, N.N. Raju, K. Malleswararao, M.M. Rahman, A critical review on design aspects and developmental status of metal hydride based thermal machines, Int. J. Hydrog. Energy 43 (2018) 17753–17779, https://doi.org/10.1016/j.ijhydene.2018.07.157.

[36] A. Singh, M.P. Maiya, S.S. Murthy, Experiments on solid state hydrogen storage device with a finned tube heat exchanger, Int. J. Hydrog. Energy 42 (2017) 15226–15235, https://doi.org/10.1016/j.ijhydene.2017.05.002.

[37] S. Mellouli, E. Abhilash, F. Askri, S. Ben Nasrallah, S. Ben Nasrallah, Integration of thermal energy storage unit in a metal hydride hydrogen storage tank, Appl. Therm. Eng. 102 (2016) 1185–1196, https://doi.org/10.1016/j.applthermaleng.2016.03.116.

[38] R. Busqué, R. Torres, J. Grau, V. Roda, R. Busqu, R. Torres, J. Grau, V. Roda, Mathematical modeling, numerical simulation and experimental comparison of the desorption process in a metal hydride hydrogen storage system, Int. J. Hydrog. Energy 43 (2018) 16929–16940, https://doi.org/10.1016/j.ijhydene.2017.12.172.

[39] H. Ben Maad, F. Askri, S. Ben Nasrallah, Heat and mass transfer in a metal hydrogen reactor equipped with a phase-change heat-exchanger, Int. J. Therm. Sci. 99 (2016) 271–278, https://doi.org/10.1016/j.ijthermalsci.2015.09.003.

[40] M. Valizadeh, M.A. Delavar, M. Farhadi, Numerical simulation of heat and mass transfer during hydrogen desorption in metal hydride storage tank by Lattice Boltzmann method, Int. J. Hydrog. Energy (2015) 1–12, https://doi.org/10.1016/j.ijhydene.2015.11.075.

[41] P. Muthukumar, M. Prakash Maiya, S.S. Murthy, M.P. Maiya, S.S. Murthy, Experiments on a metal hydride-based hydrogen storage device, Int. J. Hydrog. Energy 30 (2005) 1569–1581, https://doi.org/10.1016/j.ijhydene.2004.12.007.

[42] M. Linder, R. Mertz, E. Laurien, Experimental analysis of fast metal hydride reaction bed dynamics, Int. J. Hydrog. Energy 35 (2010) 8755–8761, https://doi.org/10.1016/j.ijhydene.2010.05.023.

[43] G. Mohan, M. Prakash Maiya, S. Srinivasa Murthy, Performance simulation of metal hydride hydrogen storage device with embedded filters and heat exchanger tubes, Int. J. Hydrog. Energy 32 (2007) 4978–4987, https://doi.org/10.1016/j.ijhydene.2007.08.007.

[44] H. Dhaou, A. Souahlia, S. Mellouli, F. Askri, A. Jemni, S. Ben Nasrallah, Experimental study of a metal hydride vessel based on a finned spiral heat exchanger, Int. J. Hydrog. Energy 35 (2010) 1674–1680, https://doi.org/10.1016/j.ijhydene.2009.11.094.

[45] S. Anbarasu, P. Muthukumar, S.C. Mishra, Thermal modeling of $LmNi_{4.91}Sn_{0.15}$ based solid state hydrogen storage device with embedded cooling tubes, Int. J. Hydrog. Energy 39 (2014) 15549–15562, https://doi.org/10.1016/j.ijhydene.2014.07.088.

[46] M. Bhouri, J. Goyette, B.J. Hardy, D.L. Anton, Numerical modeling and performance evaluation of multi-tubular sodium alanate hydride finned reactor, Int. J. Hydrog. Energy 37 (2012) 1551–1567, https://doi.org/10.1016/j.ijhydene.2011.10.044.

[47] P. Muthukumar, M. Prakash Maiya, S.S. Murthy, Parametric studies on a metal hydride based single stage hydrogen compressor, Int. J. Hydrog. Energy 27 (2002) 1083–1092, https://doi.org/10.1016/S0360-3199(02)00005-8.

[48] P. Muthukumar, A. Satheesh, U. Madhavakrishna, A. Dewan, Numerical investigation of coupled heat and mass transfer during desorption of hydrogen in metal hydride beds, Energy Convers. Manag. 50 (2009) 69–75, https://doi.org/10.1016/j.enconman.2008.08.028.

[49] E.I. Gkanas, Metal hydrides: Modeling of metal hydrides to be operated in a fuel cell, in: Portable Hydrogen Energy Systems, Elsevier Inc., 2018, pp. 67–90.

[50] M. Afzal, P. Sharma, Design of a large-scale metal hydride based hydrogen storage reactor: simulation and heat transfer optimization, Int. J. Hydrog. Energy (2018) 1–17, https://doi.org/10.1016/j.ijhydene.2018.05.084.

[51] K.B. Minko, V.I. Artemov, G.G. Yan'Kov, Numerical simulation of sorption/desorption processes in metal-hydride systems for hydrogen storage and purification. Part II: verification of the mathematical model, Int. J. Heat Mass Tran. 68 (2014) 693–702, https://doi.org/10.1016/j.ijheatmasstransfer.2013.09.057.

Chapter 20

Oleaginous microbes for biodiesel production using lignocellulosic biomass as feedstock

Falak Shaheen[1], Palvi Ravinder[2], Rahul Jadhav[3], Navanath Valekar[4], Sangchul Hwang[5], Ranjit Gurav[5] and Jyoti Jadhav[2,3]

[1]*Department of Environmental Biotechnology, Shivaji University, Kolhapur, Maharashtra, India;* [2]*Department of Biochemistry, Shivaji University, Kolhapur, Maharashtra, India;* [3]*Department of Biotechnology, Shivaji University, Kolhapur, Maharashtra, India;* [4]*Department of Chemistry, Shivaji University, Kolhapur, Maharashtra, India;* [5]*Ingram School of Engineering, Texas State University, San Marcos, TX, United States*

1. Introduction

The progress of finding innovative products for the need of the world's population is presently focused on methods that preserve a healthy environment. Biorefineries are facilities to manufacture that translate biological raw materials (such as residues from agriculture) to several goods such as chemicals, bioplastics, biochar, feed, food, fuels, and energy [1–4]. The processing of biomass is equivalent to petroleum refineries, which polish crude oil into numerous products like fuels (e.g., petrol, diesel, and kerosene) and precursors of chemicals like butanol for manufacturing various materials [5]. According to the policy of renewable energy network in 2011, highly 78% of energy expenditure of the world was from fossil fuels, nuclear energy contributed about 3%, and the left 19% contributed from renewable energy sources that are acquired from renewable resources (solar, wind, hydrothermal, geothermal, and biomass). Carbon contributes nearly 13% of the renewable energy harnessed from rich biologically originated materials obtainable on the earth either by direct burning of biomass or by the thermochemical alteration of biomass to power and heat. Presently, biofuels are produced majorly using extracted sugars from the feedstock or by conversion to sugars from starch mostly from edible grains. Then sugars obtained from both sources are fermented using yeast into ethanol [6]. Oleaginous microorganisms (OMs) are the microorganisms that gather lipid that contains 20% or more of their dry weight depending on stress conditions, i.e., low nitrogen sources and higher carbon, and may range more than 70% during the stress circumstances [7]. These microorganisms use several renewable materials and change them into microbial oil, which can be exploited further to synthesize biodiesel via the process of transesterification [8]. One new product, Single-cell oil (SCO) has been known to be a substitute for biofuels, edible oils, and oleochemicals in the last decade [9]. SCO is also known as microbial oil. The initial SCO was produced commercially by using the filamentous fungus *Mucor circinelloides* in 1985, but its application for the production of biodiesel was not believed at that time [10]. SCOs formed from OMs have analogous chemical features such as plant lipids. Though, one of the supreme advantages of OMs over plants is that they need less ground and quick growers [11]. OMs are proficient in exploiting cheap feedstocks like lignocellulosic substrates, agro-residues, and other waste raw materials for the accumulation of higher lipid content [12,13]. Also, using OMs could realize the potential for development and a circular economy of efficient and cheap processes. For example, a secondary product of the biodiesel industry, crude glycerol, can be augmented as a source of carbon for lipid accumulation [13,14]. Subsequently, these lipids can be transformed into glycerol and fatty acid methyl esters (FAMEs) that can be used again. Visualizing this process could understand the feasibility of the process and hence profitable production of biodiesel using glycerol, agro-residues, and other waste substrates [15]. OMs production scale-up is easy under controlled conditions and they accumulate higher amounts of lipids within a short incubation period. For example, oleaginous microalgae can integrate atmospheric CO_2 into the synthesis of lipid that eventually supports carbon sequestration. The overall scheme of biodiesel synthesis using OMs is culturing of microorganisms, harvesting of biomass, biomass drying, extraction of lipids, and transesterification of the gained lipid [16]. Production of biodiesel based on OMs has many advantages such as vanishing the food versus fuel disaster, rapid incubation period compared to animal and plant resources, and the individuality of lipid production from the divergence of climate, season, and topography [17]. Furthermore, the microbial oil has comparable caloric value and composition to those isolated from plant and animal sources and has a low viscosity. Moreover, OMs can be employed to translate cheap industrial and agricultural wastes and even urban wastes into microbial oil with comparable quality to many high-value lipids [18]. Low-cost wastes like lignocellulose may help as a possible source for the production of microbial oil [19]. Though, tolerance against the products obtained after and lesser lipid yield produced out of lignocellulosic pretreatment procedures are among the major blocks for cost-effective microbial oil or SCO production. OMs consist of diverse microbial families, namely, bacteria, filamentous fungi or molds, yeast, and microalgae [20]. At present, biodiesel based on OMs is the emphasis of research, as it may offer enough oil as feedstock for consumption worldwide and synthesis of biodiesel far higher than those noted every hectare from plant feedstock. Moreover, they are probable of alleviating land use and food versus fuel wars and being cultured in environments that are not advantageous for energy crops. They are also gifted to decrease the greenhouse effect via CO_2 sequestration [21]. Depending on the methods of nutrition, OMs are divided into autotrophic, mixotrophic, and heterotrophic, which indicate that they have dissimilar metabolism. In the heterotrophic approach, organic compounds would help as a source of carbon that can be employed for growth, which is beneficial in terms of higher concentration and productivity. In the autotrophic approach, carbon dioxide provides a carbon source to microalgae, which is reduced in presence of light energy, and O_2 is released. Mainly, most of the microalgae limit to this class with a slight necessity for organic compounds and vitamins for growth. The mixotrophy approach includes exploiting both CO_2 and organic carbon sources as growth substrates for cell respiration pathways and photosynthesis. The pattern of cell growth in mixotrophy is analogous to the heterotrophic approach as the final growth is the total of heterotrophic and autotrophic approaches and results in enhanced productivity [22]. The high cost of microbial-based biodiesel production is the key problem for microalgae to be employed as the right feedstock. As per previous work, the production cost of microbial-based oil is 5.3–8.0 USD/L [8,23]. The main factors affecting the high cost of production of microbial oil comprise biomass productivity, production scales, price of oil retrievals from biomass, and lipid content [24]. Mixotrophic growth and heterotrophic growth have more benefits of higher lipid productivity and biomass, but providing a low-cost suitable resource of organic carbon is one of the main boundaries and hence one of the key aspects of the high production cost of biodiesel based on the microbial source. Wastes like lignocellulose that are obtainable abundantly presented as the best choice for a cost-effective organic carbon source for the growth of microorganisms [15,25,26]. Filamentous fungi or molds are OMs that can accumulate lipids up to 80% of their total composition [8]. Several oleaginous fungi species have been studied in the last decade. High lipid storage and exploiting renewable sources of carbon for their growth preferred oleaginous fungus to be a significant candidate for biodiesel feedstock. Furthermore, in contrast to microalgae, they can be cultivated in the classical bioreactors which will result in a reduction in manufacturing (biomass and oil) prices. Regardless, the exploitation of fungal oil to yield biodiesel is still at a lab scale. Several yeast species (around 30 species) have been previously known as oleaginous, and the list is increasing nonstop [27]. Generally, these species have appropriate for fermentation at a large scale and are also genetic enhancement amenable. Oleaginous yeasts are observed to accumulate lipids up to 70% under starving conditions. They have speedy growth as compared to microalgae and comprise triglycerides as the mainstream of oil production from the lipids of yeast is observed to be more expensive than the production of vegetable oil because of the costly commercial raw material; hence, choosing species of oleaginous yeast that can cultivate using inexpensive substrate is vital [28]. Waste like lignocellulose is one of the inexpensive sources of carbon that has been exploited for producing yeast oil using several species of yeast in many studies. Newly, the capability of oleaginous yeast for the production of lipid has been studied. Though, yeast oil obtainability for profitable utilization is not yet touched. Bacteria are having the features of ease of genetic manipulation and a high growth rate which can be completely exploited for high microbial oil production. Generally, bacteria are not identified for high lipid storage excluding some species, for example, *Streptomyces*, *Rhodococcus*, and *Mycobacterium* [29].

2. Prospective oleaginous microorganisms for biodiesel production

OMs are microbes with lipid content of more than 20% accumulated inside their cells, commonly including heterogeneous fungi, bacteria, yeast, and microalgae. Stored in the form of triacylglycerols (TAGs), microbial oils also known as SCOs are used for the synthesis of biodiesel [30]. Factors such as genetic composition and different culture conditions (pH, temperature, etc.) affect the lipid contents and composition of even in-between individual strains [31]. Table 20.1 consists of a few economically important OMs.

TABLE 20.1 A list of oleaginous microorganisms and their lipid content [31–34].

	Oleaginous microorganism	Lipid content (%, w/w)
Microalgae		
1	*Scenedesmus* sp.	34.1
2	*Chlorella protothecoides*	49
3	*Tetraselmis elliptica*	14
4	*Chlorella vulgaris* NIES-227	89
5	*Auxenochlorella protothecoides*	66–63
6	*Botryococcus braunii*	28
7	*Chlamydomonas reinhardtii*, CC1010	59
Oleaginous yeast and filamentous fungi		
1	*Cryptococcus* sp. (KCTC 27583)	34
2	*Rhodosporidium kratochvilovae* HIMPA1	53–70
3	*Trichosporon fermentans* CICC 1368	36–62
4	*Rhodosporidium toruloides*	56–67
5	*Lipomyces starkeyi*	48–61
6	*Rhodotorula glutinis*	20
7	*Cryptococcus curvatus*	53–70
8	*Lipomyces starkeyi* CBS 1807	30
9	*Fusarium oxysporum*	22–53
10	*Fusarium equiseti* UMN-1	56
11	*Sarocladium kiliense* ADH17	33
12	*Mortierella alpina* LP M 301	31
13	*Mortierella isabellina*	83
14	*Mucorales* fungi	42.7–65.8
15	*Cunninghamella echinulata*	57.5
Bacteria		
1	*Rhodococcus opacus* DSM 1609	4
2	*R. opacus* PD630	14–70
3	*R. opacus* DSM 43205	66
4	*R. opacus* PD630	46
5	*Gordonia* sp. DG	13–52

2.1 Oleaginous yeast and filamentous fungi

Oleaginous yeast and fungi are the preferred microbes for the production of microbial lipids due to their high lipid content and rapid growth rate. Oleaginous species of filamentous fungi and yeasts belong to *Umbelopsis* (*Mortierella*), *Microsphaeropsis*, *Fusarium*, *Candida*, *Meyerozyma*, *Rhodotorula*, *Rhodosporidium*, *Pichia*, *Cryptococcus*, *Lipomyces*, *Trichosporon*, and *Yarrow* [35]. The micro lipid content can be as high as 80% in some species of yeast [34]. They mostly utilize glucose, sucrose, glycerol, arabinose, xylose, sodium glutamate, and sugarcane bagasse hydroxylates as carbon sources. Up to 40% of their biomass can be lipid on average, which can increase up to 80% under stressed conditions. Even with similar lipid content among species, the lipid profile can vary greatly between them, whereas with different lipid content, the profile can be similar in various species much as in *Rhodotorula*. Their environmental adaptability, less doubling time, and ease of cultivation make them preferred over microalgae for biodiesel production [30]. Fungi can store up to 70% of their dry weight as carbohydrates and lipids [10]. The nature of the carbon source usually affects the lipid content in the fungi, along with other factors such as temperature, pH of the growth medium, nitrogen source, C/N ratio, etc. Nitrogen acts as a limiting source, but its excess can create a deficiency of other nutrients increasing the lipid content [16].

2.2 Microalgae

Microalgae have high photosynthesis efficiency, are easily culturable, require less growth area, and high cell division rate accountable for higher biomass generation for biodiesel production. Microalgae can be cultured by various modes of nutrition as autotrophic, heterotrophic, and photoheterotrophic mixotrophic modes utilizing both organic and inorganic nutrient sources [36]. As a result of several enzymatic reactions, the chloroplast and endoplasmic reticulum synthesize TAGs and store as glycerolipids. The TAGs are stored in the oil bodies within the subcellular components. The TAGs are synthesized under stressed conditions and stored as a neutral source of carbon and energy [37]. The micro lipid content can vary from anywhere between 20% and 80%. Representative microalgae belong to *Neochloris, Chlorella, Scenedesmus, Chlamydomonas, Nannochloropsis, Chlorococcum, Isochrysis, Cylindrotheca, Tetraselmis*, and *Auxenochlorella* [16,30,35]. Few species of *Chlorella, Neochloris*, and *Nannochloropsis* can store up to 40%–80% of their dry weight as lipids in their cell walls. Divided into various categories some of the commonly used oleaginous algae belong to Chlorophyceae (green algae), Chrysophyceae (golden algae), Cyanophyceae (blue-green algae), and Bacillariophyceae (diatoms). Freshwater green algae usually save starch and oil as their storage material along with photosynthetic material comparable to higher plants. Blue-green algae are found in most habitats including soil, freshwater, and marine. Apart from chlorophyll, they have a higher content of phycobilin proteins which give the algae their characteristic blue-green color. Starch and oil are the main storage elements in blue-green algae. Golden algae and brown-red algae are mostly marine and have TAGs, and carbohydrates are their main storage substance. Experiments have been conducted to cultivate them in artificially created seawater. Diatoms can be usually found in the ocean as well as fresh and brackish water and contain chrysolaminarin (polymer of carbohydrates) and TAGs as their storage materials. Having the capability to switch between heterotrophic and autotrophic modes of nutrition, microalgae can also grow mixotrophically. During heterotrophic growth, it requires glucose or easily accessible sources of carbon for metabolism and proliferation. Since they do not require agricultural land for their cultivation, they do not compete with food crops and human interests, making them a favored biomass feedstock [16].

2.3 Oleaginous bacteria

Characteristics such as high growth rate make bacteria a preferred source for biodiesel; however, it is observed that in comparison to microalgae and yeast their utilization is limited. Bacteria have the potential to grow on various carbon substrates such as glucose, sodium glutamate, glycerol, sucrose, plant oils, etc. [30,37]. Bacteria belonging to the genus *Rhodococcus, Streptomyces, Nocardia, Acinetobacter*, and *Arthrobacter* are observed to accumulate high content of lipids. *Rhodococcus* sp. is the most studied bacteria in terms of biodiesel production [30,38]. Lignin is left underutilized during the use of bacteria as an OM for biodiesel production in biorefineries. Recent studies show the ability of *Rhodococcus* species to break down lignin into its monomeric units. Dextrose, sunflower oil, sesame seed oil, peanut oil, etc., as substrates are observed to give a high yield of about 50%–70% of lipids in the oleaginous bacteria [37]. Under high carbon and low nitrogen conditions, the lipid content can go up to 80%. Most of the storage occurs during the stationary phase of microbial growth.

2.4 Cocultivation of microorganisms

Various studies have been conducted to study the consortia of microbes, for example, microalgae and bacteria consortia or microalgae and yeast consortia. An increase in the lipid content was reported making it a favorable study for further development in technology for biodiesel production. However, studies showed that several factors can affect biodiesel production which may vary according to individual species due to the difference in the physiology of different microbes [18].

3. Lignocellulosic biomass as a carbon source

Considered the most abundant and economical biomass source, lignocellulosic components consist of carbohydrates, lignin, cellulose, hemicellulose, extractives, and ashes and are the building blocks of secondary plant cell walls [16,39].

The total worldwide production of cellulose and hemicellulose is about 85×10^9 ton/annum, with cereal straw estimated to exceed 2.9×10^9 ton/annum [40]. The proportion of components varies among different plant species, types, plant parts, age, and growth conditions. Carbohydrate components have higher concentrations of cellulose and hemicellulose along with a lesser amount of pectin. Extractives include terpenes, waxes, fats steroids, and phenolic compounds (lipophile and hydrophiles) and are a minor fraction of 1%—5% of biomass [16]. Lignocellulose is the most favorable and sustainable alternative to fossil resources as they are generated through biological photosynthesis. Comprising a high percentage of assimilable sugars, LCB is the preferred source to produce biofuels [35]. Even with easy availability and abundance, the conversion of LCB to biodiesel has been a technical and economical challenge [41]. Even though the biomass resource divisions are not clear, they can be roughly divided into four categories: (i) food crops, (ii) nonfood/energy crops, (iii) forest residues, and (iv) industrial process residues [41,16].

3.1 Food crops

Large area of land available on the earth has been utilized for agriculture. However, if the current scenario pertains, it is expected that in the future the world would face severe food crisis. Rice, vegetables, pulses, wheat, coconut, millets, maize, and sugarcane comprise some of the main food crops [41]. Among the food crops, some of the main energy-producing crops include wheat, barley, sorghum, maize, sugarcane juice, bagasse, etc. First-generation biofuels utilized food crops as the source, which ultimately lead to the discussion on the importance of food versus fuels [16]. The OMs efficiently utilize food crops as the source of simple sugars that helps the microbes to accumulate a good amount of lipids. Vegetable oils such as soybean, rapeseed, sunflower palm, poppy seed, sesame seeds, wheat grain, linseed, bay leaf, peanut kernel, hazelnut kernel, olive kernel, and coconut oils are some of the feedstocks for the conventional first-generation biofuels.

3.2 Nonfood/energy crops

The nonfood crops have higher energy potential, generates high biomass as compared to food crops, and can be cultivated in a soil not suitable for food crops. These are usually comprised of herbaceous and perennial, viz., Switchgrass [30]. Broadly called lignocellulosic, its conversion to biofuels is difficult due to the presence of toxins. Studies suggested growing energy crops on marginal lands, such as lands that are vulnerable to high soil erosion, or land with poor water drainage or wastewater reclaimed land [41]. Energy crops have comparatively higher productivity than their alternatives. They are observed to be more homogenous in their chemical and physical features. Among all the energy crops the most preferred are the perennial crops. Some of the common energy crops include switchgrass (*Panicum virgatum*), elephant grass (*Pennisetum purpureum* Schumach.), poplar (*Populus* spp.), willow (*Salix* spp.), mesquite (*Prosopis* spp.), Napier grass, reed canary grass, *Miscanthus, Bambusa bambos, Ricinus communis, Saccharum spontaneum*, sugarcane tops, cotton stalks, cedar, sorghum, alfalfa, pine, tobacco, etc. [16,30]. Energy crops grown on wastelands can take up heavy metals leached into the soil, thus decreasing environmental toxicity [41].

3.3 Forest trees and residues

Forest resources are the largest source of lignocellulosic biomass second to agricultural residues (Fig. 20.1) [41,42]. Biomass can be obtained from all parts of the forest wood including top, root, tree trimmings, foliage, wooden debris, etc. Trees affected by natural disasters such as floods, tsunamis, storms, drought, landslides, fire, or diseases can also be utilized along with trees grown specifically for energy production. "Disturbance wood" or the wood infected by insects is an additional source of biomass; as these are nonrenewable, they can be used supplementary to the other sources for biodiesel production [16]. Considered the most sustainable and renewable source for biofuel production, they have additional environmental benefits such as protection of soil from erosion, carbon sequestration, protection of forests, etc. [30].

3.4 Industrial process residues

With the advancement in technology, the waste of industrial residues can be reduced as various methods are applied to extract the products of our need. The residues left out after the processing of industrial crops are sources of abundant cellulosic biomass and are completely reusable. Sugarcane bagasse, rice bran, and maize cobs are some of the most common feedstocks that produce a large amount of biomass [43]. As these are residues after industrial processing, the chemical and physical characteristics such as cellulose content, and fermentable sugar content, are different; hence the technologies used for biodiesel production must vary according to the raw material used.

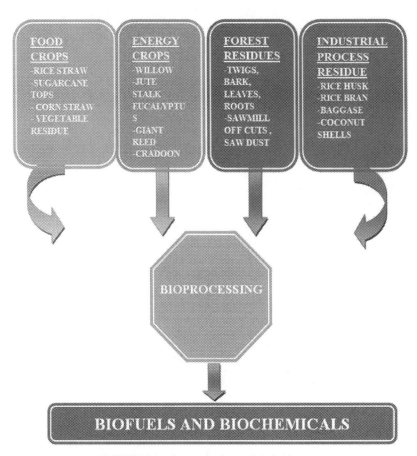

FIGURE 20.1 Sources for lignocellulosic biomass.

4. Structural features of lignocellulosic biomass and recalcitrance

In a network of cellulose, hemicellulose, lignin, and a trace amount of pectin, starch, and ashes, lignocellulose provide mechanical support and stress tolerance to plants. 30%–50% of dry biomass both in softwood and hardwood is cellulose making it the major constituent. Cellulose is a linear homopolymeric compound of more than 10000 glucose molecules linked by $\beta(1-4)$ glycoside with cellobiose as dimeric units. Hemicellulose is a amorphous and branched polysachharide formed by one main chain shorter than the cellulose chains and several side chains typically linked by $\beta(1-4)$ glycoside bonds and occasionally by $\beta(1-3)$ bonds [35]. Hexose and pentose sugars are present along with uronic acid and deoxyhexose. Hardwood usually contains glucuronoxylans, while softwoods contain galactoglucomannans, which are about 18%–35% of the biomass. The most abundant polymer after cellulose is lignin, about 19%–35% of the biomass. Mainly present in the vascular tissues, lignin provides mechanical strength to the xylem [30]. Compounds like tannins, phenols, pectin, starch, and protein are present along with lignin in little quantities. Free amino acids, alkaloid inorganic compounds (e.g., Ca, K, Mg, etc.), and trace elements (e.g., Ba, Al, Fe, Zn, etc.) are also present [35].

Lignocellulose is innately recalcitrant to enzymatic degradation significantly affecting its introduction as biomass for bioenergy industries. Lignocellulosic biomass recalcitrance is affected both by the physical and chemical properties of the bonds formed during polymerization. Consisting mainly of cellulose, hemicellulose, and lignin the chemical properties of the biomass affect the hydrolysis [44]. Embedded into the lignocellulosic matrix, cellulose creates resistance for enzymatic activity. Acting as a physical barrier, hemicellulose limits the access of enzymes for the degradation of biomass. Both cellulose and hemicellulose are bound to the cell wall by lignin making the structure rigid and hydrophobic. Cellulose and hemicellulose are tightly bound by hydrogen bonds while lignin binds to cellulose and hemicellulose covalently. Lignin enhances the mechanical properties of the plant cells as it controls the permeability in turn controlling the microbial and enzymatic degradation. It acts as the most important factor in imparting the recalcitrant properties to lignocellulosic biomass [35,45]. Physical properties such as crystallinity, particle size, accessible surface area (ASA), and accessible volume (Pore Size: Internal Surface Area) play a deciding role in the recalcitrant properties of the biomass. An increase in the crystallinity and particle size increases the lignocellulosic biomass recalcitrance, whereas an increase in the accessible surface area and accessible volume decrease [45].

5. Pretreatment

Plant biomass is innately recalcitrant to protect the invading microorganisms in nature. Natural degradation of biomass may take several months or years; however, industrial processing of biomass into biofuels is needed to be done within a few days. The recalcitrance of the biomass needs to be treated to enhance the accessibility of the enzymes for the hydrolysis of the polysaccharides (cellulose and hemicellulose) by breaking down the hemicellulose−lignin complex [46]. Consisting of physical, chemical, biological, and a combination of them, several pretreatment processes have been studied to enhance the efficiency of biodiesel production as it increases the accessibility of cellulose and hemicellulose for enzymatic hydrolysis [47]. Pretreatment is a primary requisite for the conversion of agro-industrial residues to biofuels. To enhance the reagent action, pretreatment decreases the degree of polymerization by increasing surface area and biomass porosity [35]. Pretreatment of lignocellulosic biomass is also considered the lignin depolymerization stage in biodiesel production [48]. To some extent, the biomass gets hydrolyzed during pretreatment [49]. Some of the basic characteristics of potential pretreatment techniques are [50]:

(i) It should efficiently break down cell walls and release lignin to the surface and can densify the biomass without the aid of an external binding agent.
(ii) Densified pretreated biomass in addition to a biorefinery feedstock should have multiple applications, for example, as fertilizers, animal feed, etc.
(iii) Methods that produce lesser toxic degradation products which negatively affect the downstream process and decrease yield.
(iv) Methods that can be scaled up to biorefinery demands of up to 2000 tons per day or more.
(v) Energy capital costs should be less as it decreases the overall capital costs of production.
(vi) Pretreatment techniques should use inexpensive chemicals.
(vii) Techniques that can preserve lignin, as it produces heat and energy during subsequent steps during the production of biofuel.
(viii) Processes done at moderate pressure and temperature are favored, to maintain the cost of instrumentation and safety measures to be implemented.
(ix) Use of less hazardous chemicals is preferred.
(x) During pretreatment catalyst recovery is an important step for the environment, even though it increases the production cost slightly. The catalyst should be neutralized with an acid or as a base before releasing up into the waste stream as most of the catalysts are water soluble. It is expected for the pretreatment techniques to follow most of the abovementioned requirements. Proper pretreatment can drastically increase the efficiency of enzyme hydrolysis and decrease the concentration of enzyme required. The amount of lignin present after pretreatment decides the efficiency of conversion of fermentable sugars after a pretreatment technique [50]. Various pretreatment methods are categorized as mechanical and nonmechanical methods (Fig. 20.2).

5.1 Mechanical pretreatment

Preferred to reduce the particle size and increase surface area, mechanical processes such as disk milling or grinding are used to subject the lignocellulosic biomass to shear stress and pressure under high temperatures. The stress leads to the breakdown of the fiber polymers and their shortening [50], further reducing their crystallinity. Most common mechanical treatments include grinding, chipping, and milling. The specific energy requirement for the mechanical processing of lignocellulosic biomass is affected by several biochemical and structural characteristics of biomass [16]. Solid shear forces and liquid shear forces are two main forms of mechanical forces. The solid shear forces include bead milling and high-speed homogenization, whereas the liquid shear forces consist of high-pressure homogenization and microfluidization. Mechanical treatment also includes direct energy transfer by waves (microwaves, ultrasonication, laser treatment), pulse-field application, and autoclave.

5.1.1 Oil expeller

Dry biomass is crushed using mechanical stress resulting in the extraction of lipids from the broken cell. Employed mostly for oil extraction from seeds, it has also been found useful for microalgae. On the type of biomass, different press

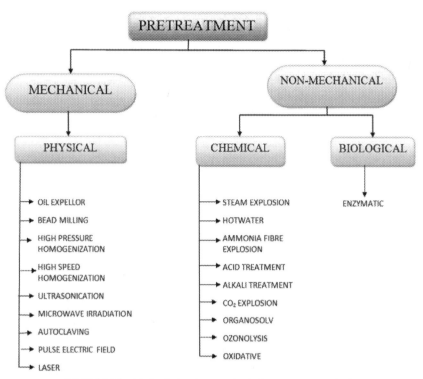

FIGURE 20.2 Mechanical and nonmechanical pretreatment methods.

configurations are used such as screw, expeller, piston, etc. However, this method is disadvantageous for low quantities of biomass and is comparatively slow.

5.1.2 Bead milling

Applied during the downstream processing of microbial cell disruption, this reduces the particle size. Biomass concentration, bead size, bead-to-biomass ratio, slurry flow rate, speed of agitator, etc., affect the disintegration of cells and are most favorable for oleaginous yeast. High operation energy consumption provides a disadvantage for bead milling.

5.1.3 High-pressure homogenization

Microbial cells are homogenized under highly pressurized fluids on the stationary valve. The efficiency of the cell disruption varies as per the valve settings. Several studies have suggested its efficiency for the disruption of wet biomass of microalgae, yeast, and bacteria. The processing fluid is passed under high pressure through a homogenization chamber, where the energy stored in the fluid increases its velocity which leads to high mechanical stress in the form of cavitation, stress and shear forces followed solvent extraction treatment with high-pressure homogenization leading to a higher yield.

5.1.4 Ultrasonication

As the ultrasound is applied to the microbial cells, cavitation and acoustic streaming are two different events that occur. Microbubbles are formed under pressure as it disrupts the cell walls and cell membranes. Several studies are conducted to find a suitable solvent for ultrasonication, such as chloroform-methanol, diethyl ether, hexane, etc. Suitable for all microorganisms, it is found highly efficient for microalgae.

5.1.5 Microwave irradiation

In the treatment of polar compounds with microwave, high energy is produced as the applied alternating current converts into electromagnetic energy. Arranged identically to the direction of current, the polar molecules rotate at high speed on the alteration of microwave direction. The frictional movement of the compounds leads to high energy and enhances the lipid yield as compared to the standard Bligh and Dyer method. It is found to be the simplest and most potent method for lipid extraction from microalgae. Appropriate only for polar compounds and inappropriate for volatile compounds are a disadvantage of microwave irradiation. Long-time use of the method leads to the generation of free radicals in the biomass as well as an increase in temperature.

5.1.6 Autoclaving

Autoclaving is a technique utilized for sterilization of laboratory instruments and microbial culture media before the growth. Cells of various microalgae are disrupted by autoclaving at 121°C and 1.5 MPa for 5—30 min. The sudies showed that *Botryococcus* sp. which had the highest yield of nutraceutical important fatty acids were obtained after autoclaving of the cell biomass. However, it is not a favorable technique when a consortium of microalgae is used. Some of the other common mechanical techniques include laser treatment, pulse-field treatment, and the physiochemical process of acid-catalyzed hot water treatment. The laser disintegrates the cell wall into its components without causing damage to the cell organelles. It is one of the fast methods as it lacks the requirement of any organic solvent and no laborious work. It is observed that the highest disruption was obtained from cells being treated with laser (96%) followed by microwave (94%) [51]. Pulse electric field works on the principle of electroporation, as an external electric field creates critical electric potential across the cell, and electric field—induced tension and electromechanical compression. With the increase in applied electric field and as it is pulsated, the membrane porosity increases proportionately leading to irreversible or reversible pore formation. Pores formed are proportional to the cell wall or cell membrane total surface area. Hot water treatment is a well-known mechanical treatment technique. High pressure under high temperature is applied during the hot water treatment, leading to the extra cost of water removal from the final mixture. Acid-catalyzed hot water resolved this advantage as the pretreatment is carried out under acidic conditions; it reduces the production cost, hence effective for wet biomass [34].

5.2 Nonmechanical pretreatment

With several disadvantages like the use of toxic solvents, low extraction yield, and long processing time, conventional mechanical techniques are less preferred after the development of technology. The energy consumption of mechanical techniques is higher when compared to mechanical techniques. Nonmechanical techniques mainly use chemical and biological or enzymatic methods for the disruption of cell walls.

5.2.1 Chemical treatment

This is usually carried out under acidic, basic, or neutral conditions.

5.2.1.1 Ammonia fiber explosion

Like the steam explosion, biomass is exposed to hot liquid ammonia under elevated pressure for a short period and the pressure is released. 1—2 kg of ammonia per kg of lignocellulosic biomass is the standard dosage for about 30 min at 90°C. Ammonia fiber explosion (AFEX) treatments are more effective for herbaceous plants and agricultural residues than hardwood and softwood. AFEX pretreatment causes structural alterations in the biomass increasing the water-holding capacity and digestibility significantly. A six-fold increase in yield is observed after treatment with AFEX when compared to untreated biomass [16].

5.2.1.2 Acid pretreatment

One of the most commonly used physiochemical techniques, acid pretreatments breaks the glycosidic bonds between lignin and hemicelluloses as they solubilize most of the hemicellulose and some parts of cellulose and lignin. The acids used can be both organic and inorganic [35]. Sulfuric acids are the most commonly used acid; some other acids also include hydrochloric acid, nitric acid, acetic acid, and phosphoric acid for the degradation of lignocellulosic biomass. Acids for the treatment are used both in dilute and concentrated form; however, the use of extremely concentrated acids may lead to corrosiveness, high toxicity, and the generation of inhibitory products such as furfural and hydroxymethyl furfural (HMF) decreasing the yield downstream. It also causes the need to extract the excess amount of acid during purification and its reuse. The use of dilute acid for acid pretreatment is preferred as it provides more suitable reaction conditions [16]. The use of concentrated acid for the pretreatment provides a higher yield of assimilable sugars and is low on energy costs as it is operated under low temperatures. However, it may cause the corrosion of instruments leading to the high cost of maintenance and safety issues. Dilute acid methods are low on cost due to overall less use of acid, less corrosive, and hence preferred over concentrated acids. Most of the assimilable sugars are obtained in the end with a low concentration of inhibitory products. Most of the biomass after treatment is easily accessible to enzymes after the treatment, therefore giving a biomass pulp with over 70% digestibility. The only disadvantage is a higher energy requirement than concentrated acid as it requires a higher temperature further increasing the energy costs. Acids first hydrolyze the hemicellulose into its monomers and depolymerize the cellulose into oligosaccharides which further break down into monomers. This leads to an increase in porosity and a decrease in lignin concentration. However, this method is not suitable for the removal of lignin from softwood because of its structural characteristics of softwood lignin [35]. Further studies show the use of dry diluted acid as an alternative to wet dilute acid pretreatment, where both raw material and product are solid. Even more than 70% of the biomass is utilized with this method. The low production of furfural and HMF gives it a major advantage over other techniques [16].

5.2.1.3 Alkali pretreatment

Bases such as calcium, ammonia, potassium, and sodium hydroxides are used to prepare solutions that soak the lignocellulosic biomass at the desired temperature. Unlike acid treatment, basic treatment techniques aim at breaking glycosidic bonds and ester linkages which alters lignin structure and causes partial decrystallization. Cellulose fibers partially swell and cause partial dissolution of hemicellulose (>40%). This leads to enhanced porosity of the biomass as the process leads to dissolution along with saponification of ester linkages. Further, the polysaccharides are broken down by the hydrolysis of glycosidic bonds and acetyl groups which enhances the subsequent enzymatic reactions [16]. Comparatively alkali treatment favors hardwood with less lignin as biomass rather than softwood with high lignin content. The low costs involved are an advantage for basic alkali pretreatment, whereas the salts formed during the treatment along with low monomer yield proves to be a disadvantage. Alkali pretreatment needs to be followed up with the removal of dissolved lignin, excess alkali to be neutralized with acid, and removal of excess salt formed due to neutralization. Potential inhibitors such as aldehydes, phenolic compounds, and furfuryl are formed during the alkali pretreatment which needs to be removed before the hydrolysis of biomass [35].

5.2.1.4 Steam pretreatment/steam explosion

Lignocellulosic biomass is exposed to hot steam up to 240°C under high pressure for steam pretreatment, after which the biomass is allowed to cool down as the steam releases. Following the process, hemicellulose solubilizes partially, and cellulose is better exposed to the enzymes and chemicals with the production of inhibitors. Autoclaving of hemicellulose takes place as its solubilization leads to the formation of enzymes that break down the soluble hemicellulose into its oligomers. A longer retention time is required in case the biomass contains a higher moisture level [35]. Steam explosion is considered one of the most effective processes for LCB from hardwood. The biomass is exposed to high pressure of water vapor (0.69–4.38 MPa) and high temperature (160–260°C) for short period and the quick depressurizing which disrupts the biomass and partial breakdown of hemicellulose and transformation of lignin. As the hemicellulose moves out of fibrils, enzymes get better access to cellulose and its further hydrolysis. Particle size, moisture content, residence time, and temperature affect the efficiency of steam explosion pretreatment. The disadvantages of steam explosion include partial breakdown of lignin and carbohydrate, breakdown of pentose sugars, and formation of inhibitory compounds. The inhibitory products must be removed before the downstream processing of lignocellulosic biomass [16].

5.2.1.5 Hot water pretreatment

Instead of steam liquid, hot water is used for the pretreatment of biomass before hydrolysis. High pressure is applied to keep the temperature up to 200–230°C. Almost 40%–60% of biomass is dissolved after hot water pretreatment, with complete dissolution of hemicellulose, 4%–20% of cellulose, and 35%–60% of lignin. Since cooking in hot water breaks down the lignocellulosic biomass, the prereduction of the size of particles is not necessary. As the temperature and residence time increase, higher yield of hemicellulose-derived sugar increases. Acetylated xylan residues generated after pretreatment decreases the digestibility of treated substances. Maintenance of pH between 4 and 7 prevents the formation of inhibitory products. The high lignin content of softwoods makes this technique ineffective for its pretreatment [16].

Some of the other commonly used chemical pretreatments include CO_2 explosion, ozonolysis, and organosolv pretreatment. CO2 acts like a superfluid with its density identical to liquid CO_2 at pressures 7.39 MPa and temperature of 31°C, its critical point. The diffusibility and penetration power remain similar to its gaseous state. At the supercritical point, CO2 is not an indefinite solid or liquid state and easily effuses through solid biomass like gas with the size of the molecules being similar to water and ammonia; it has access to all the pores. Carbonic acid is formed as the CO_2 reacts with moisture and further degrades the biomass. Strong oxidants like ozone (E0 = 2.07 V at 25°C) can degrade lignin and hemicellulose.

The terminal oxygen of ozone being electron deficient attacks electron-rich lignin molecules leaving carbohydrates unaffected. Performed under moderate temperature and pressure, ozonolysis requires a large amount of ozone thus increasing the cost. Utilization of organic solvents or organic aqueous solvents is termed organosolv pretreatment. It may include acidic (organic/inorganic) and basic compounds as catalysts, suitable for biomass with high lignin content, as it breaks the lignin and hemicellulose internal bonds. As the lignin breaks down into low-molecular-weight fragments, residual catalysts need to be extracted and recycled to reduce cost and prevent disturbance of the downstream process. The low boiling point and flammability of the solvents remain a major disadvantage for the organosolv pretreatment.

5.2.2 Biological pretreatment

The biological pretreatment methods are comparatively environment-friendly, and less expensive techniques as they do not need harsh chemicals, and high energy consuming instruments. The pretreatment is usually performed under mild conditions and low capital cost, however this pretreatment is slow and requires several days of incubation [50]. It takes advantage of saprophytic fungi which have a natural degradation mechanism for the degradation of lignocellulosic biomass preceding its hydrolysis. White, brown, and soft rot fungi are capable of delignifying LCB as they produce enzymes like laccase, lignin peroxidases, manganese-dependent peroxidases, and polyphone oxidases which degrade lignin and cellulose [16]. It is observed that the biomass treated with only biological methods has a lower rate of sugar conversion when compared to chemical methods. A combination of biological methods followed by chemical methods such as steam explosion, hot water treatment, and other physical and chemical treatments are found to be better effective as well as they require less severe hydrolysis of the lignocellulosic biomass in the next step [50]. Biological pretreatment can extensively delignify the lignocellulosic biomass. With a C: N ratio of 30:1 for fungi and 10:1 for bacteria, fungi naturally degrade lignocellulosic biomass as the C: N ratio of LCB is very high and they are less dependent on the nitrogen content. The source of biomass, moisture content, temperature, particle time, residence time, etc., affect the degree of delignification of biomass. Toxic compounds which may lead to the inhibition reactions during the down streaming process are usually not produced during the biological treatment. However, it is a rather slow process for large-scale industries because it may take several days, making it difficult and expensive to maintain sterile conditions (Table 20.2) [16].

TABLE 20.2 Advantages and disadvantages of pretreatment methods [25,34,35,50].

	Pretreatment techniques	Advantages	Disadvantages
Mechanical pretreatment			
1	Disk milling	No chemicals required Scalable	Low sugar content energy-intensive process
2	Bead milling	Solvent-free short residence time Simple and effective process	Low efficiency No residual effect
3	Hot water	Size of particles reduced, solubilization of hemicellulose lignin removed 73% Sugars obtained 95% Low inhibitory compounds, catalysts, and chemical free	The high amount of degradation products or inhibitors generated High water and energy demand
4	Ultrasonication	Less processing times Low solvent consumption Greater penetration into hard cell walls	High power consumption
5	High-pressure homogenization	Solvent-free, simple short, effective process	Energy-intensive process The increased temperature during operation
6	Pulsed electric field treatment	Simple High energy efficiency Relatively fast method	High maintenance cost High temperature during operation decomposition of fragile compounds

Continued

TABLE 20.2 Advantages and disadvantages of pretreatment methods [25,34,35,50].—cont'd

	Pretreatment techniques	Advantages	Disadvantages
7	Microwave	Cost-efficient Short reaction time Degrades lignin and hemicellulose Applied to acids and alkali and steam Eco-friendly	Increased formation of inhibitors
Nonmechanical pretreatment			
1	Steam explosion	Works well for both hardwood and softwood Cost-effective Lignin removal effectively	Expensive reactor required Inhibitory products generated
2	AFEX	Partial disruption of fiber 99% solid recovered Low inhibitor concentration 85% lignin removed 95% sugars obtained	Not suitable for softwood Ammonia dosage increases cost Residual lignin affects hydrolysis making unspecific bonds Insufficient biomass with high lignin
3	CO_2 explosion	Efficient for both hardwood and softwood Low inhibitory compounds generated Low-cost treatment	Yield lower than steam explosion and AFEX High pressure required
4	Acidic	Hydrolyzed xylose produced soluble lignin-rich hemicellulose stream Effective on softwood High yield of fermentable sugars	Corrosive Energy-intensive reactions Expensive reactors Toxic by-products Difficult to control reactions High water usage to recover neutralization salt High cost of recovery
5	Alkaline	Lesser degradation products formed Low reaction temperature	A large amount of water required Longer residence time Energy-intensive Not efficient for hardwood Catalyst recovery is expensive
6	Ionic liquids	Low carbohydrate loss Low concentration of by-products	Solvent cost is high Residual contents of ILs interfere with enzymatic hydrolysis
7	Organosolv	Pure lignin obtained Increased cellulose digestibility Mild processing conditions	Flammability Volatile substances
8	Biological	Mild treatment conditions Low energy consumption No chemicals needed	Slow process Low rate of sugar conversion Large space required Continuous monitoring needed Maintenance of sterile conditions

5.3 Importance of catalyst recovery

The catalysts used during the pretreatment process are mostly water soluble and end up in wastewater streams. In situations where the quantity of catalyst used is low, there is no need for catalyst recovery. Catalyst recovery is an expensive and energy-extensive process, done by expensive ultrafiltration or chemical precipitation. As the catalysts are neutralized by the use of acid or base, it leads to the formation of salts, which further increases the cost as it further decreases the efficiency of the downstream process and requires a large quantity of water for their removal.

6. Downstream processing of the biomass

6.1 Hydrolysis/saccharification

Saccharification/hydrolysis of the lignocellulosic biomass is the process to breakdown carbohydrate polymers into monomer that is usually performed by various pretreatment methods. Chemical and enzymatic methods are usually used for hydrolysis.

$$(C_6H_{10}O_5)n + nH_2O \rightarrow nC_6H_{12}O_6$$

6.1.1 Enzyme hydrolysis

The conversion of sugar polymers to their monomeric units is termed hydrolysis [52]. Biomass is degraded into monomeric sugars by the enzymes secreted by the microbes (including bacteria and fungi). Hydrolysis of pretreated LCB using enzymes is the last step before fermentation [35]. The enzymatic hydrolysis depends upon the cell wall composition of the OMs. Cellulase and hemicellulase are the most common cell wall degrading enzymes. Other lesser-used enzymes include xylanase, amylase, papain, and pectinase. Experiments with enzyme cocktails of up to 40–50 different enzymes are added to the microbial culture along with agricultural residues [53]. Cellulases are mainly divided into endoglucanases, exoglucanases, and β-glucosidases which break the β-1,4-D linkages of cellobiose and release glucose monomers. Hemicellulase causes a complete degradation of hemicellulose and releases a group of enzymes mainly consisting of xylanases which break down xylan to release xylose [35]. The cooperative effect of hemicellulase and cellulases increases the total yield of fermentable sugars. The enzymatic treatment of microbes is a preferred technique as it lacks the involvement of chemical solvents and severe physical stress on the cells. It is favored due to its specificity and low energy requirements. Several factors affect the yield of lipids and rate of reaction in microbes as pH, temperature, biomass constitution, and activity of cellulase enzyme during enzymatic hydrolysis. Negatively affecting factors include the high concentration of sugar, the presence of inhibitory products generated during pretreatment, and the long residence time. Various experiments observed the increased efficiency of solvent extraction of lipids after enzymatic hydrolysis and in its absence [34]. Simultaneous saccharification (enzymatic hydrolysis) and fermentation are proposed as an alternative to reduce the pretreatment costs. The synthesis of metabolites by the microbes and synthesis of sugars are combined in a single stage, reducing the reaction time, enzymatic load, contamination risk, and instruments decreasing the total production cost while giving a higher yield. Several factors such as pH, sugar and enzyme concentrations, temperature, saccharification, and fermentation (at different optimal temperatures) must be optimized to obtain a higher yield.

6.1.2 Chemical hydrolysis

Chemical hydrolysis includes acid hydrolysis and ionic liquid (IL) hydrolysis. Sulfuric acid is the most preferred acid followed by hydrochloric acid. Based on the concentration of acid it is categorized into (i) dilute acid hydrolysis and (ii) concentrated acid hydrolysis. The acid concentration in DAH is <10% (w/w) and is carried out between 100 and 240°C to hydrolyze both cellulose and hemicellulose [52]. Also considered a pretreatment technique, it usually hydrolyzes hemicellulose completely and partial degradation of cellulose and lignin [16]. Concentrated acid hydrolysis usually has an acid concentration of >10% (w/w). Sulfuric acid is used for CAH under low temperatures and atmospheric pressure; it is followed by water dilution to hydrolyze the soluble sugars. Total cost is less, as moderate working conditions allow the use of low-cost materials and energy requirement is low. About 90% of both hemicellulose and cellulose get hydrolyzed during CAH. Common acids for hydrolysis are mineral acids, e.g., sulfuric acid (H2SO4), phosphoric acid (H3PO4), nitric acid (HNO3), and hydrochloric acid (HCl), as well as organic acids such as trifluoroacetic acid (CF3COOH), oxalic acid (HOOC COOH), and acetic acid (CH3COOH) [52]. ILs are a developing category of green solvents consisting of inorganic/organic anions and organic cations. Pyridine, imidazole, triethylamine, and triethanolammonium are usually the base of most ILs [16]. ILs are dissimilar from organic solvents as they are nonvolatile and ionic. They function well with both organic and inorganic compounds and function over a large range of temperatures from ambient to over 300–400°C while being thermostable. Considered an environment-friendly substitute for organic solvents, they are effective cellulose solvents. Cellulose is efficiently dissolved by the anionic part of ILs. Acetate, chloride, and other basic components break down the hydrogen bond facilitating its dissolution. ILs are less time-consuming and highly efficient; however, the industrial-scale development of IL hydrolysis is expensive. Further studies are being conducted to prepare on-stage reactions of pretreatment, enzymatic saccharification, and fermentation to reduce the total costs [52].

6.2 Fermentation

The pretreatment delignifies the biomass, breaking down the polymers of cellulose and hemicellulose as they enter the fermentation broth. Sugars from lignocellulosic biomass are converted to biofuels and biochemical by bacteria, yeast, and microalgae by fermentation. Microbes are capable of storing up to 60%–80% of their dry weight as lipid in their cell wall. OMs store lipids in the form of fatty acids such as myristic acid, stearic acid, linoleic acid, linolenic acids, palmitic acids, and oleic acids. Acids, bases, or lipases are used to convert fatty acids to biodiesel. Reactor configuration remains one of the major factors affecting lipid accumulation along with other factors [16]. Frequently two types of fermentation are carried out for biodiesel production from lignocellulosic biomass: solid-state fermentation (SSF) and submerged fermentation (Fig. 20.3).

6.2.1 Solid-state fermentation

SSF is a comparatively simple technique as it requires a low volume of biomass, its energy cost is low, and comparatively it is inexpensive. The low-quality, natural, renewable energy crops (by-products of food, agriculture industries) can be utilized for SSF giving it a major advantage. However, feedback inhibition remains the major concern for SSF. The fermentable sugars produced if in excess can decrease the microbial growth and even its excess can hamper the culture. Either way, the yield would get affected. Another disadvantage of solid-state fermentation is the low lipid yield due to the absence of free-flowing water [41]. Economou et al. modified solid-state fermentation to semisolid-state fermentation as they increase the water content enhancing the growth of fungi and increasing the production of SCO. The lipid extraction process gets hampered as the microorganisms are grown on the substrate; during extraction, the substrate dissolves along with the microorganisms and hampers the desired product [16].

6.2.2 Submerged/liquid-state fermentation

Generally, yeast, bacteria, and molds are cultivated in solid-state fermentation, whereas microalgae are cultivated in submerged or liquid-state fermentation. It requires prior extraction of fermentable sugars from biomass to bulk liquid. This gives a major disadvantage to submerged fermentation as the procedure is time-consuming and expensive [54].

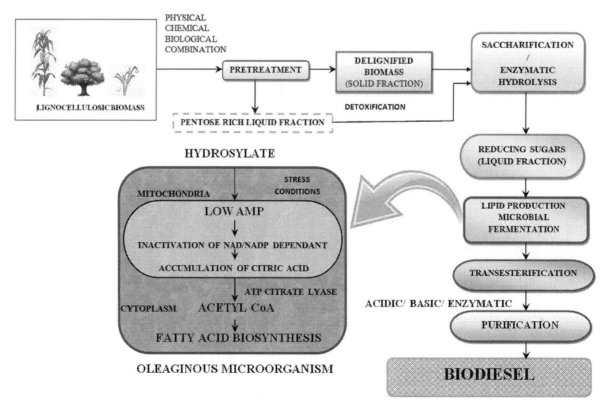

FIGURE 20.3 Production of biodiesel from oleaginous microorganisms using lignocellulosic biomass.

6.3 Transesterification

Transesterification (also called alcoholysis) is the reaction of a fat or oil with an alcohol to form esters and glycerol [52]. Rich in lipids, the biomass contains a significant amount of carbohydrates and proteins. Even microalgae with a significantly high amount of lipids contain a nonlipid component [48]. During transesterification, fatty acid alkyl esters or FAMEs are converted from the triacylglycerides obtained after the extraction of lipids. In the presence of alcohol (ethanol or methanol) and catalyst (acid or base), glycerol is obtained as a by-product [37]. Transesterification consists of three reversible reactions: conversion of triglycerides to diglycerides and further to monoglycerides. The third step leads to the formation of glycerol. The high viscosity of the biodiesel is treated well after transesterification. The process is significantly environment-friendly and nontoxic [16]. Most commonly used base catalysts include sodium and potassium hydroxide or sodium methoxide. Base catalysts are preferred over acid catalysts as they work under moderate reaction conditions and the process is faster [32]. The quantity of alcohol, catalyst concentration, and type, reaction time and temperature, reactor operating conditions, selected solvents, as well as feedstock quality significantly affect the yield of lipids [52,55]. Diversified into different classes transesterification can be classified into three basic categories: acidic, basic, and enzymatic. The transesterification process can be homogenous acidic or alkaline, or heterogeneous alkaline, acidic, or enzymatic.

6.3.1 Acid/base transesterification

The main catalysts used for acid catalysis include H_2SO_4 and HCl. Effective usually in the presence of free fatty acids (FFAs) and moisture, the catalysts have a large reaction time. It causes temperature rise during the reaction leading to corrosion. KOH, NaOH, and CH_3NaO are the primary base catalysts used due to their low reaction time, low temperature, and high reactivity. However, they are proved to be highly corrosive and difficult to recover.

6.3.1.1 Homogenous catalysis

The reversible reaction proceeds with the mix of reactants with the liquid acid or liquid base catalysts [4].

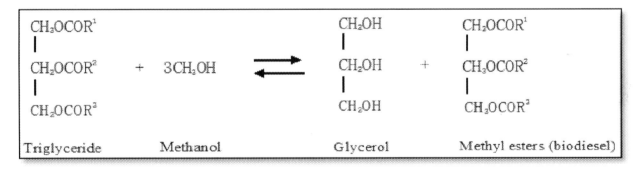

6.3.1.2 Heterogeneous catalysis

High free fatty acid (FFA) content possesses difficulty during homogenous basic catalysis as the conversion of oil decreases. The use of heterogeneous catalysts such as solid catalysts is preferred, as the transesterification of triglycerides and esterification of FFA simultaneously have the potential to replace homogenous catalysts, the use of solid acid catalysts can prevent corrosion, and they are environmentally toxic [4,56].

6.3.2 Enzymatic transesterification

Lipase enzymes are versatile in terms of their substrate specificity and chiral selectivity. Commonly used for enzymatic transesterification it has the potential to produce biodiesel of high purity. Lipase (EC 3.1.1.3) can be sourced from plant-animal as well as microbial sources. Microbial lipases are preferred for industrial processes. *Penicillium restrictum, Penicillium simplicissimum, Rhizopus homothallicus, Aspergillus niger*, etc., are some of the common microorganisms which can utilize lignocellulosic biomass to produce lipase [16]. Lipases are hydrolytic enzymes that can function under diverse environmental conditions and varied solvents. Recent studies show the use of immobilized lipases for transesterification reactions [37]. Transesterification is a diverse process and is observed that a combination of different techniques yields better results. The use of catalysts consists of a 30%–40% cost of transesterification [37] and is the main target for future endeavors to reduce the cost of production.

6.4 Purification of biodiesel

The final product quality is the major concern after the production of biodiesel. The final mixture after transesterification is not pure; it may contain enzymes, water, acid, bases soap, metal ions, or nondesirable lipids which are needed to be removed. Depending on the previous steps the purification methods are decided, as there are no predefined treatment methods and the combination of by-products in the mixture always varies [37]. During the production of biodiesel, glycerol is the main by-product that should be removed. In this case, the first step is separating nonpolar fatty acids from polar glycerol which may be achieved by centrifugation. As the number of by-products such as used catalysts, soaps, enzymes, and triglycerides increases the number of purification steps increases. Different methods such as various solvents combined with vacuum, sodium sulfate, and filtration can be used [57]. Some of the most common techniques are membrane separation, wet washing, and dry washing. Wet washing removes excess contaminates from previous steps; however, the need for a high amount of water and the need for absolute removal of water after purification give it a major disadvantage. To decrease the increased cost because of water disposal techniques, overall production time, and cost, the dry washing method is recommended. Since there is no need for water, their product is lost. Compounds used for dry washing include silica, starch, cellulose derivatives, and ion resin exchange. Apart from conventional methods, novel methods such as membrane separation are being studied. The process is based on rejection coefficients; composed of various coating materials the membranes are suitable for different processes. Polyvinylidene fluoride and polydimethylsiloxane are common materials for membrane coating along with ceramide material for organic solvents. The membranes are selected based on their thermal and chemical stability to use under desired conditions such as pH, temperature, solvents, etc. Membrane technologies have low operation costs but high product costs. The major disadvantage is biofouling of the membranes. On the other hand, the low energy use of membrane technology reduces the environmental footprint making it a method of future. The final purified biodiesel should meet the standard and quality properties of international organizations, such as the European Union and the American Society of Testing and Materials.

7. Environmental issues

It is well observed that biofuels are way more advantageous than traditional fossil fuels. With the current trend of environmental degradation, the development of environment-friendly fuels is obligatory. The main cause leading to the development of biofuels is to produce fuels with low net carbon emission and sustainable. Biofuels from LCB have significantly reduced carbon emissions and fixed CO_2 [50]. This has led to the use of crops with higher water efficiency (C_4 plants) and cultural practices with low carbon emission and low runoff of fertilizers into the water bodies, and the use of marginal lands earlier considered as wastelands. The use of perennial and woody plants or forest species as feedstock leads to lower depletion of soil carbon and nitrogen emission. The extensive root system of perennial plants prevents the escape of soil carbon into the atmosphere, prevents soil erosion, minimizes nitrogen runoff, favors nitrogen-fixing bacteria, and helps in carbon sequestration [49]. Various steps during the production of biodiesel such as biomass harvesting or grinding release particles into the atmosphere causing air pollutin, noise pollution during explosion pretreatment, and the release of pretreated chemicals into the wastewater runoff affect the environment significantly. To prevent severe impact on the environment, life cycle analysis (LCA) of each step is important as it provides the impact of each step on the environment. Emissions estimated by the LCA affect the cost of biorefinery setups (Jin et al., 2017).

8. Opportunities and challenges

Oleaginous microbes in addition to producing value-added products can be used to clean municipal and industrial wastewater, and earlier research has identified microalgae as suitable candidates for this function. Some microalgae can provide benefits such as nutrient removal, for example, macromolecules like nitrates and phosphates, which can cause eutrophication in water bodies, various multiscale harms to the receiving environment [58]. Furthermore, microalgae cultured in wastewater have high photosynthetic efficiency, high biomass productivity, and a quick growth rate [59].

8.1 Challenges

The most challenging part is the limited sources of petro-diesel and the high production cost of biodiesel using plant oils and animal fats in this era [60]. Increasing demand for vegetable oil as a biodiesel feedstock is transforming global agricultural landscapes and the ecological services they supply, highlighting several negative consequences [61]. There are various technical, environmental, and economical constraints to be faced for productive biodiesel production which include feedstock availability, food security, land-use changes, water source, and environmental constraints (deforestation, pollution, etc.) [62]. The main goal is to illustrate the technological problems and opportunities associated with commercializing biodiesels. However, industrial-scale biodiesel synthesis from lignocellulosic materials still faces obstacles. The fermentation process is the greatest impediment to large-scale biodiesel production because pentose sugars, such as arabinose, xylose, and glucuronic acids, cannot be fermented effectively [16]. Furthermore, due to carbon catabolite restriction, where glucose is used before xylose, most microorganisms eat these mixed sugars sequentially or selectively, prolonging the microbial production process and reducing carbon conversion efficiency. Other issues, such as the accumulation of harmful chemicals such as phenols, furans, organic acids, and inorganic salts during lignocellulosic pretreatment, also restrict microbial development [63,64]. Under these conditions, genetically designed, viable, and economically sound strains are required, as they can use both glucose and pentose sugars at the same time and overcome growth inhibitors [16]. There are various technical, environmental, and economic constraints to be faced for productive biodiesel production which include feedstock availability, food security, land-use changes, water source, and environmental constraints (deforestation, pollution, etc.) [62]. The main goal is to illustrate the technological problems and opportunities associated with commercializing biodiesels. To make biodiesel profitable, many technical challenges must be overcome. First, the overpriced virgin vegetable oil as a triglyceride source has a significant impact on process profitability. As a result, extra processes are required to remove any water or else the FFAs or soap from the reaction mixture. Commercial processors frequently use an acid-catalyzed esterification reactor to handle surplus FFAs [60]. A feasible process is an important component of mature technology, and it can be developed using a circular economy strategy. A circular economy differs from a linear economy and creates value in at least four different areas: (1) unprocessed material improves reutilization for higher-value biofuels; (2) mutual economic growth is facilitated by the development of liquid assets and marketplaces, where things are traded among users; (3) integration of bioprocesses that produce no waste; and (4) reusability of things that can be used for a variety of purposes [15]. During the conversion of LCB to bioethanol production in the bioethanol sector has been completed, but in terms of economic feasibility, significant questions remain unchanged [48]. Engineered strain stability and strategies for ensuring steady production in industrial microbial processes are well-known challenges. The key obstacles in changing the lipid composition are changing the degree of fatty acid unsaturation and decreasing or increasing the chain length of fatty acids. All of these are regulated by enzymes, although the majority of enzymes are membrane-bound, making purification and analysis of their function difficult [10,31]. Cloning genes of essential enzymes, transgenic expression of these genes to generate a fine high-product microbial oil recombination strain, and modification of cloned genes to engineer the expressed protein are three of the most important interdependent genetic technologies involved [31].

9. Recent advancement

Every oleaginous microbe has unique lipid synthesis capabilities, and there are numerous strategies to manipulate and improve lipid metabolism and production. Scientists can modify lipid production, storage, and profile pathways. They can also affect the pathways that microbes use to adapt to their surroundings, resulting in variations in product, production rates, and quantities. Some of these modifications will be discussed in detail which are genetic engineering of the most feasible strains, metabolic modeling, bioflocculation, cocultivation, and an approach to the biorefinery.

9.1 Genetic and metabolic engineering

Recent findings on fatty acid synthesis, turnover, and regulation, lipids and lipid particle assembly, and events leading to oleagenecity have revealed a collection of interesting access sites for genetic and metabolic engineering. These technologies have allowed changes in substrate specificities such as the use of a broader spectrum of chemicals, lipid accumulation in nonoleaginous species, fatty acid overproduction, heterologous gene expression, and direct biodiesel production to be made [32]. Oleaginous bacteria are attracting a lot of attention because of their capacity to produce target chemicals from a variety of carbon sources [16]. However, to use these OMs at an industrial scale with high lipid productivity and yield, the strains must be adjusted. Synthetic biology, ME, genome editing, and genetic elements are being developed; OMs could be changed for biofuel generation, industrial chemicals, and biorefinery by integrating these necessary components: (1) fatty acid−derived compounds [FAMEs, TAGs, fatty alcohols, FFAs]; (2) acetyl-CoA-derived products [terpenoids, poly-3-hydroxybutyrate]; (3) industrial compounds such as citric acid, succinic acid, α-ketoglutarate, itaconic acid, and erythritol [48]. The synthetic pathways of fatty acids can be enhanced by utilizing vast scope of molecular and bioinformatics techniques from metabolic engineering by understanding the process of turning bio-waste into microbial lipids. Frequently used strategies include (1) enhancing fatty acid or TAG biosynthesis pathways via overexpression of the specific enzymes involved, (2) developing genetically engineered strains with greater tolerance to elevated acid concentrations, (3) introduction and expression of exogenous genes responsible for the rate-limiting enzymes in the engineered strains, (4) raising objective flux by limiting competing routes, and (5) promoting lipid biosynthesis by regulating associated bypass [65]. The effective expression or overexpression of important enzymes is one of the metabolic engineering strategies. Overexpression of genes encoding ACC and FAS is one of the primary candidates based on lipid production pathways. In other circumstances, coexpression of multiple genes is required for a successful increase in lipid synthesis, as the pathway's later steps may limit the prior outcomes. In *Escherichia coli* cells, for example, acyl-ACP can prevent ACC overexpression. Overexpression of ACS genes in *E. coli* cells, for example, can result in increased acetate generation, which leads to increased acetyl-CoA activation and, as a result, lipid synthesis [66]. Suppression or knocking out of genes involved in lipid oxidation, degradation, and synthesis inhibition, on the other hand, is a widely employed strategy. Inactivation or malfunction of enzymes involved in oxidation, such as acyl-coenzyme oxidases (AOXs) via knock-out of their genes (POX), results in increased lipid buildup. Scientists who could disrupt the phospholipid biosynthetic route in the case of TAC synthesis improvement reviewed that combining more than one of the following strategies is becoming more common as gene metabolic engineering research progresses [66]. Recent advancements in synthetic biology research have aided the development of novel methods for genetic engineering in nontraditional yeasts like *Yarrowia lipolytica* and *Rhodosporidium toruloides*. In *R. toruloides* and *Y. lipolytica*, important genetic elements such as constitutive, inducible, targeted, and repressible promoters are being cloned. In addition, the hybrid promoter has been strengthened by engineering tandem copies relevant to upstream activation sequences for increased expresses in *Y. lipolytica* [48]. Different fatty acid profiles were seen in mutants and strains created by direct manipulation of genes that encode for elongases and desaturases compared to the wild-type strain. In addition, bacterial transcriptional factors and genetically encoded biosensors relevant to the malonyl-CoA and flavonoid pathways have been recruited in *Y. lipolytica* to improve the stability and yield of oleaginous altered strains [16]. The multigene strategy produces excellent results in all types of microbes, resulting in increased fatty acid production for nutritional or energetic purposes. Bioengineering approaches such as homologous recombination, mutagenesis, and use of short interfering RNA and microRNA exist for gene alterations, where the suppression or activation of certain genes is required. All of the changes, furthermore the best techniques to use, are reliant on the type of microbe, the strain, their genetic profile, and the desired outcome [37]. The employment of metabolic engineering technologies in conjunction with de novo and ex novo pathways could lead to even more lipid accumulation. Furthermore, advancements in metabolic methods connected to ex novo lipid synthesis may lead to the use and recycling of oil wastes, minimizing their environmental impact, and providing high-value goods [37]. Reduced lipid catabolism is a useful technique for increasing lipid accumulation through ex novo culture. It is possible to achieve this by knocking out genes involved in oxidation. The POX genes, which code for acyl-CoA oxidases, are the most frequent AOXs. The deletion of GUT2 or overexpression of GPD1 genes, which are related to the glycerol phosphate pool, along with the restriction of oxidation resulted in lipid content of 80% of dry weight. In the abovementioned microorganism, knocking down POX genes in combination with glycerol engineering can result in a 40%−70% increase in lipid accumulation [66]. It was suggested that on an industrial scale, metabolic engineering can be very useful in improving microbial strains. Compared to traditional genetic strain development procedures such as selection, mutagenesis, and hybridization, metabolic engineering has the following advantages: Strains are modified in a controlled manner without the accumulation of undesirable mutations. To enrich the microorganism with novel features, foreign genes from other organisms are introduced. The second method is particularly useful in industrial biotechnology. This can lead to alternative processes that broaden the range of industrial mediums such as LCB use and generate compounds that are not found naturally [16]. The lack of adequate validation of the acquired transcription factors for routinely employed OMs (e.g., yeasts) is currently limiting transcription factor engineering, necessitating additional research in the near future [55].

9.2 Metabolic modeling

Powerful computer modeling techniques have been developed to help the control of process parameters and the regulation of metabolic pathways to increase lipid production. As a result, more research could be conducted shortly to collect additional genomic data using sequencing techniques. In the long run, more genetic engineering techniques for a broader spectrum of OMs will be needed to meet the technical demands of genome-scale modeling [55].

9.3 Biorefineries

The findings suggest that using techniques for better genetic engineering for higher lipid accumulation in combination with biorefinery in OMs could be a potential path to commercially viable solutions [48]. Biorefineries are facilities that convert bio-based materials (such as agricultural leftovers) into a variety of products, including food, feed, fuels, chemicals, and energy. Biomass processing is similar to petroleum refineries, which convert crude oil into a variety of products such as fuels (e.g., gasoline, diesel, and kerosene) and chemical precursors such as butanol for use in various materials [50]. Future (second generation) biorefineries will need to be able to produce a broad range of products, including fuels, commodities, fine chemicals, and materials. To advance the notion of a second-generation refinery, OMs must be used to produce not only biodiesel as a commodity product but also other added-value fine chemicals for the nutraceutical and pharmaceutical industries, such as the omega-6 fatty acids—linolenic and arachidonic [50].

9.4 Bioflocculation

Auto-flocculating algae-based, bacteria-based, filamentous-fungi-based, and plant-extract-based bioflocculation are some of the microalgae bioflocculation techniques. Microalgal bioflocculation efficiency may be influenced by the type of flocculating microorganism, the strain of target microalgae, and various cocultivating conditions. According to a review by scientists, they developed a procedure to harvest the oleaginous marine alga using bioflocculation technique with an oleaginous fungus *Mortierella elongata* AG77 to overcome the serious hurdles in algal biofuel production, including high harvesting, lipid extraction, and nutrient supply costs, as well as low lipid content in algae [37]. TAG biofuel yields have been enhanced by adjusting incubation conditions. Whereas it was studied that overexpression of DGTT5, a gene encoding the type II acyl-CoA: diacylglycerol acyltransferase 5, was used to genetically engineer *Nannochloropsis oceanica* for increased TAG content in the nutrient-replete medium. When combined with bioflocculation, this method resulted in a 10% increase in TAG production in nutrient-rich circumstances compared to the wild type (6% DW) [37].

9.5 Biocatalyst

Biocatalysts are gaining popularity these days, and they have the potential to outperform chemical catalysts in the generation of biodiesel in the future. The utilization of enzymes in new biochemical methods for biodiesel generation has piqued curiosity. Rice bran oil, olive oil, soybean oil, canola oil, sunflower oil, and castor oil were employed in the majority of studies. According to a review by Vasudevan and Briggs [60], several lipases from microbial strains have been shown to have transesterification activity, including *Pseudomonas fluorescens*, *Pseudomonas cepacia*, *Rhizomucor miehei*, *Rhizopus oryzae*, *Candida rugosa*, *Thermomyces lanuginosus*, and *Candida antarctica*. Some nonenzymatic heterogeneous catalysts have been researched in an attempt to alleviate separation and soap production difficulties. Other solid catalysts which were studied during the transesterification of palm oil and coconut oil are ZrO_2, ZnO, SO^{2-}_4/SnO_2, SO^{2-}_4/ZrO_2, KNO_3/KL zeolite, and KNO_3/ZrO_2. The transesterification reaction can also be catalyzed by other sulfonated solid catalysts. Amorphous carbon was recently used to make one of the more fascinating sulfonated solid catalysts [60]. ILs are one of the best alternatives to replace alkali catalysts because of their special characteristics. However, their application has been limited due to their high viscosity and difficult recovery. Lately, ILs on solid supports have been suggested to circumvent these difficulties. These findings also point to the significant potential of IL—nanoporous material hybrids as biodiesel catalyst replacements. Nonetheless, by focusing future investigations on specific constraints, a significant step toward the employment of these catalysts in the biodiesel sector can be taken [67].

9.6 Cocultivation

During cocultivation, the connection between strains is complicated. Each strain can profit from the products generated by the other, or only one strain can benefit from the goods synthesized by the other. To overcome the constraints of single strain's ability to use substrates, a study on the cocultivation of microorganisms for lipid production has been done. Each

OM has a unique ability to utilize different substrates, and a mixed culture of them can boost lipid yield by maximizing organic substrate bioconversion [18]. Several researchers have used bacteria and yeast to cocultivate microbial strains. Microalgae and yeast cultures mixed can boost lipid production. Microalgae can absorb carbon dioxide, nutrients, and oxygen in general, but their poor development rate in wastewater has been identified as a restriction. Microalgae and cyanobacteria cocultivation can also boost lipid productivity. The scientists observed that during cocultivation, a less productive strain tends to follow a more productive strain, resulting in higher total biomass productivities than separate cultures. *Pseudokirchneriella subcapitata* and *Synechocystis salina* mixed cultures produced 51.6% more lipid productivity than a single *P. subcapitata* culture and 28.2% more lipid productivity than a single *S. salina* culture reviewed by Cho et al. [18].

9.7 Osmotic stress

According to experiments conducted by Singh et al. [68], when *R. toruloides* ATCC204091 was grown on minimal media with increased NaCl concentrations (up to 10% w/v) resulted in a slower growth rate and delayed the commencement of the stationary phase. The log phase was prolonged as expected due to salt stress, but there was no improvement in biomass or lipid production. However, in certain instances, salt stress led to enhanced lipid biosynthesis, specifically total sterol and squalene biosynthesis.

10. Applications of biodiesel

Biodiesel's fuel properties must be identical to diesel fuel for the engine to run efficiently. Flashpoint, calorific value, density, moisture content, viscosity, oxidation stability, cetane value, acid value, and cold flow qualities are among the most essential fuel properties. Different authorities have formalized the specifications for diesel and biodiesel which must be met by diesel and biodiesel fuels before use in engines, based on the varying global geographical conditions [62]. Maintaining the abovementioned specifications is one of the most crucial aspects of biodiesel applications.

10.1 On-road and off-road vehicles

Biodiesel's natural properties, its carbon footprint, and reduction in emissions make it a suitable fuel for on-road and off-road vehicles. The aim is to use biodiesel in particular every diesel-powered vehicle on the road. Natural fuel can be used in off-road vehicles, mining equipment, and farm machinery [69].

10.2 Marine vessels

Natural gasoline is completely safe to use in marine engines. The lack of any danger of waterway contamination makes marine application particularly appealing.

10.3 Solvents

Biodiesel can be used as an industrial solvent or to replace petroleum solvents with high VOC levels. With rules lowering the VOC level of solvents used in industry, biodiesel is an appealing alternative. Household cleaning products are another solvent application. Biodiesel can also be utilized as a lubricity agent or enhancer in a variety of applications. It's particularly useful in marine applications, where contamination of water with petroleum lubricity additives might cause issues. Biodiesel can be utilized as a lubricity additive in the future with low-sulfur fuel regulations. Diesel lubricity can be increased by 65% by adding 1%–2% biodiesel to the fuel. Biodiesel can also be used as a petroleum diesel additive to clean injectors, pumps, and other combustion components. For this purpose, a 1%–2% blend should suffice [69].

10.4 Automotive diesel engine

The performance of biodiesel-fueled engines is critical for biodiesel application. Corrosion, material degradation, injector coking, filter blocking, piston ring sticking, and engine deposits are some of the most common issues. By keeping all these factors in mind it is possible to use biodiesel in automotive diesel engines [70].

10.5 Drilling fluids

Biodiesel has a tolerable low kinematic viscosity, a sufficiently high flashpoint, a pour point below the most ambient temperature, nontoxicity, a reasonable acid value, adequate thermal stability, impressive biodegradability, lubricity, an environmentally friendly production process, and abundance. Biodiesel has become a viable choice in drilling operations due to all of these features. Many researchers have created various drilling mud compositions that include biodiesel as a basic fluid or as an addition such as free water biodiesel-based mud, biodiesel in water emulsion mud, and water in biodiesel invert emulsion mud. Here the mud components used were water, $CaCl_2$, lime, emulsifier, viscosifier, and barite [71].

10.6 Stationary power generation

As new power generation capacity comes online, biodiesel is an appealing option for meeting regulatory requirements. Many stationary applications are approved sources that require exhaust emission control systems, which are compatible with biodiesel but not with diesel fuel. With natural gas costs on the rise, biodiesel may simply be swapped for natural gas in boilers with just modest alterations to the burner train. Biodiesel will make good reforming fuel for hybrid vehicles, which are now required in many jurisdictions (including the fuel cell hybrid) [62].

Agriculture Adjuvants: Because biodiesel is nontoxic and biodegradable, it is utilized as a transporter for pesticides and fertilizers in agricultural sprays [62].

The advancement of biofuels does not stop with the first- and second-generation biofuels (Table 20.3). Interest in third- and fourth-generation biofuels has increased in recent years because of the food crisis of 2007–08. Third-generation

TABLE 20.3 First-generation biofuels versus second-generation biofuels.

	First-generation biofuels	Second-generation biofuels
Definition	Biofuels of the first generation are made from biomass that is normally edible.	Second-generation biofuels are defined as fuels made from a diverse range of feedstocks, including lignocellulosic Feedstocks and municipal solid waste [4].
Substrate/feedstock	• Sugar beet, sweet sorghum, and sugarcane are examples of sucrose-containing biomasses. • Grains of wheat, corn, barley, and rice are examples of starchy substances. • Rapeseed, sunflower, soybean, coconut, peanut, safflower, palm, corn, Cottonseed, and other consumable Vegetable oil crops. • Castor, seashore mallow, algae, halophytes, jatropha, Chinese tallow tree, and other nonedible vegetable oil crops. • Used cooking oil or recycled oil. • Tallow, yellow grease, chicken fat, fish oil, and other items made from animal fats [72].	Biomass used to make second-generation biofuels is commonly Divided into three types: • Homogeneous, such as white wood chips, • Quasihomogeneous, such as agricultural and forest residues, and • Nonhomogeneous, which includes low-value feedstock like municipal solid wastes; • Including lignocellulosic biomass; • Cellulosic crops: switchgrass, *miscanthus*; • Crop residues such as corn stover, rice straw, wheat straw, sugarcane bagasse, etc. [72].
End products	Biofuels include bioethanol and biodiesel or FAME, corn ethanol, and sugar alcohol.	Second-generation biofuels include lignocellulosic ethanol, butanol, bio-oil, FT oil, mixed alcohols, and hydrotreating oil [4].
Advantages	• First-generation biofuels are environmentally friendly when compared to crude fuels. • Help in reducing the carbon footprint.	• This biomass is far less expensive than vegetable oil, corn, or sugarcane, which is a significant benefit. • Many varieties of second-generation biomass are perennial crops that are not required to be replanted each year.
Disadvantages	• Production is limited due to use of edible biomass. • Production of fuels from oil crops is not an economically feasible process. • Need to blend with conventional biofuels [72].	Biomass is often more difficult to convert, and its production is reliant on innovative technologies. Second-generation bioenergy crops do not yet have well-established markets or technologies. To efficiently produce, transport, and transform these crops into energy, further research and development is needed [72].

biofuels are made from algae or engineered biofuel crops that are specially designed to convert biomass to fuel more efficiently. However, third- and fourth-generation biofuels require further research and development before becoming economically feasible. First-generation and second-generation biofuel production has a substantial economic and environmental impact globally. However, until new technology advancements unwrap the pathways for third generation of biofuels from algal biomass, agricultural commodities will face stiff competition as a source of food or fuel [72].

11. Future prospects and scope

The modern world's energy and environmental issues are prompting a reevaluation of the efficient use of natural, renewable resources and the development of alternative uses employing clean technologies. In this regard, lignocellulosic biomass has a lot of potentials to meet the world's present energy needs. More improved biotechnologies are needed for the discovery, characterization, and manufacture of novel enzymes in homologous and heterologous systems, which will eventually lead to the low-cost conversion of lignocellulosic biomasses into biofuels and biochemicals. Future advancements in lignocellulose biotechnology and genetic engineering are being directed toward improved processes and products in the current scenario. It is expected that lignocellulosic biomass along with green biotechnology will be used to solve today's energy concerns [73]. There is a need for the development of LCB hydrolysis technologies that give better sugar yields. Enzymatic hydrolysis appears to be very promising for this application, but numerous parts of the process, particularly enzyme separation and reutilization following biomass treatment, need to be improved [41]. The primary phases in the manufacturing of bioethanol are delignification, saccharification, and fermentation. The lignin destroyed mostly by enzymatic mechanisms after delignification can be employed as antioxidants, glucose biosensors, reusable adsorbents, silver nanoparticles, electric double-layer capacitors, and other applications [68]. Pentoses are largely wasted during the fermentation process. As a result, using these pentoses as a carbon source for OMs to produce lipids could be a realistic approach. These lipids can also be converted to FAMEs, such as biodiesel and glycerol, which can be further used as a carbon source for biodiesel production. The biomass produced following lipid extraction can be combined with cyanobacteria for nutrient enrichment and used as bio manure and biofertilizer, respectively [48]. Improvements to biodiesel's low-temperature qualities, as well as monitoring and preserving biodiesel quality against degradation during long-term storage (owing to its unstable double bond), are all issues. Treatment with antioxidant chemicals, such as the combination with hydrogen to reduce the double bond, is the most cost-effective way to improve biodiesel's oxidative stability. Biodiesel will be more stable in storage as a result of this approach, similar to diesel oil [70]. Several approaches can be used to reduce the commercial cost of biofuels, including the following: improving the conversion of solar energy; increasing bioreactor efficiency; lowering the overall capital cost of bioreactors; coproduction of high-value products; temperature, O_2, and CO_2 control; for generating more revenue recycling nutrients; media optimization for biomass and biofuel production—carbon; by balancing nutrition and pH; as well as minimizing contamination [74]. The optimization of growth factors for maximum biomass and metabolite yield is critical so that these conditions can be implemented at large-scale operations and the process may be productive and cost-effective [59]. Furthermore, finding microorganisms that are both more efficient and capable of using low-cost substrates such as wastewater and sewage for biomass accumulation must be extensively established [59]. To reduce expenses, an innovative growing system comprising all starch or carbon ingredients, including residential and industrial wastes, will be required. However, the economic research revealed that the cost of medium components must be reduced, and microbial systems must be developed that can yield higher lipid productivities in minimal media [30].

Changing the lipid content of oleaginous bacteria to enable them to satisfy in actual biodiesel production as well as genetic engineering strategies to modify the fatty acid have both proven effective. Thus, metabolic management and culturing of OMs under various conditions should allow for the synthesis of lipids with changing fatty acid composition [31]. Through system integration and bioengineering, high-value coproducts created from improved process chains could improve the overall economics of biofuel production, closing the financial advantage gap currently held by fossil sources [75]. Deeper insights into the underlying processes and metabolic pathways of microbial lipid accumulation in designed strains are expected to be offered with the fast evolution of metabolic and genetic engineering techniques including metagenomic sequencing [55]. Though stress manipulation approach optimization has been the primary focus, optimal bioproduction requires a combination of microalgal strain creation and process integration [59]. To begin, current hurdles in VFA generation from acidogenic fermentation of bio-waste include separation and in situ recovery of VFAs. Due to their unique qualities, emerging membrane processes such as electrodialysis, forward osmosis, and membrane distillation appear to be at the forefront of VFA recovery approaches [55]. Bioelectrochemical systems (BESs) have recently developed as a technology capable of producing biofuels. Bacteria operate as catalyzers for processes taking place at the anode and/or cathode in BESs. Microbial electrosynthesis and the coupling of BES with AD should be regarded as developing and promising technology for biofuel production. Several studies have been done concerning this technology, whereas various value-added products, ethanol, and butanol have been synthesized at the cathode of BES [75].

Future research on the optimization of methanolysis of microbial oils to boost biodiesel output is needed. Due to excessive FFAs in microbial oils, proper catalysts are required, as the usual alkaline catalyst (NaOH) may interact with FFA, resulting in soap production, lowering biodiesel output, and lowering the purity of the coproduced glycerol. The utilization of microbial lipases in the enzymatic production of biodiesel offers significant gains in process efficiency [41]. Microbial lipids are transesterified with methanol to produce biodiesel and glycerol after effective extraction [16]. There is a future need of optimizing the efficiency of transesterification reactions, which is influenced by the reactor type and operating conditions, the catalyst type and dosage, and the feedstock quality (i.e., microbial lipids) [55]. Cocultivation: It is a complex process. Experiments must be conducted to differentiate mutualism from commensalism during cocultivation to fully identify the codependent effects of cocultivation on lipid production; this task can be accomplished through appropriate experimental design. More research is needed, in particular, to optimize the strain ratios during cocultivation to better understand the potential mechanisms between cocultivated species and improve lipid yields [18]. High yield of TAG and total fatty acids were successfully produced by the process of bioflocculation [76] using oleaginous marine alga *N. oceanica* with an oleaginous fungus *M. elongata* AG77. However, future studies are needed to be done for better productivity. Membrane separation technology (e.g., microfiltration, ultrafiltration, and nanofiltration) has emerged as a feasible alternative to conventional biodiesel purification technologies in recent years, owing to lower energy usage and operational and maintenance expenses. Membrane technology works based on the membrane's selective permeation capabilities, which allows impurities to be separated from the product. Modifications to the structure and behavior of membranes have been researched and are regarded as promising solutions (e.g., nanoparticle-modified membranes) to overcome this obstacle. Future research on the scale-up of new membranes should be done in this area to cut capital and operating costs even more [55]. There are usually several by-products usable in the industry during fatty acid manufacturing. OMs contain considerable amounts of proteins, carbohydrates, and other nutrients in addition to oils. As a result, figuring out how to make use of these by-products and increase their economic worth is another option to cut biodiesel manufacturing costs. The remaining biomass from biodiesel manufacturing, for example, might be utilized as animal feed or to make methane through anaerobic digestion. As a result, leftover glycerol from the biodiesel sector might be easily converted into a valuable chemical. Furthermore, the final product's price could be significantly decreased in this manner [31].

References

[1] S.K. Bhatia, S.V. Otari, J.-M. Jeon, R. Gurav, Y.-K. Choi, R.K. Bhatia, A. Pugazhendhi, V. Kumar, J. Rajesh Banu, J.-J. Yoon, K.-Y. Choi, Y.-H. Yang, Biowaste-to-bioplastic (polyhydroxyalkanoates): conversion technologies, strategies, challenges, and perspective, Bioresour. Technol. 326 (2021b) 124733.

[2] R. Gurav, S.K. Bhatia, T.-R. Choi, Y.-K. Choi, H.J. Kim, H.-S. Song, S.M. Lee, S. Lee Park, H.S. Lee, J. Koh, J.-M. Jeon, J.-J. Yoon, Y.-H. Yang, Application of macroalgal biomass derived biochar and bioelectrochemical system with Shewanella for the adsorptive removal and biodegradation of toxic azo dye, Chemosphere 264 (2021a) 128539.

[3] R. Gurav, S.K. Bhatia, T.-R. Choi, Y.-L. Park, J.Y. Park, Y.-H. Han, G. Vyavahare, J. Jadhav, H.-S. Song, P. Yang, J.-J. Yoon, A. Bhatnagar, Y.-K. Choi, Y.-H. Yang, Treatment of furazolidone contaminated water using banana pseudostem biochar engineered with facile synthesized magnetic nanocomposites, Bioresour. Technol. (2019) 122472.

[4] S.N. Naik, V.V. Goud, P.K. Rout, A.K. Dalai, Production of first- and second-generation biofuels: a comprehensive review, Renew. Sustain. Energy Rev. 14 (2010) 578–597.

[5] D.S.J. Jones, P.R. Pujado, Handbook of Petroleum Processing, Springer, New York, NY, USA, 2006.

[6] A. Mohr, S. Raman, Lessons from first generation biofuels and implications for the sustainability appraisal of second-generation biofuels, Energy Pol. 63 (2013) 114–122.

[7] M. Rossi, A. Amaretti, S. Raimondi, Getting lipids for biodiesel production from oleaginous fungi, in: M. Stoytcheva, G. Montero (Eds.), Biodiesel—Feedstocks and Processing Technologies, InTech Publishers, Rijeka, 2011, pp. 71–92.

[8] S. Pinzi, C.D. Leiva, G.I. López, M.D. Redel, M.P. Dorado, Latest trends in feedstocks for biodiesel production, Biofuels Bioprod. Biorefin. 8 (2013) 126–143.

[9] M. Dourou, D. Aggeli, S. Papanikolaou, G. Aggelis, Critical steps in carbon metabolism affecting lipid accumulation and their regulation in oleaginous microorganisms, Appl. Microbiol. Biotechnol. 102 (2018) 2509–2523.

[10] C. Ratledge, Fatty acid biosynthesis in microorganisms being used for single-cell oil production, Biochimie 86 (2004) 807–815.

[11] J. Singh, S. Gu, Commercialization potential of microalgae for biofuels production, Renew. Sustain. Energy Rev. 14 (2010) 2596–2610.

[12] S.K. Bhatia, R. Gurav, T.-R. Choi, Y.H. Han, Y.-L. Park, H.-R. Jung, S.-Y. Yang, H.-S. Song, Y.-H. Yang, A clean and green approach for odd chain fatty acids production in *Rhodococcus* sp. YHY01 by medium engineering, Bioresour. Technol. 286 (2019a) 121383.

[13] K.K. Kumar, F. Deeba, S. Negi, Y.S. Gaur, Harnessing pongamia shell hydrolysate for triacylglycerol agglomeration by novel oleaginous yeast *Rhodotorula pacifica* INDKK, Biotechnol. Biofuels 13 (2020) 175.

[14] P. Shah Maulin, Microbial Bioremediation & Biodegradation, Springer, 2020.

[15] A.K. Chandel, V.K. Garlapati, S.P.J. Kumar, M. Hans, A.K. Singh, S. Kumar, The role of renewable chemicals and biofuels in building a bio-economy, Biofuels Bioprod. Biorefin. 14 (2020) 830–844.

[16] D. Kumar, B. Singh, J. Korstad, Utilization of lignocellulosic biomass by oleaginous yeast and bacteria for production of biodiesel and renewable diesel, Renew. Sustain. Energy Rev. 73 (2017) 654–671.

[17] L. Gujjala, K.S. Kumar, S.P.J. Talukdar, B. Dash, A. Kumar, S.,K.C.H. Sherpa, Biodiesel from oleaginous microbes: opportunities and challenges, Biofuels 10 (2017) 1–15.

[18] H.U. Cho, J.M. Park, Biodiesel production by various oleaginous microorganisms from organic wastes, Bioresour. Technol. 256 (2018) 502–508.

[19] S.K. Bhatia, R. Gurav, T.-R. Choi, Y.H. Han, Y.-L. Park, J.Y. Park, H.-R. Jung, S.-Y. Yang, H.-S. Song, S.-H. Kim, K.-Y. Choi, Y.-H. Yang, Bioconversion of barley straw lignin into biodiesel using *Rhodococcus* sp. YHY01, Bioresour. Technol. 289 (2019b) 121704.

[20] Y.L. Ma, Microbial oils and its research advance, Chin. J. Bioprocess. Eng. 4 (2006) 7–11.

[21] Y. Chisti, Biodiesel from microalgae, Bitechnol. Adv 25 (2007) 294–306.

[22] H. Aratboni, A. Rafiei, N. Garcia, R.G. Alemzadeh, J.R. Morones, Biomass and lipid induction strategies in microalgae for biofuel production and other applications, Microb. Cell Factories 18 (2019) 1–17.

[23] P. Shah Maulin, Removal of Refractory Pollutants from Wastewater Treatment Plants, CRC Press, 2021.

[24] M. Balat, Potential alternatives to edible oils for biodiesel production—a review of current work, Energy Convers. Manag. 52 (2011) 1479–1492.

[25] S.M. Lee, D.-H. Cho, H.-J. Jung, B. Kim, S.-H. Kim, S.K. Bhatia, R. Gurav, J.-M. Jeon, J.-J. Yoon, W. Kim, K.-Y. Choi, Y.-H. Yang, Finding of novel polyhydroxybutyrate producer *Loktanella* sp. SM43 capable of balanced utilization of glucose and xylose from lignocellulosic biomass, Int. J. Biol. Macromol. (2022).

[26] S.M. Lee, H.-J. Lee, S.H. Kim, M.J. Suh, J.Y. Cho, S. Ham, H.-S. Song, S.K. Bhatia, R. Gurav, J.-M. Jeon, J.-J. Yoon, K.-Y. Choi, J.-S. Kim, S.H. Lee, Y.-H. Yang, Engineering of *Shewanella marisflavi* BBL25 for biomass-based polyhydroxybutyrate production and evaluation of its performance in electricity production, Int. J. Biol. Macromol. 183 (2021b) 1669–1675.

[27] A. Luis, L.A. Garay, I.R. Sitepu, T. Cajka, I. Chandra, S. Shi, Eighteen new oleaginous yeast species, J. Ind. Microbiol. Biotechnol. 43 (2016) 887–900.

[28] M. Athenaki, C. Gardeli, P. Diamantopoulou, S.S. Tchakouteu, D. Sarris, A. Philippoussis, Lipids from yeasts and fungi: physiology, production, and analytical considerations, J. Appl. Microbiol. 124 (2018) 336–367.

[29] G. Zuccaro, D. Pirozzi, A. Yousuf, Lignocellulosic biomass to biodiesel, in: A. Yousuf, D. Pirozzi, F. Sannino (Eds.), Lignocellulosic Biomass to Liquid Biofuels, Academic Press, Cambridge, MA, 2020, pp. 127–167.

[30] S. Uthandi, A. Kaliyaperumal, N. Srinivasan, K. Thangavelu, I.K. Muniraj, X. Zhan, V.K. Gupta, Microbial biodiesel production from lignocellulosic biomass: new insights and future challenges, Crit. Rev. Environ. Sci. Technol. (2021) 1–30.

[31] X. Meng, J. Yang, X. Xu, L. Zhang, Q. Nie, M. Xian, Biodiesel production from oleaginous microorganisms, Renew. Energy 34 (1) (2009) 1–5.

[32] L.F. Bautista, G. Vicente, V. Garre, Biodiesel from microbial oil, Adv. Biodiesel Production 8 (2012) 179–203.

[33] R. Kumar, P. Kumar, Future microbial applications for bioenergy production: a perspective, Front. Microbiol. 8 (2017) 10–12.

[34] A. Patel, F. Mikes, L. Matsakas, An overview of current pretreatment methods used to improve lipid extraction from oleaginous microorganisms, Molecules 23 (7) (2018) 1562.

[35] G. Valdés, R.T. Mendonça, G. Aggelis, Lignocellulosic biomass as a substrate for oleaginous microorganisms: a review, Appl. Sci. 10 (21) (2020) 7698.

[36] N. Chandel, V. Ahuja, R. Gurav, V. Kumar, V.K. Tyagi, A. Pugazhendhi, G. Kumar, D. Kumar, Y.-H. Yang, S.K. Bhatia, Progress in microalgal mediated bioremediation systems for the removal of antibiotics and pharmaceuticals from wastewater, Sci. Total Environ. 825 (2022) 153895.

[37] A. Patel, D. Karageorgou, E. Rova, P. Katapodis, U. Rova, P. Christakopoulos, L. Matsakas, An overview of potential oleaginous microorganisms and their role in biodiesel and omega-3 fatty acid-based industries, Microorganisms 8 (2020) 434.

[38] S.K. Bhatia, R. Gurav, Y.-K. Choi, H.-J. Lee, S.H. Kim, M.J. Suh, J.Y. Cho, S. Ham, S.H. Lee, K.-Y. Choi, Y.-H. Yang, *Rhodococcus* sp. YHY01 a microbial cell factory for the valorization of waste cooking oil into lipids a feedstock for biodiesel production, Fuel 301 (2021a) 121070.

[39] R. Gurav, S.K. Bhatia, T.-R. Choi, H.J. Kim, H.-S. Song, S.-L. Park, S.-M. Lee, H.-S. Lee, S.-H. Kim, J.-J. Yoon, Y.-H. Yang, Utilization of different lignocellulosic hydrolysates as carbon source for electricity generation using novel *Shewanella marisflavi* BBL25, J. Clean. Prod. (2020) 124084.

[40] X.F. Sun, R.C. Sun, J. Tomkinson, Degradation of wheat straw lignin and hemicellulosic polymers by a totally chlorine-free method, J. Polym. Degrad. Stab. 83 (2004) 47–57.

[41] A. Yousuf, Biodiesel from lignocellulosic biomass—prospects and challenges, Waste Manag. 32 (11) (2012) 2061–2067.

[42] R. Gurav, S.K. Bhatia, T.-R. Choi, Y.-K. Choi, H.J. Kim, H.-S. Song, S.L. Park, H.S. Lee, S.M. Lee, K.-Y. Choi, Y.-H. Yang, Adsorptive removal of crude petroleum oil from water using floating pinewood biochar decorated with coconut oil-derived fatty acids, Sci. Total Environ. 781 (2021b) 146636.

[43] G.D. Vyavahare, R.G. Gurav, P.P. Jadhav, R.R. Patil, C.B. Aware, J.P. Jadhav, Response surface methodology optimization for sorption of malachite green dye on sugarcane bagasse biochar and evaluating the residual dye for phyto and cytogenotoxicity, Chemosphere 194 (2018) 306–315.

[44] R. Gurav, S.K. Bhatia, T.-R. Choi, H.J. Kim, Y.-K. Choi, H.-J. Lee, S. Ham, J.Y. Cho, S.H. Kim, S.H. Lee, J. Yun, Y.-H. Yang, Adsorptive removal of synthetic plastic components bisphenol-A and solvent black-3 dye from single and binary solutions using pristine pinecone biochar, Chemosphere 296 (2022) 134034.

[45] A. Zoghlami, G. Paës, Lignocellulosic biomass: understanding recalcitrance and predicting hydrolysis, Front. Chem. (2019) 874.

[46] S.K. Bhatia, R. Gurav, T.-R. Choi, H.-R. Jung, S.-Y. Yang, Y.-M. Moon, H.-S. Song, J.-M. Jeon, K.-Y. Choi, Y.-H. Yang, Bioconversion of plant biomass hydrolysate into bioplastic (polyhydroxyalkanoates) using *Ralstonia eutropha* 5119, Bioresour. Technol. 271 (2019c) 306–315.

[47] H.S. Lee, H.-J. Lee, S.H. Kim, J.Y. Cho, M.J. Suh, S. Ham, S.K. Bhatia, R. Gurav, Y.-G. Kim, E.Y. Lee, Y.-H. Yang, Novel phasins from the Arctic *Pseudomonas* sp. B14-6 enhance the production of polyhydroxybutyrate and increase inhibitor tolerance, Int. J. Biol. Macromol. 190 (2021a) 722–729.

[48] A.D. Chintagunta, G. Zuccaro, M. Kumar, S. Kumar, V.K. Garlapati, P.D. Postemsky, N. Kumar, A.K. Chandel, J.S. Gandara, Biodiesel production from lignocellulosic biomass using oleaginous microbes: prospects for integrated biofuel production, Front. Microbiol. 12 (2021).

[49] A. Carroll, C. Sommerville, Cellulosic biofuels, Annu. Rev. Plant Biol. 60 (2009) 165–182.

[50] V. Balan, Current challenges in commercially producing biofuels from lignocellulosic biomass, ISRN Biotechnology 2014 (2014) 3–21.

[51] J.R. McMillan, I.A. Watson, M. Ali, W. Jaafar, Evaluation and comparison of algal cell disruption methods: Microwave, waterbath, blender, ultrasonic and laser treatment, Appl. Energy 103 (2013) 128–134.

[52] A.W. Bhutto, K. Quraishi, R. Abro, K. Harijan, Z. Zhao, A.A. Bazmi, T. Abbas, G. Yu, Progress in production of biomass-to-liquid biofuels to decarbonize transport sector—prospectus and challenges, RSC Adv. 6 (2016) 3–22.

[53] M. Jin, C. Sarks, B.D. Bals, N. Posawat, C. Gunawan, B.E. Dale, V. Balan, Toward high solids loading process for lignocellulosic biofuel production at a low cost, Biotechnol. Bioeng. 114 (5) (2017) 980–989.

[54] C.N. Economou, A. Makri, G. Aggelis, S. Pavlou, D.V. Vayenas, Semisolid state fermentation of sweet sorghum for the biotechnological production of single cell oil, Bioresour. Technol. 101 (2010) 1385–1388.

[55] B. Zhang, Y. Jiang, M. Gao, T. Ma, W. Sun, H. Pan, Recent progress on hybrid electrocatalysts for efficient electrochemical CO_2 reduction, Nano Energy 80 (2021) 105504.

[56] S.K. Bhatia, R. Gurav, T.-R. Choi, H.J. Kim, S.-Y. Yang, H.-S. Song, J.Y. Park, Y.-L. Park, Y.-H. Han, Y.-K. Choi, S.-H. Kim, J.-J. Yoon, Y.-H. Yang, Conversion of waste cooking oil into biodiesel using heterogenous catalyst derived from cork biochar, Bioresour. Technol. 302 (2020) 122872.

[57] R.W. Nicol, K. Marchand, W.D. Lubitz, Bioconversion of crude glycerol by fungi, Appl. Microbiol. Biotechnol. 93 (2012) 1865–1875.

[58] T. Fazal, A. Mushtaq, F. Rehman, A.U. Khan, N. Rashid, W. Farooq, S.U.R. Muhammad, J. Xu, Bioremediation of textile wastewater and successive biodiesel production using microalgae, Renew. Sustain. Energy Rev. 82 (2018) 3107–3126.

[59] T. Mutanda, D. Naidoo, J.K. Bwapwa, A. Anandraj, Biotechnological applications of microalgal oleaginous compounds: current trends on microalgal bioprocessing of products, Front. Energy Res. 8 (2020) 1–21.

[60] P.T. Vasudevan, M. Briggs, Biodiesel production—current state of the art and challenges, J. Ind. Microbiol. Biotechnol. 35 (5) (2008) 421–430.

[61] D. Huang, H. Zhou, L. Lin, Biodiesel: an alternative to conventional fuel, Energy Proc. 16 (Part C) (2012) 1874–1885.

[62] G. Joshi, J.K. Pandey, S. Rana, D.S. Rawat, Challenges and opportunities for the application of biofuel, Renew. Sustain. Energy Rev. 79 (2017) 850–866.

[63] T.-R. Choi, H.-S. Song, Y.-H. Han, Y.-L. Park, J.Y. Park, S.-Y. Yang, S.K. Bhatia, R. Gurav, H.J. Kim, Y.K. Lee, K.Y. Choi, Y.-H. Yang, Enhanced tolerance to inhibitors of *Escherichia coli* by heterologous expression of cyclopropane-fatty acid-acyl-phospholipid synthase (cfa) from Halomonas socia, Bioproc. Biosyst. Eng. 43 (5) (2020) 909–918.

[64] H.R. Jung, J.H. Lee, Y.M. Moon, T.R. Choi, S.Y. Yang, H.S. Song, J.Y. Park, Y.L. Park, S.K. Bhatia, R. Gurav, B.J. Ko, Y.H. Yang, Increased tolerance to furfural by introduction of polyhydroxybutyrate synthetic genes to *Escherichia coli*, J. Microbiol. Biotechnol. 29 (5) (2019) 776–784.

[65] L. Zhang, K.C. Loh, A. Kuroki, Y. Dai, Y.W. Tong, Microbial biodiesel production from industrial organic wastes by oleaginous microorganisms: current status and prospects, J. Hazard Mater. 402 (2021) 123543.

[66] M.H. Liang, Jian Guo, Advancing oleaginous microorganisms to produce lipid via metabolic engineering technology, Prog. Lipid Res. 52 (4) (2013) 395–408.

[67] A. Gholami, P. Fathollah, M. Akbar, Recent advances of biodiesel production using ionic liquids supported on nanoporous materials as catalysts: a review, Front. Energy Res. 8 (2020).

[68] G. Singh, S. Sinha, K.K. Kumar, N.A. Gaur, K.K. Bandyopadhyay, D. Paul, High density cultivation of oleaginous yeast isolates in 'mandi' waste for enhanced lipid production using sugarcane molasses as feed, Fuel 276 (2020) 118073.

[69] N. Kanthavelkumaran, P. Seenikannan, Recent trends and applications of bio diesel, Int. J. Eng. Res. Afr. 2 (6) (2012) 197–203.

[70] Y. Li, T. Guohong, H. Xu, Application of biodiesel in automotive diesel engines, in: Biodiesel—Feedstocks, Production and Applications, No. May 2014, 2012.

[71] A. Ahmed, E. Salaheldin, A. Saad, Applications of biodiesel in drilling fluids, Geofluids 2021 (2021).
[72] A.H. Hirani, N. Javed, M. Asif, S.K. Basu, A. Kumar, A review on first-and second-generation biofuel productions, in: Biofuels: Greenhouse Gas Mitigation and Global Warming, 2018, pp. 141–154.
[73] Z. Anwar, G. Muhammad, I. Muhammad, Agro-industrial lignocellulosic biomass a key to unlock the future bio-energy: a brief review, J. Radiat. Res. Appl. Sci. 2 (2014) 163–173.
[74] J.K. Patra, C.N. Vishnuprasad, G. Das, Microbial biotechnology, Microb. Biotechnol. 1 (2018) 1–479.
[75] A. Callegari, S. Bolognesi, D. Cecconet, A.G. Capodaglio, Production technologies, current role, and future prospects of biofuels feedstocks: a state-of-the-art review, Crit. Rev. Environ. Sci. Technol. 50 (4) (2020) 384–436.
[76] Z. Du Yan, J. Alvaro, B. Hyden, A. Zienkiewicz, N. Benning, B. Gregory, C. Benning, Enhancing oil production and harvest by combining the marine alga *Nannochloropsis oceanica* and the oleaginous fungus *Mortierella elongata*, Biotechnol. Biofuels 11 (1) (2018) 1–16.

Chapter 21

An insight into rice straw–based biofuel production

Manswama Boro and Anil Kumar Verma
Department of Microbiology, Sikkim University, Gangtok, Sikkim, India

1. Introduction

Rice straw (RS) is an attractive feedstock for biofuel production due to several attractive reasons that include its worldwide abundant availability, its rich lignocellulose content primarily comprising of 35%–40% cellulose, and adding value to the otherwise waste crop residue. Rice is among the major food crops around the world with China and India being the top two rice-producing countries annually. Rice production techniques have been gradually modernizing with the introduction of modern technologies to farmers, yielding far more than traditional ways of farming. With higher rice yield, harvesting also leaves behind biomass residue in the form of RS and stubble parts in the fields. Unavailability of proper management techniques for these crop residues often lets farmers choose open-field burning of such valuable crop residues which not only damage the soil microbial load and quality but also are a cause of environmental pollution. Processing RS for biofuel production is one such concept that solves this problem. Biofuel production from RS helps in adding value to the biomass residue, reduces problems regarding residue management, helps in the production of fuels that are an effective alternative to fossil-based fuels, especially in the transportation sector, helps in decreasing emissions of greenhouse gases (GHGs), and reduces environmental pollution.

The recalcitrant nature of RS lignocellulose biomass is an important factor that hinders biofuel production. There are several pretreatment strategies introduced to overcome this drawback in a much more effective, economic, and environmentally friendly manner. These pretreatment methods help in degrading the complex lignocellulose structure by disintegrating the hemicellulose and lignin structures, as well as releasing the crystalline cellulose, which makes the lignocellulose biomass much more accessible to enzymatic hydrolysis to release simple sugars to be used in fermentation for biofuel production.

2. Rice straw

RS is the residue that is left behind after rice harvesting and is among the most abundant lignocellulosic crop residues in the world because of the large production of rice worldwide. China is the leading producer of rice followed by India with around 148.9 and 129 million tons in the year 2021–22, respectively. Table 21.1 shows the annual rice production in a few Asian countries (2021–22).

2.1 Rice ecosystem

Rice is one of the most important cereal crops in the world along with wheat and maize, growing on all six continents except Antarctica, including Asia, Africa, Australia, Europe, North America, and South America. Rice is celebrated as a staple by almost half of the global population, including several Asian countries.

The ecosystem in which rice is cultivated is referred to as rice ecosystem. A wide range of cultivation conditions leads to four distinct types of rice ecosystems, according to the International Rice Research Institute: upland, irrigated, rainfed lowland, and flood-prone [3]. The rice yield varies extremely in each ecosystem depending on various conditions, including water availability, droughts, flooding, soil fertility, weed conditions, pests, and plant diseases.

TABLE 21.1 Annual rice production in few Asian countries (2021–22) [1,2].

Country	Annual rice production (million tons)
China	148.9
India	129.0
Bangladesh	35.8
Indonesia	35.4
Vietnam	27.2
Thailand	19.7
Burma	12.6
Philippines	12.4
Pakistan	8.7
Japan	7.5
Cambodia	5.9
South Korea	3.8
Nepal	3.7
Sri Lanka	2.7
Malaysia	1.8
North Korea	1.3
Taiwan	1.2

Upland ecosystems consist of rainfed fields located from valleys to steep sloping mountainous land which are rarely flooded and are associated with high runoff and lateral water movement. Some of the major limitations to upland rice ecosystems include low soil fertility and aggressive growth of weeds [3]. Rice cultivation in the upland ecosystem depends on rainwater which results in varying yields every harvest season.

Irrigated rice ecosystems consist of leveled grounds with bunded fields that are facilitated such that water is irrigated according to seasonal water availability required for rice cultivation. The potential yield of rice varieties from irrigated ecosystems is high compared to other ecosystems. About half of the world's rice land has such irrigated rice ecosystems, which cover about 75% of the world's rice production [4].

Rainfed lowland ecosystems consist of slightly leveled to sloping fields with bunds. Such ecosystems depend on rainwater, which provides uncertainty and uncontrolled water levels causing floods and droughts often during the same rice cultivation season, leading to variable rice yields. The climatic conditions of these ecosystems are adverse and the soil quality is low which add to the low rice yields of such ecosystems [4].

Flood-prone rice ecosystems consist of rice fields that are subjected to uncontrolled floods. The rice fields remain submerged in water for weeks to months. Due to flood conditions, rice cultivators often prefer rice varieties with elongated stems that emerge over the floodwater. Flood-prone rice ecosystems are usually located near banks and deltas of rivers such as the Ganga and Brahmaputra rivers in India and Bangladesh, and the Mekong River in Vietnam and Cambodia. Rice yield in such an ecosystem varies greatly on the unpredictable flood levels and droughts [4].

2.2 RS availability and management

RS is the residue generated after rice grain harvesting from the rice, composed of panicle rachis, leaves, and stems [5], whereas the part of the rice plant that remains attached to the soil after harvesting is called stubble. The total biomass of RS depends on various factors including the rice variety, soil type, nutrients, as well as the rice ecosystem. It is one of the most abundant crop residues around the globe mostly because of rice being the staple food for almost 50% of the population.

After harvesting rice, RS is spread on the fields or collected from the fields and used as a raw material for paper industries, animal feed, and fertilizer, or burnt to clear the field for the next rice season. The uncut portion of the RS that is left in the fields after harvesting is usually burnt, or left to naturally decompose and mix into the soil [6]. Collection of RS is often quite expensive and tedious for the farmers, whereas incorporation into soil takes longer because of the slower rate of decomposition of RS which hinders two to three yearly cropping rounds [6]. Because of these reasons, farmers opt for open-field burning of the RS. About 50% of RS is usually burnt in the fields, with only about 20% of RS being utilized in various ways [5]. Therefore, easier and more profitable methods should be introduced to promote the cutting and collection of RS from the fields by the farmers. Balers are commonly used in Asia for collection purpose as these machines are quite efficient and require only a few skilled hands to operate the machine [7]. Farmers should also be provided with the knowledge of RS management methods such as mushroom production, composting, energy production methods, and building materials, for the maximum valorization of RS. A few RS management techniques have been tabulated in Table 21.2.

The annual RS production ranges from 100 to 140 million tons in Southeast Asia, 330–470 million tons in Asia, and 370–520 million tons over the world per year [6].

2.3 RS constituents and characteristics

The cell wall of RS is recalcitrant in nature, which has evolved based on several factors including rice variety, climatic conditions, pest management, and nutrient availability.

TABLE 21.2 Rice straw (RS) management techniques.

RS management strategy	Specification	Advantages	Disadvantages
Straw removal	Widespread in India, Bangladesh, and Nepal	Straw used as fuel for cooking, ruminant fodder, and stable bedding or as a raw material in industrial processes like paper industries	Depletion of soil K and Si reserves Some or all of the nutrients contained in straw may be lost to the rice field
Burning	Large amount of straw left behind after harvesting, high labor cost of manual straw collection, and slow decomposition rate influence farmers to burn the straw in open fields	Effective removal of large volumes of straw	Nutrient loss, depletion of soil organic matter (SOM), and reduction in the presence of beneficial soil biota, environmental pollution, greenhouse gas emissions, and release of aerosol particles
Soil incorporation	Straw is incorporated onto pits or left behind on the field to naturally decompose. Straw has a very slow decomposition rate due to which farmers usually avoid these methods especially if farmers opt for two to three rounds of cultivation per year	Natural decomposition takes place which involves natural microbial load. No labor cost for farmers. The fertility of the soil remains intact up to certain limits	Very slow decomposition process that keeps the crop fields barren for a longer time. Does not allow intensive cropping multiple times a year
Composting	Animal manure and enzymes are often added to RS and mixed by a turner and ensilage to homogenize the mixture for composting	Can be used as a natural fertilizer for growing vegetables substituting expensive inorganic fertilizers. Improves nutrient and organic matter of the soil	Lack of proper knowledge of composting is one of the major drawbacks among farmers
Mushroom production	Edible mushroom production is an effective way of RS management that requires less resources and labor	Edible mushroom are rich source of fiber, protein, and antioxidants, that are for good human health	Lack of proper knowledge of mushroom cultivation is one of the major drawbacks among farming villagers
Cattle feed	RS as a cattle feed is considered a potential feed additive for increasing the energy and protein content	Urea treatment of straw can improve consumption and digestibility of RS as fodder	RS is of poor-quality livestock feed having a low C:N ratio and high NDF and ADF, which affect its nutritive value

A major physical property of RS is its bulk density. The bulk density of RS varies largely on the length of RS, the cutting technique, cutting equipment, and the compression ratio. Loose RS from the fields has a bulk density of approximately 20 kg m^{-3}, whereas chopped RS can have a bulk density ranging from 50 to 150 kg m^{-3} depending on the straw length based on the cutting technique [6], which is comparatively low as compared to coal at approximately 600 kg m^{-3} [8]. The low density of RS is mainly because of high intrinsic and apparent porosity [9]. For bioenergy production, the low bulk density of RS results in poor mixing, longer combustion time, and poor energy distribution, which has a negative impact on the overall process decreasing the energy efficiency [6].

The bulk density of unprocessed RS is low, because of this, larger volumes of RS are required per kilogram which has a direct impact on the storage, shipping, handling, and processing costs. To address this issue, volumes of RS can be reduced using several methods to increase the density of RS. Some of the well-known techniques include pelleting to smaller lengths using pellet mills, pressing using high pressure to form different sizes and shapes using roller presses, piston presses, and briquette presses, which decrease RS volumes from 400% to 700% [7]. Increasing the RS bulk density using densification processes also affects the moisture content of the RS which is an important factor, especially for energy production techniques as it affects the combustion temperature. The higher moisture content of RS decreases the combustion temperature, decreasing the efficiency and reducing power generation [8]. Dry RS has a moisture content of 10%–20% which can increase during the wet seasons [8].

RS is lignocellulosic biomass mainly composed of cellulose, hemicellulose, and lignin as the major components, and other components such as lipids, proteins, pectins, simple sugars, ash, and silica in lesser percentages. The concentration of the components such as ash content in RS may vary depending on several factors including rice variety, rice ecosystem, and harvesting method. Ash content includes the inorganic components in RS that ranges from 18.67% to 29.1% which is quite high and is responsible for the decrease in the calorific value of RS [6]. Ultimate analysis shows that RS is composed of carbon, hydrogen, oxygen, nitrogen, and sulfur. The carbon content of RS is lower than that of fossil fuel, whereas hydrogen and oxygen contents are higher [6].

RS is highly volatile in nature as it contains high concentrations of about 60%–70% of volatile components like carbon, combined water, hydrogen, nitrogen, and sulfur [6]. RS also has a high combustion rate as compared to that of coal and wood due to which, RS as biomass has a greater advantage for bioenergy production, especially through combustion [8].

3. Biofuel production from RS

Biofuels are liquid or gaseous compounds produced from renewable lignocellulosic organic biomass, that can substitute for fossil-based fuels, and therefore, have the ability to reduce emissions of GHGs, especially in the transportation field. However, first-generation biofuel is being produced using food crops such as grains and sugarcane, which creates food vs fuel competition. This issue has been solved by introducing second-generation biofuels produced from agricultural residue such as wood, straw, husk, bagasse, etc. Lignocellulose biomass from agricultural crops can be used to produce biofuels like bioethanol and biobutanol, and this has been proved by researchers all over the world. RS is a lignocellulosic biomass and can be effectively used for the production of second-generation biofuel after proper pretreatment and hydrolysis. A few reports of biofuel production from RS have been tabulated in Table 21.3.

3.1 Types of biofuels that can be produced from RS

Bioethanol has become quite a popular alternative to gasoline in recent times. Bioethanol has a calorific value of 23.4 MJ L^{-1} compared to 35.7 MJ L^{-1} of petrol [20] and is one of the most commonly used liquid biofuel effectively used by blending with gasoline as a transport fuel in several countries, especially with blending fuel mandates established by the government. Currently, the United States is the highest bioethanol-producing country at 15 billion gallons in 2021 followed by Brazil at 7.5 billion gallons of bioethanol produced in 2021 [21]. Second-generation bioethanol production from agricultural lignocellulose biomass such as RS helps in the production of energy that is both eco-friendly and sustainable in nature. RS has several factors that make it a potential raw material for ethanol production. RS consists of a high percentage of cellulose and hemicellulose that can be degraded to simple fermentable sugars using pretreatment methods. These sugars can be further used for enzymatic hydrolysis using microorganisms or direct enzymes for the production of bioethanol [22].

Biobutanol is considered an effective alternative to fossil fuels globally, which can reduce carbon emissions for a cleaner environment. Biobutanol can be blended at very high percentage with diesel to be used in diesel engines without engine modifications and result in about 60% reduction in GHG emissions [23]. Butanol is one of the main products of

TABLE 21.3 A few reports of biofuel production from rice Straw.

Biofuel	Pretreatment method applied	Highlighted conditions	Outcome	Refs.
Bioethanol	Dilute sulfuric acid	Coculture of thermophilic anaerobic bacteria consisting of cellulolytic Clostridium sp. DBT-IOC-C19 and noncellulolytic Thermoanaerobacter sp. DBT-IOC-X2	142 mM ethanol concentration	[10]
Bioethanol	Low-cost alkali chemicals ($NH_3 \cdot H_2O$, CaO) and liquid hot water	Transgenic rice lines that overexpressed AtCesA6 gene with reduced cellulose crystallinity and polymerization were used as RS substrate	19.1 g L^{-1} ethanol concentration	[11]
Bioethanol	Na_2CO_3	Fed-batch semisimultaneous saccharification and fermentation (FB-S-SSF) with Aspergillus fumigatus	118.9 ± 3.6 g L^{-1}	[12]
Bioethanol	40% ammonia	Cellulolytic and xylanolytic enzymes by Sporotrichum thermophile was used for saccharification of RS that liberated 356.34 mg g^{-1} reducing sugars	28.88 g L^{-1}	[13]
Biobutanol	Alkaline pretreatment with 1.5% NaOH and 0.2% H_2O_2	Polyvinyl alcohol (PVA) immobilized Clostridium acetobutylicum ATCC 824 in ABE fermentation without yeast extract	13.80 g L^{-1} biobutanol titer	[14]
Biobutanol	NaOH/Urea pretreatment at −12°C	Cellulase produced from Trichoderma viride was used for saccharification of pretreated RS, and a coculture system of Clostridium beijerinckii and Saccharomyces cerevisiae was used for fermentation	10.62 g L^{-1} biobutanol concentration	[15]
Biobutanol	NaOH and H_2SO_4	Fermentation with a coculture of S. cerevisiae and Pichia sp.	13.3 g L^{-1} biobutanol concentration	[16]
Biobutanol	0.75% (w/v) of NaOH alkaline pretreatment	Simultaneous saccharification and fermentation combined with gas stripping operated in fed-batch with RS loading of 18.4% (w/v)	24.80 g L^{-1} biobutanol titer	[17]
Biomethane	Ozonolysis at ozone dosage of 0.006 gO_3/g followed by thermal (55°C, 4d) treatment	Treatment of lignocellulose by ozonolysis and temperature for higher methane production from RS was the main focus of the study	Methane yield 374 ± 6 mL CH_4/g-$VS_{substrate}$	[18]
Biomethane	Cellulase enzyme-producing bacterial strain (Bacillus sp. − accession number KX373535)	Delignification of RS using ultrasonic homogenizer prior to biological pretreatment removed 70.28% lignin content enhancing enzyme accessibility during biopretreatment	165.5 mL per g^{-1} VS methane production	[19]

acetone−butanol−ethanol (ABE) fermentation, having several applications as an industrial solvent, in various reactions as a chemical intermediate, as an extractant, and as a biofuel. Biobutanol is a colorless and flammable alcohol with 4-carbon straight chain molecules that can be burned in the presence of oxygen; has a high octane number, high heating value, low vapor pressure, low volatility, and fewer ignition problems; is highly intersoluble and flexible in fuel blends; and has high viscosity that together makes butanol a suitable alternative [24]. Second-generation biobutanol production is currently popular among researchers all over the world with several reports of biobutanol production from agricultural biomass. RS is one popular agricultural lignocellulosic biomass that has been widely researched for biobutanol production through the ABE fermentation technique after proper pretreatment of the RS. Biobutanol production often faces challenges regarding cost of biomass substrate, low production yield, and the requirement of intensive product recovery processes.

Biomethane is an easily storable and flexible fuel that can be used in both liquid and gaseous form. Biomethane can be easily used in any equipment that operates with natural gas without modifications. Liquefied biomethane is the liquefied form of biomethane that can be used as an alternative fuel for vehicles that run on liquefied natural gas and compressed natural gas [25]. A wide variety of lignocellulose biomass can be used as a feedstock for biomethane production by anaerobic digestion (AD), RS being one such commonly used biomass. During the biomethane production process, wet biomass with high moisture content can be used as a feedstock which is a major advantage over bioethanol and biobutanol production processes. However, the biomethane recovery processes after AD are highly energy consumable and expensive that requires efforts to introduce upgraded technologies [26].

3.2 Pretreatment of RS

In the second-generation biofuel production process, RS has to be first pretreated to reduce its recalcitrant nature by reducing the interactions between the cellulose, hemicellulose, and lignin components. The pretreatment steps aim to reduce cellulose crystallinity, enhance cellulose accessibility, and remove the complex lignin network. Proper pretreatment is necessary to enhance enzymatic hydrolysis to increase biofuel yield [27]. Methods of pretreatment can be categorized as physical, chemical, biological, and their combination Fig. 21.1

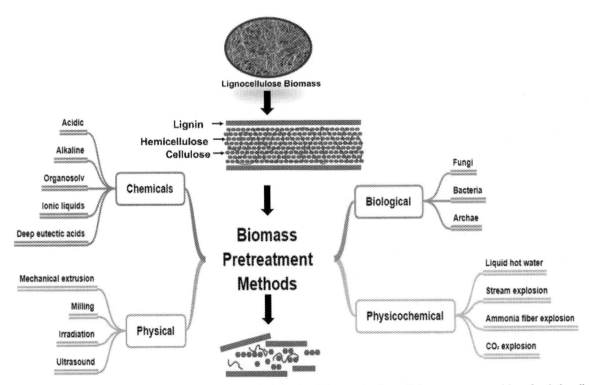

FIGURE 21.1 Different lignocellulosic biomass pretreatment methods that break the complex lignocellulose structure comprising of majorly cellulose, hemicellulose, and lignin. These pretreatment methods result in the breakage of complex lignin structure and decrease the crystallinity of cellulose.

3.2.1 Physical pretreatment

Physical pretreatment involves the use of different grinding techniques and machinery such as hammer mills [28] for the mechanical fragmentation and size reduction of RS. Physical pretreatments help in increasing the surface area of the biomass which can be achieved by various processes such as chopping, steaming, irradiation, heating, extrusion, grinding, or milling [29] to enhance enzyme accessibility in the hydrolysis process for high sugar yield. The grinding technique has a significant impact on the sugar yield by enzymatic hydrolysis of RS particles. The energy requirement for mechanical straw grinding is often overlooked by biorefineries while focusing on maximum sugar yield for biofuel production. The energy requirement is dependent on several factors including moisture content, mean particle size, cellulose crystallinity, specific surface area, and the oxygen-to-carbon surface ratio [30].

Several researchers have reported that physical pretreatment of RS by irradiation such as gamma irradiation, electron beam irradiation, and microwave irradiation are effective in lowering the lignocellulose crystallinity and degree of polymerization resulting in increased accessibility of enzymes to crystalline cellulose [29]. Microwave irradiation, for instance, changes the cellulose ultrastructure as well as degrades the hemicellulose and lignin content of the lignocellulose biomass [31]. Microwave irradiation combined with alkali treatment has been considered effective in the pretreatment of lignocellulose biomass in several reports, with a large percentage (up to 80%–90%) of hemicellulose and lignin removal [31].

3.2.2 Chemical pretreatment

Chemical pretreatment methods are considered effective low-cost methods involving acids, alkali, organosolv agents, and ionic liquid treatments at different temperature and pressure conditions. Acid pretreatment methods in which low pH is combined with a partial high temperature effectively hydrolyze the hemicellulose as well as lignin and cellulose crystals, increasing cellulose accessibility for enzymatic degradation to release monomeric glucose units. Dilute acid pretreatment involves treatment with minimum acid concentration to produce the optimum amount of glucose units. Though concentrated acids have been widely used for pretreatment, they are usually linked to the production of process inhibitors and corrosiveness [32]. Dilute nitric acid pretreatment of RS at 30 °C and 200 rpm for 72 h followed by enzymatic hydrolysis of the resultant supernatant using cellulase and laccase enzyme-producing microbial strains was found to be a novel and energy-efficient approach [32]. A high amount of glucose yield of 205 mg glucose g^{-1} of RS was reported by the method [32]. In another study, 0.6% formic acid was found effective in hemicellulose removal of RS, thereby increasing cellulase accessibility to cellulose, resulting in enhanced enzymatic hydrolysis, and was considered a promising lignocellulose pretreatment strategy for subsequent biofuel production [33]. Pretreatment of RS using a combination of dilute sulfuric acid and five organic acids including oxalic acid, acetic acid, tartaric acid, succinic acid, and fumaric acid at 121°C resulted in a significant decrease of inhibitory products and high glucose yield as compared to the pretreatment of RS with only dilute sulfuric acid [34]. Chemical pretreatment methods often result in the production of byproducts such as furans, organic acids, and phenolics that reduce microbial growth rate resulting in inhibition of fermentation and decreasing productivity [35].

Alkaline pretreatment is known for the potential for lignocellulose disruption by causing hemicellulose, lignin, and silica to partially dissolve and disintegrate. The lignocellulose structure upon alkaline pretreatment swells because of deacetylation, which enhances enzyme associability to cellulose and hemicellulose. The basic mechanisms of alkaline pretreatment methods are based on the hydrolysis of the intermolecular ester bonds in the lignocellulose biomass that crosslinks xylan, hemicellulose, and lignin [35]. Because of the comparatively lower price and the ability to solubilize lignin to a large extent, the alkali sodium hydroxide (NaOH) is commonly used for the pretreatment of lignocellulosic biomass, although it involves extensive washing, as well as the production of toxins and black liquor in vast amounts [36]. Potassium hydroxide (KOH) pretreatment is considered much more environmentally friendly as compared to the traditional methods of chemical pretreatment such as NaOH and sulfuric acid, which cause severe ecological contamination [36]. The property of selectively removing lignin from biomass and reducing the problem of wastewater production has gained KOH a lot of attention, as KOH treatment usually results in the production of potassium, which can be used in agricultural fields [36]. A combination of KOH and urea, a commonly used nitrogen (N) fertilizer that has a high N amount (46%), further improves the rate of enzymatic hydrolysis even at low temperatures, whereas at room temperatures, such combined pretreatment could significantly improve the delignification and biodegradability of RS [36].

In organosolv pretreatment, organic solvents such as methanol, acetone, formic acid, acetic acid, and glycerol are used in combination with or without a catalyst such as acid, base, or salt that hydrolyzes lignin partially or completely [37]. Organosolv pretreatment causes partial dissociation of the hydrogen ions which promotes the hydrolysis and dissolution of lignin [37]. The organic acids used in such pretreatment methods modify the lignin structure and tend to increase the accessible surface area of lignocellulosic biomass, enhancing hemicellulose and lignin solubilization [37]. Organosolv pretreatment of RS with organic acid followed by treatment with a base was found to have removed about 82% of hemicellulose and lignin, and 90.4% of silica from RS to obtain cellulose [38]. Glycerol organosolv (GO) pretreatment was found to be effective in selective disintegration of several lignocellulosic biomass, improving enzymatic hydrolysis [39]. This method has several advantages including the ability to operate in low pressure with rapid temperature fluctuation, less production of by-products that inhibit fermentation, production of valuable glyceryl glycosides, and survival of solvent that can be used as a carbon source for the microorganisms [39]. Moreover, the glycerol in these pretreatment methods protects components from unnecessary degradation caused by *trans*-glycosylation (glyceryl xylosides/glucosides) or hydroxylation grafting (glyceryl phenolics) during GO pretreatment [39].

3.2.3 Biological pretreatment

Biological pretreatment is better for the environment than other pretreatment methods, requires lesser energy input, and is much safer [29]. But these methods are quite a time-consuming and often tend to reduce overall sugar yields. In biological pretreatment processes, whole cells of bacteria and fungi, or enzymes produced by them, are used for the pretreatment of lignocellulosic biomass [29]. According to a report, *Bacillus subtilis* subsp. *subtilis* JJBS300, *Myceliophthora thermophila* BJTLRMDU3, and *Aspergillus oryzae* SBS50 released maximum of 52.41, 86.74, and 49.59 mg g^{-1} substrate reducing sugar, were found to degrade hemicellulose and lignin interaction in RS, and produced cellulolytic enzymes like CMCase, FPase, and β-glucosidase [40]. Biological pretreatment is often combined with other pretreatment methods for efficient lignocellulose degradation [48]. Pretreatment of RS with an ionic liquid (1-ethyl-3-methylimidazolium acetate) at a high substrate load of 25 wt.% followed by bacterial delignification using bacterial strain *Cupriavidus basilensis* B-8 resulted in reducing sugar and glucose yield of 510.3 and 324.5 mg g^{-1}, respectively [41].

3.3 The biofuel production process

There are extensive efforts underway to produce biofuels from biodegradable waste, including agricultural lignocellulosic biomass like RS. Agricultural biomass has low market value and helps to overcome numerous environmental issues such as GHG emissions especially caused due to burning, and fulfilling global energy requirements. Production of ethanol and butanol from lignocellulosic biomass consists of a few major steps including pretreatment, saccharification, fermentation, and distillation as well as several intermediate substeps like solid−liquid separation, detoxification of pretreated slurry, fermentation of the pretreated biomass, and sterilization before fermentation to remove unwanted microbial cultures that might inhibit the growth of desired microbial inoculum. Whereas, the major steps for the production of biomethane from lignocellulosic biomass include biomass size reduction, pretreatment, AD of the pretreated biomass followed by separation of digested slurry and separation of biogas to methane which is a much complicated process that releases a high amount of heat that can be utilized as an energy source. Pretreatment of the lignocellulosic biomass before enzymatic hydrolysis for saccharification is an important step to expose cellulose molecules to the enzymes as well as breaking down the complex lignocellulose structure consisting of lignin, hemicellulose, and crystalline cellulose with a small percentage of other components.

3.3.1 Enzymatic hydrolysis

Effective saccharification during hydrolysis of biomass is another important step in the liquid biofuel production process. Enzymatic hydrolysis of the pretreated lignocellulose biomass results in further saccharification of polysaccharides like cellulose to monomeric sugar units that are utilized during fermentation to produce biofuel. The cellulase enzymes catalyze the hydrolysis of cellulose by breaking the 1,4-β-glycosidic bonds in the cellulose chain of biomass, whereas xylan is among the crucial hemicellulase enzymes that are engaged in the enzymatic saccharification of hemicellulose in the lignocellulosic biomass. This step is often considered expensive due to the involvement of enzymes; therefore, cheaper, cost-effective enzyme production methods are necessary [42]. There are several bacteria and fungi involved in cellulase and hemicellulase production. The most examined fungi for cellulase and hemicellulase production are *Trichoderma* spp. and *Aspergillus* spp., whereas some commonly examined bacteria are *Bacillus* spp., *Cellulomonas* spp., etc.

3.3.2 Fermentation of sugars

The process of industrial fermentation can be divided into three important stages: prefermentation phase, tumultuous phase, and postfermentation complementary phase [43]. During the prefermentation phase, the process proceeds in the lag phase after the inoculation of yeast cells at a ratio of 20% v/v into the product obtained from enzymatic hydrolysis. This phase is considerably slow with gradual temperature increase and CO_2 release, aiming for a large number of cells in about 5 h with maximum fermentation ability. The tumultuous phase, for the next 10 h, proceeds with a rapid rise in the system temperature and CO_2 release, increasing the acidity of the system, with lowermost density due to alcohol formation and equivalent sugars utilization. The complementary phase runs for about 7 h slowly and steadily, decreasing the overall system temperature with continuous CO_2 release [43]. The product obtained will contain a complex mixture of liquid, solid, and gaseous components with a small amount of sugar which will be separated using several separation techniques such as distillation to separate liquid components. Yeasts, especially *Saccharomyces cerevisiae*, are commonly used in fermentation systems for the production of ethanol due to their high ethanol productivity, ethanol tolerance, and the ability to ferment a wide range of sugars [44]. Yeast fermentation has a few drawbacks that have negative impacts on the ethanol production process such as the intolerance to high temperatures and relatively high concentration of ethanol in the fermentation system, as well as the ability to ferment pentose sugars [44]. Several yeast strains have been examined and studied to be used in fermentation for ethanol production including wild-type, hybrid, and recombinant strains.

Biobutanol is commonly produced by *Clostridium acetobutylicum* or *Clostridium beijerinckii* in ABE fermentation consisting of two metabolic phases that include acidogenesis and solventogenesis. During the acidogenesis phase, bacterial cells multiply in the log phase of growth that results in production of acids like acetic acid and butyric acid, as well as hydrogen and CO_2; whereas the solvent (acetone, butanol, and ethanol) production and endospore formation take place in the solventogenesis stage [45]. Selection of active microbial strains and proper fermentation conditions are key features for higher yields by this fermentation process.

There are three fermentation processes commonly used in biofuel production: separate hydrolysis and fermentation (SHF), simultaneous saccharification and fermentation (SSF), and simultaneous saccharification and cofermentation (SSCF). In SHF systems, hydrolysis of lignocellulosic biomass is separated from the fermentation system, which allows enzymatic hydrolysis to proceed at high temperatures for better enzyme performance while allowing fermentation organisms to process at moderate temperatures for optimizing sugar utilization. The overall process duration of SSF and SSCF is comparatively shorter due to the simultaneous enzymatic hydrolysis and fermentation, keeping the concentration of glucose low. For SSF systems, glucose fermentation is separated from pentoses, while in SSCF systems, glucose and pentoses are fermented in the same reactor. Both SSF and SSCF systems are preferred over SHF systems because of the ability to perform the processes in the same tank, as well as the requirement of a lower production cost to run the processes that provide a higher ethanol yield in a shorter processing time [44]. SSF process has been found ideal for biobutanol production by ABE fermentation from microwave processed RS [46].

3.3.3 Anaerobic digestion

AD is a highly complex and dynamic process involving several biochemical steps, including hydrolysis, acidogenesis, acetogenesis, and methanogenesis in a strictly anaerobic system to degrade organic matter. Hydrolysis is the main rate-limiting step in which complex lignocellulosic molecules are broken down into simpler molecules. Hydrolysis is followed by acidogenesis, which converts the products into volatile fatty acids (VFAs), releasing H_2, CO_2, H_2S, and NH_3 as by-products. The VFAs are further converted to acetates, CO_2, and H_2 by the process of acetogenesis. These acetates, CO_2, and H_2 are processed into biomethane by methanogens in the methanogenesis process. The AD process results in the production of slurry and biogas, which consist mainly of 50%–70% CH_4, 30%–50% CO_2, and traces of NH_3, H_2S, and water vapor [47]. Fig. 21.2 shows the overview of biofuel production process from RS.

4. Economic evaluation of biofuel produced from rice straw

Biofuels such as bioethanol, biobutanol, and biomethane have been viewed over time as potential alternatives to fossil fuels. Bioethanol in particular has the potential to replace gasoline, making it a popular alternative vehicle and engine fuel worldwide. Bioethanol, which is produced from agricultural lignocellulosic biomass, is gaining recognition worldwide as it can replace or be blended with gasoline, reducing dependence on petroleum-based fuels and thus reducing pollution. Several factors play an important role in increasing the economic viability of biofuels, especially those made from lignocellulosic biomass: (1) reducing the release of GHGs normally associated with fossil fuel combustion, (2) adding value to abundant crop residues such as RS, (3) reducing the burning of crop residues that pollute the environment, (4) promoting waste recycling, (5) taxing the carbon footprint, which forces countries to shift to biofuels and reduce carbon emissions, and (6) providing energy security for the future [40]. Crop residues such as RS can be strategically used for the production of bioethanol, biobutanol, and biomethane, which are used as fuel for vehicles and industrial engines, as well as for the generation of electric power.

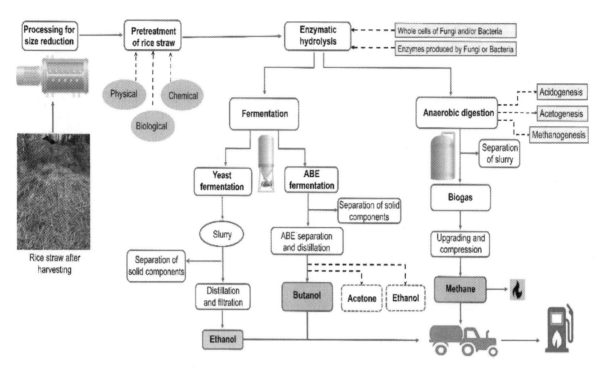

FIGURE 21.2 The basic process of biofuel production from RS. After harvesting of rice, the resulting lignocellulosic waste in the form of RS is first ground or cut for size reduction followed by pretreatment by suitable methods to break the complex lignocellulose structure for maximum enzymatic hydrolysis of the cellulose and hemicellulose. Enzymatic hydrolysis helps in releasing fermentable sugars which are fermented or anaerobically digested to produce biofuels such as bioethanol, biobutanol, and methane to be used as vehicle and machinery fuel.

Biofuel production is often debated as food or fuel. However, advances in agricultural technologies in recent years have increased crop yields and agricultural biomass residues, contributing to sustainable food and biofuel production, reducing dependence on petroleum-based fuels, and thus initiating a socioeconomic approach. There are several ways in which a biofuel economy can benefit a community, most notably by creating jobs in biomass collection, transportation, and biofuel production facilities. These jobs and the taxes collected from the operation and marketing of biofuel products can also benefit the community in the areas of health and education and lead to improved living conditions in rural areas [40]. The biofuel industry also promotes agriculture on a larger scale by introducing newer technologies due to the high demand for crop residues such as RS. This practice not only provides farmers with additional income from residues, but also increases food and grain production, reduces environmental problems associated with residue burning, and promotes the use of fallow land, leading to sustainable food and fuel production in an environmentally friendly and economical manner.

However, there are several challenges in producing biofuels from lignocellulosic biomass such as RS to make the process both economically feasible and environmentally friendly:

(1) Continuous and abundant availability of RS.
(2) The cost of collecting RS from crop fields, transporting it to biofuel production facilities, and fluctuations in the market price of RS have a major impact on production costs.
(3) The cost of establishing a biofuel production plant, operating costs, the use of economically feasible technologies, and trained personnel for the smooth operation of the biofuel production process are important for the successful operation of biofuel production plants.
(4) The use of effective technologies to pretreat RS using cost-efficient and environmentally friendly methods is an important aspect of biofuel production.
(5) The cost of transporting biofuels after production is important for economically viable biofuel production. In this regard, the availability of buyers should be considered, as transportation costs could increase if there are no possible buyers in the immediate vicinity.

5. Conclusion and future prospects

With the introduction of latest technologies in the farming sector, global rice production has been increasing at a rapid pace. RS is abundantly available as it is the major crop residue left behind after rice harvesting. RS as a lignocellulosic biomass has the potential to be used as a feedstock for biofuel production. The complex lignocellulose structure consisting of cellulose, hemicellulose, and lignin often acts as a challenge for the biofuel production process. Therefore, proper pretreatment methods that are efficient and cost-effective are to be adapted by biofuel production companies. There are several research going on all over the world to set optimal pretreatment methods for economic biofuel production. Care is also needed to be taken for effective fermentation process as well as biofuel recovery for economic biofuel production such that in spite of being a cost-effective method, no harmful substance is released to the environment.

6. Conflict of interest statement

The authors declare that the study was conducted in the absence of any commercial or financial relationships that could be construed as a potential conflict of interest.

7. List of abbreviations

ABE Acetone–butanol–ethanol
AD Anaerobic digestion
GHGs Greenhouse gases
GO Glycerol organosolv
KOH Potassium hydroxide
NaOH Sodium hydroxide
RS Rice straw
SHF Separate hydrolysis and fermentation
SSCF Simultaneous saccharification and cofermentation
SSF Simultaneous saccharification and fermentation
VFAs Volatile fatty acids

Acknowledgments

The authors would like to thank the Department of Microbiology, Sikkim University, for providing the computational infrastructure and central library facilities for procuring references and plagiarism analysis (URKUND-Ouriginal: Plagiarism Detection Software).

References

[1] World Rice Production 2021/2022, World Agricultural Production.Com, 2022. http://www.worldagriculturalproduction.com/crops/rice.aspx. accessed Mar. 29, 2022.
[2] N.W. Childs, Rice Outlook: May 2021 - USDA ERS, 2021. https://www.ers.usda.gov/webdocs/outlooks/101196/rcs-21d.pdf?v=2388.7.
[3] R. Prasad, Y.S. Shivay, D. Kumar, Current status, challenges, and opportunities in rice production, Rice Prod. Worldw. (2017) 1–32.
[4] A. Ferrero, A. Tinarelli, Chapter 1 - rice cultivation in the E.U. Ecological conditions and agronomical practices, Pestic. Risk Assess. Rice Paddies (2008) 1–24.
[5] A. Satlewal, R. Agrawal, S. Bhagia, P. Das, A.J. Ragauskas, RS as a feedstock for biofuels: availability, recalcitrance, and chemical properties, Biofuels, Bioprod. Biorefin. 12 (1) (2018) 83–107.
[6] N.V. Hung, M.C. Maguyon-Detras, M. Gummert, M.V. Migo, R. Quilloy, C. Balingbing, P. Chivenge, RS overview: availability, properties, and management practices, Sustain. RS Manag. 1 (2020).
[7] C. Balingbing, N. Van Hung, N.T. Nghi, N. Van Hieu, A.P. Roxas, C.J. Tado, E. Bautista, M. Gummert, Mechanized collection and densification of RS, Sustain. RS Manag. (2020) 15–32.
[8] J. Logeswaran, A.S. Silitonga, T.M. Mahlia, Prospect of using RS for power generation: a review, Environ. Sci. Pollut. Res. 27 (21) (2020) 25956–25969.
[9] A. Ali, A. Abakar, A.D. Mahamat, A. Donnot, J.L. Tanguier, R. Benelmir, Physical and chemical characteristics of RS, Res. J. Appl. Sci. Eng. Technol. 17 (4) (2020) 115–121.
[10] N. Singh, R.P. Gupta, S.K. Puri, A.S. Mathur, Bioethanol production from pretreated whole slurry RS by thermophilic co-culture, Fuel 301 (2021) 121074.

[11] Y. Ai, S. Feng, Y. Wang, J. Lu, M. Sun, H. Hu, Z. Hu, R. Zhang, P. Liu, H. Peng, Y. Wang, Integrated genetic and chemical modification with RS for maximum bioethanol production, Ind. Crop. Prod. 173 (2021) 114133.

[12] X. Jin, J. Ma, J. Song, G.Q. Liu, Promoted bioethanol production through fed-batch semisimultaneous saccharification and fermentation at a high biomass load of sodium carbonate-pretreated RS, Energy 226 (2021) 120353.

[13] B. Singh, A. Bala, A. Anu, V. Kumar, D. Singh, Biochemical properties of cellulolytic and xylanolytic enzymes from Sporotrichum thermophile and their utility in bioethanol production using RS, Prep. Biochem. Biotechnol. 52 (2) (2022) 197−209.

[14] T.Y. Tsai, Y.C. Lo, C.D. Dong, D. Nagarajan, J.S. Chang, D.J. Lee, Biobutanol production from lignocellulosic biomass using immobilized Clostridium acetobutylicum, Appl. Energy 277 (2020) 115531.

[15] J. Wu, L. Dong, B. Liu, D. Xing, C. Zhou, Q. Wang, X. Wu, L. Feng, G. Cao, A novel integrated process to convert cellulose and hemicellulose in RS to biobutanol, Environ. Res. 186 (2020) 109580.

[16] S. Mohapatra, R.R. Mishra, B. Nayak, B.C. Behera, P.K. Mohapatra, Development of co-culture yeast fermentation for efficient production of biobutanol from RS: a useful insight in valorization of agro industrial residues, Bioresour. Technol. 318 (2020) 124070.

[17] A. Valles, J. Álvarez-Hornos, M. Capilla, P. San-Valero, C. Gabaldón, Fed-batch simultaneous saccharification and fermentation including in-situ recovery for enhanced butanol production from RS, Bioresour. Technol. 342 (2021) 126020.

[18] R. Patil, C. Cimon, C. Eskicioglu, V. Goud, Effect of ozonolysis and thermal pre-treatment on RS hydrolysis for the enhancement of biomethane production, Renew. Energy 179 (2021) 467−474.

[19] Y. Kannah, S. Kavitha, P. Sivashanmugam, G. Kumar, Ultrasonic induced mechanoacoustic effect on delignification of RS for cost effective biopretreatment and biomethane recovery, Sustain. Energy Fuel. 5 (6) (2021) 1832−1844.

[20] B. Sharma, C. Larroche, C.G. Dussap, Comprehensive assessment of 2G bioethanol production, Bioresour. Technol. 313 (2020) 123630.

[21] N. Sönnichsen, Fuel ethanol production worldwide in 2021, by country, Statista (2022), 281606. https://www.statista.com/statistics/281606/ethanol-production-in-selected-countries/.

[22] P. Binod, R. Sindhu, R.R. Singhania, S. Vikram, L. Devi, S. Nagalakshmi, N. Kurien, R.K. Sukumaran, A. Pandey, Bioethanol production from RS: an overview, Bioresour. Technol. 101 (13) (2010) 4767−4774.

[23] D. Fernández-Rodríguez, M. Lapuerta, L. German, Progress in the use of biobutanol blends in diesel engines, Energies 14 (11) (2021) 3215.

[24] N. Vivek, L.M. Nair, B. Mohan, S.C. Nair, R. Sindhu, A. Pandey, N. Shurpali, Binod P "Bio-butanol production from RS−recent trends, possibilities, and challenges", Bioresour. Technol. Rep. 7 (2019) 100224.

[25] S.C. Bhatia, Advanced Renewable Energy Systems (Part 1 and 2), 17-Biogas, 2014.

[26] M.U. Khan, J.T. Lee, M.A. Bashir, P.D. Dissanayake, Y.S. Ok, Y.W. Tong, M.A. Shariati, S. Wu, B.K. Ahring, Current status of biogas upgrading for direct biomethane use: a review, Renew. Sustain. Energy Rev. 149 (2021) 111343.

[27] A. Sharma, G. Singh, S.K. Arya, Biofuel from RS, J. Clean. Prod. 277 (2020) 124101.

[28] O. Kaddour, Manufacturing an integrated hammer mill unit for grinding RS intended for multiple applications, J. Soil Sci. Agric. Eng. 10 (6) (2019) 325−335.

[29] S. Rakesh, K. Subburamu, N. Arunkumar, Pretreatment of paddy straw for sustainable bioethanol production, Sustain. Bioprocess. Clean Green Environ. (2021) 93−102, no. CRC Press.

[30] G. Ji, W. Xiao, C. Gao, Y. Cao, Y. Zhang, L. Han, Mechanical fragmentation of wheat and RS at different scales: energy requirement in relation to microstructure properties and enzymatic hydrolysis, Energy Convers. Manag. 171 (2018) 38−47.

[31] M.R. Swain, A. Singh, A.K. Sharma, D.K. Tuli, Bioethanol production from rice-and wheat straw: an overview, Bioethanol Prod. Food Crop. (2019) 213−231.

[32] M. Chownk, R.S. Sangwan, S.K. Yadav, A novel approach to produce glucose from the supernatant obtained upon the dilute acid pre-treatment of RS and synergistic action of hydrolytic enzymes producing microbes, Braz. J. Microbiol. 50 (2) (2019) 395−404.

[33] J. Zhao, X. Tao, J. Li, Y. Jia, T. Shao, Enhancement of biomass conservation and enzymatic hydrolysis of RS by dilute acid-assisted ensiling pretreatment, Bioresour. Technol. 320 (2021) 124341.

[34] Y. Li, B. Qi, J. Feng, Y. Zhang, Y. Wan, Effect of combined inorganic with organic acids pretreatment of RS on its structure properties and enzymatic hydrolysis, Environ. Prog. Sustain. Energy 37 (2) (2018) 808−814.

[35] E.C. van der Pol, R.R. Bakker, P. Baets, G. Eggink, By-products resulting from lignocellulose pretreatment and their inhibitory effect on fermentations for (bio) chemicals and fuels, Appl. Microbiol. Biotechnol. 98 (23) (2014) 9579−9593.

[36] W. Wang, X. Tan, M. Imtiaz, Q. Wang, C. Miao, Z. Yuan, X. Zhuang, RS pretreatment with KOH/urea for enhancing sugar yield and ethanol production at low temperature, Ind. Crop. Prod. 170 (2021) 113776.

[37] B. Tsegaye, P. Gupta, C. Balomajumder, P. Roy, Optimization of Organosolv pretreatment conditions and hydrolysis by Bacillus sp. BMP01 for effective depolymerization of wheat straw biomass, Biomass Convers. Biorefin. 11 (6) (2021) 2747−2761.

[38] N. Aggarwal, P. Pal, N. Sharma, S. Saravanamurugan, Consecutive organosolv and alkaline pretreatment: an efficient approach toward the production of cellulose from RS, ACS Omega 6 (41) (2021) 27247−27258.

[39] C. Sun, H. Ren, F. Sun, Y. Hu, Q. Liu, G. Song, A. Abdulkhani, P.L. Show, Glycerol organosolv pretreatment can unlock lignocellulosic biomass for production of fermentable sugars: present situation and challenges, Bioresour. Technol. 344 (2022) 126264.

[40] V. Kumar, D. Singh, B. Singh, A greener, mild, and efficient bioprocess for the pretreatment and saccharification of RS, Biomass Convers. Biorefin. (2021) 1−3.

[41] Y. Liu, Z. Yan, Q. He, W. Deng, M. Zhou, Y. Chen, Bacterial delignification promotes the pretreatment of RS by ionic liquid at high biomass loading, Process Biochem. 111 (2021) 95−101.

[42] M. Madadi, Y. Tu, A. Abbas, Recent status on enzymatic saccharification of lignocellulosic biomass for bioethanol production, Electron J. Biol. 13 (2) (2017) 135–143.
[43] R.P. Brexó, A.S. Sant'Ana, Impact and significance of microbial contamination during fermentation for bioethanol production, Renew. Sustain. Energy Rev. 73 (2017) 423–434.
[44] S.H. Azhar, R. Abdulla, S.A. Jambo, H. Marbawi, J.A. Gansau, A.A. Faik, K.F. Rodrigues, Yeasts in sustainable bioethanol production: a review, Biochem. Biophys. Rep. 10 (2017) 52–61.
[45] A. Pugazhendhi, T. Mathimani, S. Varjani, E.R. Rene, G. Kumar, S.H. Kim, V.K. Ponnusamy, J.J. Yoon, Biobutanol as a promising liquid fuel for the future-recent updates and perspectives, Fuel 253 (2019) 637–646.
[46] A. Valles, F.J. Álvarez-Hornos, V. Martínez-Soria, P. Marzal, Gabaldón C "Comparison of simultaneous saccharification and fermentation and separate hydrolysis and fermentation processes for butanol production from RS", Fuel 282 (2020) 118831.
[47] S. Kumar, T.C. D'Silva, R. Chandra, A. Malik, V.K. Vijay, A. Misra, Strategies for boosting biomethane production from RS: a systematic review, Bioresour. Technol. Rep. 15 (2021) 100813.
[48] M. Shah (Ed.), Microbial Bioremediation & Biodegradation, Springer, Singapore, 2020.

Chapter 22

Sources and techniques for biofuel generation

S.A. Aransiola[1], M.O. Victor-Ekwebelem[2], S.S. Leh-Togi Zobeashia[3] and Naga Raju Maddela[4,5]

[1]*Bioresources Development Centre, National Biotechnology Development Agency, Ogbomoso, Nigeria;* [2]*Department of Microbiology, Alex Ekwueme Federal University, Ndufu-Alike, Abakaliki, Nigeria;* [3]*National Biotechnology Development Agency, Abuja, Nigeria;* [4]*Departamento de Ciencias Biológicas, Facultad la Ciencias de la Salud, Universidad Técnica de Manabí, Portoviejo, Manabí, Ecuador;* [5]*Instituto de Investigación, Universidad Técnica de Manabí, Portoviejo, Manabí, Ecuador*

1. Introduction

An increase in population and high technology development increases the need for energy and is overingested, thus resulting in environmental issues [1]. Fossil fuel—derived energy poses not only a serious concern to the environment as a result of the emission of greenhouse gas (GHG) and climate change but also ecological development of the living biota [2]. Fossil-based energy limitation and its elevated demand require the need to find an alternative source of energy to complement the fossil-derived fuel. The need to redirect energy from fossil fuel requires biofuel, a renewable source of organic energy derived from the breakdown of biological resources (biomass) sourced mainly from biological origin [3]. Biofuel is a sustainable energy that globally impacts positively on the economy, social, and environment of a country. Biofuel is an alternative to conventional fossil fuel because it is nontoxic, biodegradable, and sulfur-free. Biofuel can be used in its unadulterated form or as fuel additives in transportation and heating process; for example, biofuel can be blended into gasoline to improve the octane rating and reduce harmful emissions.

Biofuels are referred to first, second, third, and fourth generation depending on the raw material used for generation. The first-generation biofuel known as conventional biofuel utilizes well-organized process and technology (transesterification, fermentation, distillation) for the manufacture of these biofuels from food and oil crops [4].

Second-generation biofuels are obtained from waste products, lignocellulosic materials, energy crops, and agricultural and forest residues (nonfood feedstock). This generation biofuel involves enhanced process method for biofuel recovery and secondary metabolites production [4] compared to the first generation; the production process is more complex but more sustainable and low GHG emission. The third-generation biofuels are known as advanced biofuels because they are manufactured in large scale, are easy to refine, and can absorb carbon dioxide. They are further from commercialization and are still undergoing research but have attracted a lot of credit while the fourth generation is recorded as advanced research that uses techniques in modern biotechnology for biofuel production. The feedstocks used for this generation of biofuel are economically less expensive, available, and unlimited [5].

Biofuel can be classified as solid, liquid, and gaseous biofuel. The solid biofuel (biochar) contains organic solid biomass of nonfossil origin produced from wood residue, animal wastes, and charcoal, among others. The biomass used for solid biofuel is important for generation of electricity and production of heat [6,7]. Liquid biofuels (bioethanol, bio-oil, and biodiesel) are generated from biodegradable wastes and are mainly used in transportation because of its high density. Liquid biofuel can be used as substitute to conventional fuel, denoting 18% of the energy consumed [8]. 80% of liquid biofuel is produced in bioethanol and 20% for biodiesel [9]. Liquid fuels are utilized as blend for conventional fuel. The leading country consuming, manufacturing, and importing biodiesel is EU, while Brazil, the United States, China, India, and EU are the world's largest countries producing and exporting bioethanol [10]. Gaseous biofuels produced via anaerobic digestion (AD) of organic wastes are gaseous in nature with low density and are produced to generate heat and electricity. Examples of such biofuels are biosyngas, biohydrogen, and biogas [11].

2. Biofuel and its sources

Biofuel is generated by processing materials from various animals and plants. The feedstocks used to create biofuel are as follows.

2.1 Bioethanol

Bioethanol known as liquid alcohol is produced through the fermentation of variety of renewable feedstocks containing carbohydrate or sugar. Bioethanol is derived from edible first-generation agricultural products such as potatoes, wheat, rice, sugarcane, corn, barley, and plants oils (rapeseed oil, sunflower oil, mustard oil, palm oil, and soybean oil, among others) [12]. Though production of bioethanol from the first-generation feedstock was simple, economically feasible, and advanced in most countries, for example, in European Union, bioethanol is produced from grains and sugar beet, in North America, producing about 53 billion liters of bioethanol from maize crop by 200 production plants. India and China also utilize maize and sugarcane as main source of feedstocks, while sweet potatoes and sorghum utilized are less popular raw material for bioethanol in China and Asia [10]. However, this generation source of biomass still had a major setback of rivalry with food-based product supply and land use and also poses an environmental risk from the application of fertilizer and pesticide to grow the crops [13]. Strategic approaches were developed to generate bioethanol from non−food-based biomass sources such as lignocellulosic and starches (second generation), algal (third generation), and captured carbon dioxide.

Lignocellulosic biomass used for bioethanol consists of lignin, hemicellulose, cellulose, and inorganic materials [14−16]. This biomass is fermentable by microorganisms to produce bioethanol and is derived from agro-industrial waste, municipal waste, agricultural waste, forest residues, nonedible energy crops such as carrot grass, switch grass, phragmites, wheat straw, rice straw, wood waste, sugarcane bagasse, palm kernel, corn straw, cocoa pod, molasses, and Kans grass [14].

The use of lignocellulosic sources of feedstock from both plant and nonplant products has increased globally. Agricultural activities contribute about 342 million tons per year for lignocellulosic material for bioethanol production while that of forest waste is about 325 tons in a year [17]. The bioconversion of lignocellulose to bioethanol poses several benefits; it does not affect the food chain supply, reduced GHG (CO_2 2.85 kg kg^{-1}) emission compared to the first generation, and requires lesser cost of production. The major drawback of this source of feedstock is, it requires land, water, and the risk of deforestation [13].

Starchy feedstocks (agricultural residues, agro-industrial wastes, and municipal solid wastes) are mainly biomass wastes high in carbohydrates and protein. The sources of biomass are available, less costly, and easily hydrolyzed by enzymatic saccharification to sugars, and also more than 95% of ethanol creation originates from sugar and starch materials based commercial enterprises and it is contributing 42% and 58% to the aggregate generation, individually [18]. Unlike lignocellulosic they are easily hydrolyzed and therefore require less processing time. The production of starchy biomass depends on availability of land, productivity, and fertile soil.

Algae are autotrophic microorganisms that produce their food through photosynthesis in sunlight, water, and carbon dioxide [19]. They are a group of chlorophyll-containing eukaryotic organisms found in marine and freshwater habitats. Algal biomass (micro- and macroalgae) represents a promising source for bioethanol production compared to the first- and second-generation source of biomass. Algal biomass is renewable and contains high energy mainly due to high composition of carbohydrate and non-lignin content which reduces cost of production resulting from pretreatment process of lignin materials. Algae energy from oil extraction or as material for the process of fermentation can be obtained from less expensive sources such as sea water, food oil, and waste [20]. Some algal species such as *Schizochytrium limacinum*, *Botryococcus braunii*, *Chlorella*, *Isochrysis galbana*, and *Chaetoceros calcitrans* have been examined for bioethanol production [21].

The cultivation process is less expensive, utilizes waste and nonportable water/saline for generation of bioethanol, has high energy density, and provides conversion efficiency; however, the bioethanol of algae origin is less stable compared to the other sources of bioethanol production [22]. Bioethanol can also be produced from genetically modified algae and captured carbon dioxide using oxide electrolysis, electrochemical synthesis, petroleum hydroprocessing, and advanced biotechnology. This source of feedstock is regarded as the fourth-generation feedstock [23] and is considered as carbon negative since most of carbon produced is less compared to the carbon captured.

2.2 Biodiesel

Biodiesel is another important liquid fuel of first-generation biofuel made from seeds, used cooking oil, algae, sewage sludge, short alcohol, and animal fats using transesterification processes [8]. Biodiesel is used as a fuel for vehicles in its pure form and as additives to fuel to reduce pollutant from diesel-powered vehicles [24,25]. Biodiesel has low toxicity, reduces the emission of carbon monoxide, and produces polycyclic aromatic hydrocarbons and contaminants. The potential source of biodiesel is from seed oil such as rapeseed, soybean, palm, flax, peanuts, castor, sunflower, safflower, jatropha [24], jojoba, avocado, cotton, and tung (Fig. 22.1). Biodiesel extraction from seed sources includes decortification of seed coats followed with oil extraction, free fatty acid analysis, and production of biodiesel using sodium hydroxide and methanol. The seed oils used globally for manufacturing biodiesel are rapeseed, soybean, and palm. In European Union countries, rapeseed is mainly used; in Central America and Asian countries, palm seeds are used as feedstock; while the United States and Argentina use soybean for biodiesel production [26]. The use of seed oil (nonedible oil) is of great benefit for biodiesel production. The cost of production is low due to the fact that, foodstuff is not consumed during the cost of production [27].

Used cooking oil (vegetable oil, animal fat, and Tallow) has gained prominent as biomass source for biodiesel production. Used cooking oil is converted into esters of fatty acids that produce the oil and the fatty acid methyl esters (FAMEs) are purified to obtain biodiesel that is alike to fossil-derived diesel [28].

Algae are a good source for biodiesel production and their oil yield is double the amount produced by other feedstocks. Algae use microalgae and macroalgae as biomass source for biodiesel generation. Algae biomass used for production depends on the content of the lipid and the oil content contains within the algae which is within 20%–50% but some species contain higher percentage of lipid, for example, *Chlorella protothecoides* contain about 70% of lipid and generate about 7.4 g per liter per day (g L^{-1} d) of biodiesel [29]. However, some microalgae are not sufficient for diesel production. For massive growth of algae and oil content, access to mineral salt, water, nutrient carbon dioxide, and light is required. Examples of algae that have been used for production of biodiesel include microalgae (*Dunaliella, Pleurochrysis carterae, Chlorella, B. braunii*) and macroalgae such as *Gracilaria, Sargassum*, and *Ulva* [30,31].

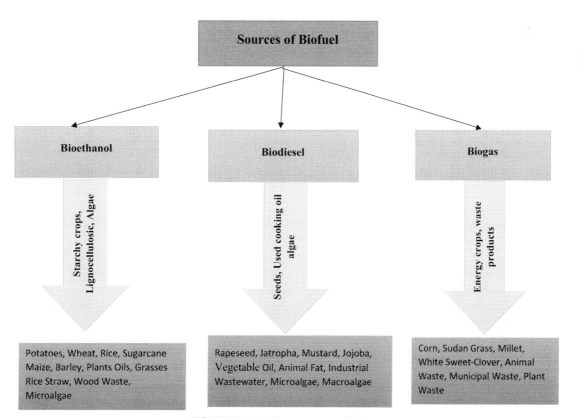

FIGURE 22.1 Different sources of biofuel.

2.3 Biogas

Biogas produced from organic material through AD is composed mainly of methane (55%—75%), carbon dioxide (30%—45%), oxygen, hydrogen, hydrogen sulfide, a few hydrocarbon traces, residues of organic, and few traces of nitrogen [32]. Similar to natural gas, it is a versatile fuel that is used directly to generate electricity, to provide heat at low or high temperatures, or to power vehicles [33]. The biochemical degradation of polymer to biogas is carried out by a variety of microorganisms. Biogas production offers an effective means of pollution reduction, superior to that achieved via conventional aerobic processes. Although practiced for decades, interest in anaerobic fermentation has only recently focused on its use in the economic recovery of fuel gas from industrial and agricultural surpluses. The input biomass source for generation depends on the amount of biogas produced [34]. The raw material used for biogas contains carbohydrate, fats, protein, and cellulose. The potential substrate for biogas includes energy crops such as corn, Sudan grass, millet, white sweet clover, cellulose-rich biomass, algae and seaweed (water based), by-products of ethanol, and biodiesel production [35]. Organic manure such as cattle dung, poultry manure, excreta of other animals, rural and urban composts, other animal wastes, crop residues and green manure, agro-industrial waste, municipal waste, plant, and cellulose-rich crop (maize, wheat, white sweet clover, barley) can also generate biogas. In Germany, most biogas uses a large amount of fodder-maize as biomass to produce biogas. This biogas is usually burnt for local heat and electricity, but is now increasingly being upgraded to biogas with natural-gas-pipeline quality (typically 97% methane). Biogas can be compressed for transport fuel compared to other biofuels; the energy yield of biogas production is higher.

3. Environmental benefits of biofuel: global perspectives

Biofuel production does not only benefit the socioeconomy growth of the economy, its global attention is not restricted to net energy merit but also on the environmental benefits such as clean environment, reduction in GHG such as sulfur oxide and carbon monoxide, producible in large amount, reduced risk of global warming, decrease in damage to ecosystem, and depletion of natural resources. The impact and potential benefits to the environment solely depend on the biomass sources, type of biofuel, and the life cycle of the biofuel.

The use of fossil-derived fuel is a major source of emission of GHG and climate change. Biofuel development and its use can reduce fossil fuels consumption and thus reduce emission of GHG (mostly carbon dioxide). Twenty-three percent of the estimated global energy-related carbon dioxide emission and 15% of the global GHG emission are from road transportation [36]. Transportation has tremendously increased over the past 40 years and is projected to continuously rise mostly especially in developing countries, middle-income countries experiencing rapid economic growth, middle-class expansion, and urbanization. In European Union, 25% of the GHG release is from transportation; reducing emission from transportation is important to achieve the 2050 EU target for climate neutral environment. A target of 6% was set to reduce the GHG emission by 2020 compared to 2010 by Fuel Quality Directive. A reduction of 4.3% was achieved in 2019 which is not up to the 2020 target. The decrease in GHG emission achieved between 2019 is attributed to increase in biofuel used. Biofuels have a lower emission intensity than fossil fuels. Across the member state, Sweden and Finland exceeded the 6% target, while Netherland had a reduction of 5.8%. The reduction of GHG emission is linked to the mixture of biofuel with transport fuel; for example, Netherland had a high mix proportion of 22% of biofuel, Finland was 9%, and Sweden 7%. GHG emission depends on the feedstocks used to produce biofuel. The decrease between 2018 and 2019 was from decreased use of oil crops substituted by sugars. Oil crops generally have higher emission intensity compared to other feedstocks in biofuel. Biofuels consumption will decrease the emission of GHG pollution by close to 50%. Incentives from State Government to subside fuel prices meet the expectations of automobiles factories and agricultural field and to achieve a cleaner environment will encourage and support renewable biofuels from crop biomass [37].

Biofuel production requires growing energy crops and using the biomass as feedstock for the fuel production. Cultivation of energy crops also generates GHG; the use of biofuel has no negative impact on land use by displacing the production of feed and food crops and driving the land conversion which is known as indirect land use change. To reduce GHG, feedstock with low GHG emission is utilized, for example, bioethanol made from rye, potatoes, and soy and biodiesel produced from Brazilian soy [38] had low GHG emission but can negatively impact on the environment. Lipid crops for biofuel have low carbon intensities as feedstocks; the life cycle emissions for lipid biomass are low because lipids feedstock were used initially for other purpose; emissions related to transport of these biofuels feedstocks account for emissions that occur after the oil waste is collected. Due to the lower carbon potential of lipid feedstock State Government supports its usage to reduce GHG emission in the environment. In California, lipids crop is used as feedstocks for bioethanol and for credits under California low carbon fuel standard.

An increase in biofuel product puts pressure on food supply and price, the risk of increase in GHG emission via direct and indirect land use change from biomass use for biofuel production, land degradation, water, ecosystem, and forest [39]. Biofuel feedstock from first generation is of major concern due to the competition with production of food crop and the worry of using agricultural land for biofuel production which can involve the use of land with high biodiversity value, use of pesticide, fertilizer, increased deforestation, and food waste [40], thereby having negative impact on the environment. This can be prevented by using second- and third-generation feedstock which poses low environmental risk. The use of algae biomass in the third generation reduces emission of GHG by converting CO_2 into oxygen for the atmosphere during photosynthesis, which is beneficial to the environment by creating a carbon neutral environment where the growth cycle absorbs the gases.

The biofuel with high net energy balance decreases GHG emission more frequently however, the net balances for bioethanol produced from corn are small but more significantly for biodiesel from soybeans and bioethanol from cellulose and sugarcane. The assessment of biodiesel exhaust emission is important to evaluate the impact on the environment and human health. Biodiesel consists of various components of FAMEs and the precise constitutes depend on the feedstock which is different from the hydrocarbon content of fossil-derived diesel. The difference in the chemical composition affects physical properties and in turn the tailpipe emission in a way that is different from conventional fuel. As fuel for vehicle, biodiesel benefits from GHG over fossil-based diesel and petrol. Carbon monoxide, particulate matter, and hydrocarbons reduce when biodiesel is used as fuel and the carbon dioxide (CO_2) emitted during combustion of biodiesel fuel is offset by CO_2 captured by the plant's biomass serving as feedstock for production of biodiesel. An estimation from Environmental Protection Agency indicates that algae-derived biodiesel produced via fatty acid methyl transesterification emits less particulate and can reduce the emission of GHG emissions by more than 60% compared to petroleum diesel [41].

Biofuel feedstocks are biodegradable and contribute to sustainability. Bioethanol and biodiesel are less toxic and degradable; when spilt in the environment they are broken to harmless substances. The impact of biofuel production and use on ecosystem (biodiversity) are reduced contrast to fossil fuel and depend on the raw material used [42]. Compared to first-generation biofuel feedstock, the second-generation agricultural and forest biomass have lower negative impact on biodiversity than the energy crops [43]. Biofuel life cycle from planting to cultivation to conversion of feedstock to biofuel and combustion reduces climatic effects on biodiversity and maintenance of species dependent on cultural habitat and ecosystem services [44]. Algae biofuel minimizes waste; for example, the growth of algae is beneficial to the environment by preventing contaminated water mixing with river or lake water. The properties in the water such as nitrate, ammonia, and phosphate that render the water unsafe serve as nutrients for algae [8,45].

4. Biomass feedstocks in the production of biofuel

In addition to raw materials, capital, and labor, energy is the workhorse of development. Fossil fuels, which are nonrenewable, not consistently present on the earth [37,46], and environmentally harmful, account for a large amount of the energy we consume today. As a result, it's critical to look for a long-term energy source that can be conveniently supplied at a minimal cost while still being environmentally beneficial. The resource generation from solid waste, known as biomass, is one of several alternative energy sources developed to replace fossil fuels. Municipal solid waste, agricultural solid waste, and forestry trash are all examples of solid waste [47]. These wastes have been successfully used to make biofuels (biogas and bio-liquids) [48]. Biomass is a natural organic matter such as wood, trees, manure, sewage, lumber waste, grass cuttings, rice husk, and other natural organic matter that are utilized as an excellent starting material for biofuel production due to its long-term sustainability. Biofuel is a type of fuel that gets its energy from biological carbon fixation; it includes biomass conversion fuels [49].

Biofuels made from biomass are divided into four categories based on the method of production and the type of biomass utilized to make the biofuel: first-generation, second-generation, third-generation, and fourth-generation biofuels [50]. Biofuels are made from food crops in the first generation, agricultural leftovers in the second, algal biomass in the third, and solar energy in the fourth generation by algae and cyanobacteria [48,51–53]. Biofuel production from the second and third generations is currently proven to be advantageous due to plentiful biomass availability [54].

4.1 Biomass with some of their application

4.1.1 Algae

Algae are harvested from stagnant ponds and algae farms for the sole purpose of producing biodiesel. Algae is a self-generating biomass that can produce up to 300 times more oil per acre than conventional crops while emitting no CO_2. Algae have been tested as a new type of green jet fuel for commercial travel, among other applications.

4.1.2 Oil-rich biomaterial

Food crops such as rapeseed (canola), sunflower, corn, and others produce oil-rich biomass. It can be used as a waste vegetable oil after it has been used for other purposes, such as food preparation, or it can be consumed right from the can (straight vegetable oil). They are resistant to microbial destruction and have a high availability and reused material. It is used in the creation of biodiesel fuel for automobiles, home heating, and experimentally as a pure fuel itself.

4.1.3 Rice straw: two types of residues are obtained from rice cultivation: straw and husk

Rice straw (RS) is one of the wastes derived from rice cultivation and is a plentiful lignocellulosic agricultural waste material in the globe, as rice is the world's third most important grain crop, behind wheat and corn, in terms of overall production [55]. Every kilogram of rice grain harvested is accompanied by the production of 1–1.5 kg of straw, resulting in a total of 650–975 million tons of RS, the majority of which is used as cow feed and the remainder stays wasted [54].

Field burning is currently the most common method of eliminating RS, although it pollutes the air and has a negative impact on human health [56,57]. RS has a low feedstock quality due to its high ash content (10%–17%) and high silica content in ash (SiO_2 is 75% in rice) [54]. RS, on the other hand, has the advantage of having a low overall alkali concentration (sodium oxide and potassium oxide typically make up less than 15% of total ash) [58].

4.1.4 Woody or lignocellulosic biomass

Carbohydrate polymers (cellulose and hemicellulose), lignin, and a smaller portion (extractives, acids, salts, and minerals) make up lignocellulosic or woody biomass. The polysaccharides cellulose and hemicellulose, which make up around two-thirds of the dry mass, can be hydrolyzed to sugars and then fermented to ethanol. Lignin cannot be used for alcohol production [59].

Several parameters, including cellulose crystallinity, moisture content, polymerization rate, lignin concentration, and surface area, limit the breakdown of lignocellulosic biomass into biofuels [60,61]. Pretreatments are feasible for lignocellulosic biomass wastes in order to overcome these challenges and achieve efficient biofuel generation with minimal retention time.

Physical, chemical, physicochemical, biological, and combination pretreatment procedures have all been explored and implemented in various ways [61]. The effectiveness of any lignocellulosic biomass pretreatment is determined by three key variables: (1) chemical or reagent concentration, (2) retention or residence period, and (3) temperature. Fluctuating any of these parameters, the hydrolysis rate of lignocellulosic biomass can be modified [54].

Others biomasses include: Crop residues: Residues from the harvest of wheat, maize, rice, other coarse grains, sugar beet, sugarcane, soybean, and other oilseeds, including sequential crops between two harvested crops as a soil management strategy that helps to keep soil fertility, retain soil carbon, and prevent erosion; these crops do not compete for agricultural land with food or feed crops.

Animal manure: From livestock including cattle, pigs, poultry, and sheep.

Organic fraction of municipal solid waste (MSW): Food and green waste (e.g., leaves and grass), paper and cardboard, and wood that are not otherwise utilized (e.g., for composting or recycling). MSW also includes some industrial waste from the food-processing industry.

Wastewater sludge: Semisolid organic matter recovered in the form of sewage gas from municipal wastewater treatment plants [62].

5. Extraction and determination of different biofuels

In the extraction of biofuel there are various techniques that exist based on fermentation, enzymatic (biological), chemical, and thermochemical (pyrolysis, gasification, liquefaction) routes to convert the lignocellulosic biomass into biofuels and chemicals [59,63] (Fig. 22.2) as well as the type (generation) of biomass. However, among the existing technologies, thermochemical processes such as fast pyrolysis and gasification are highly preferred due to shorter processing time and value of products obtained from them [59].

5.1 Fast pyrolysis

In fast pyrolysis, biomass is heated at high rates (>1000 Cs^{-1}) to moderate temperatures (400–700°C) in an inert environment. It produces a 70% liquid product called 'pyrolysis oil' (organics + water) in addition to solid (char + coke) and gaseous products [48]. The pyrolysis oil is a complex mixture of aldehyde/ketone, carboxylic acids, furan and pyran

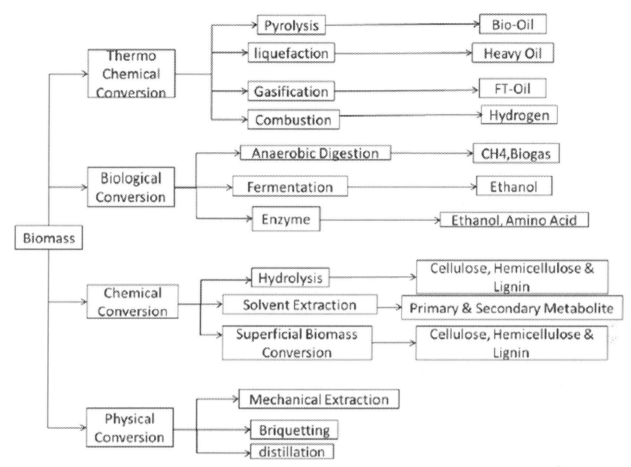

FIGURE 22.2 Biomass conversion processes [49,70].

derivatives, anhydrosugars, and phenol derivatives [59,64,65]. The pyrolysis oil is highly unstable, is highly acidic, and has a lower calorific value than petroleum crude oil due to its high oxygen concentration. The pyrolysis oil's quality is substantially determined by the feedstock genotype and pyrolysis conditions. The presence of nonbiological contaminants such as ash and minerals, on the other hand, could have a significant impact on the process's development and the production of various products [66]. The ideal time scale for quick pyrolysis is on the order of a few seconds. However, in real-world settings where heat and mass transport limits are significant, pyrolyzing the sample and sweeping off the vapors take longer. Krumm et al. [67] and Maduskar et al. [68] used the pulse-heated analysis of solid reactions (PHASRs) approach to visualize the time scale of fast pyrolysis. Their research revealed that the pyrolysis reaction might be completed in less than 1000 ms. Maduskar et al. [68] found that the critical characteristic length for operating pyrolysis in an isothermal, reaction-controlled environment should be smaller than 10 m. It was projected that cellulose conversion at 500°C can be completed in 1000 ms using a microkinetic model [59].

5.1.1 Extraction methods for biodiesel, bioethanol, and biogas (biomethane)

Bioethanol: Ethanol is a renewable fuel that is created by fermenting starch or sugar. It is one of the most important renewable fuels because of its economic and environmental benefits. Sugary, starchy, and lignocellulosic biomasses are the three types of biomass used in bioethanol synthesis [69]. Forest residue, agricultural residue, yard waste, wood products, animal and human wastes, and other lignocellulosic biomass are examples of renewable resources that store energy from sunlight in chemical bonds. Carbohydrate polymers comprising 5C and 6C sugar units make up 50–60°C of lignocellulosic biomass. Waste wood and other lignocellulosic biomasses are the most promising feedstocks for bioethanol production.

Pretreatment (acid or enzyme hydrolysis), fermentation, and distillation are the three steps in traditional bioethanol extraction from lignocellulosic biomasses (Fig. 22.3) [49,70]. Pretreatment is the chemical reaction that changes

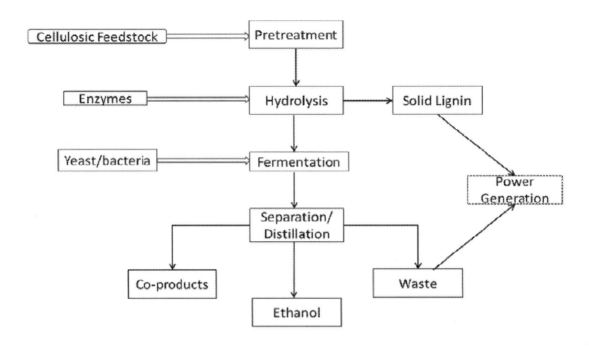

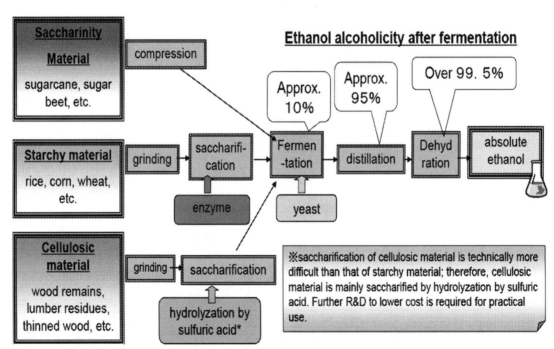

FIGURE 22.3 (A) Lignocellulosic biomass to bioethanol [49,70]. (B) Production of absolute ethanol from saccharinity, starch, and cellulosic materials [69]. (C) The distillation process for ethanol production [69].

complicated polysaccharides to simple sugar; prior to enzymatic hydrolysis of the polysaccharides, the surrounding matrix of lignin and hemicellulose must always be removed and modified. This process changes the state of lignocellulosic biomass from its natural state. The biomass structure is broken down into fermentable sugars in this stage. *Saccharomyces cerevisiae*, yeast or bacteria that feeds on sugar, causes the fermentation reaction. Following that, the pure ethanol is separated from the combination using a distiller, which boils the mixture and evaporates it to condensate at the top of the apparatus to produce the ethanol.

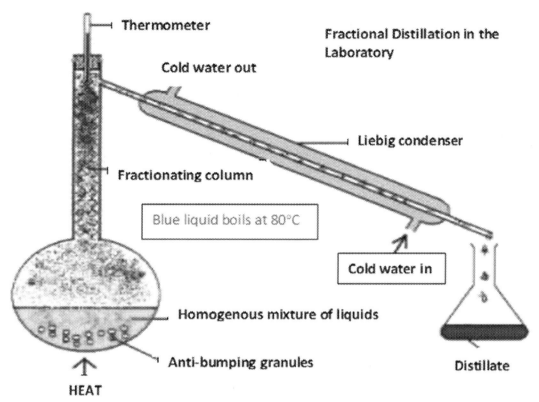

FIGURE 22.3 cont'd

5.1.2 Quantitative determination of ethanol

The quantity of extracted ethanol can be determined by:

i. Dichromate oxidation method
ii. Distillation-hydrometric method
iii. Direct injected GC/MS method

Biodiesel: Biodiesel is a type of diesel fuel made from vegetable oils, seeds, and animal fats that burn cleanly. In its pure form, biodiesel can be used as a vehicle fuel, but it is most commonly employed as a diesel additive to reduce particulates, carbon monoxide, and hydrocarbon emissions from diesel engines. Transesterification of triglycerides in the presence of a catalyst (alkali, acid, enzyme) yields glycerine as a primary by-product, which is used to produce biodiesel from oils or fats [47]). The glycerine is separated after the reaction by settling or centrifugation. The removal of glycerine from biodiesel is critical since glycerine is one of the most essential precursors for biodiesel quality.

5.2 Biodiesel extraction methods

5.2.1 Transesterification

This step involves the use of oil sample, methanol, and potassium hydroxide (as a catalyst). Methanol and potassium hydroxide are premixed to prepare potassium methoxide, and then added to oil in the reactor with a mixing speed of 400 rpm for 2 h at 50°C. Finally, the mixture was left overnight to settle forming two layers, namely: biodiesel phase (upper layer) and the glycerin-rich phase (Fig. 22.4).

For biofuel production, you need to remove as much of oxygen to gain energy density. At the same time, the cost for oxygen removal needs to be minimized. Consequently, it was found that metals like palladium can be used along with iron to yield higher rates [71]. Pretreatment and deconstruction: AFEX, an ammonia-based method, is currently being developed for pretreatment processes. The ammonia breaks the lignin structure which leads to an improvement in the efficiency of the enzyme action. Multifunctional catalysts: The most common issue with using catalysts for biodiesel production is the higher cost associated with the involute nature of the process, which includes separation techniques and

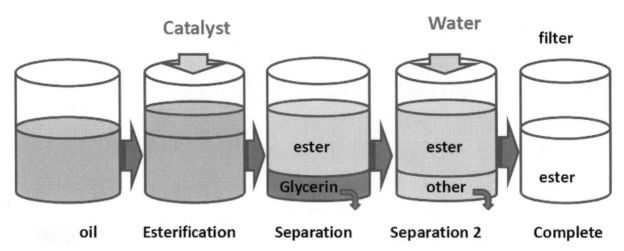

FIGURE 22.4 The biodiesel extraction process (steps) [69].

regular upgrades. As a result, the development of heterogeneous catalysts for biomass to fuel conversion is critical. As potential biomass catalysts, zeolite catalysts have proven to be groundbreaking. The low molecular weight, low coking tendency, and high deoxygenation activity of ZSM-5 make it a must-have material. The addition of certain base metals to the zeolite has found to increase the selectivity. For example, addition of nickel has increased the conversion of oxygenates [49].

Therefore, there is an urgent need for the development of heterogeneous catalysts to convert biomass to fuel. Zeolite catalysts have proven to be breakthrough in this field as potential biomass catalysts. ZSM-5 is a quintessential material due to low molecular weight, low coking tendency, and high deoxygenation activity. The addition of certain base metals to the zeolite has found to increase the selectivity. For example, addition of nickel has increased the conversion of oxygenates [49].

5.2.2 Qualitative determination of biodiesel

The Borax/phth test is special test for detection. It involves the use of 1 mL of glycerol layer mix with 1 mL of Borax/phth (red color), if the red color disappear in cold and appearing after heating confirms the presence of biodiesel fuel. Alternate method is the use of Fourier transform infrared spectroscopy analysis.

5.2.3 Biogas (biomethane)

Biomethane, a constituent of biogas, can be produced by AD process of lignocellulosic biomass. Various agricultural residues along with sewage sludge have been used by researchers for methane production [61]. A number of industrial effluents had also been used for methane production abiotically or by fermentation process [72]. RS can also be fermented to produce methane by AD process [73].

AD for biogas production from biomass is one of the best [74] as it is economically feasible due to its smaller requirements and using waste as feedstock. AD is a biological process where bacteria breaks down organic matter in the absence of oxygen to form CH_4, CO_2, and inorganic nutrients and compost with evolution of the hydrogen gas [74]. Usually four major reactions occur during the complete process of the AD to methane: hydrolysis, acidogenesis, acetogenesis, and methanogenesis. All these reactions occur concurrently and are interdependent.

6. Conclusion

Biofuels are carbon-fixing fuels that get their energy from biological processes. They are biofuels or energy fuels made from organic materials. Because biofuels are nontoxic, sulfur-free, and biodegradable, and they come from renewable sources, they are being developed as a replacement for petroleum. Based on the feedstock, biofuels are split into four categories: first generation (oil-based plants, sugar, and starch), second generation (agricultural and woodland leftovers), third generation (algae), and fourth generation (biogas) (engineered cyanobacterial). Transesterification of triglycerides produces biofuel, which must be refined before being utilized as a fuel. Dichromate oxidation, direct injected gas chromatography, and distillation-hydrometric methods are used to extract biogas (biomethane).

References

[1] S. Hafeez, S.M. Al-Salem, G. Manos, A. Constantinou, Fuel production using membrane reactors: a review, Environ. Chem. Lett. 18 (2020) 1477–1490, https://doi.org/10.1007/s10311-020-01024-7.

[2] K. Zaman, M.A. el Moemen, Energy consumption, carbon dioxide emissions and economic development: evaluating alternative and plausible environmental hypothesis for sustainable growth, Renew. Sustain. Energy Rev. 74 (2017) 1119–1130.

[3] E. Iacovidou, D.G. Ohandja, N. Voulvoulis, Food waste co-digestion with sewage sludge–realising its potential in the UK, J. Environ. Manag. 112 (2012) 267–274.

[4] H.K. Jeswani, A. Chilvers, A. Azapagic, Environmental sustainablility of biofuels: a review, Royal Soc. Pub. 476 (2243) (2020) 2020035, https://doi.org/10.1098/rspa.2020.0351.

[5] D. Sharma, D. Singh, S.L. Soni, S. Sharma, P. Kumar Sharma, A. Jhalani, A review on feedstocks, production processes, and yield for different generations of biodiesel, Fuel (2019), https://doi.org/10.1016/j.fuel.2019.116553. ISSN 0016-23611165539.

[6] Y. Dahman, K. Syed, S. Begum, P. Roy, B. Mohtasebi, Biofuels, Biomass, Biopolymer- Based Materials, and Bioenergy, 2019, pp. 277–325.

[7] P. Shah Maulin, Microbial Bioremediation & Biodegradation, Springer, 2020.

[8] A. Demirbas, A. Bafail, W. Ahmad, M. Sheikh, Biodiesel production from non-edible plant oils, Energy Explor. Exploit. 34 (2) (2016) 290–318.

[9] M.K.K. Figueiredo, G.A. Romeiro, L.A. dAvila, R.N. Damasceno, A.P. Franco, The isolation of pyrolysis oil from castor seeds via a low temperature conversion (LTC) process and its use in a pyrolysis oil–diesel blend, Fuel 88 (2009) 2193–2198, https://doi.org/10.1016/j.fuel.2009.05.025.

[10] International Energy Agency, IEA World Energy Outlook 2012, 2012 (Paris).

[11] S.Y. No, Production of Liquid Biofuels from Biomass, Application of Liquid Biofuels to Internal Combustion Engines, Green Energy and Technology, Springer, Singapore, 2019, https://doi.org/10.1007/978-981-13-6737-3_1.

[12] A.E. Ghaly, A review of: osamu kitani and carl, in: W. Hall (Ed.), Biomass Handbook, Gordon and Breach Science Publishers, New York, 1991, pp. 409–410, 1989). Energy Sources, 13(3).

[13] R.A. Lee, J.M. Lavoie, From first- to third-generation biofuels: challenges of pro- ducing a commodity from a biomass of increasing complexity, Animal Front. 3 (2) (2013) 6–11.

[14] P. Purohit, D. Subash, Lignocellulosic biofuels in India: current perspectives, potential issues and future prospects, AIMS Energy 6 (2018) 453–486.

[15] P. Shah Maulin, Removal of Refractory Pollutants from Wastewater Treatment Plants, CRC Press, 2021.

[16] M.J. Taherzadeh, K. Karimi, Pretreatment of lignocellulosic wastes to improve ethanol and biogas production: a review, Int. J. Mol. Sci. 9 (2008) 1621–1651.

[17] S. Biom, Vision for 1 Billion Dry Tonnes Lignocellulosic Biomass as a Contribution to Biobased Economy by 2030 in Europe, 2016.

[18] S.A. Aransiola, N.O. Falade, M.P. Obagunwa, B.R. Babaniyi, Environmental management and uses of vinasse-review, Asian J. Curr. Res. 1 (2) (2016). http://www.ikpress.org/abstract/5797.

[19] B. Kamm, M. Kamm, Principles of biorefineries, Appl. Microbiol. Biotechnol. 64 (2004) 137–145, https://doi.org/10.1007/s00253-003-1537-7.

[20] R.C. Rajak, S. Jacob, & B.S. A. Kim, Holistic zero waste biorefinery approach for macroalgal biomass utilization: a review, Sci. Total Environ. 716 (2020) 137067.

[21] S. Behera, S.K. Kumar, M.S. Jab, Potential and prospects of biobutanol production from agricultural residues, in: A. Rastegari, A.G.A. Yadav (Eds.), Biofuel Biorefinery Technology, SpringerCham, 2019, pp. 285–318, https://doi.org/10.1002/9781119459866, ch9.

[22] J. Singh, S. Gu, J. Singh, S. Gu, Commercialization potential of microalgae for biofuels production, Renew. Sustain. Energy Rev. 14 (2010) 2596–2610.

[23] B. Abdullah, S.A. Faua'ad, S. Muhammad, Z. Shokravi, S. Ismail, K. Anuar Kassim, A. Nik Mahmood, M. Maniruzzaman, A. Aziz, Fourth generation biofuel: a review on risks and mitigation strategies, Renew. Sustain. Energy Rev. 107 (2019) 37–50, https://doi.org/10.1016/j.fuel.2019.116553.

[24] O.P. Abioye, U.J.J. Ijah, S.A. Aransiola, Phytoremediation of soil contaminants by biodiesel plant *Jatropha curcas*, chapter 4, in: K. Bauddh, et al. (Eds.), Phytoremediation Potential of Bioenergy Plants, Springer Nature Singapore Pte Ltd, 2017, https://doi.org/10.1007/978-981-10-3084-0_4, 2017, https://link.springer.com/book/10.1007%2F978-981-10-3084-0.

[25] O.P. Abioye, P.F. Aina, U.J.J. Ijah, S.A. Aransiola, Effects of cadmium and lead on the biodegradation of diesel-contaminated soil, J. Taibah Univ. Sci. 13 (1) (2019) 628–638. https://www.tandfonline.com/doi/full/10.1080/16583655.2019.1616395.

[26] International Energy Agency, IEA Technology Roadmap: Biofuels forThe Impact of Biofuels Transport, 2011. http://www.iea.org/publications/freepublications/publication/name,3976,en.html.

[27] S.D. Romano, E. González Suárez, M.A. Laborde, Biodiesel, in: Combustibles Alternativos, second ed., Ediciones Cooperativas, Buenos Aires, 2006.

[28] RECOIL, Assessment of Best Practices in UCO Processing and Biodiesel Distribution, 2013.

[29] C.Y. Chen, K.L. Yeh, R. Aisyah, D.J. Lee, J.S. Chang, Cultivation, photobioreactor design, and harvesting of microalgae for biodiesel production: a critical review, Bioresour. Technol. 102 (2011) 71–81.

[30] T. Shirvani, X. Yan, O.R. Inderwildi, P.P. Edwards, D.A. King, Life cycle energy and greenhouse gas analysis for algae-derived biodiesel, Energy Environ. Sci. 4 (10) (2011) 3773–3778.

[31] B. Thangaraj, P.R. Solomon, Scope of biodiesel from oils of woody plants: a review, Clean Energy 4 (2020) 89–106.

[32] A.J. Ward, D.M. Lewis, F.B. Green, Anaerobic digestion of algae biomass: a review, Algal Res. 5 (2014) 204–214.

[33] R. Ruan, Y. Zhang, P. Chen, S. Liu, L. Fan, N. Zhou, K. Ding, P. Peng, M. Addy, Y. Cheng, E. Anderson, Y. Wang, Y. Liu, H. Lei, B. Li, Biofuels: alternative feedstocks and conversion processes for the production of liquid and gaseous biofuels, in: Biomass, Biofuels, Biochemicals, Elsevier, 2019, pp. 3–43.

[34] O.M. Ilori, A.S. Adebusoye, A.K. Lawal, A.O. Awotiwon, Production of biogas from banana and plantain peels, Adv. Environ. Biol. 1 (2007) 33–38.

[35] P. Demetriades, Thermal Pre-treatment of Cellulose Rich Biomass for Biogas Production, Swedish University of Agricultural Sciences, Uppsala, 2008.

[36] R. Sims, et al., Transport, in: O. Edenhofer, et al. (Eds.), Climate Change 2014: Mitigation of Climate Change Contribution of Working Group III to the Fifth Assessment Report of the Intergovernmental Panel on Climate Change, Cambridge University Press, Cambridge, UK and New York, NY, 2014.

[37] U. Lucia, G. Grisolia, Time:a Constructal viewpoint & its consequences, Sci. Rep. 9 (2019) 1–7.

[38] R. Zah, H. Boni, M. Gauch, et al., Empa Report. Life Cycle Assessment of Energy Products: Environmental Assessment of Biofuels, 2007. Executive summary available from: http://www.bfe.admin.ch/themen/00490/00496/index.html?lang=de&dossier_id=01273.

[39] UNEP, Towards Sustainable Production and Use of Resources: Assessing Biofuels, United Nations Environment Programme, 2009. https://www.resourcepanel.org/file/560/download?token=04PkF6fe.

[40] E. Moioli, F. Salvati, M. Chiesa, R.T. Siecha, F. Manenti, F. Laio, M.C. Rulli, Analysis of the current world biofuel production under a water–food–energy nexus perspec- tive, Adv. Water Resour. 121 (2018) 22–31, https://doi.org/10.1016/j.advwatres.2018.07.007.

[41] A. Hayder, Alalwan, H. Alaa, H.A. Alminshid, S. Aljaafari, Promising evolution of biofuel generations, Subject, Rev. Renew. Energy Focus 28 (2019) 127–139, https://doi.org/10.1016/j.ref.2018.12.006.

[42] D.F. Correa, H.L. Beyer, H.P. Possingham, S.R. Thomas-Hall, P.M. Schenk, Biodiversity impacts of bioenergy production: microalgae vs. first generation biofuels, Renew. Sustain. Energy Rev. 74 (2017) 1131–1146, https://doi.org/10.1016/j.rser.2017.02.068.

[43] IEA, Sustainable Production of Second-Generation Biofuels—Potential and Perspectives in Major Economies and Developing Countries, International Energy Agency, Paris, France, 2010.

[44] A. Gasparatos, P. Stromberg, K. Takeuchi, Biofuels, ecosystem services and human wellbeing: putting biofuels in the ecosystem services narrative, Agric. Ecosyst. Environ. 142 (3–4) (2011) 111–128.

[45] T.M. Mata, A.A. Martins, N.S. Caetano, Microalgae for biodiesel production and other applications: a review, Renew. Sustain. Energy Rev. 14 (1) (2010) 217–232.

[46] C.H. Liao, C.W. Huang, C.S.W. Jeffrey, Hydrogen production from semiconductor-based photo catalysis via water splitting, Catalyst (2012) 490–516.

[47] S. Rezania, B. Oryani, J. Cho, A. Talaiekhozani, F. Sabbagh, B. Hashemi, P.F. Rupani, A.A. Mohammadi, Different pretreatment technologies of lignocellulosic biomass for bioethanol production: an overview, Energy 199 (2020) 117457.

[48] A. Jeihanipour, R. Bashiri, Perspectives of Biofuels from Wastes: A Review, Institute of Chemical Technology and Polymer Chemistry, Department of Chemistry and Biosciences, Karlsruhe Institute of Technology (KIT), Germany, 2015, 2015.

[49] A. Godbole, S. Awate, O.N. Varadana, S. Suresha, Biomass to biofuel: a review of technologies of production and future prospects, Int. J. Eng. Res. Technol. 7 (2018) 24–32.

[50] P. Awasthi, S. Shrivastava, A.C. Kharkwal, A. Varma, Biofuel from agricultural waste: a review, Int. J. Curr. Microbiol. App. Sci. 4 (1) (2015) 470–477.

[51] E.M. Aro, From first generation biofuels to advanced solar biofuels, Ambio 45 (S1) (2016) 24–31.

[52] L. Chaudhary, P. Pradhan, N. Soni, P. Singh, A. Tiwari, Algae as a feedstock for bioethanol production: new entrance in biofuel world, Int. J. ChemTech Res. 6 (2) (2014) 1381–1389.

[53] S.N. Naik, V.V. Goud, P.K. Rout, A.K. Dalai, Production of first- and second-generation biofuels: a comprehensive review, Renew. Sustain. Energy Rev. 14 (2) (2010) 578–597.

[54] N. Mosier, C. Wyman, B. Dale, R. Elander, Y.Y. Lee, M. Holtzapple, et al., Features of promising technologies for pretreatment of lignocellulosic biomass, Bioresour. Technol. 96 (6) (2005) 673–686.

[55] A. Goel, L. Wati, Ethanol production from rice (*Oryza sativa*) straw by simultaneous saccharification and cofermentation, Ind. J. Exp. Biolog. 54 (2016) 525–529.

[56] J. Gabhane, A. Tripathi, S. Athar, P. William, A.N. Vaidya, S.R. Wate, Assessment of Bioenergy potential of agricultural wastes: a case study cum template, J. Biofuel. Bioenergy 2 (2) (2016) 122–131.

[57] M.O. Victor-Ekwebelem, U.J.J. Ijah, O.P. Abioye, Y.B. Alkali, Physicochemical and microbiological properties of soil in ahoko: a suspected petroleum bearing site (SPBS), Sci. Res. J. (SCIRJ) VIII (V) (2020) 15–31.

[58] C.N. Hamelinck, G.V. Hooijdonk, A.P.C. Faaij, Ethanol from lignocellulosic biomass: techno-economic performance in short-, middle- and long-term, Biomass Bioenergy 28 (4) (2005) 384–410.

[59] D.K. Ojha, D. Viju, R. Vinu, Fast pyrolysis kinetics of lignocellulosic biomass of varying compositions, Energy Convers. Manag. 10 (2021) 1–10.

[60] Z. Sapci, The effect of microwave pretreatment on biogas production from agricultural straws, Bioresour. Technol. 128 (2013) 487–494.

[61] D. Kumari, R. Singh, Pretreatment of lignocellulosic wastes for biofuel production: a critical review, Renew. Sustain. Energy Rev. 90 (2018) 877–891.

[62] IEA, Outlook for Biogas and Biomethane: Prospects for Organic Growth, IEA, Paris, 2020. https://www.iea.org/reports/outlook-for-biogas-and-biomethane-prospects-for-organic-growth.

[63] K. Kohli, R. Prajapati, B.K. Sharma, Bio-based chemicals from renewable biomass for integrated biorefineries, Energies 12 (2019) 233–273.
[64] P.R. Patwardhan, R.C. Brown, B.H. Shanks, Understanding the fast pyrolysis of lignin, ChemSusChem 4 (11) (2011) 1629–1636.
[65] P.R. Patwardhan, J.A. Satrio, R.C. Brown, B.H. Shanks, Product distribution from fast pyrolysis of glucose-based carbohydrates, J. Anal. Appl. Pyrol. 86 (2009) 323–330.
[66] N. Gomez, S.W. Banks, D.J. Nowakowski, J.G. Rosas, J. Cara, M.E. Sanchez, et al., Effect of temperature on product performance of a high ash biomass during fast pyrolysis and its bio-oil storage evaluation, Fuel Process. Technol. 171 (2018) 97–105.
[67] C. Krumm, J. Pfaendtner, P.J. Dauenhauer, Millisecond pulsed film unify the mechanism of cellulose fragmentation, ACS Chem. Mat. 28 (2016) 3108–3114.
[68] S. Maduskar, G.G. Facas, C. Papageorgiou, C.L. Williams, P.J. Dauenhauer, Five rules for measuring biomass pyrolysis rates: pulse-heated analysis of solid reaction kinetics of lignocellulosic biomass, ACS Sustain. Chem. Eng. 6 (2018) 1387–1399.
[69] E.A. Shalaby, Biofuel: Sources, Extraction and Determination, 2013.
[70] S. Chakraborty, D.M. Aggarwal, K. Andras, Biomass to Biofuel: A Review on Production Technology, 2012, pp. 254–262.
[71] O.B. Ayodele, H.F. Abass, A.W.D. Wan Mohd, Hydrodeoxygenation of stearic acid into normal and iso-octadecane biofuel with zeolite supported palladium-oxalate catalyst, Energy Fuel. (2014) 5872–5881.
[72] D. Kumari, R. Singh, Recent advances in bioremediation of wastewater for sustainable energy products, in: A.K. Rathoure (Ed.), Zero Waste: Management Practices for Sustainability, CRC Press, Taylor & Francis Group, Boca Raton, 2019, ISBN 978-0-367-18039-3, pp. 247–276, 2019.
[73] R. Chandra, H. Takeuchi, T. Hasegawa, Methane production from lignocellulosic agricultural crop wastes: a review in context to second generation of biofuel production, Renew. Sustain. Energy Rev. 16 (3) (2012) 1462–1476.
[74] N.H.S. Ray, M.K. Mohanty, R.C. Mohanty, Anaerobic digestion of kitchen wastes: biogas production and pretreatment of wastes, A Rev. Int. J. Sci. Res. 3 (11) (2013) 2250–3152.

Chapter 23

Biomass waste and feedstock as a source of renewable energy

Kondakindi Venkateswar Reddy[1], Nalam Renuka Satya Sree[1], Pabbati Ranjit[1] and Naga Raju Maddela[2,3]

[1]Center for Biotechnology, Institute of Science and Technology, JNTUH, Hyderabad, Telangana, India; [2]Departamento de Ciencias Biológicas, Facultad la Ciencias de la Salud, Universidad Técnica de Manabí, Portoviejo, Manabí, Ecuador; [3]Instituto de Investigación, Universidad Técnica de Manabí, Portoviejo, Manabí, Ecuador

1. Introduction

Years ago, the contemporary world developed as an outcome of low cost and simply accessed fossil fuels as an energy source. On the basis of fossil fuels, new technologies were evolved to convert fuels into electricity, mechanical power, and heat [1]. Present world population is double that of 1960 and estimated further raise to 9 billion by 2050 [2]. Currently, we are tackling a situation where fossil fuel is limited and the oil price is at rise. Though coal is relatively cheap, it is the principal cause for global warming and sulfur dispersion. From past few years we are hearing about frequent natural disasters like heavy storms—a major issue in East Asia and the United States—and unpredicted and unstable weather patterns, and the arctic ice cap has been reduced compared to several years ago. Such issues are probably due to global warming caused by CO_2 emission from combustion of fossil fuels [1]. Besides, oil reserves are also not renewable and are not available worldwide [3]. Yearly human activities dispose carbon nearly 8 billion metric tons, 1.5 billion resulting from deforestation and 6.5 billion tons resulting from fossil fuels. This huge fossil fuel consumption also affects our ecological cycle [4]. As fossil fuels dwindle eventually, the whole world is in serious search of clean and sustainable energy which can replace fossil fuels [5]. Also, natural resources exhaustion hastened the claim of conventional energy which ultimately forced policymakers and planners rush to alternate sources [6]. Problems with power stocks and employment are not only associated with global warming but also with ecological affairs like acid precipitation, jungle damage, air contamination, and ozone depletion. Renewable energy sources appear to be efficient and dynamic energy source [7] for abovementioned problems.

Renewable energy is derived from regenerative resources and they do not diminish over time. It offers our planet an opportunity to lower carbon emissions and tidy up air, and by single investment, we can gather energy required for decades without causing trouble to environment. It also presents a chance to the countries round the world to enhance economic development and spur energy security. By opting renewable energy sources, in the chemistry zone, petroleum chemistry can be swapped by green chemistry; in agriculture, natural methods can be implemented to facilitate soil as carbon dioxide sinks [4,6,8]. Oceans contribute two-thirds of the Earth's surface which is a massive pool of renewable hydro power. Biomass is used traditionally for cooking and heating. Solar energy contributes highest possibility to provide safe, clean, and reliable power. Wind, another means of gathering energy, is ultimately propelled by atmospheric air. The heat raised from the earth is geothermal energy. All these renewable energies can be substitutes of fossil fuels without emitting greenhouse gases and air pollutants. Thus renewable energy usage is a worth trying promising hope in the future as an alternative green approach [4,6]. As there is demand and increase in pressure to replace fossil fuels to renewable energy, achieving this kind of shift requires swift advance of novel methods and technologies for implementing at large scale [8].

Once upon a time besides sunshine, biomass was the principal energy source for human kind. Later on due to industrialization, coal and oil became major energy source. Biomass sources are not being used effectively instead decomposed or combusted [8]. Biomass represents all the organic matter present in the biosphere. It includes straw, sawdust, wood, household waste, manure, sewage waste, wastewater, paper waste, etc. Biomass, a carbon-neutral energy source, represents rising renewable energy source with prominent growth potential [2]. By the approach of green chemistry, there is a stipulation to generate alternate fuels and energy source from biomass [9]. In the past few years, biomass as an alternative renewable energy source gained renewed interest worldwide due to its cost effectiveness and zero risk to environment as the amount of carbon dioxide released by biomass conversion is same as that of absorption during plant growth [10].

Through this chapter, we will ride a journey from biomass sources, their types, how biomass is converted to energy, different conversion methods, bio-based products, their applications, and till latest advancements.

2. Biomass

2.1 Biomass—various forms available

If we could wait for the replenishment of depleted natural gas or petroleum deposits by natural process which takes several million years, there would be no scarcity of organic fuels. But, there is no such anticipated society. Consequently, carbon-containing compounds which renew themselves in a short time and are constantly available in huge quantities are required to sustain and substitute present energy supplies. Biomass is one such huge source of carbon that meets the instant needs of fuel [11]. Biomass can be broadly classified into four generations. First generation of biomass comprises of traditional plant-based biomass like starch crops, sugar, soya bean, straw, vegetable oil, and animal fat. The products obtained from first-gen biomass are sugar alcohol, biodiesel, and corn ethanol. Second-generation biomass includes agricultural waste, wood, animal manure, municipal solid waste, landfills, and grass. The products like bio-oil, hydrotreating oil, FT oil, etc., are produced from second-gen biomass. Third and fourth generation of biomass include seaweeds, algae-derived fuels like biodiesel produced from microalgae oil, bioethanol produced from microalgae, and H_2 produced from green microalgae and from microbes. Further, these extracted fuels are converted into other forms of energy through different conversion routes. "Drop-in" fuels such as green diesel, green gasoline, and green aviation fuel which are also produced from biomass are considered under fourth generation of biomass [9]. Biomass sources can also be categorized as food and nonfood basis as mentioned in Fig. 23.1.

2.1.1 Food crop as biomass

1. **Primary food crop/Starch sugar crop:** Few crops are classified as chief food crops for human population. Among them cereals are prioritized according to human food perspective. Hydrocarbons production can be increased by high yield from the crops including—Rice: a tropical crop with high production in Australia, Egypt, and China. It is considered as the high calorie food for the Asians, Africans, etc., across the globe. Soybean: another chief food crop which contains good amount of protein that supplements animal protein. Some other food crops with importance are corn, wheat, rye, barley, sugarcane, and oats.
2. **Energy crop:** Majorly there are three types of energy pathways in plants to perform photosynthesis. The most common pathway, C_3 system generally occurs in temperate and colder climates, which yields 3-phosphoglyceric acid. The second pathway—C_4 system is observed in warm climates, that yields malate and aspartate. The third pathway—CAM system yields 4-carbon organic acids. One or other systems are observed in potential energy crops. Few interesting energy crops are: Giant King Grass—high yielding; Switch Grass—feedstock for bioethanol; Hybrid poplar, Quoran—high protein content.
3. **Aquatic crop:** Seaweeds, hyacinth, and algae are considered as aquatic biomass source for bioenergy. Algae are an indirect resource for several animals. Green algae and kelp produce huge amounts of fatty oil and fatty acids which interestingly can be converted to biofuel.
4. **Oil seed crop:** Plants like jatropha, trees like palm, and crops like soybean contribute for biofuel production [8,9].

2.1.2 Nonfood biomass

1. **Agricultural waste (AW):** This includes sugarcane bagasse, cornstalk, grass fodder, etc.; it is the typical categorization of carbon-rich biomass which contains huge amount of cellulose and lignocellulose. Compared to food waste, AW has more bio-methane potentials. AW by different conversion methods can be converted to bioenergy.
2. **Forest waste/residue (FR):** It is an important lignocellulosic raw form for generating bioenergy. FR turns into hazardous flaming wooden biomass. It can be converted to biochar or bio-oil by pyrolysis.
3. **Manure and sewage sludge:** Manure from livestock is a by-product of cattle yard. Besides producing methane gas, manure also contains heavy metals, bacteria, and vermin. Biological, physical, and chemical treatment of wastewater

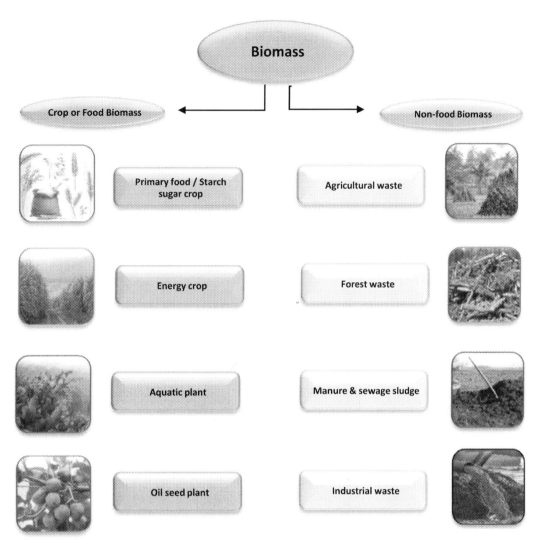

FIGURE 23.1 Biomass sources categorized on food and nonfood sources for bioenergy. *Modified from H. Andreas, Transformation of Biomass, Wiley, 2014.*

results in sewage sludge. Both manure and sewage sludge are converted to biogas and have the uppermost position as biogas manufacturers.
4. **Industrial waste:** Due to industrialization, from few years, world community has been suffering from huge environmental damage. Industrial residues are widely generated from pharmaceutical, petrochemical, food, and agricultural industries.

All the above mentioned wastes are converted to solid fuel, liquid fuel, and gaseous fuel by different conversion technologies [7].

2.2 Biomass composition and their energy content

Biomass, being a complex fuel, varies from species to species and is composed of various components. Every species differs in relative abundance of carbon (C), oxygen (O), hydrogen (H), sulfur (S), and nitrogen (N) which ultimately reflects energy value of fuels. Biomass nature is generally hygroscopic and presence of water is crucial for biomass as it stimulates microbial colonization which helps in nutrient uptake. The biochemical content of biomass is generally denoted as the percentage of dry weight. The dry matter of biomass is dried and expressed using oven—based methods at 105°C. Volatile organic compounds at this temperature vaporize, and the dry matter contains both organic and inorganic matter.

The inorganic composition of biomass is expressed as mass of ash remnants after combustion. The major organic components of biomass are cellulose, lignin, hemicellulose, extractives, sugars, pectin, starch, and proteins.

2.2.1 Cellulose
Cellulose, an abundant glucose polymer, hugely contributes to plant biomass. It has both amorphous and crystalline regions. $C_6H_{10}O_5$, the repeating units, are strung by β-glycosidic linkages which make cellulose highly stable.

2.2.2 Lignin
Lignin, an unusual biopolymer, has aromatic, complex, random, and cross-linked undefined structure. In cell walls, lignin always bonds with hemicelluloses covalently. It is the second abundant biopolymer in nature, and its subunits are phenylpropenyl. It contributes 17%−33% of mass of hardwoods, grasses, and softwoods. Being amorphous in nature it acts as glue and maintains structural integrity of proteins.

2.2.3 Hemicellulose
Hemicellulose is a mixed branched polymer with pentoses and hexoses. It is in amorphous form and is highly reactive. When plant undergoes stress, static forces or dynamic forces, hemicellulose absorbs energy and acts like a connecting link between lignin and cellulose.

2.2.4 Extractives
Extractives are compounds that can be extracted using nonpolar or polar solvents. Waxes, fats, oils, resins, lipids, etc., are plant extractives and typically have high energy.

2.2.5 Sugar
In reference to biomass, "sugar" is typically considered to sucrose, a combination of glucose and fructose. Best example of plant with high sugar content is sugarcane.

2.2.6 Proteins
Proteins are composed of nitrogen. Lentils and beans are the highest form of protein source in plants.

2.2.7 Starch
Starch is a combination of amylopectin and amylose and is often observed in grains like wheat, maize, and rice and underground tubers like cassava and potato [8,9].

Besides the organic matter, the inorganic content present in biomass is expressed in terms of ash mass which remains after combustion. During the combustion, all the components of biomass undergo oxidation and generate gases including nitrogen, sulfur, carbon dioxide, water in the form of vapor, and fly and bottom ash. Ash elements present in dry biomass may cause few operational troubles during energy conversion, hence need to be focused. The inorganic matter present in biomass is mainly due to plant uptake of macro- and micronutrients from soil. Plants specifically tend to contain Mg, Na, Ca, Si, Mn, Cu, N, S, Mo, Co, Cl, Cr, Ni, etc., and have a crucial role in biological functions besides being nutritious [8].

2.2.8 Energy content
Bioenergy's most crucial character is its energy. The heat value or energy value of biomass is the volume of heat generated per unit mass during combustion and the energy generated is related to covalent bonds and valency for C, O, H, S, and N atoms in biomass. The order of calorific value of covalent bonds in different gases is:

$$O-H < C-O < C=O < N-H < C-N < C-C < C-H < S-H < C=C$$

The energy value of dry content in biomass is described as gross calorific value (GCV) or gross energy. It is also known as higher calorific value, upper heating value, or higher heating value. This thermal energy can be detected by completely combusting biomass to CO_2, SO_2, H_2O, and NO_2 in a bomb calorimeter in presence of high concentration of oxygen. The heat generated in between initial and final temperature is considered as GCV. It takes latent heat of vaporization of water into consideration during combustion. Plant extractives like wax, oil, etc., have calorific value nearly 38 kJ g^{-1}; α-cellulose has GCV of 17.492 kJ g^{-1} [8].

2.3 Measuring and analysis of biomass

Biomass components and characteristics can be analyzed by proximate and ultimate analysis. Proximate analysis classifies the products into four clusters: (1) moisture, (2) fixed carbon, the fraction of nonvolatiles in biomass, (3) volatile matter, containing vapors and gases released during pyrolysis or torrefaction, and (4) the inorganic residue—ash which remains after complete combustion. Ultimate analysis determines the C and H content in the substance that is observed in gaseous product obtained after complete combustion. By this analysis molecular weight of that substance can be easily calculated [9].

2.3.1 Volume, weight, and moisture content

Volume of biomass can be measured and verified easily, but the result of measurement depends on the material compactness. In recent days, weight is the leading and objective measure as the weight is less altered by transport. Hence, weight is mostly accepted for shipment regulations. In order to determine density, volume and weight are combined and the density obtained can be used to calculate the moisture content (MC) of specific material approximately, but the result is affected by degree of compression, type of material, and distribution of fraction. Another parameter to analyze biomass is the measuring MC as it detects the amount of dry mass. It is important to calculate MC as it affects the combustion properties, energy content, and price.

$$\text{Dry mass} = \text{Total mass} \times (1 - \text{MC})$$

2.3.2 Moisture measurement

2.3.2.1 Gravimetric moisture measurement

The samples are merged into single sample, from which a smaller part of the sample is selected and observed for weight in a tray. The tray is kept in oven for drying till 24 h. After drying the sample is weighed by preanalyzing the empty tray weight; MC is determined. For gravimetric method, indicative precision is nearly 2%—units of moisture of same samples in two trays are simultaneously compared. The result depends on the material heterogeneity.

2.3.2.2 Instrumental methods

Gravimetric methods may subject to sampling errors and time taking for analyzing complex mixtures. Nowadays, novel technologies such as—microwave spectroscopy, near-infrared spectroscopy (NIR), and X-ray spectroscopy methods have arisen for advanced applications.

2.3.2.2.1 Microwave spectroscopy
Microwave spectroscopy is mostly referred as radiofrequency spectroscopy. It detects the absorption of microwaves. Based on the amount of moisture, the microwave pulse is subjected to various attenuations and time intervals. It can analyze numerous properties of a material even in closed and huge compartments as radio waves can travel entirely through the material which is to be measured.

2.3.2.2.2 Near-Infrared Spectroscopy
NIR uses 800–2500 nm, a near-infrared zone of electromagnetic spectrum. This method considers the data in combination with vibrations and molecular overtone. Bulk materials are analyzed in intersections as the depth of penetration differs with material. NIR can provide the data of organic matrix of different fuels and can estimate energy content and ash content.

2.3.2.2.3 X-ray spectroscopy
X-ray spectroscopy determines the photons released when electrons of inner shell are replaced with atoms of outer shell when undergo excitation. Information about ample elemental composition is considered to analyze the MC. It can determine huge part of bulk in a material or sample. By utilizing two different levels of energy, two different samples can be identified [8].

3. Available biomass conversion technologies

Biomass waste can be turned into gaseous, liquid, and solid fuels by various conversion technologies such as thermal, chemical, physical, and biochemical conversions. Common reactions that occur during these conversions are isomerization, dehydration, oxidation, hydrolysis, hydrogenation, and dehydrogenation, and the actual methods that involve these reactions are pyrolysis, combustion, alcoholic fermentation, gasification, liquefaction, and so on. The products obtained during these conversions are mentioned in Table 23.1.

3.1 Pyrolysis

There are various methods followed in pyrolysis. Among them fast pyrolysis, intermediate pyrolysis, flash pyrolysis, and slow pyrolysis have major significance in converting biomass to bioenergy.

3.1.1 Fast pyrolysis

Fast pyrolysis generally targets liquid production. Its main features are short vapor residence time and high heating rates. It requires sample with low particle size. The reactors commonly used for fast pyrolysis are: fluidized beds, vacuum pyrolysis

TABLE 23.1 Different conversion methods to convert biomass to bioenergy.

		Biomass		
Thermochemical conversion	Liquefaction	Direct	Heavy oil	
		Indirect		
	Pyrolysis (catalytic/ noncatalytic)	Fast	Biogas Bio-oil Tar Char	
		Vacuum		
		Slow		
		Flash		
		Intermediate		
	Gasification	Torrefaction	Syngas FT oil Bio-hydrogen	
		Direct		
	Combustion	—	—	
Biochemical conversion	Anaerobic	—	—	
	Fermentation	Batch	Gobar gas Ethanol	
		Fed batch		
		Continuous		
		Semi		
	Enzyme	Partly microbial	Amino acid Ethanol Hydrogen and protein-based chemicals	
		Various anaerobes		
		Cyanobacteria		
		Photosynthetic bacteria		
		Klebsiella and *Clostridium*		
Chemical conversion	Hydrolysis	Acid hydrolysis	Primary and secondary metabolites Lignin Cellulose Sugar Hemicellulose	
		Enzymatic hydrolysis		
	Solvent extraction	Leaching Liquid–liquid extraction		
	Supercritical conversion			
Physical conversion	Mechanical extraction	—	—	
	Briquetting	—	—	
	Distillation	—	—	

Modified from H. Andreas, Transformation of Biomass, Wiley, 2014.

systems, ablative systems, and stirred/moving beds. Reaction time for fast pyrolysis is just a few seconds or even less. In this process along with chemical reaction kinetics, mass and heat transfer methods and phase transition process play a crucial role. But the main issue in this process is to bring down the reacting biomass particle to optimum temperature. Hence, the particle size in sample must be small as biomass undergoes decomposition to release aerosols, charcoal, and vapors. After lowering the temperature and condensing, dark-brown-colored liquid bio-oil is obtained.

Fast pyrolysis is a complex process which needs attentive regulation of parameters to obtain higher yield of biofuel. Essentials of this process are:

- Absolute high heating (450–600°C) and transfer of heat at reaction interface
- Finely grounded biomass

3.1.2 Intermediate pyrolysis

In this process the reactor is maintained at 400–500°C, and it consists of inner screw, two conveyor coaxial screws, and a covering screw which commonly termed as pyrolyzer. With the help of outer screw and pyrolyzer, biochar travels through the reactor and acts as heat carrier with formation of bed. This process results in yield with less viscosity tar and also provides moderate residence time. Intermediate pyrolysis avoids formation of tars with dry, brittle chars and high molecular weight. In this process mechanical briquetting is not needed for feedstock processing.

3.1.3 Slow pyrolysis

Slow pyrolysis, a carbonization process, has more residence time and requires low temperature. The product targeted in this process is char and tar. The form and character of product totally depends on oxygen concentration, temperature, and time used in processing.

3.2 Torrefaction

It is a thermochemical conversion of biomass in presence of 200–300°C. This process is carried out in the absence of oxygen, water, and volatiles. The final product obtained is the remnant of slight decomposition and appears dry, solid, black-colored substance which is referred as terrified biomass/bio-coal. It has high energy density than raw biomass. It is hydrophobic, hence easy to store and handy to grind, crush, or pulverize.

3.3 Gasification

An alternate thermochemical conversion method to treat municipal solid waste or any other organic biomass functions by slight combustion of biomass in presence of controlled oxygen level and high temperature ranging between 500 and 800°C. The final product obtained by this process is syngas and char. The gas released during this process is a blend of hydrogen, carbon dioxide, nitrogen, and methane. The gas produced is then utilized to run high-efficiency gas turbine.

3.4 Combustion

In combustion, fuel undergoes rapid oxidation that results thermal energy. Biomass is the major substance used in combustion and takes place in presence of air. It is the simplest process of generating energy from available biomass. It is generally performed to release steam that runs turbines which ultimately produce electricity [9,12].

3.5 Biological conversion

- **Anaerobic digestion:** It is a biological phenomenon resulting in biogas. Biomass waste is the sample and the agents are bacteria. This process happens in two steps:
  (i) Degradation of complex organic matter by acid-forming bacteria into simple substances like volatile acids.
  (ii) The obtained volatile acids are then converted to CO_2 and CH_4—biogas by methane-producing bacteria.

  For a combination of $60:35:5 = CH_4: CO_2:$ inerts, obtained heat value of biogas is nearly 22,350 kJ m^{-3}.

- **Fermentation:** Biomass can also be subjected to fermentation to generate alcohol liquid fuels like ethanol [1].

4. Products and their applications

4.1 Biochar

Biochar also called black/dark earth is usually obtained as a by-product while generating coal from wood by pyrolysis. The natural ancient method to add carbon to soil to protect and make soil fertile can be achieved by adding biochar to soil which is capable to allow soil to absorb and preserve water. In this way the organic content of the soil also increases. By this approach agriculture sector also gets benefited. Biochar can be used in cocombustion and for generating activated carbon. It is the only way to store and reduce carbon dioxide in the form of char [9].

4.2 Nanocellulose

Nanocellulose is obtained from minute cellulosic material with nanometer range and the source of availability is from animals, plants, and bacteria. It mainly includes three classes:

(i) CNCs—Cellulosic nanocrystals are obtained by acid hydrolyzing lignocellulosic substances, by separating amorphous cellulose from rod like cellulose nanocrystals.
(ii) CNFs—Cellulose nanofibers or microfibrillated cellulose are obtained by mechanical pretreatment for cellulosic fibers—defibrillation.
(iii) BC—Bacterial nanocellulose or biocellulose are mainly produced by various Acetobacteraceae. BCs are used in cosmetics and wound dressing.

Chemical structure of nanocellulose is a combination of hydroxyl groups, aldehydes, and carboxyl groups. Cellulose in nanoform has high surface area, mechanical strength, active functional groups, and crystallinity. This makes nanocellulose high-end product in different fields. Nanocellulose provides a stable microliving surrounding for cell. They are utilized in drug delivery, tissue engineering, wound healing, and bionanocomposite tubes. Nanocomposites, a combination of nanocrystals and hyaluronic acid hydrogels, enhance the proliferation of human adipose tissue—derived stem cells. Nanocomposites are also applicable in aerospace models, electrochemical cells, and binders in wastewater treatment [13].

4.3 Biofuel

Bioproducts from biomass waste are value-added substances which may compete with fossil fuels as drop-in substitutes. Biomass can be directly converted to liquid fuels—biofuels. Most commonly used biofuels in recent days are ethanol and biodiesel.

4.3.1 Ethanol

Ethanol, a renewable fuel made from biomass waste, is an alcohol used in combination of gasoline to reduce carbon monoxide and smog emissions. Most commonly used conventional blend is E_{10} (10% of ethanol and 90% of gasoline) and E_{15} (15% of ethanol and 85% of gasoline).

4.3.2 Biodiesel

Biodiesel is produced from vegetable oil or by animal fat. Like petroleum, biodiesel can be used as fuel for diesel engine. Biodiesel can be combined with petroleum in any range of combination.

4.3.3 Renewable hydrocarbon fuels—drop-in fuels

Fossil fuels like gasoline, jet fuel, and diesel contain a combination of hydrocarbons which can also be synthesized from biomass through variety of conversion technologies [14]. Bio-hydrogen is produced by fermentation process utilizing nonsulfur bacteria in presence of energy source—light [15].

For aviation, particulate and noise emissions are issues for safety and health of public. Without considering an alternate for aviation fuel, it must rely on drop-in fuels. Aviation fuel should be able to operate at extreme conditions and meet the required standards. All fuels for aviation are a combination of different hydrocarbons. Generally two basic versions of jet fuel are available which only differs in hydrocarbons proportion. Though biodiesel from methyl esters can be used as fuel to some extent, their low cold weather sustainability makes them less preferred for aviation and further research is needed [8].

5. Implementation

Even though the United States contributes only 5% of overall population in the world, it is responsible for energy demand in the first place. Hence the markets for bioenergy in the United States have already been established. The cost of fossil fuels in Europe is still higher than other countries, but the commercial scale-up of biofuels has been better enhanced than before [16].

In India, recently energy consumption is rapidly increasing due to high population rate and industrialization. The demand has increased and Indian Govt. now targeted sustainable development. Ministry of New and Renewable Energy of India (MNRE) has made many policies and project to meet the demand of bioenergy. Indian Govt. in the 12th 5-year plan has included—power distribution based on biomass gasifier and 100% biofuel engines [17].

6. Recent advances

Few advanced biomass—supported technologies are mentioned below:

- Designs of advanced biorefinery system for multiple production of bioproducts.
- Designs of advanced plantation for biomass multicropping integrated with conversion technology.
- Development of hybrid trees, herbaceous biomass species for energy crops.
- Biomass combustion with zero emissions by energy recycle and recovery.
- Genetically engineered microbes to convert sugars from biomass to ethanol.
- Production of steam and hot water by close—coupled biomass combustion system for commercial bocks and schools.
- Production of "super cetane" from triglycerides and tall oils [16].

7. Environmental impacts

- Numerous environmental impacts are connected to biomass directly or indirectly. Environment is benefited if fossil fuels are replaced with biofuel which ultimately reduces the harsh effects caused by using fossil fuels.
- Besides, using biomass and fossil fuels in combination to generate electricity from coal and wood in dual-fuel combustion can reduce unnecessary emissions.
- Another greater factor for environment results from blended application of biomass waste and energy-retrieving technologies. One such example is biogas recovery from treating municipal wastewater by using anaerobic digestion conversion method.
- Another environmental concern is to grow energy crops and harvest biomass. If high amount of biomass is harvested than planted such design is not operative as energy source. Such systems may cause negative impact on environment because balance of carbon in soil and carbon dioxide in environment gets disturbed. In such cases biomass will not be a renewable factor as it ends up releasing carbon dioxide.
- Power plants which are based on fossil fuels must be geographically apart.
- Few state that the CO_2 in the atmosphere is mostly contributed by biomass combustion than fossil fuels. It is demonstrated in such a way due to portraying terrestrial biomass as larger CO_2 sink.
- At least, biomass which is harvested to get energy and fuel must be substituted with newly planted biomass which equals the harvested. New biomass regions like forests must be extended to reduce carbon dioxide pile up the atmosphere [16].
- To produce electricity large amount of biomass residues are burnt in power plant. It is demonstrated that, all the ash of burnt biomass and point source CO_2 can be blended together in carbonated product production [18] which reduces the impact on environment.

8. Challenges

- Challenges faced during handling biomass—storage, pretreatment, consistency, diversity, availability, and constant supply.
- Challenges faced during processing or converting biomass—enzymatic conversion, energy demand, water demand, capital investment, competition with fossil fuels, and financial support.
- Intensive research and development of conversion process and sustainability must be done [3].

- The difference in transportation of biofuels and climatic factors is all time factor that remains as a challenge and most discussed in scientific communities [19–22].

9. Conclusion

A full-fledged analysis of biomass waste management is a must requirement to meet up the depletion of fossil fuels. By doing so, we in this generation can turn the depleting fossil fuel scenario into surplus biomass-based bioenergy and can design chemical-free environment to next generations. Nonfood biofuels should be encouraged. In automobile industry, while developing motor vehicles, biodiesel/biofuel friendly engine designs should be prioritized.

References

[1] D. Erik, Technologies for Converting Biomass to Useful energy, in: Series: Sustainable Energy Developments, vol. 4, 2013.
[2] P.-M. Miguel-Angel, S.-M. Esther, P.-M. Alberto-Jesus, Biomass as renewable energy: world wide research trends, Sustainability (11) (2019) 863, https://doi.org/10.3390/su11030863.
[3] M. Nicolas Clauser, G. Giselle, M. Carolina Mendieta, K. Julia, C. Maria Area, E. Maria Vallejos, Biomass waste as sustainable raw material for energy and fuels, Sustainability (13) (2021) 794, https://doi.org/10.3390/sus13020794.
[4] K.H.M.A. Nada Alrikabi, Renewable energy types, J. Clean Energy Technol. 2 (2014), https://doi.org/10.7763/JOCET.2014.V2.92.
[5] M. Guozhu, Research on biomass energy and environment from the past to the future: a bibliometric analysis, Sci. Total Environ. (635) (2018) 1081–1090, https://doi.org/10.1016/j.scitotenv.2018.04.173.
[6] K. Ashwani, K. Kapil, K. Naresh, S. Satyawati, M. Saroj, Renewable energy in India: current status and future potentials, Renew. Sustain. Energy Rev. (14) (2010) 2434–2442, https://doi.org/10.1016/j.rser.2010.04.003.
[7] S.S. Samarjeet, Z. Qibo, D. Nishu, K.S. Adesh, S. Vipin, P. Bhawna, G. Sergejs, T. Vijay Kumar, Renew. Sustain. Energy Rev. (150) (2021) 111483, https://doi.org/10.1016/j.rser.2021.111483.
[8] D. Erik, Biomass as Energy Source—Resources, Systems and Applications, in: Series: Sustainable Energy Developments, vol. 3, 2012.
[9] H. Andreas, Transformation of Biomass, Wiley, 2014.
[10] S. Jasvinder, G. Sai, Biomass conversion to energy in India—a critique, Renew. Sustain. Energy Rev. (14) (2010) 1367–1378, https://doi.org/10.1016/j.rser.2010.01.013.
[11] L. Donald Klass, Biomass for Renewable Energy, Fuels and Chemicals, Academic Press, Elsevier, California, 1998.
[12] K. Sharukh, P. Vivek, V.P. Vikrant, K. Vijay, Biomass as renewable energy, Int. Adv. Res. J. Sci. Eng. Technol. 2 (2015) (Special issue).
[13] Y. Sujie, S. Jianzhong, S. Yifei, W. Qianquian, W. Jian, L. Jun, Nano cellulose from various biomass wastes: it's preparation and potential usages towards the high value added products, Environ. Sci. Ecotechnol. (5) (2021) 100077, https://doi.org/10.1016/j.ese.2020.1000777.
[14] Office of Energy Efficiency and Renewable Energy, https://www.energy.gov/eere/bioenergy/biofuel-basics.
[15] S. Neha, S. Manish, P.K. Mishra, S.N. Upadhyay, W. Pramod Ramteke, G. Vijay Kumar, Sustainable approaches for biofuels production technologies—from current status to practical implementation, in: Series: Biofuel and Biorefinery Technologies, vol. 7, Springer, 2019.
[16] L. Donald Klass, Biomass for renewable energy and fuels, in: Encyclopedia of Energy, vol. 1, Elsevier, 2004.
[17] K. Anil, K. Nitin, B. Prashant, S. Ashish, A review on biomass energy resources, potential, conversion and policy in India, Renew. Sustain. Energy Rev. (45) (2015) 530–539, https://doi.org/10.1016/jrser.2015.02.007.
[18] T. Nimisha, D. Colin Hills, S. Raj Singh, Biomass waste utilisation in low–carbon products: harnessing a major potential resource, NPJ Clim. Atmos. Sci. (2) (2019) 35, https://doi.org/10.1038/s41612-019-0093-5.
[19] N. Yadhu Guragain, V. Praveen Vadlani, Renewable biomass utilisation: a way forward to establish sustainable chemical and processing industries, Cleanroom Technol. (3) (2021) 243–259, https://doi.org/10.3390/cleantechnol3010014.
[20] Maulin Shah (Ed.), Microbial Bioremediation & Biodegradation, Springer, Singapore, 2020.
[21] Maulin Shah (Ed.), Removal of Refractory Pollutants from Wastewater Treatment Plants, CRC Press, USA, 2021.
[22] Maulin Shah (Ed.), Removal of Emerging Contaminants through Microbial Processes, Springer, Singapore, 2021.

Chapter 24

Bioenergy and beyond: biorefinery process sources, research, and advances

Thamarys Scapini[1,2], Sérgio Luiz Alves Júnior[3,4], Aline Viancelli[5], William Michelon[5], Aline Frumi Camargo[2,4], Angela Alves dos Santos[6], Laura Helena dos Santos[2] and Helen Treichel[2,4]

[1]*Bioprocess Engineering and Biotechnology Department, Federal University of Paraná, Curitiba, Parana, Brazil;* [2]*Laboratory of Microbiology and Bioprocess (LAMIBI), Federal University of Fronteira Sul, Erechim, Rio Grande do Sul, Brazil;* [3]*Laboratory of Biochemistry and Genetics, Federal University of Fronteira Sul, Chapecó, Santa Catarina, Brazil;* [4]*Department of Biological Science, Graduate Program in Biotechnology and Bioscience, Federal University of Santa Catarina, Florianópolis, Santa Catarina, Brazil;* [5]*University of Contestado, Concórdia, Santa Catarina, Brazil;* [6]*Department of Biochemistry, Federal University of Santa Catarina, Florianópolis, Santa Catarina, Brazil*

1. Introduction

Most chemicals and fuels are produced by petrochemical-based refineries [1]. Petroleum extraction and refining processes have high efficiency, and high market dependence is linked to these systems to meet the energy demands of many products. On the other hand, environmental impacts and the finitude of this nonrenewable resource are widely discussed problems. Associated issues include increased greenhouse gas (GHG) emissions (mainly carbon dioxide (CO_2)) and waste generation. The demand for energy and chemical products associated with cost-efficient bioprocesses which have fewer environmental impacts and are associated with residual management with converted products is why residual biomasses have been identified as an exciting strategy as feedstock for integrated biorefineries to obtain multiple products.

A biorefinery is a sustainable installation integrating various bioconversion technologies to produce biofuels and high-value-added chemicals, potentially reducing some of the overdependence on nonrenewable feedstock. Managing residues by combining capital generation through marketable products and the mitigation of GHG emissions into the atmosphere is aligned with sustainable development goals and the transition from a linear-based economy focused on producing and discarding to a circular economy with a holistic view of feedstock, processes, and products that ideally reduces end-of-process residues and results in higher conversion efficiencies [2,3].

Residual biomass biorefineries are commonly used for residue conversion through diverse approaches, including thermal, chemical, physical, enzymatic, and microbial growth processes. In addition, there is a considerable desire to make residual biomass refining a critical part of the world economy as a source of energy and value-added biomaterials, to improve consumption and production indicators, and to promote the development of innovative technologies through more sustainable routes that avoid residue generation [3,4].

The residual feedstocks for use in biorefineries include lignocellulosic, algal, pectin-rich, and starch-rich biomasses. They are particularly beneficial due to their low prices, high bioavailability, and status as residual biomasses, which avoids using arable land for bioenergy conversion. The potential for these biomasses in integrated biorefineries comes from their structural composition, which may comprise cellulose, hemicellulose, lignin, proteins, starch, free sugars, and lipids. These materials have vast potential for conversion into functional sugars, nanomaterials, phenolic compounds, animal feed, and a range of bioenergies, such as ethanol, butanol, biogas, and biohydrogen [5].

However, the complexity and structural heterogeneity of residual biomass pose significant challenges to bioconversion efficiency and limit its cost-effectiveness. The flexible coproduction of distinct bioproducts for greater conversion efficiency must address these challenges. It is necessary to fully minimize these technological and economic risks for process implementation [6]. To overcome these barriers, biomass must be maximized by structural fractionation approaches that allow full use of the components in the conversion into value-added bioproducts [2,7]. In addition, technologies associated with flexible processes (facilitated conversion into various products), efficient fractionation unit operations for different biomasses, and genetic engineering of microorganisms to maximize substrate utilization may be essential routes to feasibility and may reduce technical and economic risks.

Currently, research and industrial projects are being designed and optimized to advance the technological developments in this sector, considering aspects of natural resources, economic impacts, biomass improvement, GHG emission reduction, and environmental, social, and economic perspectives as important indicators of the benefits of biorefineries for a more sustainable economy [2,4]. The complexity of residual biomass biorefinery operations includes analysis of biomass characteristics to define the range of products to be obtained. Moreover, unit operations require integration and optimization of energy, heat, and reagent supplies. These variables should be considered when analyzing the complex independence of a project [4].

Considering the significant research advances in bioenergy and biomaterials using residual biomasses. With a focus on integrated biorefinery processes, this chapter discusses residual biomass based on structural composition, emphasizing the relevance of bioproducts and bioenergy from residual biomass-based platforms. Additionally, the key barriers to integrating biorefining processes are discussed, highlighting the challenges of fractionation for process performance. Also included is a comprehensive discussion of advances in biomass bioenergy production, focusing on value-added products. The genetic engineering of microorganisms is addressed relative to its relevance for bioprocesses.

2. Biomass sources for integrated biorefineries: bioenergy and beyond

In recent years, different processes have been explored for residual biomass valorization with the aim of aggregating value to structural fractions in integrated recovery plants to reach commercial viability. Various strategies and conversion routes aimed at transforming biomass into new products have been explored to develop renewable processes as solutions that reduce the impacts of fossil fuels.

Transformation routes depend on biomass chemical characteristics, which may vary because of source, geographical origin of the material, crop residue, soil characteristics, and seasonality, all of which can interfere with the structural elements and conversion processes of these biomasses. Because no unique classification exists for biomasses, they have been grouped according to their purpose, origin, function, generated product, or predominant chemical constitution. Biomasses are presented here according to the principal structural composition (lignocellulosic, pectin, starch) that directly influences the products obtained from the feedstock. Additionally, microalgae biomasses are discussed relative to biorefining. Table 24.1 overviews recent studies on integrated processes for obtaining multiple products based on the residual biomasses discussed here. Finally, microorganism engineering for application enhancement is also addressed.

TABLE 24.1 Overview of recent studies on residual biomass within a multiproduct biorefinery context.

Residual biomasses	Conditions process	Efficiency	References
Lignocellulosic			
Poplar wood Pinewood Wheat straw Rice straw Mōsō bamboo	*Step 1*: Biomass pretreatment was performed in two steps. First, hemicellulose was removed by hydrothermal acid-catalyzed pretreatment (separated liquid fraction) and sequenced by deep eutectic solvent (DES) pretreatment (choline chlorine and lactic acid). A cellulose-rich solid fraction and a lignin liquid fraction were recovered. *Step 2*: Activated nanocarbons (ANCs) were produced with the liquid fraction of the hydrothermal pretreatment. The experimental procedure was composed of precipitation, drying, chemical mixing (KOH), and carbonization under a nitrogen atmosphere. *Step 3*: Lignin-containing cellulose nanofibers (LCNFs) were produced from the solid fraction of the DES pretreatment. The samples were ultrasonicated under specific experimental conditions.	Poplar wood ANC: 0.8% w w_{waste}^{-1} LCNF: 40.3% w w_{waste}^{-1} LNS: 11.3% w w_{waste}^{-1} Pinewood ANC: 0.9% w w_{waste}^{-1} LCNF: 54.3% w w_{waste}^{-1} LNS: 5.3% w w_{waste}^{-1} Wheat straw ANC: 2.1% w w_{waste}^{-1} LCNF: 45.0% w w_{waste}^{-1} LNS: 6.7% w w_{waste}^{-1} Rice straw ANC: 2.2% w w_{waste}^{-1} LCNF: 45.8% w w_{waste}^{-1} LNS: 4.7% w w_{waste}^{-1} Mōsō bamboo	[8]

TABLE 24.1 Overview of recent studies on residual biomass within a multiproduct biorefinery context.—cont'd

Residual biomasses	Conditions process	Efficiency	References
	Step 4: Lignin nanospheres (LNSs) were produced from the liquid fraction from the DES pretreatment by a dialysis method using dimethyl sulfoxide.	ANC: 1.5% w w_{waste}^{-1} LCNF: 33.5% w w_{waste}^{-1} LNS: 10.1% w w_{waste}^{-1}	
Sugarcane bagasse	Step 1: Biomass was pretreated with H_2SO_4 in an autoclave for 30 min. A liquid fraction rich in hemicellulosic sugars (majority xylose) was used for fed media in fed-batch operation and was a substrate for xylitol conversion. The solid fraction was rich in cellulose and used as a substrate for ethanol fermentation. Step 2: Xylitol fermentation was performed by *Candida tropicalis* yeast in a fed-batch operation pulse feed started at 36 h of a 96 h batch, with feeding every 12 h at 25% of the total batch volume. Step 3: SSF was conducted with *Saccharomyces cerevisiae* (0.035 $g_{cells}\, g_{bagasse}^{-1}$) and cellulase enzymes (7.5 FPU $g_{bagasse}^{-1}$) in a fed-batch operation.	Xylitol: 0.50 g g_{xylose}^{-1} Ethanol: 0.44 g $g_{glucose}^{-1}$	[9]
Starchy			
Food waste	Step 1: Enzymatic hydrolysis was performed with a mixture of cellulases (10 FPU g_{waste}^{-1}) and α-amylases (20 AGU g_{waste}^{-1}) for 24 h at 50°C. The medium was then separated into three phases: (1) upper oil—biodiesel production; (2) middle aqueous—algae cultivation; (3) lower solid. Step 2: Characterization of food waste oils and transesterification was conducted by extraction using hexane. After, oil was transesterified in two steps: (1) methanol:H_2SO_4 followed by hexane extraction and (2) transesterification by methanol and KOH. Step 3: Cultivation of *Auxenochlorella protothecoides* was carried out in food waste hydrolysate and glycerol from transesterification. Compound concentrations were adjusted based on previous assays, and cells' dry weights and total lipids were measured. The steps are integrated into the study for additional yields in various experimental strategies.	Microalgae biomass yield in food waste: 0.346 g g_{sugar}^{-1} (183.55 g kg_{waste}^{-1}) Lipid yield from microalgae cells: 0.216 g g_{sugar}^{-1} (114.77 g kg_{waste}^{-1}) Fatty acid methyl ester (FAME): 109.03 g kg_{waste}^{-1} (majority oleic and linoleic acid) Microalgae biomass yield in glycerol from transesterification: 0.31 g $g_{glicerol}^{-1}$ (7.76 g kg_{waste}^{-1}) Lipid yield from microalgae cultivated in glycerol from transesterification: 0.15 g $g_{glicerol}^{-1}$ (3.61 g kg_{waste}^{-1}) Overall FAME: 248.21 g kg g kg_{waste}^{-1}	[10]
Potato peel wastes	Step 1: Organosolv pretreatment was conducted with ethanol and H_2SO_4 at 120°C for 1 h. Step 2: Cellulases (20 FPU g_{waste}^{-1}) were applied in solid residue to enzymatic hydrolysis. After, the second hydrolysis step was performed for liquefaction and saccharification of starch with α-amylases (2 g kg^{-1}) and glucoamylase (2 g kg^{-1}), respectively. Step 3: Alcoholic fermentation was performed with a liquid fraction of enzymatic hydrolysis with *S. cerevisiae* at 32°C for 48 h. Step 4: Anaerobic digestion was performed with solid fraction enzymatic hydrolysis and liquid fraction of pretreatment. The process was conducted in a mesophilic condition (37°C) for 45 days, with a mesophilic inoculum of an anaerobic reactor from a wastewater treatment plant.	Bioethanol: 284.1 L kg_{waste}^{-1} Biomethane: 57.9 L kg_{waste}^{-1}	[11]

Continued

TABLE 24.1 Overview of recent studies on residual biomass within a multiproduct biorefinery context.—cont'd

Residual biomasses	Conditions process	Efficiency	References
Pectin			
Orange peels	Step 1: Dried orange peels and sieved, followed by hydrothermal pretreatment (140°C, 30 min) Step 2: Enzymatic hydrolysis with CellicCtec2 and CellicHTec2 (9:1 ratio). Cellulase loading was 15 FPU $g_{biomass}^{-1}$, pH 4.8, 45°C. Step 3: Anaerobic digestion with solid enzymatic hydrolysis and conducted in a mesophilic condition (37°C) for 45 days with a mesophilic inoculum of an anaerobic reactor from a wastewater treatment plant. Step 4: Pretreated hydrolysate was detoxified with $Ca(OH)_2$ solution (100 g L^{-1}) for 24 h at 30°C after acetone–butanol–ethanol (ABE) fermentation. Step 5: ABE fermentation was performed with enzymatic hydrolysate or pretreatment with liquor-detoxicated and microorganism-inoculated *Clostridium acetobutylicum* NRRL B-591 (6% w w^{-1}) at 150 rpm for 72 h.	*Biohydrogen*: 9.4 g kg_{waste}^{-1} *Biobutanol*: 42.3 g kg^{-1} *Bioethanol*: 3.4 g kg^{-1} *Biomethane*: 28.4 g kg^{-1} *Acetone*: 33.1 g kg^{-1} *Acetic acid*: 19.4 g kg^{-1} *Butyric acid*: 20 g kg^{-1}	[12]
Citrus pulp of floater (orange)	Step 1: Hesperidin extraction was conducted with methanol (3 h) and precipitation of HCl (6% v v^{-1}). Step 2: Enzymatic hydrolysis was conducted with solid residue from hesperidin extraction, with a cocktail of cellulase (0.118 U mL^{-1}), pectinase (58 U g^{-1}), and β-galactosidase (1.33 U mg^{-1}) at 45°C for 24 h. Step 3: Alcoholic fermentation was conducted with *Candida parapsilosis* 48375IFM and *S. cerevisiae* for 24 h at 35°C. Step 4: Nanocellulose extraction was performed in solid residues of enzymatic hydrolysis in chemical reactions (NaOH and $NaClO_2$) and sonification.	*Hesperidin*: 1.2% (w w^{-1}) with a purity of 92.9% *Ethanol*: ~ 0.509 g g^{-1} *Nanocellulose*: 1.4% (w w^{-1}) with a purity of 98%	[13]
Microalgae			
Scenedesmus dimorphus	Step 1: Crude biomass. Lipid extraction (chloroform:methanol — 1:2 v v^{-1}) was performed, followed by acid-catalyzed transesterification (HCl and methanol). Step 2: Lipid-extracted biomass. SSF was performed based on starch conversion into ethanol—amyloglucosidase enzyme (60 U mL^{-1}), pH 5, 36°C, 3 g L^{-1} of *S. cerevisiae*.	*Lipid yield* (extracted): 14 ± 0.6% (w w^{-1})—the majority was palmitic and oleic acid *Starch* (solid): 54% (w w^{-1}) *Bioethanol*: 0.26 $g_{ethanol}$ $g_{biomass}^{-1}$	[14]
Chlamydomonas sp. KNM0029C	Step 1: Crude biomass. Lipid extraction (chloroform:methanol — 1:2 v v^{-1}) was performed, followed by sonification and chloroform addition. Step 2: Lipid-extracted biomass. Pretreatment (sonification + enzyme) and ethanol fermentation by *S. cerevisiae* were performed for 48 h at 30°C.	*Biodiesel*: 0.16 $g_{biodiesel}$ $g_{dry\ cells}^{-1}$ *Carbohydrate* (solid): 0.39 g $g_{biomass}^{-1}$ *Bioethanol*: 0.22 $g_{ethanol}$ $g_{biomass}^{-1}$	[15]

2.1 Lignocellulosic biomass

Lignocellulosic biomasses (LCBs) are residues generated primarily from crops, livestock, forest harvesting, and wood processing residues (e.g., rice straw, corn stover, sugarcane bagasse, grasses, and wheat straw). The material is formed primarily by three complexly organized polymers: cellulose (30%–60%), hemicellulose (20%–40%), and lignin (15%–25%). LCBs may also include minor amounts of pectin, protein, extractive, and ash [3].

Cellulose presents an excellent potential for industrial applications and has been explored mostly in monosaccharide conversion processes and bioenergy production; it has more recently been applied in nanomaterials production [16,17]. Cellulose is characterized by a primarily crystalline structure, where β-1,4-glycosidic bonds link D-glucose molecules. The structure is linear, with high tensile strength, making cellulose insoluble in most solvents. Unlike cellulose, hemicellulose is an amorphous structure, with a random composition of pentoses (e.g., D-xylose and L-arabinose) and hexoses (e.g., D-glucose, D-galactose, and D-galactose), with acid contents such as acetic and uronic. Compared with cellulose, the hemicellulose structure is more easily hydrolyzed, and among the possible uses are the recovery of oligosaccharides with potential application in the food industry and the bioenergy pathways using the hydrolysate as a substrate for biofuel production (the pathway is dependent on pentose fermenting microorganisms—the majority monosaccharides in the hemicellulose structure which is still a challenge) [16].

Lignin consists of a structure composed mainly of phenolic heteropolymers of amorphous characteristics, interacting with LCB polysaccharides in a compacted and resistant matrix. The structures of these polysaccharides (cellulose and hemicellulose) are interlaced by lignin (the non-polysaccharide fraction of LCB); despite being the second most abundant polymer in nature, it is still underutilized by industrial processes. In general, the amorphous organization of the structure is a challenge for the breakdown into specific compounds of commercial interest during biorefining processes. It can hinder the efficient fractionation of the structures and the recovery of the other macromolecules [18].

LCB has been exploited for integrated processes because it is the most abundant biological source globally, renewable, and low cost. Due to material complexity, its utilization involves depolymerization of feedstock, enabling the structures to be broken for conversion into functional biomaterials. Most of the LCB generated is used for heat generation, representing 97% of all renewable heat produced in the world [1,19]. However, processes are already full-scale implemented, where the LCB has been used for a range of biofuels, chemicals, and biomaterials (see Table 24.1). Although LCBs have been explored for biofuel production, studies that evaluate the impact of integrated processes and focus on obtaining more than one product from the same feedstock are recent.

An integrated biorefinery concept using LCBs as feedstock was evaluated by Ref. [9], who studied ethanol and xylitol production from sugarcane bagasse. After conducting acid pretreatment (H_2SO_4) in an autoclave, the solid fraction (rich in cellulose) was applied in ethanol production by *Saccharomyces cerevisiae* in a combined saccharification and fermentation process. In contrast, the liquid fraction (hemicellulosic fraction, rich in xylose) for xylitol production by *Candida tropicalis*. In the fed-batch configuration, the yield results were 0.44 $g_{ethanol}\, g_{glucose}^{-1}$ for ethanol and 0.50 $g_{xylitol}\, g_{xylose}^{-1}$ for xylitol in 48 h. In the batch configuration, the yield results were 0.43 $g_{ethanol}\, g_{glucose}^{-1}$ for ethanol and 0.36 $g_{xylitol}\, g_{xylose}^{-1}$ for xylitol in 96 h. The fed batch increases xylitol production over the batch configuration. The authors demonstrated that the economic impact of the process would be negative if the xylitol coproduction in the process were low. The integration was highly dependent on the product title obtained [9].

Nanomaterial production was recently evaluated by Ref. [8] using five LCB fractionation processes (Table 24.1) to produce activated nanocarbon, lignin-containing cellulose nanofibers, and lignin nanospheres. Employing a sequential fractionation strategy, the researchers used a hydrothermal process to recover a solid fraction with higher cellulose and lignin content and a liquid fraction with hemicellulose. The hydrothermal pretreatment's hemicellulose removal from the solid fraction was 100% in all biomasses. The second fractionation used deep eutectic solvents, which presented varying yields for the hydrothermally pretreated solid but with yields more significant than 80% for cellulose recovery. The authors expose that the strategy was promising to improve the cellulose characteristics for the nanofibrillation process. The nanomaterial presented dependence on the biomass source. The applied technological route enabled the lignin spheres to be more uniform and stable with higher yields. The mass balance demonstrated a complete process for converting LCBs into value-added nanomaterials [8].

The processes can be configured in a range of pathways that can be separate or simultaneous, applying enzymatic, thermal, physical, and chemical techniques. In a general scenario for obtaining bioenergy from LCB, pretreatment, hydrolysis, and fermentation steps are commonly considered. However, LCB also has the potential to be converted into value-added products (nanocomposites, functional foods, bioplastics, biofuels), which can be designed simultaneously with heat generation. Biofuels and biomaterials production from LCB thus has been a critical strategy to increase the efficient use of residual biomass and the economic value of processes by expanding the products in different sectors mainly focused on bioenergy demand.

2.2 Starchy biomass

The industrial conversion of starch-based biomass has been occurring for more than 200 years. The industry is growing worldwide, producing more than 150 million tons of starch and derivatives [20]. Starch production is generally based on cereal grains and tubers, with wheat, corn, rice, potatoes, and cassava as the primary sources. By 2021, the European starch industry was responsible for a turnover of US$ 3.25 billion for feedstock modification via enzymatic, physical, and chemical routes for application in food products, pharmaceuticals, and paper [21]. Starch is a polysaccharide composed of glucose units in α-1,4 and α-1,6-glycosidic bonds, with amylose and amylopectin as the organizational structures of the compound. This compound is among the most important dietary components, accounting for about 50%−60% of the calories needed by humans [20]. In addition to primary food crops, starch is present in residual biomass, such as crop residues, potato peels, and discarded food.

Food waste generation has become a highly discussed topic for research addressing food safety, environmental protection, and waste valorization, resulting in advances in recent decades related to treatment and disposal methods and energy production. In a biorefinery context, starch-rich residual biomass has potential application in a wide range of chemical products via chemical and biological conversion (see Table 24.1), with sectors that extend beyond food and pharmaceuticals, such as in the paper and adhesives industry [20]. It is important to note that corn starch is the primary source of bioethanol production in the USA, which in 2016 was responsible for more than 50% of the global production of this biofuel [22]. Biogas production from food waste is already operating at full-scale. In 2019, Europe accounted for 64% of energy from this waste and was also highlighted in biogas production [1].

The human diet is a rich source of energy, and food waste is one of the most exploited starch-rich residual biomasses for energy production. It is generated in large quantities, can be obtained at a low economical cost to obtain, and has high levels of biodegradable compounds (in terms of complexity, these residues are often composed of starch, lignocellulose, proteins, lipids, and pectin) [23]. Poe et al. [24] reported on starchy food wastes for butanol production via acetone−butanol−ethanol fermentation by *Clostridium* species. The authors presented a butanol production of 8.2 g L^{-1} with a yield of 7.7 $g_{butanol}$ 100 $g_{waste^{-1}}$ and a ratio of about 6:3:1 for butanol, acetone, and ethanol, respectively. The butanol yield was positively and strongly correlated to the starch content of the waste [24].

Starchy residues can also be used to feed microorganisms for lipid production to biodiesel. Obtaining lipids from starch usually involves the enzymatic degradation of a polysaccharide into fermentable sugars and bioconversion into microbial lipids. Consolidated biodiesel production processes from starch-based lipids are already being explored. Chaturvedi et al. [25] reported on lipid and enzyme production using starchy wastes and oleaginous fungus under submerged fermentation. The food waste was used without any pretreatment. Residues from wheat bran and banana peel presented the best results for *Clostridium curvatus,* with a lipid yield of 16 and 14.3 $g_{lipid}\ g_{starch^{-1}}$, respectively [25].

In a biorefinery context, starch-rich biomasses can also be applied in itaconic acid, succinic acid, lactic acid, butanol, and ethanol production. They can also be an important feedstock to explore for industrial bioplastics production, particularly considering that demand for biodegradable packaging tends to generate increased interest in using polyhydroxyalkanoates in starch-based plastics [23].

2.3 Pectin biomass

Pectin-rich biomasses also have enormous potential for conversion into high-value-added products. Despite being generated in large quantities, mainly by juice and juice by-products from the food processing industry, these biomasses are underused and destined for use in animal feed, soil coverage, or landfills. They include fruit and vegetable residues (e.g., orange and apple peels) and some other agricultural residues (e.g., beet waste pulp).

Compared with LCBs, these biomasses have a structure with high pectin and free sugar content. The lignocellulosic structure has significant cellulose and hemicellulose content and low lignin content that can be interesting for biofuel conversion [26]. Consequently, biofuel production from pectin-rich biomass has emerged as a valorization strategy, mainly due to the abundance of fermentable sugars and low lignin content. A recent bibliometric review [27] introduced research trends on valorizing pectin-rich biomass in a biorefinery context and emphasized that these biomasses are currently being evaluated for biofuels production (bioethanol; biogas; solid biofuels), chemical and physical transformation for biocomposite recovery and lignocellulosic fractions (besides pectin), butyric, lactic, mucic, and succinic acid production via biotechnological valorization, and agricultural uses (e.g., fertilizer).

The pectin used as a feedstock for biofuel production via biotechnological routes is a challenge due to the presence of galacturonic acid and pentoses, which are resistant to fermentation by most yeasts. For example, *S. cerevisiae* yeast (used mainly for ethanol production), when their strains are not genetically modified, cannot metabolize the sugars present within the pectin structure [26,27]. Some strategies, such as genetic engineering and microorganism isolation, can be used. A unit operation for pectin extraction may be more feasible and straightforward, focusing on commercialization as a product of the integrated process and converting other structures (e.g., cellulose) to biofuels [27]. Therefore, greater economic value could be derived from the biorefinery by recovering other bioactive compounds before biofuel production [28].

The valorization strategies in biorefineries for pectin-rich biomasses are several and should be based on the characteristics and generation of the target biomass. In general, citrus residues (such as orange peels) are usually first managed in integrated processes by essential oil extraction, which has broad applicability in pharmaceutical and self-care products. The second step is typically the pectin extraction, which results in a rich LCB that can be routed through a third step to obtain bioproducts or biofuels, such as bioethanol, biogas, nanocellulose, and hesperidin, such as the study development by Cypriano et al. [13] (Table 24.1).

An integrated biorefinery combining hydrothermal processing of waste pomegranate peels was developed for simultaneous pectin and phenolics recovery and ethanol production. Hydrothermal treatment (115°C for 40 min) yielded a maximum pectin recovery of 207 g_{pectin} kg_{waste}^{-1} and phenolic compounds of 118 g_{GAE} kg_{waste}^{-1}; punicalagin was found in 57.2% ($g_{punicalagin}$ g_{GAE}^{-1}) of the recovered phenolics (67.5 mg g_{waste}^{-1}). The hydrolysis yielded 177 g_{sugar} kg_{waste}^{-1}, and fermentation of waste pomegranate hydrolysates yielded 80 $g_{ethanol}$ kg_{waste}^{-1} and 88% efficiency [29]. Cypriano et al. [13] evaluated orange waste (citrus pulp of floater) in a biorefinery concept. Three bioproducts were isolated from the pectin waste biomass: nanocellulose, hesperidin, and ethanol. Nanocellulose and hesperidin were obtained with high purity (98% and 92.6%, respectively), demonstrating a strategy to integrate unit operations with a flexible approach to biomaterials and biofuel production [13].

The seasonality of pectin-rich residual biomass generation is a challenge for integrated biorefinery management. However, the potential for obtaining high-value-added products from these biomasses is excellent (see Table 24.1). To solve this barrier, Fazzino et al. [30] recently proposed ensiling processes (a forage preservation method under anaerobic conditions) to maximize high-value-added products from orange peel residual biomass, such as ethanol and lactic and acetic acids. Approaches to biorefineries of pectin-rich residue biomass have recently been reviewed with estimates of waste generation and valorization strategies by Manhongo et al. [28].

2.4 Microalgae

Microalgae are unicellular photosynthesizing microorganisms that use light as an energy source to convert CO_2 into biomass and produce O_2. These microorganisms play a vital role in aquatic ecosystems and are crucial to the planet due to their capacity to modulate atmospheric CO_2 [31]. Microalgae cultivation requires simple nutritional composts for development—macronutrients (C, N, and P) and micronutrients (Fe, Mg, Ca, S, Na, Cl, Zn, Cu, Mo, Mn, and Br) [32]. These compounds can be found in various wastewaters, such as municipal, agricultural, and industrial, and can be used as cultivation media to grow microalgae [32]. This association of wastewater treatment with microalgae cultivation is called phycoremediation and is a promising alternative.

Using microalgae as a tertiary treatment to remove nutrients (N and P) from wastewater is gaining interest in the bioenergy scenario due to higher growth rates, fast life cycles, and higher productivity per area (7–13 times higher) than traditional biofuel production using feedstocks such as corn or soybean [32]. The simultaneous process of microalgae biomass production and wastewater treatment is fundamental to supporting the economic viability of microalgae-based industries by reducing the need for additional nutrient input. Microalgae cultivation with wastewater is under the circular economy concept. The integration of wastewater treatment with biomass production results in biomass that could be harvested and used as raw material for biofuels production, such as biomethane, biodiesel, bioethanol, and biohydrogen [33].

The significant possibilities of wastewater and microalgae interaction and the bioproducts application are summarized in Fig. 24.1. The obtention of these materials starts with a chemical or mechanical process to harvest the biomass. After that, a cell rupture step is followed by specific extraction procedures, which depend on the intended bioproduct, such as transesterification used to obtain fatty acid methyl ester for biodiesel production. When lipids are extracted, the residual biomass could be used in anaerobic digestion to produce biomethane or convert it into additional biofuels such as bioethanol via hydrolysis and fermentation. A biomethane purification system could be coupled to the microalgae to remove the CO_2 present in this gas. The digestate (liquid fraction) produced by anaerobic digestion is a rich source of nutrients that could be reused in microalgae cultivation. Finally, the solid fraction resulting after anaerobic digestion could be applied as a fertilizer by-product.

Considering the range of possibilities, strategies for integration processes have been studied worldwide. One study reported the application of anaerobic codigestion of defatted microalgae residue (from two distinct species) and rice straw. The resulting cumulative biomethane yield was 382 mL g_{VS}^{-1}, obtained from *Chlorella* CG12 + rice straw. Additionally,

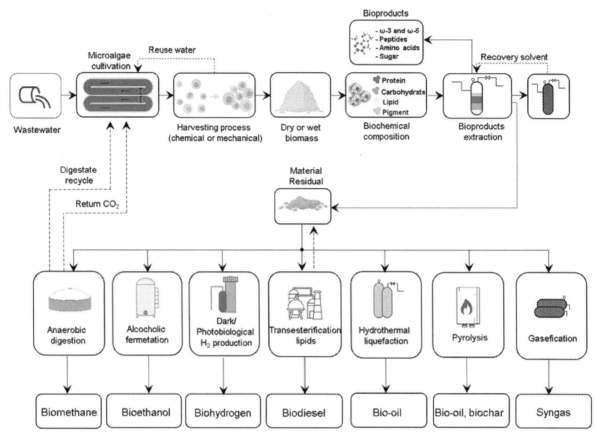

FIGURE 24.1 Schematic flow sheet representation of microalgae multiapplication possibilities in the biorefinery concept.

311 mL g_{VS}^{-1} was obtained from the codigestion of *Desmodesmus* GS12 + rice straw, an increase of 49.8% and 22.2%, respectively, over the control [34]. The defatted microalgae biomass showed a low C:N ratio due to high protein content after lipid extraction. This can become an inhibitory condition for methane production due to the excessive ammonia production after breaking down these proteins, which can be avoided by codigestion with other biomasses to increase the C:N ratio and consumption through various technological routes, e.g., fertilizer production [35].

Carbohydrates accumulated in microalgae biomass can be hydrolyzed to glucose, an efficient substrate for fermentative biohydrogen. In this sense, bacteria present in the anaerobic degradation process, such as *Clostridium* sp. and other mixed anaerobic bacteria, are efficient in using starch to produce hydrogen [36]. Microalgae biomass pretreated with acid-hydrothermally was used as the substrate to produce hydrogen and methane through integrated dark fermentation + photofermentation and photofermentation + anaerobic digestion processes. The study demonstrated that integrating photofermentation and anaerobic digestion processes is the most effective way for energy recovery, with an energy yield of 5.98 kJ g_{VS}^{-1} [37].

Fetyan et al. [38] cultivated *Nannochloropsis oculata* microalgae on sugarcane bagasse aqueous extract. They then submitted the harvested biomass to pretreatment with acid hydrolysis (maximum concentration of 5.0% v $v_{H_2SO_4}^{-1}$) or acid enzyme hydrolysis. The ultrasound pretreatment of defatted microalgae biomass improved ethanol yield and produced a maximum of 86.70 ± 0.52 $mg_{bioethanol}$ $g_{biomass^{-1}}$ [39]. The highest concentration of sugars (232.39 mg $g_{defatted\ biomass^{-1}}$) was obtained with biomass pretreated with 3.0% (v v^{-1}) of H_2SO_4 associated with the enzymatic process, resulting in a maximum ethanol concentration of 6.17 ± 0.47 g L^{-1} (0.26 $g_{ethanol}$ $g_{sugar^{-1}}$), that is higher than the concentration obtained with the biomass submitted only to the acid pretreatment that results in 1.17 g L^{-1} (0.007 $g_{ethanol}$ $g_{sugar^{-1}}$) [38].

In biodiesel synthesis, the addition of alcohol acts as an extraction solvent and subsequently as a transesterification reagent. Direct biodiesel production eliminates the loss of lipids once all the lipids present in the biomass are converted into biodiesel and the production of value-added coproducts, such as glycerol carbonate, diethyl ether, ethyl levulinate, and ethyl formate [40].

A process combining wet in situ transesterification and hydrothermal liquefaction for fatty acid ethyl ester production from microalgae biomass was applied. The yield of fatty acid ethyl ester in biocrude was 91.85% [41]. To fully use microalgae biomass, the residual material can be transformed into bio-oil through pyrolysis [42].

3. Genetically engineered microorganisms for CO_2 fixation

Metabolic and genetic engineering can improve the performance of residual biomass biorefinery processes, whether applied to the organisms responsible for biomass conversion or applied to the feedstock to be converted into products. Gene overexpression can be a strategy to develop a consolidated process where genetic engineering tools can maximize the CO_2 fixation potential and increase the biomass production of an organism. These changes can culminate in significant advances for biorefineries, expanding the possibilities of feedstocks and products obtained.

Naturally, all green plants and autotrophic microorganisms can fix CO_2 through various biochemical cycles. Autotrophic microorganisms can absorb CO_2 during biomass production using solar energy (photosynthesis) or chemical energy from the oxidation of inorganic substances (chemosynthesis). Using the natural ability to fix CO_2 in these organisms, metabolic engineering has been applied to remodel the intracellular pathways, capture atmospheric CO_2, and convert it into chemicals or biofuels. Biorefineries based on microalgae as feedstock aim to use these microorganisms as cell factories to convert atmospheric CO_2 into fuels and chemicals. The role of the photosynthetic apparatus is the main factor contributing to the productivity of the process. For CO_2 assimilation, the engineering of ribulose 1,5-biphosphate carboxylase/oxygenase (RuBisCO) has received attention for demonstrating interesting results in improving the performance of biological systems [43].

In autotrophic microorganisms and plants, the Calvin–Benson–Bassham (CBB) cycle is the most critical mechanism for CO_2 fixation into biomass, with RuBisCO being the key enzyme of the pathway [44]. As the primary representative of autotrophic microorganisms, photosynthetic cyanobacteria are the most studied for metabolic modification of the CBB pathway for CO_2 bioconversion into products. The overexpression of RuBisCO in cyanobacteria can result in increased biomass, ethanol, and lipids production, among other products of commercial interest [43].

More recent advances have developed artificial autotrophic microorganisms from naturally heterotrophic organisms. As heterotrophic models, the bacterium *Escherichia coli* and the yeast *S. cerevisiae* has been widely used as expression platforms for natural CO_2 fixation pathways [45]. *S. cerevisiae* is the microorganism most used for the fermentation of sugars into ethanol, mainly due to its high productivity and strong alcohol tolerance [46]. Numerous industrial processes are conducted using this yeast, especially in alcoholic fermentation, during which ethanol production is associated with CO_2 release, causing atmospheric emission [47]. Therefore, among the microorganisms developed for CO_2 capture, *S. cerevisiae* may be an attractive option to take advantage of the yeast's industrial processes, becoming more sustainable by aiming at carbon neutrality.

CO_2 fixation in *S. cerevisiae* has been accomplished by the heterologous introduction of two key enzymes: RuBisCO and phosphoribosyl kinase (PRK). The PRK enzyme converts ribulose-5-phosphate, an intermediate of the pentose–phosphate pathway, to ribulose 1,5-biphosphate, which in turn is converted to two molecules of 3-phosphoglycerate in a reaction catalyzed by RuBisCO using CO_2 [47]. In 2013, researchers co-expressed spinach PRK and RuBisCO from the chemoautotrophic bacterium *Thiobacillus denitrificans* in *S. cerevisiae* to capture the CO_2 generated by the yeast during the fermentation process and to use CO_2 as an electron acceptor for oxidation of the cofactor NADH [48]. The reason is that, in anaerobic fermentation processes using this yeast, the unbalance of redox cofactors usually restricts product yields, often seen with the glycerol production as an undesired by-product of bioethanol production, a consequence of the necessity to reoxidize excess NADH. The strain modified with the CBB pathway demonstrated a 90% reduction in glycerol by-product accumulation and a 10% increase in ethanol production in media with glucose and galactose [49]. Another study used a strategy for low RuBisCO expression that integrated nine copies of the gene into the *S. cerevisiae* genome, increasing the expression of the enzyme [50].

A study reported the introduction of a heterologous CBB pathway in *S. cerevisiae* SR8 (a strain modified to ferment xylose) for simultaneous lignocellulosic ethanol production and CO_2 recycling. Specifically, the RuBisCO of *Rhodospirillum rubrum* and a PRK of *Spinacia oleracea* were introduced in *S. cerevisiae* SR8. The result was higher ethanol yields and lower by-product yields than the control strain [51]. Besides *S. cerevisiae*, CO_2 fixation has also been explored in nonconventional yeasts, such as *Kluyveromyces marxianus* [52] and *Pichia pastoris* [53], and by the expression of the CBB pathway.

Although the natural CBB pathway is an attractive CO_2 fixation mechanism for expression in heterotrophic microorganisms, an inherent problem of RuBisCO is the low carboxylation efficiency [54]. Therefore, the emerging field of synthetic biology facilitates the development of artificial CO_2 fixation pathways, aiming to increase the catalytic efficiency of the process and further target the desired products [47]. For example, Schwander et al. [54] have developed an optimized synthetic carbon fixation pathway in vitro using 17 enzymes, including three modified enzymes, from nine organisms from all three life domains. The developed path was five times more efficient than the in vivo rates of the natural CBB pathway. In the future, in vivo assays of this artificial pathway or the development of new routes using synthetic biology approaches may improve biological CO_2 fixation [54].

4. Key barriers to converting biomass into bioenergy

Developing biorefineries to manufacture chemicals and bioenergy is an exciting strategy to reduce dependence on nonrenewable resources. Compared with traditional petrochemical-based refineries, however, biorefinery processes are more complex and associated with higher technical and market risks. For large-scale feasibility, the singularities of these processes must be understood to reduce their complexity and minimize technical—economic risks [6]. Biorefinery research and industries have been searching for strategies to achieve economic competitiveness, and for this, bioconversion barriers must be explored.

Efficient use of biomass with minimal generation of liquid and solid wastes, reduced energy consumption, and flexible coproduction of products based on market demand are significant challenges for the residual biomass-based biorefinery industries [2,6]. The selection of appropriate feedstock, seasonality of biomass characteristics, feasible process design to achieve the most efficient route integrating the operations unit, and technological aspects, are some of the strategies necessary for possible implementation [55]. Despite their many challenges, biorefineries are the key to transitioning from the traditional economic model to a circular economy with less environmental impact.

Overcoming the challenge of operating biorefineries with the efficient use of residual biomass is associated with several factors, with biomass characteristics being one of the most important. The definition of the products and the capital value in the market will define the economic viability of the biorefinery. Moreover, the characterization of the residual biomass components and the operations unit is critical for the industrial system definition. Moreover, logistics systems for biomass transportation to the bioconversion plant can increase bioproduct production costs and affect the entire technical and economic process chain [55]. Detailed analysis of the whole integration process, from feedstock supply to market expansion of the recovered bioproducts, is crucial for the viability of residual biomass biorefineries.

To obtain multiple products from the same feedstock, a fractionation process to separate the structures of various technological bioconversion routes is essential. When there is a focus on bioenergy biorefineries, polysaccharide structures in residual biomass are converted to monomeric sugars through several steps: obtaining the feedstock; homogenization milling; pretreatment fractioning the polymers that compose the biomass; saccharification; fermentation; and finally, product recovery. This process changes depending on the other products obtained and the biomass composition. An example is in biomasses rich in pectin, where pectin and essential oil extraction can occur before the LCB fractionation process. In this regard, using a technological route to obtain chemicals will consequently change the arrangement of an operations unit. Indeed, the energy and equipment optimization of the process poses a great challenge.

The heterogeneity of the structural components of residual biomass results in biomass recalcitrance and resistance to fractionation. In an integrated system, biomass fractionation implies higher capital costs due to pretreatment and distillation equipment installations, which demand increased economic investment in acquisition, maintenance, and energy. For a switchgrass biorefinery producing furfural, lignin, and ethanol, the structure of the biomass fractionation equipment corresponded to 17.1% of the capital cost [6].

When defining the most appropriate fractionation method, the structural characteristics of the biomass must be considered, and the properties involved in each technique must be evaluated. The choice of a fractionation method comes with challenges. A predominant concern is that fractionation methods based on chemicals must incorporate downstream operating units such as separation, recovery, or neutralization. The processes must be performed in a corrosion-resistant reactor, significantly increasing the fractionation cost [56]. Energy consumption must be highlighted for methods requiring multiple conditions over long periods and generating liquid and solid effluents. Fractionations often generate toxic compounds, limiting the practical product applications of biomass bioconversion [57]. The technical combination, pretreatment process development, and reuse of biological agents (enzymes or microorganisms) and chemical products can be promising alternatives to reduce energy and operational costs [58].

The low productivity of microalgae biomass can generate a high specific energy input for the biomass pretreatment process, leading to an inefficient energy ratio. When associated with microalgal biomass, the seasonal variation in microalgae productivity becomes a significant barrier to economic biorefinery production, with winter productivity that may be one-fifth to one-third of its summer level [59]. Thus, microalgae growth must be improved with advances and greater efficiency in biomass thickening and dewatering [58].

Biorefinery progress will only be possible with integrated technologies and a holistic vision of biomass fractionation to ensure optimum structural use and unit operations integration [2,60]. Despite recent advances in the multiproduct biorefinery concept and holistic biomass use, most studies on biomass fractionation have focused on obtaining a unique product, usually derived from carbohydrates within the structure. This approach, which ignores coproducts that can be obtained in parallel, causes a loss in economic viability, with processes becoming disadvantageous [5].

Effective use of residual biomass after extraction of the target compound is essential to improve the residual biomass biorefinery process economics. For this purpose, all biomolecules (carbohydrates, fatty acids, proteins, pigments, peptides, and minerals) present in residual biomass should be fractionated and converted into various bioproducts [33]. Other value-added products can also be developed, including animal feed and food additives. It is worth noting that valuable biomaterials, such as nanoparticles, can be produced from microalgae and other biomasses (lignocellulosic; rich in pectin and starch) using enzymatic, catalytic, and thermochemical processes [35]. Replacing toxic solvents with bio-based solvents, supercritical fluids, ionic liquids, or a mixture of organic salts (e.g., quaternary ammonium salt and phosphonium salt) is a green alternative for extracting compounds from biomasses [61].

Efforts to fill the gaps in the viability of multiproduct biorefineries have been proposed worldwide. Biomass utilization has been described as a fundamental pillar of a biobased economy. However, considering only the application of residual biomass as sufficient for sustainable processes is an unstable path because biorefineries are complex systems, often dependent on chemical reagents, large volumes of water, electricity, and side flows that, if not accounted for, could cause negative impacts. Therefore, residual biomass as a feedstock is necessary but insufficient to guarantee that the process will be economically and environmentally sustainable. Other aspects, such as natural resources, generated effluents, and market expansion of recovered products, should be considered for the successful performance of the biorefinery [4].

5. Final considerations

Residual biomasses have focused on integrated processes to obtain multiple products to minimize environmental impacts and as a strategy for transition from the traditional economy to a circular economy. The use of residual biomass as a feedstock is due to the range of products that can be obtained from the heterogeneous structure that composes it, this being a feature of great value in two contexts: first, by enabling the recovery of different structures for conversion into a range of products, and second, challenging industrial design by the complexity of developing integrated and flexible unit operations to maintain efficiency and profitability in the face of the heterogeneity of these structures.

The essential part (and at the same time, the great challenge) of these systems is the fractionation processes of the residual biomasses that aim at the whole deconstruction of the bonds between the biomass components minimizing the loss of carbohydrates or other compounds of interest. The efficient fractionation of biomass components defines the quality and efficiency of downstream bioprocesses and is crucial for the technical and economic feasibility of the processes. Due to advances in the integrated biorefinery concept, biomass fractionation processes have been evaluated in sequential systems to achieve different feedstock structures because of the higher yield. Reducing the consumption of chemical reagents, recycling effluents, evaluating potable water use, and optimizing operational conditions are challenges to overcome.

Research advances have demonstrated a variety of possible technological routes for the most diverse residual biomasses, with conversion efficiencies for a range of products. However, integration processes are still scarce or have low efficiency, demonstrating a gap between academic research and large-scale development. Despite this, it is clear that substantial effort has been made to move toward more sustainable processes for obtaining bioproducts through integrated methodologies, the minimization of residue and effluent generation, and the optimization of economic costs, which will guide advances in biorefinery implementation in the present and near future.

References

[1] WBA, WBA Global Bioenergy Statistic 2021, 2021. https://www.worldbioenergy.org/uploads/211214 WBA GBS 2021.pdf. Accessed 4 April 2022.

[2] J.R. Banu, S.P. Kavitha, V.K. Tyagi, et al., Lignocellulosic biomass based biorefinery: a successful platform towards circular bioeconomy, Fuel 302 (2021) 121086, https://doi.org/10.1016/j.fuel.2021.121086.

[3] A. Wagle, M.J. Angove, A. Mahara, et al., Multi-stage pretreatment of lignocellulosic biomass for multiproduct biorefinery: a review, Sustain. Energy Technol. Assess. 49 (2022) 101702, https://doi.org/10.1016/j.seta.2021.101702.

[4] J.C. Solarte-Toro, C.A.C. Alzate, Biorefineries as the base for accomplishing the sustainable development goals (SDGs) and the transition to bioeconomy: technical aspects, challenges and perspectives, Bioresour. Technol. 340 (2021) 125626, https://doi.org/10.1016/j.biortech.2021.125626.

[5] E. Scopel, C.A. Rezende, Biorefinery on-demand: modulating pretreatments to recover lignin, hemicellulose, and extractives as co-products during ethanol production, Ind. Crops Prod. 163 (2021) 113336, https://doi.org/10.1016/j.indcrop.2021.113336.

[6] G. Zang, A. Shah, C. Wan, Techno-economic analysis of an integrated biorefinery strategy based on one-pot biomass fractionation and furfural production, J. Clean. Prod. 260 (2020) 120837, https://doi.org/10.1016/j.jclepro.2020.120837.

[7] P. Shah Maulin, Microbial Bioremediation & Biodegradation, Springer, 2020.

[8] D. Tian, F. Shen, J. Hu, et al., Complete conversion of lignocellulosic biomass into three high-value nanomaterials through a versatile integrated technical platform, Chem. Eng. J. 428 (2022) 131373, https://doi.org/10.1016/j.cej.2021.131373.

[9] P. Unrean, N. Ketsub, Integrated lignocellulosic bioprocess for coproduction of ethanol and xylitol from sugarcane bagasse, Ind. Crop. Prod. 123 (2018) 238–246, https://doi.org/10.1016/j.indcrop.2018.06.071.

[10] A. Patel, K. Hrůzová, U. Rova, et al., Sustainable biorefinery concept for biofuel production through holistic volarization of food waste, Bioresour. Technol. 294 (2019) 122247, https://doi.org/10.1016/j.biortech.2019.122247.

[11] A. Soltaninejad, M. Jazini, K. Karimi, Sustainable bioconversion of potato peel wastes into ethanol and biogas using organosolv pretreatment, Chemosphere 291 (2022) 133003, https://doi.org/10.1016/j.chemosphere.2021.133003.

[12] F. Saadatinavaz, K. Karimi, J.F.M. Denayer, Hydrothermal pretreatment: an efficient process for improvement of biobutanol, biohydrogen, and biogas production from orange waste via a biorefinery approach, Bioresour. Technol. 341 (2021) 125834, https://doi.org/10.1016/j.biortech.2021.125834.

[13] D.Z. Cypriano, L.L. da Silva, L. Tasic, High value-added products from the orange juice industry waste, Waste Manag. 79 (2018) 71–78, https://doi.org/10.1016/j.wasman.2018.07.028.

[14] L.M. Chng, D.J.C. Chan, K.T. Lee, Sustainable production of bioethanol using lipid-extracted biomass from *Scenedesmus dimorphus*, J. Clean. Prod. 130 (2016) 68–73, https://doi.org/10.1016/j.jclepro.2016.02.016.

[15] E.J. Kim, S. Kim, H.-G. Choi, S.J. Han, Coproduction of biodiesel and bioethanol using psychrophilic microalga *Chlamydomonas* sp. KNM0029C isolated from Arctic sea ice, Biotechnol. Biofuel. 13 (2020) 20, https://doi.org/10.1186/s13068-020-1660-z.

[16] D. Pradhan, A.K. Jaiswal, S. Jaiswal, Emerging technologies for the production of nanocellulose from lignocellulosic biomass, Carbohydr. Polym. 285 (2022) 119258, https://doi.org/10.1016/j.carbpol.2022.119258.

[17] P. Shah Maulin, Removal of Refractory Pollutants From Wastewater Treatment Plants, CRC Press, 2021.

[18] Y. Zhao, U. Shakeel, M. Saif Ur Rehman, et al., Lignin-carbohydrate complexes (LCCs) and its role in biorefinery, J. Clean. Prod. 253 (2020) 120076, https://doi.org/10.1016/j.jclepro.2020.120076.

[19] T.M. Le, U.P. Tran, Y.H. Duong, et al., Development of a paddy-based biorefinery approach toward improvement of biomass utilization for more bioproducts, Chemosphere 289 (2022) 133249, https://doi.org/10.1016/j.chemosphere.2021.133249.

[20] S. Marques, A.D. Moreno, M. Ballesteros, F. Gírio, Starch biomass for biofuels, biomaterials, and chemicals, in: S. Vaz Jr. (Ed.), Biomass and Green Chemistry, first ed., Springer International Publishing, Cham, 2018, pp. 69–94.

[21] Market Data Forest, Europe Starch Market, 2022. https://www.marketdataforecast.com/market-reports/europe-modified-starch-market. Accessed 4 Apr 2022.

[22] C. Bonatto, A.F. Camargo, T. Scapini, et al., Biomass to bioenergy research: current and future trends for biofuels, in: V. Gupta, H. Treichel, R. Kuhad, S. Rodriguez-Couto (Eds.), Recent Developments in Bioenergy Research, first ed., Elsevier, 2020, pp. 1–17.

[23] V. Kumar, P. Longhurst, Recycling of food waste into chemical building blocks, Curr. Opin. Green Sustain. Chem. 13 (2018) 118–122, https://doi.org/10.1016/j.cogsc.2018.05.012.

[24] N.E. Poe, D. Yu, Q. Jin, et al., Compositional variability of food wastes and its effects on acetone-butanol-ethanol fermentation, Waste Manag. 107 (2020) 150–158, https://doi.org/10.1016/j.wasman.2020.03.035.

[25] S. Chaturvedi, A. Bhattacharya, L. Nain, et al., Valorization of agro-starchy wastes as substrates for oleaginous microbes, Biomass Bioenergy 127 (2019) 105294, https://doi.org/10.1016/j.biombioe.2019.105294.

[26] L.C. Martins, C.C. Monteiro, P.M. Semedo, I. Sá-Correia, Valorisation of pectin-rich agro-industrial residues by yeasts: potential and challenges, Appl. Microbiol. Biotechnol. 104 (2020) 6527–6547, https://doi.org/10.1007/s00253-020-10697-7.

[27] C. Sabater, M. Villamiel, A. Montilla, Integral use of pectin-rich by-products in a biorefinery context: a holistic approach, Food Hydrocoll. 128 (2022) 107564, https://doi.org/10.1016/j.foodhyd.2022.107564.

[28] T.T. Manhongo, A.F.A. Chimphango, P. Thornley, M. Röder, Current status and opportunities for fruit processing waste biorefineries, Renew. Sustain. Energy Rev. 155 (2022) 111823, https://doi.org/10.1016/j.rser.2021.111823.

[29] S. Talekar, A.F. Patti, R. Vijayraghavan, A. Arora, An integrated green biorefinery approach towards simultaneous recovery of pectin and polyphenols coupled with bioethanol production from waste pomegranate peels, Bioresour. Technol. 266 (2018) 322–334, https://doi.org/10.1016/j.biortech.2018.06.072.

[30] F. Fazzino, F. Mauriello, E. Paone, et al., Integral valorization of orange peel waste through optimized ensiling: lactic acid and bioethanol production, Chemosphere 271 (2021) 129602, https://doi.org/10.1016/j.chemosphere.2021.129602.

[31] B.O. Abo, E.A. Odey, M. Bakayoko, L. Kalakodio, Microalgae to biofuels production: a review on cultivation, application and renewable energy, Rev. Environ. Health 34 (2019) 91–99, https://doi.org/10.1515/reveh-2018-0052.

[32] D. Nagarajan, A. Kusmayadi, H.-W. Yen, et al., Current advances in biological swine wastewater treatment using microalgae-based processes, Bioresour. Technol. 289 (2019) 121718, https://doi.org/10.1016/j.biortech.2019.121718.

[33] M. Moon, W.-K. Park, S.Y. Lee, et al., Utilization of whole microalgal biomass for advanced biofuel and biorefinery applications, Renew. Sustain. Energy Rev. 160 (2022) 112269, https://doi.org/10.1016/j.rser.2022.112269.

[34] G. Srivastava, V. Kumar, R. Tiwari, et al., Anaerobic co-digestion of defatted microalgae residue and rice straw as an emerging trend for waste utilization and sustainable biorefinery development, Biomass Convers. Biorefin. 12 (2022) 1193–1202, https://doi.org/10.1007/s13399-020-00736-8.

[35] R. Maurya, C. Paliwal, T. Ghosh, et al., Applications of de-oiled microalgal biomass towards development of sustainable biorefinery, Bioresour. Technol. 214 (2016) 787–796, https://doi.org/10.1016/j.biortech.2016.04.115.

[36] S.K. Mandotra, C. Sharma, N. Srivastava, et al., Current prospects and future developments in algal bio-hydrogen production: a review, Biomass Convers. Biorefin. (2021), https://doi.org/10.1007/s13399-021-01414-z.

[37] O. Phanduang, S. Lunprom, A. Salakkam, et al., Improvement in energy recovery from *Chlorella* sp. biomass by integrated dark-photo biohydrogen production and dark fermentation-anaerobic digestion processes, Int. J. Hydrog. Energy 44 (2019) 23899–23911, https://doi.org/10.1016/j.ijhydene.2019.07.103.

[38] N.A.H. Fetyan, A.E.-K.B. El-Sayed, F.M. Ibrahim, et al., Bioethanol production from defatted biomass of *Nannochloropsis oculata* microalgae grown under mixotrophic conditions, Environ. Sci. Pollut. Res. 29 (2022) 2588−2597, https://doi.org/10.1007/s11356-021-15758-6.

[39] K. Dhandayuthapani, V. Sarumathi, P. Selvakumar, et al., Study on the ethanol production from hydrolysate derived by ultrasonic pretreated defatted biomass of *Chlorella sorokiniana* NITTS3, Chem. Data Collect. 31 (2021) 100641, https://doi.org/10.1016/j.cdc.2020.100641.

[40] B.H.H. Goh, H.C. Ong, M.Y. Cheah, et al., Sustainability of direct biodiesel synthesis from microalgae biomass: a critical review, Renew. Sustain. Energy Rev. 107 (2019) 59−74, https://doi.org/10.1016/j.rser.2019.02.012.

[41] B. Kim, J. Park, J. Son, J.W. Lee, Catalyst-free production of alkyl esters from microalgae via combined wet in situ transesterification and hydrothermal liquefaction (iTHL), Bioresour. Technol. 244 (2017) 423−432, https://doi.org/10.1016/j.biortech.2017.07.129.

[42] Z. Huang, J. Zhang, M. Pan, et al., Valorisation of microalgae residues after lipid extraction: pyrolysis characteristics for biofuel production, Biochem. Eng. J. 179 (2022) 108330, https://doi.org/10.1016/j.bej.2021.108330.

[43] R. Velmurugan, A. Incharoensakdi, Metabolic transformation of cyanobacteria for biofuel production, Chemosphere 299 (2022) 134342, https://doi.org/10.1016/j.chemosphere.2022.134342.

[44] I. Andersson, A. Backlund, Structure and function of rubisco, Plant Physiol. Biochem. 46 (2008) 275−291, https://doi.org/10.1016/j.plaphy.2008.01.001.

[45] N. Antonovsky, S. Gleizer, R. Milo, Engineering carbon fixation in *E. coli*: from heterologous RuBisCO expression to the Calvin−Benson−Bassham cycle, Curr. Opin. Biotechnol. 47 (2017) 83−91, https://doi.org/10.1016/j.copbio.2017.06.006.

[46] H. Alper, J. Moxley, E. Nevoigt, et al., Engineering yeast transcription machinery for improved ethanol tolerance and production, Science 314 (2006) 1565−1568, https://doi.org/10.1126/science.1131969.

[47] B. Liang, Y. Zhao, J. Yang, Recent advances in developing artificial autotrophic microorganism for reinforcing CO_2 fixation, Front. Microbiol. 11 (2020), https://doi.org/10.3389/fmicb.2020.592631.

[48] X. Pan, J.N. Saddler, Effect of replacing polyol by organosolv and kraft lignin on the property and structure of rigid polyurethane foam, Biotechnol. Biofuel. 6 (2013) 12, https://doi.org/10.1186/1754-6834-6-12.

[49] V. Guadalupe-Medina, H.W. Wisselink, M.A. Luttik, et al., Carbon dioxide fixation by Calvin-Cycle enzymes improves ethanol yield in yeast, Biotechnol. Biofuel. 6 (2013) 125, https://doi.org/10.1186/1754-6834-6-125.

[50] I. Papapetridis, M. Goudriaan, M. Vázquez Vitali, et al., Optimizing anaerobic growth rate and fermentation kinetics in *Saccharomyces cerevisiae* strains expressing Calvin-cycle enzymes for improved ethanol yield, Biotechnol. Biofuel. 11 (2018) 17, https://doi.org/10.1186/s13068-017-1001-z.

[51] P.-F. Xia, G.-C. Zhang, B. Walker, et al., Recycling carbon dioxide during xylose fermentation by engineered *Saccharomyces cerevisiae*, ACS Synth. Biol. 6 (2017) 276−283, https://doi.org/10.1021/acssynbio.6b00167.

[52] D.M. Ha-Tran, R.-Y. Lai, T.T.M. Nguyen, et al., Construction of engineered RuBisCO *Kluyveromyces marxianus* for a dual microbial bioethanol production system, PLoS One 16 (2021) e0247135, https://doi.org/10.1371/journal.pone.0247135.

[53] T. Gassler, M. Sauer, B. Gasser, et al., The industrial yeast *Pichia pastoris* is converted from a heterotroph into an autotroph capable of growth on CO_2, Nat. Biotechnol. 38 (2020) 210−216, https://doi.org/10.1038/s41587-019-0363-0.

[54] T. Schwander, L. Schada von Borzyskowski, S. Burgener, et al., A synthetic pathway for the fixation of carbon dioxide in vitro, Science 354 (80) (2016) 900−904, https://doi.org/10.1126/science.aah5237.

[55] R. Reshmy, E. Philip, A. Madhavan, et al., Lignocellulose in future biorefineries: strategies for cost-effective production of biomaterials and bioenergy, Bioresour. Technol. 344 (2022) 126241, https://doi.org/10.1016/j.biortech.2021.126241.

[56] H.A. Ruiz, M. Galbe, G. Garrote, et al., Severity factor kinetic model as a strategic parameter of hydrothermal processing (steam explosion and liquid hot water) for biomass fractionation under biorefinery concept, Bioresour. Technol. 342 (2021) 125961, https://doi.org/10.1016/j.biortech.2021.125961.

[57] M. Zhou, O.A. Fakayode, A.E. Ahmed Yagoub, et al., Lignin fractionation from lignocellulosic biomass using deep eutectic solvents and its valorization, Renew. Sustain. Energy Rev. 156 (2022) 111986, https://doi.org/10.1016/j.rser.2021.111986.

[58] A. Salakkam, S. Sittijunda, C. Mamimin, et al., Valorization of microalgal biomass for biohydrogen generation: a review, Bioresour. Technol. 322 (2021) 124533, https://doi.org/10.1016/j.biortech.2020.124533.

[59] B.D. Wahlen, L.M. Wendt, A. Murphy, et al., Preservation of microalgae, lignocellulosic biomass blends by ensiling to enable consistent year-round feedstock supply for thermochemical conversion to biofuels, Front. Bioeng. Biotechnol. 8 (2020), https://doi.org/10.3389/fbioe.2020.00316.

[60] H.A. Ruiz, M. Conrad, S.-N. Sun, et al., Engineering aspects of hydrothermal pretreatment: from batch to continuous operation, scale-up and pilot reactor under biorefinery concept, Bioresour. Technol. 299 (2020) 122685, https://doi.org/10.1016/j.biortech.2019.122685.

[61] G. Muhammad, M.A. Alam, M. Mofijur, et al., Modern developmental aspects in the field of economical harvesting and biodiesel production from microalgae biomass, Renew. Sustain. Energy Rev. 135 (2021) 110209, https://doi.org/10.1016/j.rser.2020.110209.

Chapter 25

Endemic microalgae biomass for biorefinery concept and valorization

Samanta Machado-Cepeda, Rosa M. Rodríguez-Jasso and Héctor A. Ruiz

Biorefinery Group, Food Research Department, School of Chemistry, Autonomous University of Coahuila, Saltillo, Coahuila, Mexico

1. Introduction

Recognizing the breakthroughs and patterns left by first-generation biorefineries, where plants were cultivated to produce biodiesels, such as palm and soybean, and bioethanols, such as corn and sugarcane, a necessary comment is that currently, competition with food crops for land and nutrients (nitrogen and phosphorous) raises serious sustainability concerns about food vs. fuel, as well as the deforestation of native woods for those crops, which negative affects the ecosystem. Additionally, the process generates substantial solid, liquid, and gaseous wastes without applying the concept of a double purpose.

In this sense, it is important to find new options for the energy industry that incorporate the concept of a circular bioeconomy. With the arrival of the second-generation biorefineries, or so-called lignocellulosic materials biorefineries, the agriculture residues, e.g., corn stoves, corn fiber, and wheat straw, were processed with various treatments, such as hydrolysis, saccharification (enzymatic hydrolysis), and other hydrothermal processes, to make the cellulosic material an optimal substrate for biofuels fermentation and production. The third-generation biorefineries, based on aquatic biomass (micro–macro algae) [1], are the subject of this chapter. Fig. 25.1 shows a graphical summary of the history of biorefineries described above.

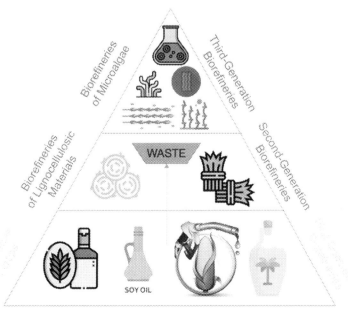

FIGURE 25.1 Development of biorefinery biomass.

Microalgae are single-cell photosynthetic eukaryote organisms that utilize solar light energy, inorganic nutrients, and CO_2 to generate biomass [2] for use as a feedstock for foods and other value-added products. To date, more than 15,000 novel compounds derived from microalgae have been chemically characterized (between them lipids and carbohydrates) and used for the biological generation of raw materials for the energy industry and food and cosmetic products [2–4]. They vary in size from nanometers to millimeters and exist as independent organisms or in chains or groups suspended in water. They include green algae, red algae, and diatoms [4].

Furthermore, there is an advantage to this approach because microalgae have simplicity in their nutritional requirements [5], having the alternative of using several sources of wastewater, brackish saline water, and other low-cost culture media, such as lignocellulosic hydrolysates from residues of first- and second-generation biorefineries. In addition, they can tolerate a range of abiotic conditions, such as temperature, salinity, and pH, and can consume organic carbon for microalgal culture [2–4,6].

Due to the versatility of microalgae, as described above, their high-value compounds with relatively low-cost production costs, and their easy cultivation alternatives, such as using wastewater with higher growth rates, their biomass has experienced important economic growth [2–4,6,7]. This is a very important argument in the era of the circular bioeconomy, closing the cycle of waste from other agro-industrial processes. For example, residues from first-generation biorefineries, such as those of vinasse, have substantial amounts of nitrogen and phosphorus, but their high turbidity limits nutrient availability. However, this factor can be reduced if it is mixed with carbon dioxide, taking advantage of their components from both subproducts of the ethanol from sugarcane to provide important support to microalgae production [8].

Once again, highlighting the importance of third-generation biorefineries, bioethanol and biodiesel have the highest biofuel demand worldwide and still have oil crops as their main raw materials [9]. Conversely, biodiesel is based on triacylglycerols (TAGs), which represent mostly a concentration of lipids that can be obtained for microalgae, making them an attractive source for biooil production [10]. The highest biodiesel production efficiency is based on land use (12,000–98,500 L ha^{-1} year), up to 220 times the productivity of oil crops (soybean: 446; sunflower: 952; rapeseed: 1190; jatropha: 1892; and oil palm: 5950) [1]. Additionally, the low water consumption required to produce microalgae biomass is an important factor because its production only needs about 0.75 L of water per kilogram of biomass [11]. Therefore, this alternative could help save water for use in crop irrigation.

Consequently, it is important to find new species of microalgae that can contribute to this dynamic of circular bioeconomy. For that reason, the concept of endemic microalgae is becoming increasingly relevant so that each geographic region in nations with microalgae availability can take advantage of this valuable resource and explore the potential of microalgae for its region. The applications have already been mentioned, but it is imperative to note that to be environmentally responsible, extracted microalgae must be subsequently cultivated at the laboratory level, which means no alteration of the natural environment, guaranteeing sustainable development of the region and satisfying the needs of the present without compromising the resources of future generations.

The production of bioethanol from microalgae is carried out in three stages: first, recovery of the fermentable fraction present in the cell in the form of starch and other carbohydrates or sugars; second, hydrolysis of carbohydrates by enzymatic or chemical treatments; and third, fermentation of carbohydrates using microorganisms such as yeasts and bacteria to produce bioethanol [12]. This process provides the potential to optimize the microalgae biomass as an important source of compounds such as lipids and carbohydrates for use in the energy industry. This chapter analyzes the fractionation of microalgal biomass to extract these valuable compounds while considering the identification and cultivation of microalgae.

2. Methodologies to extract and isolate microalgae

Most microalgae extractions are aleatory [13] in places with visible populations, where they can be realized at the surface or at a depth of the water body [3]. However, some microalgae populations are not suspended in the water, such as diatoms found in sludge, which can be identified as brown foam on its surface. These are collected by taking a sample of the sludge. After a few hours, the sludge settles, and the diatoms come to the surface. Then, with a pipette or spoon, the supernatant containing the microalgae is collected.

It is important to identify the single place of the extraction. For this, it is necessary to write the name of the body of water and its coordinates. When several samplings are made at the same point, the cardinal points of each should be indicated; for example, whether it was in the north, south, east, west, northeast, southwest, etc. If possible, it is always better to make a map for better visualization, either a sketch of the water body or a freehand drawing, as shown in Fig. 25.2, where the red dots (gray in print version) represent the exact sampling location in the visible microalgae populations.

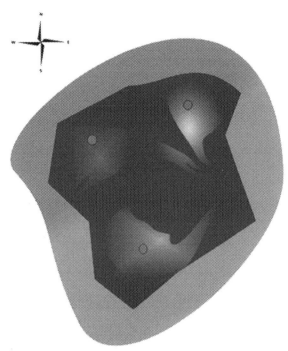

FIGURE 25.2 Mapping of sampling area.

For each sampling and before storage, is necessary to ensure that the samples have a microalgae population. To do this, the sample must first be filtered to preserve only the microalgae that may exist and avoid the passage of other objects or organisms that also inhabit the area, such as crustaceans, since they may consume the collected microalgae or cause their deterioration. Then, with a visualization instrument such as microscopy, microalgae are sought according to morphology, as discussed in the next section.

Field extraction work can be complicated because sterility must be maintained in the process to not contaminate the samples while they are taken to the laboratory for isolation. Transportation must be realized in refrigerated and dark storage. For isolation, it is essential to sterilize all materials and instruments, not only to quickly achieve axenic cultures but also to avoid the contamination and subsequent death of the collected microalgae.

To isolate single microalgal species from field water samples, studies indicate that the samples must be diluted several times in a standard medium to make them grow under parameters because several physicochemical factors affect microalgae growth, such as light intensity, photoperiod, temperature, and nutrient concentrations in the medium, such as carbon (glucose or glycerol), phosphates, and nitrogen (ammonium or nitrate) [14], that must be controlled to achieve an axenic culture. For more, see Fig. 25.3.

As mentioned, the production of microalgae is relatively easy. They only demand that the medium has the necessary nutrients, independently of which, for example, they can grow in wastewater. But in isolation, it is important to ensure fast-growing colonies according to the preferences of the different species of microalgae in the media. Table 25.1 shows various methods of growing microalgae with standard media according to the samples taken and the culture dilutions.

It is always important to know the physicochemical properties of the water sources from which the microalgae are isolated, either by literature or by sampling, analyzing at least its pH, dissolved oxygen, salinity, conductivity, total suspended solids, temperature, and nitrate content, which is determined by the cadmium reduction method [Method 8171] [16,18]. This last parameter is quite relevant since it facilitates the selection of the best growth medium for the microalgae and favors isolation and preservation.

However, to produce high-value components, replicating the growth conditions of microalgae in their habitat is not something that should be done strictly. The growth conditions should be optimized; this subject is discussed in Section 4.

It is important to note that the nutrient media must be prepared with distilled water because water with high conductivity, i.e., due to high levels of minerals, the medium's ingredients are precipitated and do not ensure correct mixing. On the other hand, any wastewater to be used must be filtered to remove unwanted particles and, together with the medium and the recipients selected for the process, must be sterilized in an autoclave at 121°C for 15 min. Later, they must be allowed to cool to room temperature before inoculation. In addition to the above, to achieve a microalgal cell culture free of

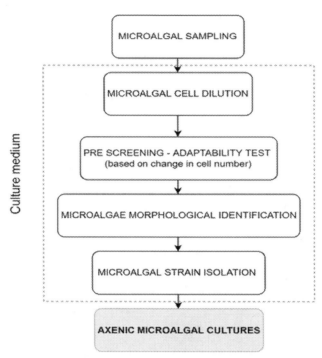

FIGURE 25.3 Isolation of microalgae.

bacteria, fungi, and other microorganisms, the cultures could be supplemented with ampicillin (100 μg mL^{-1}) and chloramphenicol (50 μg mL^{-1}) [5].

After visible mixed growth on the liquid media of the microalgae, they must be spread evenly onto the surface of the agar plate using the same liquid medium but with 1% of agar to solidify. The most common medium is BG11 agar [5] or bacteriology agar [15] under the same conditions of temperature and light previously mentioned, as well as the same time to see growth. Through microscopic observation, we can determine whether there is an axenic culture. If plate growth is uniform and different microalgae colonies are observed, the microalgae can streaked and then inoculated in a liquid medium to accelerate isolation under sterile conditions. The above procedure must be repeated until the cellular morphology of an axenic culture can be identified in the plate by microscopy.

The isolated mother cultures should be stored at 4°C in an agar plate for 1 month. After this period, replating is necessary [3,5]. Another way to conserve microalgae is by using a cryoprotectant solution of sterilized skim milk powder at 20% (w/v) [19]. That approach requires the deposit of 500 μL broth culture medium and 500 μL^1 skim milk powder into 1.5 mL^1 vials. The vials are frozen at −35°C for 24 h and later freeze-dried at −80°C in a condenser at a vacuum pressure of 0.030 mbar for 24 h.

3. Review of morphological microalgae identification

This general classification method is used to distinguish colonies for isolation. It is based on the morphology of the individual cells by microscopic examination [3]. As mentioned in a previous chapter, it is also important to identify the microalgae population in the first step of the process, which is extraction the easiest way to do this is through visualization according to morphology. For example, Fig. 25.4 shows microscopic visualization photographs of microalgae to provide a guide for identifying them according to their different forms, colors, aggrupations, etc. The underlined species are well known to accumulate significant amounts of lipids; specifically, up to 31% of the weight of the *Scenedesmus* species can be made up of lipids [5,20].

However, for strict classification in the literature, dichotomous keys allow for the identification of organisms at the level of species, genus, family, or any other taxonomic category. These keys are based on whether microalgae have a particular morphological character; the process is repeated until the organism in question has been identified. In addition, if identification is not achieved according to microalgal morphology, it is necessary to resort to molecular identification based on the extraction and amplification of multiple DNA and ARN fragments to obtain their sequences and compare them with existing database sequences. For *Chlorella* sp. and *Scenedesmus* sp., this process has worked with the 18S part of ribosomal DNA and RNA [5,17].

TABLE 25.1 Liquid media references for growing microalgae species.

Medium[a]	Conditions[b]	Growing time	Microalgal isolates	References
40 mL of F/2 Gaillard's medium, Jaworski's medium, Stosch's medium, bold basal medium, or blue-Green Medium (BG-11) for 1 mL sample	20–25°C Constant aeration 24 µE m^{-1} s of illumination	14 days	Coelastrum Chlorella Scenedesmus Dictyochloropsis Cosmarium Cladophora Ulothrix Oscillatoria Phormidium	[3]
150 mL of bold basal medium, blue-Green Medium (BG-11), or F/2 Gaillard's medium for 150 mL sample	25 ± 3°C 14 W of white fluorescent light/TL5 8:16 h light-dark cycle	14 days	Chlorella Microcystis Scenedesmus Oscillatoria	[15]
10 mL sample with 100 mL of medium (BG -11)	28 ± 2°C Illuminated by cool white fluorescent light 80 µmol photons m^{-2} s^{-1}	12 days	Desmodesmus sp. Chlorella sp. Scenedesmus sp. Monoraphidium	[5]
Plate incubation with conway media and i/2 of samples from marine water. Bold basal media and WC media from freshwater	25°C 300–500 µmol photons m^{-2} s^{-1}, with 12 h light:12 h dark	15 days	Chlorella Nannochloropsis, Desmodesmus	[16]
Agar plate streaking with F/2 medium for isolation To cultivation: 50 mL of inoculated microalgal cultures with 450 mL medium	Orbital shaking at 100 rpm 25 ± 2°C 16 h:8 h light:dark cycle, 16 h:8 h from white LED lights), at 150 µmol photons m^{-2} s^{-1}	15 days	Scenedesmus dimorphus	[17]

[a] Does not work with all recipient capacities; at least 10% of them should be freed.
[b] It should be mentioned that aeration is for mixing effects, because in all cases, agitation is necessary.

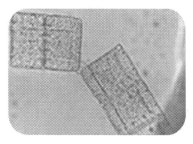
Diathomea Terpsinoe Musica

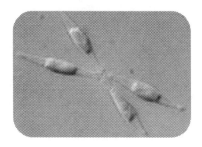
Diathomea Phaeodactylum Tricornutum

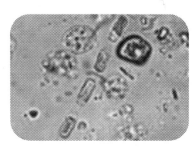
Diathomea Skeltomema Potamos

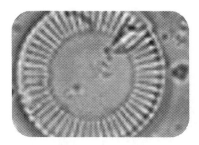
Diathomea Cyclotella Meneghiniana

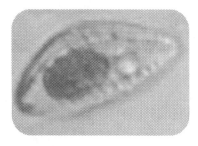
Diathomea Cymbella sp.

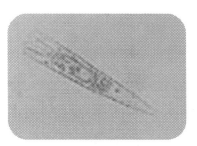
Diathomea Licmophora sp.

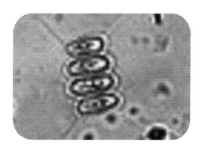
<u>Scenedesmus Quadricauda</u>

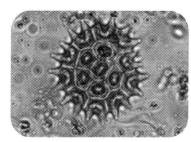
Pediastrum Borynun

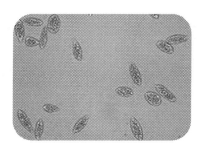
Euglena Gracilis

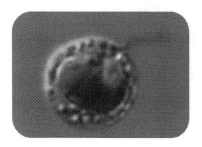
Ochromona

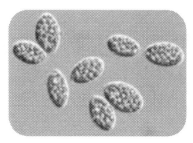

Polytoma

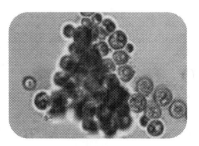
<u>Chlorella</u>

FIGURE 25.4 Some microalgae morphology.

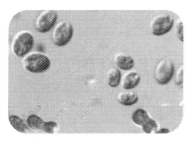 *Chlamydomonas Reinhardii*

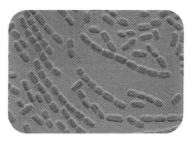 *Stichococcus*

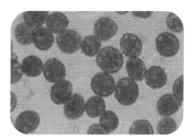 *Dictyochloropsis*

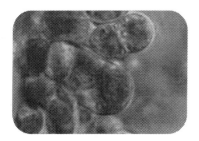 *Leptosira*

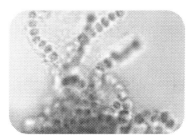 *Coccobotrys*

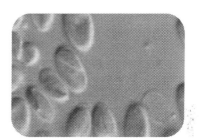 *Coccomyxa*

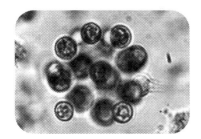 *Gloeocystis*

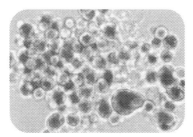 *Trebouxia*

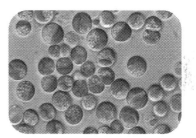 *Elliptochloris*

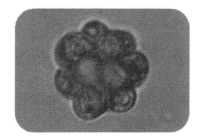 *Coenobium of Coelastrum sp.*

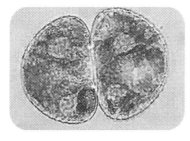 *Cosmarium*

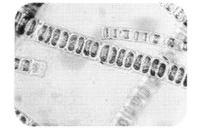 *Ulothrix*

FIGURE 25.4 cont'd

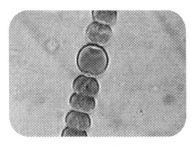
Anabaena cylindrica

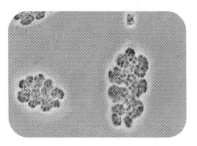
Dunaliella salina

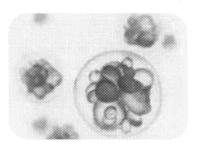
Porphyridium cruentum

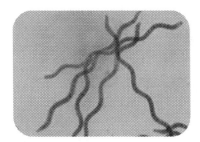
Spirulina maxima

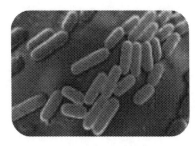
Synechococcus sp.

Haematococcus pluvialis

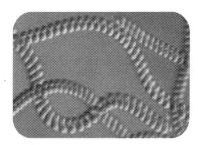
Arthrospira (Spirulina) platensis

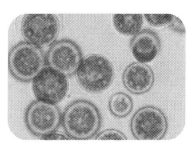
Nannochloropsis Sp.

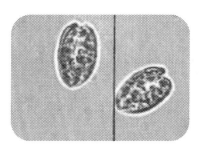
Tetraselmis

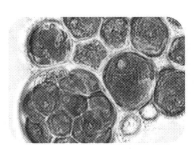
Chlorococcum

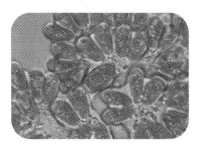
Botryococcus braunii

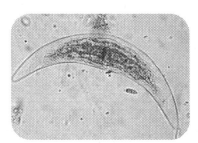
Monoraphidium

FIGURE 25.4 cont'd

4. Optimal culture media for laboratory-level photobioreactor biomass production, depending on microalgae species

First, it is important to know the microalgae metabolism. For that, Fig. 25.5 illustrates where microalgae take their nutrients and factors naturally or in cultivation to grow. It should be noted that chlorophyll formation is due to the reduction reaction of nitrate (NO_3^-) as a common source of nitrogen, essential for its growth, to nitrite ion (NO_2^-) and finally to ammonium by the enzyme *Nitrate Reductase* EC 1.7.99.4 inside a chloroplast of the microalgae cell [5]. Chlorophyll is common to all microalgae, so some authors propose a wavelength (λ) for all measures, such as 680 nm [3].

The yields of compounds obtained from microalgae species depend on inducing stress and adapting certain cultivation strategies [2,3]. Section 2 mentioned the methodologies for extracting and isolating microalgae and discussed the standard media for achieving microalgae growth, starting from samples extracted from its natural habitat. Scientific studies have looked at various culture media to obtain greater biomass yields, such as mixing the same standard methods and variations in illumination conditions depending on the high-value compounds of interest. This is explained posteriorly in Section 4. In any case, it is important to maintain aeration for correct mixing. On the other hand, the culture can be scaled to work with larger volumes than in isolation to achieve higher yields.

Additionally, microalgae can be cultivated in both open and closed systems [2,6]. The first option is cheaper because it does not require strict control of conditions related to nutrients, pH, temperature, aeration rate, CO_2 concentration, and light regime, as required in closed systems [6,21].

For this same reason, open systems like lakes and ponds have a lower biomass production yield, or at least not the same quality, due to significant changes in the ionic composition of the culture medium, temperature fluctuations caused by diurnal cycles and seasonal variations, potential CO_2 deficiencies derived from its diffusion into the atmosphere, and poor mixing by inefficient stirring mechanisms. These conditions may result in poor mass CO_2 transfer rates and light limitations due to top layer thickness causing low biomass productivity. If the optical path is less than 0.01 m, however, photosynthetic efficiency increases by up to 8%, thereby increasing the amount of biomass produced. On the other hand,

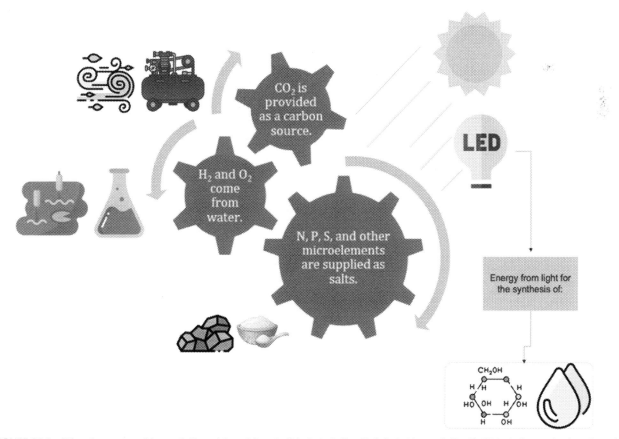

FIGURE 25.5 Microalgae autotrophic metabolism. *Adapted from I. Gifuni, A. Pollio, C. Safi, A. Marzocchella, G. Olivieri, Current bottlenecks and challenges of the microalgal biorefinery, Trends Biotechnol 37(3) (2019) 242–252. https://doi.org/10.1016/J.TIBTECH.2018.09.006.*

open systems have disadvantages such as pollution and contamination risks that negatively impact the extraction and quality of high-value compounds for use in industry [20].

Nevertheless, species such as *Scenedesmus* spp. and *Chlorella* sp. have good yields in open systems. This last one can achieve biomass productivity of 40 g per liter, but other species, such as *Spirulina platensis, Anabaena* sp., and *Haematococcus pluvialis*, have reported growth of only 1.60, 0.23, and 0.20 g per liter of culture, respectively, for an open pond production system. This contrasts with closed pond systems, with the biomass produced from *H. pluvialis* achieving a yield of up to 7 g per liter of culture. Therefore, the correct choice of production system depends on the microalgal species [20,22].

Additionally, microalgae can be grown in closed systems photoautotrophically, mixotrophically, or heterotrophically [6,23].

4.1 Closed system production

4.1.1 Heterotrophic cultivation production

Heterotrophic production takes place on acetate or glucose as carbon sources. Nevertheless, not all microalgae can live in this kind of culture medium because of the chemical composition of the no facultative autotrophic microalgae that often change under these conditions [24].

4.1.2 Closed photoautotrophic production

This closed photoautotrophic culture system is widely used in the aquaculture industry, although most of these systems are operated in batch or semicontinuous mode. However, it appears not to be commercial production because the system must be operated indoors with large-diameter transparent bags to supply the light energy needs of the microalgae. Nevertheless, photoautotrophic production has been developed with smaller closed photobioreactor (PBR) technologies [2].

4.1.3 Mixotrophic production

Many microalgal organisms can use either autotrophic or heterotrophic metabolic processes for their growth, meaning they can photosynthesize as well as ingest organic materials. Therefore, light energy is not an absolute limiting factor for growth it is influenced by media supplementation with glucose during the light and dark phases. Hence, there is less biomass loss during the light phase.

Growth rates of mixotrophic microalgae are better than those from the cultivation of photoautotrophic microalgae in closed PBRs. In addition, these rates are higher than those obtained in open pond cultivation, although they are considerably lower than the results of heterotrophic production [6].

In conclusion, the successful production of mixotrophic microalgae in closed systems by PBRs allows the photosynthetic and heterotrophic components to be integrated during the diurnal cycle. It reduces the impact of biomass loss during dark respiration and decreases the number of organic substances utilized during growth because it facilitates control and optimizes growth conditions. Depending on the bioreactor design, it can improve production yields, as shown in Table 25.2. For example, it can reduce the light path, thus increasing the amount of light available to each cell.

As mentioned, PBRs are controlled closed systems; they can be aligned horizontally, vertically, with an incline, or as a helix. They are made of transparent materials for maximum solar or artificial light capture because light availability strongly influences microalgae growth. For that reason, they generally consist of an array of straight glass or plastic; however, it is important to establish an optimal light/dark cycle [21].

Another advantage is that microalgae cultures have proper mixing or agitation besides that provided mechanically with helices or through the location of the PBR on a moving platform. Mixing and agitation can also be provided by gas aeration by a pump or airlift system with O_2 or CO_2 to accomplish the dual purposes of agitation and nutrition [22].

4.2 Recovery of microalgal biomass

Microalgal biomass recovery generally requires one or more solid−liquid separation steps. Harvesting technology selection is crucial to the low-cost production of microalgal biomass but depends on characteristics of the microalgae species, e.g., size, density, and the values of target products. Generally, microalgae harvesting is a two-stage process of bulk harvesting and thickening [6].

TABLE 25.2 Comparison of photobioreactors and ponds in microalgae cultivation.

System	Advantages	Limitations
Raceway pond	• Easy to clean • Low energy inputs because it utilizes sun energy	• Poor biomass productivity • Large land area required • Limited to a few strains of algae • Poor mixing, light, and CO_2 utilization • Cultures are easily contaminated • Some biomasses grow in the wall • Difficult to sterilize
Cascade	• Low culture thickness (<6 mm) to improve the light path • Substantial biomass production	• Culture is prone to contamination • Light intensity, temperature, and released CO_2 are not controllable • Does not suitable to high-value products unless it improves indoor
Tubular photobioreactor	• Large illumination surface area • Also suitable for outdoor cultures • Good biomass productivities • Reduces contamination • High level of control over culture parameters	• Some biomasses grow in the wall, but due to the small entrance of the tube, it is difficult to remove • Requires large land space • Gradients of pH, dissolved oxygen and CO_2 along the tubes
Flat plate photobioreactor	• High biomass productivity • Easy to sterilize • Good light transfer and large illumination surface area that can be improved with inclinations • Suitable for outdoor cultures • Easy to scale and operate • Reduces contamination • High level of control over culture parameters	• Difficult temperature control • Low mixing by bubbling

Continued

TABLE 25.2 Comparison of photobioreactors and ponds in microalgae cultivation.—cont'd

System	Advantages	Limitations
Column photobioreactor	• Good mixing with low shear stress • Easy to sterilize • Reduced photoinhibition and photooxidation • Reduces contamination • High level of control over culture parameters	• Sophisticated construction • Small illumination area

Adapted from I. Gifuni, A. Pollio, C. Safi, A. Marzocchella, C. Olivieri, Current bottlenecks and challenges of the microalgal biorefinery. Trends Biotechnol., 37(3) (2019) 242–252. https://doi.org/10.1016/j.TIBTECH.2018.09.006 and L. Brennan, P. Owende. Biofuels from microalgae—a review of technologies for production, processing, and extractions of biofuels and co-products. Renew. Sustain. Energy Rev., 14(2) (2010) 557–577. https://doi.org/10.1016/J.RSER.2009.10.009.

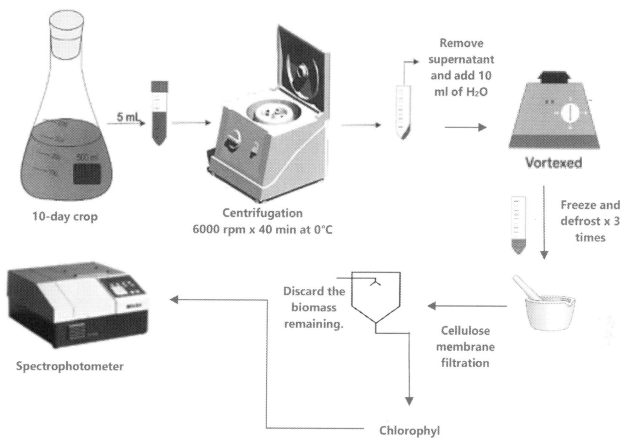

FIGURE 25.6 Methodology to find correct wavelength (λ) to measure optical density.

4.2.1 Bulk harvesting

Bulk harvesting involves directly separating biomass from the bulk suspension culture system through various technologies, including flocculation, flotation, or gravity sedimentation [6].

4.2.1.1 Flocculation

Flocculation is intended to aggregate a substance to the microalgal cells to increase the effective "particle" size. Flocculation is a preparatory step before other harvesting methods, such as filtration, flotation, or gravity sedimentation. The microalgae cells carry a negative charge that prevents the natural aggregation of cells in suspension. Adding flocculants such as multivalent cations and cationic polymers (2 mg of chitosan per liter of microalgal culture) neutralizes or reduces the negative charge [22].

4.2.1.2 Flotation

Flotation methods are based on trapping algae cells using dispersed micro-air bubbles; in contrast with flocculation, this technique does not require any chemical additions. Here, it is important to observe that some strains naturally float at the surface of the water if their lipid content increases [22]. Consequently, such microalgae are optimal for producing biodiesel.

4.2.2 Thickening

The target is to concentrate the slurry through techniques such as centrifugation, filtration, and ultrasonic aggregation; hence, thickening generally requires higher energy consumption than bulk harvesting [22].

TABLE 25.3 Comparison of adaptive lab evolution (ALE) in batch [29] and continuous [30] cultivation.

	ALE in batch cultivation	ALE in continuous cultivation
Concept	Adaptation is made in deep-well recipients with small culture volumes, and at regular intervals, the culture is transferred to new recipients containing fresh media.	Made in bioreactors operated under tightly regulated pH, dissolved oxygen, and nutrient availability.
Advantages	Advantages include inexpensive lab equipment and chemicals, ease of conducting multiple experiments in parallel, and well-controlled abiotic conditions, such as temperature and mixing.	Achieves constant growth rate and population density because it may more efficiently control pH, dissolved oxygen, and nutrient availability.
Disadvantages	Cultivation suffers from variable populations and fluctuating pH, dissolved oxygen, and nutrient availability that affect the growth rate of the microorganism.	Higher cost of equipment and difficulty running multiple adaptations.

4.2.2.1 Ultrasonic aggregation

Ultrasonic aggregation is an acoustic technique for optimizing the aggregation efficiency and concentration factor in the flocculation, achieving 92% separation efficiency and a concentration factor of 20 times. The main advantage of ultrasonic harvesting are that it can be operated continuously without inducing shear stress on the biomass that could destroy potentially valuable compounds [22].

4.2.2.2 Gravity and centrifugal sedimentation

The centrifugation unit operation is recommended for the recovery of compounds with high added value. The settling of parameters such as revolutions per minute or gravities, time, and temperature is determined by the density and radius of algae cells. For example, *Spirulina platensis* has been recovered at $4430 \times g$ for 15 min at 0°C. To reduce centrifugation energy costs, filtration as a first step is a good strategy [25].

In another study, the biomass was centrifuged at 5000 rpm for 10 min and then dried at 60°C until constant weight [5].

4.2.2.3 Filtration

Filtration is most appropriate for harvesting relatively large (>70 mm) microalgae. Filtration aids such as diatomaceous earth or cellulose can be used to improve efficiency. To recover smaller algae cells (<30 mm), membrane microfiltration and ultrafiltration are ways to optimize filtration using hydrostatic pressure and are technically viable alternatives to conventional filtration. It is suitable for fragile cells that require low transmembrane pressure and low cross-flow velocity conditions [22].

The harvested biomass slurry has between 5% and 15% dry solid content that is perishable and must be processed rapidly. Dehydration or drying is commonly used to extend the viability of high-value compounds. Also, sun drying, low-pressure shelf drying, spray drying, drum drying, fluidized bed drying, freeze drying, and reactance technology drying have been used [6]. For example, the biomass of *Botryococcus braunii, Synechococcus nidulans, Chlorella kessleri, Chlorella vulgaris, Neochloris oleoabundans,* and *Scenedesmus obliquus* were conditioned by withdrawing and centrifuging at 3000 g for 15 min. They were then washed once with bicarbonate buffer (0.1 M) and dried at 80°C until a constant weight was achieved [8].

To determine whether a culture medium is optimal, its growth tendency must be measured. A common technique uses optical density and absorbance values to calculate dry cell weight. This is achieved by making a lineal plot where absorbance depends on dry cell weight. For correct readings, the optimal wavelength (λ) must be known. For this, it is recommended that the chlorophyll is analyzed by a microplate reader in a range within the visible region (400–700 nm) to find a λ with major absorbance [25], using the process described in the Fig. 25.6.

For example, in an investigation, the biomass of *Desmodesmus* sp. *Chlorella* sp. and *Scenedesmus* sp. was measured spectrophotometrically with an absorbance at 680 nm^{-1} [4]. However, in a different study with *Chlorella, Nannochloropsis* and *Desmodesmus* were measured at 750 nm^{-1} [16].

It is unnecessary to strictly replicate the natural habitat in the culture medium to produce a microorganism, as the amount of biomass produced is not critically important in microalgae cultivation. For example, for the energy industry,

TABLE 25.4 Natural carbohydrate and lipid composition of various microalgae [4].

Microalgae	Carbohydrates (% D.W.)[a]	Lipids (% D.W.)[a]
Anabaena cylindrica	25–30	4–7
Chlamydomonas reinhardtii	17	21
Chlorella vulgaris	12–17	14–22
Dunaliella salina	32	6
Porphyridium cruentum	40–57	9–14
Spirulina maxima	13–16	6–7
Monoraphidium sp.	31	31
Synechococcus sp.	15	11

[a]D.W., dry weight.

metabolites (carbohydrates and lipids) are the objective. Therefore, the culture medium and growth conditions are the key factors for the productivity of these value-added compounds.

The yields of the compounds obtained from microalgae species depend on inducing stress and adapting certain cultivation strategies [2,3]. For example, to achieve greater biomass yields or high-value compounds such as lipids, carbohydrates, and other compounds, a technique called adaptive lab evolution [26] can be used. The technique consists of adapting microalgae to a specific medium in uncommon conditions but with parameters that achieve the objective of producing valuable compounds at high yields [27]. This technique can be used by batch and continuous cultivation [28]. The differences are explained in Table 25.3.

Selecting the best options depends on the resources and conditions in terms of the laboratory context or the place of production, such as the amount of culture. However, it is important to mention that the optimal passage size, which is the number of cells inoculated for gradual adaptation, is between approximately 13.5% and 20%. That is, a relation for the nutrient availability for the technique can be correctly carried out with a small passage size, the evolution rate can decrease; with a large passage size, the microorganism would die. However, this can decrease in every round of adaptation [29].

This technique requires a long period because it is necessary to attach the microorganisms slowly and not cause their death (opposite effect) [31]. Because of this, this technique is not optimal for production unless it already has the modified strain. The following discussion addresses modifications to the medium to produce certain compounds.

4.3 Production of lipids

Oleaginous microalgae naturally offer between 20% and 60% lipid content [3,10], which compares perfectly with terrestrial crops such as soybeans. Nevertheless, nitrogen limitation at the beginning of the exponential growth phase and the presence of excess carbon can cause fatty acid synthesis enzyme activation. Therefore, the intracellular lipid content can be increased until it accumulates up to ~70% lipid yield [20], becomes stressed, and accumulates lipids as an energy reserve [14,15]. The above converts the microalgal biomass into an important source for biodiesel production.

With respect to light irradiance, polyunsaturated fatty acids require low irradiance at the beginning of the stationary phase, whereas TAGs are produced during the late stationary phase at high light irradiance [20].

Lipid quality is important for the production quality of biodiesel and depends on a good balance between cold flow and stability in the oxidation of fatty acids. Oleic acid is an optimal component for producing biooils because it has a good ignition point, combustion heat, cold filter plug points, viscosity, lubricity, and oxidative stability. According to the above, *Scenedesmus* sp. reportedly has excellent potential for producing biodiesel compared with standard combustibles because its fatty acid profile includes oleic acid [5].

4.4 Production of carbohydrates

The production of carbohydrates from microalgae can be improved by up to 70% by supplying 2% CO_2 in the growth medium, thus enhancing the recovery of nitrogen and phosphorus nutrients [32]. In addition, under light saturation

TABLE 25.5 Lipid quantification.

Method	References
Extraction with methanol: chloroform 1:1 (v/v) followed by liquid–liquid extraction with hexane. Composition was determined using gas chromatography equipped with a hydrogen flame ionization detector.	[8]
Extraction with methanol: Chloroform 1:2 (v/v) and were quantified gravimetrically.	[5]
Lyophilized algal biomass (5 mg) can be hydrolyzed and methyl-esterified in 300 μL of a% (v/v) H_2SO_4 methanol solution for 2 h at 80°C with 50 μg C21:0 (heneicosanoic acid) as an internal recovery standard. A total of 300 μL of hexane and 300 μL of 0.9% (w/v) NaCl can be added to the mixture. The mixture must be vortexed for 20 s and centrifuged at 16,000×g for 3 min to facilitate phase separation. A 150 μL portion of the hexane layer must be injected into an agilent 6890 gas chromatograph connected to a mass spectrometer. A linear calibration curve is based on the C22:4(n-6) (adrenic acid) internal standard; the C21:0 internal recovery standard is used for fatty acid quantification. The mg L^{-1} day^{-1} lipids quantity is calculated and multiplied by average biomass productivity.	[17]

TABLE 25.6 Carbohydrate quantification.

Method	References
Phenol sulfuric extraction: A 2 mL sample must be mixed with 1 mL of 5% aqueous phenol solution in a test tube. Subsequently, 5 mL of concentrated sulfuric acid must be added rapidly to the mixture. After allowing the test tubes to stand for 10 min, they are vortexed for 30 s and placed in a water bath at room temperature for 20 min for color development. Then, light absorption at 490 nm is measured on a spectrophotometer. Standard calibration solutions are prepared in an identical manner as above.	[5,37]
Sulfuric acid–UV: A 1 mL sample must be rapidly mixed with 3 mL of concentrated sulfuric acid in a test tube and vortexed for 30 s. The temperature of the mixture rises rapidly, within 10–15 s, after sulfuric acid has been added. Then, the solution is cooled in ice for 2 min to bring it to room temperature. Finally, UV light absorption at 315 nm is read using a UV spectrophotometer. Standard calibration solutions are prepared in an identical manner as above.	[37]

conditions, carbohydrates such as sugars and starch are formed in greater proportions [14]. In contrast to the case of lipid production, for carbohydrate production, the nitrogen limitation should be made at the beginning of exponential growth [20].

It is important to mention that if there is no nitrogen at the beginning of exponential growth, there will be no biomass production. Therefore, if batch production is used, there must be a nitrogen concentration according to the above, with that nitrogen limitation when the end of exponential growth commences. Conversely, if production is continuous, less nitrogen should be supplied than was supplied at the beginning or vice versa, depending on the higher yields desired. The amounts and percentages of the nitrogen reduction depend on the microalgae species in production, and this is where experimentation takes center stage. When production is made separately, the recovery of the target product is simplified, but the time to cultivation may become longer with lower productivity [20]. The best way to coproduce lipids and carbohydrates is by nitrogen limitation at the end of the exponential phase, although the extraction could be complex.

Finally, Table 25.4 shows the natural compositions of carbohydrates and lipids from some microalgae species compared with soybean, which is 30% carbohydrates and 20% lipids in terms of dry biomass weight. We can observe that lipids produced from some microalgae are not viable for commercial use because more energy is spent than is economically viable, given the yields that can be achieved (low composition percentage). This is due to factors such as the high cost of agitation and aeration during cultivation and downstream processing (mainly harvesting and drying) to extract and purify lipids from biomass [33]. The oil content must be at least 40% for lipid production from microalgae to be profitable [10]. The obvious challenge is determining an optimal way of extracting or optimizing media to obtain satisfactory yields.

Practically, a study proved that yields of 47.07% w/w of carbohydrates in *Chlorella* and 31.30% w/w of lipids in *Scenedesmus* could be achieved, representing increases in the natural compositions of these components of microalgae [5].

5. Biomass fractionation

Extracting high-value compounds from microalgal biomass requires cell disruption because the products are intracellular. The process entails cell uptake of solvent molecules, which causes alterations to the cell membrane to enhance the movement of compounds toward the outside of the cell. For that process, solvents are indispensable, such as hydrochloric acid, sodium hydroxide, or alkaline lysis, for traditional methods [6]. Alternatively, emergent technologies can be used, such as high-pressure homogenizers, autoclaves, or supercritical/subcritical fluid methods; however, these approaches are associated with high capital costs [4].

It is important to mention that to optimize energy, time, and reactive consumption during extraction, the biomass quantity must be high, at around 100 g L^{-1} [20]. Such a yield is not easy to achieve even less so on a laboratory scale. For that reason, the harvested biomass should be stored, as mentioned in Section 4.2, until a sufficient biomass quantity is obtained to warrant the process for high lipid and carbohydrate yields. Authors have reported ways to optimize the methods mentioned—for example, freezing and warming the biomass before using solvents or other extraction techniques [25]. In addition, simultaneous pressurized liquid extraction and fractionation of microalgal biomass with enzymatic pretreatment can produce polar lipid yields of up to 53% [34].

Therefore, combining all these techniques with others, such as supercritical fluid extraction, three-phase partitioning, and aqueous two-phase separation [35], could be important subjects to study for continuously improving the production of high-value compounds.

It is also important to ensure the quality of those compounds, which in turn ensures the quality of the products to which they are applied. For that reason, the fractionation method should not be selected solely based on the quantity of product that can be extracted; rather, molecular bioactivity is sometimes the first factor to consider. With that in mind, ionic liquids are proposed as solvents for selective fractionation—i.e., salts with low melting points that maintain biomolecule structural integrity and activity [36].

6. Techniques to identify and quantify lipids and carbohydrates

Microalgae typically have a high carbohydrate content that is approximately 50% higher than their dry weight. They are primarily composed of glucose, starch, cellulose, galactose, rhamnose, mannose, arabinose, N-acetyl glucosamine, and N-acetyl galactosamine [6].

Tables 25.5 and 25.6 show some methods for quantifying lipids and carbohydrates from dry biomass.

Cascade extraction can be an attractive option for cost optimization [35], but yields may be low or incomplete [20]. Cascade extraction involves sequential recovery using consecutive steps. The priority of the compounds in cascade extraction generally starts with high-value products or the more sensitive or water-soluble components. In this case, carbohydrates must be extracted first, then lipids. Nevertheless, the optimal strategies to decide whether it is better to use cascade extraction depends on the strain and the target product. Eco-friendly techniques are recommended and should be considered [20].

7. Conclusions

Microalgal biomass can be produced by a batch or continuous process. The continuous mode is not an effective strategy, however, especially in outdoor cultures in addition to having to control environmental factors, culture media concentrations must be varied during microalgae growth phases to achieve higher yields of high-value constituents. Outdoor microalgae production at a large scale generates low biomass production due to weather conditions and poor light penetration to the system.

The hyperaccumulation strategy using a closed PBR with a low light path maximizes the target product yield. However, the provision of nitrogen must be increased at the end of exponential growth if more lipid production is desired. On the other hand, if higher carbohydrate yields are desired, the nitrogen limitation must be made at the beginning of exponential growth. Nitrogen should not be reduced completely, however without some nitrogen, microalgae will not grow, and no biomass production will occur.

Acknowledgments

This research project was supported by the Mexican Science and Technology Council (CONACYT, Mexico) with the Infrastructure Project — FOP02-2021—04 (Ref. 317250). The author Samanta Machado thanks the National Council for Science and Technology (CONACYT, Mexico) for her Ph.D. Fellowship support (grant number: 1154370).

References

[1] C. Larroche, M.A. Sanromán, G. Du, A. Pandey, Current developments in biotechnology and bioengineering: bioprocesses, bioreactors and controls, in: Current Developments in Biotechnology and Bioengineering: Bioprocesses, Bioreactors and Controls, 2016.
[2] F.A. Ansari, P. Singh, A. Guldhe, F. Bux, Microalgal cultivation using aquaculture wastewater: integrated biomass generation and nutrient remediation, Algal. Res. 21 (2017) 169—177, https://doi.org/10.1016/J.ALGAL.2016.11.015.
[3] M.L.E. Keesoo Lee, F. Rindi, P. Swaminathan, P.K. Nam, Isolation and screening of microalgae from natural habitats in the midwestern United States of America for biomass and biodiesel sources, J. Nat. Sci. Biol. Med. 5 (2) (2014).
[4] Y.H. Park, S.-I. Han, B. Oh, H.S. Kim, M.S. Jeon, S. Kim, Y.-E. Choi, Microalgal secondary metabolite productions as a component of biorefinery: a review, Bioresour. Technol. 344 (2022) 126206, https://doi.org/10.1016/J.BIORTECH.2021.126206.
[5] A. Pandey, S. Srivastava, S. Kumar, Isolation, screening and comprehensive characterization of candidate microalgae for biofuel feedstock production and dairy effluent treatment: a sustainable approach, Bioresour. Technol. 293 (2019) 121998, https://doi.org/10.1016/J.BIORTECH.2019.121998.
[6] M.C. Deprá, A.M. dos Santos, I.A. Severo, A.B. Santos, L.Q. Zepka, E. Jacob-Lopes, Microalgal biorefineries for bioenergy production: can we move from concept to industrial reality? Bioenergy Res. 11 (4) (2018) 727—747. Springer New York LLC, https://doi.org/10.1007/s12155-018-9934-z.
[7] B. Úbeda, J.Á. Gálvez, M. Michel, A. Bartual, Microalgae cultivation in urban wastewater: coelastrum cf. pseudomicroporum as a novel carotenoid source and a potential microalgae harvesting tool, Bioresour. Technol. 228 (2017) 210—217, https://doi.org/10.1016/J.BIORTECH.2016.12.095.
[8] E.B. Sydney, C.J.D. Neto, J.C. de Carvalho, L.P. de S. Vandenberghe, A.C.N. Sydney, L.A.J. Letti, S.G. Karp, V.T. Soccol, A.L. Woiciechowski, A.B.P. Medeiros, C.R. Soccol, Microalgal biorefineries: integrated use of liquid and gaseous effluents from bioethanol industry for efficient biomass production, Bioresour. Technol. 292 (2019) 121955, https://doi.org/10.1016/J.BIORTECH.2019.121955.
[9] Organization for Economic Co-operation and Development. (s.f.). OECD iLibrary. Recuperado el 2022, de OCDE-FAO Perspectivas Agrícolas 2020—2029: https://www.oecdilibrary.org/sites/8d79647ees/index.html?itemId=/content/component/8d79647e-es.
[10] N. Arora, H.-W. Yen, G.P. Philippidis, Sustainability harnessing the power of mutagenesis and adaptive laboratory evolution for high lipid production by oleaginous microalgae and yeasts, Sustainability 12 (12) (2020) 5125, https://doi.org/10.3390/su12125125.
[11] R.H. Wijffels, M.J. Barbosa, An outlook on microalgal biofuels, Science 329 (5993) (2010), https://doi.org/10.1126/science.1189003.
[12] J. Julich, E.B. Werlang, M.V.G. Muller, G.D.A. da Silva, F.D.F. Neves, R.D. C. de S. Schneider, Estudo da hidrólise enzimática de biomassa de microalga empregando uma sequência de enzimas, Revista Jovens Pesquisadores 9 (2) (2019) 77—84, https://doi.org/10.17058/rjp.v9i2.13519.
[13] L. Charmaine, T. Kai Heng, L. Kar Leong, V. Vimala Gana, Y.F. Sarah Mei, R.C. Teng, M.M. Hui, X.S. Wei, M. Sarah Liyana, Q.N. Joscelyn Jun, B.N. Nazurah Syazana, Md A. Nurhazlyn Bte, W.E. Zephyr Yu, M. Punithavathy, N. Jen Yan, Identification of microalgae cultured in Bold's Basal medium from freshwater samples, from a high-rise city, Informes Científicos 11 (1) (2021) 4474.
[14] M. Sacristán de Alva, V.M. Luna Pabello, M.T. Orta Ledesma, M.J. Cruz Gómez, Carbon, nitrogen, and phosphorus removal, and lipid production by three saline microalgae grown in synthetic wastewater irradiated with different photon fluxes, Algal Res. 34 (2018) 97—103, https://doi.org/10.1016/J.ALGAL.2018.07.006.
[15] T.E. Sero, N. Siziba, T. Bunhu, R. Shoko, Isolation and screening of microalgal species,native to Zimbabwe, with potential use inbiodiesel production, Life (2021) 256—264, https://doi.org/10.1080/26895293.2021.1911862.
[16] N.A. Kasan, F.S. Hashim, N. Haris, M.F. Zakaria, N.N. Mohamed, N.W. Rasdi, M.E.A. Wahid, T. Katayama, K. Takahashi, M. Jusoh, Isolation of freshwater and marine indigenous microalgae species from terengganu water bodies for potential uses as live feeds in aquaculture industry, Int. Aquat. Res. 12 (1) (2020), https://doi.org/10.22034/IAR(20).2020.671730.
[17] B. Bao, S.R. Thomas-Hall, P.M. Schenk, Fast-tracking isolation, identification and characterization of new microalgae for nutraceutical and feed applications, Phycology 2 (1) (2022) 86—107, https://doi.org/10.3390/phycology2010006.
[18] APHA, Standard Methods For The Examination Of Water And Wastewater, 2005.
[19] M.S. McGrath, P.M. Daggett, S. Dilworth, Freeze-drying of algae: chlorophyta and chrysophyta, J. Phycol. 14 (4) (1978), https://doi.org/10.1111/j.1529-8817.1978.tb02480.x.
[20] I. Gifuni, A. Pollio, C. Safi, A. Marzocchella, G. Olivieri, Current bottlenecks and challenges of the microalgal biorefinery, Trend. Biotechnol. 37 (3) (2019) 242—252, https://doi.org/10.1016/J.TIBTECH.2018.09.006.
[21] E.A. Cezare-Gomes, L. del C. Mejia-da-Silva, L.S. Pérez-Mora, M.C. Matsudo, L.S. Ferreira-Camargo, A.K. Singh, J.C.M. de Carvalho, Potential of microalgae carotenoids for industrial application, Appl. Biochem. Biotechnol. 188 (3) (2019) 602—634, https://doi.org/10.1007/s12010-018-02945-4. Humana Press Inc.
[22] L. Brennan, P. Owende, Biofuels from microalgae—a review of technologies for production, processing, and extractions of biofuels and co-products, Renew. Sustain. Energy Rev. 14 (2) (2010) 557—577, https://doi.org/10.1016/J.RSER.2009.10.009.

[23] D. Sun, Z. Zhang, X. Mao, T. Wu, Y. Jiang, J. Liu, F. Chen, Light enhanced the accumulation of total fatty acids (TFA) and docosahexaenoic acid (DHA) in a newly isolated heterotrophic microalga Crypthecodinium sp. SUN, Bioresour. Technol. 228 (2017) 227–234, https://doi.org/10.1016/J.BIORTECH.2016.12.077.

[24] M.A. Borowitzka, Commercial production of microalgae: ponds, tanks, tubes and fermenters, J. Biotechnol. 70 (1999).

[25] G. Rosero-Chasoy, R.M. Rodríguez-Jasso, C.N. Aguilar, G. Buitrón, I. Chairez, H.A. Ruiz, Growth kinetics and quantification of carbohydrate, protein, lipids, and chlorophyll of Spirulina platensis under aqueous conditions using different carbon and nitrogen sources, Bioresour. Technol. (2021) 126456, https://doi.org/10.1016/J.BIORTECH.2021.126456.

[26] V.A. Portnoy, D. Bezdan, K. Zengler, Adaptive laboratory evolution — harnessing the power of biology for metabolic engineering, Curr. Opin. Biotechnol. 22 (4) (2011) 590–594, https://doi.org/10.1016/J.COPBIO.2011.03.007.

[27] X. Li, Y. Yuan, D. Cheng, J. Gao, L. Kong, Q. Zhao, W. Wei, Y. Sun, Exploring stress tolerance mechanism of evolved freshwater strain Chlorella sp. S30 under 30 g/L salt, Bioresour. Technol. 250 (2018) 495–504, https://doi.org/10.1016/J.BIORTECH.2017.11.072.

[28] M. Dragosits, D. Mattanovich, Adaptive laboratory evolution - principles and applications for biotechnology, Microb. Cell Factor. 12 (1) (2013), https://doi.org/10.1186/1475-2859-12-64.

[29] R.A. LaCroix, B.O. Palsson, A.M. Feist, A model for designing adaptive laboratory evolution experiments, Appl. Environ. Microbiol. 83 (8) (2017), https://doi.org/10.1128/AEM.03115-16.

[30] R.L.H.,K.K.C. Winkler J, in: H. Alper (Ed.), Adaptive Laboratory Evolution for Strain Engineering, Systems Metabolic Engineering, vol. 985, 2013.

[31] D. Cheng, X. Li, Y. Yuan, C. Yang, T. Tang, Q. Zhao, Y. Sun, Adaptive evolution and carbon dioxide fixation of Chlorella sp. in simulated flue gas, Sci. Total Environ. 650 (2019) 2931–2938, https://doi.org/10.1016/J.SCITOTENV.2018.10.070.

[32] H. Zheng, J. Chen, X. Hu, F. Zhu, A. Ali Kubar, X. Zan, Y. Cui, C. Zhang, S. Huo, Biomass production of carbohydrate-rich filamentous microalgae coupled with treatment and nutrients recovery from acrylonitrile butadiene styrene based wastewater: synergistic enhancement with low carbon dioxide supply strategy, Bioresour. Technol. 349 (2022) 126829, https://doi.org/10.1016/J.BIORTECH.2022.126829.

[33] A. Valente, D. Iribarren, J. Dufour, How do methodological choices affect the carbon footprint of microalgal biodiesel? A harmonised life cycle assessment, J. Clean. Prod. 207 (2019) 560–568, https://doi.org/10.1016/J.JCLEPRO.2018.10.020.

[34] N. Castejón, F.J. Señoráns, Simultaneous extraction and fractionation of omega-3 acylglycerols and glycolipids from wet microalgal biomass of Nannochloropsis gaditana using pressurized liquids, Algal Res. 37 (2019) 74–82, https://doi.org/10.1016/J.ALGAL.2018.11.003.

[35] S. Sarkar, M.S. Manna, T.K. Bhowmick, K. Gayen, Priority-based multiple products from microalgae: review on techniques and strategies, Crit. Rev. Biotechnol. 40 (5) (2020) 590–607. Taylor and Francis Ltd, https://doi.org/10.1080/07388551.2020.1753649.

[36] M.H.M. Eppink, S.P.M. Ventura, J.A.P. Coutinho, R.H. Wijffels, Multiproduct microalgae biorefineries mediated by ionic liquids, Trends Biotechnol. 39 (11) (2021) 1131–1143. Elsevier Ltd, https://doi.org/10.1016/j.tibtech.2021.02.009.

[37] A.A. Albalasmeh, A.A. Berhe, T.A. Ghezzehei, A new method for rapid determination of carbohydrate and total carbon concentrations using UV spectrophotometry, Carbohydr. Polym. 97 (2) (2013) 253–261, https://doi.org/10.1016/j.carbpol.2013.04.072.

Chapter 26

Microalgae biomass: a model of a sustainable third-generation of biorefinery concept

Alejandra Cabello-Galindo, Rosa M. Rodríguez-Jasso and Héctor A. Ruiz

Biorefinery Group, Food Research Department, School of Chemistry, Autonomous University of Coahuila, Saltillo, Coahuila, Mexico

1. Introduction

Throughout human history, the evolution of industry and technology has brought about great advances and benefits. However, the increase in the world's population and in industrialization has generated increasing demands for resources to meet needs. This has resulted in several disadvantages such as environmental pollution, climate change, and ecological degradation associated with the high demand for resources to generate energy sources. Fossil fuels are formed under the earth's crust through fossilization, whereas fuels that are obtained by processing biomass by chemical methods and technologies instead of being generated by this slow geological process are known as biofuels.

Fossil fuels such as oil, natural gas, and coal are dominant sources of energy, satisfying about 80% of the world's energy demand [1]. Their production and consumption are attributed to high greenhouse gas emissions and pollution, and their non-renewable nature and limited reserves, coupled with their high rate of exploitation and consumption, severely compromise the availability of these resources for the future. This indicates that there is an urgent need to find an alternative energy source that meets society's expectations, is a long-term replacement for fossil fuels, and reduces greenhouse gas emissions [2]. It leads us to a crucial search for new sources of resources to produce fuels to address the growing energy crisis and environmental pollution.

With the advent of bioenergy technology, a smarter way to reduce carbon emissions and fossil fuel consumption as well as use raw materials was developed, favoring the agricultural economy and generating a sustainable process [3]. Fortunately, the concept of bio-refinery encompasses all of these concepts through the use of diverse raw materials and biomasses for their transformation into biofuels and high value-added compounds. Bio-refineries facilitate the conversion of the material by relying on technologies such as thermochemical, biochemical, and hydrothermal treatment [4,5].

Bio-refineries are classified by the type of raw material used in the technological process. There are three generations. In first-generation bio-refineries, food crops are processed as feedstocks. The most used raw materials are sugarcane, cornstarch, and vegetable oils. Their main use is to produce biofuels, but because of the current situation of scarcity of resources for human food, the use of this type of raw material has been questioned. Second-generation bio-refineries use as raw materials for agricultural and industrial processing waste that does not compete with food sources. These include lignocellulosic materials such as cane bagasse, agave bagasse, and corncobs. The third-generation bio-refineries use aquatic biomass as feedstock [6]. Algal biomass, microalgae, and macroalgae are considered good sources of feedstock owing to their rapid growth in simple conditions without competition for freshwater and land use. Furthermore, because of their composition, they are viable sources to produce bio-composites, biofuels, and bioproducts [7,8].

There has been increasing interest in the use of marine biomasses such as microalgae. Because of their growth characteristics and capacity for bioremediation, microalgae are one of the best sources of bioenergetics. This chapter addresses the use of microalgae as an alternative for bioremediation and the production of biofuels and high value-added products, emphasizing sustainable and bio-economic systems with perspectives for the maximum use of this biomass.

2. Current status of third-generation bio-refineries with microalgae

The term biorefinery came from "refinery." It consists of integrated production that transforms different raw materials and biomass into biofuels and high value-added compounds. Biorefineries provide renewable sources of energy as an alternative to reduce environmental contamination using a manufacturing process that optimizes and investigates technologies in a sustainable, circular, bio-economical way [9,10].

2.1 Biorefinery classification

Biorefineries are classified by the type of raw material used in the technological process. They can be classified into three generations.

2.2 First-generation biorefineries

In first-generation biorefineries, food crops are processed as feedstock. The most used raw materials are sugarcane, cornstarch, and vegetable oils. Their main use is to produce biofuels, but because of the scarcity of resources for human food, the use of this type of raw material has been questioned [11,12].

2.3 Second-generation biorefineries

Second-generation biorefineries use raw materials from agricultural and industrial processing waste that does not compete with food sources. These include lignocellulosic materials such as cane bagasse, agave bagasse, and corncobs [13,14].

2.4 Third-generation biorefineries

Third-generation biorefineries processes microalgae and macroalgae biomass to produce biofuels and other high added-value compounds. This type of biorefinery is considered to be promising because of the growth characteristics and composition of the raw material used.

The use of microalgae biomass in biorefineries does not compromise food agriculture and security because it does not compete with arable land and water and is a renewable source of energy and biocomposites. It has high photosynthetic efficiency, and its nutritional requirements are low.

Over the past decade, microalgae have attracted attention for producing biofuels. Compounds extracted from this biomass, such as lipids, can be used to produce biodiesel, whereas other biofuels such as bioethanol and biobutanol can be produced from carbohydrates by fermentative methods, replacing conventional carbohydrate sources such as lignocellulosic materials. Other important compounds obtained from microalgae, such as pigments, proteins, and long-chain fatty acids, can be used in industries such as pharmaceuticals and food [15].

The selection of appropriate technologies for the treatment of biomass and extraction of the compounds is a key issue. Such a selection will depend on available technologies, the type of biomass, economic and energetic considerations, and the final form of the desired product.

These biomass conversion technologies can be divided into thermochemical conversion, whose principle is the decomposition of organic matter into biomass, the main examples of which are gasification, thermal liquefaction, pyrolysis, and direct combustion; biochemical conversion, such as anaerobic digestion and alcoholic fermentation; and transesterification, the reaction of triglycerides with alcohol in the presence of catalysis for the production of fatty acid chains and glycerol [16].

Although there are many challenges to carrying out a third-generation biorefinery, the growing need to curb environmental changes generated by the exploitation and use of fuels has been proposed and technologies and processes have been improved. For a third-generation biorefinery to be profitable, factors such as the availability of feedstock, processes such as harvesting and extraction, the quality of biofuel and other compounds obtained, operations, and energy infrastructure costs must be considered (Table 26.1). Therefore, it is important to propose a multidisciplinary approach that will make such processes efficient, including system biology and strain development, scale up processes, and set up an integrated production chain.

3. Microalgae and sustainability

Microalgae are unicellular, eukaryotic, photosynthetic microorganisms. Owing to their characteristics and metabolism, they are closely related to cyanobacteria. Cyanobacteria are prokaryotic microorganisms with an internal organization involving a central region (nucleoplasm) rich in DNA and a peripheral region (chromoplast) where the photosynthetic membranes are located.

TABLE 26.1 Biomass conversion technologies.

Conversion method	Principle	Example	Compound obtained	References
Thermochemical conversion	Thermal decomposition of organic compounds	Pyrolysis Liquefaction Gasification Torrefaction	Biogas Biochar Bio-oils	[16,17]
Biochemical conversion	Conversion of biomass into biofuels in presence of microorganisms or enzymes	Anaerobic digestion Alcoholic fermentation Photobiological H_2 production	Biogas Biodiesel Bioethanol	[18]
Transesterification	Lipids or triglycerides converted into fatty acid methyl ester using methanol and catalyst	Acid catalyst Base catalyst Enzyme catalyst	Biodiesel	[19,20]

Photosynthesis is a process carried out by photoautotrophic organisms in which inorganic compounds and light energy are converted into organic matter. This process is fundamental to life on earth. All forms of life depend on it directly or indirectly. Studies indicate that 2 million years ago, prokaryotic bacteria helped to generate the earth's atmospheric oxygen.

Microalgae have the ability to transform CO_2 into O_2 using light energy to extract protons and electrons from donor molecules to reduce CO_2 to form organic molecules. Therefore they are credited with producing about 50% of atmospheric oxygen [17]. They are highly efficient in CO_2 fixation and use solar energy to produce biomass with an efficiency up to four times higher than plants [18]. Microalgae have a faster growth rate compared with other crops. Moreover, because they do not need fertile soil or freshwater, they do not compete for these resources. Microalgae can grow in saline water [19], freshwater, and wastewater. In the latter, they have been shown to have the capacity to purify it and use it for growth and biomass production [20]. Compared with crops, microalgae are not subject to seasonal harvests because they can withstand extreme weather conditions (Fig. 26.1) [21].

Microalgae require an adequate supply of carbon source and light to perform photosynthesis, but in adequate concentrations of other nutrients and other factors are also necessary. The main factors in microalgae cultivation are pH, temperature, nutrients, and light. Adequate amounts of micronutrients and macronutrients in the culture ensure the maximum biomass yield. The main macronutrients required are nitrogen, carbon, free CO_2 in the medium, phosphorus, sulfur, and potassium [22] (please see Fig. 26.2).

To achieve maximum biomass productivity, the culture must be optimized. Biomass productivity highly depends on a wide range of cultivation parameters, such as microalgae strain, nutrients, light, temperature, pH, mixing, substrate

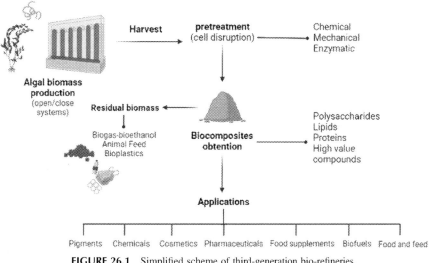

FIGURE 26.1 Simplified scheme of third-generation bio-refineries.

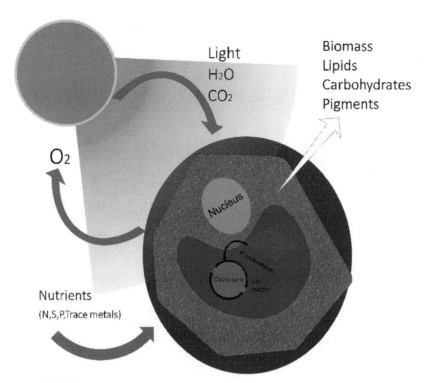

FIGURE 26.2 Microalgae photosynthesis diagram. *ATP*, adenosine triphosphate.

concentrations, and culture purity. The balance between operating parameters such as molecular oxygen, CO_2, pH, temperature, and light intensity is crucial. A low growth rate and therefore low biomass productivity are the main obstacles to microalgae cultivation.

Together with their growth rate, oxygen production, and capacity for bioremediation, microalgae offer an unlimited range of possibilities for biotechnological and economic development. Microalgae have advantages over other microbiological organisms because of their minimal nutritional requirements. In this sense, owing to their metabolic flexibility and high biomass production rates, microalgae offer almost unlimited possibilities for the development of a modern bioeconomy [23,24].

Since ancient times throughout diverse cultures, humans have used algae for various purposes. There are records of the consumption of microalgae such as *Arthrospira spp.* (spirulina). or *Nostoc spp.* as a staple in various civilizations from South America to Asia [25]. Although this practice of consumption was lost for many years because of cultural changes in eating habits, there has been growing interest in resuming it. Applications of this type of biomass are not limited to food. According to research, microalgae are also able to produce different types of bioactive compounds such as antioxidants, enzymes, lipids such as polyunsaturated acids, peptides, pigments, polysaccharides alginates, and carrageenan, among others.

This is because microalgae have endless applications in biotechnology industries such as pharmaceuticals, cosmetics, food, wastewater treatment, and raw materials to produce biofuels. According to the bioeconomy and biorefinery concept, microalgae have an extraordinary potential to obtain value compounds and take advantage of biomass for other uses such as animal feed and biofuels production (Table 26.2).

To ensure the sustainability of the process, it is important to develop cultivation technologies for microalgae in which minimal material and energy conditions are used to reduce production costs. To evaluate the potential of microalgae as a biotechnological feedstock, it is important to conduct systematic research including technoeconomic and life-cycle assessments to validate the economic viability and environmental sustainability of processing models.

4. Bioremediation capacity of microalgae

As a result of population growth, urbanization, and industrialization, environmental pollution has been increasing. The consequences are the excessive production of waste and garbage, contamination of aquifers, global warming, extinction of

TABLE 26.2 Advantages and disadvantages of microalgae as biorefinery feedstock.

Advantages	Disadvantages
• Higher rates of growth and photosynthesis than terrestrial plants • No competition for farmland • Bioremediation capacity • Competent feedstock for bioethanol production and extraction of valuable products • No need for fertilizers • Can grow in wastewater • Fewer greenhouse emissions	• Difficulties scaling up systems • Expensive to harvest microalgal biomass • Difficult to recover biomass • Risk for contamination with another microorganism

species, and seriously modified and damaged ecosystems. The consequences of these problems have great impact that compromises the sustainability of the planet for both present and future generations, so it is a great challenge for the scientific community to seek solutions to mitigate this damage (Fig. 26.3).

Microalgae are widely studied in environmental biotechnology for monitoring environmental toxicants such as pesticides and heavy metals and for bioremediation and bioassays, but the most important environmental applications include the mitigation of CO_2 emissions and wastewater treatment. These microorganisms can be used for biological treatments because of their capacity of toxic substantial biosorption.

4.1 Environmental toxicant monitoring

For centuries, the problem of water contamination has generated health issues, among others. This concern has been increasing along with industrial development. Many times, wastewater generated by industries and agriculture is returned

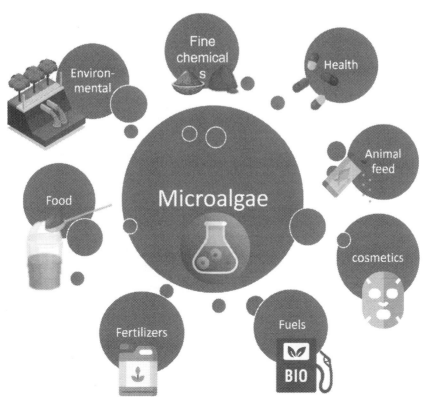

FIGURE 26.3 Main commercial areas for microalgae application.

to the environment without adequate treatment, bringing the risk for contamination of the environment. Some of these contaminants are found in small quantities, or their composition is so complex as to make their detection difficult, so it is important to develop efficient methods to deal with them.

Microalgae are considered the first rung on the food chain and are sensitive to toxic pollutants in the environment. If these microorganisms are affected by the contamination, they will negatively affect higher trophic levels.

Because of the need to monitor aquatic ecosystems, the search for the development of bioassays to detect of pollutants has been increasing. Microalgae are characterized according to their predominant role in the food chain, so they have great potential as indicators of environmental ecotoxicity. Microalgae are able to consume toxic pollutants and fix heavy metals from wastewater. Owing to this characteristic, they are considered a useful indicator of environmental quality.

One type of pollutant present in wastewater that is not easily detoxified is heavy metals such zinc (Zn), copper (Cu), chrome (Cr), nickel (Ni), lead (Pb), mercury (Hg), arsenic (As), and cobalt (Co). Trace metals are essential elements for cell metabolism and growth, but a high concentration of these compounds is toxic to organisms. Microalgae can take up and accumulate heavy metals through two mechanisms. The first phase is via rapid extracellular adsorption, and the second is slow intracellular absorption, diffusion, and bioaccumulation [26–28].

For microalgae, the effect of high heavy metal concentrations is reflected in the diminution or suppression of vital metabolism activities such as photosynthesis, cell growth, and motility [23]. These changes in the appearance and metabolism of microalgae can be used as indicators of contamination by excess heavy metals. On the other hand, an increase in phosphorus and nitrogen content in water is typical of a deficient wastewater system, but this situation is closely related to the high productivity of microalgal biomass, which takes advantage of these compounds as a source of nutrients for development. Therefore, the presence of microalgae is usually an indirect indicator of this kind of pollutant in the water.

Environmental monitoring using bioassays is becoming increasingly frequent. Several environmental agencies and commissions such as those in Europe and the United States have standardized tests that use different microalgae indicators to determine the quality of lake water. These indicators are based on the average concentration of phosphorus and chlorophyll [24]. Not all microalgae species are efficient for environmental monitoring; the most commonly used and reported are the genera *Chlamydomonas*, *Chlorella*, *Scenedesmus*, *Selenastrum*, *Dictyosphaerium*, *Raphidecelis*, and *Pseudokirchneriella* [28–31].

4.2 Use of microalgae for wastewater treatment

Globally, one of the greatest threats facing most countries is water pollution. This issue is caused by the discharge of domestic and industrial waste and pesticides from the agricultural industry that enter the marine ecosystem through discharge into rivers. Wastewater contains heavy metals, phosphorus, and nitrogen in the form of nitrate or ammonium [32]. To address these problems, the use of microalgae as wastewater treatment has emerged as an eco-friendly and cost-effective method because the wastewater loads serve as a nutrient medium for their growth, and biomass is produced as a by-product that can be used in various ways.

Traditional wastewater treatments to remove solid and chemical pollutants include coagulation, flocculation, ion exchange, filtration, precipitation, and activated sludge. However, because of their high energy consumption, carbon emissions, and consumption of nonrecyclable substrates, these processes are far from being considered sustainable.

Microalgae-based wastewater treatment has great potential owing to its metabolic flexibility and capacity for growth and adaptation in wastewater. Microalgae fix carbon dioxide by photosynthesis and bioabsorb heavy metals, using nitrogen (N) and phosphorus (P) in the media for their metabolic protein production [20]. These metabolic characteristics of microalgae mean that wastewater can be used as a growth medium to produce biomass (with high-valued compounds) simultaneously with wastewater treatment. The most studied species of microalgae for wastewater treatment are *Arthrospira* ssp., *Chlorella* ssp., *Scenedesmus* ssp., *Botryococcus* ssp., *Desmodesmus* ssp., *Nannochloropsis* ssp., and *Heterochlorella* ssp [32,33].

For wastewater treatment, microalgal cultivation can be done in suspended or attached-cell systems [34]. Suspended-cell cultivation systems use closed photobioreactors, or open bioreactors such as pounds and lakes. The closed system has advantages such as control of the parameters and growth conditions, and it poses a low risk for contamination. Disadvantages include difficulties in scaling up and construction and maintenance costs. Open systems have the advantages of low construction and maintenance cost as well as a high rate of biomass production, and are more easily scaled up, but with poor control of parameters such as agitation and aeration, culture contamination is more likely to occur with greater frequency [35,36].

Attached or immobilized-cell cultivation systems consist of cell immobilization by natural or artificial techniques. Natural immobilization occurs through the natural ability of microalgae to adhere to specific surfaces generating biofilms. Artificial immobilization uses semipermeable membranes, which employ adsorption and covalent coupling to immobilize microalgal cells in a polymeric matrix [37]. The principal advantages of these systems are the low energy cost and easy handling and operation of the harvest [38].

Microalgae can grow on wastewater and can promote microbial groups with other bacteria because of photosynthesis. This symbiosis among microorganisms may occur because during photosynthesis, microalgae synthesize CO_2, generating O_2. This oxygen is freely released in the media and is used by bacteria to assimilate elements into the organic state as phosphorus, nitrogen, and carbon. In turn, bacteria release CO_2, inorganic nitrogen, and phosphorus, which can be used by microalgae as nutrients for photosynthesis [34]. Microalgae—bacteria symbiosis can significantly improve the assimilation of nutrients with nitrate and phosphorus removal, reflected in higher biomass productivity [39,40].

Despite the promising prospects for the use of microalgae for wastewater treatment, there are limiting factors such as light and temperature. Microalgae require optimal temperature and light conditions for growth, so application of this technology is limited to countries with warm-temperate climates and appropriate periods of sunlight. The solution for areas that do not meet these conditions is to use lighting systems and bioreactor cultures with temperature controls, but using these increases the cost of treatment and makes it less profitable, so it must be compensated for by obtaining biocomposites with high added value.

4.3 Microalgae mitigation capacity of CO_2

A dominant cause of climate change is the excessive accumulation in the atmosphere of greenhouse gases, mainly CO_2. Most of this carbon dioxide is produced by the combustion of fossil fuels. The alarming increase in CO_2 atmospheric concentrations is the reason for concentrate efforts to develop efficient methods for CO_2 capture. Methods of CO_2 capture are chemical absorption, physical absorption, membrane separation, and cryogenic distillation. The widest strategies used are carbon capture and storage (CCS), carbon capture, utilization, and storage (CCUS), and carbon capture and utilization (CCU) [41,42].

The objective of CCS is the store CO_2 captured in the subsurface and deep ocean to prevent its emission into the atmosphere. Like CCS, CCUS stores the captured CO_2 and also seeks its application and reuse. The problem with CO_2 storage is that storage space is limited, and at high concentrations it can cause damage to the environment such as acidification of the sea. CCU seeks to use of the captured CO_2 immediately [41]. Captured CO_2 can be used as a precursor to producing high value-added compounds such as fuel and chemicals [43].

Biological processes focused on CO_2 sequestration for capture include reforestation, agriculture, and the use of photosynthetic microorganisms. Microalgae are highly used in CO_2 biofixation owing to their adaptability to the environment, high photosynthetic and growth rates, and ability to employ CO_2 for biomass production, from which bioethanol and bioproducts can be produced [44,45].

Flue gases released from power plants contain high amounts of CO_2 and can be used for the growth of microalgae bacteria. However, these gases also have high concentrations of CO_2 and SO_x and NO_x, which can inhibit the growth of cultures. Therefore, further gas cleaning treatments may be required to standardize the gas composition. The addition of an extra step such as gas cleaning treatments may increase the energy and economic costs, however. A possible solution suggests investigating microalgae strains that can be coupled to inhibitory compounds in industrial flue gas for direct use, without treatment [46]. There are some studies of microalgae with the capacity for growth under these extreme conditions, such as *Chlorella* sp. Also, *Chlorella* showed a 59%−63% removal rate of CO_2, NO_x, and CO_x from fuel gases [47].

For these processes to be sustainable, research is important to determine which strains of microalgae are most suitable for removing CO_2 and other components of combustion gases, as well as strategies to overcome the challenges of dissolving these compounds in culture media.

5. Integration and reuse of industrial waste for microalgae cultivation

A critical point of this chapter is the importance of the valorization of microalgae biomass and its use in producing fuels and high-value products. For the feasibility of a third-generation biorefinery, these must be considered in conjunction with optimizing harvesting processes and selecting the culture media.

The principle of biorefineries uses residues to produce biofuels and biocompounds. Microalgae can grow in wastewater, taking advantage of polluting elements such as nutrients and transforming them into biomass that can be used for valorization. This has advantages in the sense of environmental bioremediation and is also an integrated process for a third-generation biorefinery.

Considering the current energy demand and the importance of cleaner fuels, it is essential to develop biorefineries with an integrated approach to producing biofuels. The use of wastewater for microalgae cultivation can be an economical source of carbon. It allows large-scale cultivation, can provide abundant trace elements, and can cope with conventional wastewater treatment. Studies have tested different microalgae strains for treating domestic wastewater and wastewater from industries such as breweries, pharmaceutical plants, slaughterhouses, agroindustrial complexes, textile factories, and wastewater containing heavy metals, starch, palm oil, and so on. [48].

Despite the advantages of a microalgae-based biorefinery, it is not economically viable to use microalgae to produce biofuels because of the operational cost of harvesting and extraction [49] and the limited market for potential bioactive chemicals.

The selection of microalgae strain is important to the success of a biorefinery. Genetically modified strains are being investigated to improve the accumulation of compounds such as lipids and pigments and increase photosynthetic efficiency and biomass production, among other characteristics. Also important are the development of more efficient photobioreactor designs with more efficient lighting, agitation, and feeding strategies, and efficient nutrient distribution. These factors can achieve the success of a microalgae-based biorefinery, but social, economic, technological, and environmental aspects must also be integrated [50].

6. Co-production of microalgae-derived products for circular bio-economy-based bio-refinery

Let us talk about the differences between the current system of the linear economy and the circular bioeconomy. The linear economy consists of the use of raw materials that are processed after consumption. These products are discarded. On the other hand, the central axis of the circular bioeconomy encourages the reduction of environmental loss by recovering waste, to allow materials for reuse and valorization to be put back into the supply chain [51]. These processes require a consideration of the sustainability of these factors: the resource base, processes, and products to improve and implement circular processes of material fluxes. The development of circular processes emphasized the role of all of the sciences and biotechnologies. In the same way, it is necessary to promote collaboration among industry, enterprises, and research institutions to improve the reuse of biomass and waste residue and provide support for developing bio-based activities. In global development, bio-economic strategies have been developed by many countries and organizations such as the Organization for Economic Co-operation and Development, the European Union, the west Nordic Countries (Iceland, Greenland, and the Faroe Islands), Australia, Finland, France, Germany, Japan, Malaysia, South Africa, Spain, Sweden, and the United States [52].

Fortunately, the concept of bio-refinery encompasses all of these concepts through the use of diverse raw materials and biomasses for transformation into biofuels and high value-added compounds. This makes it an ideal tool for generating products in a circular bio-economy. As mentioned throughout this chapter, owing to the composition of microalgae, they can be used in a third-generation bio-refinery for the co-production of biofuels and high-value compounds. The applications of compounds extracted from microalgae focus on a variety of markets, mainly in the areas of nutrition, human health, and food. One of the most widely used species of microalgae is spirulina, which has been used throughout history in food production owing to its rich protein and vitamin content. The main industrial application of spirulina is for developing pharmaceutical and nutritional supplements because of its bioactivity as an antioxidant, principally in pigments such phycocyanin. *Chlorella* is rich in proteins, carotenoids, and vitamins. Its large-scale production began in Japan and Taiwan, and its biomass is mainly applied in health food and to produce peptides. Another strain of interest is *Dunaliella*, owing to its high content of lipids, proteins, glycerol, and β-carotene [53–56] (Table 26.3).

Several challenges need to be overcome to consider large-scale microalgae production. These include the optimization of biomass yields and their components as well as the development of more efficient technologies and methodologies for each stages, as well as strategies coupled with bioremediation and the co-production of biofuels and value-added compounds.

7. Future perspectives

Creating strategies for the co-production of high value-added products in the production of biofuels from microalgae is an essential strategy to achieve the profitability of the third-generation biorefinery process. For this, it is important to cover all bases of the bioeconomy, including technoeconomic, ecological, and social factors.

TABLE 26.3 Main species of microalgae used to extract high-value compounds.

Species	Growth strategy	Main compound	Principal producer	Global production (dry mass per year)	Principal application	References
Spirulina	Phototropical	• Antioxidants • Sterols • Pigments • Proteins	China (10,000 t year^{-1})	15,000	• Human health and nutrition • Cosmetics	
Chlorella	Phototropical, heterotrophical and combination of both	• Peptides • Proteins • Plant biostimulants	Japan Taiwan	5000	• Human health and nutrition • Peptide production • Plant biostimulant	[17,24], [53–56]
Dunaliella	Stress conditions (high salt concentration, light variations)	• Proteins • Pigments • Carotenoids	Australia Israel	2000	• Human health and nutrition • Cosmetics • Water bioremediation	
Hematococcus	Oxidative stress conditions	• Astaxanthin • Pigments	China	–	• Human health	

In technoeconomic terms, the development of technologies that simplify critical points such as the collection and extraction of compounds would significantly reduce energy and economic costs. The development of environmental bioremediation plants with microalgae, the application of biomass to obtain compounds of interest, and the generation of biofuels would contribute to mitigating pollution and reducing CO_2 emissions into the atmosphere. For the social aspect, awareness is important to facilitating the acceptance of products obtained from microalgae.

8. Conclusion

The threat of climate change and the need to use products of natural origin have led humanity to search for new resources and technologies. Microalgae have been used throughout history, but currently, people have paid special attention to them. Today, microalgae are widely used in industry to produce various products. Despite this, the cultivation systems and strategies that are employed are still under development. Although the current approach to microalgae as third-generation biorefinery feedstock has great advantages, especially when combined with bioremediation processes such as wastewater treatment, there is still a long way to go to ensure that these processes are sustainable in all aspects. Some improvements that might be applied are emerging technologies for biomass extraction and harvesting, the genetic modification of strains, and optimization of open and closed culture systems.

Acknowledgments

This research project was supported by the Mexican Science and Technology Council (CONACYT, Mexico) with the Infrastructure Project —FOP02-2021—04 (Ref. 317250). Alejandra Cabello thanks the National Council for Science and Technology (CONACYT, Mexico) for Ph.D. fellowship support (Grant No. 711463).

References

[1] M. Guo, W. Song, J. Buhain, Bioenergy and biofuels: history, status, and perspective, Renew. Sustain. Energy Rev. 42 (2015) 712–725, https://doi.org/10.1016/j.rser.2014.10.013. Elsevier Ltd.

[2] R. Ganesan, et al., A review on prospective production of biofuel from microalgae, Biotechnol. Rep. 27 (Sep. 01, 2020), https://doi.org/10.1016/j.btre.2020.e00509. Elsevier B.V.

[3] M.A. Koondhar, Z. Tan, G.M. Alam, Z.A. Khan, L. Wang, R. Kong, Bioenergy consumption, carbon emissions, and agricultural bioeconomic growth: a systematic approach to carbon neutrality in China, J. Environ. Manag. 296 (Oct. 2021), https://doi.org/10.1016/j.jenvman.2021.113242.

[4] A.T. Ubando, C.B. Felix, W.H. Chen, Biorefineries in circular bioeconomy: a comprehensive review, Bioresour. Technol. 299 (Mar. 01, 2020), https://doi.org/10.1016/j.biortech.2019.122585. Elsevier Ltd.

[5] H.A. Ruiz, et al., Engineering aspects of hydrothermal pretreatment: from batch to continuous operation, scale-up and pilot reactor under biorefinery concept, Bioresour. Technol. 299 (2020), https://doi.org/10.1016/j.biortech.2019.122685. Elsevier Ltd.

[6] L. Bhatia, R.K. Bachheti, V.K. Garlapati, A.K. Chandel, Third-generation biorefineries: a sustainable platform for food, clean energy, and nutraceuticals production, Biomass Convers. Biorefin. (2020), https://doi.org/10.1007/s13399-020-00843-6. Springer.

[7] R. Peñalver, J.M. Lorenzo, G. Ros, R. Amarowicz, M. Pateiro, G. Nieto, Seaweeds as a functional ingredient for a healthy diet, Mar. Drug. 18 (6) (2020), https://doi.org/10.3390/md18060301. MDPI AG.

[8] C.E. Solís-Salinas, G. Patlán-Juárez, P.U. Okoye, A. Guillén-Garcés, P.J. Sebastian, D.M. Arias, Long-term semi-continuous production of carbohydrate-enriched microalgae biomass cultivated in low-loaded domestic wastewater, Sci. Total Environ. 798 (2021), https://doi.org/10.1016/j.scitotenv.2021.149227.

[9] M.S. Pino, M. Michelin, R.M. Rodríguez-Jasso, A. Oliva-Taravilla, J.A. Teixeira, H.A. Ruiz, Hot compressed water pretreatment and surfactant effect on enzymatic hydrolysis using agave bagasse, Energies 14 (16) (2021), https://doi.org/10.3390/en14164746.

[10] K.D. González-Gloria, et al., Macroalgal biomass in terms of third-generation biorefinery concept: current status and techno-economic analysis — a review, Bioresour. Technol. Rep. 16 (2021) 100863, https://doi.org/10.1016/j.biteb.2021.100863.

[11] S.N. Naik, V.v. Goud, P.K. Rout, A.K. Dalai, Production of first and second generation biofuels: a comprehensive review, Renew. Sustain. Energy Rev. 14 (2) (2010) 578–597, https://doi.org/10.1016/j.rser.2009.10.003.

[12] F. Saladini, N. Patrizi, F.M. Pulselli, N. Marchettini, S. Bastianoni, Guidelines for emergy evaluation of first, second and third generation biofuels, Renew. Sustain. Energy Rev. 66 (2016) 221–227, https://doi.org/10.1016/j.rser.2016.07.073. Elsevier Ltd.

[13] A. Aguilar-Reynosa, A. Romaní, R.M. Rodríguez-Jasso, C.N. Aguilar, G. Garrote, H.A. Ruiz, Microwave heating processing as alternative of pretreatment in second-generation biorefinery: an overview, Energy Convers. Manag. 136 (2017) 50–65, https://doi.org/10.1016/j.enconman.2017.01.004. Elsevier Ltd.

[14] S.S. Hassan, G.A. Williams, A.K. Jaiswal, Moving towards the second generation of lignocellulosic biorefineries in the EU: drivers, challenges, and opportunities, Renew. Sustain. Energy Rev. 101 (2019) 590–599, https://doi.org/10.1016/j.rser.2018.11.041. Elsevier Ltd.

[15] K.K. Jaiswal, S. Dutta, I. Banerjee, C.B. Pohrmen, V. Kumar, Photosynthetic microalgae–based carbon sequestration and generation of biomass in biorefinery approach for renewable biofuels for a cleaner environment, Biomass Convers. Biorefin. (2021), https://doi.org/10.1007/s13399-021-01504-y. Springer Science and Business Media Deutschland GmbH.

[16] K.W. Chew, et al., Microalgae biorefinery: high value products perspectives, Bioresour. Technol. 229 (2017) 53–62, https://doi.org/10.1016/j.biortech.2017.01.006.

[17] M. Rizwan, G. Mujtaba, S.A. Memon, K. Lee, N. Rashid, Exploring the potential of microalgae for new biotechnology applications and beyond: a review, Renew. Sustain. Energy Rev. 92 (Sep. 01, 2018) 394–404, https://doi.org/10.1016/j.rser.2018.04.034. Elsevier Ltd.

[18] D.Y.Y. Tang, K.S. Khoo, K.W. Chew, Y. Tao, S.H. Ho, P.L. Show, Potential utilization of bioproducts from microalgae for the quality enhancement of natural products, Bioresour. Technol. 304 (May 01, 2020), https://doi.org/10.1016/j.biortech.2020.122997. Elsevier Ltd.

[19] T. Ishika, N.R. Moheimani, P.A. Bahri, Sustainable saline microalgae co-cultivation for biofuel production: a critical review, Renew. Sustain. Energy Rev. 78 (2017) 356–368, https://doi.org/10.1016/j.rser.2017.04.110. Elsevier Ltd.

[20] W.S. Chai, W.G. Tan, H.S. Halimatul Munawaroh, V.K. Gupta, S.H. Ho, P.L. Show, Multifaceted roles of microalgae in the application of wastewater biotreatment: a review, Environ. Pollut. 269 (2021), https://doi.org/10.1016/j.envpol.2020.116236.

[21] B. Sajjadi, W.Y. Chen, A.A.A. Raman, S. Ibrahim, Microalgae lipid and biomass for biofuel production: a comprehensive review on lipid enhancement strategies and their effects on fatty acid composition, Renew. Sustain. Energy Rev. 97 (2018) 200–232, https://doi.org/10.1016/j.rser.2018.07.050. Elsevier Ltd.

[22] S.Y.A. Siddiki, et al., Microalgae biomass as a sustainable source for biofuel, biochemical and biobased value-added products: an integrated biorefinery concept, Fuel 307 (2022), https://doi.org/10.1016/j.fuel.2021.121782.

[23] B. dos, S.A.F. Brasil, F.G. de Siqueira, T.F.C. Salum, C.M. Zanette, M.R. Spier, Microalgae and cyanobacteria as enzyme biofactories, Algal Res. 25 (2017) 76–89, https://doi.org/10.1016/j.algal.2017.04.035. Elsevier B.V.

[24] F.G.A. Fernández, A. Reis, R.H. Wijffels, M. Barbosa, V. Verdelho, B. Llamas, The role of microalgae in the bioeconomy, N. Biotech. 61 (2021) 99–107, https://doi.org/10.1016/j.nbt.2020.11.011.

[25] J.L. Pérez-Lloréns, Microalgae: from staple foodstuff to avant-garde cuisine, Int. J. Gastrono. Food Sci. 21 (2020), https://doi.org/10.1016/j.ijgfs.2020.100221.

[26] D. Kaplan, 32 Absorption and Adsorption of Heavy Metals by Microalgae, 2013.

[27] Y.K. Leong, J.S. Chang, Bioremediation of heavy metals using microalgae: recent advances and mechanisms, Bioresour. Technol. (2020), https://doi.org/10.1016/j.biortech.2020.122886. Elsevier Ltd.

[28] A.P. Peter, et al., Microalgae for biofuels, wastewater treatment and environmental monitoring, Environ. Chem. Lett. 19 (4) (2021) 2891–2904, https://doi.org/10.1007/s10311-021-01219-6. Springer Science and Business Media Deutschland GmbH.

[29] T. Yamagishi, H. Yamaguchi, S. Suzuki, Y. Horie, N. Tatarazako, Cell reproductive patterns in the green alga Pseudokirchneriella subcapitata (=Selenastrum capricornutum) and their variations under exposure to the typical toxicants potassium dichromate and 3,5-DCP, PLoS One 12 (2) (2017), https://doi.org/10.1371/journal.pone.0171259.

[30] N.G.A. Ekelund, D.P. Häder, Environmental monitoring using bioassays, in: Bioassays: Advanced Methods and Applications, Elsevier, 2018, pp. 419–437, https://doi.org/10.1016/B978-0-12-811861-0.00021-8.

[31] R.K. Goswami, K. Agrawal, M.P. Shah, P. Verma, Bioremediation of heavy metals from wastewater: a current perspective on microalgae-based future, Lett. Appl. Microbiol. (2021), https://doi.org/10.1111/lam.13564. John Wiley and Sons Inc.

[32] F. Hussain, et al., Microalgae an ecofriendly and sustainable wastewater treatment option: biomass application in biofuel and bio-fertilizer production. A review, Renew. Sustain. Energy Rev. 137 (2021), https://doi.org/10.1016/j.rser.2020.110603. Elsevier Ltd.

[33] M. Plöhn, et al., Wastewater treatment by microalgae, Physiol. Plant. 173 (2) (2021) 568–578, https://doi.org/10.1111/ppl.13427.

[34] A.L. Gonçalves, J.C.M. Pires, M. Simões, A review on the use of microalgal consortia for wastewater treatment, Algal Res. 24 (2017) 403–415, https://doi.org/10.1016/j.algal.2016.11.008. Elsevier B.V.

[35] K.W. Chew, S.R. Chia, P.L. Show, Y.J. Yap, T.C. Ling, J.S. Chang, Effects of water culture medium, cultivation systems and growth modes for microalgae cultivation: a review, J. Taiwan Inst. Chem. Eng. 91 (2018) 332–344, https://doi.org/10.1016/j.jtice.2018.05.039. Taiwan Institute of Chemical Engineers.

[36] G. Zuccaro, A. Yousuf, A. Pollio, J.P. Steyer, Microalgae cultivation systems, in: Microalgae Cultivation for Biofuels Production, Elsevier, 2019, pp. 11–29, https://doi.org/10.1016/B978-0-12-817536-1.00002-3.

[37] L.L. Zhuang, M. Li, H. Hao Ngo, Non-suspended microalgae cultivation for wastewater refinery and biomass production, Bioresour. Technol. 308 (2020), https://doi.org/10.1016/j.biortech.2020.123320. Elsevier Ltd.

[38] L.L. Zhuang, et al., The characteristics and influencing factors of the attached microalgae cultivation: a review, Renew. Sustain. Energy Rev. 94 (2018) 1110–1119, https://doi.org/10.1016/j.rser.2018.06.006. Elsevier Ltd.

[39] K. Li, et al., Microalgae-based wastewater treatment for nutrients recovery: a review, Bioresour. Technol. 291 (2019), https://doi.org/10.1016/j.biortech.2019.121934. Elsevier Ltd.

[40] B.B. Makut, D. Das, G. Goswami, Production of microbial biomass feedstock via co-cultivation of microalgae-bacteria consortium coupled with effective wastewater treatment: a sustainable approach, Algal Res. 37 (2019) 228–239, https://doi.org/10.1016/j.algal.2018.11.020.

[41] E. Daneshvar, R.J. Wicker, P.L. Show, A. Bhatnagar, Biologically-mediated carbon capture and utilization by microalgae towards sustainable CO_2 biofixation and biomass valorization — a review, Chem. Eng. J. 427 (2022), https://doi.org/10.1016/j.cej.2021.130884.

[42] J.B. Beigbeder, M. Sanglier, J.M. de Medeiros Dantas, J.M. Lavoie, CO_2 capture and inorganic carbon assimilation of gaseous fermentation effluents using Parachlorella kessleri microalgae, J. CO_2 Util. 50 (2021), https://doi.org/10.1016/j.jcou.2021.101581.

[43] M.K. Lam, K.T. Lee, A.R. Mohamed, Current status and challenges on microalgae-based carbon capture, Int. J. Greenh. Gas Control 10 (2012) 456–469, https://doi.org/10.1016/j.ijggc.2012.07.010. Elsevier Ltd.

[44] M.A. Kassim, T.K. Meng, Carbon dioxide (CO_2) biofixation by microalgae and its potential for biorefinery and biofuel production, Sci. Total Environ. 584–585 (2017) 1121–1129, https://doi.org/10.1016/j.scitotenv.2017.01.172.

[45] M.G. de Morais, E.G. de Morais, J.H. Duarte, K.M. Deamici, B.G. Mitchell, J.A.V. Costa, Biological CO_2 mitigation by microalgae: technological trends, future prospects and challenges, World J. Microbiol. Biotechnol. 35 (5) (2019), https://doi.org/10.1007/s11274-019-2650-9. Springer Netherlands.

[46] S.R. Chia, K.W. Chew, H.Y. Leong, S.H. Ho, H.S.H. Munawaroh, P.L. Show, CO_2 mitigation and phycoremediation of industrial flue gas and wastewater via microalgae-bacteria consortium: possibilities and challenges, Chem. Eng. J. 425 (2021), https://doi.org/10.1016/j.cej.2021.131436. Elsevier B.V.

[47] P.K. Kumar, S. Vijaya Krishna, K. Verma, K. Pooja, D. Bhagawan, V. Himabindu, Phycoremediation of sewage wastewater and industrial flue gases for biomass generation from microalgae, S. Afr. J. Chem. Eng. 25 (2018) 133–146, https://doi.org/10.1016/j.sajce.2018.04.006.

[48] S. Mehariya, R.K. Goswami, P. Verma, R. Lavecchia, A. Zuorro, Integrated approach for wastewater treatment and biofuel production in microalgae biorefineries, Energies 14 (8) (2021), https://doi.org/10.3390/en14082282. MDPI AG.

[49] L.Y. Batan, G.D. Graff, T.H. Bradley, Techno-economic and Monte Carlo probabilistic analysis of microalgae biofuel production system, Bioresour. Technol. 219 (2016) 45–52, https://doi.org/10.1016/j.biortech.2016.07.085.

[50] Z. Yin, et al., A comprehensive review on cultivation and harvesting of microalgae for biodiesel production: environmental pollution control and future directions, Bioresour. Technol. 301 (2020), https://doi.org/10.1016/j.biortech.2020.122804. Elsevier Ltd.

[51] D.A. Campos, R. Gómez-García, A.A. Vilas-Boas, A.R. Madureira, M.M. Pintado, Management of fruit industrial by-products—a case study on circular economy approach, Molecules 25 (2) (2020), https://doi.org/10.3390/molecules25020320. MDPI AG.

[52] C. Priefer, R. Meyer, One concept, many opinions: how scientists in Germany think about the concept of bioeconomy, Sustainability 11 (15) (2019), https://doi.org/10.3390/su11154253.

[53] R. Ma, et al., Comprehensive utilization of marine microalgae for enhanced co-production of multiple compounds, Mar. Drugs 18 (9) (2020), https://doi.org/10.3390/md18090467. MDPI AG.

[54] I. Barkia, N. Saari, S.R. Manning, Microalgae for high-value products towards human health and nutrition, Mar. Drugs 17 (5) (2019), https://doi.org/10.3390/md17050304. MDPI AG.

[55] M.I. Khan, J.H. Shin, J.D. Kim, The promising future of microalgae: current status, challenges, and optimization of a sustainable and renewable industry for biofuels, feed, and other products, Microb. Cell Factor. 17 (1) (2018), https://doi.org/10.1186/s12934-018-0879-x. BioMed Central Ltd.

[56] E.M. Sierra, M.C. Serrano, A. Manares, A. Guerra, Y.A. Díaz, Microalgae: potential for bioeconomy in food systems, Appl. Sci. 11 (23) (2021), https://doi.org/10.3390/app112311316. MDPI.

Chapter 27

Green hydrogen production: a critical review

Gilver Rosero-Chasoy[1], Rosa M. Rodríguez-Jasso[1], Cristóbal N. Aguilar[1], Germán Buitrón[2], Isaac Chairez[3] and Héctor A. Ruiz[1]

[1]*Biorefinery Group, Food Research Department, School of Chemistry, Autonomous University of Coahuila, Saltillo, Coahuila, Mexico;* [2]*Laboratory for Research on Advanced Processes for Water Treatment, Unidad Académica Juriquilla, Instituto de Ingeniería, Universidad Nacional Autónoma de Mexico, Queretaro, Mexico;* [3]*Unidad Profesional Interdisciplinaria de Biotecnología, UPIBI, Instituto Politécnico Nacional, Mexico City, Mexico*

1. Introduction

Hydrogen is the most abundant element in the universe, and it can be used as a clean fuel to combat global warming since its only atmosphere emission is water steam [1]. However, hydrogen is not found in its isolated form on the Earth, and its current production is obtained from natural gas through steam reforming and coal gasification [2]. These processes generate a huge carbon footprint. For each tone of hydrogen produced from the steam methane reforming process, nearly 9-10 tons of equivalent carbon dioxide are released into the atmosphere [3]. As carbon emissions result, this process is known as grey hydrogen production [4].

Carbon capture, utilization, and storage technologies can be integrated into hydrogen production to obtain blue hydrogen with a low carbon footprint. However, hydrogen production is still linked to the use of fossil fuels, which are essential to accelerate the hydrogen market in the short-term and ensure an organized energy transition based on the hydrogen economy [1,4]. Methane pyrolysis is a cleaner technology based on natural gas, known as turquoise hydrogen. This technology produces hydrogen and solid carbon without releasing carbon dioxide into the atmosphere [1]. Like blue hydrogen, turquoise hydrogen production is not sustainable in the long-term since this is dependent on fossil fuel reserves unless methane production comes from the methanogenesis of renewable biomass such as microalgae culture.

The world urgently needs carbon-free hydrogen production from renewable resources to accelerate the decarbonize the global economy and achieves the goal of net zero emissions for 2050 set up in the Paris agreement [1]. Green hydrogen production from water electrolysis utilizing renewable energy sources such as solar, wind, wave, hydroelectric, and biofuels has enormous potential in the net zero scenario [5]. Thus, many countries have focused significant efforts on developing clean power generation technologies to produce green hydrogen. Table 27.1 shows the advantage and disadvantages of current hydrogen production colors.

In this chapter, various technologies for sustainable hydrogen production are discussed. The first one is power generation from hydropower plants, in which electrolysis produces hydrogen from water. The second technology involves the thermochemical processes on microalgae biomass to obtain biofuels that can be used for power generation by steam turbines—finally, biological pathways for hydrogen production to small-scale.

2. Hydrogen production from renewable energy

Currently, the largest power generation in the world is based on thermoelectric technology from the utilization of nonrenewable resources such as natural gas, coal, and oil [1,5]. The main problem with using these sources in power generation is the excess greenhouse gas emission into the atmosphere (approximately 65 % of total greenhouse gas emission), accelerating the temperature rise on the Earth's surface and triggering climate change. Nowadays, Global warming has a significant concern in the international community. Therefore, many countries have been making great efforts to reduce greenhouse gas emissions, especially carbon dioxide emissions [6].

TABLE 27.1 Advantage and disadvantage from hydrogen production by label color.

	Grey	Blue	Turquoise	Green
Technology	Reforming/ Gasification	Reforming/ Gasification	Pyrolysis	Electrolysis
Advantages	Low cost	Cheaper than green hydrogen	No CO_2 emissions	No CO_2 emissions
	Large scale	Low carbon emissions	Black carbon formation	Sustainable to long-term
	Mature technology	Infrastructure use of grey hydrogen		Food production is decoupled from fossil fuels
Disadvantage	High CO_2 emissions	Carbon capture technology are required	Low hydrogen yield	Dependent on electricity costs and water availability
	Vulnerable to carbon taxes	Dependent on nonrenewable resources	Dependent on nonrenewable resources	Needing of new infrastructure
	Dependent on nonrenewable resources			

Adapted from D. Hjeij, Y. Biçer, M. Koç, Hydrogen strategy as an energy transition and economic transformation avenue for natural gas exporting countries: Qatar as a case study, Int. J. Hydrog. Energy (2021).

Fig. 27.1 shows a significant increase in publications about renewable energy from the 21st century. China, USA, and India are top list in worldwide because these are the primary consumers of fossil fuels due to their high industry development. In Latin America, countries like Brazil and Mexico lead the search for renewable energy because these joint with Colombia, Argentina, and Venezuela represent 72 % of the population in the Latin American and Caribbean region, and they produce 80 % of carbon dioxide emissions in the region [7].

Hydrogen can substitute all energy matrix from fossil fuels as clean renewable energy, achieving a sustainable energy transition to a circular economy based on itself [7]. However, hydrogen production directly comes from natural gas via steam methane reforming. In this process, we have a cost-effective hydrogen production, where the 50 % is consumed in ammonia synthesis by the Haber-Bosch cycle. Oil industries producing natural gas could export to produce hydrogen, but

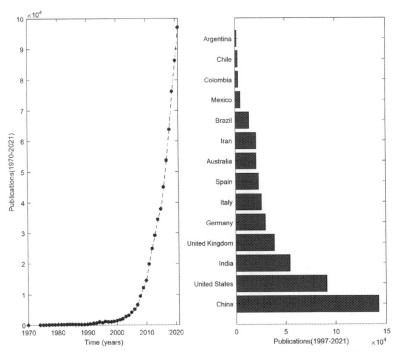
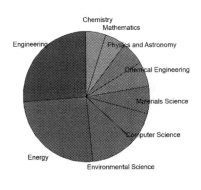

FIGURE 27.1 Renewable energy publications from 1970 to 2021 by number of publications, countries, and area of research. *Source: www.scopus.com.*

it is not sustainable long-term [7,8]. Another way of making hydrogen is using water, it is known as electrolysis (Eq. 27.1), and two British scientists, William Nicholson and Anthony Carlisle, discovered this process in 1800. For a long time, electrolysis was the leading technology for hydrogen generation with efficiency levels between 70 % to 90 % [7]. However, the increasing use of natural gas in the 20th century displaced this technology [8]. Now, water splitting with clean renewable energies such as solar, wind, wave, and hydroelectric are experiencing a renaissance [9].

The Latin American region has enormous potential for hydrogen production since Brazil has the largest hydroelectric sources in the region (42 %), followed by Colombia (16 %), Peru (10 %), Mexico (9 %), Venezuela (7 %), and Argentina (7 %) [7]. Hydrogen can be produced in this region in two ways. The first one is to assume that a percentage of the electric generation on the dams will apply to produce hydrogen by electrolysis (Eq. 27.1), but this generates an energy security problem with more vulnerable populations. The second one is based on obtaining hydrogen from excess electricity. It can be possible by reducing the energy demand or through high flow values in dams [7]. This one proposes a way for hydrogen production which allows these countries to obtain cost-effective green hydrogen that directly contributes to the production of fertilizer, natural gas, fuels, fuel cells, and other chemicals, which is economically feasible [9].

$$2H_2O \rightarrow 2H_2 + O_2 \quad \Delta H° = 286 \text{ kJ/mol} \tag{27.1}$$

3. Hydrogen production from biofuels obtained by thermochemical processes of microalgae biomass

To reach net zero goals on greenhouse gas emissions for 2050 very countries are seeking to accelerate the adoption of clean energy technologies to ensure an organized transition toward new clean energy industries. A clean energy technology to produce fuels is microalgae cultivation. 1 kg of microalgal biomass can fix 1.83 kg CO_2 [10]. Further, the fuels made from microalgae do not release SO_x and NO_x into the atmosphere; these are locally accessible, eco-friendly, not toxic, derived from renewable sources, and they do not compete with arable lands, i.e., third-generation biofuels [11].

Microalgal biomass can be converted into various forms of renewable fuels such as bio-oil (liquid fuels), biochar (solid), biogas, and biohydrogen (gas) via thermochemical processes without extracting lipids or carbohydrates as chemical and biochemical conventional processes require [12]. Thermochemical processes are categorized into direct combustion, liquefaction, gasification, and pyrolysis. These are promising technologies for biofuel production because they do not need additional chemicals, the reaction is completed in a short period, and all biofuels can convert afterward into green hydrogen directly or through power generation by steam turbines [12,13].

3.1 Direct combustion

This process breaks down the biomass with excess air (as oxidant) at a temperature range between 800 − 1000 °C [12] The direct combustion of biomass is the classical method to obtain energy in the form of heat from the calorific value of biomass, and it can be transformed into electric power through steam turbines [12,14]. This method is the easiest path for power generation from biomass with a low value of activation energy [15]. However, microalgal biomass cannot be put into direct combustion because it has a high moisture content (> 55 %) [14]. Therefore, it is necessary to reduce moisture content in microalgal biomass at values < 20 % to be introduced into the combustion chamber [12]. The cost of energy production from direct combustion is slightly higher than pyrolysis and gasification due to the need to remove the moisture in biomass [14]. However, direct combustion from microalgal dried biomass has an energy return 2.6 times higher than liquefaction because this process requires less energy input [16]. The direct combustion of the microalgal dried biomass is an expensive way for power generation owing to its low calorific value compared with coal (Fig. 27.2) and it is not suitable since inside microalgal biomass there are many high-value compounds such as pigments, vitamins, and antioxidants [17]. Therefore, it is important looking for another technologies that can take advantage of the moisture content in microalgal biomass is important.

3.2 Liquefaction

The moisture content is a critical parameter in microalgal biomass because it affects the product quality and the processing costs associated with valorization in the conversion to biofuels by using direct combustion. Therefore, reducing the moisture content in microalgal biomass before using it for conversion is crucial for avoiding fumes and wetness. This one is infeasible in biofuel production because it increases production cost making that microalgae biomass to be less economically feasible [12]. Thus, the liquefaction process is considered the most effective technology to produce biofuels since it is a biomass-to-liquid conversion route that uses the moisture content inside microalgal biomass and water to produce mainly bio-crude oil (yield of 50 − 60 %) and biochar with a higher caloric value at comparison with the raw biomass without dewatering [18]. This conversion occurs in the supercritical phase through the breakdown of the cell wall owing to the effect of the temperature (200 to 370 °C) and high pressures (5 to 20 MPa), with a reaction time of 5 − 120 min [19,20]. The bio-oil cannot be directly used as fuel, but it is expected to be a suitable renewable feedstock for co-refining in existing fossil-based refineries [18]. Fig. 27.2 shows the high heat value from bio-oil of some algal biomasses obtained by liquefaction.

3.3 Gasification

Gasification is a conversion technique that uses dry biomass to produce syngas through a partial oxidation of biomass at elevated temperatures (higher than 700 °C) and pressures between 15 to 23 MPa under a limited air supply [12,18,19]. As result it is obtained hydrogen (H_2), carbon monoxide (CO), carbon dioxide (CO_2), methane (CH_4), and other hydrocarbons such as char and tar [12]. Again, the moisture content of the algal biomass seems to be an insurmountable barrier to the development of this technology [18]. However, this thermochemical conversion can be particularly suitable for producing biofuels from dry macroalgal biomass due to the content of metals inside the cell since these can be used as catalysts in the process [19]. The composition of the syngas from gasification strongly depends on biomass type, gasification agent (air, O_2, CO_2, steam, or their combination), operating parameter, and gasifier type [12]. Fig. 27.2 shows the weight percent from main products reported by A. Raheem et al. for the gasification of C. vulgaris at 800 °C for 30 min with 10 % catalyst

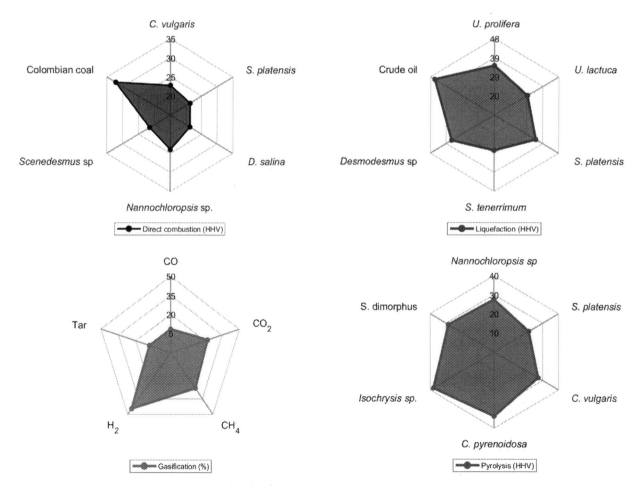

FIGURE 27.2 Comparison of the high heat value (MJ kg^{-1}) and main products of some microalgal species via direct combustion, liquefaction, gasification, and pyrolysis [12,15,18,21,22].

loading [21]. The gases, tar, and char distribution under operation conditions above were 78, 8, and 14 % at weight, respectively. Hydrogen is the highest compound, followed by methane. Methane obtained from gasification can be converted to hydrogen by steam reforming (Eq. 27.2), achieving high hydrogen production from microalgal biomass [21]. Gasification technology can be the via thermochemical most efficient to produce blue hydrogen from different microalgal biomass type through close looping for the CO_2 emissions.

$$CH_4 + 2H_2O \rightarrow CO_2 + 4H_2 \quad \Delta H° = 165 \text{ kJ/mol} \tag{27.2}$$

3.4 Pyrolysis

Pyrolysis is a thermal decomposition process in the absence of air or oxygen at high temperatures and atmospheric pressure to convert dry biomass into bio-oil, charcoal, and a gaseous fraction [14,15,18]. Again, the microalgal biomass's moisture content limits this technology's development [18]. During pyrolysis, the microalgae biomass's organic components (carbohydrate, protein, lipids, and volatile material) start to degrade at temperatures above 200 °C up to 500 °C [22]. Based on the heating rate, pyrolysis is divided into two operation settings: fast pyrolysis and slow pyrolysis [18]. Fast pyrolysis, the biomass is heated to a high heating rate (300 °C/min) with a short vapor residence time (in the order of seconds) to achieve a high bio-oil fraction, decreasing the solid fraction yield [12,14,18]. Slow pyrolysis uses a low heating rate (10 °C/min) and long residence times (5-30 min) to produce more tar and char [12,14,18]. The bio-oil produced from the pyrolysis of microalgal biomass is highly acidic, and it cannot be used directly for power generation. Therefore, bio-oil needs to be deoxygenated before it is used as gasoline or diesel [12,23]. Fig. 27.2 shows the high heat value from bio-oil of some algal biomasses obtained by pyrolysis.

4. Hydrogen production by microorganisms

Hydrogen can be produced biologically from unicellular organisms through direct and indirect bio-photolysis, dark fermentation, and photo-fermentation as a sustainable route at a small-scale [24–26].

4.1 Direct and indirect bio-photolysis

The bio-photolysis is a light-driven process where water splits into hydrogen and oxygen under controlled conditions by using two key catalysts: photosystem II (PSII) and the [FeFe]-hydrogenase enzyme [24,27]. This process occurs inside the thylakoid membrane of the chloroplast in green algae and plants through electron transfer. Light in the form of photons is directly absorbed by the PSII protein complex at 680 nm and/or the PSI protein complex at 700 nm, generating a strong oxidant that can oxidize water into protons (H^+), electrons, and oxygen as waste material [28]. The electrons generated inside the PSII protein complex are transported to PSI protein complex through the electron transport chain composed by the PQ pool and cytochrome b6f complex (Cyt b6f). Once electrons are inside the PSI protein complex, they are transported to the chloroplast ferredoxin (Fd). Reduced Fd acts as an electron donor to [FeFe]-hydrogenases which reversibly facilitates the reduction of proton (H^+) to hydrogen molecules via direct bio-photolysis as it is shown in Fig. 27.3 [25,28]. The [FeFe]-hydrogenase in microalgae is the machinery that consumes electrons to generate hydrogen, it is monomeric enzyme with a molecular weight of almost 48 KDa and it has a sole prosthetic group (H-group) in its active center, which result in 100-fold higher enzyme activity than [NiFe]-hydrogenase and [Fe]-hydrogenase [27,28]. In model organisms such as *Chlamydomonas reinhardtii* has been extensively studied the deprivation of sulfur to decrease the oxygen production in PSII and establishes anoxic conditions in the culture, which is mandatory to activate the synthesis and activity of [FeFe]-hydrogenase [24,29].

The electrons generated from water splitting in PSII produce ATP and NADPH, which will be used to fix carbon dioxide into glycogen via photosynthesis before being used to form hydrogen via indirect bio-photolysis [28]. On this point, microalgae are allowed to grow under normal cultivation conditions with light because they aim to increase the carbohydrates inside the cells since they will be used as an endogenous substrate for hydrogen production [25,27]. This alternative pathway (indirect bio-photolysis) is also light-dependent, and it is known as the PSII-independent pathway since it does not involve PSII [30]. Electrons from the glycolytic pathway and citric acid cycle of the endogenous substrate are transferred via the plastoquinone pool (PQ) to PSI, and then electrons arrive at the [FeFe]-hydrogenase resulting in hydrogen production (Fig. 27.3) [30]. Compared to direct bio-photolysis, oxygen inhibition is not significant for hydrogen production. However, both the PSII-dependent and independent pathways act as primary sources of electrons in sulfur deprivation [24,28]. Both paths are considered the most desirable route for hydrogen production since only water and light are required according to the Eq. 27.3, 27.4 and 27.5 [25,28]:

$$2H_2O \xrightarrow[\text{PS and } H_2\text{ase}]{} 2H_2 + O_2 (\text{Direct biophotolysis}) \tag{27.3}$$

$$6H_2O + CO_2 \xrightarrow[\text{PS}]{} C_6H_{12}O_6 + O_2 (\text{Indirect biophotolysis}) \tag{27.4}$$

$$6H_2O + C_6H_{12}O_6 \xrightarrow[H_2\text{ase}]{} 12H_2 + 6CO_2 (\text{Indirect biophotolysis}) \tag{27.5}$$

4.2 Dark fermentation

Dark fermentation is an alternative path for biological hydrogen production from microalgae biomass. It is a process carried out in the dark under anaerobic conditions and utilizes organic substrate as the sole energy source to generate electrons [28,31]. Several facultative/obligate anaerobic bacterial genera can use the carbohydrates, proteins, and lipids inside the microalgae as substrate to produce hydrogen, carbon dioxide, and some simple organic compounds such as volatile fatty acids and alcohol [24,32]. Electrons that are used to generate hydrogen come from two stages in fermentation: In the first stage, substrate oxidation to pyruvate in cytoplasm generates NADH, which can be used by some microorganisms that possess the FNR to generate reduced Fd. In the second stage, pyruvate can be converted to acetyl-CoA and formate by pyruvate formate lyase (PFL). Some organisms such as *Escherichia coli* and Enterobacteriaceae use formate hydrogen lyase (FHL) complex that contains a special class of [NiFe]-hydrogenase or a [FeFe]-hydrogenase to convert formate to hydrogen and carbon dioxide (Fig. 27.3) [28,32]. Also, pyruvate can be converted to acetyl-CoA and reduced Fd by pyruvate-Fd oxidoreductase (PFOR). The reduced Fd produced here and in the first stage by FNR is oxidized by [FeFe]-hydrogenase to produce hydrogen (Fig. 27.3) [32]. This type of hydrogen-producing reaction is typical of *Clostridium* species [28]. The substrate usually used to produce hydrogen is glucose, and according to stoichiometry, 12 moles of hydrogen can be generated from 1 mol of glucose (Eq. 27.5). However, the maximum yield in dark fermentation is 4 mol hydrogen/1 mol glucose, i.e., only 33 % theoretical yield [28]. Another important aspect during the hydrogen production from glucose is the accumulation of alcohol and various organic acid inside the reactor that, beyond a certain level, inhibits cell growth and decreases the hydrogen yield [28].

4.3 Photo-fermentation

As with dark fermentation, photo-fermentation utilizes short-chain organic acids as electron donors in the presence of light to produce hydrogen and carbon dioxide by employing photosynthetic bacteria under anaerobic conditions [28,33]. Generally, two types of bacteria are responsible for hydrogen production: purple non-sulfur bacteria and green sulfur bacteria. These bacteria include *Rhodobacter sphaeroides*, *Rhodopseudomonas palustris*, *Rhodobacter capsulatus*, and *Rhodospirillum rubrum*. They can be helpful in the conversion of the organic acid (acetic, butyric, propionic, malic, and lactic) formed during the anaerobic fermentation of organic wastes to hydrogen and carbon dioxide [33,34], since oxygen does not evolve during hydrogen production because purple non-sulfur bacteria do not have PS sufficiently powerful to split water. There is not inhibition of the hydrogenase by oxygen [28,33]. Thereby, 4 mol of hydrogen can be produced from 1 mol of acetic acid according to stoichiometry (Eq. 27.6). Nevertheless, nitrogenase is responsible for hydrogen production under anaerobic conditions [35]. This reaction is not spontaneous and requires external energy input in the form of light under anaerobic conditions to oxidize the substrate and producing ATP and electrons. The electrons scavenged from organic acid oxidation are transferred to oxidized ferredoxin (Fdox) through a series of membrane-bound, electron transport carrier molecules. Then, the electrons in reduced ferredoxin (Fdrd) are primarily used to reduce molecular dinitrogen (N_2) to ammonia (NH_3) by the action of nitrogenase (N_2ase) (Eq. 27.7). However, more hydrogen can be produced in the absence of N_2 since nitrogenase catalyzes the reduction of protons to produce hydrogen (Eq. 27.8) [27,33,35]. This technology is an excellent way to convert the by-products from dark fermentation to hydrogen and carbon dioxide reaching a yield coefficient of up to 80 % of the theoretical yield [35].

$$CH_3COOH + H_2O \xrightarrow[N_2\text{ase}]{} 4H_2 + 2CO_2 \tag{27.6}$$

$$N_2 + 8H^+ + 8e^- + 16ATP \xrightarrow[N_2\text{ase}]{} 2NH_3 + H_2 + 16ADP + 16P_i \tag{27.7}$$

$$8H^+ + 8e^- + 16ATP \xrightarrow[N_2\text{ase}]{} 4H_2 + 16ADP + 16P_i \tag{27.8}$$

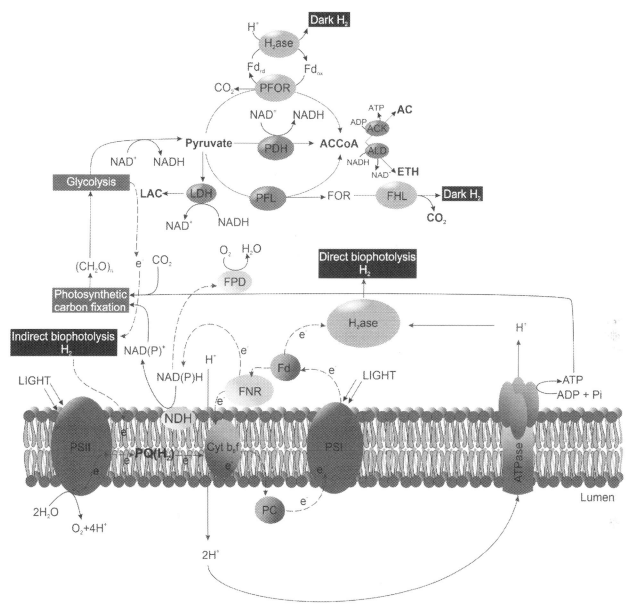

FIGURE 27.3 Hydrogen production from direct and indirect bio-photolysis processes by microalgae and hydrogen production from dark fermentation by bacteria. The electron transfer reactions by the thylakoid membrane and directions are drawn in *blue dotted arrows* (dark gray in print) for direct photolysis. *Black lines* indicate the multiple reactions. *AC*, acetate; *ACK*, acetate kinase; *ACCoA*, acetyl-CoA; *ALD*, alcohol dehydrogenase; *ATH*, ethanol; *ATPase*, ATP synthase; *Cyt b6f*, cytochrome b6f complex; *Fd*, ferredoxin; *FNR*, ferredoxin-NAD(P)$^+$ reductase; *H$_2$ase*, hydrogenase; *FOR*, formate; *FHL*, formate hydrogen lyase; *LAC*, lactate; *LDH*, lactate dehydrogenase; *NDH*, NAD(P)H, dehydrogenase; *PS*, photosystem; *PC*, plastocyanin; *PQ*, plastoquinones; *PQ(H2)*, reduced plastoquinones; *PDH*, pyruvate dehydrogenase; *PFL*, pyruvate:formate lyase; *PFOR*, pyruvate ferredoxin oxidoreductase. *This figure is adapted and modified from Y.K. Oh, S.M. Raj, G.Y. Jung, S. Park, Current status of the metabolic engineering of microorganisms for biohydrogen production, Bioresour. Technol. 102(18) (2011) 8357–8367.*

5. Benefits and challenges of hydrogen production

Hydrogen as an energy carrier is considered one of the critical resources in the energy transition to decarbonize the economy in the world because it is energy-efficient. It has a net heat value of 120 MJ/kg, higher than gasoline (46 MJ/kg), diesel (42.6 MJ/kg), natural gas (45 MJ/kg), and ethanol (27 MJ/kg), respectively. Hydrogen does not free greenhouse gas when it is burning. Hydrogen can be produced from almost all energy resources as renewable energy. It is an economically promising solution to the increasing global energy and food demand because it is safe, clean, reliable, and affordable, allowing the possibility of building a sustainable economy beyond the transportation sector [36–38]

Another critical aspect of hydrogen production is ammonia synthesis. Each year, around 180 million tons of ammonia are produced to sustain global food production [39−41], given that it is well known as indispensable raw material for soil fertilization, either as ammonium nitrate or urea [39,42]. Ammonia is produced from hydrogen obtained by steam methane reforming through the Haber-Bosch process (Eq. 27.9) by using atmospheric nitrogen at high pressure (150-250 bar) and temperature (400-500 °C) [4]. Ammonia production is one of the most energy-intensive industrial chemical processes, demanding nearly 100 GJ/tonNH3, which stands for 1-2 % of global energy consumption, and as a consequence, it is one of the biggest emitters of industrial carbon dioxide (1-1.5 % of anthropogenic carbon dioxide emissions) [41]. Therefore, hydrogen production from renewable energy is necessary to migrate toward a sustainable future where food and fossil fuels are decoupled [39].

$$N_2 + 3H_2 \rightarrow 2NH_3 \; \Delta H° = -91.8 kJ/mol \qquad (27.9)$$

Gaseous hydrogen has a density of 0.089 kg/m^3 at 25 °C and 1 atm [2]. Therefore, at ambient temperature and pressure, hydrogen's volumetric energy density is only 10 MJ/m^3 [4]. It implies that hydrogen has a very low volumetric energy density compared with methane (33 MJ/m^3), propane (87 MJ/m^3), gasoline (31.150 MJ/m^3), and methanol (15,800 MJ/m^3) in the same condition of temperature and pressure [4]. The above-mentioned means that hydrogen storage and transportation large scale are a potential challenge because it requires high pressures to be compressed (200-700 bar the volumetric energy density of hydrogen is 1800−4820 MJ/m^3) and very low temperature to be stored as liquid (-253 °C), leading to high operational cost [4,38]. An alternative for hydrogen storage and transportation is using hydrogen carriers that can absorb and release hydrogen when it is needed through chemical reaction for power generation [4]. Ammonia has been identified as a potential hydrogen carrier to store and transport hydrogen due to its high hydrogen content (17.65 % w/w) compared with methanol (12.5 % w/w) [41]. Ammonia and methanol can be stored and transported more easily. Besides, Ammonia and methanol can be liquefied easily [37].

Another challenge to overcome is the low yield that faces the hydrogen production biological methods because the dehydrogenase's activity is inhibited in oxygen presence. Also, biological methods are less reliable due to issues associated with controlling and maintaining cultivation, given that many bioreactors are expensive in their implementation. Therefore, it is least likely a scale-up [36]. Regarding thermochemical methods for hydrogen production, these are typically applied at a larger scale since various countries can take advantage of their fossil fuel-based infrastructure to produce blue hydrogen from cheap renewable feedstock that releases little greenhouse gas emissions, which can be storage by carbon capture and sequestration technologies [4,36]. The use of microalgae biomass in the thermochemical conversion to hydrogen is little likely because the cultivation yields are extremely low, and biomass harvesting still faces high operative costs.

Electrolysis is the most efficient and cost-effective method to obtain green hydrogen using energy from hydropower plants since this resource does not generate greenhouse gas emissions. This technology can be applicated only in countries rich in hydric resources. Also, it is highly beneficial to the Latin American region since it has a minimal impact on human health and can lead to new employment opportunities [36].

6. Conclusions and perspectives

This review indicates that hydrogen is the most promissory fuel to combat climate change and replace all energetic matrix derivatives from fossil fuels. Therefore, hydrogen production must be clean, safe, reliable, efficient, and cost-effective to achieve with success a sustainable energy transition that decarbonizes our economy. This implies that all countries should focus their efforts on sustainable energy systems integrating to produce green hydrogen through electrolysis. However, electrolysis does not is the unique pathway to obtain green hydrogen. It can be produced in a small-scale, safe, reliable way through microorganisms either by bio-photolysis, photo-fermentation, or dark fermentation despite being economically unviable.

According to our analysis, gasification is the method most promisor to produce blue hydrogen in thermochemical conversion in a safe way, reliable, efficient, and cost-effective by using of cheap feed stock that releases greenhouse gas emissions such as municipal solid waste. Blue hydrogen obtained by this method can accelerate the energy transition in at shorter term through carbon capture and storage, while green hydrogen becomes more affordable in the long term. On the other hand, using microalgae biomass to produce direct hydrogen by pyrolysis and gasification is not feasible since a high energy input is required to remove its moisture content and obtain hydrogen at industrial level, and a huge amount of biomass is required.

In the coming years, all clean technology for power generation, such as wind, solar, geothermal, and hydro, will converge on green hydrogen production via electrolysis for its application in transportation, power generation, heating, and chemical product synthesis (ammonia and methanol). This will allow humanity to overcome the climate crisis that faces and end its dependence on fossil fuels as the war by obtaining fossil resources. Therefore, the development of the most efficient system will be prioritized by countries to increase power generation through sustainable resources to obtain cost-effective hydrogen production.

Acknowledgments

This research project was supported by the Mexican Science and Technology Council (CONACYT, Mexico) with the Infrastructure Project FOP02-2021−04 (Ref. 317250). The author Gilver Rosero thanks the National Council for Science and Technology (CONACYT, Mexico) for his Ph.D. Fellowship support (grant number: 750752).

References

[1] N. Sánchez-Bastardo, R. Schlögl, H. Ruland, Methane pyrolysis for zero-emission hydrogen production: a potential bridge technology from fossil fuels to a renewable and sustainable hydrogen economy, Ind. Eng. Chem. Res. 60 (32) (2021) 11855−11881.
[2] N.S. Muhammed, B. Haq, D. Al Shehri, A. Al-Ahmed, M.M. Rahman, E. Zaman, A review on underground hydrogen storage: insight into geological sites, influencing factors and future outlook, Energy Rep. 8 (2022) 461−499.
[3] C. Kurien, M. Mittal, Review on the production and utilization of green ammonia as an alternate fuel in dual-fuel compression ignition engines, Energy Convers. Manag. 251 (2022) 114990.
[4] D. Hjeij, Y. Biçer, M. Koç, Hydrogen strategy as an energy transition and economic transformation avenue for natural gas exporting countries: Qatar as a case study, Int. J. Hydrog. Energy 47 (8) (2021) 4977−5009.
[5] M. Neuwirth, T. Fleiter, P. Manz, R. Hofmann, The future potential hydrogen demand in energy-intensive industries-a site-specific approach applied to Germany, Energy Convers. Manag. 252 (2022) 115052.
[6] G. Wang, Y. Chao, Y. Cao, T. Jiang, W. Han, Z. Chen, A comprehensive review of research works based on evolutionary game theory for sustainable energy development, Energy Rep. 8 (2022) 114−136.
[7] W.C. Nadaleti, E.G. de Souza, V.A. Lourenço, Green hydrogen-based pathways and alternatives: towards the renewable energy transition in South America's regions−Part B, Int. J. Hydrog. Energy 47 (1) (2022) 1−15.
[8] J.A. Turner, Sustainable hydrogen production, Science 305 (5686) (2004) 972−974.
[9] W.C. Nadaleti, V.A. Lourenço, G. Americo, Green hydrogen-based pathways and alternatives: towards the renewable energy transition in South America's regions−Part A, Int. J. Hydrog. Energy 46 (43) (2021) 22247−22255.
[10] J. Mathushika, C. Gomes, Development of microalgae-based biofuels as a viable green energy source: challenges and future perspectives, Biointerface Res. Appl. Chem. 12 (2021) 3849−3882.
[11] D.J. Farrelly, C.D. Everard, C.C. Fagan, K.P. McDonnell, Carbon sequestration and the role of biological carbon mitigation: a review, Renew. Sustain. Energy Rev. 21 (2013) 712−727.
[12] H.M.U. Ayub, A. Ahmed, S.S. Lam, J. Lee, P.L. Show, Y.K. Park, Sustainable valorization of algae biomass via thermochemical processing route: an overview, Bioresour. Technol. 344 (2022) 126399.
[13] S. Thanigaivel, A.K. Priya, K. Dutta, S. Rajendran, Y. Vasseghian, Engineering strategies and opportunities of next generation biofuel from microalgae: a perspective review on the potential bioenergy feedstock, Fuel 312 (2022) 122827.
[14] R. Sirohi, H.I. Choi, S.J. Sim, Microalgal fuels: promising energy reserves for the future, Fuel 312 (2022) 122841.
[15] R.K. Vij, D. Subramanian, S. Pandian, S. Krishna, S. Hari, A review of different technologies to produce fuel from microalgal feedstock, Environ. Technol. Innovat. 22 (2021) 101389.
[16] H.I. Choi, Y.J. Sung, M.E. Hong, J. Han, B.K. Min, S.J. Sim, Reconsidering the potential of direct microalgal biomass utilization as end-products: a review, Renew. Sustain. Energy Rev. (2021) 111930.
[17] G. Rosero-Chasoy, R.M. Rodríguez-Jasso, C.N. Aguilar, G. Buitrón, I. Chairez, H.A. Ruiz, Growth kinetics and quantification of carbohydrate, protein, lipids, and chlorophyll of Spirulina platensis under aqueous conditions using different carbon and nitrogen sources, Bioresour. Technol. 346 (2022) 126456.
[18] D.L. Barreiro, W. Prins, F. Ronsse, W. Brilman, Hydrothermal liquefaction (HTL) of microalgae for biofuel production: state of the art review and future prospects, Biomass Bioenergy 53 (2013) 113−127.
[19] B.E. Morales-Contreras, N. Flórez-Fernández, M.D. Torres, H. Domínguez, R.M. Rodríguez-Jasso, H.A. Ruiz, Hydrothermal systems to obtain high value-added compounds from macroalgae for bioeconomy and biorefineries, Bioresour. Technol. 343 (2022) 126017.
[20] G. Su, H.C. Ong, Y.Y. Gan, W.H. Chen, C.T. Chong, Y.S. Ok, Co-pyrolysis of microalgae and other biomass wastes for the production of high-quality bio-oil: progress and prospective, Bioresour. Technol. 344 (2022) 126096.
[21] A. Raheem, G. Ji, A. Memon, S. Sivasangar, W. Wang, M. Zhao, Y.H. Taufiq-Yap, Catalytic gasification of algal biomass for hydrogen-rich gas production: parametric optimization via central composite design, Energy Convers. Manag. 158 (2018) 235−245.
[22] R. Nagarajan, J. Dharmaraja, S. Shobana, A. Sermarajan, D.D. Nguyen, S. Murugavelh, A comprehensive investigation on Spirulina platensis−part I: cultivation of biomass, thermo−kinetic modelling, physico−chemical, combustion and emission analyses of bio−oil blends in compression ignition engine, J. Environ. Chem. Eng. 9 (3) (2021) 105231.
[23] N. Pragya, K.K. Pandey, P.K. Sahoo, A review on harvesting, oil extraction and biofuels production technologies from microalgae, Renew. Sustain. Energy Rev. 24 (2013) 159−171.

[24] G. Buitrón, J. Carrillo-Reyes, M. Morales, C. Faraloni, G. Torzillo, Biohydrogen production from microalgae, in: Microalgae-Based Biofuels and Bioproducts, Woodhead Publishing, 2017, pp. 209–234.
[25] J. Jimenez-Llanos, M. Ramirez-Carmona, L. Rendon-Castrillon, C. Ocampo-Lopez, Sustainable biohydrogen production by *Chlorella* sp. microalgae: a review, Int. J. Hydrog. Energy 45 (15) (2020) 8310–8328.
[26] P. Cheng, Y. Li, C. Wang, J. Guo, C. Zhou, R. Zhang, R. Ruan, Integrated marine microalgae biorefineries for improved bioactive compounds: a review, Sci. Total Environ. (2022) 152895.
[27] M.K. Lam, A.C.M. Loy, S. Yusup, K.T. Lee, Biohydrogen production from algae, in: Biohydrogen, Elsevier, 2019, pp. 219–245.
[28] Y.K. Oh, S.M. Raj, G.Y. Jung, S. Park, Current status of the metabolic engineering of microorganisms for biohydrogen production, Bioresour. Technol. 102 (18) (2011) 8357–8367.
[29] A.R. Limongi, E. Viviano, M.D. Luca, R.P. Radice, G. Bianco, G. Martelli, Biohydrogen from microalgae: production and applications, Appl. Sci. 11 (4) (2021) 1616.
[30] E. Eroglu, A. Melis, Microalgal hydrogen production research, Int. J. Hydrog. Energy 41 (30) (2016) 12772–12798.
[31] G. Antonopoulou, I. Ntaikou, K. Stamatelatou, G. Lyberatos, Biological and fermentative production of hydrogen, in: Handbook of Biofuels Production, Woodhead Publishing, 2011, pp. 305–346.
[32] C. Ding, K.L. Yang, J. He, Biological and fermentative production of hydrogen, in: Handbook of Biofuels Production, Woodhead Publishing, 2016, pp. 303–333.
[33] M.M. Habashy, E.S. Ong, O.M. Abdeldayem, E.G. Al-Sakkari, E.R. Rene, Food waste: a promising source of sustainable biohydrogen fuel, Trends Biotechnol. 39 (12) (2021) 1274–1288.
[34] E. Sağır, P.C. Hallenbeck, Photofermentative hydrogen production, in: Biohydrogen, Elsevier, 2019, pp. 141–157.
[35] F. Dalena, A. Senatore, A. Tursi, A. Basile, Bioenergy production from second-and third-generation feedstocks, in: Bioenergy Systems for the Future, Woodhead Publishing, 2017, pp. 559–599.
[36] S.F. Ahmed, M. Mofijur, S. Nuzhat, N. Rafa, A. Musharrat, S.S. Lam, A. Boretti, Sustainable hydrogen production: technological advancements and economic analysis, Int. J. Hydrog. Energy 47 (88) (2021) 37227–37255.
[37] C. Hakandai, H.S. Pramono, M. Aziz, Conversion of municipal solid waste to hydrogen and its storage to methanol, Sustain. Energy Technol. Assess. 51 (2022) 101968.
[38] V. Okoro, U. Azimov, J. Munoz, Recent advances in production of bioenergy carrying molecules, microbial fuels, and fuel design—a review, Fuel 316 (2022) 123330.
[39] P.H. Pfromm, Towards sustainable agriculture: fossil-free ammonia, J. Renew. Sustain. Energy 9 (3) (2017) 034702.
[40] J. Osorio-Tejada, N.N. Tran, V. Hessel, Techno-environmental assessment of small-scale Haber-Bosch and plasma-assisted ammonia supply chains, Sci. Total Environ. (2022) 154162.
[41] S. Chatterjee, R.K. Parsapur, K.W. Huang, Limitations of ammonia as a hydrogen energy carrier for the transportation sector, ACS Energy Lett. 6 (12) (2021) 4390–4394.
[42] P. Wang, S. Wang, B. Wang, L. Shen, T. Song, Green production of ammonia from nitrogen-rich biomass pyrolysis: evolution of fuel-N under H2-rich atmosphere, Fuel Process. Technol. 227 (2022) 107126.

Chapter 28

Algal biofuel production using wastewater: a sustainable approach

Sougata Ghosh[1,2], Bishwarup Sarkar[3] and Sirikanjana Thongmee[1]

[1]*Department of Physics, Faculty of Science, Kasetsart University, Bangkok, Thailand;* [2]*Department of Microbiology, School of Science, RK University, Rajkot, Gujarat, India;* [3]*College of Science, Northeastern University, Boston, MA, United States*

1. Introduction

The persistent increase in the global population is contributing to increased energy demands, leading to an increase in the indiscriminate use of fossil fuels [1]. Such excessive use of fossil fuels significantly contributes to greenhouse gas emissions and global warming. The increase in atmospheric CO_2 concentration also has a negative impact on biodiversity. Surveys have also shown an almost 50% increase in energy demand between 2020 and 2030, especially in developing countries [2]. Hence, alternative sources of energy being sought that are renewable, carbon-neutral, and environmentally benign.

Algae are photoautotrophs that have the ability to grow in various wastewater samples in the presence of sunlight and CO_2 to form biomass. Most algal strains store considerable energy inside the cells [3]. Therefore, various microalgae are used to prepare a considerable amount of biomass and algal oil that is efficiently employed in biofuel production [1]. This chapter provides detailed insight into the use of various wastewater samples as the nutrient source for algal cultivation followed by biofuel production. Several algal genera such as *Actinastrum*, *Chlamydomonas*, *Chlorella*, *Heynigia*, *Micractinium*, and *Scenedesmus* are discussed in this chapter for the production of biofuel (Table 28.1). These different algal strains are cultivated for biofuel production using dairy farm, industrial, domestic, piggery, aquaculture, oasis, municipal, mill, and livestock wastewater samples with efficient lipid productivity rates.

Further evaluation of the large-scale applicability, economic feasibility, and long-term environmental implications of the use of wastewater for producing algal biofuels will provide insight into this sustainable approach and help solve the energy crisis and water pollution issue.

2. Algal biofuel production using wastewater

The feasibility of microalgae-mediated biodiesel production in the presence of industrial wastewater as the nutrient source was evaluated [4]. Two freshwater microalgal strains (*Desmodesmus* sp.TAI-1 and *Chlamydomonas* sp. TAI-2) were isolated and cultivated in Bold's basal medium, and then the strains were cultured in a photobioreactor in the presence of a raw industrial wastewater sample, as shown in Fig. 28.1. The biomass as well as lipid productivity of *Chlamydomonas* sp. TAI-2 was better than that of *Desmodesmus* sp.TAI-1. The lipid productivity of *Chlamydomonas* sp. TAI-2 was 0.34 g L^{-1} under continuous illumination, compared with 0.25 g L^{-1} under a 14:10 light–dark cycle. Moreover, the ammonium consumption rate of *Chlamydomonas* sp. TAI-2 was 19.2 mg of NH_4^+-N/L/day whereas the phosphate removal rate was 33% after 10 days of reactor operation. Maximum biomass of 1.5 g L^{-1} was achieved using the industrial wastewater sample after 5 days of operation along with a maximum lipid productivity of 0.28 g L^{-1} after 10 days of photobioreactor operation. Thus, the fatty acid methyl ester (FAME) profile analysis demonstrated the presence of 54%–79% saturated fatty acids with a dominance of C16:0, which is palmitic acid. In addition, the concentration of linolenic acid was less than 12% after 8 days of operation, which is ideal for producing high-quality biodiesel. Similarly, the oleic acid content was 31.6%; it is an optimal fatty acid for biofuel generation that has oxidative stability.

TABLE 28.1 Algae-mediated biofuel production using wastewater as substrate.

Algae	Type of wastewater	Total lipid productivity	Biofuel production rate	References
Chlamydomonas sp. TAI-2	Industrial wastewater	0.34 g L^{-1}	–	[4]
Chlorella pyernoidosa	Dairy wastewater	–	–	[5]
Chlorella sp.	Sewage wastewater	29.6%	4.9 mg L^{-1} day	[6]
Chlorella sp., *Chlamydomonas* sp., *Scenedesmus* sp., and *Gloeocystis* sp.	Carpet and rug mill wastewater and municipal sewage	6.82%	9.2–17.8 tons/ha year	[7]
Chlorella sp., *Heynigia* sp., *Hindakia* sp., *Micractinium* sp., and *Scenedesmus* sp.	Concentrated municipal sewage	17.41–33.53%	94.8 mg L^{-1} day	[8]
Chlorella sp., *Micractinium* sp., *Scenedesmus* sp., and *Actinastrum* sp.	Municipal wastewater and anaerobically treated dairy wastewater	4.9–11.3% and 10–29%	24 mg L^{-1} day and 17 mg L^{-1} day	[9]
C. vulgaris	Oasis wastewater	30.5%	–	[10]
C. vulgaris, Chlorella sorokiniana, and *Scenedesmus simris002*	Domestic wastewater	44.1%	0.57 g L^{-1} day	[11]
Chlorella zofingiensis	Piggery wastewater	45.81%	110.56 mg L^{-1} day	[12]
Desmodesmus sp.	Raw wastewater	26.3%	–	[20]
Desmodesmus sp. EJ8-10	Livestock wastewater	19.4–28%	4.8 ± 0.3 mg L^{-1} day	[13]
Mixed microalgal community	Municipal wastewater	44.5 ± 4.7%	–	[14]
Mixed microalgal community	Municipal wastewater	52.2%	–	[15]
Mixed microalgal community	Dairy farm wastewater	23.62 ± 1.19%	33.38 thousand liters per hectare per year	[16]
Nannochloropsis oculata	Anaerobically and aerobically treated swine wastewater	–	0.035–0.177 g L^{-1} day	[17]
Platymonas subcordiformis	Aquaculture wastewater	–	–	[18]
Scenedesmus sp. ZTY1	Domestic wastewater	148 mg L^{-1}	–	[19]

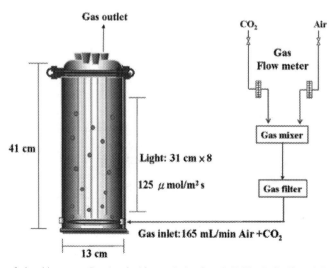

FIGURE 28.1 Schematic diagram of photobioreactor. *Reprinted with permission from L.F. Wu, P.C. Chen, A.P. Huang, C.M. Lee, The feasibility of biodiesel production by microalgae using industrial wastewater, Bioresour. Technol. 113 (2012) 14–18, Copyright ©2011 Elsevier Ltd.*

Chlorella pyernoidosa was used to integrate wastewater treatment and biofuel production [5]. For this study, a dairy wastewater sample was used as the nutrient source for algal growth. The content of nitrate and phosphate was reduced by 49% and 83%, respectively, in the effluent. Then, the algal biomass was quantified after 15 days of growth, in which 18.8 g L^{-1} of the fresh weight of *Chlorella* was attained in the presence of 75% influent whereas 14.1 g L^{-1} of biomass was obtained using 75% of effluent wastewater as the growth medium. Likewise, the dry weight of algae was 4-fold higher when influent wastewater samples were used. After 15 days of algae growth, n-hexane and methanol were used to extract the algal oil, which provided 3.5 and 2.0 g of extracted oil from the influent and effluent, respectively, as the substrate. In addition, 6.7 and 4.9 mL of biodiesel were produced using the influent and effluent wastewater samples, respectively.

Sewage wastewater was used for microalga-mediated CO_2 fixation and biodiesel production [6]. *Chlorella* sp. was isolated from a wastewater sample of a sewage treatment plant that was enriched in Blue-Green 11 (BG-11) medium. *Chlorella* sp. isolate exhibited a maximum growth of 475 mg L^{-1} in the presence of 3500 flux of light intensity. The optimal temperature and pH for isolate were 25°C and 9.0, respectively. *Chlorella* sp. displayed a maximum lipid yield of 29.6% along with biomass of 330 mg L^{-1} in the presence of 1 mM $NaNO_3$, 0.04 mM KH_2PO_4, and 50 mM $NaHCO_3$ which provided the nitrogen, phosphorus, and carbon sources, respectively. The lipid productivity rate of *Chlorella* sp. in the presence of 50 mM of $NaHCO_3$ was 4.9 mg L^{-1} day. The FAME profile analysis of *Chlorella* sp. displayed 29.80% saturated and 70.20% unsaturated fatty acids.

In another, similar study, Chinnasamy et al. [7] demonstrated the potential of microalgae cultivation along with biofuel production in the presence of wastewater samples. Wastewater effluents from carpet and rug mills in conjunction with municipal sewage were used in this study, in which 27 different green algal species were identified. *Chlorella, Chlamydomonas, Scenedesmus,* and *Gloeocystis* were isolated to create a microalgal consortium that was used for wastewater treatment and biomass production. Along with the prepared consortium, two freshwater algal strains (*Botryococcus braunii* UTEX 572 and *Chlorella saccharophila* var. *saccharophila* UTEX 2469) and two marine algal species (*Dunaliella tertiolecta* UTEX LB999, and *Pleurochrysis carterae* CCMP 647) were considered for comparative analysis with the prepared consortium. More than 96% of nutrients in the wastewater were removed by the microalgal consortium within 72 h of reactor operation at 15°C. The biomass production rate and total lipid content were around 9.2—17.8 tons/ha year and 6.82%, respectively. Around 63.9% of algal oil that was produced during wastewater treatment was successfully converted into biodiesel. Therefore, biofuel production is possible using native algal cultures present in the wastewater samples. Energy recovery from such native microalgal species can provide more enhanced biofuel production rates.

Zhou et al. [8] demonstrated the use of concentrated municipal wastewater for algal biofuel production. The enrichment of algae from the concentration of municipal waste was carried out using a BG-11 medium; 17 isolates were obtained that had stable growth in terms of stable biomass and lipid productivities. The 18S rRNA gene sequencing results identified the strains as *Chlorella* sp., *Heynigia* sp., *Hindakia* sp., *Micractinium* sp., and *Scenedesmus* sp. A high biomass concentration range of 0.48—1.08 g L^{-1} of total volatile suspended solids (VSS) and with the highest growth rate of 0.492/day^{-1} were achieved after 4 days of operation in the presence of concentrated municipal waste. The total lipid content of the isolated algal species ranged from 17.41% to 33.53% of total VSS. The highest lipid productivity rate of 94.8 mg L^{-1} day was observed for *Chlorella* sp. The total organic carbon removal efficiency ranged from 82.27% to 96.18%, highlighting the photoheterotrophic growth of algal species using concentrated municipal waste as the medium. Gas chromatography-mass spectrometry results further demonstrated the presence of C16—C18 fatty acids as the dominant lipid type of total fatty acid.

Dairy and municipal wastewater samples were used to remove nutrients as well as algal growth followed by biofuel production [9]. The semicontinuous flow experiments using municipal wastewater samples showed a dominance of *Chlorella, Micractinium,* and *Actinastrum* algal genera with a steady-state biomass after 11 days of reactor operation. The lipid content of algae cultivated from municipal wastewater samples ranged from 4.9% to 11.3% of VSS by weight and had a maximum lipid productivity rate of 24 mg L^{-1} day. In the case of dairy wastewater samples, the biomass was increased to a maximum of 900 mg L^{-1} of VSS after 13 days of reactor operation. *Scenedesmus, Micractinium, Chlorella,* and *Actinastrum* were the dominant algal genera in the dairy wastewater. The total lipid content in 25% diluted dairy wastewater sample ranged from 10% to 29% by weight with a maximum lipid productivity of 17 mg L^{-1} day that was attained after 5 days of reactor operation. The removal of nutrients from wastewater samples was also investigated, in which more than 99% of ammonium and orthophosphate was removed when treated with CO_2 sparging, with a hydraulic retention time of 3 and 4 days. After 6 days of operation, the total ammonia nitrogen and orthophosphate content from 25% diluted dairy wastewater samples was reduced from 30 and 2.6 mg L^{-1} to less than 5 and 0.6 mg L^{-1} after 6 and 9 days of operation, respectively. This study highlights the waste-to-biofuel production potential using algae.

The potential of algal biofuel production using wastewater was evaluated by Fathi et al. [10]. *Chlorella vulgaris* was used in this study; it was isolated from a lake water sample. The inoculation of *C. vulgaris* in the wastewater led to an increase in the pH value that was attributed to the increased photosynthetic activity of the planktonic algae. No lag phase was seen with regard to the growth of *C. vulgaris*, which highlights the highly adaptive behavior of the algae. The wastewater sample with higher nitrogen, phosphorus, and chemical oxygen demand (COD) levels facilitated improved algal growth. The inorganic phosphate and nitrate contents of the wastewater samples were reduced with considerable growth of the algae. On the contrary, the dissolved oxygen increased after the wastewater samples were treated with algae, which was attributed to algal photosynthesis, whereas the increase in alkalinity was suggested to result from bacteria-mediated organic matter decomposition. The COD value also increased consistently because the algae used CO_2 as the carbon source in the absence of organic substrates. The chemical composition analysis of *C. vulgaris* was 54% and 30.5% dry weights of protein and lipid contents, respectively, which implied the high biofuel production ability of the algal strain from the wastewater sample. Hence, 72.54% dry weight biodiesel production was achieved in this study along with 32.52% sediment formation.

Khalekuzzaman et al. [11] reported producing algal biofuels using a hybrid anaerobic baffled reactor (HABR) and photobioreactor system. The uninsulated HABR system removed $93 \pm 7\%$ and $91 \pm 7\%$ of COD and total suspended solids (TSS), respectively, from the wastewater. The organic loading and removal rate of the insulated HABR system were 0.67 ± 0.31 and 0.61 ± 0.31 kg of COD/m^3 day, respectively. The turbidity of the wastewater samples was also reduced by $98 \pm 1\%$ and $97 \pm 2\%$ for uninsulated and insulated HABR systems, respectively. A maximum of $96 \pm 5\%$ of VSS and $95 \pm 9\%$ of TSS was removed by the insulated HABR system, which facilitated the production of a healthy feedstock for cultivating microalgae with a 3:1 ratio for nitrogen–phosphorus. A coculture of three microalgal strains (*C. vulgaris*, *Chlorella sorokiniana*, and *Scenedesmus simris002*) exhibited the significant production of lipid (44.1%), whereas gas chromatography with flame ionization detection results highlighted that 87.9% of the biodiesel produced was composed of FAME. The highest biofuel productivity of 0.57 g L^{-1} day was achieved in this study in the presence of a low light intensity of 110 ± 31 μmol m^{-2} s in the morning, which gradually increased to 644 ± 485 μmol m^{-2} s in the afternoon.

In another study, Zhu et al. [12] investigated the biodiesel production of algae using piggery wastewater along with the removal of nutrients that could be useful for wastewater treatment, as schematically depicted in Fig. 28.2. *Chlorella zofingiensis* was used in this study. It was initially grown in BG-11 medium, after which the algae was cultivated in the

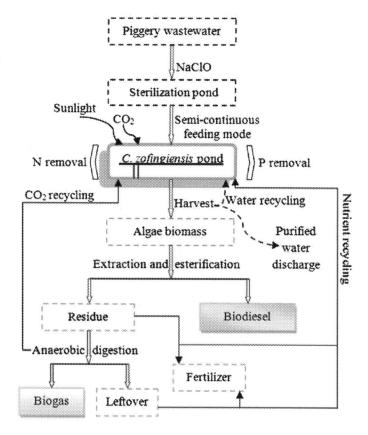

FIGURE 28.2 Proposed scale-up scheme for *Chlorella zofingiensis*-based biofuel production using piggery wastewater. Although this image is meant to illustrate connections between each process, the model does not reflect the physical layout of the integrated approach, because algal ponds must be located close to the pig farm. *Reprinted with permission from L. Zhu, Z. Wang, Q. Shu, J. Takala, E. Hiltunen, P. Feng, Z. Yuan, Nutrient removal and biodiesel production by integration of freshwater algae cultivation with piggery wastewater treatment. Water Res. 47 (2013) 4294–4302, Copyright ©2013 Elsevier Ltd.*

piggery wastewater sample using a tubular bubble column photobioreactor. An initial COD of 1900 mg L^{-1} displayed the highest biomass concentration of 2.962 ± 0.192 g L^{-1} along with a biomass productivity of 296.16 ± 19.16 mg L^{-1} day. A considerable reduction in COD concentrations of wastewater samples was observed within 2 days of operation, in which 79.84% of COD reduction was attained from cultures with a COD of 1900 mg L^{-1}. Moreover, 82.70% of total nitrogen and 98.17% of total phosphorus were efficiently removed from the wastewater sample with 1900 mg L^{-1} of COD. The maximum algal lipid content of 45.81% and 42.16% was achieved with a wastewater sample that had initial COD concentrations of 400 and 800 mg L^{-1}, respectively. The highest lipid productivity rate of 110.56 mg L^{-1} day was observed using the wastewater sample with 1900 mg L^{-1} of COD. FAME analysis revealed the presence of C16:0 (palmitic acid methyl ester), C18:2 (octadecadienoic acid methyl ester), and C18:3 (octadecatrienoic acid methyl ester) as the most abundant fatty acids.

In another study, Li et al. [13] demonstrated efficient microalgal biodiesel production using anaerobically digested livestock wastewater. The *Desmodesmus* sp. EJ8-10 algal strain was used in this study. It was isolated from a freshwater sample and cultured in BG-11 medium. The average NH_4^+-N removal rate of *Desmodesmus* sp. from digested livestock wastewater ranged from 1.61 to 4.84 mg L^{-1} day, which was lower than the removal rate of 1.76–7.92 mg L^{-1} day from the undigested wastewater sample. The total nitrogen removal efficiencies were 89% and 84% when 20% of undigested wastewater and 10% of digested wastewater samples were used, respectively. The phosphorus removal efficiency reached almost 100% after 14 days of the process operation. The dry microalgal biomass concentration increased from 0.01 to 0.33–0.39 g L^{-1} after 14 days of incubation in the presence of the undigested livestock wastewater sample, whereas the final biomass concentration in the presence of the anaerobically digested wastewater sample ranged from 0.15 to 0.35 mg L^{-1} under similar conditions. The presence of a higher concentration of macronutrients such as phosphorus along with trace elements and other micronutrients in the untreated wastewater sample was attributed to the increased biomass. The digested wastewater samples facilitated the production of a higher lipid content of 19.4%–28%, compared with 18.7% –22.3% of lipids using undigested wastewater samples. The lipid productivity of 10% digested livestock wastewater was 4.8 ± 0.3 mg L^{-1} day along with biomass productivity, whereas total lipid yields were 24.7 ± 0.1 and 67.1 ± 03.8 mg L^{-1}, respectively.

Indigenous microalgae were also grown in raw wastewater and evaluated for their biodiesel production ability [20]. The microalgae used in this study were isolated from an artificial lake that was cultivated in open batch reactors containing the wastewater sample to simulate the open pond system. *Desmodesmus* sp. and two mixed algal cultures were successfully cultivated in this system, resulting in the production of maximum *Desmodesmus* sp. biomass (0.58 g L^{-1}). After 22 days of operation, 83.9% of total nitrogen was removed from the wastewater by the mixed microalgal culture that was dominated by *Desmodesmus* sp. In addition, 99.8% of total coliform present in the wastewater was successfully removed from the wastewater by *Desmodesmus* sp. The biomass obtained was then separated using the ozone-floatation method, which improved lipid extractability. A maximum lipid productivity of 26.3% was obtained using the mixed algal cultures, whereas 13.3% biomass of *Desmodesmus* sp. was attained. FAME profile analysis further revealed the dominance of C16:0 (palmitic acid) in the harvested biomass of *Desmodesmus* sp. The total saturated FAME had a higher abundance compared with monounsaturated and polyunsaturated FAME in the ozone-separated microalgal biomass. The ozone treatment facilitated the saturation of FAME, which was proposed to be desirable for producing biodiesel.

In another study, Oberts et al. [14] demonstrated a feasible strategy for algal biofuel production using municipal wastewater samples. An integrated wastewater algae-to-biocrude process was carried out using pilot-scale algal cultivation ponds in the presence of wastewater as the nutrient source for algae. A mixed microalgal community was efficiently cultivated using wastewater composed of 18 algal species such as *Cladophora* sp., *Cosmarium* sp., *Golenkinia radiata*, *Micractinium pusillum*, *Pediastrum boryanum*, *Scenedesmus bijuga*, *Scenedesmus quadricauda*, and *Selenastrum* sp.. The obtained microalgal community had 48.9%, 37.5%, and 14.0% of ash-free dry weight (afdw) carbon, oxygen, and lipid, respectively. Moreover, 3 g of freeze-dried harvested algae and 50 mL of water were used to carry out hydrothermal liquefaction at 350°C for 3 h. The biocrude yield of 44.5 ± 4.7% afdw was achieved, in which 78.7%, 10.1%, 4.4%, and 5.5% of elemental weights were carbon, hydrogen, nitrogen, and oxygen, respectively, along with an energy content of 39 MJ kg^{-1}. Around 18.4 ± 4.6% afdw and 45.0 ± 5.9% dry weight of aqueous coproducts (ACPs) and solid biochar were obtained after hydrothermal processing, in which organic carbon, total nitrogen, and total phosphorus in the ACPs were 4550 ± 460, 1640 ± 250, and 3.5 g L^{-1}, respectively. In addition, the energy density of the solid biochar product ranged from 8 to 10 MJ kg^{-1} in the presence of more than 20% dry weight of carbon.

Zhou et al. [15] demonstrated the treatment of wastewater using algae with the simultaneous production of biofuel. That study reported a novel system for the hydrothermal liquefaction of wastewater biosolids as well as algal strains to generate biocrude oil, named environment-enhancing energy. A mixed algal culture was used in the study, which was obtained from clarifier outlets of a wastewater treatment plant. Preliminary microscopic observations revealed the presence of diverse algal species such as *Chlorella* sp., *Scenedesmus obliquus*, and cyanobacteria, in which the algal mixture was bioaugmented with other algal strains such as *Botryococcus braunii*, *Chlamydomonas reinhardtii*, *Chlorella protothecoides*, *C. vulgaris*, *Nannochloropsis oculata*, and *Scenedesmus dimorphus*. A maximum cellular dry weight of 541 mg L^{-1} was achieved using 0.5% of posthydrothermal liquefaction wastewater (PHWW) wherein 94% of biomass was attributed to autotrophic microorganisms such as algae. The total biomass of 2500 mg L^{-1} was obtained which showed effective organic and nutrient removal ability from the 1% PHWW sample. Hence, a total of 86%, 100%, and 95% of total soluble nitrogen, ammonium nitrogen, and soluble phosphorus removal was removed from the wastewater, whereas 63% of organics was removed under similar conditions. A maximum biocrude oil yield of 52.2% was observed after hydrothermal liquefaction using biomass cultivated in carboy batch jars, in which the lipid content ranged from only 15% to 20%. It was suggested that this novel system could be viable for providing sustainable biofuel.

Similarly, Hena et al. [16] cultivated an algal consortium in the presence of dairy farm wastewater and evaluated its biodiesel production ability. The consortium was composed of two pure strains of *Chlorella sachharophila* and *Scenedesmus* sp. along with other wastewater isolates that were cultivated in BG-11 medium. The maximum biomass production of the consortium was 219.87 tons with around 51.37 thousand liters of algal oil production per hectare per year in the presence of treated wastewater, whereas biomass production and algal oil productivity using untreated wastewater were 137.68 tons and 33.38 thousand liters of algal oil production per hectare per year, respectively. The lipid production of the prepared consortium was 23.62 ± 1.19% and 21.34 ± 1.26% in the presence of treated and untreated dairy farm wastewater, respectively. The lipid productivity of the native algal strains was also investigated in high-rate algae ponds, revealing an average productivity of 0.213 g L^{-1} day after 10 days of incubation. The stowed energy in the consortium was reduced by 25.18% after the extraction of lipids. The dynamic hexane method was employed to convert algal biomass to algal oil, after which acid esterification was carried out to increase the yield of biodiesel. Gas chromatography analysis displayed the presence of a mixture of fatty acids with an abundance of C16—C18 fatty acids. Almost 63% of algal oil was composed of unsaturated fatty acids.

Wu et al. [17] cultivated *Nannochloropsis oculata* using swine wastewater for enhanced biofuel production. A comparative study of anaerobically and aerobically treated swine wastewater (AnATSW) against an artificial 3 × f/2 medium was carried out to analyze the growth rate and amount of oil produced by the algal strain. A biomass range of 0.94—3.22 g L^{-1} was attained within 5 days of cultivating *N. oculata* in the presence of 0%—50% (v/v) AnATSW. An optimal lipid productivity rate of 0.035—0.177 g L^{-1} day was achieved using 0%—50% (v/v) AnATSW, whereas a maximum lipid productivity of 0.122 g L^{-1} day was observed using 3 × f/2 medium when it was supplemented with vitamins along with 48.9% of oil content. A total nitrogen removal rate of 86.4 mg L^{-1} day was observed along with a biomass productivity rate of 170 mg L^{-1} day in the presence of 50% (v/v) AnATSW. The ammonia and nitrate use rates of *N. oculata* in the presence of 50% (v/v) AnATSW were 74.7% and 69.4%, respectively. The unsaturated oils were thought to have more oxidatively stable compositions after vitamin supplementation. The bis-allylic-position-equivalent and iodine value ranges of 30.64—46.13 and 90.5—117.46 showed similarity in their structure of the oil content to that of rapeseed oil. A low-heating value of 17.6—22.9 MJ kg^{-1} indicated the similarity of the oils to coal.

Microalgae were also cultivated by Guo et al. [18] using aquaculture wastewater, which facilitated biomass production. A wild-type strain of *Platymonas subcordiformis* was initially cultivated in natural seawater medium, after which the isolate was inoculated in pretreated flounder aquaculture wastewater. The total nitrogen concentration of the wastewater samples was reduced from 2.39 to 0.32 mg L^{-1} after 14 days of incubation. The phosphate removal efficiency was 99% in the 14-day culture, in which the metabolic uptake and precipitation were thought to be the underlying mechanisms for such high removal efficiency. The algal growth kinetics was rapidly increased during the initial phase of 4 days, after which it remained constant. The biomass of the algae increased 8.9 times after 14 days of incubation. Therefore, this study showed the simultaneous removal of nitrogen and phosphorus from wastewater and an increase in algal biomass that could be beneficial for biofuel production.

Zhang et al. [19] reported the application of *Scenedesmus* sp. ZTY1 microalgal strain for simultaneous wastewater treatment and biofuel production. The microalgal strain was isolated from an effluent sample obtained from a wastewater treatment plant and cultured in BG-11 medium. Both primary and secondary domestic wastewater effluents (2.5% v/v) were used as the nutrient source for the growth of the microalgal strain. After 21 days of cultivation, a maximum density of 3.6×10^6 and 1.9×10^6 cells/mL was achieved using the primary and secondary effluents, respectively. Almost 97% of total phosphorus present in both effluents was efficiently removed after sufficient algal growth. Around 90% of the total nitrogen rate was successfully removed from the samples. The dissolved organic carbon was removed by 60%, which was attributed to the synergistic action of algae and bacteria. The lipid content of the algae was 32.2% after 21 days of growth in the presence of the wastewater, whereas the concentration of triacylglycerol increased from 32 to 148 mg L^{-1} after 45 days of growth in the presence of the primary effluent.

3. Conclusion and future perspectives

Algal biofuel is a promising alternate energy source that can address problems such as the rapid depletion of fossil fuel reserves, climate change, rising crude oil prices, and environmental deterioration. Effluents from industries associated with poultry, slaughterhouses, distilleries, paper, food, tanneries, textiles, and wineries can serve as potential sources of substrate to support algal growth and subsequent biofuel production. Attempts should be made to use agricultural and domestic wastewater to develop cost-effective and sustainable means of biofuel production. Coupling microalgae-mediated biofuel production with wastewater treatment can be a powerful strategy to convert waste to wealth. This can provide a novel and economical route to produce bioenergy from biobased by-products.

The thorough optimization of reaction parameters such as the inoculum density, volume of wastewater, substrate concentration, duration of the reaction, temperature, pH, aeration, and agitation will help to achieve maximal biomass production. Similarly, harvesting conditions should be standardized for large-scale biomass production. The development of commercial processes to harvest algal biomass with subsequent conversion to biofuel will help to overcome technoeconomic barriers and address serious sustainability concerns.

Algae-mediated biofuel production using wastewater will certainly help to move us forward toward ensuring a clean environment.

Acknowledgment

Dr. Sougata Ghosh acknowledges Kasetsart University, Bangkok, Thailand for the postdoctoral fellowship and funding under the Reinventing University Program (Ref. No. 6501.0207/10870, November 9, 2021 and Ref. No. 6501.0207/9219 dated 14th September, 2022).

References

[1] J.K. Pittman, A.P. Dean, O. Osundeko, The potential of sustainable algal biofuel production using wastewater resources, Bioresour. Technol. 102 (2011) 17–25.

[2] S. Li, X. Li, S.H. Ho, Microalgae as a solution of third world energy crisis for biofuels production from wastewater toward carbon neutrality: an updated review, Chemosphere 291 (2022) 132863.

[3] S. Mehariya, R.K. Goswami, P. Verma, R. Lavecchia, A. Zuorro, Integrated approach for wastewater treatment and biofuel production in microalgae biorefineries, Energies 14 (2021) 2282.

[4] L.F. Wu, P.C. Chen, A.P. Huang, C.M. Lee, The feasibility of biodiesel production by microalgae using industrial wastewater, Bioresour. Technol. 113 (2012) 14–18.

[5] R. Kothari, V.V. Pathak, V. Kumar, D.P. Singh, Experimental study for growth potential of unicellular alga *Chlorella pyrenoidosa* on dairy waste water: an integrated approach for treatment and biofuel production, Bioresour. Technol. 116 (2012) 466–470.

[6] N. Maheshwari, P.K. Krishna, I.S. Thakur, S. Srivastava, Biological fixation of carbon dioxide and biodiesel production using microalgae isolated from sewage waste water, Environ. Sci. Pollut. Res. 27 (2020) 27319–27329.

[7] S. Chinnasamy, A. Bhatnagar, R.W. Hunt, K.C. Das, Microalgae cultivation in a wastewater dominated by carpet mill effluents for biofuel applications, Bioresour. Technol. 101 (2010) 3097–3105.

[8] W. Zhou, Y. Li, M. Min, B. Hu, P. Chen, R. Ruan, Local bioprospecting for high-lipid producing microalgal strains to be grown on concentrated municipal wastewater for biofuel production, Bioresour. Technol. 102 (2011) 6909–6919.

[9] I. Woertz, A. Feffer, T. Lundquist, Y. Nelson, Algae grown on dairy and municipal wastewater for simultaneous nutrient removal and lipid production for biofuel feedstock, J. Environ. Eng. 135 (11) (2009) 1115–1122.

[10] A.A. Fathi, M.M. Azooz, M.A. Al-Fredan, Phycoremediation and the potential of sustainable algal biofuel production using wastewater, Am. J. Appl. Sci. 10 (2) (2013) 189–194.

[11] M. Khalekuzzaman, M. Alamgir, M.B. Islam, M. Hasan, A simplistic approach of algal biofuels production from wastewater using a Hybrid Anaerobic Baffled Reactor and Photobioreactor (HABR-PBR) System, PLoS One 14 (12) (2019) e0225458.

[12] L. Zhu, Z. Wang, Q. Shu, J. Takala, E. Hiltunen, P. Feng, Z. Yuan, Nutrient removal and biodiesel production by integration of freshwater algae cultivation with piggery wastewater treatment, Water Res. 47 (2013) 4294–4302.

[13] G. Li, J. Zhang, H. Li, R. Hu, X. Yao, Y. Liu, Y. Zhou, T. Lyu, Towards high-quality biodiesel production from microalgae using original and anaerobically-digested livestock wastewater, Chemosphere 273 (2021) 128578.

[14] G.W. Roberts, M.O.P. Fortier, B.S. Sturm, S.M. Stagg-Williams, Promising pathway for algal biofuels through wastewater cultivation and hydrothermal conversion, Energy Fuels 27 (2013) 857–867.

[15] Y. Zhou, L. Schideman, G. Yu, Y. Zhang, A synergistic combination of algal wastewater treatment and hydrothermal biofuel production maximized by nutrient and carbon recycling, Energy Environ. Sci. 6 (2013) 3765–3779.

[16] S. Hena, S. Fatimah, S. Tabassum, Cultivation of algae consortium in a dairy farm wastewater for biodiesel production, Water Resour. Ind. 10 (2015) 1–14.

[17] P.F. Wu, J.C. Teng, Y.H. Lin, S.C.J. Hwang, Increasing algal biofuel production using *Nannocholropsis oculata* cultivated with anaerobically and aerobically treated swine wastewater, Bioresour. Technol. 133 (2013) 102−108.

[18] Z. Guo, Y. Liu, H. Guo, S. Yan, J. Mu, Microalgae cultivation using an aquaculture wastewater as growth medium for biomass and biofuel production, J. Environ. Sci. 25 (2013) S85−S88.

[19] T.Y. Zhang, Y.H. Wu, H.Y. Hu, Domestic wastewater treatment and biofuel production by using microalga Scenedesmus sp, ZTY1. Water Sci. Technol. 69 (12) (2014) 2492−2496.

[20] O. Komolafe, S.B.V. Orta, I. Monje-Ramirez, I.Y. Noguez, A.P. Harvey, M.T.O. Ledesma, Biodiesel production from indigenous microalgae grown in wastewater, Bioresour. Technol. 154 (2014) 297−304.

Chapter 29

Rice straw: a potential substrate for bioethanol production

Quratulain[1], Ali Nawaz[1], Hamid Mukhtar[1], Ikram ul Haq[1] and Vasudeo Zambare[2,3]

[1]Institute of Industrial Biotechnology, Government College University Lahore, Lahore, Punjab, Pakistan; [2]Om Biotechnologies, Nashik, Maharashtra, India; [3]Aesthetika Eco Research Pvt Ltd, Nashik, Maharashtra, India

1. Introduction

Ethanol derived from renewable energy sources has piqued the curiosity of researchers throughout the world for centuries as a potential replacement fuel for existing carbon fuels. Lignocellulose biomass, such as wood, as well as field crops waste, such as straw, and sugar beet pulp, are possible basic resources for the manufacture of a variety of improved products such as fuel bioethanol and biodiesel. Polysaccharides make up to 80% of lignocelluloses [1]. These naturally available resources are be attractive alternatives to replace ecologically unfavorable fossil hydrocarbon raw materials as well as producing "green" goods. In contrast to conventional fuels, fermentation ethanol is a sustainable and environmentally friendly resource that does not contribute to the greenhouse gas effect.

Rice husk is among the most common lignocellulosic waste materials on the planet. Rice is the world's third most important cereal grain, behind only wheat and maize. As according to FAO data, the annual global rice output in 2007 was over 650 million tonnes, with 1–1.5 kilos of straw being produced for every kilogram of grain produced [2]. Each year, around 650–975 million tonnes of rice straw are produced globally, with the bulk of this being utilized as cow fodder and the remainder being discarded [3,4].

Due to the relatively low volume fraction, slow soil deterioration, the prevalence of rice stalk diseases, and high mineral composition, rice straw management options are restricted. Crop burning is presently the most common method of removing rice straw, even though it impacts air quality and thus impacts global health [5]. As climate change is increasingly acknowledged as a danger to the future, there is increased curiosity about alternative energy uses for agro-industrial by-products. Therefore, rice straw could be an excellent option for meeting our future energy needs. The aim of this study is to provide a summary of the techniques available for manufacturing bioethanol utilizing rice straw [6,7]. Bioethanol, a prospective near-term output that has been intensively explored over a couple of decades, is a viable liquid fuel that could be used to solve the challenge of energy constraints. Lignocellulosic biomass is among the most enticing and important prospective raw resource for the production of bioethanol among renewable power sources because of its simple accessibility, cheap cost, as well as excessive sugar concentration. Cellulose, hemicellulose, and lignin, are the three key components of lignocellulosic materials. Through noncovalent as well as covalent connections, these constituents produce an efficient system that can limit enzyme availability, decreasing the efficiency of hydrolytic enzymes [8–10]. Rice straw, on the other hand, is largely composed of cellulose, hemicellulose, and lignin. Hydrogen bonds connect cellulose with hemicellulose. Because of its complicated system, rice straw is difficult to degrade due to ether links among lignin and cellulose, including ester bonds, ether bonds, glycoside bonds, and hydrogen bonds, including acetals among hemicellulose and lignin. While hemicellulose and cellulose can be transformed into monosaccharides, the restrictive organization of rice straw must be overcome [11,12]. As a result, it is vital to execute proper rice straw pretreatment. Several pretreatment techniques for lignocelluloses have been devised, all of which work by increasing the inner surface area. This is performed in part by hemicellulose solubilization and in part by lignin degradation [4,13–15].

Studies are currently being conducted on the conversion of residues to useable products such as enzymes and sugar syrups by microbial and enzymatic degradation, which can be used in a variety of applications [5,16]. Cellulolytic enzymes are crucial in the spontaneous breakdown of organic matter, during which cellulolytic fungi, as well as bacteria, degrade plant lignocellulosic components effectively. These enzymes have found new uses in industry, such as the generation of fermentable sugars and ethanol. Fungal cellulases are naturally occurring inducible proteins, some of which are exposed to the environment and are dependent on the type of cellulose acting on the organism (amorphous or crystalline) [17–19].

2. Rice straw potential for bioethanol production

Bio-based ethanol is becoming an increasingly interesting choice for fuel. Nevertheless, bioethanol production using crops including such as cereals (first-generation biofuels) has also culminated in an unfavorable direct rivalry with food production. The shift to more prevalent inedible organic materials should assist in alleviating this strain on crop production. The majority of these plant components are composed of complex carbohydrates like hemicellulose and cellulose, which may be transformed into fermentable sugars. Such sugars may be used by ethanol-fermenting bacteria to produce ethanol.

Straw, being a primary source for bioethanol production, is a significant possibility. Although straw contains lignocellulose, it can be used as a form of renewable power due to its high carbohydrate content. Polysaccharides encased in lignin with strong enough linkages are the major components of lignocellulose. It is simpler to use straw biomass for ethanol conversion after these carbohydrates have been digested into monosaccharides such as xylose, sugar, arabinose, and sucrose. Cellulose biomass is made up of three major components: hemicellulose, cellulose, and lignin. Cellulose is the most abundant element in plant cell walls, accounting for up to 50% of the plant's dry mass. Rice straw has a high cellulose concentration of 34.2% dry mass, a 24.5% hemicellulose ratio, and a lignin level of more than 23.4%. Because the assembly of cellulosic biomass is complicated, it is found in components that are more challenging to break down and convert to carbohydrate biomass [20–22].

The composition of feedstock has a significant impact on bioenergy generating efficiency. Rice straw's poor feedstock integrity is mostly influenced by its relatively high concentration (10%–17%) in comparison to wheat straw (about 3%), as well as its high concentration of silica in ash (55% in wheat and 75% SiO_2 in rice). Rice straw, on the other hand, has the added value of installing a comparatively low overall alkali content (K_2O as well as Na_2O typically compose 15% of whole ash), but wheat straw can often have >25% alkali content in residue [23,24].

3. Production of bioethanol from rice straw

3.1 Basic concepts of rice straw

Rice straw, which is largely made up of cellulose and hemicellulose, has long been touted as a reliable source of bioethanol, animal fodder, and organic compounds. The techniques for converting this feedstock to ethanol have been developed on two systems: the sugar basis and the syngas platform. Fig. 29.1 depicts the major phases of these platforms.

In the sugar platform, cellulose, as well as hemicellulose, are fermentable sugars, which are fermented to produce ethanol. Arabinose, glucose, galactose, xylose, and mannose are examples of fermentable sugars. To create these sugars, cellulose and hemicellulose can be hydrolyzed utilizing acids or enzymes [25]. Silicification and chemical composition are important aspects that influence the utilization proficiency and nutritional excellence of lignocellulosic materials like rice straw, thus predicting the chemical composition of rice straw quickly and accurately would be beneficial [26]. The production of bioethanol through cellulose, on the other hand, has still not been widely used. Fig. 29.1 shows the production process of bioethanol using rice straw. Chemical techniques and HPLC are two common analytical methods for determining the configuration of rice straw. The primary biochemical makeup of rice straw is after machine-driven high-

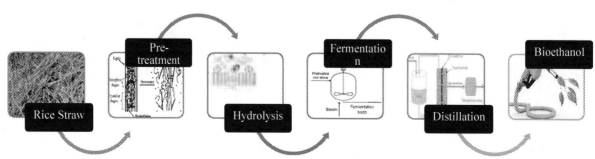

FIGURE 29.1 Flowchart demonstrating the procedure for manufacturing bioethanol using rice straw.

pressure steam treatment. The chemical arrangement was discovered to be considerably impacted by the time required. As the treatment time progresses, the amount of hemicellulose and lignin decreases, while the amount of cellulose increases. The cellulose concentration increases from 44.3% to 69.2% after 54 h of treatment, while hemicellulose and lignin contents decrease to 8.2% and 14.1%, respectively [27,28].

This suggests that high-pressure mist improves the solubilization of hemicellulose and lignin, resulting in intimate contact between lignin and water, and consequently improved lignin solubilization. Rice straw that has been processed with power-driven high-pressure steam has more cellulose, fewer hemicelluloses, as well as less lignin. This chemical alteration in the rice straw fibers leads to a higher percentage of crystalline cellulose, which advances the fibers' strength thermal and constancy [29,30].

3.2 Cellulolytic strain culture and enzyme activity testing

Throughout the biological preliminary treatment, microbes including *Bacillus* sp., *Cellulomonas* sp., as well as a few fungi, for example, *Phanerochaete chrysosporium* and *Aspergillus niger*, and also related enzymes, have been often used. Manganese peroxidases, laccases, lignin peroxidases, and some other flexible peroxidases participate in lignin breakdown by diverse oxidation mechanisms [31]. The cellulolytic properties of bacterial strains in naturalistic environments are affected by the origins and quantities of biowaste. The six fungal strains mostly used for enzymatic hydrolysis are: *T. reesei* ITCC4025, *Trichoderma viride* ITCC6413, *Aspergillus aculeatus* ITCC5078, *A. fumigatus* ITCC4768, *Fusarium solani* ITCC6397, and *A. niger* ITCC302 from ITCC. All stock cultures were subcultured on potato dextrose agar (PDA) slants (Hi−Media, India) at 30°C for 7 days and then subcultured again. After 7 days, fully sporulated cultures were subcultured onto fresh PDA slants. To find an effective strain for enzymatic hydrolysis of rice straw, researchers screened nine fungal strains. In 250 mL, 1.0 mL (about 2×10^6 spores/mL) spore suspension of appropriate fungal strains was added to create the enzyme extract. The fungal spore suspension was made by adding 5 mL distilled water to the slants and dislodging the spores with an inoculation needle under aseptic conditions. The suspension was diluted correctly and utilized as the inoculum. Using a hemocytometer, the spore count of the fungal suspensions was set to roughly 2×10^6 spores/mL. Filter paper (exo-b-glucanase) and CMCase (endo-b-1, 4-glucanase) activity is usually carried out in sodium citrate buffer at pH 4.8 [32,33].

3.3 Pretreatment techniques

Rice straw has a dense loading of lignin, hemicelluloses, and cellulose, which creates bioethanol production challenges. To remove lignin, destabilize the lignocellulosic complex, diminish cellulose crystallinity, and also improve the permeability of the constituents, a preprocessing step is required. The resulting cellulosic biomass is more receptive to hydrolysis after a preprocessing step. Rice straw pretreatment improves both the production rate and the overall sugars yield [34,35]. It also improves the compounds' digestibility for enzymatic hydrolysis. Physical, chemical, biological, and combination pretreatment procedures are among the different kinds of pretreatment technologies [36,37].

There are three ways for pretreatment of LCB (lignocellulosic biomass): biological, physical, and chemical. Pretreatment using chemicals is the most successful and commonly used approach for removing the lignin content. Nevertheless, due to the increasing price of the neutralizing and chemical agents, this technique is not always viable for commercial use. In addition to the expense of pretreatment, the expense of LCB enzymatic hydrolysis through cellulase enzymes is an issue for its economic and industrial use. Furthermore, these cellulolytic enzymes are now created utilizing artificial substrate by adopting modified strains (via genetically engineered) individually, which raises the overall price of the biomass for the biohydrogen manufacturing technique. In this context, the application of cheap sustainable methodologies for efficient pretreatment and cheap enzymatic activities may be beneficial in greatly reducing the key problems. With the integration of proper pretreatment methods and minimal substrate-based effective cellulase synthesis from mesophilic or thermotolerant microorganisms, the total manufacturing cost of biohydrogen employing LCB may be reduced. Several experiments have investigated whether chemical pretreatment and thermotolerant fungus might be possible factors to lower the cost of the LCB to the biohydrogen manufacturing process [38]. Several pretreatment procedures, including alkali treatment and ammonia explosion, have emerged in recent years. Many techniques have been demonstrated to provide large sugar production, exceeding 90% of the theoretical production for lignocellulosic biomasses, for instance forests, grasses, maize, and many more. It should be noted that it is not at always feasible for transference of the outcomes of pretreatment through one type of material to another. Moreover, a method that works well for one type of biomass material may not work well for another.

The pretreatment method utilized for certain biomass is determined by its content and the by-products formed as an outcome of pretreatment. These elements have a significant impact on the costs associated with a pretreatment process. Several studies have been published that compare various pretreatment techniques for biomass [39–41]. Rosgaard [42] tested the efficacy of three alternative pretreatment processes on barley and wheat straw, namely water or acid impregnation accompanied by steam explosion using hot water. Following enzyme action with a cellulase enzyme method, the pretreatments were evaluated. Because of the consequent glucose concentration in the liquescent crude extract after enzymatic hydrolysis, acid, as well as water adsorption accompanied by barley straw thermal cracking, was the best preparation. Silverstein [43] investigated the efficacy of pretreatments using sodium hydroxide, sulfuric acid, ozone, and hydrogen peroxide aimed at cotton straws for ethanol conversion. Sulfuric acid pretreatment led to a tremendous xylan drop (95.2% for 2% acid, 121°C/15 psi, 90 min), however the lowest cellulose to glucose transformation was through hydrolysis (23.9%). The highest possible level of delignification and cellulose conversion was obtained after pretreatment with sodium hydroxide (65% for 2% NaOH, 121°C/15 psi, 90 min) (60.8%). In comparison to hydrogen peroxide pretreatment, sodium hydroxide pretreatment resulted in considerably lower delignification (max of 29.5% for 2%, 121°C/15 psi, 30 min) and cellulose reconfiguration (49.8%). Over time, ozone did not affect lignin, xylan, or glucan substances.

As a result, biomass pretreatment is a significant step in the biosynthetic pathways of bioenergy from lignocellulosic biomasses, although there is a vital requirement to recognize the basics of various methods that can contribute to making the appropriate choice based on the framework of the biomass material as well as the hydrolyzing operator.

3.3.1 Physical pretreatments

Irradiation, grinding and milling, heating, pressure, and temperature are all examples of physical preparation. Microwave irradiation has been shown to alter the ultrastructure of cellulose, destroy hemicelluloses, and also lignin in lignocellulosic materials, and increase lignocellulosic materials' enzymatic vulnerability. Wet disc milling outperforms ball-milling in terms of recovery of glucose and power savings, according to Hideno et al. [44,45].

3.3.2 Chemical pretreatments

Chemical pretreatment methods for biomass have been developed, with steam explosion, including lime, acid, sulfur dioxide explosion, ionic liquid, ammonia fiber explosion, and others. Chemicals such as organic acids, alkali, peroxides, acids, and others are used in the chemical pretreatment process. For alkali pretreatment, calcium hydroxide ($Ca(OH)_2$) potassium, sodium hydroxide (NaOH), and hydroxide (KOH) are commonly utilized. Sodium hydroxide raises the internal surface of cellulose and crystallinity, and reduces polymerization and disrupts the lignin structure as a result [34,46].

3.3.3 Biological pretreatment

Rice straw biological pretreatment is an environment-friendly and safe method. It features a reduced energy and chemical demand, a better target product yield, and enhanced substrates, as well as reaction precision. White rot fungus from the Basidiomycetes class is commonly used for pretreatment methods. Endophytic fungi generate ligninolytic enzymes including laccase, manganese peroxidase (MnP), and lignin peroxidase (LiP), which are oxidized by H_2O_2 in the presence of veratryl alcohol (VA) and Mn (III). The form of oxidized enzymes induces lignin oxidation. Laccase includes the oxidation of lignin phenolic materials in the presence of lignin-oxidizers such as laccase-oxidizing precursors and 2,20-azinobis-(3)-ethyl-benzyl thiazoline-6-sulfonate (ABTS) or 3-hydroxyanthranilic acid (HA) such as O_2 [46]. Various studies have found that white rot fungi like *Ceriporiopsis subvermispora*, *Phanerochaete chrysosporium*, *Pleurotus ostreatus*, and *Phlebia subserialis* may effectively degrade lignin. *Phlebia subserialis*, *Ceriporiopsis subvermispora*, *Phanerochaete chrysosporium*, and *Pleurotus ostreatus* are white rot fungi that can metabolize lignin. Similarly, many actinomycetes produce enzymes that aid in the degradation of lignocelluloses. Zhang [47] used *Streptomyces griseorubens* JSD-1 to biologically pretreat rice straw. After a 10-day incubation period, *S. griseorubens* JSD-1 cellulolytic enzyme enabled the alteration of cellulose to form reducing sugars, with an ideal saccharification efficacy of 88.13%. According to Zeng [48], adding inorganic salts to the SSF (solid state fermentation) of wheat straw increased the biomass breakdown proportion by *P. chrysosporium* considerably. After 21 days of pretreatment with ligninolytic fungi (*Irpex lacteus* and *Pleurotus eryngii*), mild alkali-treated wheat straw exhibited the highest digestibility [35]. For pretreatment of wheat straw, five distinct fungi were identified by screening [49].

3.3.4 Combined pretreatment

Kun [50] showed that processing of rice straw using alkali aided by photo-catalysis enhanced the physical properties as well as the morphology of rice straw by reducing the lignin material. In the absence of H_2O_2, rice straw alkali treatment enabled the dissolution of lower molecular size hemicelluloses productive in glucose, most likely derived by glucan, though alkaline peroxide treatment favored the suspension of larger molecular sized hemicelluloses rich in xylose [51]. Kim [52] described a two-step process for obtaining fermentable sugar from rice straw using liquid ammonia and also dilute sulfuric acid. The combined processing of rice straw with aqueous ammonia and dilute sulfuric acid produced fermentable sugar with a crystallinity index that was nearly identical to that of cellulose. Numerous mixtures of microwave preprocessing of rice straw with alkali and acid were described by Zhu [53], which removed lignin and hemicelluloses. Yang and Fang [54] investigated the effects of ultrasonic irradiation and ILs on rice straw pretreatment in several processes. ILs and rice straw that had been processed with ultrasound had the best enzymatic efficiency of the hydrolysis tests. Jin and Chen [55] investigated rice straw enzymatic hydrolysis using a blend of steam explosion and superfine grinding [56,57].

3.3.5 Enzymatic hydrolysis of pretreated straw

Following rice straw pretreatment, degradation or enzymatic saccharification is carried out and includes the cleavage of hemicellulose and cellulose polymer along with the assistance of enzymes, and instead the sugar produced through hydrolysis is transformed into ethanol by fermentation [51,58]. The volume of ethanol formed is determined by the proportion of sugar recovered, the kind of simple sugar used (pentose/hexose), as well as the number of inhibitors produced [59]. The hydrolysis of hemicellulose yields numerous pentose as well as hexose, whereas the hydrolysis of cellulose yields glucose [60]. The content of the substrates, the kind of pretreatment process, the dose, and the effectiveness of the enzymes utilized for hydrolysis all have an impact on enzyme hydrolysis. The amount of polymerization and crystallinity of cellulose are critical considerations in affecting the rate of enzymatic hydrolysis. The hemicellulose concentration influences the hydrolysis reaction because removing hemicelluloses increases the mean pore size of the substrate, which enhances enzyme permeability and hence increases the degradation efficiency.

Rice straw was initially treated with sodium sulfite—formaldehyde catalytic enzymatic hydrolysis, and the maximum sugar yields of 71.1%, 79.0%, 88.8%, and for total glucan, xylan, and sugar in both, were acquired at 160°C [61] by such an enzyme concentration of 40 FPU/g-substrate and 12% sodium sulfite. Yao et al. [62] investigated the SO_3 collaborative environment diluted alkali process, in which SO_3 in gaseous form was dispersed into the interior arrangement and responded along with the water within the straw, resulting in an inner microthermal explosion, and then a dilute alkali solution was utilized to wash out the lignin, resulting in an enzymatic hydrolysis rate of 91% based on the preprocessed rice straw along SO_3 for 4 h followed by a 7 h 1% w/v NaOH treatment [63,64]. Sulbaran-de-Ferrer et al. [65] discovered a 61% increase in the amount of fermentable sugar in ammonia-treated rice husk samples that were similar to raw rice straw (11%).

3.4 Fermentation of rice straw hydrolysates

Fermentation is an important phase in the production of bioethanol as ethanol is primarily created through the metabolic activity of the fermenting agent. The fermentation method is particularly important because of the general complex structure of lignocellulosic hydrolysate (Fig. 29.2). The classic approach for fermenting biomass hydrolyzates entails a systematic method during which cellulose breakdown, as well as fermentation, occurs in discrete units. This alignment is known as SHF, which stands for separate hydrolysis and fermentation. The fermentation and hydrolysis are accomplished in a particular unit in the alternate option, SSF (simultaneous-saccharification, and fermentation). *S. cerevisiae* is the most commonly used bacterium aimed at fermenting lignocellulosic hydrolyzates as it ferments the hexoses but not the pentose [66–68].

Fermentation necessitates certain parameters for the bacteria utilized, including pH and temperature range. Usually, fermenting agents are mesophilic microbes that thrive at temperatures ranging from 30 to 35°C. A pH of 6.5–7.5 is typically required to maintain the progress of various kinds of fermentation microorganisms. Fungi may survive in more acidic environments (pH 3.5–5.0) than bacteria [33,69,70]. Other significant elements in the fermentation method include the bacteria's genetic stability and growth rate; tolerance to regulators, permeability, and ethanol; efficiency, as well as ethanol output. There have been several bacteria found that are suited as fermenting microbes. *Zymomonas mobilis* is a common fermenting bacterium that transforms glucose, sucrose, and fructose into ethanol. *S. cerevisiae*, as *Z. mobilis*, consumes hexose sugar normally (e.g., glucose, fructose) [71–73]. Despite being the most widely used microorganisms,

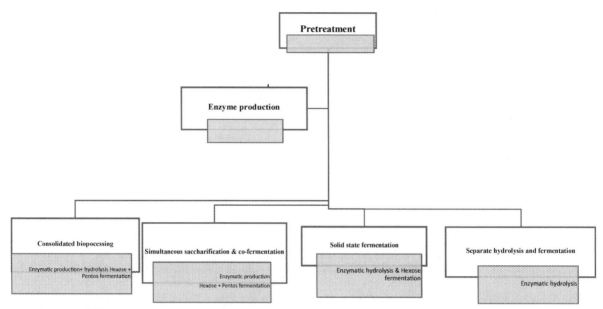

FIGURE 29.2 Illustration of various possible ethanol production methods using lignocellulosic biomass.

Z. mobilis and *S. cerevisiae* are both inefficient at fermenting pentose carbohydrates. The capacity of *Candida shehatae*, *Pachysolen tannophilus*, and *Pichia stipitis* to transform pentose sugars is well known (xylose) [74,75]. Several studies are being conducted to change the behavior as well as the function of natural types of yeasts and bacteria due to the constraints of various wild-type microorganisms. More fermenting potential in yeasts can be acquired by genetic manipulation, allowing them to ferment more than one kind of sugar. Buaban et al. (2010) [76] found that utilizing the modified fermentation agent *Pi. stipitis* BCC15191 results in 8.4 g L^{-1} of ethanol during 24 h of hydrolyzed bagasse ash fermentation [56].

3.5 Bioethanol production's techno-economic worthiness

The overall cost of producing straw-based ethanol is susceptible to important criteria including feedstock type, structure, as well as cost, ethanol plant capacity, power efficiency, and also asset base. The gross cost of production of biofuel is classified into four categories: (1) investment costs, (2) fixed operational costs (including incomes, overall operational costs, general liability, taxes, as well as general upkeep), (3) variable operational costs (such as consumable purchases and surplus power sales), and (4) raw material costs. Bioethanol has a good octane number, a relatively higher heat of evaporation, as well as a low cetane quantity, making it an ideal fuel for mixing into gasoline. Biofuel is utilized as an oxygenated component because it includes 35% oxygen, which decreases particulate matter as well as nitrogen oxide emissions. It is combined with gasoline in various ratios, the most common of which is E85, which includes 15% gasoline and 85% bioethanol. In Brazil, bioethanol is utilized either in its basic state or as a combination with gasoline known as gasohol (24% bioethanol, 76% gasoline) [77]. The bioethanol generated is transferred from the unit to a regional storage facility before being delivered to consumer fueling stations. It is mixed with gasoline and used in automobiles. In Fig. 29.3, the demand for petrol and bioethanol from 2015 to 2021 is provided in a graphical presentation. In most cases, a diesel truck is employed for travel. Wojnar [78] estimated an entire trip transference distance of 160 km for biofuel supply. The average mileage of automobiles that ran on petrol, E85, and E15 was 11.91 km L^{-1}, 8.29 km L^{-1}, and 11 km L^{-1}, respectively [79]. Depending on the petrochemical industry data source, the estimated conveyance cost per liter is Rs. 3.5 per liter of ethanol [80].

An increase in production results (from 200 to 400 mL year^{-1}) resulted in a 10% decline in the net cost of production (from US$0.73 L^{-1} to US$0.67 L^{-1}). A 50% reduction in production capability (between 200 and 100 mL year^{-1}) resulted in a 15% increase in the net production rate (from US$0.73 L^{-1} to US$0.84 L^{-1}). Besides assessing the environmental effect of bioethanol production, difference between the sizes of the plants and transport distance can contribute distinctly in terms of product price.

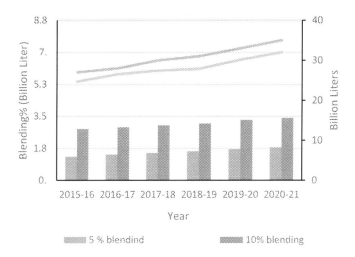

FIGURE 29.3 Future demands for petrol and bioethanol.

4. Conclusions and future recommendations

The consumption of lignocellulosic biomass for bioethanol production needs an environmentally friendly and cost-effective production process. Rice straw seems to be an effective and promising method for bioethanol synthesis, given the development and demand for second-generation biofuels, due to its plentiful availability and appealing composition. Due to environmental considerations, the biological conversion of rice straw into fermentable sugars using hydrolyzing enzymes is currently the most appealing option. However, there are numerous obstacles on the way to creating economically viable techniques due to its high-lignin, multifaceted nature, and ash content, and numerous efforts are being made to develop an effective preprocessing technique to eliminate unwanted portions in order to obtain readily accessible sugars, with significant accomplishments to date.

However, the pretreatment procedure should be upgraded in terms of energy productivity and environmental friendliness. As a result, future research should focus on addressing several concerns, such as the growth of large-scale commercialization through economic analysis and the adoption of an appropriate technique based on various LCB architectures. Finally, it is suggested that an appropriate study be designed to catch an effective combination of prevailing methodologies. The use of lignocellulosic biomass for bioethanol production needs a cost-effective and eco-friendly production process. Rice straw seems to be an effective and promising candidate aimed at bioethanol synthesis, given the evolution and demand for second-generation biofuels, due to its plentiful availability and appealing composition.

Due to environmental considerations, the biological transfiguration of rice straw into fermentable sugars using hydrolyzing enzymes is currently the most appealing option. Nevertheless, there remain a few obstacles to creating an economically viable technology because of its complex nature, ash and high-lignin contents, and numerous efforts are being made to develop an efficient preprocessing technique to eliminate undesirable portions in order to obtain readily accessible sugars, with significant success to date. According to existing data, rice straw might meet the demand for bioethanol in the transportation industry. To overcome the challenges of glucose cofermentation and xylose and improve system efficiency, approaches in the process of strain engineering must be used. For maximal process performance, an intelligent and well-balanced combination of pretreatment, fermentation, and hydrolysis must be used.

References

[1] P. Kaparaju, M. Serrano, A.B. Thomsen, P. Kongjan, I. Angelidaki, Bioethanol, biohydrogen and biogas production from wheat straw in a biorefinery concept, Bioresour. Technol. 100 (9) (2009) 2562–2568.
[2] B.L. Maiorella, Ethanol, Comprehen. Biotechnol. 3 (1985) 861–914.
[3] R. Sindhu, R.K. Sukumaran, R.R. Singhania, P. Binod, N. Kurien, S. Vikram, et al., Bioethanol Production from Rice Straw: An Overview, 2010.
[4] X. Jin, J. Song, G.Q. Liu, Bioethanol production from rice straw through an enzymatic route mediated by enzymes developed in-house from Aspergillus fumigatus, Energy 190 (2020) 116395.
[5] S.I. Mussatto, I.C. Roberto, Alternatives for detoxification of diluted-acid lignocellulosic hydrolyzates for use in fermentative processes: a review, Bioresour. Technol. 93 (1) (2004) 1–10.

[6] P. Binod, R. Sindhu, R.R. Singhania, S. Vikram, L. Devi, S. Nagalakshmi, et al., Bioethanol production from rice straw: an overview, Bioresour. Technol. 101 (13) (2010) 4767–4774.

[7] J. Hou, X. Zhang, S. Zhang, K. Wang, Q. Zhang, Enhancement of bioethanol production by a waste biomass-based adsorbent from enzymatic hydrolysis, J. Clean. Prod. 291 (2021) 125933.

[8] Q. Zhang, H. Huang, H. Han, Z. Qiu, V. Achal, Stimulatory effect of in-situ detoxification on bioethanol production by rice straw, Energy 135 (2017) 32–39.

[9] J. Hou, J. Tang, J. Chen, J. Deng, J. Wang, Q. Zhang, Evaluation of inhibition of lignocellulose-derived by-products on bioethanol production by using the QSAR method and mechanism study, Biochem. Eng. J. 147 (2019) 153–162.

[10] A. Goshadrou, K. Karimi, M.J. Taherzadeh, Bioethanol production from sweet sorghum bagasse by Mucor hiemalis, Ind. Crop. Prod. 34 (1) (2011) 1219–1225.

[11] S. Jin, H. Chen, Near-infrared analysis of the chemical composition of rice straw, Ind. Crop. Prod. 26 (2) (2007) 207–211.

[12] A. Agbagla-Dohnani, P. Nozière, G. Clément, M. Doreau, Sacco degradability, chemical and morphological composition of 15 varieties of European rice straw, Anim. Feed Sci. Technol. 94 (1–2) (2001) 15–27.

[13] N. Yoswathana, P. Phuriphipat, P. Treyawutthiwat, M.N. Eshtiaghi, Bioethanol production from rice straw, Energy Res. J. 1 (1) (2010) 26.

[14] E. Palmqvist, B. Hahn-Hägerdal, Fermentation of lignocellulosic hydrolysates. II: inhibitors and mechanisms of inhibition, Bioresour. Technol. 74 (1) (2000) 25–33.

[15] H.F. Hörmeyer, P. Tailliez, J. Millet, H. Girard, G. Bonn, O. Bobleter, J.P. Aubert, Ethanol production by Clostridium thermocellum grown on hydrothermally and organosolv-pretreated lignocellulosic materials, Appl. Microbiol. Biotechnol. 29 (6) (1988) 528–535.

[16] R.K. Andren, R.J. Erickson, J.E. Medeiros, Cellulosic substrates for enzymatic saccharification, in: Biotechnol. Bioeng. Symp.;(United States), vol 6, Army Natick Development Center, MA, January 1976.

[17] E.B. Belal, Bioethanol production from rice straw residues, Braz. J. Microbiol. 44 (2013) 225–234.

[18] J. Hou, J. Tang, J. Chen, Q. Zhang, Quantitative Structure-Toxicity Relationship analysis of combined toxic effects of lignocellulose-derived inhibitors on bioethanol production, Bioresour. Technol. 289 (2019) 121724.

[19] X. Liming, S. Xueliang, High-yield cellulase production by Trichoderma reesei ZU-02 on corn cob residue, Bioresour. Technol. 91 (3) (2004) 259–262.

[20] K.L. Kadam, L.H. Forrest, W.A. Jacobson, Rice straw as a lignocellulosic resource: collection, processing, transportation, Environ. Aspect. Biomass Bioenergy 18 (5) (2000) 369–389.

[21] H. Chen, Chemical Composition and Structure of Natural Lignocellulose in Biotechnology of Lignocellulose, Springer, Dordrecht, 2014, pp. 25–71.

[22] S. Manju, B.S. Chadha, Production of hemicellulolytic enzymes for hydrolysis of lignocellulosic biomass, in: Biofuels, Academic Press, 2011, pp. 203–228.

[23] M. Zevenhoven, The prediction of deposit formation in combustion and gasification of biomass fuels – fractionation and thermodynamic multiphase multi-component equilibrium (TPCE) calculations, in: Combustion and Materials Chemistry, Lemminkäinengatan, Finland, 2000, p. 38.

[24] L.L. Baxter, T.R. Miles, T.R. Miles Jr., B.M. Jenkins, D.C. Dayton, T.A. Milne, R.W. Bryers, L.L. Oden, The Behavior of Inorganic Material in Biomass-Fired Power Boilers—Field and Laboratory Experiences: Volume II of Alkali Deposits Found in Biomass Power Plants, National Renewable Energy Laboratory, Golden, CO, 1996. Report: NREL/TP-433-8142.

[25] C.M. Drapcho, N.P. Nhuan, T.H. Walker, Biofuels Engineering Process Technology, Mc Graw Hill Companies, Inc, 2008.

[26] B. Tsegaye, C. Balomajumder, P. Roy, Biodelignification and hydrolysis of rice straw by novel bacteria isolated from wood feeding termite, 3 Biotech 8 (10) (2018) 1–11.

[27] J. Vadiveloo, The effect of agronomic improvement and urea treatment on the nutritional value of Malaysian rice straw varieties, Anim. Feed Sci. Technol. 108 (1–4) (2003) 133–146.

[28] K. Malik, J. Tokkas, R.C. Anand, N. Kumari, Pretreated rice straw as an improved fodder for ruminants-An overview, J. Appl. Natural Sci. 7 (1) (2015) 514–520.

[29] J. Vadiveloo, Nutritional properties of the leaf and stem of rice straw, Anim. Feed Sci. Technol. 83 (1) (2000) 57–65.

[30] J. Vadiveloo, B. Nurfariza, J.G. Fadel, Nutritional improvement of rice husks, Anim. Feed Sci. Technol. 151 (3–4) (2009) 299–305.

[31] M. Rout, B. Sardar, P.K. Singh, R. Pattnaik, S. Mishra, Utilization of microbial potential for bioethanol production from lignocellulosic waste, Biotechnol. Zero Waste: Emerg. Waste Manag. Techniques (2022) 263–282.

[32] S. Prasad, S. Kumar, K.K. Yadav, J. Choudhry, H. Kamyab, Q.V. Bach, et al., Screening and evaluation of cellulytic fungal strains for saccharification and bioethanol production from rice residue, Energy 190 (2020) 116422.

[33] J. Hou, Z. Qiu, H. Han, Q. Zhang, Toxicity evaluation of lignocellulose-derived phenolic inhibitors on *Saccharomyces cerevisiae* growth by using the QSTR method, Chemosphere 201 (2018) 286–293.

[34] C. Sarnklong, J.W. Cone, W. Pellikaan, W.H. Hendriks, Utilization of rice straw and different treatments to improve its feed value for ruminants: a review, Asian-Australas. J. Anim. Sci. 23 (5) (2010) 680–692.

[35] M. López-Abelairas, T.A. Lu-Chau, J.M. Lema, Fermentation of biologically pretreated wheat straw for ethanol production: comparison of fermentative microorganisms and process configurations, Appl. Biochem. Biotechnol. 170 (8) (2013) 1838–1852.

[36] W.H. Chen, B.L. Pen, C.T. Yu, W.S. Hwang, Pretreatment efficiency and structural characterization of rice straw by an integrated process of diluteacid and steam explosion for bioethanol production, Bioresour. Technol. 102 (3) (2011) 2916–2924.

[37] J. Wu, S.R. Collins, A. Elliston, N. Wellner, J. Dicks, I.N. Roberts, K.W. Waldron, Release of cell wall phenolic esters during hydrothermal pretreatment of rice husk and rice straw, Biotechnol. Biofuel. 11 (1) (2018) 1–12.

[38] K.Y. Show, Z.P. Zhang, J.H. Tay, D.T. Liang, D.J. Lee, N. Ren, A. Wang, Critical assessment of anaerobic processes for continuous biohydrogen production from organic wastewater, Int. J. Hydrogen Energy 35 (24) (2010) 13350–13355.

[39] Z.P. Zhang, K.Y. Show, J.H. Tay, D.T. Liang, D.J. Lee, W.J. Jiang, Effect of hydraulic retention time on biohydrogen production and anaerobic microbial community, Process Biochem. 41 (10) (2006) 2118–2123.

[40] H.H. Fang, H. Liu, Effect of pH on hydrogen production from glucose by a mixed culture, Bioresour. Technol. 82 (1) (2002) 87–93.

[41] S.Y. Wu, C.H. Hung, C.N. Lin, H.W. Chen, A.S. Lee, J.S. Chang, Fermentative hydrogen production and bacterial community structure in high-rate anaerobic bioreactors containing silicone-immobilized and self-flocculated sludge, Biotechnol. Bioeng. 93 (5) (2006) 934–946.

[42] L. Rosgaard, S. Pedersen, A.S. Meyer, Comparison of different pretreatment strategies for enzymatic hydrolysis of wheat and barley straw, Appl. Biochem. Biotechnol. 143 (3) (2007) 284–296.

[43] R.A. Silverstein, Y. Chen, R.R. Sharma-Shivappa, M.D. Boyette, J. Osborne, A comparison of chemical pretreatment methods for improving saccharification of cotton stalks, Bioresour. Technol. 98 (16) (2007) 3000–3011.

[44] N. Singh, R.P. Gupta, S.K. Puri, A.S. Mathur, Bioethanol production from pretreated whole slurry rice straw by thermophilic co-culture, Fuel 301 (2021) 121074.

[45] M. García-Torreiro, M. López-Abelairas, T.A. Lu-Chau, J.M. Lema, Fungal pretreatment of agricultural residues for bioethanol production, Ind. Crop. Prod. 89 (2016) 486–492.

[46] R. Singh, M. Srivastava, A. Shukla, Environmental sustainability of bioethanol production from rice straw in India: a review, Renew. Sustain. Energy Rev. 54 (2016) 202–216.

[47] D. Zhang, Y. Luo, S. Chu, Y. Zhi, B. Wang, P. Zhou, Biological pretreatment of rice straw with Streptomyces griseorubens JSD-1 and its optimized production of cellulase and xylanase for improved enzymatic saccharification efficiency, Prep. Biochem. Biotechnol. 46 (6) (2016) 575–585.

[48] J. Zeng, D. Singh, S. Chen, Biological pretreatment of wheat straw by Phanerochaete chrysosporium supplemented with inorganic salts, Bioresour. Technol. 102 (3) (2011) 3206–3214.

[49] T.A. Lu-Chau, M. García-Torreiro, M. López-Abelairas, N.A. Gómez-Vanegas, B. Gullón, J.M. Lema, G. Eibes, Application of fungal pretreatment in the production of ethanol from crop residues, in: Bioethanol Production from Food Crops, Academic Press, 2019, pp. 267–292.

[50] K. Niu, P. Chen, X. Zhang, W.S. Tan, Enhanced enzymatic hydrolysis of rice straw pretreated by alkali assisted with photocatalysis technology, J. Chem. Technol. Biotechnol: Int. Res. Proc., Environ. Clean Technol. 84 (8) (2009) 1240–1245.

[51] V.S. Chang, M.T. Holtzapple, Fundamental factors affecting biomass enzymatic reactivity, in: Twenty-first Symposium on Biotechnology for Fuels and Chemicals, Humana Press, Totowa, NJ, 2000, pp. 5–37.

[52] S.B. Kim, S.J. Lee, J.H. Lee, Y.R. Jung, L.P. Thapa, J.S. Kim, et al., Pretreatment of rice straw with combined process using dilute sulfuric acid and aqueous ammonia, Biotechnol. Biofuels 6 (1) (2013) 1–11.

[53] S. Zhu, Y. Wu, Z. Yu, C. Wang, F. Yu, S. Jin, et al., Comparison of three microwave/chemical pretreatment processes for enzymatic hydrolysis of rice straw, Biosyst. Eng. 93 (3) (2006) 279–283.

[54] C.Y. Yang, T.J. Fang, Combination of ultrasonic irradiation with ionic liquid pretreatment for enzymatic hydrolysis of rice straw, Bioresour. Technol. 164 (2014) 198–202.

[55] S. Jin, H. Chen, Superfine grinding of steam-exploded rice straw and its enzymatic hydrolysis, Biochem. Eng. J. 30 (3) (2006) 225–230.

[56] M.R. Swain, A. Singh, A.K. Sharma, D.K. Tuli, Bioethanol production from rice-and wheat straw: an overview, Bioethanol Prod. Food Crop. (2019) 213–231.

[57] R. Sindhu, M. Kuttiraja, T.P. Prabisha, P. Binod, R.K. Sukumaran, A. Pandey, Development of a combined pretreatment and hydrolysis strategy of rice straw for the production of bioethanol and biopolymer, Bioresour. Technol. 215 (2016) 110–116.

[58] M.J. Taherzadeh, C. Niklasson, Ethanol from Lignocellulosic Materials: Pretreatment, Acid and Enzymatic Hydrolyses, and Fermentation, 2004.

[59] R. Singh, A. Shukla, S. Tiwari, M. Srivastava, A review on delignification of lignocellulosic biomass for enhancement of ethanol production potential, Renew. Sustain. Energy Rev. 32 (2014) 713–728.

[60] R.P. Chandra, R. Bura, W.E. Mabee, D.A. Berlin, X. Pan, J.N. Saddler, Substrate pretreatment: the key to effective enzymatic hydrolysis of lignocellulosics? Biofuels (2007) 67–93.

[61] F. Gu, W. Wang, L. Jing, Y. Jin, Sulfite–formaldehyde pretreatment on rice straw for the improvement of enzymatic saccharification, Bioresour. Technol. 142 (2013) 218–224.

[62] R.S. Yao, H.J. Hu, S.S. Deng, H. Wang, H.X. Zhu, Structure and saccharification of rice straw pretreated with sulfur trioxide micro-thermal explosion collaborative dilutes alkali, Bioresour. Technol. 102 (10) (2011) 6340–6343.

[63] R. Singh, S. Tiwari, M. Srivastava, A. Shukla, Performance study of combined microwave and acid pretreatment method for enhancing enzymatic digestibility of rice straw for bioethanol production, Plant Knowledge J. 2 (4) (2013) 157–162.

[64] R. Singh, S. Tiwari, M. Srivastava, A. Shukla, Microwave assisted alkali pretreatment of rice straw for enhancing enzymatic digestibility, J. Energy (2014), 2014.

[65] B. Sulbarán-de-Ferrer, M. Aristiguieta, B.E. Dale, A. Ferrer, G. Ojeda-de-Rodriguez, Enzymatic hydrolysis of ammonia-treated rice straw, Appl. Biochem. Biotechnol. 105 (1) (2003) 155–164.

[66] O.J. Sanchez, C.A. Cardona, Trends in biotechnological production of fuel ethanol from different feedstocks, Bioresour. Technol. 99 (13) (2008) 5270–5295.

[67] R.C. Sun, Detoxification and separation of lignocellulosic biomass prior to fermentation for bioethanol production by removal of lignin and hemicelluloses, Bioresources 4 (2) (2009) 452–455.

[68] J. Zaldivar, A. Martinez, L.O. Ingram, Effect of selected aldehydes on the growth and fermentation of ethanologenic *Escherichia coli*, Biotechnol. Bioeng. 65 (1) (1999) 24–33.
[69] J. Hou, C. Ding, Z. Qiu, Q. Zhang, W.N. Xiang, Inhibition efficiency evaluation of lignocellulose-derived compounds for bioethanol production, J. Clean. Prod. 165 (2017) 1107–1114.
[70] F.K. Agbogbo, G. Coward-Kelly, Cellulosic ethanol production using the naturally occurring xylose-fermenting yeast, Pichia stipitis, Biotechnol. Lett. 30 (9) (2008) 1515–1524.
[71] M. Sonderegger, M. Jeppsson, C. Larsson, M.F. Gorwa-Grauslund, E. Boles, L. Olsson, et al., Fermentation performance of engineered and evolved xylose-fermenting *Saccharomyces cerevisiae* strains, Biotechnol. Bioeng. 87 (1) (2004) 90–98.
[72] J. Zaldivar, J. Nielsen, L. Olsson, Fuel ethanol production from lignocellulose: a challenge for metabolic engineering and process integration, Appl. Microbiol. Biotechnol. 56 (1) (2001) 17–34.
[73] S.G. Wi, I.S. Choi, K.H. Kim, H.M. Kim, H.J. Bae, Bioethanol production from rice straw by popping pretreatment, Biotechnol. Biofuels 6 (1) (2013) 1–7.
[74] B. Basak, B.H. Jeon, T.H. Kim, J.C. Lee, P.K. Chatterjee, H. Lim, Dark fermentative hydrogen production from pretreated lignocellulosic biomass: effects of inhibitory byproducts and recent trends in mitigation strategies, Renew. Sustain. Energy Rev. 133 (2020) 110338.
[75] S.M. Shevchenko, K. Chang, J. Robinson, J.N. Saddler, Optimization of monosaccharide recovery by post-hydrolysis of the water-soluble hemicellulose component after steam explosion of softwood chips, Bioresour. Technol. 72 (3) (2000) 207–211.
[76] B. Buaban, H. Inoue, S. Yano, S. Tanapongpipat, V. Ruanglek, V. Champreda, R. Pichyangkura, S. Rengpipat, L. Eurwilaichitr, Bioethanol production from ball milled bagasse using an on-site produced fungal enzyme cocktail and xylose-fermenting Pichia stipitis, J. Biosci. Bioeng 110 (1) (2010) 18–25.
[77] M.E. Dias De Oliveira, B.E. Vaughan, E.J. Rykiel, Ethanol as fuel: energy, carbon dioxide balances, and ecological footprint, Bioscience 55 (7) (2005) 593–602.
[78] Z. Wojnar, T. Banach, A.M. Hirschberger, Renewable fuels roadmap and sustainable biomass feedstock supply for New York, Forest 4 (8.2) (2010) 2–8.
[79] S. González-García, C.M. Gasol, X. Gabarrell, J. Rieradevall, M.T. Moreira, G. Feijoo, Environmental profile of ethanol from poplar biomass as transport fuel in Southern Europe, Renew. Energy 35 (5) (2010) 1014–1023.
[80] Maulin Shah (Ed.), Removal of Emerging Contaminants through Microbial Processes, Springer, Singapore, 2021.

Chapter 30

Algal bioenergy: the fuel for tomorrow

Elham M. Ali[1], Mostafa Elshobary[2] and Mostafa M. El-Sheekh[2]

[1]Department of Aquatic Environment, Faculty of Fish Resources, Suez University, Suez, Egypt; [2]Department of Botany, Faculty of Science, Tanta University, Tanta, Egypt

1. Introduction

An assumption for population growth is made at the high end of the current UN forecasts with 9.6–12.3 billion capita by the year 2100 [46]. Most of this growth occurs in Africa. Urbanization is very rapid everywhere, with a greater rate in the Mediterranean region, achieving about 156 million new residents between 1970 and 2015, representing 82% of the total new population during this period. By 2050, it is expected that urban population will increase by 45% compared with 2015, amounting to 476 million people. Urbanization in the MENA area to date has been much faster compared with the Mediterranean Europe, and this shift from rural to urban areas is expected to continue, with 69% of the MENA population living in cities by 2050, compared with only 42% in 1970. The urban areas in Libya, Algeria, Turkey, and Tunisia have been the fastest growing in the region, with annual rates of 3.5%–4.6%. The building stock in almost all Mediterranean EU countries is rather old, with 31%–55% of dwellings built before 1969, while the use of space cooling in residential dwellings varies significantly (i.e., with a range of 5%–81%) between countries [42].

Due to growing energy demands all over the world, the ongoing growth of the human population poses several problems to the global economy in general, with a specific focus on environmental conservation and energy security [4]. For example, present global petroleum usage is more than 100 times greater than what nature can produce [100,116] According to the Energy Information Administration research (EIA, 2014), this rate will increase by nearly 56% by 2040, potentially resulting in an energy catastrophe. The global economy is mostly dependent of the nonrenewable fossil fuels [68,101,117] with an overuse status affecting both the environment and the energy security as these fuels are greatly contributing to greenhouse gas (GHG) emissions with a CO_2 level of nearly 400 ppm [27,92], with an obvious increase of about 51% during the last 20 years, mainly due to fossil fuel combustion and hence resulting in global warming [58]. These threatening facts exert pressures and have forced both governments and research sectors to look for other renewable alternatives of clean resources replacing the fossil fuels. Using biologic resources for renewable fuels became very important and necessary for the global fuel market and are known as biofuels.

In 2014, global biofuel output was predicted to be 127.7 billion liters, with biodiesel accounting for 23% of that total. This estimate was increased 13%–29% in 2014 with the United States, Brazil, Germany, China, and Argentina accounting for the top countries producing 60, 29.9, 4.3, 3.9, and 3.6 billion liters, respectively [60–62]. Recently, biodiesel became the main alternative for fossil fuel that is receiving much attention worldwide. Adopting biodiesel as a sustainable substituent to fossil fuel is vital in reducing unburned hydrocarbon emissions in addition to the fact of not contributing to atmospheric CO_2 emissions being derived from biologic origin [146]. Biodiesel is also of minimal toxicity due to its high biodegradability [15,93].

1.1 Climatic change consequences on energy

The future evolution of the relevant infrastructure and its spatial distribution will greatly influence the potential exposure to the climate-related impacts on electricity supply and transmission. The Mediterranean Network of Transmission System Operators (Med-TSO) has predicted an increase of the Mediterranean production capacity from 250 to 400 GW in 2030 and estimated the contribution from renewable energy sources at approximately 40%–60% [12]. It also made a prediction of almost 18,000 MW of new interconnection capacity, of which 6600 MW new capacity is estimated in Turkey and

4550 MW in Egypt. The share of renewable generation within the Mediterranean area is expected to rise from 25% in 2016 (mainly from hydropower) to up to 32%–42% by 2030. The concept of a "green economy" has emerged as a new tool achieving sustainable development goals during the Rio+20 conference in 2012 [128,147].

1.2 Climate impacts on mediterranean countries

Most of the global population (67%) lives in urban areas, with about 329 million capita in 2015 [127]; of these, about 54% (i.e., 177 million) are inhabiting large cities with about 300 1000 inhabitants or more. Urbanization exists in the MENA region to date, and this shift from rural to urban areas is expected to continue, with 69% of the MENA population living in coastal cities by 2050, compared with only 42% in 1970. The urban areas in Libya, Algeria, Turkey, and Tunisia were determined as the fastest growing areas in the region, with annual rates of 3.5%–4.6%. The building stock in all Mediterranean European countries is old, with 31%–55% of dwellings being built before 1970. The use of space cooling in residential dwellings significantly varies between countries (5%–81%) [42] as well as the climate-related impacts [111].

The Mediterranean area is characterized by very high human mobility with a particular effect of climate change [25]. Coastal cities along the Mediterranean are particularly sensitive to climate change because of their precise spatial and topographical features, which are defined by proximity to the shore, high water demand, and a dry summer [103]. In addition, Climate change has a detrimental impact on Mediterranean soil resources, such as intensive erosion processes that have significant on-site and off-site consequences, posing a threat to the region's long-term development [148]. It might also have a significant impact on coastal infrastructure, housing, industrial facilities, energy and sanitation systems, transportation and communication networks, and tourism and cultural sites. All of the above effects will destabilize towns and the dense population that lives in this area.

One of the specific examples of the regional risks along the Mediterranean region, particularly in the southern part, is the great challenge to meet the sustainable development goals (SDGs) [133]. The Mediterranean countries were described as one of the global "hotspots" with regard to risks related to climate change. Reduced summer river flows together with increased water temperatures in the region would generate water-cooling constraints for thermoelectric power plants [125,143]. In addition, the hydropower in most of the Mediterranean region is also subjected to significant impact with reductions ranging between 1% and 18% with a temperature increase of 1.5–3.0°C, as estimated by Tobin et al. [125], for the Mediterranean EU countries (except Italy). In the rest of the Mediterranean countries, the hydropower reduction rate will increase to over 20% annually by mid-century but decreases when warming levels are less severe [125,140]. A small decrease of up to 5% of the mean annual wind energy is potentially expected over the Southern European countries after mid-century [34,125,135], but a probability for low wind speed of <3 m s^{-1} is estimated over the Maghreb countries by the end of the century. For solar energy, there is a level of disagreement on the signs for photovoltaic production and future cloud coverage under climate change over Southern Europe, e.g., Refs. [94,125], while some studies document some small changes (i.e., ±2%) for the other Mediterranean countries [94]. Impacts of climate change–related risks are significantly affecting energy with its various types and/or sources. In addition, the increase of cooling degree days and other societal needs will drive the electricity and energy demand in the Mediterranean region.

The urban heat island is an additional effect that occurs in many Mediterranean cities and will also strengthen the increase of energy demand [115,142]. Accordingly, most economic and social sectors in the region are significantly impacted by climate change including tourism, agriculture trade, urbanization, and transport, creating many challenges and opportunities with the predicted increase in energy demand, particularly in the Southern Mediterranean countries with up to 98% by 2040 [82,95], with a zero-emission energy sources scenario being projected by 2040, which would make up to 60% of installed capacity [95]. Those impacts activated the regional initiatives and integral role that those countries would play for energy security not in conventional energy sources only but in renewable sources too. Further regional cooperation is needed, particularly in technology transfer for clean energy.

1.2.1 Pollution problems of coal energy and the need for "clean" alternatives

Several environmental challenges would emerge thorough coal mining, including mining accidents, land subsidence, and impacts on water environment [141]. Researchers hypothesized that the habitual use of fire began nearly 300–400 thousands years ago [51,113] using wooded and other organic materials (e.g., animal bones, seaweed, and straw) as the first used fuels [57] that significantly influencing human biology [51] and human behavior and also transformed the developing societies and shaped their culture [109]. Such information confirms a relatively recent shift from using biomass for energy to fossil fuels (e.g., petroleum, coal, and natural gas) during the technologic progress and has vigorously played a critical role in the recent, current, and future development, industry, and civilizations, creating the need for modern technologies and clean energy sources to sustain the environment and meet the SDGs.

1.3 Potentiality of algae as a biologic source of energy

Algae are photosynthetic prokaryotic or eukaryotic creatures that have evolved to thrive in a wide range of environments due to their diverse morphologies, which range from unicellular and basic to multicellular and complicated. They include various species/strains with various forms and structure and are potential able to convert CO_2 to produce different bio-products, such as fuel, food, feed, and other bioactive materials [86]. Moreover, they are efficient as bioremediation agents [91] and as nitrogen-fixing fertilizers [131]. Algae are the oxygen precursors producing approximately half of the atmospheric oxygen needed for living organisms on earth [81,21]. With regard to their abundance, the annual world production of biomass of microalgal species was estimated as 5×10^3 to 7.5×10^3 t, which can provide about US$1.25 billion as annual income [121,24].

As a matter of fact, there are many reasons and algal basics that led the researchers and entrepreneurs to think of biofuel production from algae such as (1) their high productivity and biomass yields; (2) their ability to grow in wastewater, recycled water, and sea water, to secure freshwater sources especially in countries with limited freshwater supplies; (3) the minimal competition with arable/agriculture land as well as nutrients; and (4) their ability to recycle CO_2 from stationary gas emissions.

Algae can efficiently grow in any water source, within a wide variety of climate conditions and through varied methods of production, such as natural (i.e., open ponds) or artificial (i.e., photobioreactors or fermenters) habitats. Such great ability for cultivation works as a "job creation engine" opening venues for a wide variety of jobs including researchers and engineers who contributed to the cultivation processes. As a real estimate, in the United States about 220,000 jobs were created by 2020 in this sector through "The Algal Biomass Organization" projects [144].

1.4 Types of microalgae

Microalgae can be found in a wide range of environments with light and water, including the ocean, lakes, soils, ice, and rivers [37]. Microalgae have a high diversity (between 200,000 and several millions of species) that can be separated into categories based on pigmentation, biologic structure, and metabolism [99].

1.5 Size classification

Microalgae are small organisms that are classified into four sizes: microplankton (20–1000 m), nanoplankton (2–100 m), ultraplankton (0.5–15 m), and picoplankton (0.2–2 m) [22,50]. Their small size allows them to perform efficient photosynthesis, turning solar energy into lipids and carbohydrates from CO_2 dissolved in water.

2. Species

Microalgae (organisms capable of photosynthesis with a diameter of less than 0.4 mm, such as diatoms and cyanobacteria) rather than macroalgae, such as seaweed, are the focus of research into algae for mass-production of oil. Microalgae have gained popularity due to their less complex structure, rapid growth rates, and high oil content (for some species). Nonetheless, due to the high availability of this resource, some study is being done into using seaweed for biofuels.

3. Potential strains

Despite the fact that microalgae are a diverse group of microorganisms, not all species are ideal for biodiesel production. [54] listed the criteria to consider: (1) carbon dioxide tolerance and absorption; (2) temperature tolerance; (3) stability for cultivation in specific bioreactors; (4) secondary value products; (5) specific growth requirements and competing algae; (6) vulnerability to and herbivory potential; (7) excretion of auto inhibitor; (8) harvesting and downstream processing; and (9) manipulation potential of genetic engineering.

3.1 History of biofuel research

Coal is the largest worldwide source of solid fuel [88], which significantly assisted launching the recent revolution and supported the evolution of several heavy industrial activities and general manufacturing, for example, iron and steel production, power generation, and railways [44,69]. The industry passed through a level of crisis during the past decade due to the decline in demand and hence in productivity creating many negative consequences affected by the quick change in coal industry, while such a decline in coal dependency presents positive outcomes in terms of environmental and for

TABLE 30.1 A comparative statistic among the three common biofuel generations generated from different biofeedstocks.

	First-generation biofuel		Second-generation biofuel		Third-generation biofuel	
	Value	Feedstock source	Value	Feedstock source	Value	Feedstock source
Yield of oil produced in relation to generated biodiesel (liters/hectare): (Kg/hectare/year) The later values are in "bold"	5366: 4747	Palm	1307: 1156	Castor	58,700: 51,927	Microalgae (WB)
	1190: 862	Canola	741: 656	Jatropha	136,900: 121,104	Microalgae (DB)
	1070: 946	Sunflower				
	446: 562	Soybean				
	172: 152	Corn				

DB, Dry biomass; *WB*, wet biomass.

human health. This creates the need for clean energy sources and initiates the global wave of using algae as a renewable environmentally friendly source for energy. Utilization of algae as energy producing sources started in the late 1950s, and this was coincident with the use of algal carbohydrate to produce methane gas [104].

4. What is biofuel?

Biofuel is the fuel derived from lipids derived from living organisms (oils and fats), animals or plants. Most biodiesels used in industry are mainly from the triglycerides derived from plants, e.g., sunflower, soybean, etc. See Table 30.1.

4.1 First-generation biofuel

The first generation of biofuels are the types that depend on plant crops as raw materials, such as beet seeds, corn sugar, and cane, which are mainly food crops [14] that produce the ethanol fuel during the process of sugar fermentation or the vegetable oil fuels [75] from oil plants (e.g., palm) as well as the biogases emitted from landfills [98]. At both environmental and societal levels, several limits exist in using food-based crops for biofuels production. These include environmental pollution and competition with agricultural land, as well as the deforestation phenomenon [49]. If most of the lands available for oil plants, in different countries in Africa and Europe, were devoted for biofuels production, this would lead to lack of available land needed for biofuel production, creating problems for soil poverty [26].

4.2 Second-generation biofuel

The second generation of biofuels are those depending on cellulose obtained from raw materials of crops that are not used as food, such as wood and/or straw. Wood-based diesel, bio-alcohols, and other bio-oils are forms of the second-generation biofuels [114].

4.3 Third-generation biofuel

The third-generation biofuels are those that rely on microorganisms as the raw materials or feedstock, including yeast, fungi, and microalgae. These include vegetable oils, biodiesel, jet fuels, biohydrogen, and many other bio-oils [102,149]. It is anticipated that, compared with the first generation biofuels, the second- and third-generation biofuels are favored with regard to sustainable development due to the ability to reduce the atmospheric CO_2 [98]. Fossil fuels are hugely contributing to GHG emissions with high CO_2 levels reaching about 400 ppm [27,92]. In comparison, plant-based biofuels, e.g., soybean (first-generation type), would reduce GHG emissions by about 41%, but a reduction rate of about 180% would be induced when use third-generation types [56]. Authors (e.g., Ref. [30]) have estimated that about 1.8 tons of CO_2 would be consumed for each ton of microalgal biomass produced. In addition, algal cultivation for biofuel is known to compete less with agriculture compared with land-based biofuels that are based on agricultural feedstocks.

5. Algal biofuel

A wave of increasing interest in algal cultivation has increasingly been considered due to their potential abilities to produce biofuels (e.g., algal-based biofuel; ABB) ,which could be used for transportation, substituting the expensive and unrenewable fossil fuel. Microalgae are able to produce several types of biofuels, including methane [150], biodiesel [17,45], and biohydrogen [43,67]. Although, using microalgae as a source of energy is not a new approach [29,97], its production as well as applications are more developed nowadays; this is mainly attributed to the petroleum prices increase. The global warming phenomena and the climatic changes associated with burning of the fossil fuels are also significant reasons for the worldwide consideration of algal/microalgal biofuel [45].

Currently, several companies and government agencies are considering the conversion to bio-oils to make algae fuel production commercially viable to reduce capital and operating costs. The high productivity of algae and the year-round production together with the direct utilization of combustion gas and their potential for wastewater treatment are among the aspects that increase the future dependence on microalgae for biofuel production. Algal fuel derived from the algal energy-rich oils is an alternative to liquid fossil fuels as well as to other common sources for biofuel, e.g., sugarcane and corn.

Algal-based fuels could help reduce the national dependence on fossil fuel in some countries due to its market potentiality, enabling the transfer from importing fossil fuels from other countries. It is anticipated that a higher fraction of algal biomass could be converted to oil when compared with food crops (e.g., soybeans) with an estimate of nearly 60% compared with 2%–3%, respectively. Several researches showed that algae (some species) would generate oil yield of about 60×10^3 or 136×10^3 L for each cultivated hectare per year of wet and dry weights, respectively, giving more than 60% of their dry/wet weight [80,85]. The yield of oil per unit area of algae is estimated to be from 5.9×10^4 to 13.7×10^4 L ha^{-1} year. These estimates are mainly depended on the lipid content of the algal cell, which is 10–23 times higher than the oil palms that only produce $<6 \times 10^3$ L ha year [30].

In addition, algal cells can easily grow in aqueous solutions when cultivated either in open ponds or in photobioreactors with an efficient access to water, CO_2, and nutrients producing biomass with large yield. The algal oil produced could then be turned into sellable biodiesels and can be utilized for transportation. In addition, algae do not need herbicides or pesticides on cultivation and hence could potentially avoid land agriculture problems [21]. Microalgae are also recognized for their metabolic flexibility, which allows for simple changes to biochemical pathways (such as protein, carbohydrate, or oil synthesis) and cellular composition [126].

This indicates that modifying the growing conditions can vary, modify, and improve the biochemical and oil content of the cultivated algae. There is a report published by FAO in 2009 including all different methodologies for algal cultivation and all kinds of bio-oils and other co-products that can be produced from algae. The report also gave detailed information on the environmental benefits of transferring into biofuels as well as the expected threats that might emerge with the production process.

5.1 Advantages of algae as a biofuel source

Algae are becoming one of the most promising, sustainable sources of cultivated biomass that produce considerable yields and oils that can be used for the production of biofuel, food, feed, and other useful products. There are several reasons that make them much favored, such as the easy methods for cultivation, the large biomass produced, as well as the various environmental and socioeconomic benefits associated with their cultivation process. It is known that algae, particularly microalgae, have been naturally existing for billions of years, growing, reproducing, and storing energy within their cells as oil; therefore, it becomes more crucial to encourage the researches to study "algae" and strengthen our understanding to be much aware of all its numerous and unlimited benefits.

For a better understanding of the mechanisms of algal-based biofuel production, it is important to know some basic information about algae:

- *Nature and diverse habitat*

Microalgae are very diverse and can live within diversified environments where only light and water exist. Hence, microalgae can grow and flourish in oceans, lakes, rivers, ice, humid soils, etc. [37]. Microalgae show a high level of biodiversity among aquatic organisms with up to several millions of species, to which there are many thousands of species that are continuously being discovered. Algal populations are greatly diversified with respect to size including microscopic species that are so-called microalgae and to giant species or the seaweed (macroalgae) having species with more than 100 feet in length. Algal species/strains are also varied according to their developmental level including the organism's shape, cell structure, cell components, particularly pigments, and also internal physiologic processes [99]. Microalgae are classified into four size classes: (1) microplankton (20–1000 μm), (2) nanoplankton (2–100 μm), (3) ultraplankton (0.5–15 μm), and (4) picoplankton (0.2–2 μm) [22,50]. The smaller the size of the species is, the more effective is the photosynthesis and hence more lipids generated by converting light energy and atmospheric CO_2.

- *Diversity*

Most of the researches on algae of mass-production for oil focus mainly on microalgae due to their high photosynthesis (e.g., organisms <0.4 mm, including diatoms and cyanobacteria). Microalgae are more advantageous due to their less complex structure, rapid development rate, and high oil content (some species have substantially higher oil content than others). It is worth noting that, although being diverse communities with a diverse range of species and strains, not all algae species are ideal for biofuel generation [54].

- *Fast growing*

Algae can reproduce very rapidly, and they are almost faster than any other plants and also can be harvested daily, as the population may reach double or three times of its original size within 1 day or a few hours [71]. This means that a good volume of biomass could be continuously produced with a rate greater than any other productive crops, increasing the potentiality of algae for biofuel production. Microalgae can also be used in food, cosmetics, fertilizers, and a variety of other industries [63].

- *Easy cultivation*

One of the favored characteristics of algae is that they can grow within various climatic conditions and with many different cultivation and production methods; they can grow in ponds, photobioreactors, or/and fermenters. Algal cultivation can be achieved under different culturing conditions, such as photoautotrophic, mixotrophic, or heterotrophic. If they grow photoautotrophically, they utilize light to reproduce and provide new cells/biomass, but in heterotrophic cultivation method, algae only require a carbon source for energy (e.g., sugars), instead of light [66]. In mixotrophic cultivation, algae are able to use both or any of the energy sources (light or carbon) to grow and provide biomass yield. When relying on any carbon source, algae achieve high biomass yield, and this is cost-effective as it will reduce the cost of the overall process [136]. The heterotrophic and mixotrophic cultivation strategies are therefore more advantaged compared with photoautotrophic methods.

5.2 Sources of algal biomass

5.2.1 Natural standing crop

Tropical or subtropical regions with available solar light throughout the year and fresh, saline, or even wastewater are commonly favorable conditions for mass cultivation of algae called algal blooms [20,53]. Environmental conditions at specific locations can greatly affect microalgae populations and their growth dynamics [40]. Moreover, excess nutrients such as nitrogen and phosphorus compounds in untreated wastewater lead to algal blooms. At least 30,000 microalgae species have been identified, and they can be found in seas, freshwater lakes, and other bodies of water. Cultivation is the most efficient approach to generate biomass from microalgae due to their small size, which makes successive harvests more difficult.

5.3 Artificial cultured biomass

5.3.1 Microalgal cultures

Because algae are unevenly distributed, the density of algae in nature is very low [23]. To a certain extent, it brings some difficulties to the large-scale production of biofuel. To meet the fuel needs of explosive populations and industrial production, algae biomass must be supplied throughout the year [64]. Culture systems vary widely between macroalgae (seaweed) and microalgae. Microalgae have to be cultivated in a specific system placed on land or floating on water to be easily harvested in contrary to seaweed that can be grown directly on open sea surfaces [70].

Closed and open microalgae growing systems are generally characterized based on their configuration parameters. Microalgae are grown in open systems such as ponds, lagoons, channels, shallow circulating units, and other open environments. Microalgae are grown in closed systems in transparent vessels that are exposed to sunlight or artificial radiation to facilitate photosynthesis [32]. The advantages and drawbacks of various cultivation systems are tabulated in Table 30.2.

TABLE 30.2 Advantages and limitations of open and closed systems [21,30,35,65,87,110].

Cultivation methods	Advantages	Drawbacks
Open pond	Low investment cost Simple system Easy to clean Utilizes nonarable land Low energy consuming Easy maintenance Easy to scale up in pilot scale	Low biomass productivity Need large area setup Not suitable for all microalgae Low light and CO_2 consumption efficiency Cultures are simply contaminated High evaporation and water loss rate Lack of temperature control The cost of harvesting is comparatively higher Low mixing efficiency Low nutrient diffusion Flotation of dead and living algae in static open pond
Tubular photobioreactor	Appropriate for outdoor cultures High illumination efficacy Large biomass productivity Suitable for axenic culture Easy to control and operate Easy to harvest due to large biomass Higher photosynthetic rate due to higher surface to volume ratio	Expensive Large land area needs PH, dissolved O_2 and CO_2 gradients in the tubes Supply of air needed for operation using airlift pumps Photosynthesis is inhibited by high concentrations of O_2. Restricts the maximum continuous run tube length
Stirring tank photobioreactor	High productivity Low consumption of energy High mixing efficacy Easy to operate Easy to work under antiseptic conditions Suitable for microalgae immobilization Low photo-inhibition and photo-oxidation issues	Relatively costly High price of repair Area of small illumination Increased shear pressure on cultures of microalgae Sophisticated material for building Difficult to scale up Complex pattern of flow Low ratio of surface and volume Large unit for building due to limited diameter and height adjustability
Flat-plate photobioreactor	Comparatively cheap High biomass productions easy to operate and clean Low oxygen concentration High illumination efficiency Contamination is low, suitable for immobilized microalgae Suitable for outdoor and indoor cultivations	Challenging large scale-up Difficulties with temperature control Limited amount of growth in the near-wall area Difficulties with regulating the diffusion rate of CO_2 Photo-inhibition can happen. There may be hydrodynamic stress

6. Lipids yield of microalgae

Lipids are one of microalgae's most important main metabolites (e.g., 15% and 60% of each dry cell weight is lipid content), and its lipids are classified as polar (structural) or nonpolar or neutral (storage) lipids based on their polarity, of which the polar type is further subdivided into phospholipids and glycolipids. However, the nonpolar lipids exist as triacylglycerides (TAGs) and work as a store for energy. The stored lipids are transesterified to produce biodiesel. Microalgae lipids and their composition vary by species; for example, some microalgae have a higher proportion of neutral lipid than others [151]. A microalga modifies its metabolic pathway to store neutral lipids, primarily in the form of TAGs, when it is starved or deprived of nutrients. Controlled development of microalgae lipids, which produce saturated and unsaturated fatty acids, is critical for nutritional value. Polyunsaturated fatty acids, which contain important fatty acids, are employed in animal feed and human food.

7. Growth requirements for optimal microalgal growth

Temperature, light intensity, pH, and nutrient concentrations are only a few of the environmental characteristics that must be kept within physiologic limits. To provide and maintain a good growing environment, the reactor system is essential [108]. As a result, reactor design necessitates an understanding of algae physiology, such as morphology, nutritional requirements, and stress tolerance.

7.1 Temperature

Temperature is one of the vital attributes for algal growth and also is among the most difficult characteristics to keep at optimum, particularly in large-scale and outdoor growing systems. Any fluctuations in temperature level can lead to significant losses in produced yields. Many algal species can tolerate temperature variations of up to 15°C below their optimum with reduced growth rates; however, a few degrees above this optimum level can cause cell deaths [84]. It is worth mentioning that ideal temperature for microalgae growth is mostly between 20 and 30°C with some variations among species [31]. Closed reactor systems almost always require some form of temperature control. They often suffer from overheating during hot days when temperatures inside the reactor can reach in excess of 50°C. The culture system can also be placed inside a greenhouse or contacted with water to minimize temperature fluctuations [30].

7.2 Light

The intensity and utilization efficiency of the light supply are crucial in reactor design since light is the primary limiting element in the growing of photosynthetic organisms [108]. In three separate locations, the photosynthetic activity of microalgae fluctuates in response to light intensity [77]. Cells are light limited at low light intensities, and photosynthetic rate increases with increasing irradiance. When cells become light saturated, photon absorption exceeds electron turnover in photosystem II (PS II), and the photosynthetic rate does not rise any further with increased light intensity. After a certain level of irradiance, cells become photo-inhibited due to damage to the photosynthetic apparatus, and the photosynthetic rate begins to fall.

7.3 Nutrient provision

High growth rates require an optimal supply of nutrients, primarily carbon, nitrogen, and phosphorus, as well as a variety of additional macro- and micronutrients essential for algal growth. Any nutritional deficiency causes metabolic abnormalities, physiologic alterations, and diminished production [108]. The provision of nutrients to the culture is rather straightforward, but the delivery of nutrients to individual cells is contingent on effective mass transfer, which is linked to mixing and gas sparing [54]. In microalgal cultivation, nutrients are also a substantial cost; consequently, the reactor system must be designed to allow for effective culture medium recycling [52].

Certain algae species can grow heterotrophically or mixotrophically, through which they can get their needs of carbon and energy from organic carbon sources (e.g., glucose or acetate) [79]. Using heterotrophic species can minimize the need for light and CO_2, and this would remove some of the major design constraints on reactors [108]. Although biomass productivity has yet to reach that of yeast and other heterotrophic organisms, heterotrophic growth of microalgae in serializable fermenters has shown some commercial success [79].

7.4 Mixing

In microalgal culture systems, mixing is critical because it is linked to other critical parameters including light, gas movement, and nutrient provision. Good mixing maintains cell suspension, removes temperature stratification, establishes the light-dark regime by moving cells via an optical gradient, assures effective nutrition distribution, and enhances gas exchange, minimizes photo-inhibition at the surface, and reduces mutual shading at the reactor's center [130]. Secure, continuous mixing prevents the biomass settling [89], and mixing rates constitute a trade-off between increased growth rate, cell damage, and a higher risk of infection. Shear stress can be imposed on microalgal cells by high mixing rates, especially in filamentous species or those with fragile morphology [52].

Mixing in open ponds is typically provided by a paddlewheel or rotating arm; however, it can be achieved by mechanically tools (e.g., pumping) or by gas transfer systems (e.g., pipes, propellers, jet aerators).

8. Closed system

8.1 Horizontal tube photobioreactor

The horizontal tube photobioreactor was the first constructed design of a closed reactor for microalgae cultivation. It consisted of long horizontal tubes that can be configured in designs, including panels, walls, or helices (Fig. 30.1A). This design needs a large area of land owing to the high surface area required. In this model, the diameter of the tube may be smaller owing to the long and large distance covered to permit light to penetrate the culture fluid. This reactor type is only appropriate for microalgae that can survive under sunlight. In this type of reactor, the culture medium is circulated within the long tubes to exposure to the light source (sun or artificial light) and returned to the reservoir by an electric pump with a highly turbulent flow regime to avoid flocculation of microalgae cells [74]. After the medium circulates, a certain portion of the medium must be collected to allow the continuing operation of the system. The use of horizontal photobioreactors can better reduce dissolved oxygen content than vertical reactors, as the increase in reactor height increases tank pressure leading to increase oxygen content in the medium [138].

8.2 Vertical tube photobioreactor

A vertical photobioreactor, also known as a tubular reactor, consists of a vertical tube supplied with air pump located at the bottom of the bioreactor to mix the algal culture by generating air bubbles. The bioreactor diameter is half its height and is provided with an external light source. There are two types of vertical tube photobioreactor: bubble column and airlift column (Fig. 30.1B). A bubble column is a cylindrical tube without any internal configuration where the liquid flow is primarily driven by the air bubbles produced. Using airlift reactors is more advanced and cost-effective for cultivating several microalgal species. It is called "airlift" due to the process in which the air bubbles push the fluid from a riser area

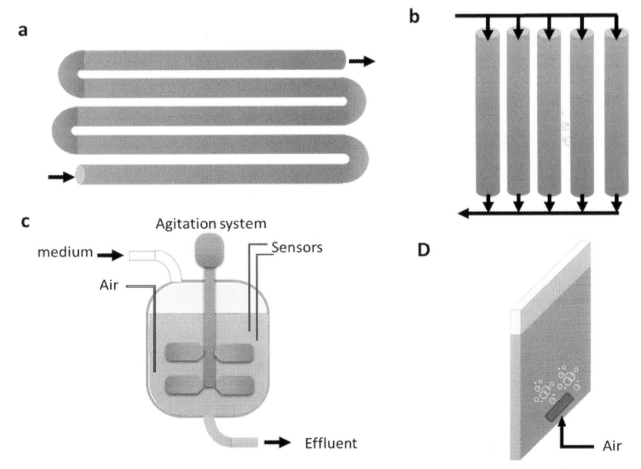

FIGURE 30.1 Examples of different photobioreactor designs. Basic designs: (A) horizontal tubular, (B) vertical tubular, (C) stirring tank photobioreactor, (D) flat panel.

within the concentric tube to an irradiated area outside, allowing circulation of liquid between dark and light areas, enhancing growth of microalgae. Airlift design adds more advantages, good mixing, efficient mass transfer, and is perfect for algae immobilization [28].

8.3 Stirring photobioreactor

The stirring tank design is mainly coming from the fermentation tank design. The models are similar, while the microalgal tank should be transparent and provided with a light source to allow microalgae to grow photosynthetically. For this bioreactor, an agitator is used to mix the algal culture and to ensure optimal temperature and mass transfer (Fig. 30.1C). A stirring tank is shown to perform well because it provides better mixing and transfers in indoor cultivation, where providing aeration may help to optimize solubility [124]. The use of various media is likely to influence overall fuel productivity; for example, a medium with an increased salt concentration will lessen biomass growth while increasing lipid accumulation. The medium type used must be optimized to achieve the highest lipid production [28].

8.4 Flat-panel photobioreactor

This type of photobioreactor is the most common, the photobioreactor (PBR), and it is suitable for use for either outdoor or indoor cultivations. The flat-panel photobioreactor consists of two transparent or semitransparent glass sheets forming a transparent cuboid arranged in a cascade toward the light source (Fig. 30.1D). The very short light path allows for significantly easier light penetration to culture, maximizing the light intensity received from the light source [59]. An air sparger could be also applied in the PBR for mixing and circulation.

9. Open system

9.1 Static pond

It is the first and open system for microalgae cultivation with a simple approach with a shallow pond without stirring. Static ponds are set up in natural lakes or lagoon ponds with depths of 50 cm allowing light penetration without mixing (Fig. 30.2A).

9.2 Raceway pond

This is one of the most common open systems used for algal cultivation, and it is suitable for large-scale commercial production (Fig. 30.2B). Microalgae species such as *Spirulina, Dunaliella, Chlorella,* and *Haematococcus* have been extensively cultivated in this type of system. Raceway ponds are typically constructed as a single or set of tracks with 15–50 cm depth [59]. Paddlewheels are used for mixing microalgae cells. Turbulent mixing may be used to maintain a uniform density of the algal cells.

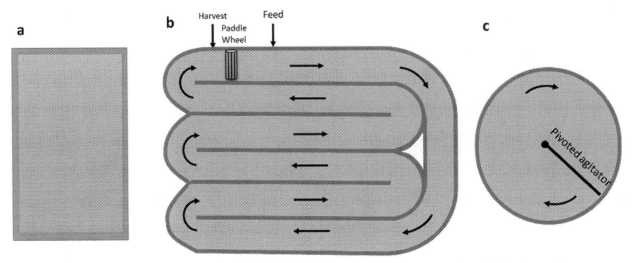

FIGURE 30.2 Examples of different open pond systems: (A) static pond, (B) raceway pond, and (C) circular pond.

9.3 Circular ponds

The design of circular ponds originated from ponds used for treating wastewater with a circular bioreactor with a rotating arm placed at the center and moving in the axial direction to mix the culture [83]. This pond usually has a diameter not exceeding 50 cm and a depth of 20–30 cm (Fig. 30.2C).

9.4 Hybrid photobioreactors

For different growth phases, hybrid systems alternate between closed photobioreactors and open systems. Microalgae are cultivated in a closed system in the first stage to avoid contamination from the environment. To increase biomass production, the microalgae will be grown in nutrient-rich open pond systems in the second stage. For example, hybrid systems are used to treat *Haematococcus pluvialis* with yearly production rates of 10–24 tons per hectare [21].

Closed PBRs, either complete or just the solar collector, are occasionally floating in a temperature-modulating water bath. Although all such modifications raise the cost of manufacturing [130], built double-walled reactors with a portion of the liquid volume are used for heating and cooling. Temperature and the amount of light available have a relationship. When the temperature is below ideal (as it is in early-morning outdoor cultures), a fast increase in light intensity can create photo-inhibitory stress, in which cells become too cold to process incoming photons, decreasing photosynthetic efficiency for a large portion of the morning [132]. As a result of low temperatures, they are particularly inefficient in the early morning, so any efforts to use heat reactors should be focused just before sunrise.

10. Macroalgal cultures

Approximately 95% of seaweed used by human beings is cultivated [47], and nearly 93% of this cultivated seaweed belongs to four genera: *Porphyra*, *Undaria*, *Laminaria*, and *Gracilaria* [139]. Only 6% of the world's seaweed originates from natural sources [47], with the rest being cultivated.

11. Types of algal bioenergy

11.1 Bioethanol production

Due to the scarcity of nonrenewable fuel resources and their environmental consequences, the world faces an energy problem [76]. As a result, researchers have turned their attention to the topic of bioenergy. Edible vegetation, agricultural wastes, and algae are the most common biomass materials for bioethanol production today. A substantial phycology research has recently been conducted to investigate the process of generating energy as bioethanol from macroalgal biomass [137]. Algae contain significant amounts of structural polysaccharides. The type and concentration of carbohydrates, on the other hand, vary across algal classes.

It necessitates developing new technologies to facilitate the efficient bioconversion of algal biomass into bioethanol [3,33]. Algae can produce bioethanol with about 0.43 g per each gram of substrate, which is more than what terrestrial feedstock can provide, while algae have a shorter production time [33].

Because the lignin content of macroalgal biomass is negligible, it does not necessitate an additional delignification step during bioprocessing [73]. In comparison to microalgae, macroalgal biomass contains a higher proportion of holocellulose than microalgae. Additionally, macroalgae can be grown in greater quantities in the marine environment. Bioethanol can be produced from algal biomass via three main methods: hydrolysis followed by fermentation of the hydrolyzed biomass via dark or photofermentation. Each pathway or process has distinct characteristics [3].

11.2 Biodiesel production

Biodiesel is made up of fatty acid methyl esters that are produced by transesterifying renewable oil feedstocks with alcohol and an acid or base catalyst. Vegetable oils, animal fats, waste cooking oils, and microalgal lipids can all be used to make biodiesel [4,119]. The net benefit to society of biodiesel feedstock is based on a number of aspects, including its impact on net energy supply, GHG emissions, water and air quality, and global food impact [5,6]. Biodiesel is a diesel fuel made from the lipids of animals or plants (oils and fats). According to studies, some algae species can create 60% or more of their dry weight in the form of oil. Microalgae may produce enormous amounts of biomass and useable oil in either high-rate algal ponds or photobioreactors because the cells grow in aqueous suspension, where they have better access to water, CO_2, and dissolved nutrients. This oil can then be converted into biodiesel, which can subsequently be marketed for use in cars.

Regional microalgae cultivation and biofuel processing will benefit rural communities financially. Microalgae can develop at a faster rate than terrestrial crops because they do not need to manufacture structural materials like cellulose for leaves, stems, or roots, and they can be grown floating in a rich nutritional solution. They can also convert a far higher percentage of their biomass to oil than traditional crops, such as 60% compared with 2%–3% for soybeans. Oil from algae has a per unit area production of 58,700 to 136,900 L ha^{-1} year, depending on lipid content, which is 10–23 times more than the next best yielding crop, oil palm, which has a yield of 5950 L ha^{-1} year [30,37].

Using cultivated plants for biodiesel production is currently a substitute fuel, but it affects huge areas of land to cultivate fuel plants instead of plants for human food; at the same time, those plants require massive areas for cultivation. For example, to cultivate palm oil, which is one of the largest oil-producing crops, to produce the amount of biodiesel to satisfy the need for the sector of transportation in the United Kingdom (22.7 million tons) [19], it requires >200 times of the total UK area. This implies the need for alternative sources, and algal cultivation is a possible biologic fuel source.

12. Difference between diesel and biodiesel

Because of its high energy density, diesel, while not as popular as high-octane gasoline these days, is more suitable to high-commodity diesel-engine cars. Both biodiesel and diesel are limited in terms of availability, with biodiesel derived from plant oils and animal fats and diesel derived from the rapidly depleting nonrenewable energy sources found in the earth's crust. Diesel and biodiesel are two carbon-based organic compounds with differing sources of carbon. The former is known to be extracted from animal fats and plant oils, whereas the latter is known to be extracted from petroleum crude oil cracking. The biggest difference among both diesel types is that biodiesel is a renewable energy source, whereas diesel is a depleting one.

When compared with gasoline-powered engines, diesel engines were once thought to be exceptionally high-tech and efficient machinery. Diesel requires less refining than gasoline due to its thick composition, making it a less expensive option. The damaging emission of carbon, sulfurous, and nitrogenous oxides into the environment is caused by incomplete combustion of diesel. These are the primary compounds/gases that cause most of the problems with the earth's atmosphere and the ozone layer's deterioration. Acid rain and other climate changes can result from the presence of such contaminants in the atmosphere, causing harm to the natural order, the ecosystem. Inhalation can produce headaches and nausea, and long-term exposure can result in brain damage or death. Because the earth's natural resources are rapidly depleting, diesel, a nonrenewable form of energy, is now available in less quantities as one of Earth's natural resources Fig. 30.3.

Nonenvironmentally friendly diesel supplies are nearly depleted, and it will only be a matter of time before they are fully depleted. Biodiesel is quite feasible and has minimal environmental impact. Biodiesel was developed to increase engine efficiency by consuming less fuel while producing a lot of energy. As a result, it has a number of environmental advantages over diesel.

FIGURE 30.3 Combustion of biodiesel and petroleum diesel. *Source: https://en.wikipedia.org/wiki/Biodiesel.*

13. Microalga species for biodiesel

The production of biodiesel is mainly achieved from "microalgae" so let's briefly describe what microalgae are. Microalgae are sunlight-driven cell factories that convert carbon dioxide to potential biofuels, foods, feeds, and high-value bioactive materials [86].

Lipids can make up to 80% of the dry weight of some microalgae species including *Botryococcus braunii* and *Schizochytrium* sp [37]. These plants may produce up to 770 times more lipids per acre than oleaginous plants (colza, sunflower, etc.), and their large-scale production enables for the development of high-yield biodiesel [30]. Another benefit of using microalgae to make biodiesel is that it can double in size from one to three times in 24 h [71]. As a result, the biomass of microalgae can be harvested more than once a year. Microalgae has the potential to be employed in a variety of industries, including food, cosmetics, fertilizer, and many others [63].

14. Cultivation systems

For microalgal cultivation, a wide range of open and closed reactor systems have been developed, possibly reflecting the variability in the physiology and requirements of different algae species. Finally, the overall purpose is to maintain a desired algal culture in ideal productivity-producing circumstances. High volumetric and areal outputs lower costs by reducing reactor volume and the needed reactor volume and land area, correspondingly.

According to Ref. [112], there are some important factors that achieve optimal yield, these factors include the following:

- Ensuring adequate light and thick algal culture
- Optimum rate of mixing and mass transfer
- Avoiding damage to cells by shear stress
- Ainimization of temperature variation from ideal
- Minimization of dissolved oxygen tension
- Simple to clean and maintain
- Energy input requirements are kept to a minimum
- Minimization of water use (e.g., evaporation from ponds, evaporative cooling use)
- Low initial and ongoing capital and operating expenditures per unit of harvested output

14.1 Photoautotrophic versus heterotrophic

Cultivation of algae can be achieved via photoautotrophic, heterotrophic, or mixotrophic methods, which also vary in their challenges and advantages (Table 30.1). In photoautotrophic cultivation, algae require light to grow and create new biomass. In heterotrophic cultivation, algae are grown without light and are fed a carbon source, such as sugars, to generate new biomass.

Mixotrophic cultivation harnesses both the photoautotrophic and heterotrophic ability of algae. Heterotrophic and mixotrophic cultivation strategies present a different set of advantages and challenges compared with photoautotrophic methods. Optimal conditions for production and contamination prevention are often easier to maintain, and there is the potential to utilize lignocellulosic sugars or carbon-rich wastewater for algal growth. Growth on a carbon source also achieves high biomass concentrations that can reduce the extent and cost of the infrastructure required to grow the algae [136]. However, the primary challenges with these approaches are the cost and availability of suitable feedstocks such as lignocellulosic sugars. Because these systems rely on primary productivity from other sources, they could compete for feedstocks with other biofuel technologies.

14.2 Biohydrogen production

Hydrogen is usually recognized as a clean fuel, an environmentally safe, renewable energy resource, and an excellent substitute for fossil fuels, as well as a possible candidate with the highest energy density, as well as many other technologic, socioeconomic, and environmental advantages (143 GJ per ton). It is the only known fuel that does not emit carbon dioxide as a by-product when used in fuel cells to generate electricity [16,24]. In certain physiologic conditions, such as hypoxia and high light intensity, algae are metabolically able to produce hydrogen. This ability to generate hydrogen from microalgae has been investigated, and various cultivation strategies have been used to improve hydrogen production by microalgae. In addition, carbohydrate-rich microalgae or macroalgae have been used as a raw material for anaerobic bacteria to produce fermentative hydrogen. Consequently, two main processes are used to produce biohydrogen from algae: (1) light-dependent reaction and (2) light-independent reaction. The light-dependent process depends on the photolysis of water molecules during the photosynthesis process; however, the light-independent reaction is based on the utilization of algal biomass as a feedstock of the dark fermentation process [41,96].

15. Methodologies of production

Selecting the desired algae, either micro- or macroalgae, and which strain is the key step in the production process of all biofuel types. This step depends on many factors such as the biochemical composition of the algal strain, harvesting methods, cultivation methods, and response to physiologic stress.

15.1 Steps of bioethanol production from algae biomass

There are three main steps in bioethanol production: pretreatment, fermentation processes, and bioethanol recovery (Fig. 30.4).

15.2 Pretreatment process

Appropriate biomass pretreatment is a critical process in bioethanol production because it decomposes polymer structure like cellulose and starch to form simple fermentative sugars, resulting in rapid hydrolysis and higher yields [90,123]. It was discovered in this context that using intact algal cells results in low bioethanol productivity due to lower conversion rates [134]. Furthermore, the starch content of microalgal biomass can be transformed directly into biofuel via dark and anaerobic conditions, but the biofuel production rate and yield are very poor [3,129]. Consequently, algal biomass requires pretreatment to increase the production yield. As a result, pretreatment of algal biomass is required to increase production yield. Pretreatment or hydrolysis/saccharification can be accomplished through one of four methods: enzymatic, chemical, mechanical, or combined [106]. The cost of these pretreatment saccharification methods is determined by a number of factors, including (1) type of alkaline or acid, (2) the power cost, (3) the length of thermal pretreatment, (4) required temperature, (5) catalysts in enzymatic hydrolysis, (6) the type of hydrolytic enzymes, and (7) the type of raw material used [55].

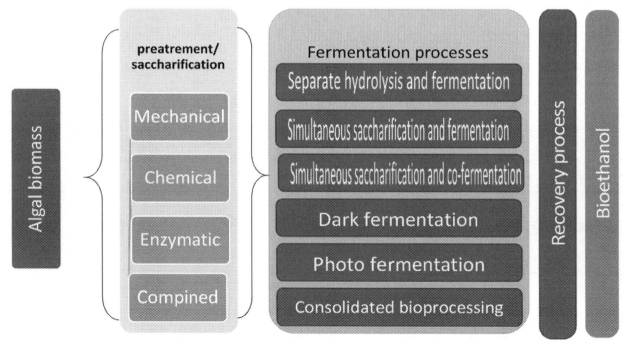

FIGURE 30.4 Standard description of bioethanol production from algal biomass.

15.3 Fermentation process

Fermentation is a metabolic activity that converts monosaccharides to bioethanol and other by-products in the presence of fermentative microorganisms under favorable conditions, such as temperature and pH [39]. The most common fermentative microorganisms are yeast and bacteria, including *Saccharomyces cerevisiae, Pichia stipitis, Kluyveromyces fragilis, K. marxianus, Escherichia coli, Klebsiella oxytoca*, and *Zymomonas mobilis* [36]. Some starch-rich microalgae, such as *Chlamydomonas reinhardtii, C. moewusii, Chlorella vulgaris, Chlorococcum littorale*, and *Spirulina* sp., can synthesis ethanol via an internal mechanism in anaerobic conditions in the absence of light [106]. Fermentation using carbohydrate-rich microalgae as bioethanol feedstock can be classified into six main routes, summarized in Table 30.3.

TABLE 30.3 Remarks, advantages, and drawbacks of different fermentation processes used for bioethanol production.

Fermentation processes	Remarks	Advantages	Drawbacks
Separate hydrolysis and fermentation (SHF)	SHF is a fermentation technology where both hydrolysis and fermentation are done separately and sequentially using two different reactors to continually breakdown the feedstock into simple sugars via hydrolysis followed by converting fermentable sugars to ethanol process using fermentative microorganisms [106a]	— low cost of chemicals — short residence duration — simple equipment system [106b]	— Accumulation of glucose and cellobiose during hydrolysis causes complete process inhibition [106a, 152]. — to avoid the formation of undesirable by-products during the hydrolysis process, further neutralizing and purifying procedures are required [106c]
Simultaneous saccharification and fermentation (SSF)	In this technique, hydrolysis and fermentation occur simultaneously in the same step at the same time in the same bioreactor	— The inhibitory effect of glucose and cellobiose is reduced because they immediately are converted to bioethanol by microorganisms [106a] — SSF needs lower enzyme loading and higher bioethanol yield compared with SHF — The presence of ethanol in the medium prevents undesirable microbes — short processing time [106]	Not suitable for hydrolysis of pentose sugars
Simultaneous saccharification and co-fermentation (SSCF)	SSCF entails simultaneous hydrolysis and fermentation of pentose and hexose sugars in a single bioreactor	— High bioethanol yield over SSF route and SHF route [154] — SSCF can ferment both glucose and pentoses in the same bioreactor — low cost	

Continued

TABLE 30.3 Remarks, advantages, and drawbacks of different fermentation processes used for bioethanol production.—cont'd

Fermentation processes	Remarks	Advantages	Drawbacks
Dark fermentation (df)	DF is a metabolic process where complex organic polymers found in fermentative and hydrolytic microorganisms will be hydrolyzed into monomers, which will subsequently be transformed into a combination of low molecular weight organic acids and alcohols primarily acetic and ethanol [155]	No need for a light source, making it a cost-effective process.	DF is still less efficient due to the impact of light and oxygen [64]
Photofermentation (PF)	The term "photofermentative" refers to a natural mechanism that converts solar energy into a fermentation product via an ethanol synthesis metabolic route [155]	The inhibitory effect of light and oxygen were neglected	Low yield compared with SSF and SSCF process
Consolidated bioprocessing (CBP)	Consolidated bioprocessing is an integrated method that combines the production of saccharolytic enzymes, pretreatment, saccharification, and fermentation processes in a single reactor via a single microbe or microbial consortia [157,158]	− low operational costs − high process efficiency [156]	No natural microorganism has been discovered that can fulfil all the requirements of CBP

15.4 Bioethanol recovery

In the commercial manufacturing of bioethanol, the product recovery procedure is critical. The bioethanol can be retrieved by a distillation method in which the bioethanol condensate is concentrated (approximately 37%) in an azeotrope distillation column. The lower product is sent to a stripping column to remove excess water, and the residual bioethanol is drained and converted to a liquid form that can be used as a supplement or replacement for gasoline in automobiles [105]. The bioethanol recovery rate could be as high as 99.6%. However, solid waste is removed via centrifugation and then dried in a rotary dryer.

16. Steps of biodiesel production from algae biomass

16.1 Steps of biohydrogen production from algae biomass

Biohydrogen production using carbohydrate-rich biomass as a renewable energy source is one of the alternative techniques, where photosynthetic processes takes place (light-dependent reaction) and is anaerobic (light-independent reaction).

16.1.1 Light-dependent reaction (biophotolysis)

The biophotolysis process can occur in various species of cyanobacteria and microalgae that depend on the photosynthesis process. This process is divided into two main categories direct and indirect biophotolysis.

16.1.1.1 Direct biophotolysis

The mechanism of biohydrogen production through biophotolysis is dependent on the biosynthesis of hydrogen gas from the water by using sunlight and CO_2 as the sole sources for energy through the process of hydrogenase enzyme by cyanobacteria and microalgae [48]. Light energy captured by PSI and PSII transports electrons linearly from water to ferredoxin (FDX1) in the absence of oxygen [72].

$$2\ H_2O + \text{light energy} \longrightarrow 2H_2 + O_2$$

Reduced ferredoxin serves as an electron donor, allowing protons to generate hydrogen gas. For algal hydrogenases, the ferredoxin enzyme FDX1 or PETF is the only natural electron source.

$$2\ H_2 + 2\ FD \longleftrightarrow H_2 + 2\ FD$$

Chlamydomonas reinhardtii, *Chlorella fusca*, *Scenedesmus obliquus*, *Chlorococcum littorale*, and *Platymonas subcordiformis* are some of the microalgae that produce hydrogen. These microalgae are naturally capable of producing hydrogen and have hydrogenase encoded in their genomes [41]. The advantage of this mechanism is that it allows green algae in anaerobic circumstances to convert about 22% of light energy by using hydrogen as an electron donor in the CO_2 fixation process, even in low light intensity [16]. However, because the water-splitting activity of photosystem II (PSII) and the hydrogenase system are completely inhibited by oxygen evolution, hydrogen production in green microalgae is not continual [96]. A very creative hydrogen production method was invented to obtain continual hydrogen production from green algae: a two-stage strategy for growth and hydrogen generation. *Chlamydomonas reinhardtii* is cultivated in a photosynthetic growth medium and then transferred to sulfur depletion medium nongrowth phase of hydrogen evolution [159]. This causes PSII to degenerate rapidly, severely preventing electron flow at PSII, but PSI activity remains unchanged.

16.1.1.2 Indirect biophotolysis

This hydrogen production process by microalgae is also known as anaerobic dark fermentation or respiration-assisted dark fermentation of a fixed carbon during the dark growth phase. The oxygen generation and hydrogen evolution are naturally separated by the daily fluctuation in light intensity. The photosynthetic microalgae fix carbon dioxide and release oxygen during the light period. During dark times, the fixed carbon in the form of protein and carbohydrates is consumed, resulting in excess electrons that must be discarded. The generated electrons are transported from NADPH to the plasto-quinone pool via passing PSII. Under hypoxia, oxygen-sensitive hydrogenase will be activated to produce hydrogen from the generated electrons [120].

In cyanobacteria, the problem of oxygen susceptibility of hydrogenase in cyanobacteria is solved by spatially separating CO_2 fixation-based oxygen evolution and nitrogenase-based hydrogen production. Cyanobacteria can fix atmospheric nitrogen in addition to CO_2, which is done in different cells. Nitrogen is fixed in heterocysts under anoxic conditions, while CO_2 fixation occurs in normal vegetative cells [18]. Cyanobacteria have a variety of hydrogen-evolving enzymes, including uptake [NiFe] hydrogenases and a bidirectional [NiFe] hydrogenase that works in tandem with a [MoFe] nitrogenase for nitrogen fixation and hydrogen evolution [122]. The following equation [41] describes how hydrogen is produced during nitrogen fixation by nitrogenase:

$$N_2 + 8\ e^- + 8\ H^+ + 16\ ATP \longrightarrow 2\ NH_3 + H_2 + 16\ ADP + 16\ Pi$$

In the absence of nitrogen or in nonnitrogen-fixing cyanobacteria, nitrogenase acts as an ATP-powered hydrogenase generating hydrogen at a considerably greater rate, as described by the following equation [72]:

$$8\ H^+ + 8\ e^- + 4\ ATP \longrightarrow 4\ H_2 + 4\ ADP + 4\ Pi$$

16.1.2 Light-independent reaction (dark fermentation)

Fermentative bacteria are known to produce hydrogen as a by-product, which has been used for biohydrogen generation for years. Such bacteria use protons as final electron acceptors producing hydrogen to maintain the electrical neutrality of the cell [78]. Fermentative hydrogen generation is a promising waste biorefinery method, and many industrial waste effluents can be employed as a dark fermentation substrate. Algal biomass, either macro- or microalgae, has cellulose and hemicellulose in the relatively simple cell walls that are completely absent of lignin, making it easier for pretreatment. Microalgae can accumulate a large quantity of lipids or carbohydrates, which may differ from one strain to another. The storage carbohydrates of algae are mainly simple glucans like starch and glycogen, which can be easily transformed to simple fermentable sugar by an effective pretreatment. On the other hand, lipids and proteins accumulating algae cannot be fermented by hydrogen-producing bacteria, so the defatted algal biomass achieved after lipid extraction in the biodiesel production process could also be utilized as a feedstock for biohydrogen dark fermentation production.

17. Algal resources in Egypt

Water systems are now being studied not just as water supplies, but also as systems rich in other valuable resources in terms of the "blue economy." Algae is one of those resources, serving as a viable alternative to fossil fuels and a reliable source of renewable energy. Researchers in Egypt are not only interested in the algae that live in the aquatic ecosystem, but they are also interested in the algae that live on land: the arid and desert lands as habitats that are rich in algal biodiversity [1]. Microalgae species (i.e., phytoplankton) survive suspended in the water column, whereas macroalgae species (i.e., seaweed) are plant-like organisms that range in size from a few centimeters to several meters in length [160]. Gigantic algae species (such as kelp) grow from the seabed to form massive underwater forests.

Egypt's coasts stretch for more than 3500 km along the Eastern Mediterranean and the Red Sea, with distinct coastal zones [2]. Egypt's Mediterranean coast is separated into four primary subareas, according to Frihy and El-Sayed [161], including (1) the coastal field extends northwest from Sallum to Alexandria, (2) Alexandria's coastal area extends further east, from Hammam to Abu Qir, (3) the coastal sector of the Nile Delta extends eastward all the way to Port Said, and (4) the easternmost region of Egypt's Mediterranean coast is the coastal region of North Sinai, which extends from Port Said to Rafah.

From Rafah (Sinai Peninsula) to Salloum (east of the Libyan border), the Egyptian Mediterranean coast stretches for 970 km, including five natural lakes stretching from northern Sinai to Alexandria [118]. There are also the northern ones Fig. 30.5.

Mediterranean coastal lakes, such as Mariut, Edku, Manzala, Burullus, and Bardawil, are natural resources for diverse algae populations in Egypt, especially microalgae/phytoplankton as well as of fish production.

The Red Sea is the world's northernmost tropical sea, and academics were particularly interested in the Red Sea coastal plain of Egypt. The Egyptian Red Sea and Gulf of Aqaba have an estimated 1500 km of coral reef along the coastline and island edges, with 800 km of fringing reef stretching only to the Egyptian-Sudanese boundary in the east [38], creating great environments for several algal species to live and flourish.

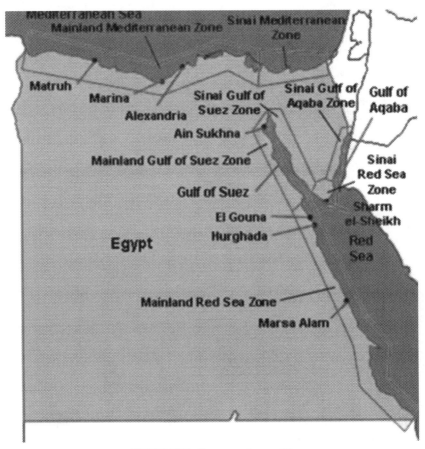

FIGURE 30.5 Egypt costal zones [2].

Algae are also found in abundance around the Nile's springs, including Bahariya Oasis (Western Desert), Ain Helwan (near Cairo), Ain El-Sokhna (near Suez town), El-Temsah Lake (Ismailia town), Hammam Musa Faroun (South Sinai), Wadi El-Natrun Lakes (Western Desert), and Farafra Oasis.

18. Future perspective in Egypt: initiatives toward a safe environment with clean energy

Given the importance of Egypt's northern lakes, it is critical to consider practical and effective solutions to prevent further damage. As a result, there is a growing desire for a national water framework directive to be proposed. Before a water body is designated as having good surface water status, elements of surface water status (ecologic and chemical status) must achieve an acceptable quality, according to a 2010 European Commission report for the Water Framework Directive [107].

Applying the known environmental indicators and integrating environmental and human systems into a single conceptual framework is a good and relevant strategy, as well as an important decision-making tool for lake restoration. There are a number of conceptual frameworks that might be used for this, the most popular of which is the "Drivers Pressure State Impact Response" framework (DPSIR), which is the most common one used in indicator-based studies for lakes worldwide as well as along Egyptian lakes [11] and other aquatic systems [162]. The DPSIR framework's application to aquatic ecosystems is related to the influence of human activities that exert pressure on the environment and are primarily driven by social and economic development, which would negatively affect and/or deteriorate the lake ecosystem's original state. As a result, it is critical to improve the ecosystem of lakes by mimicking standard circumstances and offering appropriate mitigation measures along with viable restoration options. According to various restoration studies (e.g., Refs. [7–10,13]), the "ideal" strategy would involve lowering the loading into the lakes (to a minimal of tons) and using the least expensive approach (cost-effective) methods. For aquatic ecosystems, an operational management plan should be implemented, which necessitates an effective technique that provides exact data on water quality cycles, trends, and condition. Without a monitoring strategy for lakes, geomorphological features, water quality variables, and the adoption of a water quality indexing system, such information could not be obtained.

Simulating standard circumstances and offering appropriate mitigation measures with viable alternatives would be an effective strategy for improving Egypt's lake ecosystems. This includes using techniques like *Potamogeton pectinatus* and *Ceratophyllum* to reduce effluent loading while also regulating the existing submerged vegetation in Lake Idku [10] and [7–9], such as *Otamogeton pectinatus* and *Ceratophyllum demersum*, using the grass carp (*Ctenopharyngodon idella*). Such combined management plan would enhance the lakes' ecosystems and fishing capabilities by cutting down on health problems within the northern lakes.

References

[1] H. Abbas, Biodiversity and distribution of algae in Egypt, in: Environmental Pollution, Biodiversity and Sustainable Development: Issues and Remediation, 2019, https://doi.org/10.1201/9780429265013-14.

[2] T. AbdeL-Latif, S.T. Ramadan, A.M. Galal, Egyptian coastal regions development through economic diversity for its coastal cities, HBRC J. 8 (3) (2012) 252–262.

[3] A.E. Abomohra, M.E. Elshobary, Biodiesel, bioethanol and biobutanol production from microalgae biomass, in: M.A. Alam, Z. Wang (Eds.), In Microalgae Biotechnology for Development of Biofuel and Waste Water Treatment, Springer Singapore Press, Singapore, 2019, pp. 293–321.

[4] A. Abomohra, M. El-Sheekh, D. Hanelt, Pilotcultivationofthechlorophyte microalga Scenedesmus obliquus as a promising feedstock for biofuel, Biomass Bioenergy 64 (2014) 237–244.

[5] J. Ahmad, D. Alam, M.S. Haseen, Impact of climate change on agriculture and food security in India, Int. J. Agric. Environ. Biotechnol. 4 (2) (2011) 129–137.

[6] A. Ahmad, N. Yasin, C. Derek, J. Lim, Microalgae as as ustainable energy source for biodieselproduction: areview, Renew. Sustain. Energy Rev. 15 (2011) 584–693.

[7] E.M. Ali, H.M. Khairy, Environmental assessment of drainage water impacts on water quality and eutrophication level of Lake Idku, Egypt, Environ. Pollut. 216 (2016) 437–449.

[8] E.M. Ali, H.M. Khairy, Environmental assessment of drainage water impacts on water quality and eutrophication level of Lake Idku, Egypt, J. Environ. Pollut. 27 (2016) 437–449, https://doi.org/10.1016/j.envpol.2016.05.064.

[9] E.M. Ali, H. Khairy, Variations in phytoplankton carbon biomass, community assemblages and species succession along lake Burullus, northern Egypt, Int. J. Environ. Biol. 33 (2012).

[10] E.M. Ali, Impact of drain water on water quality and eutrophication status of Lake Burullus, Egypt, a southern Mediterranean lagoon, African J. Aquatic Sci. 36 (3) (2011).

[11] E.M. Ali, DPSIR framemork and geoinfrmatics twords sustainable development for a portion of the south western coast of the Red Sea, Egypt, in: The 11th International Conference of the African Association of Remote Sensing of the Environment (AARSE) in Kampala, Uganda, October 2016 (Accepted, 2016).

[12] A. Ameyoud, A. Benbellil, Application of the cost–benefit analysis to electrical interconnection projects, in: 2019 Algerian Large Electrical Network Conference (CAGRE), IEEE, February 2019, pp. 1–6.

[13] A.A. Ansari, F.A. Khan, Household detergents causing eutrophication in freshwater ecosystems, in: A.A. Ansari, S.S. Gill (Eds.), Eutrophication: Causes, Consequences and Control, vol 2, Springer, The Netherlands, 2014, p. 139e164.

[14] B. Antizar-Ladislao, J.L. Turrion-Gomez, Second-generation biofuels and local bioenergy systems, Biofuel., Bioprod. Biorefin.: Innovat. sustain. Econ. 2 (5) (2008) 455–469.

[15] M. Aresta, A. Dibenedetto, M. Carone, T. Colonna, C. Fagale, Production of biodiesel from macroalgae by supercritical CO_2 extraction and thermochemical liquifaction, Environ. Chem. Lett. 3 (2005) 136–139.

[16] M.Y. Azwar, M.A. Hussain, A.K. Abdul-Wahab, Development of biohydrogen production by photobiological, fermentation and electrochemical processes: a review, Renew. Sustain. Energy Rev. 31 (2014) 158–173, https://doi.org/10.1016/j.rser.2013.11.022.

[17] A. Banerjee, R. Sharma, Y. Chisti, U.C. Banerjee, Botryococcus braunii: a renewable source of hydrocarbons and other chemicals, Crit. Rev. Biotechnol. 22 (3) (2002) 245–279.

[18] J.R. Banerjee Benemann, Hydrogen production by microalgae, J. Appl. Phycol. 12 (2000) 291–300.

[19] N. Boonwong, C. Wattanachant, S. Wattanasit, Effects of crude glycerin from palm oil biodiesel production as a feedstuff for broiler diet on growth performance and carcass quality, Pertanika J. Trop. Agric. Sci. 41 (3) (2018).

[20] M.A. Borowitzka, Commercial production of microalgae: ponds, tanks, tubes and fermenters, J. Biotechnol. 70 (1–3) (1999) 313–321.

[21] L. Brennan, P. Owende, Biofuels from microalgae—a review of technologies for production, processing, and extractions of biofuels and co-products, Renew. Sustain. Energy Rev. 14 (2) (2010) 557–577.

[22] C. Callieri, J.G. Stockner, Freshwater autotrophic picoplankton: a review, J. Limnol. 61 (1) (2002) 1–14.

[23] A.S. Carlsson, J.B. Beilen, R. Moller, D. Clayton, in: D. Bowles (Ed.), Micro-algae and Macro-Algae: Utility for Industrials Applications, 2007, pp. 9–33.

[24] M.C. Chang, A. Pralle, E.Y. Isacoff, C.J. Chang, A selective, cell-permeable optical probe for hydrogen peroxide in living cells, J. Am. Chem. Soc. 126 (47) (2004) 15392–15393.

[25] M. Charef, K. Doraï, Human migration and climate change in the Mediterranean region, in: S. Thiébault, J.P. Moatti (Eds.), The Mediterranean Region under Climate Change. A Scientific Update, (Marseille, France: Institut de Recherche pour le Développement), 2016, pp. 439–444.

[26] C. Charles, I. Gerasimchuk, R. Bridle, T. Moerenhout, E. Asmelash, T. Laan, Biofuels—At what Cost? A Review of Costs and Benefits of EU Biofuel Policies, International Institute for Sustainable Development (IISD), 2013, pp. 1–129.

[27] W.Y. Cheah, P.L. Show, J.-S. Chang, T.C. Ling, J.C. Juan, Biosequestration of atmospheric CO_2 and flue gas-containing CO_2 by microalgae, Bioresour. Technol. 184 (2014) 190–201.

[28] K.W. Chew, S.R. Chia, P.L. Show, et al., Effects of water culture medium, cultivation systems and growth modes for microalgae cultivation: a review, J. Taiwan Inst. Chem. Eng. 91 (2018) 332–344.

[29] Y. Chisti, An unusual hidrocarbon, Ramsay Soc. 27–28 (1981) 24–26.

[30] Y. Chisti, Biodiesel from microalgae, Biotechnol. Adv. 25 (2007) 294–306.

[31] Y. Chisti, Biodiesel from microalgae beats bioethanol, Trend. Biotechnol. 26 (3) (2008) 126–131, https://doi.org/10.1016/j.tibtech.2007.12.002.

[32] R. Craggs, Advanced integrated wastewater ponds, in: A. Shilton (Ed.), Pond Treatment Technology, IWA Scientific and Technical Report Series, IWA, London, 2005, pp. 282–310.

[33] N. Dave, R. Selvaraj, T. Varadavenkatesan, R. Vinayagam, A critical review on production of bioethanol from macroalgal biomass, Algal Res. 42 (2019) 101606.

[34] R. Davy, N. Gnatiuk, L. Pettersson, L. Bobylev, Climate change impacts on wind energy potential in the European domain with a focus on the Black Sea, Renew. Sustain. Energy Rev. 81 (2018) 1652–1659.

[35] A. Demirbas, Use of algae as biofuel sources, Energy Convers. Manag. 51 (2010) 2738–2749.

[36] A. Demirbas, Competitive liquid biofuels from biomass, Appl. Energy 88 (2011) 17–28.

[37] X. Deng, Y. Li, X. Fei, Microalgae: a promising feedstock for biodiesel, Afr. J. Microbiol. Res. 3 (13) (2009) 1008–1014.

[38] H.M. El-Asmar, M.H. Ahmed, S.B. El-Kafrawy, A.H. Oubid-Allah, T.A. Mohamed, M.A. Khaled, Monitoring and assessing the coastal ecosystem at Hurghada, Red Sea coast, Egypt, J. Environ. Earth Sci. 5 (6) (2015) 144–160.

[39] M.M. El-Dalatony, M.B. Kurade, R.A.I. Abou-Shanab, et al., Long-term production of bioethanol in repeated-batch fermentation of microalgal biomass using immobilized *Saccharomyces cerevisiae*, Bioresour. Technol. 219 (2016) 98–105.

[40] M.E. Elshobary, D.I. Essa, A.M. Attiah, et al., Algal community and pollution indicators for the assessment of water quality of Ismailia canal, Egypt, Stoch. Environ. Res. Risk Assess. 34 (2020) 1089–1103.

[41] E. Eroglu, A. Melis, Microalgal hydrogen production research, Int. J. Hydro. Energy 41 (2016) 12772–12798, https://doi.org/10.1016/j.ijhydene.2016.05.115.

[42] Eurostat, Asylum Applications (Non-EU) in the EU-27 Member States, 2008 - 2019, 2019. Last Accessed July 2020. Retrieved from: https://ec.europa.eu/eurostat/statistics-explained/index.php/Asylum_statistics.

[43] A.S. Fedorov, S. Kosourov, M.L. Ghirardi, M. Seibert, Continuous hydrogen photoproduction by Chlamydomonas reinhardtii, Appl. Biochem. Biotechnol. 121 (1) (2005) 403–412.

[44] R. Fouquet, P.J. Pearson, A thousand years of energy use in the United Kingdom, Energy J. 19 (4) (1998).

[45] M. Gavrilescu, Y. Chisti, Biotechnology—a sustainable alternative for chemical industry, Biotechnol. Adv. 23 (2005) 471−499.

[46] P. Gerland, A.E. Raftery, H. Ševčíková, N. Li, D. Gu, T. Spoorenberg, L. Alkema, B.K. Fosdick, J. Chunn, N. Lalic, G. Bay, T. Buettner, G.K. Heilig, J. Wilmoth, Science 346 (6206) (2014) 234−237, 10 Oct 2014.

[47] M. Ghadiryanfar, K.A. Rosentrater, A. Keyhani, M. Omid, A review of macroalgae production, with potential applications in biofuels and bioenergy, Renew. Sustain. Energy Rev. 54 (2016) 473−481.

[48] M. Ghirardi, Microalgae: a green source of renewable H2, Trend. Biotechnol. 18 (2000) 506−511.

[49] J. Goldemberg, P. Guardabassi, Are biofuels a feasible option? Energy Pol. 37 (1) (2009) 10−14.

[50] C.P. Gopinathan, Marine Microalgae, 2004.

[51] J.A. Gowlett, The discovery of fire by humans: a long and convoluted process, Phil. Trans. Biol. Sci. 371 (1696) (2016) 20150164.

[52] H.C. Greenwell, L.M.L. Laurens, R.J. Shields, R.W. Lovitt, K.J. Flynn, Placing microalgae on the biofuels priority list: a review of the technological challenges, J. R. Soc. Interface 7 (46) (2010) 703−726.

[53] J.U. Grobbelaar, C.H. Bornman, Algal biotechnology: real opportunities for Africa, South Afr. J. Bot. 70 (1) (2004) 140−144.

[54] J.U. Grobbelaar, Factors governing algal growth in photobioreactors: the "open" versus "closed" debate, J. Appl. Phycol. 21 (5) (2009) 489−492.

[55] D. Hernández, B. Riaño, M. Coca, M.C. García-González, Saccharification of carbohydrates in microalgal biomass by physical, chemical and enzymatic pre-treatments as a previous step for bioethanol production, Chem. Eng. J. 262 (2015) 939−945.

[56] J. Hill, E. Nelson, D. Tilman, S. Polasky, D. Tiffany, Environmental, economic, and energetic costs and benefits of biodiesel and ethanol biofuels, Proc. Natl. Acad. Sci. U. S. A. 103 (30) (2006) 11206−11210.

[57] C.S. Hirst, How Has the Morphology of the Human Mandible Varied in Response to the Dietary Changes that Have Occurred in Britain between the Neolithic and Post-Medieval Periods? Doctoral dissertation, UCL University College London, 2019.

[58] W. Hoppe, S. Bringezu, N. Thonemann, Comparison of global warming potential between conventionally produced and CO_2-based natural gas used in transport versus chemical production, J. Clean. Prod. 121 (2016) 231−237.

[59] G. Huang, F. Chen, D. Wei, et al., Biodiesel production by microalgal biotechnology, Appl. Energy 87 (2010) 38−46.

[60] IEA− International Energy Agency, Advanced Motor Fuels Report 2013, OECD publishing, Paris, 2014.

[61] IEA− International Energy Agency, World Energy Outlook 2014, OECD publishing, Paris, 2014.

[62] IEA− International Energy Agency, Renewable Energy Medium-Term Market Report 2014 Market Analysis and Forecasts to 2020, OECD publishing, Paris, 2014.

[63] E. Jacob-Lopes, T.T. Franco, Microalgae-based Systems for Carbon Dioxide Sequestration and Industrial Biorefineries, vol. 2, Biomass. Croatia, Sciyo, 2010, pp. 135−146.

[64] R.P. John, G.S. Anisha, K.M. Nampoothiri, A. Pandey, Micro and macroalgal biomass: a renewable source for bioethanol, Bioresour. Technol. 102 (2011) 186−193.

[65] Z.H.U. Junying, R. Junfeng, Z. Baoning, Factors in mass cultivation of microalgae for biodiesel, Chin. J. Catal. 34 (2013) 80−100.

[66] M. Kamalanathan, P. Chaisutyakorn, R. Gleadow, J. Beardall, A comparison of photoautotrophic, heterotrophic, and mixotrophic growth for biomass production by the green alga Scenedesmus sp. (Chlorophyceae), Phycologia 57 (3) (2018) 309−317, https://doi.org/10.2216/17-82.1.

[67] I.K. Kapdan, F. Kargi, Bio-hydrogen production from waste materials, Enzym. Microb. Technol. 38 (5) (2006) 569−582.

[68] M.A. Kassim, T.K. Meng, Carbon dioxide (CO_2) biofixation by microalgae and its potential for biorefinery and biofuel production, Sci. Total Environ. 2017 (2017) 1−9.

[69] C. Kennedy, The energy embodied in the first and second industrial revolutions, J. Ind. Ecol. 24 (4) (2020) 887−898.

[70] L. Kexun, L. Shun, L. Xianhua, An overview of algae bioethanol production, Int. J. Energy Res. 38 (2014) 965−977.

[71] S.A. Khan, M.Z. Hussain, S. Prasad, U.C. Banerjee, Prospects of biodiesel production from microalgae in India, Renew. Sustain. Energy Rev. 13 (9) (2009) 2361−2372.

[72] D.H. Kim, M.S. Kim, Hydrogenases for biological hydrogen production, Bioresour. Technol. 102 (2011) 8423−8431.

[73] G.-S. Kim, Manufacturing technology of bioenergy using algae, in: Handbook of Marine Macroalgae, John Wiley & Sons, Ltd, Chichester, UK, 2011, pp. 451−460.

[74] W. Klinthong, Y.-H. Yang, C.-H. Huang, C.-S. Tan, A review: microalgae and their applications in CO_2 capture and renewable energy, Aerosol Air Qual. Res. 15 (2015) 712−742.

[75] G. Knothe, Biodiesel and renewable diesel: a comparison, Prog. Energy Combust. Sci. 36 (3) (2010) 364−373.

[76] S. Kraan, Algal Polysaccharides, Novel Applications and Outlook, IntechOpen, 2012.

[77] A. Kumar, S. Ergas, X. Yuan, A. Sahu, Q. Zhang, J. Dewulf, F.X. Malcata, H. Van Langenhove, Enhanced CO2 fixation and biofuel production via microalgae: recent developments and future directions, Trend. Biotechnol. 28 (7) (2010) 371−380.

[78] H.Y. Lee, S.E. Lee, K.H. Jung, et al., Repeated-batch operation of surface-aerated fermentor for bioethanol production from the hydrolysate of seaweed Sargassum sagamianum, J. Microbiol. Biotechnol. 21 (2011) 323−331.

[79] Y.K. Lee, Microalgal mass culture systems and methods: their limitation and potential, J. Appl. Phycol. 13 (4) (2001) 307−315.

[80] S. Jayakumar, M.M. Yusoff, M.H.A. Rahim, G.P. Maniam, N. Govindan, The prospect of microalgal biodiesel using agro-industrial and industrial wastes in Malaysia, Renew. Sustain. Energy Rev. 72 (2017) 33−47.

[81] R.A. Kerr, The Story of O_2, 2005.

[82] Mediterranean Energy Perspectives, Observatoire Méditerranéen de l'Energie, 2015 (MEP, 2015).

[83] J. Masojídek, G. Torzillo, Mass cultivation of freshwater microalgae, in: Encyclopedia of Ecology, Elsevier, 2008, pp. 2226−2235.

[84] T.M. Mata, A.A. Martins, N.S. Caetano, Microalgae for biodiesel production and other applications: a review, Renew. Sustain. Energy Rev. 14 (1) (2010) 217–232.

[85] Y.N. Mata, E. Torres, M.L. Blazquez, A. Ballester, F.M.J.A. González, J.A. Munoz, Gold (III) biosorption and bioreduction with the brown alga Fucus vesiculosus, J. Hazard Mater. 166 (2–3) (2009) 612–618.

[86] B. Metting, J.W. Pyne, Biologically active compounds from microalgae, Enzym. Microb. Technol. 8 (7) (1986) 386–394.

[87] J. Milano, H.C. Ong, H.H. Masjuki, et al., Microalgae biofuels as an alternative to fossil fuel for power generation, Renew. Sustain. Energy Rev. 58 (2016) 180–197.

[88] D.N. Miller, J.T. Irvine, B-site doping of lanthanum strontium titanate for solid oxide fuel cell anodes, J. Power Sources 196 (17) (2011) 7323–7327.

[89] E. Molina Grima, F.G. Acien Fernandez, F. Garcia Camacho, Y. Chisti, Photobioreactors: light regime, man transfer and scaleup, J. Biotechnol. 70 (1999) 231–247.

[90] N. Mosier, C. Wyman, B. Dale, et al., Features of promising technologies for pretreatment of lignocellulosic biomass, Bioresour. Technol. 96 (2005) 673–686.

[91] R. Munoz, B. Guieysse, Algal–bacterial processes for the treatment of hazardous contaminants: a review, Water Res. 40 (15) (2006) 2799–2815.

[92] H. Mikulčić, I.R. Skov, D.F. Dominković, S.R.W. Alwi, Z.A. Manan, R. Tan, N. Duić, S.N.H. Mohamad, X. Wang, Flexible Carbon Capture and Utilization technologies in future energy systems and the utilization pathways of captured CO_2, Renew. Sustain. Energy Rev. 114 (2019) 109338. Aghajani A, Richter F, Somsen C, Fries SG, Steinbach I, Eggeler G, et al. Accepted Manuscript, 2009.

[93] M.M. Musthafa, Synthetic lubrication oil influences on performance and emission characteristic of coated diesel engine fuelled by biodiesel blends, Appl. Therm. Eng. 96 (2016) 607–612.

[94] J. Müller, D. Folini, M. Wild, S. Pfenninger, CMIP-5 models project photovoltaics are a no-regrets investment in Europe irrespective of climate change, Energy 171 (2019) 135–148.

[95] New Energy Outlook (NEO, 2016) 2016, Bloomberg.

[96] D. Nagarajan, D.-J. Lee, A. Kondo, J.-S. Chang, Recent insights into biohydrogen production by microalgae—From biophotolysis to dark fermentation, Bioresour. Technol. 227 (2017a) 373–387.

[97] N. Nagle, P. Lemke, Production of methyl ester fuel from microalgae, Appl. Biochem. Biotechnol. 24 (1) (1990) 355–361.

[98] S.N. Naik, V.V. Goud, P.K. Rout, A.K. Dalai, Production of first and second generation biofuels: a comprehensive review, Renew. Sustain. Energy Rev. 14 (2) (2010) 578–597.

[99] F.M.I. Natrah, F.M. Yusoff, M. Shariff, F. Abas, N.S. Mariana, Screening of Malaysian indigenous microalgae for antioxidant properties and nutritional value, J. Appl. Phycol. 19 (6) (2007) 711–718.

[100] A.N. Netravali, S. Chabba, Composites get greener, Mater. Today 4 (6) (2003) 22–29.

[101] M.S. Nicolò, S.P.P. Guglielmino, V. Solinas, A. Salis, Consequences of Microbial Interactions with Hydrocarbons, Oils, and Lipids: Production of Fuels and Chemicals, 2016, pp. 1–20.

[102] P.S. Nigam, A. Singh, Production of liquid biofuels from renewable resources, Prog. Energy Combust. Sci. 37 (1) (2011) 52–68.

[103] B. Önol, Effects of coastal topography on climate: high-resolution simulation with a regional climate model, Clim. Res. 52 (2012) 159–174.

[104] W.J. Oswald, C.G. Golueke, Biological transformation of solar energy, Adv. Appl. Microbiol. 2 (1960) 223–262. Academic Press.

[105] V. Patil, K.-Q. Tran, H.R. Giselrød, Towards sustainable production of biofuels from microalgae, Int. J. Mol. Sci. 9 (2008) 1188–1195.

[106] C.K. Phwan, H.C. Ong, W.H. Chen, et al., Overview: comparison of pretreatment technologies and fermentation processes of bioethanol from microalgae, Energy Convers. Manag. 173 (2018) 81–94.

[106a] C. Xiros, E. Topakas, P. Christakopoulos, Hydrolysis and fermentation for cellulosic ethanol production, WIREs Energy Environ. 2 (2013) 633–654, https://doi.org/10.1002/wene.49.

[106b] K. Li, S. Liu, X. Liu, et al., An overview of algae bioethanol production, Int. J. Energy Res. 38 (2014) 965–977, https://doi.org/10.1002/er.3164.

[106c] M.K. Lam. K.T. Lee. Bioethanol production from microalgae, in: Handbook of marine microalgae, Elsevier, 2015, pp 197–208.

[107] P. Pollard, M. Huxham, The European Water Framework Directive: a new era in the management of aquatic ecosystem health? Aquat. Conserv. Mar. Freshw. Ecosyst. 8 (6) (1998) 773–792.

[108] O. Pulz, Photobioreactors: production systems for phototrophic microorganisms, Appl. Microbiol. Biotechnol. 57 (3) (2001) 287–293.

[109] S.J. Pyne, Fire in the mind: changing understandings of fire in Western civilization, Phil. Trans. Biol. Sci. 371 (1696) (2016) 20150166.

[110] S.A. Razzak, M.M. Hossain, R.A. Lucky, et al., Integrated CO2 capture, wastewater treatment and biofuel production by microalgae culturing—a review, Renew. Sustain. Energy Rev. 27 (2013) 622–653.

[111] L. Reimann, A.T. Vafeidis, S. Brown, J. Hinkel, R.S. Tol, Mediterranean UNESCO World Heritage at risk from coastal flooding and erosion due to sea-level rise, Nat. Commun. 9 (1) (2018) 1–11.

[112] A. Richmond, Biological principles of mass cultivation. Handbook of Microalgal Culture: Biotechnology and Applied Phycology, 2004, pp. 125–177.

[113] W. Roebroeks, P. Villa, On the earliest evidence for habitual use of fire in Europe, Proc. Natl. Acad. Sci. U. S. A. 108 (13) (2011) 5209–5214.

[114] Y. Román-Leshkov, C.J. Barrett, Z.Y. Liu, J.A. Dumesic, Production of dimethylfuran for liquid fuels from biomass-derived carbohydrates, Nature 447 (7147) (2007) 982–985.

[115] A. Salvati, H.C. Roura, C. Cecere, Assessing the urban heat island and its energy impact on residential buildings in Mediterranean climate: barcelona case study, Energy Build. 146 (2017) 38–54.

[116] P. Shah Maulin, Microbial Bioremediation & Biodegradation, Springer, 2020.

[117] P. Shah Maulin, Removal of Refractory Pollutants from Wastewater Treatment Plants, CRC Press, 2021.
[118] S.H. Shabaka, Checklist of seaweed and seagrasses of Egypt (Mediterranean Sea): a review, Egyptian J. Aquatic Res. 44 (3) (2018) 203–212.
[119] K.G. Satyanarayana, A.B. Mariano, J.V.C. Vargas, A review on microalgae, a versatile source for sustainable energy and materials, Int. J. Energy Res. 35 (4) (2011) 291–311.
[120] K.-Y. Show, D.-J. Lee, J.-S. Chang, Bioreactor and process design for biohydrogen production, Bioresour. Technol. 102 (2011) 8524–8533.
[121] K. Skjånes, P. Lindblad, J. Muller, BioCO2—A multidisciplinary, biological approach using solar energy to capture CO2 while producing H2 and high value products, Biomol. Eng. 24 (4) (2007) 405–413.
[122] K. Srirangan, M.E. Pyne, C.P. Chou, Biochemical and genetic engineering strategies to enhance hydrogen production in photosynthetic algae and cyanobacteria, Bioresour. Technol. 102 (2011) 8589–8604.
[123] L. Tao, A. Aden, R.T. Elander, et al., Process and technoeconomic analysis of leading pretreatment technologies for lignocellulosic ethanol production using switchgrass, Bioresour. Technol. 102 (2011) 11105–11114.
[124] H. Ting, L. Haifeng, M. Shanshan, et al., Progress in microalgae cultivation photobioreactors and applications in wastewater treatment: a review, Int. J. Agric. Biol. Eng. 10 (2017) 1–29.
[125] I. Tobin, W. Greuell, S. Jerez, F. Ludwig, R. Vautard, M.T.H. Van Vliet, F.M. Breón, Vulnerabilities and resilience of European power generation to 1.5 C, 2 C and 3 C warming, Environ. Res. Lett. 13 (4) (2018) 044024.
[126] M.R. Tredici, Photobiology of microalgae mass cultures: understanding the tools for the next green revolution, Biofuels 1 (1) (2010) 143–162.
[127] D.E.S.A. UN, World Population Prospects 2017, United Nations, Department of Economic and Social Affairs, Population Dynamics, 2017.
[128] United Nations Development Programme (UNDP), Human Development Report 2011: Sustainability and Equity — A Better Future for All, UNDP, New York, 2011.
[129] Y. Ueno, N. Kurano, S. Miyachi, Ethanol production by dark fermentation in the marine green alga, Chlorococcum littorale, J. Ferment. Bioeng. 86 (1998) 38–43.
[130] C.U. Ugwu, H. Aoyagi, H. Uchiyama, Photobioreactors for mass cultivation of algae, Bioresour. Technol. 99 (10) (2008) 4021–4028.
[131] A. Vaishampayan, R.P. Sinha, D.P. Hader, T. Dey, A.K. Gupta, U. Bhan, A.L. Rao, Cyanobacterial biofertilizers in rice agriculture, Bot. Rev. 67 (4) (2001) 453–516.
[132] A.V.I. Vonshak, Outdoor Mass Production of Spirulina: The Basic Concept, Spirulina platensis, 1997, pp. 79–99.
[133] WHO, World Health Statistics 2016: Monitoring Health for the SDGs, Sustainable Development Goals, World Health Organization, 2016.
[134] Y. Wang, W. Guo, C. Cheng, et al., Enhancing bio-butanol production from biomass of Chlorella vulgaris JSC-6 with sequential alkali pretreatment and acid hydrolysis, Bioresour. Technol. 200 (2016) 557–564.
[135] J. Weber, F. Gotzens, D. Witthaut, Impact of strong climate change on the statistics of wind power generation in Europe, Energy Proc. 153 (2018) 22–28.
[136] H. Xu, X. Miao, Q. Wu, High quality biodiesel production from a microalga Chlorella protothecoides by heterotrophic growth in fermenters, J. Biotechnol. 126 (4) (2006) 499–507.
[137] M. Yanagisawa, S. Kawai, K. Murata, Strategies for the production of high concentrations of bioethanol from seaweed, Bioengineered 4 (2013) 224–235.
[138] H.-W. Yen, W.-C. Chiang, Effects of mutual shading, pressurization and oxygen partial pressure on the autotrophical cultivation of Scenedesmus obliquus, J. Taiwan Inst. Chem. Eng. 43 (2012) 820–824.
[139] W.D. Zemke-White, M. Ohno, World seaweed utilisation: an end-of-century summary, J. Appl. Phycol. 11 (4) (1999) 369–376.
[140] X. Zhang, H.Y. Li, Z.D. Deng, C. Ringler, Y. Gao, M.I. Hejazi, L.R. Leung, Impacts of climate change, policy and Water-Energy-Food nexus on hydropower development, Renew. Energy 116 (2018) 827–834.
[141] B.I.A.N. Zhengfu, H.I. Inyang, J.L. Daniels, O.T.T.O. Frank, S. Struthers, Environmental issues from coal mining and their solutions, Min. Sci. Technol. 20 (2) (2010) 215–223.
[142] M. Zinzi, E. Carnielo, Global energy performance of residential buildings: the role of the urban climate, Energy Proc. 142 (2017) 2877–2883.
[143] M.T.H. Van Vliet, L.P.H. Van Beek, S. Eisner, M. Flörke, Y. Wada, M.F.P. Bierkens, Multi-model assessment of global hydropower and cooling water discharge potential under climate change, Global Environ. Change 40 (2016) 156–170.
[144] https://algaebiomass.org/.
[145] https://en.wikipedia.org/wiki/Biodiesel.
[146] J. Ren, D. An, H. Liang, L. Dong, Z. Gao, Y. Geng, Q. Zhu, S. Song, W. Zhao, Life cycle energy and CO_2 emission optimization for biofuel supply chain planning under uncertainties, Energy 103 (2016) 151–166.
[147] OECD Family Database. www.oecd.org/els/social/family/database
[148] D. Raclot, Y. Le Bissonnais, M. Annabi, M. Sabir, Challenges for mitigating Mediterranean soil erosion under global change, Mediterr. Reg. Under Clim. Change 311 (2016).
[149] M.F. Demirbas, Biorefineries for biofuel upgrading: a critical review, Appl. Energy 86 (2009) S151–S161.
[150] P. Spolaore, C. Joannis-Cassan, E. Duran, A. Isambert, Commercial applications of microalgae, J. Biosci. Bioeng. 101 (2) (2006) 87–96.
[151] J.M. Lv, L.H. Cheng, X.H. Xu, L. Zhang, H.L. Chen, Enhanced lipid production of Chlorella vulgaris by adjustment of cultivation conditions, Bioresour. Technol. 101 (17) (2010) 6797–6804.
[152] S.A. Jambo, R. Abdulla, S.H.M. Azhar, H. Marbawi, J.A. Gansau, P. Ravindra, A review on third generation bioethanol feedstock, Renew. Sust. Energ. Rev. 65 (2016) 756–769.
[154] Y. Peralta-Ruíz, Y. Pardo, Á. González-Delgado, V. Kafarov, Simulation of bioethanol production process from residual microalgae biomass, in: Computer Aided Chemical Engineering, 30, Elsevier, 2012, pp. 1048–1052.

[155] de Farias Silva, A. Bertucco, Bioethanol from microalgae and cyanobacteria: a review and technological outlook, Process Biochem. 51 (11) (2016) 1833–1842.

[156] M. Devarapalli, H.K. Atiyeh, A review of conversion processes for bioethanol production with a focus on syngas fermentation, Biofuel Res. J. 2 (3) (2015) 268–280.

[157] W.H. Van Zyl, L.R. Lynd, den Haan, J.E. McBride, Consolidated bioprocessing for bioethanol production using Saccharomyces cerevisiae, Biofuels (2007) 205–235.

[158] V. Mbaneme-Smith, M.S. Chinn, Consolidated bioprocessing for biofuel production: recent advances, Energy Emiss. Control Technol. 3 (2015) 23.

[159] A. Melis, L. Zhang, M. Forestier, M.L. Ghirardi, M. Seibert, Sustained photobiological hydrogen gas production upon reversible inactivation of oxygen evolution in the green alga Chlamydomonas reinhardtii, Plant Physiol. 122 (1) (2000) 127–136.

[160] D.A. Esquivel-Hernández, I.P. Ibarra-Garza, J. Rodríguez-Rodríguez, S.P. Cuéllar-Bermúdez, M.D.J. Rostro-Alanis, G.S. Alemán-Nava, J.S. García-Pérez, R. Parra-Saldívar, Green extraction technologies for high-value metabolites from algae: a review, Biofuel Bioprod. Biorefin. 11 (1) (2017) 215–231.

[161] O.E. Frihy, M.K. El-Sayed, Vulnerability risk assessment and adaptation to climate change induced sea level rise along the Mediterranean coast of Egypt, Mitig. Adapt. Strateg. Glob. Chang. 18 (8) (2013) 1215–1237.

[162] E. EEA, Environmental indicators: typology and overview, European Environmental (1999).

Further reading

[1] M. Ashour, M.E. Elshobary, R. El-Shenody, et al., Evaluation of a native oleaginous marine microalga Nannochloropsis oceanica for dual use in biodiesel production and aquaculture feed, Biomass Bioenergy 120 (2019) 439–447.

[2] N. Ali, Z. Tingb, Y. Khan, M. Athar, V. Ahmad, M. Idrees, Making biofuels from microalgae —a review of technologies, J. Food Sci. Technol. 1 (2014) 7–14.

[3] D. Hernández, M. Solana, B. Riaño, et al., Biofuels from microalgae: lipid extraction and methane production from the residual biomass in a biorefinery approach, Bioresour. Technol. 170 (2014) 370–378.

[4] L.G. Castellanos-Barriga, F. Santacruz-Ruvalcaba, G. Hernández-Carmona, E. Ramírez-Briones, R.M. Hernández-Herrera, Effect of seaweed liquid extracts from Ulva lactuca on seedling growth of mung bean (Vigna radiata), J. Appl. Phycol. 29 (5) (2017) 2479–2488.

[5] L. Otjen, R.A. Blanchette, Selective delignification of birch wood (Betula papyrifera) by Hirschioporus pargamenus in the field and laboratory, in: Organisation for Economic Cooperation and Development (OECD) (2011), towards Green Growth, OECD, Paris, 1986.

[6] D. Raclot, Y. Le Bissonnais, M. Annabi, M. Sabir, A. Smetanova, Main issues for preserving Mediterranean soil resources from water erosion under global change, Land Degrad. Dev. 29 (3) (2018) 789–799.

Chapter 31

Plant-based biofuels: an overview

Soumya Singh[1] and Shalini Singh[2]
[1]Shri Ramswaroop Memorial University, Lucknow, Uttar Pradesh, India; [2]Amity Institute of Microbial Technology, Amity University, Jaipur, Rajasthan, India

1. Introduction

With the advent of technology and focus on sustainability, biotechnology industry is growing in leaps and bounds. Areas like biomedical equipment, biochemical products, drugs, textiles, agriculture, and many other fields have been making major advances and changes keeping sustainability in mind. Talking of sustainability, coal, oil, and petroleum sectors cannot be ignored as these sectors signficantly contribute to nation-building. Coal, petroleum, and crude oil fall under a large category of fossil fuels which are derived from dead biomass of millions of years. These fossil fuels are exhaustible resources which are depleting at a rate faster than the world's population. As per the IEA data of India Energy Outlook 2021, the domestic production of oil and gas continues to fall behind consumption trends and net dependence on imported oil is expected to rise above 90% by 2040, from current dependency of 75%. This continued reliance on imported fuels creates vulnerabilities to price cycles and volatility as well as possible disruptions to supply. Energy security hazards could arise in India's domestic market as well, notably in the electricity sector, if the necessary flexibility in power system operation does not materialize [1]. As per 2022 data submitted by Ministry of Petroleum and Natural Gas, the import dependency of crude oil is around 85% and gross petroleum import is around 23% to balance the consumption rate [2]. This heavy dependency on imported petroleum products forces India to examine the feasibility of using alternative energy sources and a great opportunity exists for a biotechnological advancement to bring major changes in oil and petroleum industry of the country.

The need of the hour is to produce biofuels to meet rising consumption requirments of petroleum products, tackle the problem of high import cost and also bring sustainability in the oil and gas market. Biofuels are being looked upon as one-stop solution to the ever-increasing energy crisis. Most of the industrialized nations are taking advantage of this opportunity and producing biological-based fuels for meeting the requirement of population and to remove the excessive dependence on imported fossil fuels.

1.1 Biofuels and their economics

Substituting fossil fuels with biofuels has the potential to reduce some adverse effects of fossil fuel production and use, including conventional and greenhouse gas (GHG) pollutant emissions, exhaustible resource depletion, and reliance on unsteady foreign suppliers. At the same time, an increase in farm income is achieved.

The production of biofuels results in GHG emissions at several stages of the process; EPA's (2010) analysis of the Renewable Fuel Standard (RFS) projected that several types of biofuels could yield lower life cycle GHG emissions than gasoline over a 30-year time horizon. Academic studies using other economic models have also found that biofuels can lead to reductions in life cycle GHG emissions relative to conventional fuels [3].

Biofuels can be produced locally, which could lead to lower fossil fuel imports [4]. If biofuel production and use reduces our consumption of imported fossil fuels, we may become less vulnerable to the adverse impacts of supply disruptions [5].

Biofuels also decrease the emission of certain pollutants. It is important to understand that biofuel production and consumption alone will not reduce GHG or conventional pollutant emissions, lessen petroleum imports, or improve pressure on exhaustible resources. Biofuel production and use must match with reductions in the production and use of fossil fuels for these benefits to accumulate. These benefits would be mitigated if biofuel emissions and resource demands expand, rather than dislocate, those of fossil fuels.

The Indian Government has been actively exploring its biofuel prospective since 2001 (GOI 2005 and 2006). Bioethanol in India is currently being produced from molasses and the possibilities of using sugarcane juice to increase bioethanol production are also being searched. In the middle of food versus fuel issues, India also hopes to increase energy security by launching one of the world's biggest nonedible oilseed-based biodiesel industries. *Jatropha* and *Pongamia* are the two prominent oilseed plants undergoing experimentation for biodiesel production [6]. It has clearly mentioned in its biofuel policy not to promote comestible feedstock or diversion of conventional farmland to biofuel production. Therefore, the biodiesel strategy considers use of waste or fallow lands, estimated at 55 million hectares nationwide [7].

India initiated biofuel production nearly a decade ago to reduce its dependence on foreign oil and began 5% ethanol blending (E5) pilot programs in 2001 and formulated a National Mission on Biodiesel in 2003 to achieve 20% biodiesel blends by 2011–12 (GOI 2002 and 2003). To strengthen, formalize, and promote a sustainable biofuels industry, India implemented a National Policy on Biofuels in December 2009. This policy document outlines a broad strategy for the biofuel program and briefly catalogs policy measures being considered to support the program. Although the policy contains limited details on how the program will be implemented, the country's intention to avoid conflicts with food security is firmly stated throughout the policy document. The policy specifically requires the use of nonfood feedstocks grown on marginal lands unsuitable for agricultural production. However, no details on how this requirement will be enforced are contained in the policy.

1.2 Generations of biofuels

Biofuels are the fuels produced by the transformation of biomass; so they can also be termed as "biomass-based fuels" including both liquid and gaseous biofuels. Biofuels can be classified as first-, second-, and third-generation biofuels, depending on their feedstock and production technology [8]. The classification of biofuels is shown in Fig. 31.1.

The first-generation biofuels are manufactured from sugar, starch, and vegetable oil, which are derived primarily from food crops. A vast majority of biofuels currently produced in commercial quantities fall into this category. The second-generation biofuels are those that are manufactured from nonfood crops and lignocellulosic wastes and their manufacturing process requires enzymatic digestion and fermentation. The third generation of biofuel is referred to as algal biofuel and is manufactured from photosynthetic algae. At the level of basic research, the fourth generation of biofuels has also appeared, using an efficient, solar-based biofuel production pathway with algae and cyanobacterial activity [10].

1.3 Plant biofuels: an introduction

Biofuels and biofuel crops have long been in use as a replacement for fossil fuels. Biofuel is a low-carbon fuel that is produced from biomass as compared to fossil fuels. Also known as energy crops, biofuels include crop materials like corn, wheat, sugarcane, and soybean but these can also be made by utilizing waste materials. Biofuels generally fall into two categories: bioalcohol and biodiesel. A bioalcohol is a product of fermentation, using yeast and bacteria to break down starch and sugar. Technologies that would allow bioalcohol to be made from cellulose and hemicelluloses are also being researched. In fact, several commercial-scale cellulosic ethanol biorefineries are already in operation. On the other hand, biodiesel is produced in refineries that use new and used vegetable oils and animal fats such as, recycled cooking oil. These

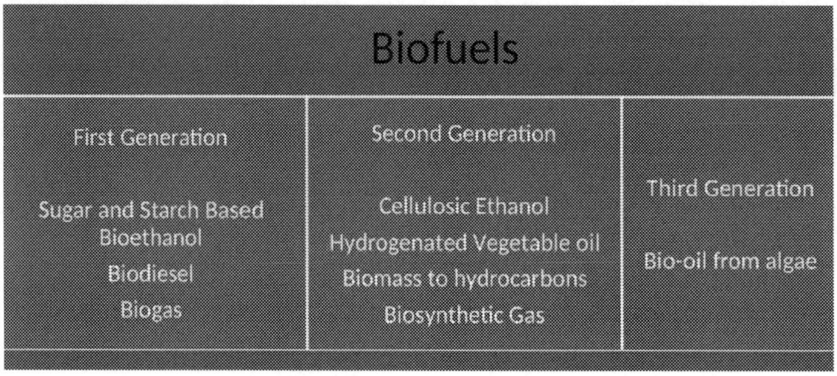

Biofuels		
First Generation	Second Generation	Third Generation
Sugar and Starch Based Bioethanol Biodiesel Biogas	Cellulosic Ethanol Hydrogenated Vegetable oil Biomass to hydrocarbons Biosynthetic Gas	Bio-oil from algae

FIGURE 31.1 Different generations of biofuels [8].

oils are then treated with alcohol and converted to biodiesel. Many plants and plant crops are being utilized for the production of these alternative fuels, some of the important ones will be discussed in this chapter. But before discussing biofuel production from plant and plant crops, it is also important to understand the underlying policies and standards governing the production of these alternative fuels.

1.4 The renewable fuel standards (RFS) II

RFS was established by US Congress as a modification to the Clean Air Act; the RFS mandates that US transportation fuels contain a minimum volume of renewable fuel. The mandated minimum volume increases annually and generally has been met using both conventional biofuel (e.g., corn starch ethanol) and advanced biofuel (e.g., cellulosic ethanol). For a renewable fuel to be applied towards the mandate, it must be used for certain purposes (i.e., road transportation fuel, jet fuel, or heating oil) and meet certain environmental and biomass feedstock criteria [11].

RFS II is the motivation for increasing the production of renewable fuels from green plants. This standard was set in 2005 and revised in 2007 to mandate quantities of renewable fuels to be incorporated into the transportation industry in the United States. The goal for 2022 is set at 36 billion gallons of renewable fuels, with 16 billion gallons required to be from lignocellulosic feedstocks, and 1 billion gallons per year of biodiesel. Additionally, 58% of the fuels produced by 2022 should be "advanced biofuels," e.g., non-starch ethanol or other types of fuels such as long-chain hydrocarbons or butanol that achieve a 50% reduction in GHG emissions. The feedstocks for these fuels are lignocellulosics and oils. However, intense research is required to make the production cost-effective. Advancements are being made in feedstock structure, ease of processing, efficiency of conversion, coproduct manufacture, and sustainability. As these discoveries come together, they can be incorporated into new industrial applications.

The statute centers on four renewable fuel categories—total renewable fuel, advanced biofuel, cellulosic biofuel, and biomass-based diesel—each with its own target volume. A key part of the statutory definition of each fuel category is whether the fuel achieves certain GHG reductions relative to gasoline and diesel fuel. Each fuel is assigned a life cycle GHG emission threshold (in proportion to baseline life cycle GHG emissions for gasoline and diesel) [12].

The statute defines the four renewable fuels. Conventional biofuel is corn starch ethanol. Advanced biofuel is renewable fuel, other than corn starch ethanol, with life cycle GHG emissions of at least 50% less than life cycle GHG emissions from its gasoline or diesel counterpart. Cellulosic biofuel is a renewable fuel derived from cellulose, hemicellulose, or lignin that is derived from renewable biomass, with life cycle GHG emissions of at least 60% less than life cycle GHG emissions from its gasoline or diesel counterpart. Biomass-based diesel is biodiesel or other renewable diesel with life cycle GHG emissions of at least 50% less than life cycle GHG emissions from its diesel counterpart. Additionally, biofuel from new facilities—those built after enactment of the 2007 law—must achieve at least a 20% GHG reduction to qualify as a conventional renewable fuel [11].

The 2020, 2021, and 2022 Proposed Rule: EPA released the proposed rule for the 2020, 2021, and 2022 RFS volume requirements on December 7, 2021, and published the proposed rule in the Federal Register on December 21, 2021. EPA proposes to reduce the volume requirements from what is required by statute for 2020, 2021, and 2022. EPA reports it is "proposing to reduce the applicable statutory volumes for 2020, 2021 and 2022 utilizing both the cellulosic waiver and reset authorities." EPA reports that, for 2022, it is proposing to set volumes of advanced biofuel and total renewable fuel that "represent growth compared to historical volumes and compared to the volumes proposed for 2020 and 2021." EPA reports the retroactive reduction of the 2020 volume requirements is necessary because "several significant and unanticipated events" occurred following the release of the 2020 final rule. These events include the COVID-19 pandemic—which led to a drop in transportation fuel demand—and lower than expected gasoline and diesel volume exemptions from the 2020 RFS obligations stemming from small refinery exemptions [11].

The RFS After 2022: Contrary to popular myth, the RFS does not "phase out" or "sunset" at the end of 2022. When Congress passed the **Energy Independence and Security Act of 2007**, it specified RFS volume requirements through the year 2022 for total renewable fuels, advanced biofuels, cellulosic biofuels, and biomass-based diesel. For years after 2022, the law clearly states that required volumes of each renewable fuel shall be determined by the EPA Administrator, in coordination with the Secretary of Energy and the Secretary of Agriculture. In other words, the law requires EPA to set RFS volumes for 2023 and beyond, according to certain criteria defined in the statute.

1.5 Feedstocks for biofuel production

1.5.1 Corn

Corn is generally considered the king of ethanol-based biofuels. Sugar-rich corn is turned into ethanol in a similar fashion to beer brewing. The kernels are ground up and mixed with warm water and yeast. The yeast ferments the mixture to produce ethanol. This ethanol is then blended with gasoline to use in existing car engines. This mixture releases less carbon monoxide, nitrogen oxide, and sulfur than run-of-the-mill gasoline and reduces smog in cities. The reason the kernels alone are used is that the main body of the plant contains cellulose, which is more difficult and expensive to break down.

1.5.2 Rapeseed/canola

Rapeseed oil has been used to cook food and in lamps for centuries. Today, it's an important biodiesel source. The most important type is canola because, compared to other rapeseeds, it is low in erucic acid, which makes it healthier for animals and humans to eat. It's an interesting fact that biodiesels tend not to fare well in cold climates. This is because vegetable oils tend to be high in saturated fats, which allow ice crystals to form at low temperatures. This is obviously not good for combustion engines. Canola just so happens to be low in saturated fats, which clearly gives it an advantage in colder environments than its alternatives. Rapeseeds are also relatively high in oil content when compared to most plants, which makes them great crops for making fuels.

1.5.3 Sugarcane

Brazil has been working tirelessly to reduce its dependence on fossil fuels over the years. This South American country has been growing energy crops since the 1970s as a direct consequence of the Middle East Oil Embargo. When oil prices leaped, the Brazilian government encouraged its farmers to plant sugarcane. Sugarcane is used to produce bioethanol, not unlike corn. Brazil has invested billions of dollars into this industry, to such an extent that it is now cheaper than gasoline. Interestingly, in the 1980s, most cars in Brazil were ethanol-powered, but today most utilize flexible fuel engines. Producing ethanol from sugarcane is six times cheaper than using corn. However, farmers burn their fields during sugarcane harvesting, releasing massive amounts of GHGs into the atmosphere and somewhat negating the carbon benefits from using bioethanol.

1.5.4 Palm oil

Palm oil is extracted from the fruit of palm trees and it is one of the energy-efficient biodiesel fuels on the market. Diesel engines also do not need to be converted to run on palm oil either. Palm oil biodiesel is less polluting than gasoline as well. Palm oil has helped develop the economies of Malaysia and Indonesia in particular. However, the growing of palm trees for biodiesel in these countries has led to the burning of thousands of acres of rainforest each day for land to grow the crops. This destruction threatens the fragile ecosystem and endangers thousands of plant and animal species.

1.5.5 Jatropha

This poisonous weed is a big player in the biofuel market. The bushes grow quickly and do not require a large amount of water, and their seeds have around 40% oil content. India is currently the world's largest Jatropha producer, and their biodiesel industry is centered on this crop. This has allowed the country to bring economic benefits to rural farmers who can grow this crop on agricultural land too poor for other crops. Jatropha plants can live for 50 years and do very well on land devastated by drought and pests. The seeds of the plant are crushed to release the oil for biodiesel production. But the seed cases and vegetable matter are not wasted. They can also be used as biomass fuel.

1.5.6 Soybean plants

Not just used for tofu, sauce, crayons, and shampoos, soybeans can also be used as a fuel source. Most biodiesel in the United States is, as it turns out, made from soybeans. Motor vehicles, heavy equipment, and even buses can run on pure soybean biodiesel or by blending the soy biodiesel with more traditional diesel fuels. Soybean diesel is reported to yield more energy than corn ethanol [13]. One bushel of soybeans can yield 5.68 L (about 1.5 gallons) of biodiesel. However, while soybeans have an oil content of around 20%, other fuels, like Canola and sunflower seeds, have double that at 40% and 43%, respectively [14].

2. Switchgrass

This plant has great potential as an alternative to fossil fuels. Unlike corn, switchgrass has a form of cellulose that uses less energy to convert to ethanol than from processing fossil fuels. Ethanol from Switchgrass's cellulose contains more energy than corn ethanol. Although there are not currently large plantations of these crops, scientists are currently working on methods to exploit this plant in the future. Researchers at the Auburn University, Alabama have grown test plots of

Switchgrass to yield 15 tons of biomass per acre [14,15]. Unlike other biofuel crops, switchgrass grows in poor soil on land not normally used for farming, so no cropland is taken away from food production. And because it is a perennial, farmers only have to plant it once, meaning less work to grow.

3. Feedstock processing technologies

3.1 Deconstruction

Producing advanced biofuels (e.g., cellulosic ethanol and renewable hydrocarbon fuels) typically involves a multistep process. First, the tough rigid structure of the plant cell wall—which includes the biological molecules cellulose, hemicellulose, and lignin bound tightly together—must be broken down. This can be accomplished by one of the two ways: high-temperature deconstruction or low-temperature deconstruction.

3.1.1 High-temperature deconstruction

High-temperature deconstruction makes use of extreme heat and pressure to break down solid biomass into liquid or gaseous intermediates. There are three primary routes used in this pathway:

Pyrolysis: During pyrolysis, biomass is heated rapidly at high temperatures (500–700°C) in an oxygen-free environment. The heat breaks down biomass into pyrolysis vapor, gas, and char. Once the char is removed, the vapors are cooled and condensed into a liquid "bio-crude" oil.

Gasification: Gasification follows a slightly similar process; however, biomass is exposed to a higher temperature range (>700°C) with some oxygen present to produce synthesis gas (or syngas)—a mixture that consists mostly of carbon monoxide and hydrogen.

Hydrothermal liquefaction: When working with wet feedstocks like algae, hydrothermal liquefaction is the preferred thermal process. This process uses water under moderate temperatures (200–350°C) and elevated pressures to convert biomass into liquid bio-crude oil.

3.1.2 Low-temperature deconstruction

Low-temperature deconstruction typically makes use of biological catalysts called enzymes or chemicals to break down feedstocks into intermediates. First, biomass undergoes a pretreatment step that opens up the physical structure of plant and algae cell walls, making sugar polymers like cellulose and hemicellulose more accessible. These polymers are then broken down enzymatically or chemically into simple sugar building blocks during a process known as hydrolysis.

3.2 Upgrading

After deconstruction, intermediates such as crude bio-oils, syngas, sugars, and other chemical building blocks must be upgraded to produce a finished product. This step can involve either biological or chemical processing. Microorganisms, such as bacteria, yeast, and cyanobacteria, can ferment sugar or gaseous intermediates into fuel blendstocks and chemicals. Alternatively, sugars and other intermediate streams, such as bio-oil and syngas, may be processed using a catalyst to remove any unwanted or reactive compounds in order to improve storage and handling properties. The finished products from upgrading may be fuels or bioproducts ready to sell into the commercial market or stabilized intermediates suitable for finishing in a petroleum refinery or chemical manufacturing plant.

4. Production strategies

4.1 Bioethanol production

4.1.1 Ethanol production from sugar and starch

Both raw juice and molasses from sugarcane and sugar beets can be used for ethanol production. Brazil leads the world in production of ethanol using sugar feedstocks. In Brazil, both sugarcane juice and molasses are used for ethanol production. The juice is extracted from sugarcane by either squeezing (roll mills) or diffusion (diffuser). Part of the juice is used for sugar manufacture and the remaining is used for ethanol production. Molasses, which is a low-value by-product, also are used for ethanol production. The solid residue from the extraction step, which is referred to as bagasse, is burned to generate energy for use in the plant. Ethanol normally is obtained by fermentation of cane juice or a mixture of cane

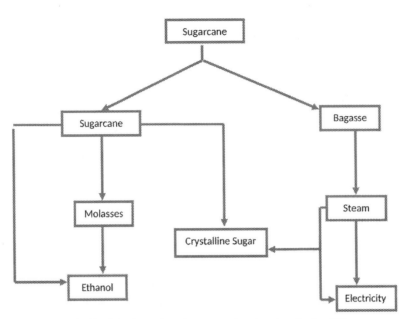

FIGURE 31.2 Processing of sugarcane for ethanol production [12].

molasses and juice. Before going to the fermentors, the sugar solution must go through purification and pasteurization. Purification normally involves treatment with lime, heating, and later decantation, similar to the treatment used in sugar manufacture. Pasteurization involves heating and immediate cooling. The cooling typically includes two stages. In the first stage the hot sugar solution is passed through a heat exchanger in countercurrent flow to the cold solution. At the end of this stage the hot solution is cooled to about 60°C. In the second stage, the sugar solution is cooled further to 30°C using water as the cooling fluid. The sugar concentration normally is adjusted to approximately 19 degrees Brix. The fermentation can be either batch/fed-batch or continuous (Fig. 31.2) [12].

Among potential starch feedstocks for ethanol production, corn is used most widely, especially in the United States. Other grains, such as wheat and barley, also are used as feedstocks for ethanol production. The process technology developed for corn can easily be adapted to other grains. Corn ethanol plants need only minor modifications to handle other grain feedstocks. Ethanol is produced from corn by either the wet milling or dry milling process. The dry milling is also referred to as the dry-grind process. The main difference between the two processes is that in the dry milling process the whole corn is ground and fed to the fermentor for ethanol fermentation, whereas in the wet milling process the corn components are fractionated first and then only the starch fraction is used in ethanol fermentation. As a result the wet milling process requires much higher capital investment, and ethanol plants using this process are much larger than those using the dry milling process. The large size of the wet milling ethanol plants is needed to justify the rate of return on capital investment. Both processes generate a number of ethanol-related coproducts. These coproducts include distillers dried grains with solubles and carbon dioxide in the dry milling process, and corn oil, corn gluten meal, corn gluten feed, and carbon dioxide in the wet milling process. All of these coproducts are of relatively moderate values. However, the wet milling plants can easily be modified to produce other products such as corn syrups and high fructose corn syrups, which can be produced independent of ethanol production. These independent coproducts are not just economically beneficial but also can be strategically important, especially during the times of reduced ethanol market demand (Fig. 31.3) [12].

Recovery of Ethanol: Ethanol from the fermentation broth is recovered by distillation followed by a dehydration step. Ethanol and water form an azeotropic mixture of 95% ethanol and 5% water by volume. In the past, ternary azeotropic distillation using an agent such as benzene, cyclohexane, diethyl ether, and *n*-pentane was employed to produce anhydrous ethanol. Molecular sieves currently are used for dehydration of 95% ethanol [12].

4.1.2 Ethanol production from lignocellulose

Lignocellulosic feedstocks consist of three main components: cellulose, hemicellulose, and lignin. Technologies for conversion of these feedstocks to ethanol have been developed on two platforms, which can be referred to as the sugar platform and the synthesis gas (or syngas) platform.

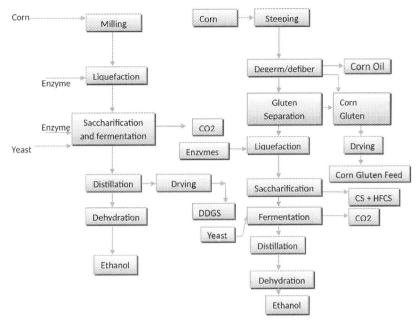

FIGURE 31.3 Concept of dry and wet milling for ethanol production [12].

In the sugar platform, cellulose and hemicellulose are first converted to fermentable sugars, which then are fermented to produce ethanol. The fermentable sugars include glucose, xylose, arabinose, galactose, and mannose. Hydrolysis of cellulose and hemicellulose to generate these sugars can be carried out by using either acids or enzymes. Pretreatments of the biomass are needed prior to hydrolysis.

In the syngas platform, the biomass is taken through a process called *gasification*. In this process, the biomass is heated with no oxygen or only about one-third the oxygen normally required for complete combustion. The biomass subsequently is converted to a gaseous product, which contains mostly carbon monoxide and hydrogen. The gas, which is called synthesis gas or syngas, then can be fermented by specific microorganisms or converted catalytically to ethanol (Fig. 31.4).

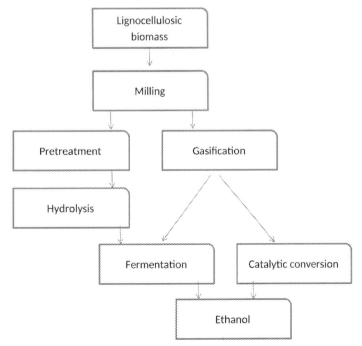

FIGURE 31.4 Production of ethanol from lignocellulosic feedstocks [12].

4.2 Biodiesel production [12]

The basic chemistry for biodiesel production is relatively simple and occurs primarily through a transesterification step rendering this process to fairly straightforward commercialization. The reactions associated most commonly to biodiesel production include transesterification and esterification, but with potential competing reactions including hydrolysis and saponification.

The general transesterification reaction for acid- or base-catalyzed conversion of oils consists primarily of neutral triacylglycerols (TAGs) in reaction with alcohols to form alkyl esters. The alkyl esters produced depend on the alcohol used where methanol ($R = CH_3$) and ethanol ($R = CH_2CH_3$) are most common. The catalysts for transesterification include KOH, NaOH, or H_2SO_4. Excess alcohol with adequate catalyst generally forces the reaction equilibrium toward the products of biodiesel esters and glycerol. The reaction for base-catalyzed systems will occur rapidly at room temperature, although higher temperatures of 50°C are often employed to reduce initial viscosity of oils while increasing reaction rates. Acid-catalyzed transesterification is often reacted at higher temperatures from just below the boiling point of the alcohol to 120°C in pressurized vessels. The transesterification reaction of TAGs takes place in three steps where TAG is first converted to a diacylglycerol (DAG) and one fatty acid ester. Then the DAG is converted to monoacyl glycerol (MAG) liberating an additional fatty acid ester, and finally the MAG is converted to glycerol liberating the final fatty acid ester. Interesterification and transesterification reactions via enzyme catalysis are biologically significant mechanisms for producing fatty acid esters and may be accomplished efficiently with enzyme catalysis using lipases. The reactant commonly used as the acyl acceptor for the interesterification of triglycerides is ethyl acetate rather than ethanol or methanol (as used in transesterification reactions). This reaction results in the production of triacetin and long-chain fatty acid methyl or ethyl esters, but not glycerol as in the case of esterification and transesterification with alcohols. Most plant oils have a promising fatty acid ester profiles and produce cetane values near 60 [12]. The most prevalent plant oil for production of biodiesel worldwide is from rapeseed or commonly called *canola* in the United States, which yields oil as much as 127 gal/acre/year [12]. Soybeans are grown for their high protein and lipid content, and being a legume, they have the advantage of introducing some nitrogen back into the soil naturally, making this plant more sustainable than intense nitrogen consumers. Because of the value of the products and ability to be used in crop rotation with nitrogen-intensive crops such as corn, soybean oil has become the largest source of food oils and consequently biodiesel. Considerable work toward production of oils by algal and microbial strains with oleaginous traits has been conducted with main goals of producing biodiesel. These species not only produce the typical fatty acids produced by higher order plants, but typically also produce long-chain polyunsaturated fatty acids, important to metabolism in animals, including nerve cell function and brain cell development. Algae have been shown to accumulate an impressive amount of lipids of over 80% of their dry weight and commonly produce levels of 20%–50% [12]. The three most prevalent groups of algae targeted for biodiesel production include the diatoms (Bacillariophyceae) containing nearly 100,000 species that make up a majority of phytoplankton in salt and brackish waters: green algae (Chlorophyceae) common in many freshwater systems, blue-green algae (Cyanophyceae), which are actually bacteria that contain chloroplasts and are important to nitrogen fixation in aquatic systems, and finally the golden algae (Chrysophyceae) with about 1000 known species able to store carbon as oil and complex carbohydrates.

4.3 Biomethane production [12]

Anaerobic digestion of animal manures, municipal wastewater or sludge, and food processing wastewater has been practiced in various forms for many decades. Anaerobic digestion decreases the organic matter content in the waste through the biological conversion of organic carbon and produces a biogas containing primarily methane and carbon dioxide. Due to the development of sophisticated treatment technologies, thousands of anaerobic treatment processes are now installed and operating worldwide in addition to the millions of small low-tech systems operating throughout less industrialized nations.

4.3.1 Methanogenic environments

The microbiological production of methane occurs in engineered systems such as anaerobic digesters used for municipal and animal waste treatment, in the gastro-intestinal tract of ruminants and non-ruminant herbivores, and in many natural environments such as freshwater, estuarine, and marine sediments. In the late 1770s, the Italian physicist, Alessandro Volta, most well-known for developing an early battery, performed experiments with what he termed *combustible air*. Volta collected gas bubbles from the disturbed sediment of a shallow lake and described the combustion of the gas that was later called methane. True anaerobic conditions, devoid of inorganic terminal electron acceptors such as oxygen, nitrate, ferrous iron, and sulfate, are required for methane production.

4.3.2 Methane process description

Methane is produced by a metabolically diverse community of bacteria and archaea that act as an integrated metabolic unit to produce methane and carbon dioxide through a series of sequential and concurrent reactions. The end-products of one group's metabolism are used as substrate by the next group. In general, the biological production of methane from complex organic compounds contained in biomass and waste sources involves four main phases: hydrolysis, fermentation (acidogenesis), acetogenesis, and methanogenesis.

4.3.2.1 Hydrolysis

Many of the potential biomass sources for methane production are high-molecular-weight, insoluble polymers such as polysaccharides, proteins, and fats that are too large to be transported across bacterial cell membranes. The initial conversion reactions may require several different types of enzymes. One unifying characteristic of these enzymes is that they are synthesized within the bacterial cells in small amounts and are secreted into the environment surrounding the bacterium until they contact the polymers. These enzymes catalyze hydrolysis reactions that cleave polymers and incorporate water, thus producing soluble monomers. Polysaccharides such as cellulose and hemicellulose are hydrolyzed to glucose and xylose by cellulase and hemicellulase enzymes. Proteins and lipids are hydrolyzed to their constituent amino acids and long-chain fatty acids by proteases and lipases, respectively. These compounds then enter the cell through active transport, and once an increase in the specific degradation products is sensed by the bacteria, genes that produce these enzymes are upregulated to increase the amount of these enzymes being secreted into the environment. In this way, the bacteria do not expend cellular energy producing these enzymes at high rates when not needed. This interconnectedness of the genetic regulation and microbial degradation process is fundamental to the ability to direct specific reactions. The rate of hydrolysis is a function of several factors, such as pH, substrate composition, and particle size.

4.3.2.2 Fermentation (acidogenesis)

The second phase of the overall process is fermentation that begins with the conversion of the sugar monomers to pyruvate ($C_3H_4O_3$), ATP, and the electron carrier molecule NADH by central metabolic pathways. The central metabolic pathways found within most bacteria are the Embden–Meyerhof pathway (glycolysis) and the pentose phosphate pathway. Next, these fermentative bacteria convert pyruvate and amino acids to a variety of short-chain organic acids—primarily acetate, propionate, butyrate, and succinate—and alcohols, CO_2, and H_2 through various fermentation pathways. During the fermentation reactions, NADH is oxidized to regenerate NAD+, while organic intermediates of the fermentation pathways are reduced. Because the fermentation process results in the formation of various short-chain organic acids, this stage of the methane fermentation is also referred to as the *acid-forming stage* or *acidogenesis*.

4.3.2.3 Acetogenesis

The short-chain organic acids produced by fermentation and the fatty acids produced from the hydrolysis of lipids are fermented to acetic acid, H_2, and CO_2 by acetogenic bacteria. Syntrophic bacteria that oxidize organic acids to acetate, H_2, and CO_2 are reliant on the subsequent oxidation of H_2 by the next group, the methanogens, to lower the H_2 concentration and prevent end-product inhibition.

4.3.2.4 Methanogenesis

In the final phase, methane is produced through two distinct routes by two different microbial groups. One route is by the action of the lithotrophic H_2-oxidizing methanogens that use H_2 as electron donor and reduce CO_2 to produce methane. In the second route, the organotrophic acetoclastic methanogens ferment acetic acid to methane and carbon dioxide. Approximately two-thirds of the methane produced derive from the acetoclastic methanogens. The complete conversion of a carbohydrate such as glucose to methane in anaerobic digestion results in 85–90% recovery of energy.

5. Future and scope of biofuels

The benefits of biofuels whether globally or to a single country are as follows:

1. Reduction in crude oil use. Liquid biofuels can supplement or replace petrol and diesel, and at low levels of blending, little engine modification is required. Biodiesel can be used up to 100% in a conventional diesel engine but higher blends of ethanol (85%) require either modifications or a flexible fuel engine. Biomass and biogas can reduce fossil fuel use for electricity generation.

2. Improvements in engine performance. Ethanol has a very high octane number and has been used to improve the octane levels of petrol. It is also a possible replacement for methyl tertiary butyl ether which is being phased out as an octane enhancer. Biodiesel addition will enhance diesel lubricity and raise the cetane number.
3. Air quality. Biofuels can improve air quality by reducing the emission of carbon monoxide (CO) from engines, sulfur dioxide, and particulates (PM) when used pure or in blends.
4. Reduction in the emission of the GHGs: carbon dioxide and methane. The replacement of fossil fuels with biofuels can reduce significantly the production of carbon dioxide, and the use of biogas reduces methane emissions.
5. Toxicity. Biofuels are less toxic than conventional fuels, sulfur-free, and easily biodegradable.
6. Production from waste. Some biofuels can also be made from wastes, for example, used cooking oil can be used to make biodiesel.
7. Agricultural benefits. Biofuel crops of all types will provide the rural economy with an alternative nonfood crop and product market.
8. Reduction of fuel imports. By producing fuels in the country, imports will be reduced and the security of energy supply will be increased.
9. Infrastructure. No new infrastructure is required for the first- and second-generation liquid biofuels and some of the solid and gaseous biofuels.
10. Sustainability and renewability. Biofuels are sustainable and renewable, as they are produced from plants and animals.

However, there are shortcomings to the use of biological materials to replace fossil fuels which are as follows:

1. Biological material may not be able to produce enough fuel to replace fossil fuels completely, and extensive cultivation of biofuel crops will compete with food crops, perhaps driving up prices.
2. Large amounts of energy are required to produce some biofuels, giving them a low net energy gain.
3. Some of the second- and third-generation biofuels will require the introduction of a completely new infrastructure, for example, hydrogen.

References

[1] India Energy Outlook Report, International Energy Agency, 2021.
[2] PPAC's Snapshot of India's Oil and Gas Data, Petroleum Planning and Analysis Cell, Ministry of Petroleum and Natural Gas, Govt of India, January 2022.
[3] https://www.epa.gov/environmental-economics/economics-biofuels.
[4] H. Huang, M. Khanna, H. Onal, X. Chen, Stacking low carbon policies on the renewable fuels standard: economic and greenhouse gas implications, Energy Pol. 56 (May 2013) (2013) 5−15.
[5] US Environmental Protection Agency, Renewable Fuel Standard Program (RFS2) Regulatory Impact Analysis, 2010. (Accessed 10 September 2013).
[6] T. Altenburg, H. Dietz, M. Hahl, N. Nikolidakis, C. Rosendahl, K. Seelige, Biodiesel in India: Value Chain Organisation and Policy Options for Rural Development Studies, Deutsches Institut fuer Entwicklungspolitik, Deutsches Institut fuer Entwicklungspolitik, Bonn, 2009.
[7] H. Gunatilake, Food Security, Energy Security and Inclusive Growth in India: The Role of Biofuels, Asian Development Bank, Manila, 2011. Processed.
[8] M.F. Demirbas, Biofuels from algae for sustainable development, Appl. Energy 88 (2011) 3473−3480.
[9] Magda, et al., The role of using bioalcohol fuels in sustainable development, in: Bio-economy and Agri-production, 2021, pp. 133−146.
[10] S.G. Hays, D.C. Ducat, Engineering cyanobacteria as photosynthetic feedstock factories, Photosynth. Res. 123 (3) (2015) 285−295, https://doi.org/10.1007/s11120-014-9980-0. Epub 2014 Feb 14.
[11] Congressional Research Service, The Renewable Fuel Standard (RFS): An Overview, January 2022, p. R43325. https://crsreports.congress.gov.
[12] C. Drapcho, J. Nghiem, C. Walker, Biofuel Engineering Process Technology, Mc Graw Hill Publishers, 2008.
[13] J. Hill, E. Nelson, S. Polasky, D. Tiffany, Environmental, economic, and energetic costs and benefits of biodiesel and ethanol biofuels, Proc. Natl. Acad. Sci. USA 103 (30) (2006) 11206−11210, https://doi.org/10.1073/pnas.0604600103.
[14] C. McFadden. https://interestingengineering.com/innovation/seven-biofuel-crops-use-fuel-production, 2021. (Accessed 15 December 2022).
[15] M. Shah (Ed.), Microbial Bioremediation & Biodegradation, Springer, Singapore, 2020.

Chapter 32

Bioenergy: biomass sources, production, and applications

Tanvi Taneja[1], Muskaan Chopra[1] and Indu Sharma[2]

[1]Department of Bio-Sciences and Technology, Maharishi Markendeshwar (Deemed-to-be University) Mullana, Ambala, Haryana, India; [2]Nims Institute of Allied Medical Science and Technology, NIMS University Rajasthan, Jaipur, Rajasthan, India

1. Introduction

Energy stored in biomass can be released to produce renewable electricity or heat. Biopower can be generated through the combustion or gasification of dry biomass or biogas (methane) captured through controlled anaerobic digestion. Cofiring of biomass and fossil fuels (usually coal) is an inexpensive way to reduce greenhouse gas emissions, improve cost-effectiveness, and reduce air pollutants in existing power plants. Thermal energy (heating and cooling) is often produced at the scale of the individual building, through the direct combustion of wood pellets, wood chips, and other sources of dry biomass.

1.1 Bioenergy

Bioenergy is a renewable energy produced from biomass. Bioenergy is one of many diverse resources available to meet our demand for energy. It is a form of renewable energy derived from recently living organic materials known as biomass, which can be used to produce transportation fuels, heat, electricity, and products [12].

Bioenergy has attracted great interest owing to rapidly increasing fuel prices, alterations in the global climate, the fast depletion of fossil fuels, and environmental degradation by fossil fuels. The manufacturing of another products for international marketing in substantial volumes of the feedstock material refers as a base of the product from food and feed crops, such as wastes and agricultural residues (fruits and vegetable waste, forestry, fishing wheat straw, rice straw, municipal waste), non-food crops (weeds, litter, miscanthus and short rotation coppice) and algae etc. Their key inputs serve as biomass feedstock, the production of bioenergy. Biomass feedstock is available from various sources such as agricultural or energy crops and waste fuel. This feedstock is sourced from organic matter [18].

1.2 Biomass sources

Biomass sources are renewable sources that can be used to generate biomass in their native or modified form. Biomass is versatile and can be converted into liquid transportation fuels equivalent to fossil-based fuels, such as gasoline, jet, and diesel fuel. Bioenergy technologies enable the reuse of carbon from biomass and waste streams into reduced-emissions fuels for cars, trucks, jets, and ships; bioproducts; and renewable power [12]. Renewable biomass resources that are available and used directly as a fuel or converted to other forms or energy products are commonly referred to as feedstock. This feedstock includes energy crops, agricultural crop residue, forestry residue, algae crops, and municipal solid waste.

(1) Energy crops: Energy crops include bamboo, switchgrass, wheatgrass, hybrid willow, hybrid poplar, and sycamore. These are nonfood crops. A single bamboo pole can be used to power a rural household for a month. Energy crops are divided into herbaceous and woody types [10]. Energy crops include some plant biomass, which is planted exclusively for the use of their biomass as raw materials for bioenergy generation. Energy crops include maize or sweet sorghum, and herbaceous and woody plants such as miscanthus (*Miscanthus giganteus*), switchgrass (*Panicum virgatum*), kenaf (*Hibiscus cannabinus*), reed canary grass (*Phalaris arundinacea*), giant reed (*Arundo donax*), and eucalyptus

(*Eucalyptus globulus*) are regarded as appropriate plants for the formation of renewable bioenergy such as biohydrogen. Mixed or single microbial culture can be employed to transform energy crops into biohydrogen. For the production of biohydrogen from energy crops, batch mode bioreactors have been widely used; however, continuous-mode biohydrogen production is also used extensively. Utilization of microorganisms, microalgae, plant biomass is frequently utilized as energy crops for biohydrogen production. These are nonfood crops that can grow on marginal land specifically to provide biomass. They are of two categories: herbaceous and woody. Herbaceous crops are perennial and are harvested annually after 2–3 years, when they reach full productive level. They include switchgrass, bamboo, sweet sorghum, miscanthus, wheatgrass, and kochia. Short rotation woody crops are harvested 5–8 years after planting. These are fast-growing hardwood trees. They include silver maple, eastern cottonwood, black walnut, green ash, sweet gum, sycamore, hybrid poplar, and hybrid willow. These plant species help to improve water (capillary water, electrical conductivity), and soil quality (organic matter, microbial load, moisture, water holding capacity), or wildlife habitat relates to annual crops, diversify sources of income and improve overall farm productivity.

(2) Agricultural crop residue: Throughout the world, there is abundant agricultural waste that can be used for biomass generation. Such waste includes stalks, leaves, husks, cobs, wheat straw, oat straw, barley straw, and rice straw. Farmers can sell this waste for biomass generation, and it will aid farmers in obtaining an additional income. This brings opportunities for agricultural resources on existing land without interfering with the production of feed, food, forest products, and fiber. The stalks and leaves are abundant, diverse, and widely distributed across the United States in agricultural crop residues. Examples include wheat straw, corn straw, and corn stover (stalks, leaves, husks, and cobs). Residue sold to a local biorefinery represents an opportunity for farmers to generate additional income.

(3) Forestry residue: Forest-related biomass sources include dead, poorly formed, unmerchantable or whole trees specifically planted for biomass generation. This biomass could be harvested without having a negative impact on the environment [7].

Forest biomass feedstock falls into one of two categories: forest residue left after logging timber (including limbs, tops, and culled trees and tree components that would be otherwise unmerchantable) and whole-tree biomass harvested explicitly for biomass. Dead, diseased, poorly formed, and other unmerchantable trees are often left in the woods after timber harvest. This woody debris can be collected for use in bioenergy while leaving enough behind to provide habitats and maintain proper nutrients and hydrologic features. There are also opportunities to use excess biomass in millions of acres of forests. Harvesting excessive woody biomass can reduce the risk for fire and pests as well as aid in forest restoration, productivity, vitality, and resilience. This biomass could be harvested for bioenergy without negatively affecting the health and stability of forest ecological structures and functions.

(4) Algae: Algae such as microalgae, microalgae, and cyanobacteria can use sunlight and nutrients to generate biomass that can be used to produce biofuels. They can be grown in wastewater such as agricultural and municipal wastewater.

(5) Sorted municipal waste: This includes yard clippings, vegetable waste, leather, rubber, and textile waste [4]. Resources in municipal solid waste include residential and commercial garbage such as paper and paperboard, rubber, textiles, and food waste. Bioenergy municipal solid waste represents an opportunity to reduce commercial and residential waste by diverting volume from landfills

1.2.1 Biomass feedstock

Every region has its own locally generated biomass feedstock from agricultural, forest, and urban sources [16].

A wide variety of biomass feedstock is available and biomass can be produced anywhere plants or animals can live. Furthermore, most feedstock can be made into liquid fuel, heat, electric power, and/or biobased products. This makes biomass a flexible and widespread resource that can be adapted to meet local needs and objectives.

Biomass includes terrestrial lignocelluloses (i.e., plants and plant-based materials not used for food or feed), which is a renewable, abundant, and sustainable resource for producing biofuels, bioproducts, and biopower. Biomass and other renewable or reusable carbon sources commonly used for bioenergy applications.

Some of the most common (and most promising) biomass feedstocks are:

- **Grains and starch crops** – sugarcane, corn, wheat, sugar beets, industrial sweet potatoes, and so on.
- **Agricultural residue** – Corn stover, wheat straw, rice straw, orchard prunings, and so on.
- **Food waste** – Waste produce, food processing waste, and so on.
- **Forestry materials** – Logging residues, forest thinnings, and so on.
- **Animal by-products** – Tallow, fish oil, manure, and so on.
- **Energy crops** – Switchgrass, miscanthus, hybrid poplar, willow, algae, and so on.
- **Urban and suburban waste** – municipal solid waste, lawn waste, wastewater treatment sludge, urban wood waste, disaster debris, trap grease, yellow grease, waste cooking oil, and so on (Fig. 32.1).

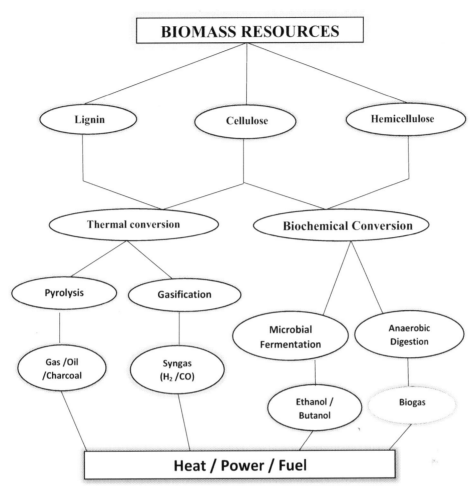

FIGURE 32.1 Different types of biomass resources.

1.3 Bioenergy production techniques

Bioenergy is energy produced from biomass or organic matter. There are many techniques for bioenergy production. They can be selected based on the biomass material used and the type of bioenergy one wants to produce. Some techniques are simple; others are relatively complex and may require specific environmental conditions and processing procedures. A number of conversion procedures can be used to convert biomass into bioenergy such as biofuel, biopower, bioelectricity, heat, biofuel, and other bioproducts.

2. Biomass conversion techniques to produce electricity and heat

- Combustion techniques

  Characteristics of combustion and emission of different flame types, including nonpremixed/premixed, moderate or intense low-oxygen dilution flameless combustion flames; colorless distributed combustion flames, and low-swirl injector combustion flames are presented with their limitations for applications [9]:

- Direct combustion: This is one of the simplest techniques for generating heat. Heat generated by this process can be used to heat water or space or to produce electricity. The electrical efficiency of these systems varies. It ranges from 25% to 30% for combustion, whereas it could be around 85% for cogeneration techniques.

  The two main combustion techniques are:

(a) Fixed bed combustion: Combusting materials are placed on a fixed or moving grate with air passing through it. It is also called the combustion of coal.

(b) Fluidized bed combustion: In this technique, biomass materials are mixed with sand, which promotes the even burning of materials. Rapid mixing ensures the uniformity of temperature.
- Cofiring: This involves burning biomass materials with another base fuel such as coal or liquid petroleum gas. This technique can reduce greenhouse emissions. It does not involve major processing procedures, and the process is simple.
- Cogeneration: This is known as combined heat and power. The special nature of this method is that it traps waste heat from electricity generation procedures. This technique is used where both heating and cooling are required
- Combined cycle electricity generation: In this method, exhaust gases are trapped and used to heat water and produce steam that can be used to drive a steam turbine to generate more electricity.

2.1 Gasification techniques

Gasification is a thermochemical method that entails heating strong biomass to 800–1000°C in a gasifier with a constrained delivery of oxygen. Under those conditions, fuel is partially burned and basically transformed to syngas, which includes a combination of methane, hydrogen, carbon monoxide, carbon dioxide, and nitrogen. Small quantities of char are produced through gasification. Syngas may be used at once for warmth or electricity applications, such as to run gas engines, gas mills, or combined cycle power systems. It can also be upgraded for biofuel manufacturing through some current and rising technologies. Gasification is usually better than combustion in terms of energy generation.

Plasma gasification is a technique that uses an arc gasifier to create excessive temperatures to break down biomass and inorganic remains into syngas to generate electric energy. Plasma gasification is particularly deployed as a waste treatment for inorganic waste streams, which include domestic waste.

2.2 Pyrolysis

Like gasification, this technique involves heating the biomass at very high temperatures in the absence of air or oxygen or a very little amount. Depending on the temperature and speed of the process, the ratio of by-products varies. These by-products can be solids, liquids, or gases, which can be used to generate bioenergy. Types of pyrolysis include:

- Slow pyrolysis: This involves heating the biomass to around 500°C. It is the slow heating of organic material in the absence of oxygen.
- Fast pyrolysis: This is done at much higher temperatures. It is an efficient process and gives a larger yield of bio-oil, in which biomass is rapidly heated to a high temperature in the absence of air.

2.3 Fermentation

In the context of industrial biotechnology, the term fermentation refers to the growth of large quantities of microbes under anaerobic or aerobic conditions within a container, called a fermenter or bioreactor. Fermented products have various applications in the food, textile, and transportation industries. The basic function of a fermenter is to provide an environment suitable for the controlled growth of a pure culture or a defined mixture of organisms. The main process steps for typical fermentation involve media preparation, sterilization, inoculum development, fermentation or production, and downstream processing for product purification.

Fermentation is basically the metabolism of microorganisms on carbon as a substrate. According to the requirements of air and oxygen, some types of fermentation are classified as aerobic and some are anaerobic. We will briefly discuss these two processes.

2.3.1 Anaerobic digestion

Anaerobic digestion breaks down biomass in the absence of oxygen. It occurs naturally in landfills and anaerobic lagoons built for the biodigestion of waste.

It is particularly used for biomass materials that lack lignin.

Some commonly used materials for anaerobic digestion include sewage, wet agricultural residues, and straw.

Anaerobic digestion of biomass results in the production of a mixture of gases such as methane, carbon dioxide, and hydrogen sulfide, called biogas. Biogas can be used to generate heat; it may also be used to drive a gas turbine to generate electricity. It may be used as fuel in vehicles to facilitate transportation.

The undigested sludge obtained from biodigesters can be dehydrated and burned to produce even more bioenergy.

2.4 Bioethanol production from biomass created by agricultural waste

Raw material — wheat straw, rice straw, corn straw, bagasse.

Rice straw is the most abundant of these raw materials. It has the potential to produce billions of liters of bioethanol per year. These materials are rich in lignocellulose, is a complex carbohydrate polymer of cellulose, hemicellulose, and lignin.

There are three processes for bioethanol production:

(1) **Pretreatment**: This technique is used to free cellulose and hemicellulose from lignocellulose.
(2) **Hydrolysis** of cellulose and hemicellulose so that fermentable sugars are produced, such as glucose, xylose, arabinose, galactose, and mannose.
(3) **Fermentation** of reducing sugars.

2.4.1 Pretreatment

2.4.1.1 Physical pretreatment

This is a technique in which raw material is treated physically.

(A) Mechanical size reduction: This includes mechanical size reduction through comminution, milling, grinding, or chipping.

This kind of pretreatment can enhance the quality of downstream processing. The method employed depends on the type of raw material used.

(B) Pyrolysis: In this method, materials are treated at 300°C and cellulose decomposes to produce gaseous products such as H_2, CO_2, and residual char. Residual char is further subjected to leaching with water or mild acid. This water contains enough carbon source to support microbial growth for bioethanol.
(C) Steam explosion: This makes biomass more accessible to cellulose attack. It involves heating using high-pressure steam for a few minutes. The reaction is stopped by sudden decompression to atmospheric pressure. The purpose of this reaction is to separate individual fibers.

2.4.1.2 Chemical pretreatment

(A) Acid treatment — Acid treatment has the advantage of giving high yields of sugars from lignocellulose. It is usually carried out by treatment with concentrated or diluted acids.
(B) Alkaline treatment — It digests the lignin matrix and makes cellulose and hemicellulose available for enzymatic degradation. The end residue can be used to produce paper or cellulose derivatives.
(C) Wet oxidation — The feedstock material is treated with water and by air or oxygen above 120°C. The product of hemicellulose hydrolysis is sugar polymers.

2.4.1.3 Biological pretreatment

The lignocellulose complex to liberate cellulose can be degraded with brown rot, white rot, and soft rot fungi. The aim of this process is to release sugars from the lignocellulose matrix [1,5].

3. Enzymatic hydrolysis

In this step for bioethanol production, complex carbohydrates are converted to simple monomers. This step is known as saccharification.

The advantage of this technique is that it requires less energy and mild environmental conditions [6].

Cellulose is converted to glucose whereas hemicellulose is converted to pentoses and hexoses by the hydrolytic action of enzymes [3,15]. Bacterial species of *Clostridium*, *Cellulomonas*, *Thermolospora*, *Bacillus*, and so on produce cellulase enzymes. Moreover, fungi species such as *Trichoderma*, *Penicillium*, and *Fusarium* have been found to produce cellulase enzyme [11,14,17]. Enzymes such as endo-1,4-β-xylanase and endo-1,4-β-mannanase have been known to break hemicellulose [8].

3.1 Fermentation

The biomass is subjected to fermentation through several microbes. The process usually employed in the fermentation of lignocellulosic hydrolysate is simultaneous saccharification and fermentation and separate hydrolysis and fermentation [4,7,10]. Fermentation is a food processing technology. It takes place through the lack of oxygen that produces adenosine triphosphate (ATP) (energy) (Fig. 32.2).

Microorganisms used in the fermentation are *S. cerevisiae*, *Escherichia coli*, *Zymomonas mobilis*, *Pachysolen tannophilus*, and so on [2,4,13,19].

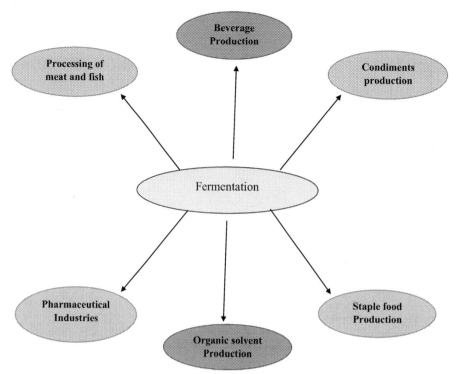

FIGURE 32.2 Applications of fermentation.

Principles of fermentation:

- In the absence of oxygen, fermentation derives energy from carbohydrates.
- Through glycolysis, glucose is first partially oxidized to pyruvate.
- Next, pyruvate is converted into alcohol or acid along with the regeneration of NAD^+, which takes part in glycolysis to produce more ATP.
- Fermentation yields about 5% energy obtained by aerobic respiration.

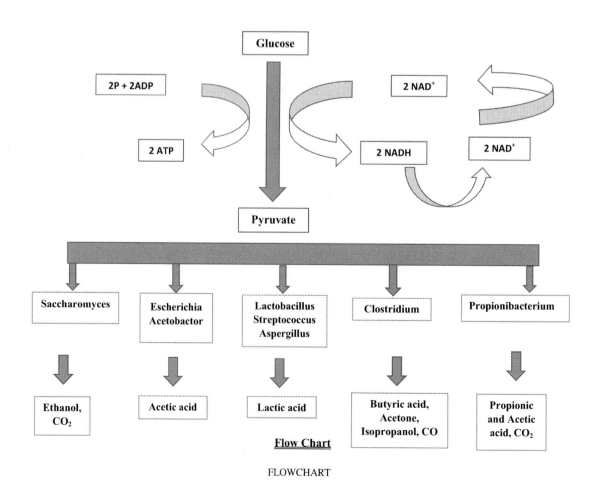

Flow Chart

FLOWCHART

Flowchart: Pathways represent the production of some fermentation end products from glucose by various organisms (Fig. 32.2).

- Fermentation is an anaerobic biochemical process used to produce energy from the partial oxidation of glucose or other carbon sources.
- Oxidation of the substrate, which occurs through the Embden–Meyerhof or Entner–Doudoroff pathway, results in the production of pyruvate, ATP, and NAD (P) H.
- In the absence of external electron acceptors, pyruvate undergoes reduction with the regeneration of $NAD^+(P)$.
- This step is essential for fermentation to progress and leads to the production of products (ethanol and organic acids).
- ATP is the main product of fermentation. It is generated by phosphorylation at the substrate level.
- NADH is then reoxidized, reverting to NAD^+ in the second phase of fermentation, which reduces pyruvate to a fermentation product such as ethanol and lactate.
- For example, in the fermentation of glucose by *Streptococcus lactis*, pyruvate is converted to lactic acid to reform NAD^+ coenzymes so two ATP molecules are produced.
- In yeasts such as *Saccharomyces*, when pyruvate is converted to ethyl alcohol (ethanol), NAD^+ is reformed.

Types of fermentation:

1. Lactic acid homofermentation
   - Glucose → lactic acid
2. Lactic acid heterofermentation
   - Glucose → lactic acid + acetic acid + ethyl alcohol + $2CO_2$ + H_2O
3. Propionic acid fermentation
   - Glucose → lactic acid + propionic acid + acetic acid + CO_2 + H_2O
4. Diacetyl and 2,3-butylene glycol fermentation
5. Alcoholic fermentation
   - Glucose → ethyl alcohol

Applications in fermentation:
Applications in medicine:

- Production of antibiotics
- Production of insulin
- Production of growth hormones
- Production of vaccines
- Production of interferon

Applications in food industry:

- Production of fermented foods such as cheese, wine, beer, and bread for high-value products.
- Food-grade biopreservatives
- Functional food/neutraceuticals
- Production of single-cell protein

Other applications:

1. Reducing greenhouse gas emissions
   - Many studies have proved that using bioenergy results in much less CO_2 emissions compared with fossil fuels.
   - Using bioenergy can reduce greenhouse gas emissions. However, the reduction in greenhouse gas emissions can vary depending on the type of biomass used and the type and efficiency of conversion technologies employed. Bioenergy is far more effective in reducing greenhouse gas emissions compared with other renewable energy resources.
   - For example, instead of burning stubble or straw burned in fields, it can be harvested and burned in bioenergy plant with facilities of controlled emissions.
2. Generates heat and electricity
   - Biomass is a source able to generate both heat and electricity. A combined heat power plant can be used to form both heat and electricity.

3. Improved air quality
   - When biomass such as straw, tree prunings, forest waste, and so on is burned in bioenergy power plants that control emissions it will improve air quality. It is definitely an advantage over open burning.
4. Biofuels are biodegradable.
   - One of the most important advantages of biofuels is that they are biodegradable. Biofuels such as bioethanol and biodiesel are degradable and do not have toxic effects like those of petroleum-based fuels and petrochemicals.
5. Regional and rural economic development and employment opportunities
   - New jobs can be created relating to the growth or harvest of biomass, its transport to different bioenergy plants, and its handling and the procurement of infrastructure for bioenergy plants.
   - Farmers can sell agricultural waste to bioenergy plants, which will provide them with additional income.
6. Less landfill
   - Using waste material to create bioenergy can reduce the pressure of dumping this waste in landfills.
7. Production of biochar
   - Wood or agricultural waste can be converted into biochar, a stable form of charcoal, by subjecting it to low oxygen conditions. This technique is known as pyrolysis and gasification. This process results in the formation of bioenergy in the form of heat and bio-oil [20,21].

Limitation of fermentation:

(1) Low-scale production requires high cost and high energy.
(2) Possibility of contamination is higher.
(3) Over time, natural variations occur.
(4) Product needs further treatment that is impure.
(5) Unexpected or undesirable end products.

References

[1] V. Aristizábal-Marulanda, J.C. Solarte-Toro, C.A. Cardona Alzate, Study of biorefneries based on experimental data: production of bioethanol, biogas, syngas, and electricity using cofee-cut stems as raw material, Environ. Sci. Pollut. Res. 28 (2021) 24590–24604.

[2] M. Balat, H. Balat, C. Oz, Progress in bioethanol processing, Prog. Energy Combust. Sci. 34 (2008) 551–573.

[3] S. Banerjee, S. Mudliar, R. Sen, B. Giri, D. Satpute, T. Chakrabarti, et al., Commercializing lignocellulosic bioethanol: technology bottlenecks and possible remedies, Biofuels, Bioprod. Biorefining 4 (2010) 77–93.

[4] A.B. Bjerre, A.B. olesen, T. Fernqvist, Pretreatment of wheat straw using combined wet oxidation & alkaline hydrolysis resulting in convertible cellulose & hemicellulose, Biotechnol. Bioeng. 49 (1996) 586–677.

[5] J.E.G. Dam, P.F.H. Harmsen, Cofee Residues Utilization, Wageningen UR - Food & Biobased Research, 2010.

[6] S. Ferreira, A.P. Durate, M.H.L. Ribeiro, J.A. Queiroz, F.C. Domingues, Response surface optimization of enzymatic hydrolysis of Cistus ladanifer and Cytisus striatus for bioethanol production, Biochem. Eng. J. 45 (2009) 192–200.

[7] C.N. Hamelinck, G.V. Hooijdonk, A.P.C. Faaij, Ethanol from lignocellulosic biomass, techno- economic performance in short-,middle-,and long-term, Biomass Bioenergy 28 (2005) 384–410.

[8] H. Jorgensen, J.P. Kutter, L. Olsson, Separation and quantification of cellulases and hemicellulases by capillary electrophoresis, Anal. Biochem. 317 (1) (2003) 85–93.

[9] M. Nemitallah, S.S. Rashwan, I. Mansir, A. Abdelhafez, Review of novel combustion techniques for clean power production in gas turbines, Energy Fuel. 32 (2) (2018), https://doi.org/10.1021/acs.energyfuels.7b03607.

[10] M.A. Neves, T. Kimura, N. Shimizu, M. Nakajima, State of the Art and Future Trends of Bioethanol Production, Dynamic Biochemistry, Process Biotechnology & Molecular Biology, Global Science books, 2007, pp. 1–13.

[11] M.L. Rabinovich, M.S. Melnik, A.V. Boloboba, Microbial cellulases (review), Appl. Biochem. Microbiol. 38 (4) (2002) 305–321.

[12] J.N. Rogers, B. Stokes, J. Dunn, H. Cai, M. Wu, Z. Haq, H. Baumes, An assessment of the potential products and economic and environmental impacts resulting from a billion ton bioeconomy, Biofuels, Bioproducts, and Biorefin. 11 (2016) 110–128, https://doi.org/10.1002/bbb.1728.

[13] Ó.J. Sanchez, C.A. Cardona, Trends in biotechnological production of fuel ethanol from different feedstocks, Bioresour. Technol. 99 (2008) 5270–5295.

[14] Y. Sun, J. Cheng, Hydrolysis of lignocellulosic material for ethanol production: a review, Bioresour. Technol. 96 (2002) 673–686.

[15] M.J. Taherzadeh, K. Karimi, Enzyme-based hydrolysis processes for ethanol from lignocellulosic materials: a review, Bioresources 2 (4) (2007) 707–738.

[16] W.Y. Wong, S. Lim, Y.L. Pang, S.H. Shuit, W.H. Chen, K.T. Lee, Synthesis of renewable heterogeneous acid catalyst from oil palm empty fruit bunch for glycerol-free biodiesel production, Sci. Total Environ. 727 (2020) 138534.

[17] J. Xu, N. Takakuwa, M. Nogawa, H. Okada, Y. Morikawa, A third xylanase from Trichoderma reesei PC-3-7, Appl. Microbiol. Biotechnol. 49 (1998) 18–724.

[18] J.S. Yuan, K.H. Tiller, H. Al-Ahmad, N.R. Stewart, C.N. Stewart Jr., Plants to power: bioenergy to fuel the future, Trends Plant Sci. 13 (8) (2008) 421−429.
[19] M. Shah (Ed.), Microbial Bioremediation & Biodegradation, Springer, Singapore, 2020.
[20] M. Shah (Ed.), Removal of Refractory Pollutants from Wastewater Treatment Plants, CRC Press, USA, 2021.
[21] M. Shah (Ed.), Removal of Emerging Contaminants through Microbial Processes, Springer, Singapore, 2021.

Further reading

[1] M. Admassie, A review on food fermentation and the biotechnology of lactic acid bacteria, World J. Food Sci. Technol. 2 (1) (2018) 19, https://doi.org/10.11648/j.wjfst.20180201.13.
[2] A.V. Bridgwater, Renewable fuels and chemicals by thermal processing of biomass, Chem. Eng. J. 91 (2003) 87−102.
[3] H. Chum, A.P.C. Faaij, J. Moreira, O. Edenhofer, Renewable energy sources and climate change mitigation, in: L.O. Girardin, M. Roman (Eds.), Special Report of the Intergovernmental Panel on Climate Change, Cambridge University Press, Cambridge, 2012, pp. 209−321.
[4] M. Ciani, F. Comitini, I. Mannazzu, Fermentation, Encyclop. Ecol. (2018) 310−321, https://doi.org/10.1016/B978-0-12-409548-9.00693-X. *June*.
[5] B. Ghosh, D. Bhattacharya, M. Mukhopadhyay, Use of fermentation technology for value-added industrial research, Princ. Applicat. Ferment. Technol. August (2018) 141−161, https://doi.org/10.1002/9781119460381.ch8.
[6] H.L. Hind, F.E. Day, Fermentation industries, J. Inst. Brew. 36 (6) (1930) 1−29, https://doi.org/10.1002/j.2050-0416.1930.tb05286.x.
[7] R. andine, C. De Garie, A. Cocci, Fermentation process, Biotechnol. Adv. 15 (3−4) (1997) 702, https://doi.org/10.1016/s0734-9750(97)87650-1.
[8] R.M. Martínez-Espinosa, Introductory chapter: a brief overview on fermentation and challenges for the Next future, New Adv. Fermentation Proc. (2020), https://doi.org/10.5772/INTECHOPEN.89418.
[9] F. Microbiology, Basic principles of food fermentation, Food Microbiol.: Princip. into Pract. (2016) 228−252, https://doi.org/10.1002/9781119237860.ch39.
[10] Principles and Applications of Fermentation Technology, in: Principles and Applications of Fermentation Technology, 2018, https://doi.org/10.1002/9781119460381.
[11] Principles of Fermentation Technology second ed.. (n.d.).
[12] R. Sharma, P. Garg, P. Kumar, S.K. Bhatia, S. Kulshrestha, Microbial fermentation and its role in quality improvement of fermented foods, Fermentation 6 (4) (2020) 1−20, https://doi.org/10.3390/fermentation6040106.
[13] F. Talebnia, D. Karakashev, I. Angelidaki, Production of bioethanol from wheat straw: an overview on pretreatment, hydrolysis and fermentation, Bioresour. Technol. 101 (13) (2010) 4744−4753.

Chapter 33

Bioprospecting microalgae for biofuel synthesis: a gateway to sustainable energy

Nahid Akhtar[1], Atif Khurshid Wani[1], Reena Singh[1], Chirag Chopra[1], Sikandar I. Mulla[2], Farooq Sher[3] and Juliana Heloisa Pinê Américo-Pinheiro[4,5]

[1]*Department of Biotechnology, School of Bioengineering and Biosciences, Lovely Professional University, Phagwara, Punjab, India;* [2]*Department of Biochemistry, School of Allied Health Sciences, REVA University, Bengaluru, Karnataka, India;* [3]*Department of Engineering, School of Science and Technology, Nottingham Trent University, Nottingham, United Kingdom;* [4]*Department of Forest Science, Soils and Environment, School of Agronomic Sciences, São Paulo State University (UNESP), Ave. Universitária, Botucatu, SP, Brazil;* [5]*Graduate Program in Environmental Sciences, Brazil University, São Paulo, SP, Brazil*

1. Introduction

It has been estimated that the global fossil fuel production will be on the rise to at least 2040 [1], and the consumption of the fossil fuels is also expected to increase, especially in developing economies such as India, China, and Indonesia [2]. The production and consumption of fossil fuels release harmful greenhouse gases that will cause global warming and will have adverse effects on human life and the environment [3]. The fossil fuel emissions are responsible for 65% additional mortality, and the phaseout of fossil fuel use can be helpful in avoiding the additional mortality of 3.6 million individuals every year globally [4]. Furthermore, it is also important to look for other alternatives to fossil fuels to cover present and future energy demands globally as the fossil fuels are nonrenewable and have a limited supply [5]. In this regard, the biofuels produced using microalgae, as feedstock, can be very helpful to tackle the environmental issues arising due to fossil fuel consumption and to maintain energy security. Microalgae are a group of metabolically, biochemically, and genetically diverse, unicellular photosynthetic microbes that can grow in both aquatic and terrestrial ecosystems [3,6]. They can grow in both freshwater and sea water as well as in wastewater, and they play an important role in oxygen production, energy flux, and nutrient cycling in different ecosystems [7]. A recent study suggests that there are more than 120,000 known species of microalgae, and more than one million species of microalgae are yet to be discovered [6]. Microalgae can be a beneficial source for the production of various types of biofuels because they can grow rapidly, adapt and grow in different environmental conditions, require simple nutrients for growth, have better growth and productivity in comparison to other plants and crops, and are easy and inexpensive to cultivate [3]. Moreover, they can grow in wastewater, seawater, and saline ground water, thus decreasing the intake of fresh water and making the biofuel production process more sustainable [8]. Furthermore, they can also act as substrate for different types of biofuels such as biodiesel, bioethanol, methane, and biohydrogen [3,9] and help in the bioremediation of hydrocarbons, dyes, heavy metals, and pesticides [10]. Recently, the feasibility of using microalgae as a source of aviation fuel was explored [11]. Along with the production of biofuels, the microalgae can also help in the carbon dioxide fixation in a more efficient manner than terrestrial plants [12]. Moreover, microalgae can be grown throughout the year and are not a major food source for humans and animals, and they do not compete with important food crops for land and resources as they do not require freshwater and cultivable land for their growth [9,13]. Microalgae are rich in lipids and different carbohydrates that can be converted to biodiesel and bioethanol [13]. The microalgae oil yield per hectare of cultivation is higher than various commercial oil plants such as soya bean and palm [8]. Due to these benefits, microalgae can be a sustainable, economical, and renewable

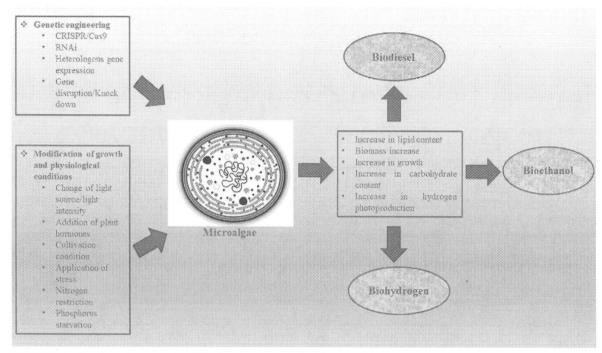

FIGURE 33.1 Genetic engineering and modification of growth and physiologic conditions to improve microalgae-derived biofuel production.

feedstock for biofuel production and a better alternative to the fossil fuels as they emit lesser amounts of harmful pollutants in the air in comparison to fossil fuels [14].

Although studies have shown that microalgae can be a potential source of biofuels, there are several limitations associated with their viable commercial use [14,15]. The biodiesel production using traditional methods of microalgae cultivation cannot be financially beneficial due to high costs associated with the microalgae biomass cultivation, harvesting, and drying [15]. Furthermore, the oil extraction from the microalgae biomass is also an expensive process [14]. Hence, to make the biofuel production more economically efficient, it is important to increase algal growth rate, enhance the microalgae biomass production, increase algal lipid content (for biodiesel production), increase carbohydrate content (for production of bioethanol), and develop viable technologies for the separation of algal biomass and oil extraction [8]. In the following sections, methods such as modification of growth and physiologic conditions and genetic engineering for improving algal biomass cultivation and increasing lipid content in the microalgae are discussed (Fig. 33.1).

2. Modification of growth and physiologic conditions for enhancing microalgae biomass and biofuel components

It is important to increase microalgae biomass concentration because low biomass production can reduce biocomponent content (lipid and carbohydrate) for biofuel production, increasing harvesting cost and utilization of water resources [16]. The growth and cellular content such as lipid and carbohydrates of microalgae are dependent on various factors such as carbon dioxide levels, pH, light intensity, and nutrient content of the algal growth medium [17]. Hence, the monitoring, control, and optimization of these parameters can be beneficial in improving microalgae growth and algal lipid content. The use of a combination of blue and red light emitting diodes (LEDs) at the intensity of 500 lux increased the biomass in *Chlorella* sp. [17]. Similarly, the combination of blue and red LEDs has also increased the biomass in four different microalgae: *Phaeodactylum tricornutum, Isochrysis galbana, Nannochloropsis salina,* and *Nannochloropsis oceanica* [18]. The use of organic dye such as rhodamine101, which can improve the efficiency of light utilization by converting useless light wavelength for algal cultivation to be useful, has also been reported to increase algal biomass [19]. The use of light sources having high red or high blue photon content have been reported to increase the biomass production of *Tetraselmis chuii* and *Nannochloropsis oculata* by 24% [20]. The cultivation conditions can also affect the biomass production of microalgae. The mixotrophic culture (energy source: light and organic carbon; carbon source: inorganic and organic carbon) and photoheterotrophic culture (energy source: light; carbon source: organic carbon) have been reported to increase biomass production in *Chlorella vulgaris* ESP-31 in comparison to the heterotrophic culture (organic compounds as source of carbon and energy) and photoautotrophic cultivation (energy source: light; carbon source: inorganic carbon) [21]. Mixotrophic culture has also been reported to increase biomass production in *C. vulgaris*, *C. protothecoides*, and *C. sorokiniana* in comparison to heterotrophic and photoautotrophic cultivation systems [22–24]. The supplementation of microalgae growth media with plant hormones and their analogs such as indoleacetic acid (IAA), diethyl aminoethyl hexanoate, 6-benzylaminopurine, and brassinosteroids has been reported to enhance the biomass production of various microalgae species such as *C. vulgaris*, *Scenedesmus obliquus*, *Chlorella pyrenoidosa*, *Chlamydomonas reinhardtii*, and *Ourococcus multisporus* [16].

Apart from modification of the growth conditions, the application of different stresses such as ultraviolet (UV) irradiation, gamma irradiation, and low-dose cold atmospheric pressure plasma (CAPP) have been reported to improve the growth and biomass production of different microalgae [25–27]. UV irradiation increased biomass productivity in *Scenedesmus* sp. [27]. However, the negative impacts of UV mutagenesis on microalgae have also been reported, which is discussed in the future prospects section of this chapter. Gamma irradiation (300–900 Gy of ^{137}Se) has increased the growth and biomass yield of microalgae belonging to *Chlorella* sp. [26]. Similarly, the gamma irradiation of microalgae belonging to the *Botryococcus* sp. by ^{60}Co at the dose of 10 Gy increased the biomass, but in *Arthrospira platensis* the gamma irradiation had a negative effect on the growth and biomass production [28,29]. In *Arthrospira platensis* the gamma irradiation (0.5–2.5 kGy) inhibited the growth, and the low irradiation doses did not decrease the biomass production significantly, but the high irradiation dose (2.0–2.5 kGy) significantly reduced the biomass production [28]. Oxidative stress caused by hydrogen peroxide has decreased biomass and growth of *Scenedesmus* sp. and *C. vulgaris* [27,30]. The pretreatment of *C. vulgaris* inoculum with low-dose CAPP for 30 s improved the growth and biomass productivity in comparison to the control microalgae [25]. The heavy metal stress caused by high concentration of cobalt nitrate (2 μM) and manganese chloride (12 μM) caused a minor increase in biomass productivity of 4% and 3% respectively in *C. vulgaris* [30].

3. Modification of growth and physiologic conditions for enhancing lipid content of microalgae

Depending upon the microalgae strains, the lipid content in the cells can vary in the range of 20%–50% of the algal dry biomass [31]. Biodiesel production from microalgae can be made more economically viable by enhancing the lipid content in the algal cells. The lipid content in microalgae can be increased by altering the carbon dioxide levels, optimizing the temperature, and controlling the light supply and nitrogen compound concentration in the microalgae growth medium [14,31]. The lipid content in various microalgae grown under different conditions is given in Table 33.1. The use of organic dye 9,10-diphenylanthracene in the photobioreactor for growing microalgae *C. vulgaris* has been reported to increase the lipid content accumulation by 30% in the microalgae in comparison to microalgae grown without the organic dye [19]. The use of red LED light at the intensity of 200Lux doubled the lipid content in *Chlorella* sp. in comparison to the cells grown in white light [17]. Similarly, the illumination with blue and red LEDs has also increased total lipid content in *C. vulgaris* and *S. obliquus* in comparison to cells illuminated with white, green, or yellow LED illumination [41]. The lipid content of the microalgae can also be affected by the culture conditions. The mixotrophic cultivation has enhanced lipid content more than photoautotrophic and heterotrophic culture in *C. vulgaris* and *C. sorokiniana* [21,23,24]. The increasing dose of NaCl (concentration up to 50 mM) has been reported to increase lipid content in plethora of microalgae such as *Chlorella* sp., *Chlamydomonas mexicana*, *S. obliquus*, and *Prymnesium saltans* growing in different synthetic media [16]. However, the increasing salinity has decreased lipid content in *Botryococcus* sp. and *Botryococcus braunii* [16,42]. Nitrogen restriction is one of the major strategies to increase lipid content in microalgae as it is able to upregulate the enzymes involved in lipid biosynthesis [43,44]. The inhibition of nitrogen assimilation to mimic nitrogen starvation condition by using chemicals like methionine sulfoximine has been able to increase lipid content in *Chlamydomonas reinhardtii* [45]. Furthermore, it has been reported that phosphorus can also play a significant role in lipid biosynthesis under nitrogen starvation conditions. The combined deficiency of both nitrogen and phosphorus has increased lipid content in *C. vulgaris* [37]. Besides the nitrogen and phosphorus stresses, other stresses are caused by heavy metals, UV irradiation, hydrogen peroxide, atmospheric and room temperature plasma (ARTP), and CAPP [25,27,30,46]. UV-C irradiation of a microalgae belonging to *Scenedesmus* sp. increased the lipid content from 40% to 55% of algal dry weight [27]. Similarly, the gamma irradiation at low doses has also been reported to increase lipid content in microalgae belonging to *Chlorella* sp., *Botryococcus* sp., and *A. platensis* [26,28,29]. The increasing concentration of magnesium chloride, ranging between 2 and 12 μM, significantly increased the lipid content in *C. vulgaris* [30]. Similarly, the heavy metal stress caused

TABLE 33.1 Lipid content in microalgae grown under different conditions.

Microalgae	Growth condition	Lipid content by dry weight (%)	References
C. vulgaris	Grown in municipal wastewater supplied with 10% CO_2	58.48%	[32]
Chlorella vulgaris CCALA 256 strain	Grown on sterilized nutrient medium 1/2 SŠ in a photobioreactor at 28°C supplied with 2% CO_2 and irradiance of 500 µmol m^{-2} s^{-1}	30%	[33]
Chlorella zofingiensis	Grown on BG11 medium in an airlift photobioreactor at 25°C with 1% CO_2 and irradiance of 150 µmol m^{-2} s^{-1}	24.5%	[34]
C. vulgaris NIES-227	Heterotrophic cultivation of microalgae supplemented with glucose at 24 ± 2°C under nitrogen starvation	89%	[35]
Spirulina platensis	Cultivated under nitrogen-limited condition in photobioreactor	29.51%	[36]
C. vulgaris	Microalgae were under nitrogen- and phosphorus-deficient conditions	54.88%	[37]
Chlorococcum sp.	Mixotrophic cultivation of microalga in medium containing waste unhydrolyzed molasses syrup, 2% nitrogen at 28 ± 2°C, with light intensity of 75.5 µmol m^{-2} s^{-1}	80.34%	[38]
Parachlorella kessleri	Grown in nutrient-depleted condition (diluted mineral medium) in photobioreactor supplied with 2% CO_2 and irradiance of 1200 µmol m^{-2} s^{-1} at 30°C	30%	[39]
Chlorella sp.	Cells treated with gamma irradiation were grown in f/2 seawater culture medium	54.9%	[26]
Scenedesmus sp.	Grown in modified BG11 medium at 25°C with light intensity of 55–60 µmol m^{-2} s^{-1} under phosphorus limitation	53%	[40]

by the addition of high concentration of cobalt nitrate caused a significant increase in lipid productivity relative to the control in *C. vulgaris* [30]. The oxidative stress caused by hydrogen peroxide has increased lipid content in *C. vulgaris* and *Scenedesmus* sp. [27,30]. The hydrogen peroxide–mediated oxidative stress can be combined with other stress to further increase the lipid content in microalgae [27,47]. Further increase of lipid content in UV-mutagenized *Scenedesmus* sp. has been reported after hydrogen peroxide –mediated oxidative stress [27]. Similarly, the combined effects of nitrogen depletion and hydrogen peroxide treatment enriched *Phaeodactylum tricornutum* with neutral lipids [47]. The pretreatment of *C. vulgaris* by CAPP for a short duration of time (30 and 60s) elevated the algal lipid content by 7.5% and 6.9% respectively [25]. ARTP-mediated mutagenesis increased the lipid productivity of *Chlorella pyrenoidosa* by 16.85% in comparison with the nonmutated strain [46].

4. Modification of growth and physiologic conditions for enhancing carbohydrate content of microalgae

Furthermore, it is important for microalgae to increase their carbohydrate content so that they can be converted to bioethanol. The restriction of nitrogen while cultivating microalgae can increase the carbohydrate accumulation in the microalgae [48]. The nitrogen restriction can lead to microalgae with carbohydrate content of 50%–55% of the dry cell weight [48]. Under nitrogen starvation condition, the starch content of *C. zofingiensis* increased from 7.6% to 43.4% within 24 h [34]. The blue LED illumination has also been reported to increase the total carbohydrate content in three microalgae: *C. vulgaris*, *S. obliquus*, and *A. platensis* [41]. Similarly, the use of LED lights having emission peaks at 465, 630, and 660 nm has increased the carbohydrate level by 29% in *T. chuii* and *N. oculata* in comparison with cells illuminated by white LEDs or LEDs with a peak at 405 nm [20]. The use of glucose as only carbon source during heterotrophic and photoheterotrophic conditions has been reported to increase carbohydrate content in *C. vulgaris* ESP-31 [21]. Growing cell under stress conditions such as phosphorus and nitrogen starvation and UV irradiation can also enhance the carbohydrate accumulation [13]. Furthermore, increasing light intensity and changing salinity and pH of growth media can also enhance carbohydrate content and productivity in microalgae [13,49]. The irradiation of microalgae with UV and gamma rays can also affect the carbohydrate content. The gamma irradiation (1 kGy) increased carbohydrate content by 248% in A. platensis, whereas UV irradiation has caused the decrease of carbohydrate content in C. sorokiniana UUIND6 [28,50].

5. Genetic engineering for enhancing microalgae biomass and biofuel components

Recently, the interest in genetically modifying microalgae to increase the biomass and the lipid content has increased [51–53]. The genetic modification of the microalgae can also help in improving the quality and decrease the cost of biofuel production [53]. Furthermore, the advanced genetic engineering tools and the current knowledge about microalgae growth and lipid metabolism can be combined for targeted metabolic engineering to increase microalgae biomass, growth, and lipid content [52]. The application of advanced gene editing tools such as TALEN (transcription activator-like effector nucleases) and CRISPR/Cas9 can further help in more efficient and effective microalgae genetic engineering [54]. Mostly, the genes associated with the lipid synthesis and lipid metabolism pathways of microalgae and the genes that control the type of fatty acid being synthesized are targeted for genetically engineering the microalgae for increasing the biofuel production [8]. Apart from the genes involved in lipid biosynthesis, the genes involved in the carbohydrate metabolic pathways such as pentose phosphate pathway can also be targeted for enhanced lipid production [55]. The genetic engineering tools can also be used to enhance the lipid production by blocking the metabolic and biosynthetic pathways that compete with lipid biogenesis pathways [56]. The disruption/knock down of the genes involved in the process of lipid catabolism using gene editing tools can also be useful in enhancing the microalgal lipid content [8]. The genetic engineering tools used for the genetic manipulation of various microalgae have been discussed by Kumar et al. [57]. In Table 33.2 various genetic manipulations in different microalgae to enhance the lipid content are listed. Regarding the enhancement of growth and biomass productivity in microalgae, different genes such as *RuBisCO, fructose–1,6-bisphosphatase*, and *sedoheptulose 1,7-bisphosphatase* and *fructose 1,6-bisphosphate aldolase* have been targeted [57]. The overexpression of these genes has shown to increase the growth and biomass productivity in different microalgae such as *Nannochloropsis oceanica* and *C. vulgaris* [57]. The production of hydrogen as biofuel from microalgae can also be improved by genetic engineering approaches. The genetic engineering tools can be used to target key enzymes necessary for the production of hydrogen by microalgae such as hydrogenase [71]. Moreover, genetic engineering can be used to elevate the activity of hydrogenase enzyme and increase the flux of electron toward hydrogen synthesis pathways by

TABLE 33.2 Genetic engineering is used to enhance lipid content in microalgae for biodiesel production.

Microalgae	Manipulated gene	Role of the gene	Lipid content	References
Tetraselmis sp.	CRISPR/Cas9-mediated knockout of ADP-glucose pyrophosphorylase	Regulation of starch biosynthesis	3.1-fold increase in lipid content in comparison to wild type	[58]
Phaeodactylum tricornutum	Overexpression of glucose-6-phosphate dehydrogenase	Involved in pentose phosphate pathway	55.7% of dry weight	[55]
Nannochloropsis salina	Disruption of trehalose-6-phosphate synthase	Synthesis of trehalose	34% more fatty acid methyl ester content than the wild type	[59]
P. tricornutum	Overexpression of plastidial pyruvate transporter	Regulate the influx of pyruvate into plastids	30% higher than the wild type	[54]
Scenedesmus quadricauda	Overexpression of acetyl-CoA carboxylase	Biosynthesis of fatty acids	1.6-fold increase in the total fatty acid content	[60]
Chlamydomonas reinhardtii	Heterologous expression of fatty acyl-ACP thioesterase gene from Dunaliella tertiolecta	Biosynthesis of fatty acids	56% increase in total lipids	[61]
C. reinhardtii	Silencing of cullin-RING E3 ubiquitin ligases by RNA interference (RNAi)	Part of ubiquitin proteasome pathway	28% increase of total lipid content	[62]

Continued

TABLE 33.2 Genetic engineering is used to enhance lipid content in microalgae for biodiesel production.—cont'd

Microalgae	Manipulated gene	Role of the gene	Lipid content	References
C. reinhardtii	Heterologous expression of glycerol-3-phosphate acyltransferase from Lobosphaera incisa	Involved in triacylglycerol synthesis pathway	50% increase in the content of triacylglycerol	[63]
N. salina	Heterologous expression of AP2 type transcription factor Wrinkled1 from Arabidopsis thaliana	Regulator of lipid biosynthesis in plants	36.5% increase of total lipid content in comparison to wild type	[64]
C. reinhardtii	Overexpression of lysophosphatidic acyltransferase (LPAAT) gene and glycerol-3-phosphate dehydrogenase (GPD) gene	Involved in triacylglycerol synthesis pathway	44.5% (LPAAT) and 67.5% (GPD) increase of lipid content	[65]
N. oceanica	Disruption of APETALA2-like transcription factor gene	Regulation of photosynthesis and lipid synthesis	40% increase of neutral lipid content	[66]
N. salina	RNAi-mediated knockdown of pyruvate dehydrogenase kinase	Regulation of fatty acid synthesis	86% increase of triacylglycerols	[67]
Neochloris oleoabundans	Overexpression of lysophosphatidic acid acyltransferase	Plastidial triacylglycerol biosynthesis	78.99% of dry cell weight (1.9-fold increase)	[68]
Synechocystis sp.	Recombinant overexpression of the acetyl-CoA carboxylase gene from Escherichia coli	Regulation of lipogenesis pathway	3.6-fold increase in lipid content	[69]
Chlorella sp. HS2	Overexpression of endogenous bZIP transcription factor	Transcriptional regulation of triacylglycerol synthesis	113% increase of fatty acid methyl ester in comparison to wild type	[70]

preventing the electron consumption by other metabolic pathways [71]. Jokel et al. knocked out the flavodiiron electron sink (competes for electron with hydrogen production) from the microalgae C. reinhardtii, and they reported an increase in hydrogen photoproduction in comparison to the wild-type microalgae [72].

6. Future prospects

There is a difference between the optimal culture conditions required for the higher biomass production and accumulation of biofuel components such as lipids and carbohydrates [73]. Thus, optimization of culture conditions for enhancing both the biomass and biofuel components accumulation has become a major limitation in the commercial microalgae-based biofuel production. This limitation can be overcome by designing two-phase systems, where in the first phase the division and growth of the microalgae can be enhanced by adding phytohormones or providing nutrient-rich conditions; and in the second phase, stress conditions such as increased salinity or nutrient starvation can be applied to increase the content of biofuel components such as lipids and carbohydrates [8]. This two-phase system for optimization of culture conditions for increased biomass production and lipid accumulation has already been studied in Dunaliella tertiolecta [73]. The use of 2,4-dichlorophenoxyacetic acid, an analog of the auxin IAA, in the first phase and increase of salinity in the later phase enhanced the biomass accumulation and lipid content in Dunaliella tertiolecta [73]. Another study, in which the microalgae Scenedesmus obliquus was grown in glucose-supplemented condition in the first phase and then subjected to phosphorus and nitrogen limited condition in second phase, showed 10-fold increase in lipid content [8,74]. In the future, the two-phase strategy of culture condition optimization can be explored in other microalgae species for sustainable, efficient, and economical microalgae-based biofuel production.

To make the microalgae-based biofuel production more economical, the cultivation of microalgae for biofuel production can be merged with the coproduction of value-added products. A plethora of value-added products such as dyes, vitamins, amino acids, omega fatty acids, antioxidants, antimicrobials, cosmetics, nutraceuticals, and various other pharmaceuticals can be obtained from microalgae [44,75–78].

UV mutagenesis has been explored for the purpose of microalgae strain improvement to increase the lipid content and biomass production. UV irradiation has been shown to enhance the biomass production and lipid content in *Scenedesmus* sp. in comparison to the wild-type microalgae [27]. However, the UV irradiation of another microalgae *C. sorokiniana* UUIND6 has been reported to affect the lipid content, growth, and biomass production negatively [50]. These contrasting effects of the UV mutagenesis can be due to the difference in the type of UV irradiation used. The mutagenesis of *Scenedesmus* sp. was caused by UV-C lamp, whereas the mutagenesis of *C. sorokiniana* UUIND6 was caused by UV-B (280–320 nm) [27,50]. Furthermore, the effects of UV irradiation on the growth and lipid content of microalgae may also vary from one species to another. Hence, in the future, using UV mutagenesis to improve a microalgae strain suitable for biodiesel production can be further explored. The effects of different ranges of UV light on different types of microalgae can be studied.

Most of the studies involving genetically engineered microalgae have been limited to the lab scale. Further scale up of these studies can be performed in the future along with the studies to monitor and assess the ecologic and environmental risks of the genetically engineered microalgae such as lateral transfer of genes to other microbes and genetic contamination of wild-type microalgae [8]. The application of advanced genetic engineering tools such as CRISPR/cas9, ZNF, and TALEN in microalgae is still underexplored. Hence, more studies are required to overcome the limitations like off-target mutations. Furthermore, there is paucity of researches related to genetic manipulation of the microalgae to enhance the carbohydrate content in microalgae. The researches in this direction will help in the development of novel microalgae strains that can be targeted for the production of bioethanol.

7. Conclusion

Due to ever-increasing energy consumption and demands for sustainable alternatives to conventional fuels, it is imperative to look for different energy sources that can meet these demands. Biofuels such as biodiesel, bioethanol, and hydrogen derived from microalgae could be an alternative to conventional energy sources in the long term. A plethora of studies conducted so far have shown that different nutritional and physiologic modifications along with genetic engineering can be used to improve the growth, biomass productivity, and lipid and carbohydrate contents in several species of microalgae, which will make the microalgae-based biofuel production process more efficient and economical. However, the development of various economical, ecofriendly, efficient, and sustainable processes are still needed to make microalgae-based biofuel business more feasible. Countries all over the world should focus on investing resources to improve the microalgae-based biofuels production process, which will make them self-sufficient and help to reduce the import of non-renewable fossil fuels.

References

[1] United Nations, Fossil Fuel Production 'dangerously Out of Sync' with Climate Change Targets, UN News, 2021. https://news.un.org/en/story/2021/10/1103472 (accessed December 31, 2021).

[2] R. Rapier, Fossil Fuels Still Supply 84 Percent of World Energy — and Other Eye Openers from BP's Annual Review, Forbes., 2020. https://www.forbes.com/sites/rrapier/2020/06/20/bp-review-new-highs-in-global-energy-consumption-and-carbon-emissions-in-2019/ (accessed December 31, 2021).

[3] T.M. Mata, A.A. Martins, N.S. Caetano, Microalgae for biodiesel production and other applications: a review, Renew. Sustain. Energy Rev. 14 (2010) 217–232, https://doi.org/10.1016/j.rser.2009.07.020.

[4] J. Lelieveld, K. Klingmüller, A. Pozzer, R.T. Burnett, A. Haines, V. Ramanathan, Effects of fossil fuel and total anthropogenic emission removal on public health and climate, Proc. Natl. Acad. Sci. U. S. A. 116 (2019) 7192–7197, https://doi.org/10.1073/pnas.1819989116.

[5] N. Akhtar, A. Karnwal, A.K. Upadhyay, S. Paul, M.A. Mannan, *Saccharomyces cerevisiae* bio-ethanol production, a sustainable energy alternative, Asian J. Microbiol. Biotechnol. Environ. Sci. 20 (2018) S202–S206.

[6] S.R. Manning, Microalgal lipids: biochemistry and biotechnology, Curr. Opin. Biotechnol. 74 (2022) 1–7, https://doi.org/10.1016/j.copbio.2021.10.018.

[7] A.M. dos Santos, K.R. Vieira, L.Q. Zepka, E. Jacob-Lopes, Environmental applications of microalgae/cyanobacteria, in: J.S. Singh (Ed.), New Future Dev. Microb. Biotechnol. Bioeng, Elsevier, 2019, pp. 47–62, https://doi.org/10.1016/B978-0-12-818258-1.00003-0.

[8] W.-L. Chu, Strategies to enhance production of microalgal biomass and lipids for biofuel feedstock, Eur. J. Phycol. 52 (2017) 419–437, https://doi.org/10.1080/09670262.2017.1379100.

[9] A. Tiwari, T. Kiran, Biofuels from Microalgae, IntechOpen, 2018, https://doi.org/10.5772/intechopen.73012.

[10] B. Koul, K. Sharma, M.P. Shah, Phycoremediation: a sustainable alternative in wastewater treatment (WWT) regime, Environ. Technol. Innovat. 25 (2022) 102040, https://doi.org/10.1016/j.eti.2021.102040.

[11] M. Mofijur, S.M. Ashrafur Rahman, L.N. Nguyen, T.M.I. Mahlia, L.D. Nghiem, Selection of microalgae strains for sustainable production of aviation biofuel, Bioresour. Technol. 345 (2021) 126408, https://doi.org/10.1016/j.biortech.2021.126408.

[12] S.S.M. Mostafa, Microalgal Biotechnology: Prospects and Applications, IntechOpen, 2012, https://doi.org/10.5772/53694.

[13] M.I. Khan, J.H. Shin, J.D. Kim, The promising future of microalgae: current status, challenges, and optimization of a sustainable and renewable industry for biofuels, feed, and other products, Microb. Cell Factor. 17 (2018) 36, https://doi.org/10.1186/s12934-018-0879-x.

[14] M. Debowski, M. Zieliński, J. Kazimierowicz, N. Kujawska, S. Talbierz, Microalgae cultivation technologies as an opportunity for bioenergetic system development—advantages and limitations, Sustainability 12 (2020) 9980, https://doi.org/10.3390/su12239980.

[15] L. Lardon, A. Hélias, B. Sialve, J.-P. Steyer, O. Bernard, Life-cycle assessment of biodiesel production from microalgae, Environ. Sci. Technol. 43 (2009) 6475–6481, https://doi.org/10.1021/es900705j.

[16] E.-S. Salama, J.-H. Hwang, M.M. El-Dalatony, M.B. Kurade, A.N. Kabra, R.A.I. Abou-Shanab, K.-H. Kim, I.-S. Yang, S.P. Govindwar, S. Kim, B.-H. Jeon, Enhancement of microalgal growth and biocomponent-based transformations for improved biofuel recovery: a review, Bioresour. Technol. 258 (2018) 365–375, https://doi.org/10.1016/j.biortech.2018.02.006.

[17] A. Severes, S. Hegde, L. D'Souza, S. Hegde, Use of light emitting diodes (LEDs) for enhanced lipid production in micro-algae based biofuels, J. Photochem. Photobiol., B 170 (2017) 235–240, https://doi.org/10.1016/j.jphotobiol.2017.04.023.

[18] C.H. Ra, P. Sirisuk, J.-H. Jung, G.-T. Jeong, S.-K. Kim, Effects of light-emitting diode (LED) with a mixture of wavelengths on the growth and lipid content of microalgae, Bioproc. Biosyst. Eng. 41 (2018) 457–465, https://doi.org/10.1007/s00449-017-1880-1.

[19] Y.H. Seo, Y. Lee, D.Y. Jeon, J.-I. Han, Enhancing the light utilization efficiency of microalgae using organic dyes, Bioresour. Technol. 181 (2015) 355–359, https://doi.org/10.1016/j.biortech.2015.01.031.

[20] P.S.C. Schulze, H.G.C. Pereira, T.F.C. Santos, L. Schueler, R. Guerra, L.A. Barreira, J.A. Perales, J.C.S. Varela, Effect of light quality supplied by light emitting diodes (LEDs) on growth and biochemical profiles of Nannochloropsis oculata and Tetraselmis chuii, Algal Res. 16 (2016) 387–398, https://doi.org/10.1016/j.algal.2016.03.034.

[21] K.-L. Yeh, J.-S. Chang, Effects of cultivation conditions and media composition on cell growth and lipid productivity of indigenous microalga Chlorella vulgaris ESP-31, Bioresour. Technol. 105 (2012) 120–127, https://doi.org/10.1016/j.biortech.2011.11.103.

[22] M.P. Caporgno, I. Haberkorn, L. Böcker, A. Mathys, Cultivation of Chlorella protothecoides under different growth modes and its utilisation in oil/water emulsions, Bioresour. Technol. 288 (2019) 121476, https://doi.org/10.1016/j.biortech.2019.121476.

[23] T. Li, Y. Zheng, L. Yu, S. Chen, Mixotrophic cultivation of a Chlorella sorokiniana strain for enhanced biomass and lipid production, Biomass Bioenergy 66 (2014) 204–213, https://doi.org/10.1016/j.biombioe.2014.04.010.

[24] X.-F. Shen, Q.-W. Qin, S.-K. Yan, J.-L. Huang, K. Liu, S.-B. Zhou, Biodiesel production from Chlorella vulgaris under nitrogen starvation in autotrophic, heterotrophic, and mixotrophic cultures, J. Appl. Phycol. 31 (2019) 1589–1596, https://doi.org/10.1007/s10811-019-01765-1.

[25] J.Q.M. Almarashi, S.E. El-Zohary, M.A. Ellabban, A.E.-F. Abomohra, Enhancement of lipid production and energy recovery from the green microalga Chlorella vulgaris by inoculum pretreatment with low-dose cold atmospheric pressure plasma (CAPP), Energy Convers. Manag. 204 (2020) 112314, https://doi.org/10.1016/j.enconman.2019.112314.

[26] J. Cheng, H. Lu, Y. Huang, K. Li, R. Huang, J. Zhou, K. Cen, Enhancing growth rate and lipid yield of Chlorella with nuclear irradiation under high salt and CO2 stress, Bioresour. Technol. 203 (2016) 220–227, https://doi.org/10.1016/j.biortech.2015.12.032.

[27] R. Sivaramakrishnan, A. Incharoensakdi, Enhancement of lipid production in Scenedesmus sp. by UV mutagenesis and hydrogen peroxide treatment, Bioresour. Technol. 235 (2017) 366–370, https://doi.org/10.1016/j.biortech.2017.03.102.

[28] A.E.-F. Abomohra, W. El-Shouny, M. Sharaf, M. Abo-Eleneen, Effect of gamma radiation on growth and metabolic activities of Arthrospira platensis, Braz. Arch. Biol. Technol. 59 (2016) e16150476, https://doi.org/10.1590/1678-4324-2016150476.

[29] D. Ermavitalini, N. Yuliansari, E.N. Prasetyo, T.B. Saputro, Efect of gamma 60Co irradiation on the growth, lipid content and fatty acid composition of Botryococcus sp. microalgae, Biosaintifika J. Biol. Biol. Educ. 9 (2017) 58–65, https://doi.org/10.15294/biosaintifika.v9i1.6783.

[30] M. Battah, Y. El-Ayoty, A.E.-F. Abomohra, S.A. El-Ghany, A. Esmael, Effect of Mn^{2+}, Co^{2+} and H_2O_2 on biomass and lipids of the green microalga Chlorella vulgaris as a potential candidate for biodiesel production, Ann. Microbiol. 65 (2015) 155–162, https://doi.org/10.1007/s13213-014-0846-7.

[31] R.A. Patil, S.B. Kausley, S.M. Joshi, A.B. Pandit, Process intensification applied to microalgae-based processes and products, in: E. Jacob-Lopes, M.M. Maronexe, M.I. Queiroz, L.Q. Zepka (Eds.), Handb. Microalgae-Based Process. Prod, Academic Press, 2020, pp. 737–769, https://doi.org/10.1016/B978-0-12-818536-0.00027-0.

[32] X. Hu, J. Zhou, G. Liu, B. Gui, Selection of microalgae for high CO2 fixation efficiency and lipid accumulation from ten Chlorella strains using municipal wastewater, J. Environ. Sci. China. 46 (2016) 83–91, https://doi.org/10.1016/j.jes.2015.08.030.

[33] P. Přibyl, V. Cepák, V. Zachleder, Production of lipids in 10 strains of Chlorella and Parachlorella, and enhanced lipid productivity in Chlorella vulgaris, Appl. Microbiol. Biotechnol. 94 (2012) 549–561, https://doi.org/10.1007/s00253-012-3915-5.

[34] S. Zhu, W. Huang, J. Xu, Z. Wang, J. Xu, Z. Yuan, Metabolic changes of starch and lipid triggered by nitrogen starvation in the microalga Chlorella zofingiensis, Bioresour. Technol. 152 (2014) 292–298, https://doi.org/10.1016/j.biortech.2013.10.092.

[35] X.-F. Shen, F.-F. Chu, P.K.S. Lam, R.J. Zeng, Biosynthesis of high yield fatty acids from Chlorella vulgaris NIES-227 under nitrogen starvation stress during heterotrophic cultivation, Water Res. 81 (2015) 294–300, https://doi.org/10.1016/j.watres.2015.06.003.

[36] H.H.A. El Baky, G.S. El Baroty, E.M. Mostafa, Optimization growth of spirulina (Arthrospira) platensis in photobioreactor under varied nitrogen concentration for maximized biomass, carotenoids and lipid contents, recent pat, Food Nutr. Agric. 11 (2020) 40–48, https://doi.org/10.2174/2212798410666181227125229.

[37] F.-F. Chu, P.-N. Chu, P.-J. Cai, W.-W. Li, P.K.S. Lam, R.J. Zeng, Phosphorus plays an important role in enhancing biodiesel productivity of Chlorella vulgaris under nitrogen deficiency, Bioresour. Technol. 134 (2013) 341–346, https://doi.org/10.1016/j.biortech.2013.01.131.

[38] A. Khanra, S. Vasistha, S. Kumar, M.P. Rai, Cultivation of microalgae on unhydrolysed waste molasses syrup using mass cultivation strategy for improved biodiesel, 3 Biotech 11 (2021) 287, https://doi.org/10.1007/s13205-021-02823-7.

[39] B. Fernandes, J. Teixeira, G. Dragone, A.A. Vicente, S. Kawano, K. Bišová, P. Přibyl, V. Zachleder, M. Vítová, Relationship between starch and lipid accumulation induced by nutrient depletion and replenishment in the microalga Parachlorella kessleri, Bioresour. Technol. 144 (2013) 268–274, https://doi.org/10.1016/j.biortech.2013.06.096.

[40] L. Xin, H. Hu, G. Ke, Y. Sun, Effects of different nitrogen and phosphorus concentrations on the growth, nutrient uptake, and lipid accumulation of a freshwater microalga Scenedesmus sp, Bioresour. Technol. 101 (2010) 5494–5500, https://doi.org/10.1016/j.biortech.2010.02.016.

[41] R. Habibi, S. G, Light emitting diode (LED) illumination for enhanced growth and cellular composition in three microalgae, Adv. Microbiol. Res. 3 (2019) 1–6, https://doi.org/10.24966/AMR-694X/100007.

[42] C. Yeesang, B. Cheirsilp, Effect of nitrogen, salt, and iron content in the growth medium and light intensity on lipid production by microalgae isolated from freshwater sources in Thailand, Bioresour. Technol. 102 (2011) 3034–3040, https://doi.org/10.1016/j.biortech.2010.10.013.

[43] L. Fu, Q. Li, G. Yan, D. Zhou, J.C. Crittenden, Hormesis effects of phosphorus on the viability of Chlorella regularis cells under nitrogen limitation, Biotechnol. Biofuels 12 (2019) 121, https://doi.org/10.1186/s13068-019-1458-z.

[44] P.K. Sharma, M. Saharia, R. Srivstava, S. Kumar, L. Sahoo, Tailoring microalgae for efficient biofuel production, Front. Mar. Sci. 5 (2018) 382, https://doi.org/10.3389/fmars.2018.00382.

[45] M. Kamalanathan, R. Gleadow, J. Beardall, Use of a chemical inhibitor as an alternative approach to enhance lipid production in Chlamydomonas reinhardtii (Chlorophyceae), Phycologia 56 (2017) 159–166, https://doi.org/10.2216/16-49.1.

[46] S. Cao, X. Zhou, W. Jin, F. Wang, R. Tu, S. Han, H. Chen, C. Chen, G.-J. Xie, F. Ma, Improving of lipid productivity of the oleaginous microalgae Chlorella pyrenoidosa via atmospheric and room temperature plasma (ARTP), Bioresour. Technol. 244 (2017) 1400–1406, https://doi.org/10.1016/j.biortech.2017.05.039.

[47] A.R. Burch, A.K. Franz, Combined nitrogen limitation and hydrogen peroxide treatment enhances neutral lipid accumulation in the marine diatom Phaeodactylum tricornutum, Bioresour. Technol. 219 (2016) 559–565, https://doi.org/10.1016/j.biortech.2016.08.010.

[48] C.E. de F. Silva, A. Bertucco, Bioethanol from microalgal biomass: a promising approach in biorefinery, Braz. Arch. Biol. Technol. 62 (2019), https://doi.org/10.1590/1678-4324-2019160816.

[49] S.-H. Ho, C.-Y. Chen, J.-S. Chang, Effect of light intensity and nitrogen starvation on CO2 fixation and lipid/carbohydrate production of an indigenous microalga Scenedesmus obliquus CNW-N, Bioresour. Technol. 113 (2012) 244–252, https://doi.org/10.1016/j.biortech.2011.11.133.

[50] V. Kumar, M. Nanda, S. Kumar, P.K. Chauhan, The effects of ultraviolet radiation on growth, biomass, lipid accumulation and biodiesel properties of microalgae, Energy Sources Part Recovery Util. Environ. Eff. 40 (2018) 787–793, https://doi.org/10.1080/15567036.2018.1463310.

[51] W. Chungjatupornchai, S. Fa-Aroonsawat, Enhanced triacylglycerol production in oleaginous microalga Neochloris oleoabundans by co-overexpression of lipogenic genes: plastidial LPAAT1 and ER-located DGAT2, J. Biosci. Bioeng. 131 (2021) 124–130, https://doi.org/10.1016/j.jbiosc.2020.09.012.

[52] C.F. Muñoz, C. Südfeld, M.I.S. Naduthodi, R.A. Weusthuis, M.J. Barbosa, R.H. Wijffels, S. D'Adamo, Genetic engineering of microalgae for enhanced lipid production, Biotechnol. Adv. 52 (2021) 107836, https://doi.org/10.1016/j.biotechadv.2021.107836.

[53] H. Shokravi, Z. Shokravi, M. Heidarrezaei, H.C. Ong, S.S. Rahimian Koloor, M. Petrů, W.J. Lau, A.F. Ismail, Fourth generation biofuel from genetically modified algal biomass: challenges and future directions, Chemosphere 285 (2021) 131535, https://doi.org/10.1016/j.chemosphere.2021.131535.

[54] S. Seo, J. Kim, J.-W. Lee, O. Nam, K.S. Chang, E. Jin, Enhanced pyruvate metabolism in plastids by overexpression of putative plastidial pyruvate transporter in Phaeodactylum tricornutum, Biotechnol. Biofuels 13 (2020) 120, https://doi.org/10.1186/s13068-020-01760-6.

[55] J. Xue, S. Balamurugan, D.-W. Li, Y.-H. Liu, H. Zeng, L. Wang, W.-D. Yang, J.-S. Liu, H.-Y. Li, Glucose-6-phosphate dehydrogenase as a target for highly efficient fatty acid biosynthesis in microalgae by enhancing NADPH supply, Metab. Eng. 41 (2017) 212–221, https://doi.org/10.1016/j.ymben.2017.04.008.

[56] X.-M. Sun, L.-J. Ren, Q.-Y. Zhao, X.-J. Ji, H. Huang, Enhancement of lipid accumulation in microalgae by metabolic engineering, Biochim. Biophys. Acta Mol. Cell Biol. Lipids. 1864 (2019) 552–566, https://doi.org/10.1016/j.bbalip.2018.10.004.

[57] G. Kumar, A. Shekh, S. Jakhu, Y. Sharma, R. Kapoor, T.R. Sharma, Bioengineering of microalgae: recent advances, perspectives, and regulatory challenges for industrial application, Front. Bioeng. Biotechnol. 8 (2020) 914. https://www.frontiersin.org/article/10.3389/fbioe.2020.00914 (accessed January 24, 2022).

[58] K.S. Chang, J. Kim, H. Park, S.-J. Hong, C.-G. Lee, E. Jin, Enhanced lipid productivity in AGP knockout marine microalga Tetraselmis sp. using a DNA-free CRISPR-Cas9 RNP method, Bioresour. Technol. 303 (2020) 122932, https://doi.org/10.1016/j.biortech.2020.122932.

[59] A.J. Ryu, N.K. Kang, S. Jeon, D.H. Hur, E.M. Lee, D.Y. Lee, B.-R. Jeong, Y.K. Chang, K.J. Jeong, Development and characterization of a Nannochloropsis mutant with simultaneously enhanced growth and lipid production, Biotechnol. Biofuels 13 (2020) 38, https://doi.org/10.1186/s13068-020-01681-4.

[60] A.E. Gomma, S.-K. Lee, S.M. Sun, S.H. Yang, G. Chung, Improvement in oil production by increasing malonyl-CoA and glycerol-3-phosphate pools in Scenedesmus quadricauda, Indian J. Microbiol. 55 (2015) 447–455, https://doi.org/10.1007/s12088-015-0546-4.

[61] K.W.M. Tan, Y.K. Lee, Expression of the heterologous Dunaliella tertiolecta fatty acyl-ACP thioesterase leads to increased lipid production in Chlamydomonas reinhardtii, J. Biotechnol. 247 (2017) 60–67, https://doi.org/10.1016/j.jbiotec.2017.03.004.

[62] Q. Luo, X. Zou, C. Wang, Y. Li, Z. Hu, The roles of cullins E3 ubiquitin ligases in the lipid biosynthesis of the green microalgae Chlamydomonas reinhardtii, Int. J. Mol. Sci. 22 (2021) 4695, https://doi.org/10.3390/ijms22094695.

[63] U. Iskandarov, S. Sitnik, N. Shtaida, S. Didi-Cohen, S. Leu, I. Khozin-Goldberg, Z. Cohen, S. Boussiba, Cloning and characterization of a GPAT-like gene from the microalga Lobosphaera incisa (Trebouxiophyceae): overexpression in Chlamydomonas reinhardtii enhances TAG production, J. Appl. Phycol. 28 (2016) 907–919, https://doi.org/10.1007/s10811-015-0634-1.

[64] N.K. Kang, E.K. Kim, Y.U. Kim, B. Lee, W.-J. Jeong, B.-R. Jeong, Y.K. Chang, Increased lipid production by heterologous expression of AtWRI1 transcription factor in Nannochloropsis salina, Biotechnol. Biofuels 10 (2017) 231, https://doi.org/10.1186/s13068-017-0919-5.

[65] C. Wang, Y. Li, J. Lu, X. Deng, H. Li, Z. Hu, Effect of overexpression of LPAAT and GPD1 on lipid synthesis and composition in green microalga Chlamydomonas reinhardtii, J. Appl. Phycol. 30 (2018) 1711–1719, https://doi.org/10.1007/s10811-017-1349-2.

[66] C. Südfeld, M. Hubáček, D. Figueiredo, M.I.S. Naduthodi, J. van der Oost, R.H. Wijffels, M.J. Barbosa, S. D'Adamo, High-throughput insertional mutagenesis reveals novel targets for enhancing lipid accumulation in Nannochloropsis oceanica, Metab. Eng. 66 (2021) 239–258, https://doi.org/10.1016/j.ymben.2021.04.012.

[67] X. Ma, L. Yao, B. Yang, Y.K. Lee, F. Chen, J. Liu, RNAi-mediated silencing of a pyruvate dehydrogenase kinase enhances triacylglycerol biosynthesis in the oleaginous marine alga Nannochloropsis salina, Sci. Rep. 7 (2017) 11485, https://doi.org/10.1038/s41598-017-11932-4.

[68] W. Chungjatupornchai, K. Areerat, S. Fa-Aroonsawat, Increased triacylglycerol production in oleaginous microalga Neochloris oleoabundans by overexpression of plastidial lysophosphatidic acid acyltransferase, Microb. Cell Factories 18 (2019) 53, https://doi.org/10.1186/s12934-019-1104-2.

[69] W. Fathy, E. Essawy, E. Tawfik, M. Khedr, M.S. Abdelhameed, O. Hammouda, K. Elsayed, Recombinant overexpression of the *Escherichia coli* acetyl-CoA carboxylase gene in Synechocystis sp. boosts lipid production, J. Basic Microbiol. 61 (2021) 330–338, https://doi.org/10.1002/jobm.202000656.

[70] H. Lee, W.-S. Shin, Y.U. Kim, S. Jeon, M. Kim, N.K. Kang, Y.K. Chang, Enhancement of lipid production under heterotrophic conditions by overexpression of an endogenous bZIP transcription factor in Chlorella sp, HS2, J. Microbiol. Biotechnol. 30 (2020) 1597–1606, https://doi.org/10.4014/jmb.2005.05048.

[71] M.P. Naghshbandi, M. Tabatabaei, M. Aghbashlo, M.N. Aftab, I. Iqbal, Metabolic engineering of microalgae for biofuel production, Methods Mol. Biol. Clifton NJ 1980 (2020) 153–172, https://doi.org/10.1007/7651_2018_205.

[72] M. Jokel, V. Nagy, S.Z. Tóth, S. Kosourov, Y. Allahverdiyeva, Elimination of the flavodiiron electron sink facilitates long-term H2 photoproduction in green algae, Biotechnol, Biofuels 12 (2019) 280, https://doi.org/10.1186/s13068-019-1618-1.

[73] H. El Arroussi, R. Benhima, I. Bennis, N. El Mernissi, I. Wahby, Improvement of the potential of Dunaliella tertiolecta as a source of biodiesel by auxin treatment coupled to salt stress, Renew. Energy 77 (2015) 15–19, https://doi.org/10.1016/j.renene.2014.12.010.

[74] S. Mandal, N. Mallick, Microalga Scenedesmus obliquus as a potential source for biodiesel production, Appl. Microbiol. Biotechnol. 84 (2009) 281–291, https://doi.org/10.1007/s00253-009-1935-6.

[75] P. Shah Maulin, Microbial Bioremediation & Biodegradation, Springer, 2020.

[76] P. Shah Maulin, Removal of Refractory Pollutants from Wastewater Treatment Plants, CRC Press, 2021.

[77] A. Sarkar, N. Akhtar, M.A. Mannan, Antimicrobial property of cell wall lysed Chlorella, an edible alga, Res. J. Pharm. Technol. 14 (2021) 3695–3699, https://doi.org/10.52711/0974-360X.2021.00639.

[78] A.K. Wani, N. Akhtar, B. Datta, J. Pandey, M. Amin-ul Mannan, Cyanobacteria-derived small molecules: a new class of drugs, in: A. Kumar, J. Singh, J. Samuel (Eds.), Volatiles Metab. Microbes, Academic Press, 2021, pp. 283–303, https://doi.org/10.1016/B978-0-12-824523-1.00003-1.

Chapter 34

Technoeconomic analysis of biofuel production from agricultural residues through pyrolysis

Kondragunta Prasanna Kumar and Neelancherry Remya
Indian Institute of Technology, Bhubaneswar, Odisha, India

1. Introduction

Waste generation and accumulation are serious problems worldwide. The agricultural sector generates an enormous quantity of agricultural waste annually. Nearly 140 billion metric tons of agricultural residue are generated each year around the globe [1]. Open-air burning of agricultural residues after crop harvesting is increasing uncontrollably, leading to air pollution due to excessive particulate emissions. It is also a major contributor to global warming. This process typically occurs under uncontrolled conditions that result in incomplete combustion products and greater polycyclic aromatic hydrocarbon (PAH) generation than optimized heating at higher temperatures. Some PAHs have acute and chronic health effects and possess high carcinogenic potential [2]. In India, the practice of agricultural residue burning results in nearly 824 Gg of $PM_{2.5}$ and 211 Tg of CO_2-equivalent greenhouse gases and emissions [3]. Lignocellulosic agricultural residue, a major agricultural residue resource, is considered a sustainable feedstock for generating biomaterials and biofuels. Safe disposal and potential energy recovery that are environmentally sustainable, are often challenges in valorizing underutilized crop residues. The waste-to-energy (WtE) technique or thermochemical or biochemical conversion is attractive for converting agricultural residue into biofuel. The conversion of the lignocellulosic agricultural residue by biological processes is a difficult and cost-intensive process as the crop residues have complex structural and chemical mechanisms, thus making them recalcitrant toward the assault on their structural sugars from the microbial and animal kingdoms [4]. The thermochemical process can convert a wide range of agricultural residues into biofuels. Combustion, gasification, torrefaction, pyrolysis, hydrothermal carbonization, and hydrothermal liquefaction are existing thermochemical techniques. Conversion of the agricultural residue by thermochemical method involves the thermal decomposition of agricultural residue into various solid, liquid, and gaseous products. Temperature, heating rate, reaction time, use of catalysts, pressure, and the absence or presence of oxygen are different parameters generated by product distribution and quality [5]. Among these, pyrolysis is one of the most promising thermochemical technologies for converting agricultural residue into biofuels [6]. Pyrolysis results in the lowest emission levels, and all end products can be used as energy sources or marketable nonenergy sources. An essential element of commercial scale-up is technoeconomic analysis (TEA) to determine economic viability. The present chapter focuses on stepwise methodologies to technically analyze biofuel production and its production cost based on the data obtained from the lab-scale study of pyrolysis and TEA were discussed to evaluate the viability of commercializing biofuel production from agricultural residues. Capital cost, variable cost estimation for commercialized biochar plants, and cash flow analyses like internal rate of return (IRR) and payback period based on the economic return of biofuel sales are also emphasized.

2. Properties of agricultural residues

The suitability of agricultural residue for thermochemical conversion generally depends upon its physical properties, chemical composition and thermal properties. It also influences the thermochemical conversion by affecting the reaction time required in the reactor. Fig. 34.1 and Table 34.1 shows different agricultural residue properties.

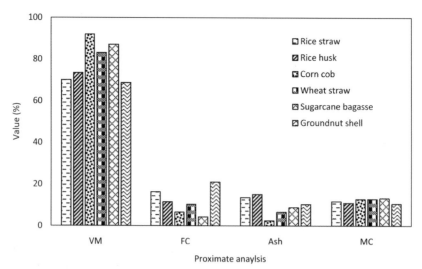

FIGURE 34.1 Proximate analysis of agricultural residues [4,6,7,8,24].

TABLE 34.1 Ultimate analysis of agricultural residues.

Agricultural residue	Ultimate analysis (%)					Higher heating value (MJ kg^{-1})	References
	C	H	O	N	S		
Rice straw	37.10	5.20	43.50	0.50	0.10	12.10	[6]
Rice husk	35.00	4.80	22.30	1.20	0.20	12.90	[4]
Corncob	42.10	5.90	51.02	0.50	0.48	16.00	[7]
Wheat straw	38.34	5.47	55.22	0.6	0.37	14.70	[7]
Sugarcane bagasse	48.20	5.40	44.00	1.60	0.80	16.30	[8]
Groundnut shell	41.00	6.30	41.20	1.00	0.20	15.00	[9]

The bulk density of agricultural residue can vary depending on its form and the processing equipment used. For instance, the dry bulk density of loose rice straw directly obtained from the field varies between 13 and 18 kg m^{-3}, whereas chopped straw (2−10 mm length) has a dry bulk density varying from 50 to 120 kg m^{-3} [10]. The average bulk densities of wheat straw, corncob, and sugarcane bagasse are 150−250, 160−210, and 120 kg m^{-3}, respectively [11]. A low bulk density can result in unfavorable operating conditions, such as poor mixing and nonuniform temperature distribution, thus reducing energy efficiency.

The proximate analysis is the measurement of the moisture content (MC), fixed carbon (FC), volatile matter (VM), and ash. MC is a property that significantly affects agricultural residue processing and application. Agricultural residues can have an MC ranging from 10.5% to 13.21%, as demonstrated in Fig. 34.1. The figure depicts a proximate analysis of rice straw, rice husk, corncob, groundnut shell, wheat straw, and sugarcane bagasse. VM is maximum in corncob at 92% [6], followed by sugarcane bagasse at 87% [9]. In thermochemical conversion, it is advantageous to have high VM content in the feedstock. FC is the amount of carbon left after the VM is expelled. Among agricultural residues, groundnut shell contains the highest FC content, at 21%, followed by rice straw at 16.3%, as shown in Fig. 34.1.

Ultimate analysis provides the elemental composition of agricultural residue as percentages of carbon (C), hydrogen (H), nitrogen (N), oxygen (O), and sulfur (S). Table 34.1 shows an ultimate analysis of various agricultural residues. Sugarcane bagasse contains maximum carbon content of 48.20%, followed by corncob with 42.10% and rice husk with the least, 35.00%. High nitrogen and sulfur content in agricultural residues may cause SO_x and NO_x formation during thermochemical conversion. The oil obtained after pyrolysis with high sulfur and nitrogenous compound content may require upgrading for application as a transport fuel [6]. Table 34.1 shows that rice straw contains minimal sulfur and nitrogen content compared with other types of biomass. Higher heating value (HHV) is the quantity of heat released by biomass during combustion. Sugarcane bagasse has the highest HHV of the biomass types shown, at 16.30 MJ kg^{-1}, followed by corncob at 16.00 MJ kg^{-1}, and rice straw at an HHV of at least 12.10 MJ kg^{-1} due to low carbon content w.r.t. other residues (Table 34.1).

3. Thermochemical conversion process for agricultural residue

Thermochemical processes degrade agricultural residue into value-added end products. These processes are categorized as combustion, gasification, and pyrolysis. Table 34.2 lists various thermochemical processes, operating conditions, and end-product distributions.

3.1 Combustion

Combustion is the thermochemical conversion method of burning organic substances in the presence of an ample quantity of air at temperatures generally ranging between 700 and 1350°C [10,15]. Various types of equipment, such as furnaces and boilers, are employed in this process to generate electrical/mechanical energy and heat. This process reduces the waste volume by nearly 90%, depending on the feedstock type, process conditions, and reactor configuration. However, flue gases contain various air pollutants, such as NO_x, SO_x, and dioxins, that must be effectively removed with the help of air pollution control equipment [16]. The inorganic part of the fuel produces fly ash and bottom ash [25], which can result in various issues in the boiler, such as accumulation, fouling, slagging, and corrosion [26].

3.2 Gasification

Gasification is a complex process involving the conversion of agricultural residue to combustible gas, generally known as synthetic (syn) gas. In this process, the agricultural residue first undergoes partial combustion in the presence of nearly 25% externally supplied oxygen in stoichiometric amounts resulting in the formation of char and gas. Later, this char then reduces the gases resulting from the partial oxidation such as CO_2 and H_2O to form a gaseous mixture of H_2 and CO, which take up nearly 15%–18% of the composition of producer gas. The favorable conditions for gasification involve temperatures in the range of 800–1000°C and pressures equal to or above atmospheric pressure. Syngas is mainly comprised of combustible gases, such as CO, and H_2, which can be further used as an energy source for biofuel generation, electricity cogeneration systems, cooking, etc. Gasification of rice straw can produce syngas with 61% hot gas efficiency and 52% cold gas efficiency, as well as HHV of 5.1 MJN m^{-3} [17,18]. Gasification of rice straw with potassium carbonate (K_2CO_3) shows an increase in H_2 and CO_2, and when K_2CO_3 was added to 9%, the H_2 yield was highest at 59.8% [19]. The largest problem in gasification is the presence of tar in the syngas, which requires further processing units to deal with tar removal using scrubbers, condensers, filters, or in situ conversion through the catalytic cracking or reforming of tar [20].

3.3 Pyrolysis

Pyrolysis is a thermochemical conversion process for the degradation of agricultural residue into its constituent elements by heat application in the absence of oxygen to produce end products such as biochar, bio-oil, and non-condensable gaseous mixture. Various operating parameters of this process, such as reactor configuration, heating rate, operating

TABLE 34.2 Pyrolysis process.

Agricultural residue	Condition	Product distribution (%)			References
		Char	Liquid	Gas	
Rice straw	400°C, 1 h	38.10	28.40	33.50	[7]
Sawdust	–	15–25	30–35	40–45	[12]
Rice straw	500°C, 24 g min^{-1} Continuous fast MAP	32.44	31.86	35.7	[13]
Wheat straw	600 W, 450°C	10.2	35.4	54.4	[14]

temperature, and agricultural residue particle size, significantly affect the composition of end products. Depending on the heat application, it can be differentiated from conventional and microwave-assisted pyrolysis. Conventional pyrolysis methods like gas heater, electric heater, and electric arc are employed to meet high-temperature requirements. On the other hand, microwave-assisted pyrolysis (MAP), in which microwave energy is used for heat rather than conventional sources, has received significant attention nowadays due to the selective, faster, and volumetric heating characteristics of microwaves [21]. Biochar is obtained with a lower operating temperature and more vapor reaction time; biogas yield is enhanced with a longer vapor reaction time and higher operating temperature, whereas bio-oil is generated when the operating temperature is moderate and vapor reaction time is shorter [27]. Table 34.2 shows various product formations during pyrolysis for different operating conditions. In conventional pyrolysis of rice straw at 300°C for 1 h, 35.1% of biochar was obtained. XRD and FTIR analysis of the biochar showed that agricultural residues were disintegrated during pyrolysis and converted into end products [7]. A continuous fast MAP system has been developed for producing biochar, bio-oil, and gas from rice straw [13]. Pyrolysis of sugarcane bagasse at 400°C results in 60% biochar [8], a yield higher than that of rice straw pyrolysis [7]. This variation in biochar yield was due to higher FC in sugarcane than in rice straw, as shown in Fig. 34.1.

4. Technoeconomic analysis

TEA is an effective tool to assess the economic and technical viability of any process or technology: the required capital cost, operating cost, and product selling price. With this information, IRR, payback period, and net present value (NPV) can be estimated [22]. Finally, the obtained data are used to decide whether the project should proceed or be abandoned. This chapter uses TEA to analyze various thermochemical conversion processes for converting agricultural residue into biofuels, including estimating the costs of commercializing the lab-scale setup into a large-scale plant [12]. TEA involves several steps:

1. Evaluate the project's economic feasibility.
2. Recognize various finance-related issues arising over the project's duration.
3. Study the various available technologies for obtaining the required product.
4. Compare the costs of available technologies.

Capital cost is the one-time investment at the start of a project and mainly includes site development and equipment costs. It refers to the major process unit equipment, i.e., everything in contact with the process stream. In the thermochemical process, capital cost includes equipment or reactor costs, consisting of drying, shredding, condensing, and storage units. Capital cost varies according to reactor capacity [22].

Operating costs are recurring company expenses incurred for normal day-to-day plant operations. Operating costs include all variable and fixed costs. They include operating labor, raw materials, maintenance, and energy charges. The total operating cost can vary by plant type and production level.

The economic feasibility of thermochemical conversion of agricultural residue plant was assessed by the following measures [9]:

1. Payback period
2. IRR

The payback period is the duration, or time in years, required for investment cash flows sufficient to recover the initial capital cost. Simply, it is the duration equal to an investment's breakeven point. For the feasibility of a pyrolysis plant, a shorter payback period is expected. IRR is a metric for evaluating a project's economic feasibility and investment profitability. Basically, the IRR is the discount rate that results in a discounted-cash-flow NPV equal to zero. It is the annual return that results in a zero NPV. Generally, a pyrolysis plant with a higher IRR is desirable, and IRR can be applied as a metric to compare plants or projects.

5. Technoeconomic analysis of pyrolysis process of agricultural residue

Numerous studies are being conducted for the commercialization of biofuel production, with interest in the thermochemical conversion of agricultural residues gaining significantly. Researchers have performed various studies on TEA for various pathways for converting agricultural residue to biofuels. Table 34.3 enlists TEA for different thermochemical methods of generating agricultural residue for biofuels.

TABLE 34.3 Technoeconomic analysis of pyrolysis process for agricultural residues.

Agricultural residue feedstock	Plant capacity (kg h^{-1})	Capital cost (USD)	Operating cost as % of capital cost	Biofuels selling cost	Payback period (years)	IRR (%)	References
Sawdust	100	142,800	7	Bio-oil = USD 0.6/L	7.5	10	[12]
Pinewood	100,000	525 million	4	Methanol = USD 0.42/kg Biochar = USD 0.25/kg	–	14.2	[28]
Rice husk	0.3	3416	6.8	Bio-oil = USD 0.8/kg	–	–	[22]
	100	111,000	5.5				
	1000	444,152	6.8				
Rice husk	4000	4.10 million	10	Bio-oil = USD 0.32/kg	6	13	[23]
Groundnut shell	30	11,656.82	10	Biochar = USD 1.16/kg	0.91	121	[9]
Wheat straw	5	8626	8.75	Bio-oil = USD 0.5/L Biochar = USD 1.07/kg	1.57	63.12	[1]

An economic analysis was performed to study the cost viability of agricultural residue to bio-oil conversion and commercial feasibility. A pyrolysis plant with 100 kg h^{-1} capacity had a capital cost of USD 142,800, with a 10% IRR specified for a 20 year plant life. The estimated bio-oil price was USD 0.6 L^{-1}, and the plant payback period was about 7.5 years [12]. TEA for coproducing methanol and biochar from pyrolysis was performed for a plant capacity of 100 tonnes h^{-1} of pinewood [28]. Biochar and methanol yields were estimated to be 0.26 and 0.25 kg kg^{-1} of pinewood, the capital cost was USD 525 million, and the operating cost was 4% of the capital cost. Revenue was obtained by selling biochar and methanol at prices of USD 0.25 kg^{-1} and USD 0.42 kg^{-1}, resulting in an IRR of 14.2%. TEA was made of the three rice husk pyrolysis plant capacities [22]. For the bench-scale plant (0.3 kg h^{-1}), operating labor costs appeared to influence cost items that affected unit production costs, whereas for the scale-up plant (100 and 1000 kg h^{-1}), equipment and operating labor costs were found to be dominating influences on unit production costs. The higher the plant capacity is, the lower the production cost of bio-oil—a 1000 kg h^{-1} capacity plant was found to be the most promising and lowest-cost option. A biofuel plant with a capacity of 4000 kg h^{-1} required a total project investment of USD 4.1 million [23]. The estimated production and selling costs were USD 0.26 kg^{-1} and USD 0.32 kg^{-1}. The sensitivity analysis results indicate that the bio-oil production cost was mainly affected by production yield and was moderately sensitive to labor costs and project investment. Based on financial analysis results, the payback period of the plant was 6 years with an IRR of 13%.

The screw-auger pyrolysis reactor was developed to process groundnut shells and has a capacity of 30 kg h^{-1} [9]. The pyrolysis was performed for 4 min at 450°C and a screw-auger speed of 20 rpm. The initial capital cost for the entire manufacturing system was USD 11,656, and the repair and maintenance costs equated to 10% of the capital cost per year. The primary revenue source was the sale of biochar at a price of USD 1.16 kg^{-1}. The production cost of biochar per annum, including 10% repair and maintenance, is USD 11,066 kg^{-1}. Considering all the expenses and revenues, the payback period was 11 months with an IRR of 121%.

6. Summary

Open burning of the agricultural residues generated by crop harvesting often causes excessive particulate matter emission, generates PAHs, and causes acute and chronic health effects. It is estimated that 211 Tg of CO_2-equivalent greenhouse gases and 824 Gg of $PM_{2.5}$ emissions result from crop residue burning. Rather than open burning, the WtE technique is an appropriate substitute for converting agricultural residues into biofuel. Of all the conversion techniques, thermochemical conversion is prominent as a means to convert agricultural residue. The pyrolysis process is favorable for converting agricultural residue end products—i.e., biochar and bio-oil can be used for energy, soil nourishment, and pollutant absorption. Laboratory-scale studies have shown that pyrolysis is a promising technique for converting agricultural residue to biofuels. For scale-up and commercialization from lab-scale studies, TEA is a critical step that combines process modeling and economic evaluation to assess the viability of obtained plant outcomes. Biofuels, i.e., bio-oil and biochar, have selling prices of USD 0.25–1.16/kg and USD 0.32–0.8/kg, depending on quality. Depending on plant capacity, the payback period and IRR can vary from 0.91 to 7.5 years and 10%–121%, respectively.

References

[1] A.S. Paul, N.L. Panwar, B.L. Salvi, S. Jain, D. Sharma, Experimental investigation on the production of bio-oil from wheat straw, Energy Sourc., Part A Recover., Util. Environ. Eff. 00 (00) (2020) 1–16, https://doi.org/10.1080/15567036.2020.1779416.

[2] K.S. Chen, H.K. Wang, Y.P. Peng, W.C. Wang, C.H. Chen, C.H. Lai, Effects of open burning of rice straw on concentrations of atmospheric polycyclic aromatic hydrocarbons in Central Taiwan, J. Air Waste Manag. Assoc. 58 (10) (2008) 1318–1327, https://doi.org/10.3155/1047-3289.58.10.1318.

[3] K. Ravindra, T. Singh, S. Mor, Short Research and Discussion Article Covid - 19 Pandemic and Sudden Rise in Crop Residue Burning in India : Issues and Prospects for Sustainable Crop Residue Management, 2021, p. 0123456789.

[4] S.L. Lo, Y.F. Huang, P.T. Chiueh, W.H. Kuan, Microwave pyrolysis of lignocellulosic biomass, Energy Proc. 105 (2017) 41–46, https://doi.org/10.1016/j.egypro.2017.03.277.

[5] F. Motasemi, M.T. Afzal, A review on the microwave-assisted pyrolysis technique, Renew. Sustain. Energy Rev. 28 (2013) 317–330, https://doi.org/10.1016/j.rser.2013.08.008.

[6] D.V. Suriapparao, R. Vinu, A. Shukla, S. Haldar, Effective deoxygenation for the production of liquid biofuels via microwave assisted co-pyrolysis of agro residues and waste plastics combined with catalytic upgradation, Bioresour. Technol. 302 (January) (2020) 122775, https://doi.org/10.1016/j.biortech.2020.122775.

[7] B. Biswas, N. Pandey, Y. Bisht, R. Singh, J. Kumar, T. Bhaskar, Pyrolysis of agricultural biomass residues: comparative study of corn cob, wheat straw, rice straw and rice husk, Bioresour. Technol. 237 (2017) 57–63, https://doi.org/10.1016/j.biortech.2017.02.046.

[8] S.D. Rabiu, M. Auta, A.S. Kovo, An upgraded bio-oil produced from sugarcane bagasse via the use of HZSM-5 zeolite catalyst, Egyptian J. Petrol. 27 (4) (2018) 589–594, https://doi.org/10.1016/j.ejpe.2017.09.001.

[9] A. Pawar, N.L. Panwar, Experimental Investigation on Biochar From Groundnut Shell in a Continuous Production System, 2020.

[10] Z. Liu, A. Xu, B. Long, Energy from combustion of rice straw: status and challenges to China, Energy Power Eng. 03 (03) (2011) 325–331, https://doi.org/10.4236/epe.2011.33040.

[11] J.M. Makavana, V.V. Agravat, P.R. Balas, P.J. Makwana, V.G. Vyas, Engineering properties of various agricultural residue, Int. J. Curr. Microbiol. Appl. Sci. 7 (06) (2018) 2362–2367, https://doi.org/10.20546/ijcmas.2018.706.282.

[12] C. Jaroenkhasemmeesuk, N. Tippayawong, Technical and economic analysis of A biomass pyrolysis, in: Energy Procedia, vol. 79, Elsevier B.V, 2015, https://doi.org/10.1016/j.egypro.2015.11.592.

[13] Y. Wang, Z. Zeng, X. Tian, L. Dai, L. Jiang, S. Zhang, Q. Wu, P. Wen, G. Fu, Y. Liu, R. Ruan, Production of bio-oil from agricultural waste by using a continuous fast microwave pyrolysis system, Bioresour. Technol. 269 (2018) 162–168, https://doi.org/10.1016/j.biortech.2018.08.067.

[14] D.V. Suriapparao, A. Yerrayya, G. Nagababu, R.K. Guduru, T.H. Kumar, Recovery of renewable aromatic and aliphatic hydrocarbon resources from microwave pyrolysis/co-pyrolysis of agro-residues and plastics wastes, Bioresour. Technol. 318 (September) (2020) 124277, https://doi.org/10.1016/j.biortech.2020.124277.

[15] P. Shah Maulin, Microbial Bioremediation & Biodegradation, Springer, 2020.

[16] S.N. Naik, V.V. Goud, P.K. Rout, A.K. Dalai, Production of first and second generation biofuels: a comprehensive review, Renew. Sustain. Energy Rev. 14 (2) (2010) 578–597, https://doi.org/10.1016/j.rser.2009.10.003.

[17] L.F. Calvo, M.V. Gil, M. Otero, A. Morán, A.I. García, Bioresource Technology Gasification of rice straw in a fluidized-bed gasifier for syngas application in close-coupled boiler-gasifier systems, Bioresour. Technol. 109 (2012) 206–214, https://doi.org/10.1016/j.biortech.2012.01.027.

[18] P. Shah Maulin, Removal of Refractory Pollutants From Wastewater Treatment Plants, CRC Press, 2021.

[19] H.A. Baloch, T. Yang, H. Sun, J. Li, S. Nizamuddin, R. Li, Z. Kou, Y. Sun, A.W. Bhutto, Parametric study of pyrolysis and steam gasification of rice straw in presence of K_2CO_3, Korean J. Chem. Eng. 32 (4) (2016) 1–8, https://doi.org/10.1007/s11814-016-0121-7.

[20] J. Brandin, M. Tunér, I. Odenbrand, V. Lund, Swedish Energy Agency Report Small Scale Gasification : Gas Engine CHP for Biofuels, 2011.

[21] Y.F. Huang, P.T. Chiueh, S.L. Lo, A review on microwave pyrolysis of lignocellulosic biomass, Sustain. Environ. Res. 26 (3) (2016) 103–109, https://doi.org/10.1016/j.serj.2016.04.012.

[22] M.N. Islam, F.N. Ani, Techno-economics of rice husk pyrolysis, conversion with catalytic treatment to produce liquid fuel, Bioresour. Technol. 73 (1) (2000) 67–75, https://doi.org/10.1016/S0960-8524(99)00085-1.

[23] L. Ji, C. Zhang, J. Fang, crossmark, Renew. Sustain. Energy Rev. 70 (October 2016) (2017) 224–229, https://doi.org/10.1016/j.rser.2016.11.189.

[24] Omoriyekomwan, et al., Production of phenol-rich bio-oil during catalytic fixed-bed and microwave pyrolysis of palm kernel shell, Bioresour. Technol. (2016), https://doi.org/10.1016/j.biortech.2016.02.002.

[25] Sadaka, Johnson. https://www.uaex.uada.edu/publications/PDF/FSA-1056.pdf, 2017.
[26] Zafar, Waste management and recycling in oil palm cultivation Salman Zafar, BioEnergy Consult, India. Achieving Sustainable Cultivation of Oil Palm 2, Burleigh Dodds Science Publishing, 2018.
[27] Bridgwater, et al., A techno-economic comparison of power production by biomass fast pyrolysis with gasification and combustion, Renew. Sust. Energ. Rev. (2002), https://doi.org/10.1016/S1364-0321(01)00010-7.
[28] Shabangu, et al., Techno-economic assessment of biomass slow pyrolysis into different biochar and methanol concepts, Fuel (2014), https://doi.org/10.1016/j.fuel.2013.08.053.

Chapter 35

Different methods to synthesize biodiesel

José Manuel Martínez Gil[1,5], Ricardo Vivas Reyes[2], Marlón José Bastidas Barranco[3], Liliana Giraldo[4] and Juan Carlos Moreno-Piraján[5]

[1]*Grupo de Investigación Catálisis y Materiales, Facultad de Ciencias Básicas y Aplicadas, Universidad de La Guajira, Colombia;* [2]*Grupo de Investigación Química Cuántica y Teórica, Facultad de Ciencias Exactas y Naturales, Universidad de Cartagena, Colombia;* [3]*Grupo de Investigación Desarrollo de Estudios y Tecnologías Ambientales del Carbono (DESTACAR), Facultad de Ingeniería, Universidad de La Guajira, Colombia;* [4]*Facultad de Ciencias, Departamento de Química, Grupo de Calorimetría Universidad Nacional de Colombia, Sede Bogotá, Bogotá, Colombia;* [5]*Facultad de Ciencias, Departamento de Química, Grupo de Investigación en Sólidos Porosos y Calorimetría, Universidad de los Andes, Sede Bogotá, Bogotá, Colombia*

This chapter presents a description of the benefits, impacts and production of biodiesel. The following topics are included:

1. Biodiesel: definition, variations, properties, and comparison with the diesel

In the universe of bioenergy, there are many liquid biofuels that have similar features, as in the case of green diesel, environmentally-friendly diesel, and the biodiesel. The first one relates to a mixture of high molecular weight aliphatic hydrocarbons (paraffins), whose molecular structure is different from that of biodiesel, which is a methyl ester. Those structural differences lead to differences in their properties as well; for example, the cetane number of green diesel is between 70 and 90, while the number of green diesel is between 50 and 65 [1]. The second biofuel results from the partial transesterification of oils and fats, resulting in a mixture of monoglycerides and methyl esters of fatty acids in relation 1:2, and in this mixture, it is not necessary to separate it because glycerol is soluble. Unlike biodiesel, ecodiesel has an atomic efficiency of 100% and high lubricating power, due to high presence of monoglycerides [2]. Biodiesel can be described from various perspectives, compositional, functional, and applicability. From the compositional point of view, it can be defined as a mono-alkyl ester of long-chain fatty acids [3]; however, biodiesel is not considered a single type of compound; in fact, it is a solution, so we can define it as a complex mixture of many hydrocarbons, whose components are fatty acid methyl esters (FAME), monoacylglycerols (MG), diacylglycerols (DG), and triacyclglycerols (TG) (Fig. 35.1A–D), among other substances [4]. The percentage of each constituent within the energy solution brings with it variations that affect the functionality of the final product, i.e., in its physical properties (cloud point, density, flash point, pour point, surface tension and viscosity) and in its chemical properties (cetane index and calorific value).

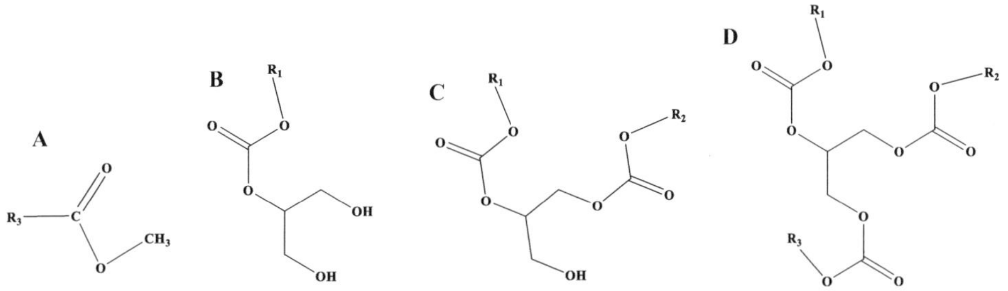

FIGURE 35.1 (A) Fatty acid methyl ester (FAME); (B) Monoacylglycerol (MG); (C) Diacylglycerols (DGs); (D) Triacyclglycerols (TGs). *Made with the software ChembioDraw 2D professional18.1. PerkinElmer.*

Green Approach to Alternative Fuel for a Sustainable Future. https://doi.org/10.1016/B978-0-12-824318-3.00035-7

The MGs are the constituents that most deteriorate the properties of biodiesel, so they are considered one of the most important impurities [5]. Under these considerations, it is possible to observe biodiesel under the following categories: (1) high-purity biodiesel, made up of a high percentage of FAME molecules (around of 99%) and a low percentage of MG molecules, and (2) low purity biodiesel, those in which the solubility limit exceeds 0.3% of MG molecules [6].

Because MGs solidify earlier than saturated FAMEs at low temperatures, biodiesel with concentrations of undesirable substances above the solubility limit bring as a consequence alteration in their properties, in particular the presence of MG molecules, which leads to alterations in their properties such as the cloud point and the precipitation of solids at low temperature.

The properties of biodiesel are also conditioned by the raw materials, so the biodiesel is classified in four generations, according to the relationship established between the raw materials, food safety, economic impact, and technologies used in the production [7]. Among the effects of raw material on the properties of biodiesel, we find the preferable use of sunflower or rapeseed oil because the biodiesel obtained has similar properties to petroleum diesel like costs, carbon footprint, water use, fertilizer use, stress tolerance, ease of conversion, and preference over oil palm, sugarcane, and sweet sorghum [8]. But, these biodiesel sources compete with food safety, so crops like jatropha, pongamia, tobacco, mahua, neem, cotton, and sesame are preferred [9]. It is also known that the properties and quality of biodiesel are directly influenced by the composition and quantity of the lipids that make up the raw materials; in such a way the percentage of saturated or unsaturated fatty acids is responsible for parameters like viscosity, density, acid number, flash point, fire point, cloud point, caloric value, cetane number, iodine number, and oxidation stability. For example, high percentages of unsaturated fatty acids, especially polyunsaturated ones, produce chemically unstable biodiesel, causing low oxidation stability, degradation, polymerization, leading to the formation of solid waste [10].

Saturated fatty acids, especially long-chain fatty acids, improve oxidation stability and cetane number, reducing emissions of NOx. [11]; however, the high presence of saturated fatty acids considerably affects the cold flow properties, like cold filter plugging point, freezing point, and cloud point [12]. In addition to these considerations, it is important to take into account that for the use of certain raw material, the life cycle must be evaluated, as well as economic value, effects on air quality, land availability, volumes of water required for cultivation, oil content, yield per hectare, energy supply, use of pesticides, loss of biodiversity, job creation, cultivation practices, effects on the soil, logistics cost, among others [13]. In addition to the above, it is necessary to modify the properties of biodiesel, so that it can be applied directly to internal combustion engines. Therefore, they must be mixed with other fuels such as oils, gasoline, diesel, or ethanol to modify their properties.

In such a way, binary or ternary mixtures are originated. Among the binary mixtures we find oils-biodiesel; diesel-biodiesel, gasoline-biodiesel, and ethanol-biodiesel. Among all different ternary mixtures, we find oil-ethanol-biodiesel, diesel-ethanol-biodiesel, and oil-diesel-biodiesel. In the middle of these combinations, the binary biodiesel-diesel mixture has great importance because it can be applied to diesel, without having to make any changes [14]; in addition, it has greater efficiency compared with other fuel mixtures, especially with oil-biodiesel mixtures [15].

Another consideration that affects biodiesel quality of its mixtures is the impurities presence or undesirable substances; for example, oxidative degradation is the main factor that affects the durability and stability of biodiesel and is caused by these impurities. A solution to this issue has been the use of additives such as antioxidants to improve the stability and durability of biodiesel (see Table 35.1). In addition, the current limitations in terms of the operating conditions of the automotive industry make the use of biodiesel or its mixtures present some restrictions. In this scenario, it is pertinent to join forces in the development of engines with a high affinity for biodiesel, taking into account that diesel or petrodiesel is responsible for most of the environmental problems, insofar as releasing polluting gases into the air such as hydrocarbons, carbon dioxide, nitrogen oxides, carbon monoxide, and particulate matter [33].

The importance of generating new technologies that seek to improve performance in the production of biodiesel and the generation of engines that can apply biodiesel B100 is evident. It is very convenient to change the vehicle fleet, and improving the quality of biodiesel lies in the fact that its use promotes the diversification of energy sources and allows over time to reduce dependence on fossil fuels. Among biofuels, biodiesel is the most important alternative because its production requires the use and transformation of inputs of biologic origin, which gives it many environmental and social advantages in relation to petrodiesel, and biodiesel has many physical properties and chemicals that make it similar to petrodiesel, so it is miscible with it, with the advantage that it is nontoxic and biodegradable [34].

Some advantages can be highlighted: high flash points, high cetane number, better lubrication, better engine performance, low sulfur content, reduced air pollution emissions, low content of aromatic compounds, and containing approximately 10% oxygen [35]. Broadly speaking, the combustion products of biodiesel are cleaner compared with those of petrodiesel [34]. Among the chemical properties, the cetane number is one of the most important, as it describes the ignition characteristics and affects the power of the engine and its polluting emissions [36]. When a small volume of

TABLE 35.1 Different antioxidants used as additives to improve the stability and durability of biodiesel.

Antioxidants	References
Catechol	[14]
Pd supported in Al_2O_3	[16]
Chalcones	[17]
N, N'-diphenyl-p-phenylenediamine	[18]
Tannins	[19]
Butylated hydroxytoluene, citric acid, ascorbic acid, and alizarin	[20]
Tertbutyl hydroquinone, 3,5-di-butil-4-hiyroxytoluene, butyl hydroxydone, pyrogallol, and propyl gallate	[21]
n-Butanol	[22]
Blackberry extracts, oregano extract, basil extract, and quercetin	[23]
Natural ethanolic extract of turmeric, synthetic butylated hydroxyanisole, and propyl gallate	[24]
Platymiscium floribundums extracts, ethanolic and chloroformic	[25]
Moringa oleifera Lam antioxidant powder	[26]
Functionalized anionic ionic liquid [MI][C6H2(OH)3COO]	[27]
Green tea extract (*Camellia assamica*)	[28]
MG: methyl gallate, TBHQ: Terc-butyl hydroxyquinone, PA: L-ascorbyl palmitate, GA: gallic acid	[29]
HT: butylated hydroxytoluene, BHA: butylated hydroxyanisole	[30]
Used coffee extract	[31]
1,4-Dihydroxyanthraquinone	[32]
2,2,4-Trimethyl-1,2-dihydroquinoline	
2.2'-methylenebis (4-methyl-6-terc-butyphenol)	
Acid hydrazide 2,4,6-tris-isopropylbenzoic	
2,5-Di-terc-butylhydroquinone	
2,6-DI-terc-butylphenol	
Nitrate de 2-ethylhexyl	
2-Terc-butyl-4-hydroxyanisole	
3.3′,5.5′-Tetramethoxy (1.1′-biphenyl)-4.4′-diol	
4.4′-dioctyldiphenylamine	
4-octyl-N-phenylaniline	
Ascorbyl palmitate	
Butylated hydroxytoluene	
Butyl hydroxyanisole	
Gallic acid	
L-ascorbic acid	
Methyl gallate	
N,N'-diphenyl-p-phenylenediamine	
N,N'-Di-sec-butyl-p-phenylenediamine	
N-isopropyl-N'-phenyl-p-phenylenediamine	
Octyl gallate	
Pentaerythritoltetrakis (3,5-di-terc-butyl-4-hydroxihydro cinnamate)	
p-Phenylenediamine	
Propyl gallate	
Pyrogallol	
Terc-butyl hydruinone	

biodiesel is available or a test engine is not available, the cetane number is used to estimate the cetane number [37]. As biodiesel has a higher cetane number in relation to diesel, it gives it greater combustion efficiency and increases the compression ratio of the engine, which translates into higher ignition speed, less noise, and less pollution [38].

The environmental damage caused by the use of technologies based on fossil energy has led to the acceleration of the effects of climate change and the overexploitation of the resource, which has consequently generated a global energy crisis. In this scenario, it is a priority for the application and search of renewal energies that overcome these challenges, without forgetting that it is necessary to make an energy transition, and that in turn deserves a very in-depth study so that the application of clean technologies does not cause any social or environmental damage at any stage of its generation. Procurement of inputs, processing of raw materials, preparation, and application due to the use of these alternative energies can hide pollutants that can be more harmful to the environment and society than those generated by the use of traditional energies. In global terms, the biodiesel manufacturing process can be improved in such a way that the environmental and social impacts generated are minimal, making it an excellent candidate to progressively overcome the demands of the energy transition.

1.1 Biodiesel production by transesterification

Therefore, because the properties of fats and oils do not allow direct use as combustibles, it is necessary that they are subjected to processes that allow their transformation to biodiesel; four methods are known for the conversion of oils or fats to biodiesel: dilution, microemulsification, pyrolysis, and transesterification, [39]. Among these methods, transesterification is the most effective and economical to produce high-heat biodiesel, and it is a process of converting heavy lipids or esters called triacylglycerides into light esters, and these lipids generally come from natural sources.

Natural sources are preferred because they do not require the participation of anthropogenic processes for their production. Oils from vegetables, fungi, bacteria, and protists, as well as fats from animals, supply a wide variety of triacylglycerols and other glycerols with a high conversion capacity to biologic diesel or biodiesel. Among the oils and fats components used for the production of biodiesel, we find triglycerides, triglyceride monomers, monoglycerides, oxides, free fatty acids, complex lipids, antioxidants, and vitamins [40] among others.

Vegetables are the most used resource to obtain the raw material to produce biodiesel, due to the ease of growing them and requiring less investment in relation to other sources like animals, algae, microalgae, fungi, and bacteria. When triacyglycerols are made, they react specifically with an alicyclic mono-alcohol such as methanol (CH_3OH), a FAME, which is usually the major component of biodiesel (see Fig. 35.2). The speed of the transesterification reaction depends on whether it is carried out in the presence or absence of a catalyst, and the catalyst used can be in the same phase (homogeneous catalysis) or a different phase of catalysis (heterogeneous catalysis) with the substances that interact and in turn can have an acidic or basic character.

Oils and fats can have a high percentage of free fatty acids, so the conversion reaction is usually carries out in two steps, although it is feasible to apply methods in a single step. Two-step methods involve the application of acid catalysis, which is intended to esterify free fatty acids, followed by base catalysis, which is intended to transesterify triacylgycetide and diacylglyceride. For both esterification and transesterification, homogeneous or heterogeneous catalysis can be used in both steps, but heterogeneous catalysis is preferred.

For the one-step esterification/transesterification reaction, the use of heterogeneous acid catalysis is preferred. The techniques used for the production of biodiesel must overcome drawbacks such as long reaction time, increase in the amount of alcohol and amount of catalyst, separation of water, purification, and recovery of catalysts, which incur high energy costs and result in low production efficiency [41]. The characteristics of the catalysts, in particular their activity, cost, availability, and affinity with the raw material, are required for the conversion process to be satisfactory [42].

For economy and greater affinity with raw materials of plant origin, basic or homogeneous alkaline catalysts are the most used for the production of biodiesel; among these catalysts we find potassium hydroxide (KOH) and sodium hydroxide (NaOH), and basic catalysis presents high catalytic activity, applying moderate temperature, pressures, and low conversion times [43]; however, this catalysis type presents a big disadvantage in its high sensitivity and requirements for high-purity reagents and separation of water content. In addition, if the source has a high percentage of free fatty acids, the catalyst is neutralized and produces soap; this problem not only poisons the catalyst but also contaminates biodiesel, forming biodiesel/glycerol emulsions that are very difficult to remove.

Opposite to the homogeneous basic catalysts, the homogeneous acid catalysts present low activity, without the drawbacks of the basic catalysts; however, to achieve acceptable yields, it is necessary to increase the molar amount of alcohol, and they have as disadvantages that they are corrosive and expensive [44]. As advantages, homogeneous acid catalysts are resistant to poisoning by the presence of free fatty acids, and as an attribute that stands out, they present

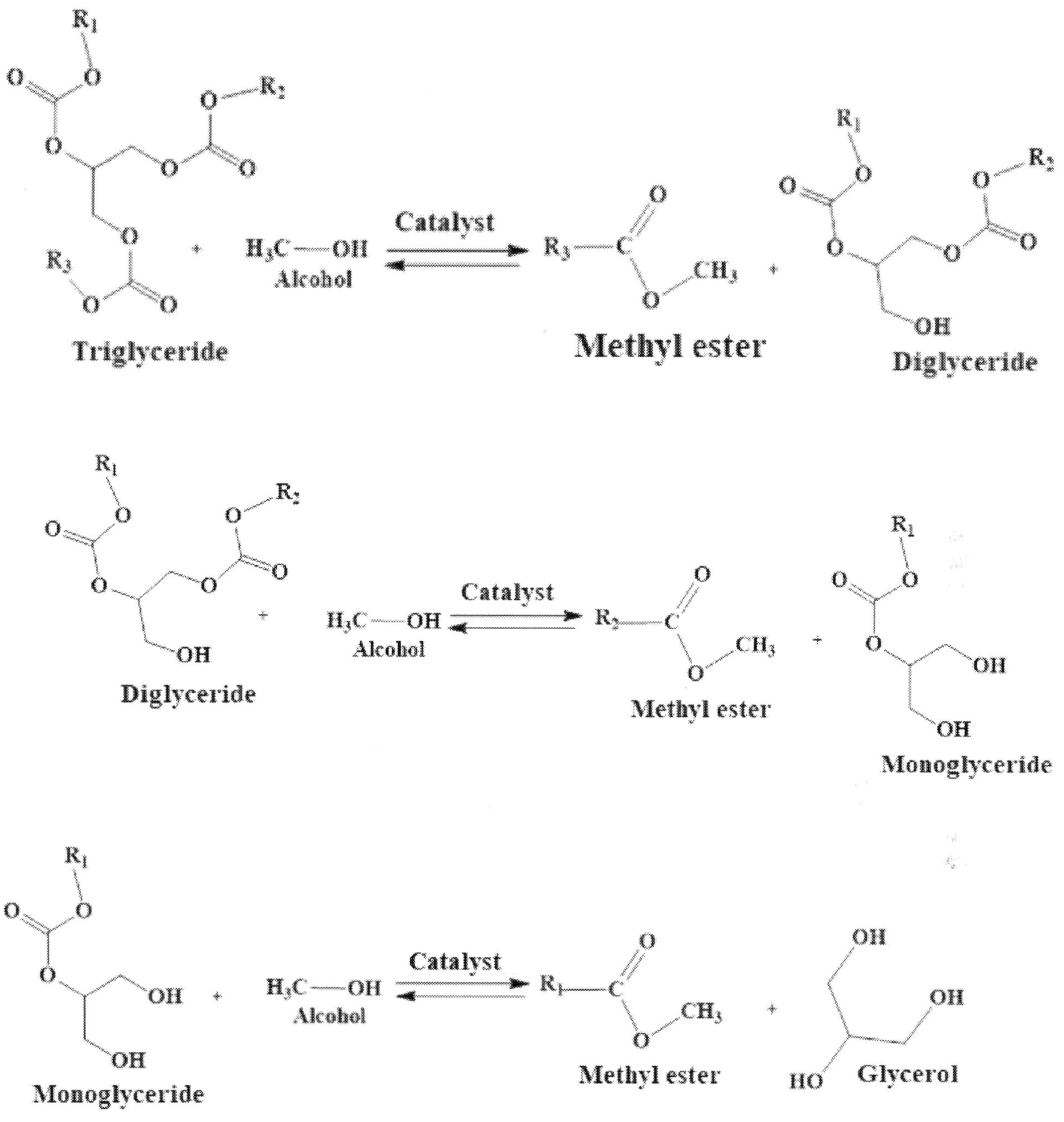

FIGURE 35.2 Transesterification reaction. *Resource: made by ChembioDraw 2D professional 18.1 Software PerkinElmer.*

duality by simultaneously catalyzing esterification and transesterification reactions, with the difficulty that transesterification reactions are limited by the presence of water [45].

Sulfuric acid (H_2SO_4), hydrochloric acid (HCl), and sulfonic acid (HSO_2OH) are the homogeneous acid catalysts generally used because they present high yields, the separation of the reaction products is simple, and they produce high purity glycerin [46]. In less proportion, trifluoroacetic acid has been used for the conversion of soybean oil to biodiesel with a yield of 98.4% [47]; however, for the use of this kind of catalyst, the following are necessary: high concentration of catalysts, high temperature, higher amount of alcohol, and specially designed reactor; likewise, traces of the catalyst may remain in the biodiesel obtained, and therefore, when applied, it causes corrosion damage to the engine [48].

Although the highest yields in biodiesel production occur in the presence of homogeneous acid or basic catalysts, the transesterification reaction takes place in the condensed phase (medium liquid), which prevents the recovery of the catalyst [49].

Once the biodiesel is separated, residues persist (sludge) and pollutants that together with glycerin represent a strong challenge for their final disposal. To clear the vision, heterogeneous catalysis, which can also be acidic or alkaline, is presented as an alternative to solve the problems presented by basic catalysis, if and only if the catalyst is released into the reaction medium [50]. The release of the catalyst guarantees its separation from the products obtained at the end of the process, which prevents the catalyst from having effects on the product and its application, and also allows its reuse [48].

Among the widely used heterogeneous alkaline catalyst we find hydrotalcites, magnesium oxide (MgO), or calcium oxide (CaO), where the CaO is the preferred solid for heterogeneous catalysis due to its many advantages, such us long shelf life, high activity, lower solubility in methanol, and requiring moderate reaction conditions, which translates into low environmental impacts. It is also economical because it can be found as a component of limestone, animal bones, and eggshells among other resources [51]. In the same line of ideas, we have found heterogeneous acid catalysts, where the most used are Nafion-NR50, sulfated zirconia, and tungsten zirconia [52]. Catalytic membranes made with polyvinyl alcohol have also been used due to its ease of being functionalized with acid groups; however, certain catalysts based on polyvinyl alcohol impregnated with Nafion have better characteristics in relation to heterogeneous basic catalysts [53]. Some Mg-MCM-41 catalysts have been used for the conversion of rapeseed oil to biodiesel, but with lower efficiency compared with KOH [54].

Most of these problems can be overcome by the application of acid-functionalized solid catalysts, but they should show catalysts activity at least comparable to their homogeneous analogs. In this sense, the efficiency of heterogeneous catalysts can be significantly improved by increasing the resistance of the supported acid sites, which leads these catalysts to be considered a viable alternative to replace conventional homogeneous catalysts systems. In this sense, the application of zeolites has been a very active area of research, but the low range of pore sizes available (<1 nm) is considered a very important issue [55]. Due to these considerations, M41S-type mesostructured materials have been studied, because their structure in the form of silica confers long-range order, large pore diameter, and high surface area [56].

Despite the benefits of these heterogeneous catalysts, they present difficulty in their diffusion, possible leaching of the active component in the reaction mixture, and poisoning of the catalyst when it adsorbs CO_2 and H_2O upon contact with air [57]. For these reasons, efforts are directed toward the preparation of solid catalysts that have marked acid characteristics (superacids) or strong basic properties; however, mass catalysts represent difficulties of a different nature: nature and concentration of active sites, particle sizes, arrangement of active phases, and leaching; therefore, the approaches are directed toward the preparation by different methods (mechanical mixtures, impregnation, ion exchange, precipitation, and graphing, of solid supported catalysts). It is then that some components of the catalyst such as the support, the promoters, and the other additives are of special importance.

In the previous scenario, metal-organic frameworks (MOFs) appear, which are a class of porous ordered materials that combine some of the best characteristics of both homogeneous and heterogeneous catalysts, discovered by Prof. Omar Yaghi [58]. They can provide structures with different amounts of basic and acidic active sites. Unlike other materials made up of coordination networks with single-metal nuclei, MOFs are massive materials made up of polynuclear secondary building units (SBUs) that confer rigidity, stability, and versatility to this type of structure [59] and allow cation exchange by partial or complete substitution of one metal ion for another, while conserving the framework [60]. The MOFs are grouped by families, and their location within a certain family depends on several considerations: the configuration of the SBU, the consequence of the properties of the metal core (preferably transition elements) anchored by organic ligands whose donors can be oxygen or nitrogen, for example, MOF-74 (Zn_2DOT), MOF-101 [$Cu_2(BDC-Br)_2(H_2O)_2$] [59], and the three-dimensional configuration of its structure, for example, the IRMOF family, which is characterized by the fact that its members present an IsoReticular MOF [60]. They are named in relation to the institution where its discovery or manufacture was made, for example, the UiO family in honor of the University of Oslo, the MIL family in honor of the Lavoisier Institute, the family HKUST in honor of the University of Science and Technology of Hong Kong, and the family LIC in honor of Leiden Institute of Chemistry [58]. In honor of research groups, for example CPL, F-MOF-1 y MOP-1, there are also families like ZIF (zeolytic imidazolate framework) whose arrangement corresponds to zeolite materials, in which the tetrahedrons are connected to imidazole rings through the nitrogen atoms [61]. Among these families of MOFs, we found a series made up of A100 (MIL-53), C300 (HKUST-1 o C-BTC), F300 (Fe-BTC), Z1200 (ZIF-8), and Basosiv M050 [62] called basolites because they are manufactured by the company BASF, and some of them, MOF-5, ZIF-8, ZIF-67, and MOF-808, have been used in transesterification reactions [63]. Among these catalysts, the MOF-199 and the ZIF-8 have drawn attention, due to their ability to act in esterification and/or transesterification reactions without or with modifications.

The MOF-199 or HKUST-1 is the first highly porous open-frame metal coordination polymer developed by Hong Kong University of Science and Technology. The MOF-199 has been used for gas storage, catalysis, and product capture, CO_2 adsorption, air purification, removal of rubidium ions from water, desulfurization and denitrogenating of hydrocarbon fuels, and as a bioreactor in the form of hollow capsules with selective permeability [60]. The basolite C_{300} or MOF-199 is usually prepared by combining Cu^{2+} with benzene-1,3,5-tricarboxylic acid ligand, using DMF (N,N-dimetylmetanamide/ethanol/H_2O) as a solvent to 85°C, between 8 and 15 h [59]. Under these conditions, the $Cu_3(BTC)_2$ (see Fig. 35.3) is produced, which crystallizes in the cubic system with binodal topology (3.4)-c tbo [64], with BET surface area of about 1800 $m^2\,g^{-1}$, and its structure is characterized by the fact that it presents two types of large cavities with diameters of ca. 1.0 and 1.3 nm and accessible porous windows ca. 9.3 and 1.1 nm, respectively [65]. The MOF-199 was first synthesized by Williams et al. [66]; since then it has been one of the most studied MOFs due to its large surface area, good thermal stability, and durability, which is why both the metal core (node) and the ligands have been replicated and modified, giving rise to variants, for example, the MOF-399 and the MOF-14 [67]. In this order of ideas, ZIF-8 basolite is generally prepared by combining Zn^{2+} salts with $Zn(NO_3)_2$ acid ligand, using CH_3OH (methanol), DMF (N, N-dimethyl methanamide) as solvent, or H_2O at 140°C in a time of 24 h [68]. This material presents a polyhedral crystalline system in the form of a Zeolite framework structure of sodalite (SOD) (see Fig. 35.4) with a connection window size of 0.34 nm, and its configuration offers great stability, its production is low cost, and its synthesis is very simple, which is why there are more than 150 ZIFs [61]. Its surface area BET is found around 1700 $m^2\,g^{-1}$ due to its structure presenting nano/micropores, high thermic stability (400°C), high porosity structure, and low density, and the synthesis conditions are smooth and fast [70].

Despite the versatility of MOF-199 and el ZIF-8, there are few investigations in which these frameworks have been used in the production of biodiesel. Table 35.2 shows some works where MOF-199 and ZIF-8 have been used in transesterification and esterification reactions.

In addition to the organic-inorganic framework MOFs synthesized with homometallic SBUs, MOFs with more than one metal in their nodes or the immobilization of MM-MOF metal ions have also been prepared, to improve their properties [74].

In this sense, works have been reported in which metallic nanoparticles, metallic complexes, and organometallic compounds have been used as MOF cavities as hosts with or without interaction with the organometallic framework [63]. These MOFs offer opportunities in terms of multifunctionality and adjustment of the properties of the material to a given application, as a result of the synergistic effects that could arise from the presence of two or more metals; however, the synthesis of homometallic MOFs and/or MM-MOF anchored to other structures such as mesostructured materials have not been reported [75]. This functionalization with MOF centers can generate materials that exhibit better performance in heterogeneous catalysis and detection and construction of materials that overcome current limitations in transesterification reactions of oils and/or fats in biodiesel production processes. Combining what was described above, both acid and catalysis consume a lot of energy and produce many by-products from poor treatment of materials.

To forcefully solve all the inconveniences associated with the production of biodiesel, it is necessary to direct studies in the search for high-quality lipid raw materials and thermally resistant, recoverable catalysts that have high activity and a high rate of conversion.

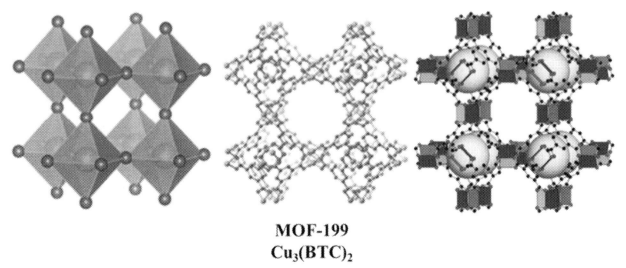

FIGURE 35.3 Structure of MOF-199. *Taken from A. Schoedel, Secondary building units of MOFs. In Metal-Organic Frameworks for Biomedical Applications (2020) 11–44. Woodhead Publishing.*

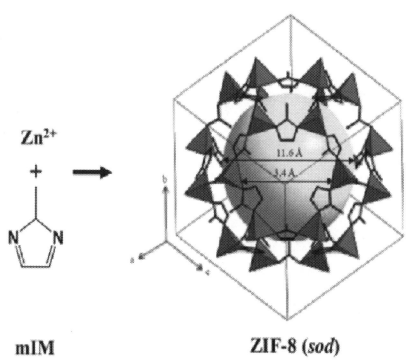

FIGURE 35.4 Structures of different ZIF. *Taken from M. Bergaoui, M. Khalfaoui, A. Awadallah-F, & S. Al-Muhtaseb, A review of the features and applications of ZIF-8 and its derivatives for separating CO_2 and isomers of C3-and C4-hydrocarbons, J. Nat. Gas Sci. Eng. 96 (2021) 104289.*

TABLE 35.2 Different reactions catalyzed with MOF-199 and modified ZIF-8.

MOF	Reaction	Resource	Modified by	Conversion	References
MOF-199	Transesterification	Palm oil	No modifier	86.0	[71]
	Esterification	Stearic acid	Acid 1,2-tungstophosphoric	94.1%	[65]
ZIF-8	Transesterification	Soy oil	*Aspergillus niger*	88%	[68]
	Transesterification	Soybean and sunflower oils	NaOH	90%	[72]
	Transesterification	Soy oil	Fe_3O_4	92.6%	[73]
	Esterification	Oleic acid	TiO_2	80%	[44]

1.2 Enzyme-catalyzed transesterification for biodiesel production

The described panorama by chemical catalysis is not very promising; however, efforts are being made in the search for viable alternatives that overcome the disadvantages of these technologies. Under this context, enzymes appear as promising materials for the development of new technologies leading to the production of biodiesel, if we take that into account enzymes reduce energy consumption, because the reactions mediated by these biocatalysts occur at low conditions and also can be applied to a wide range of substrates [76]. Additionally, enzyme catalysis does not require high-quality raw materials, since enzymes do not react with impurities [77].

Enzymes are proteins or nucleic acids that act as catalysts produced by living things to mediate most cellular processes. Enzymes are extremely selective to a certain substrate, which makes them very specific for a certain reaction; in this context, there are enzymes with high affinity for lipid substrates, and these enzymes are called lipolytic, which in turn are divided into lipases and esterases [78]. Esterases hydrolyze short-chain acylglycerols. Lipases can hydrolyze long-chain fatty acid acylglycerols into glycerol and fatty acids step by step and in turn can transform fatty acids into esters. Since lipases are soluble in water, the best catalytic activity can be achieved at the oil—water interface.

Thirteen families of lipases have been described: true lipases, family II or GDSL, family III, family IV or HSL (hormone-sensitive lipase), family V, family VI, family VII, family VIII, family IX, family X, family XI, family XII, and family XIII [79]. The lipases (triacylglycerol; EC 3.1.1.3) catalyze the hydrolysis of ester bonds present in acylglycerols in vivo [80]. Although lipases have been found in a wide variety of species: animals, plants, fungi, protists, and bacteria [81], microbial lipases are much more versatile and present interesting characteristics such as stability in organic solvents, activity under various conditions, high substrate specificity, as well as regio- and enantioselectivity [82]. Lipases have been used in conventional reactions such as hydrolysis of fats and oils, alcoholysis, aminolysis, esterification, interesterification, transesterification [83].

Although enzyme catalysis offers a variety of advantages, lipases are very expensive, difficult to recover, and unstable at temperatures above 40°C, which makes their use restricted. To overcome these drawbacks, low-cost lipase immobilization methods have gained strength in solid supports to stabilize the enzymes and reuse them [84]. Lipase immobilization still has to overcome limitations such as high cost and low mass transfer of some catalysis. To overcome these drawbacks, liquid lipases have recently been used. [85]; Chang, and the results with this type of catalysis are between 77% and 85.1% [86].

Among the liquid lipases, those produced by the species *Candida antartica*, *Pseudomona cepacian*, and *Sus* (porcine lipase "smooth and rough"), due to their low cost in relation to other enzymes and tolerance to methanol, can esterify free fatty acids and the kinetic model, and they present high conversion of biodiesel [87]. The main challenges facing the industrialization of the enzymatic process are the high resistance to mass transfer, the tendency to adsorb the glycerol byproduct on the support matrix, and the low operational stability [88]. Those problems can be solved through a good selection of supports with favorable surface characteristics and pore sizes [89]. In this aspect, the immobilization of lipases seeks to solve problems in terms of its stability under drastic conditions, activity, selectivity, and purity [90]. MOFs are considered the best candidates to support lipases, as they produce a greater amount of biodiesel without affecting the catalytic activity and at the same time increase the stability of the catalyst and its reuse [87]. Table 35.3 shows several studies reporting the use of lipases immobilized on various supports.

It is notable in Table 35.3, few studies are reported where organometallic structure support MOFs are used. Similarly, this table reports the use of modified biochars as supports for lipases, despite the fact that their surface and application properties make them excellent supports if they need to be modified. Perhaps, they have not been taken into account because their properties depend mainly on the starting material and are therefore conditioned by the biomass used and the carbonization procedure [69]. Biochars have been used us catalysts [109], because they are cheaper, have a large surface area, have a stable structure, good thermal and mechanical stability, are chemically inert, have a higher acid density, and are friendly to the environment [110]. Biochars have a basic and acid character, where the basic character is due to the presence of elements such as K, Na, Ca, Mg, Fe, P, Si, and S, and the acid character is related to the presence of P, Si, S, and CO [111]. Properties that make it ideal for the production of biodiesel, however, mean that the biochars produced have poor physicochemical properties. Additionally, during the production process, especially at low or high agitation, small carbon particles can form that have to be removed from the biodiesel, and for this reason, they become suitable materials as supports, especially for the immobilization of lipases. In summary, the design, preparation, and physicochemical characterization of hybrid lipase–MOFs and lipase–biocarbon catalysts configure promising materials for transesterification/esterification, with light alcohols, of acyl glycerides present in conventional sources of fats and oils and glycerol as a byproduct, on the way to obtain biodiesel components.

2. Sources for the synthesis of biodiesel

The use of alternative raw materials to obtain biodiesel is currently a necessity, due to the negative impacts generated by crops such as oil palm and soybeans, at a social and ecologic level [112]. It is necessary to use alternative raw materials that do not represent the use of large hectares of land and also that in its processes have minimal effects on the environmental and can somehow mitigate the effects of climate change [113].

The properties of the raw materials, their effect on the biodiesel production process, as well as the social and environmental effects that they cause are the determining criteria for the classification of biofuels, namely first, second, third, and fourth generations [114]. In the first generation, we find biodiesel made from edible oils such as palm oil, rapeseed, and tomato among other sources [115]. The advantage of using these energy sources lies in the ease of their achievement because they are used to feed the human species and fattening animals. The disadvantages of using these materials lies in the presence of unsaturated fatty acids, since these increase or reduce the cold flow properties of biodiesel; similarly, the presence of unsaturated fatty acids can also generate a high density to biodiesel; similarly, the presence of saturated fatty acids can also generate a high density to biodiesel [114] and reduction in its viscosity, its flash point, in addition to the

TABLE 35.3 Immobilized lipases with supports of different configurations.

Lipases	Support	Resource	Alcohol	Conversion	References
Candida antartica CALA	Free	Palm frying oil	Methanol	94.6%	[91]
	FeCl₃,6H₂O	Palm oil	Methanol	70.2%	[92]
Candida rugosa CRL	Multifunctional magnetic polydopamine nanospheres (PDA)	Aceite Olive oil	Methanol	87.9%	[93]
Candida antartica CALB and Thermomyces lanuginosa TLL	Epoxy functionalized silica gel	Palm oil	Methanol	94%	[94]
Pseudomona cepacia PCL	Paranitrophenyl Palmitate (p-NPP)	Argan oil	2-Propanol	74.5%	[95]
	Cellulosic polyurethane	Glyceryl trioctanoate	Ethanol	70%	[96]
	PVA/AlgNa	Beaver oil	Methanol	78%	[76]
	Polyvinyl alcohol/AlgNa	Residual glycerol	Methanol	66.4%	[108]
	Free	Olive oil	Methanol	85%	[97]
	PVA/AlgNa	Karanja oil	2-Propanol	84%	[98]
	Activated biochar	Serratia lipids sp.	Methanol	92.23%	[99]
Porcine lipase PPL	Xanthate de p-nitrobenzyl cellulose (NBXCel)	Soy oil	Methanol	96.5%	[100]
Rhizopus oryzae (ROL)	FeCl₃ 6H₂O and FeSO₄ 7H₂O	Olive oil	Methanol	88.4%.	[101]
	Silica magnetic nanoparticles (MNP)	Olive oil	Methanol	84.6%	[102]
	ZIF-8	Soy oil	Methanol	94%	[103]
	Nanofibrous polyacrylonitrile membrane electrophile (PAN)	Palm oil	Methanol	95%	[104]
Rhizomucor miehei RML	Fe₃O₄@COF-OMe	Jatropha curcas oil	Methanol	80%	[105]
	Free	Rapeseed oil	Methanol	99.84%	[106]
Rhizomucor miehei (RML) and Thermomyces lanuginosa (TLL)	Fe₃O₄@SiO₂	Used cooking oil	Methanol	82.9%	[74]
QLM de Alcaligenes sp.	Zinc acetate	Sunflower oil	Methanol	80.9%	[107]
Aspergillus niger	Magnetic nanoparticles of FeSO₄ 7H₂O and FeCl₃ 6H₂O	Lipids of A. niger	Methanol	85.3%	[84]

expansion of crops of these raw materials for biodiesel, which have generated negative impacts at a social and environmental level [116], for example, the risk of limitation in food supply and adaptability to environmental conditions caused by monocultures and land use change [117].

Among the first-generation raw materials, we find rapeseed oil, soybean oil, coconut oil, corn oil, palm oil, mustard, oil, olive oil, rice oil, among others [36]. The second generation of biodiesel is characterized by the fact that the raw materials used for the production of biodiesel do not compete with human food, and its production is more economically viable [118]. Among these raw materials, we find babassu oils, nuts, jatropha, rubber [36], and agricultural residues such as straw, stalks, husks, corn cobs, husk, and leaf litter [119], among other oils. The disadvantage of second-generation biodiesel lies in its reduced availability, so the amount of product obtained is reduced [120]. Third-generation biodiesels correspond to those made from used oils and algae, where the oils used, especially cooking oils, are the main raw material, and in second place we find the oils extracted from microalgae [121].

Used oils are relatively easy to obtain, and their reuse directly minimizes the environmental impacts caused by their incorrect disposal [122]. In addition, used oils have a high conversion rate due to the large amounts of free fatty acids they have. The disadvantage of this generation of biodiesel is the need to subject these oils to a pretreatment process before synthesis, which considerably increases production costs [123]. Nowadays, many studies are focused on the cultivation of microalgae, since the use of soil is not necessary, and they also fix much more CO_2 than other crops [124]. However, the disadvantage of the use of microalgae for the production of biodiesel is to reach a critical mass in the crops that allows the production of biodiesel to be obtained in a sustainable way [125]. Lastly, we find fourth-generation biodiesel, which is characterized by the use of synthetic and photoelectric inputs, such as synthetic algae and cyanobacteria [36]. This generation of biodiesel differs from those previously described because they do not need a geographic space, which facilitates their production on a large scale and globally, and they also have a high energy potential [126]. However, this generation has the disadvantage that there still are no technologic improvements that allow the development of synthetic raw material; in addition, research for this type of material is still in its initial phase of development [127]. According to the above explanation, the need to satisfy the great current energy demand is evident, with new sources for fuels that help to minimize the inconveniences that both biologic and fossil fuels present today.

2.1 Vegetable oils for biodiesel

Plants are the source of supply for other organisms such as fungi and animals because they have a wide variety of substances in their tissues that supply most of their metabolic needs. Among the substances that vegetables provide, we find proteins, lipids, and carbohydrates, among others. The human being is one of the few terrestrial living beings capable of understanding, manipulating, and transforming the conditions of the planet in shorter times in relation to other species. In an effort to condition the environment to their needs, human beings have used and generated energy to maintain and supply their daily life and industrial processes.

Lipids are the main source of energy for living beings, and vegetables store lipids in the form of oils in cellular structures called oligoplasts, and these can be found in various parts of the plant, leaves, stems, roots, fruits and seeds, and their concentration depends on the species, which is why they are used to meet the energy needs of today's society. However, oils of biologic origin are poor fuels for nonliving machinery; for this reason, oils of fossil origin are preferred. To overcome this drawback, biologic oils are transformed into biodiesel, where the similarity of the properties of this fuel with petrodiesel depends a lot on the processed vegetable species. Many vegetable oils are used for biodiesel; among them we have sunflower oil, African palm, safflower, coconut, barley, peanut, wheat, corn, sorghum, rice bran, soybean, sesame, palm kernel, rapeseed, hazelnut, and walnut [128] among others.

The composition of the oils is essential for their selection as fatty acid methyl ester precursors (FAME); in order of similarity between biodiesel and petroleum diesel, sunflower oil, palm oil, rapeseed oil, and soybean oil are preferred [3]. The chemical composition, conversion, and properties of biodiesel depend on the plant species chosen as the source. The properties of oils are determined by factors such as availability, cost, acyl glyceride content, and physical and chemical properties [129]. Economically, the profitability of biodiesel production is associated with the production costs of the raw material, which in turn is connected to food safety, and for this reason, oils for biodiesel are classified as edible and nonedible.

Due to effects of competition with human food and fattening of animals, the use of nonedible oils is preferred. Among the nonedible oils used for biodiesel we find jatropha, jojoba, flaxseed, cottonseed, karanja, neem, castor, rubber seed, and tobacco seed [36], among others. Vegetable oils are the most viable sources for the synthesis of biodiesel, under considerations such as land use, expansion of crops, loss of biodiversity, and low environmental pollution.

2.2 Animal fat for biodiesel

Animals as heterotrophic beings are unable to synthesize their own food. The main source of food for animals is vegetables, which provide the raw materials for them to process proteins, carbohydrates, and lipids. Despite the direct or indirect dependence of animals on vegetables, biodiesel obtained from fat has greater advantages in relation to biodiesel obtained from oils, since animal fat is much cheaper due to the fact that it is generated as a waste in the sacrifice of the animal [130]; additionally, as it contains a greater amount of unsaturated fatty acids in relation to vegetable oils, the resulting biodiesel emits less CO_2 [131]; likewise, the tallow or fat, once used, can be used for the synthesis of biodiesel.

Among the lipids present in animal tissue, we find saturated, monounsaturated, and polyunsaturated fatty acids, which can be converted into biodiesel. Several lipids of animal origin have been used for biodiesel, such as duck tallow, fish oil, castor oil, lard, and bovine tallow [132]. Despite the advantages of fats over oils, the high presence of free fatty acids means that they must be subjected to pretreatment for transesterification [128]. Moreover, the production of the different types of livestock involves the use of large extensions of land. In this scenario, there are possibilities that have not yet been deeply explored as raw material for the production of biodiesel. Insects as an energy source are made up of a great variety of species and are very abundant and are generally considered crop pests [133]. Insects have promising characteristics due to their high content of fatty acids, which can make them a good alternative for the development of third-generation biofuels.

2.3 Biodiesel from used cooking oil

Worldwide, human populations use oils in a number of culinary and industrial activities, which generate oily waste, which, when purified, becomes an excellent source for the production of biodiesel. Used cooking oils are raw materials that solve problems presented by raw materials such as edible or nonedible oils and fats, such as costs, does not compete with food security, and they do not contaminate water sources since they are not eliminated, and their supply is constant [134]. Despite the low quality of used cooking oils, they are readily available and produced in large quantities [135]. However, the properties of used cooking oils do not have the necessary properties to guarantee a high conversion to biodiesel, due to the high content of free fatty acids, which translates into low yield, as the presence of free fatty acids affects the acidity index and the water content, factors that do not favor the synthesis of biodiesel; on the contrary, it favors the formation of soap and makes it impossible to recover the catalyst in a heterogeneous phase. Under this scenario, it is necessary to mix oils and modify the process for obtaining biodiesel [136].

To optimize the transesterification reaction of used cooking oils, response surface methodologies have been applied, among which the following have been applied: reactive, supercritical, microwave, ultrasound, and membrane distillation, [137], an aspect that improves biodiesel production at the expense of increased operating costs. Cooking oils have been mixed with chicken oil, fish oil, and jatropha, and it has been found that the mixture between used cooking oil and fish oil considerably improves the properties of biodiesel [138]. Likewise, the conversion operation of used cooking oil by injecting methanol and 2-propanol under superheated semicontinuous conditions shows that the biodiesel product obtained has better performance [139]. It is evident that used oils are the promising raw materials in the generation of eco-energies, with some aspects to be improved in terms of increasing their quality and the search for related catalysts that contribute to and promote the generation of processes and methodologies that respond to the questions and challenges of the transition from fossil energies.

3. Biodiesel from microalgae oils

In the race to overcome the challenges of the energy transition, sources have been sought that overcome inconveniences such as the use of large extensions of land, competition with human food and fattening of animals, have a high content of lipids, and are bioecological.

Microalgae are the source for bioenergy that meet these demands, because they have a high content of lipids, they are easily cultivated, their growth is fast, their production can be controlled by manipulating the growth conditions, they do not generate social impacts and negative environmental effects, they increase water production, do not modify the composition and structure of soils, do not alter the abundance and composition of the fauna and flora of ecosystems, and they are not the basis of substance for human populations [140].

Compared with the use of raw materials of animal and vegetable origin, obtaining lipids from microalgae cultures has the following advantages: short life cycle, less work required for their cultivation, less effect on the place of cultivation, their crops are not affected by weather season, and additionally produce higher lipid content [141]. Regarding the production costs of microalgae, the economic expenses decrease because the use of farm and processing machinery is not necessary.

These reasons make microalgae viable as ideal candidates to produce bioenergy in quantities high enough to replace fossil fuels gradually and substantially. It has been found that on average microalgae have approximately 25% polyunsaturated fatty acids, but this percentage increases when their cells are subjected to stress [142].

Expenses incurred to cultivate microalgae are contingent on carbon supply (C) and nitrogen (N), because the C/N relation is one of the important parameters for them to produce unsaturated fatty acids. Other factors that have an important influence on the cultivation of microalgae are light, temperature, carbon source, nutrients, pH, evaporation, pollution, salinity, and culture medium, which do not involve large investments. For the production of biodiesel, microalgae of the species *Nannochloropsis oculata*, *Chlorella pyrenoidosa*, *Chlorella protothecoides*, *Chlorella* sp., *Chlorella vulgaris*, and *Nannochloropsis oceánica* have been used. Among these, microalgae *Chlorella* sp. presents the highest conversion by acid catalysis [95]. Microalgae are a potential source of renewable biodiesel that may be able to meet the global demand for fuels for transportation and hydrogen production as the energy source of the future, but they must meet some limitations such as scale for production and high production costs of operation [143]. According to the biologic properties of microalgae and the quantity and chemical characteristics of the lipids they produce, they can become the main source of raw material for biodiesel production, without negatively affecting the environment and society.

4. Production of biodiesel from oleaginous microorganisms with organic waste as raw materials

In relation to the high operating costs of microalgae cultures, currently there has been an incursion into the search for raw materials with similar characteristics, but that do not present these limitations. Oleaginous microorganisms such as bacteria and fungi are a promising alternative as they accumulate a large amount of lipids and their use in the production of biodiesel would be conditioning the pathogenic effects of these microorganisms [144]. These microorganisms have a wide variety of food substrates, so one way to meet their energy demand is to use biowaste as culture media such as food waste, sludge, wastewater, and lignocellulosic waste, which have advantages that generate costs in their production, and their poor disposal generates environmental and health issues [145]. In addition, these precursors for biodiesel reduce costs in relation to the technologies currently applied for the production of biodiesel. The high conversion of biowaste to polyunsaturated fatty acids, especially dihomo-gamma-linolenic acid, arachidonic and eicosapentaenoic acids by these microorganisms, especially by fungi, is an important factor in the production of biodiesel with low or zero CO_2 emissions [146]. Fungal, bacterial, and yeast cultures can be produced in aerobic or anaerobic media. Aerobic media limit the biowaste used as substrates, so anaerobic culture media are preferable, because they are compatible with a wide variety of biowaste, in which the microorganism transforms into biomass through several consecutive steps such as hydrolysis, acidogenesis, acetogenesis, and methanogenesis, generating biogas as by-products [147].

These chemoproductive benefits place anaerobic methods as the most effective strategy in the production of oils for biodiesel, but for the start-up of these bioprocesses, it is necessary to supply sugar to the food environment, which involves competition with human food and industrial processes. Additionally, especially for the biosynthesis of oils in lignocellulosic bed, it presents a low rate of use of available carbon, and its hydrolyzation produces furfural and phenols, which inhibit the growth of microorganisms, so the percentage of conversion of microbial lipids to biodiesel is low [148]. Applying technologies or processes can overcome the current drawbacks presented by the production of microbial biodiesel. In general terms, the production of biodiesel from microbial oils using biowaste as a food substrate is consolidated as one of the cleanest technologies for energy generation.

5. Application of nanotechnologies in the production of biodiesel

Although producing nanotechnologies involves high costs, at the level of basic research, applied research, and scaling up, nanocatalysis is one of the most advanced methods, with the greatest futuristic projection because nanocatalysts can be designed and programmed to measure, so their performance and reusability are guaranteed. In this sense, the use of nanotechnologies in biodiesel production processes is relevant. Nanometals, nanotubes, and nanofibers are the nanocatalysts that have currently been used for the synthesis of biodiesel [149], and the advantages of its application are extremely great. Among them, we find a high ratio between the surface area and the volume, which translates into a greater quantity and distribution of active sites, and they configure a large number of morphologies and have good catalytic activity, greater capacity for adsorption, higher percentage of crystallinity, and good chemical stability [150]. Nanocatalysts overcome the drawbacks of the anaerobic methods used for the production of biodiesel; insofar as when they are spread in porous materials, they very easily convert cellulose into FAMEs. Nanocatalysts based on semiconductor materials, metals, polymeric nanocomposites, and hydrogels, among others, have been designed. Nanocatalysts improve biodiesel conversion in transesterification reactions up to 85% regardless of the impurities present in the source. Among the most effective biocatalysts are carbon nanotubes, and the immobilization of enzymes on carbon nanotube supports improves their catalytic activity when immobilized, and this increase is due to nanoscale fixation and the large surface area of biologic material, which is reflected in higher production of biodiesel. In this context, the functionalization and nano conjugation of nanomaterials for transesterification reactions can also increase the conversion of biodiesel in percentages greater than 95% depending on the source used. Nanocatalysts are the materials called to generate clean energy of high quality and sustainability [151–153].

6. Conclusions

A brief description was made regarding the benefits and drawbacks of the generation of biodiesel type biofuels from the perspective of establishing more efficient alternatives in the future that help meet the growing demand for clean energy and gradually reduce dependence on fuels. A brief description was made regarding the benefits and drawbacks of the generation of biodiesel type biofuels from the perspective of establishing more efficient alternatives in the future that help meet the growing demand for clean energy and gradually reduce dependence on fuels, without forgetting that the primary objective is the implementation of clean technologies, whose start-up and production are more ecologic.

In the race for the search for better and more effective biodiesel type biofuels, raw materials such as sunflower oil, palm oil, rapeseed oil, soybean oil, and palm oil have been used, which are the key raw materials in biodiesel production worldwide. However, the high cost of raw materials and questionable sustainability have led to the search for low-cost raw materials, such as nonedible vegetable oils, microalgae oils, animal fats, used cooking oils, transgenic plants, microbial oils, and biowaste, among other sources and application of more efficient catalysts. Nowadays, biodiesel production is not profitable, because even more work is required to concentrate on raw materials and catalysts that are up to sustainability requirements, including low cost and environmental friendliness.

7. Perspectives and challenges for the future of biodiesel

Biodiesel is considered a technologic energy solution that involves the use and transformation of inputs of biologic origin, and it has many environmental and social advantages in relation to petrodiesel. However, biodiesel production has inherent problems associated with the raw materials and catalysts used. The growing global demand for sustainable energy resources is causing a rapid increase in biodiesel consumption. In this regard, studies have been continuously conducted to address process barriers in biodiesel production.

The production of biofuels has had an increase as they are much friendlier in relation to petroleum diesel, for which they have been considered a clean and priority source of renewable energy.

Despite the advantages that these biofuels present, the socioeconomic and environmental effects that their development and application may have are worrying, among which we can mention the replacement of food growing areas by crops to the production of raw material for the production of biodiesel, the increase in deforestation of forests, and therefore the loss of biologic diversity due to the expansion of these crops, and the contamination of surface waters due to the different processes in their production. In addition the use of species of subsistence produce for the production of biofuels has resulted in the flourishing of economic and social problems for the most vulnerable populations. For this reason, it is essential to propose and apply alternatives for the production of biofuels that are friendlier to the environment and that do not affect food security and the integrity of ecosystems in such a forceful way.

In this sense, several types of biofuels have been developed, which are classified according to the input or raw material and the technology used to produce them. According to advances in developed technology, biodiesel is classified by generations, in which first-generation biofuels are currently predominant. The production of new catalytic materials that in turn have relevant applicability in new processes and technology that contribute to obtaining biofuels from more profitable and environmentally friendly sources and processes is preponderant.

Acknowledgments

José Manuel Martínez Gil thanks the Universidad de la Guajira (Colombia) for the commission of doctoral studies granted. The Professors Liliana Giraldo and Juan Carlos Moreno-Piraján authors thank the framework agreement between Universidad Nacional de Colombia and Universidad de los Andes (Bogotá, Colombia) under which this work was carried out. Professor Juan Carlos Moreno-Piraján also is thanks by grant awarded the Facultad de Ciencias de Universidad de los Andes, number INV-2021−128-2257 under which this work was carried out.

References

[1] V.D. Chaudhari, V.S. Jagdale, D. Chorey, D. y Deshmukh, Características de combustión y ruptura por aspersión de biodiesel para aplicaciones de arranque en frío, Ingeniería y tecnología más limpias 5 (2021) 100285.

[2] J. Calero, D. Luna, C. Luna, F.M. Bautista, A.A. Romero, A. Posadillo, R. Estevez, Optimización por metodología de superficie de respuesta de las condiciones de reacción en la transesterificación 1,3-selectiva de aceite de girasol, utilizando CaO como catalizador heterogéneo, Catálisis Molecular 484 (2020) 110804 (Diferencias biodiesel Vs Ecodiesel).

[3] P.R. Yaashikaa, P.S. Kumar, S. y Karishma, Catalizadores bioderivados para la producción de biodiesel: una revisión de materias primas, metodologías de extracción de petróleo, reactores y evaluación del ciclo de vida del biodiesel, Combustible 316 (2022) 123379 (Mecanismo de la transesterificación).

[4] L. Seniorita, E. Minami, H. y Kawamoto, Predicción del comportamiento de solidificación de biodiesel que contiene monoacilgliceroles por encima del límite de solubilidad, Combustible 315 (2022) 123204 (Variaciones).

[5] R.W. Heiden, S. Schober, M. Mittelbach, Solubility limitations of residual steryl glucosides, saturated monoglycerides and glycerol in commercial biodiesel fuels as determinants of filter blockages, J. Am. Oil Chem. Soc. 98 (12) (2021) 1143−1165.

[6] A. Krishnasamy, K.R. y Bukkarapu, Una revisión exhaustiva de los modelos de predicción de propiedades del biodiésel para estudios de modelos de combustión, Combustible 302 (2021) 121085.

[7] G.M. Mathew, D. Raina, V. Narisetty, V. Kumar, S. Saran, A. Pugazhendi, et al., Recent advances in biodiesel production: challenges and solutions, Sci. Total Environ. 794 (2021) 148751.

[8] L. Zhang, W. y Bai, Sostenibilidad del biodiésel a base de cultivos para el transporte en China: análisis de barreras y cálculos de la huella ecológica del ciclo de vida, Pronóstico tecnológico y cambio Soc. 164 (2021) 120526.

[9] T. Badawy, M.S. Mansour, A.M. Daabo, M.M.A. Aziz, A.A. Othman, F. Barsoum, et al., Selección de cultivo de segunda generación para la extracción de biodiesel y prueba de su impacto con nano aditivos en el rendimiento y emisiones de motores diesel, Energía 237 (2021) 121605.

[10] M.J. Zarrinmehr, E. Daneshvar, S. Nigam, K.P. Gopinath, J.K. Biswas, E.E. Kwon, et al., El efecto de la polaridad de los solventes y las condiciones de extracción en el rendimiento de lípidos de microalgas, el perfil de ácidos grasos y las propiedades del biodiésel, Tecnología de biorecursos 344 (2022) 126303.

[11] M. Suvarna, M.I. Jahirul, W.H. Aaron-Yeap, C.V. Augustine, A. Umesh, M.G. Rasul, et al., Predicción de las propiedades del biodiésel y su perfil óptimo de ácidos grasos a través del aprendizaje automático explicable, Energía renovable, 2022.

[12] S.P. Yeong, Y. San Chan, M.C. Law, J.K.U. y Ling, Mejora de las propiedades de flujo en frío del biodiésel destilado de ácidos grasos de palma a través de la destilación al vacío, Diario de biorecursos y bioproductos 7 (1) (2022) 43−51.

[13] B.A. Ziyad, M. Yousfi, Y. Vander Heyden, Effects of growing region and maturity stages on oil yield, fatty acid profile and tocopherols of Pistacia atlantica Desf. Fruit and their implications on resulting biodiesel, Renew. Energy 181 (2022) 167−181.

[14] S.M. Karishma, U. Rajak, B.K. Naik, A. Dasore, R. y Konijeti, Evaluación de las características de rendimiento y emisión de un motor de encendido por compresión alimentado con mezclas de biodiésel antioxidante catecol-daok novedoso, Energía (2022) 123304.

[15] B. Najafi, F. Haghighatshoar, S. Ardabili, S. S. Band, K.W. Chau, A. y Mosavi, Efectos del hidroxi de bajo nivel como aditivo gaseoso en el rendimiento y las características de emisión de un motor diésel de combustible dual alimentado con mezclas de diésel/biodiésel, Aplicaciones de ingeniería de la mecánica de fluidos computacional 15 (1) (2021) 236−250.

[16] M. Hossain, S.S. Israt, N. Muntaha, M.S. y Jamal, Efecto de los antioxidantes y la mezcla con diesel en biodiesel de aceite de pescado parcialmente hidrogenado para mejorar la estabilidad oxidativa, Informes de tecnología de biorecursos 17 (2022) 100938.

[17] C.A. Moreira, E.C. Faria, J.E. Queiroz, V.S. Duarte, M.D.N. Gomes, A.M. da Silva, et al., Información estructural y análisis antioxidante de una trimetoxi chalcona con potencial como aditivo de mezcla de diésel y biodiésel, Tecnología de procesamiento de combustible 227 (2022) 107122.

[18] K.K. Pandey, J. Paparao, S. y Murugan, Estudios experimentales de un motor diésel DI en modo LHR que funciona con biodiésel dopado con antioxidantes, Combustible 313 (2022) 123028.

[19] L.S. Schaumlöffel, L.A. Fontoura, S.J. Santos, L.F. Pontes, M. Gutterres, Vegetable tannins-based additive as antioxidant for biodiesel, Fuel 292 (2021) 120198.

[20] A.C. Roveda, I.P. de Oliveira, A.R.L. Caires, D. Rinaldo, V.S. Ferreira, M.A.G. Trindade, Mejora de la actividad del butilhidroxitolueno con antioxidantes secundarios alternativos: alto efecto sinérgico en la estabilización de mezclas de biodiésel/combustible diésel en presencia de metales prooxidantes, Cultivos y Prod. Indus. 178 (2022) 114558.

[21] T.C.P.M. Ramos, E.P.S. Santos, M. Ventura, J.C. Pina, A.A. Cavalheiro, A.R. Fiorucci, M.S. Silva, Acciones antioxidantes de Eugenol y TBHQ en biodiesel comercial obtenido a partir de aceite de soja y grasa animal, Combustible 286 (2021) 119374.

[22] P.V. Elumalai, M. Parthasarathy, J.S.C.I.J. Lalvani, H. Mehboob, O.D. Samuel, C.C. Enweremadu, et al., Efecto de la sincronización de la inyección en la reducción de los contaminantes nocivos emitidos por el motor de combustión interna utilizando biodiésel Mahua ecológico mezclado con antioxidantes de N-butanol, Informes de energía 7 (2021) 6205−6221.

[23] L.S. De Sousa, C.V.R. de Moura, E.M. y de Moura, Acción de los antioxidantes naturales sobre la estabilidad oxidativa del biodiesel de soja durante el almacenamiento, Combustible 288 (2021) 119632.

[24] J.S. Rodrigues, C.P. do Valle, A.F.J. Uchoa, D.M. Ramos, da Ponte, M.A. de Sousa Rios, et al., Estudio comparativo de antioxidantes sintéticos y naturales sobre la estabilidad oxidativa del biodiesel de aceite de Tilapia, Energía renovable 156 (2020) 1100−1106.

[25] T.R. Nogueira, I. de Mesquita Figueredo, F.M.T. Luna, C.L. Cavalcante Jr., dos Santos, Á. JED, M.A.S. Lima, et al., Evaluación de la estabilidad oxidativa del biodiesel de soya usando extractos etanólicos y clorofórmicos de Platymiscium floribundum como antioxidante, Energía Renovable 159 (2020) 767−774.

[26] N. Jeyakumar, B. Narayanasamy, D. Balasubramanian, K. y Viswanathan, Caracterización y efecto de Moringa Oleifera Lam. Aditivo antioxidante sobre la estabilidad de almacenamiento del biodiesel de Jatropha, Combustible 281 (2020) 118614.

[27] M. Sui, F. Li, S. y Wang, Estudio del mecanismo de antioxidación del antioxidante líquido iónico [MI] [$C_6H_2(OH)_3COO$] del biodiésel, Energía Renovable 165 (2021) 565–572.

[28] R. Bharti, B. y Singh, Extracto de té verde (*Camellia assamica*) como aditivo antioxidante para mejorar la estabilidad a la oxidación del biodiésel sintetizado a partir de aceite de cocina usado, Combustible 262 (2020) 116658.

[29] Z.H. Ni, F.S. Li, H. Wang, S. Wang, S.Y. Gao, L. y Zhou, Desempeño antioxidante y propiedades solubles en aceite de antioxidantes convencionales en biodiesel de aceite de semilla de caucho, Energía Renov. 145 (2020) 93–98.

[30] Y. Devarajan, B. Nagappan, G. Mageshwaran, M.S. Kumar, R.B. y Durairaj, Estudio de viabilidad del empleo de diversos antioxidantes como aditivo en motores diésel de investigación que funcionan con mezclas de diésel y biodiésel, Combustible 277 (2020) 118161.

[31] J.R. Banu, S. Kavitha, R.Y. Kannah, M.D. Kumar, A.E. Atabani, G. y Kumar, Biorrefinería de residuos de posos de café: camino viable hacia la bioeconomía circular, Tecnología de biorecursos 302 (2020) 122821.

[32] C.J. Romola, M. Meganaharshini, S.P. Rigby, I.G. Moorthy, R.S. Kumar, S. y Karthikumar, Una revisión exhaustiva de la selección de antioxidantes naturales y sintéticos para mejorar la estabilidad oxidativa del biodiesel, Revisiones de energía renovable y sostenible 145 (2021) 111109.

[33] K. Velásquez Velásquez, Implementación de indicadores de emisión en estudios de movilidad sostenible a través del aplicativo AppiMotion, 2020, 10, Issue 2, 2022, 107265, ISSN 2213–3437. http://doi.org/10.1016/j.jece.2022.107265.

[34] E. Gutierrez Ballesteros, Study and analysis of obtaining biodiesel from jatropha curcas and application to alternative internal combustion engines, 2020.

[35] J.M.M. Quispe Huanca, Determinación de los tiempos de la biodegradación del hidrocarburo (Diésel B5) a diferentes concentraciones en biorreactores de polietileno con un inóculo de bacterias nativas de suelo contaminado con hidrocarburos, 2020.

[36] D. Singh, D. Sharma, S.L. Soni, S. Sharma, P.K. Sharma, A. Jhalani, A review on feedstocks, production processes, and yield for different generations of biodiesel, Fuel 262 (2020) 116553.

[37] M. Mostafaei, Modelos ANFIS para la predicción del número de cetano de los combustibles biodiésel utilizando la función de deseabilidad, Combustible 216 (2018) 665–672.

[38] S.F. Aguayza Pillco, J.R. Quintuña Córdova, Análisis del funcionamiento de un motor de encendido por compresión mediante diferentes porcentajes de diésel-butanol para la determinación del comportamiento de potencia y emisiones contaminantes en la ciudad de Cuenca (Bachelor's thesis), 2020.

[39] Brahma Sujata, Nath Biswajit, Basumatary Bidangshri, Das Bipul, Saikia Pankaj, Patir Khemnath, Basumatary Sanjay, Biodiesel production from mixed oils: A sustainable approach towards industrial biofuel production, Chem. Eng. J. Adv. (2022) 100284. ISSN 2666–8211. http://doi.org/10.1016/j.ceja.2022.100284.

[40] P. Siudem, A. Zielińska, K. y Paradowska, Aplicación de la RMN 1H en el estudio de la composición de ácidos grasos de los aceites vegetales, Revista de Análisis Farmacéutico y Biomédico, 2022, p. 114658 (Composición de aceites).

[41] G. Vaidya, B.T. Nalla, D.K. Sharma, J. Thangaraja, Y. Devarajan, V.S. y Ponnappan, Producción de biodiesel a partir de aceite de fénix sylvestris: técnica de optimización de procesos, Química y Farmacia Sostenible 26 (2022) 100636.

[42] M.E. Chiosso, Desarrollo de catalizadores sólidos ácidos para la eterificación de glicerol obtenido en la producción de biodiésel, Doctoral dissertation, Universidad Nacional de La Plata, 2020.

[43] I.M. Rizwanul Fattah, H.C. Ong, T.M.I. Mahlia, M. Mofijur, A.S. Silitonga, S.M. Rahman, A. y Ahmad, Estado del arte de los catalizadores para la producción de biodiesel, Fronteras en la investigación energética 8 (2020) 101.

[44] A.M. Sabzevar, M. Ghahramaninezhad, M.N. Shahrak, Enhanced biodiesel production from oleic acid using TiO_2-decorated magnetic ZIF-8 nanocomposite catalyst and its utilization for used frying oil conversion to valuable product, Fuel 288 (2021) 119586.

[45] Y. Tian, F. Zhang, J. Wang, L. Cao, Q. Han, A review on solid acid catalysis for sustainable production of levulinic acid and levulinate esters from biomass derivatives, Bioresource Technol. 342 (2021a) 125977.

[46] N.M. Marzouk, A.O.A. El Naga, S.A. Younis, S.A. Shaban, A.M. El Torgoman, F.Y. y El Kady, Optimización de procesos de producción de biodiesel vía esterificación de ácido oleico utilizando ZSM-5 mesoporoso jerárquico sulfonado como catalizador heterogéneo eficiente, Revista de ingeniería química ambiental 9 (2) (2021) 105035.

[47] F. Rajabi, R. y Luque, Estructura de líquidos iónicos próticos de piridinio funcionalizados mesoporosos altamente ordenados como sistema eficiente en reacciones de esterificación para la producción de biocombustibles, Catálisis Molecular 498 (2020) 111238.

[48] M. Miceli, P. Frontera, A. Macario, A. Malara, Recuperación/reutilización de catalizadores usados heterogéneos soportados, Catalizadores 11 (5) (2021) 591.

[49] W. Xie, H. y Wang, Líquido iónico sulfonado polimérico inmovilizado en compuestos estructurados de núcleo-carcasa de Fe_3O_4/SiO_2: un catalizador reciclable magnéticamente para transesterificación y esterificaciones simultáneas de aceites de bajo costo a biodiesel, Energía renovable 145 (2020) 1709–1719.

[50] C.M. Mayo, Estudio de la composición y estabilidad de biodiesel obtenido a partir de aceites vegetales limpios y procedentes de aceites de fritura, Universidad de La Laguna, (Canary Islands, Spain), 2020.

[51] Y.G. Blanco Vásquez, M.C. Perez Chahua, Elaboración de mármol sintético a partir de la cáscara de huevo, 2021.

[52] M.O. Faruque, S.A. Razzak, M.M. y Hossain, Aplicación de catalizadores heterogéneos para la producción de biodiesel a partir de aceite de microalgas: una revisión, Catalizadores 10 (9) (2020) 1025.

[53] I. Ahangar, F.Q. y Mir, Desarrollo de una membrana de intercambio iónico compuesta de tungstato de circonio (ZrW/PVA) soportada por alcohol polivinílico (PVA), Revista internacional de energía de hidrógeno 45 (56) (2020) 32433–32441.

[54] S. Dehghani, M. y Haghighi, Dispersión de CaO mejorada con sono sobre nanocatalizador bifuncional MCM-41 dopado con Zr con varias proporciones de Si/Zr para la conversión de aceite de cocina usado en biodiesel, Energía Renov. 153 (2020) 801–812.

[55] H. Pang, G. Yang, L. Li, J. y Yu, Transesterificación eficiente sobre zeolitas bidimensionales para la producción sostenible de biodiesel, Energía verde y medio ambiente 5 (4) (2020) 405–413.

[56] J.A.S. Costa, R.A. de Jesus, D.O. Santos, J.B. Neris, R.T. Figueiredo, C.M. Paranhos, Síntesis, funcionalización y aplicación ambiental de materiales mesoporosos a base de sílice de las familias M41S y SBA-n: una revisión, Revista de ingeniería química ambiental 9 (3) (2021) 105259.

[57] G.T. Kadja, M.M. Ilmi, N.J. Azhari, M. Khalil, A.T. Fajar, I.G.B.N. Makertihartha, et al., Avances recientes en los catalizadores nanoporosos para la generación de combustibles renovables, Revista de Investigación y Tecnología de Materiales, 2022.

[58] A. Al Obeidli, H.B. Salah, M. Al Murisi, R. Sabouni, Recent advancements in MOFs synthesis and their green applications, Int. J. Hydro. Energy (2021).

[59] A. Schoedel, Secondary building units of MOFs, in: Metal-Organic Frameworks for Biomedical Applications, Woodhead Publishing, 2020, pp. 11–44.

[60] J. Ha, J.H. Lee, H.R. Moon, Alterations to secondary building units of metal–organic frameworks for the development of new functions, Inorgan. Chem. Front. 7 (1) (2020) 12–27.

[61] V. Hoseinpour, Z. Shariatinia, Applications of zeolitic imidazolate framework-8 (ZIF-8) in bone tissue engineering: A review, Tissue Cell 72 (2021) 101588.

[62] U. Uzoh, Síntesis y caracterización de estructuras metal-orgánicas (MOF) para la captura y conversión de CO_2 (Tesis de Máster, Itä-Suomen y liopisto), 2021.

[63] C.W. Chang, Z.J. Gong, N.C. Huang, C.Y. Wang, W.Y. y Yu, Nanopartículas de MgO confinadas en ZIF-8 como catalizadores bifuncionales ácido-base para mejorar la producción de carbonato de glicerol a partir de la transesterificación de glicerol y carbonato de dimetilo, Catálisis hoy 351 (2020) 21–29.

[64] A. Schoedel, S. Rajeh, Why design matters: From decorated metal-oxide clusters to functional metal-organic frameworks, Metal-Organ. Frame. (2020) 1–55.

[65] A.S. Khder, M. Morad, H.M. Altass, A.A. Ibrahim, S.A. Ahmed, Unprecedented green chemistry approach: tungstophosphoric acid encapsulated in MOF 199 as competent acid catalyst for some significant organic transformations, J. Por. Mater. 28 (1) (2021) 129–142.

[66] T.T. Minh, N.T.T. Tu, T.T. Van Thi, L.T. Hoa, H.T. Long, N.H. Phong, et al., Synthesis of porous octahedral ZnO/CuO composites from Zn/Cu-based MOF-199 and their applications in visible-light-driven photocatalytic degradation of dyes, J. Nanomat. (2019), 2019.

[67] S. Gautam, H. Agrawal, M. Thakur, A. Akbari, H. Sharda, R. Kaur, M. Amini, Metal oxides and metal organic frameworks for the photocatalytic degradation: a review, J. Environ Chem Eng. 8 (3) (2020) 103726.

[68] Y. Hu, L. Dai, D. Liu, W. y Du, Espacio poroso hidrofóbico constituido en ZIF-8 macroporoso para la inmovilización de lipasa mejorando en gran medida el rendimiento catalítico de la lipasa en la preparación de biodiesel, Biotecnología para biocombustibles 13 (1) (2020) 1–9.

[69] M. Bergaoui, M. Khalfaoui, A. Awadallah-F, S. Al-Muhtaseb, A review of the features and applications of ZIF-8 and its derivatives for separating CO2 and isomers of C3-and C4-hydrocarbons, J. Nat. Gas Sci. Eng. 96 (2021) 104289.

[70] M. Taheri, D. Ashok, T. Sen, T.G. Enge, N.K. Verma, A. Tricoli, et al., Stability of ZIF-8 nanopowders in bacterial culture media and its implication for antibacterial properties, Chem. Eng. J. 413 (2021) 127511.

[71] T. Pangestu, Y. Kurniawan, F.E. Soetaredjo, S.P. Santoso, M. Irawaty, M. Yuliana, et al., The synthesis of biodiesel using copper based metal-organic framework as a catalyst, J. Environ. Chem. Eng. 7 (4) (2019) 103277.

[72] M.O. Abdelmigeed, E.G. Al-Sakkari, M.S. Hefney, F.M. Ismail, T.S. Ahmed, I.M. Ismail, Biodiesel production catalyzed by NaOH/Magnetized ZIF-8: Yield improvement using methanolysis and catalyst reusability enhancement, Renew. Energy 174 (2021) 253–261.

[73] W. Xie, C. Gao, J. Li, Sustainable biodiesel production from low-quantity oils utilizing H6PV3MoW8O40 supported on magnetic Fe3O4/ZIF-8 composites, Renew. Energy 168 (2021) 927–937.

[74] M. Ashjari, M. Garmroodi, F.A. Asl, M. Emampour, M. Yousefi, M.P. Lish, et al., Aplicación de una reacción multicomponente para la inmovilización covalente de dos lipasas sobre nanopartículas magnéticas funcionalizadas con aldehído; producción de biodiesel a partir de aceite de cocina usado, Bioquímica de procesos 90 (2020) 156–167.

[75] N. Wang, Q. Sun, J. Yu, Ultrasmall metal nanoparticles confined within crystalline nanoporous materials: a fascinating class of nanocatalysts, Advan. Mater. 31 (1) (2019) 1803966.

[76] A. Kumar, R. Gudiukaite, A. Gricajeva, M. Sadauskas, V. Malunavicius, H. Kamyab, et al., Enzimas lipolíticas microbianas: prometedores biocatalizadores energéticamente eficientes en biorremediación, Energía 192 (2020) 116674.

[77] B. Angulo, J.M. Fraile, L. Gil, C.I. Herrerías, Comparación de métodos químicos y enzimáticos para la transesterificación de ésteres etílicos grasos de aceite de pescado residual con diferentes alcoholes, ACS Omega 5 (3) (2020) 1479–1487.

[78] F. Kovacic, N. Babic, U. Krauss, K. y Jaeger, Clasificación de las enzimas lipolíticas de bacterias, Utilización aeróbica de hidrocarburos, aceites y lípidos 24 (2019) 255–289.

[79] L.A. Salazar Carranza, M.M. Hinojoza Guerrero, M.P. Acosta Gaibor, A.F. Escobar Torres, A.J. Scrich Vázquez, Caracterización, clasificación y usos de las enzimas lipasas en la producción industrial, Revista Cubana de Investigaciones Biomédicas 39 (4) (2020).

[80] A. Castilla, S.R. Giordano, G. Irazoqui, Extremophilic lipases and esterases: Characteristics and industrial applications, in: Microbial Extremozymes, Academic Press, 2022, pp. 207–222.

[81] F.T.T. Cavalcante, F.S. Neto, I.R. de Aguiar Falcão, J.E. da Silva Souza, L.S. de Moura Junior, P. da Silva Sousa, et al., Oportunidades para mejorar la producción de biodiesel vía catálisis de lipasa, Combustible 288 (2021) 119577.

[82] F. Rafiee, M. y Rezaee, Diferentes estrategias para la inmovilización de lipasas sobre los soportes a base de quitosano y sus aplicaciones, Revista internacional de macromoléculas biológicas 179 (2021) 170−195.

[83] P. Chandra, R. Singh, P.K. y Arora, Lipasas microbianas y sus aplicaciones industriales: una revisión exhaustiva, Fábricas de células microbianas 19 (1) (2020) 1−42.

[84] R. Jambulingam, M. Shalma, V. y Shankar, Producción de biodiesel utilizando nanocatalizadores magnéticos funcionalizados inmovilizados con lipasa a partir de lípidos de hongos oleaginosos, Revista de Producción más Limpia 215 (2019) 245−258.

[85] J.M. Loh, W. Gourich, C.L. Chew, C.P. Song, E.S. Chan, Improved biodiesel production from sludge palm oil catalyzed by a low-cost liquid lipase under low-input process conditions, Renew. Energy 177 (2021) 348−358.

[86] K.C.N.R. Pedro, I.E.P. Ferreira, C.A. Henriques, M.A.P. Langone, Enzymatic fatty acid ethyl esters synthesis using acid soybean oil and liquid lipase formulation, Chem. Eng. Commun. 207 (1) (2020) 43−55.

[87] R. Shomal, W. Du, S. Al Zuhair, Immobilization of Lipase on Metal-Organic frameworks for biodiesel production, J. of Environ. Chem. Eng. (2022) 107265.

[88] R. Shomal, B. Ogubadejo, T. Shittu, E. Mahmoud, W. Du, S. Al-Zuhair, Advances in Enzyme and Ionic Liquid Immobilization for Enhanced in MOFs for Biodiesel Production, Molecules 26 (12) (2021) 3512.

[89] H. Xia, N. Li, X. Zhong, Y. Jiang, Metal-organic frameworks: a potential platform for enzyme immobilization and related applications, Front. Bioeng. Biotechnol. 8 (2020) 695.

[90] R.R. Monteiro, J.J. Virgen-Ortíz, A. Berenguer-Murcia, T.N. da Rocha, J.C. dos Santos, A.R. Alcantara, R. Fernandez-Lafuente, Biotechnological relevance of the lipase A from Candida antarctica, Catal. Today 362 (2021) 141−154.

[91] J. Guo, S. Sun, J. Liu, Conversion of waste frying palm oil into biodiesel using free lipase A from Candida antarctica as a novel catalyst, Fuel 267 (2020) 117323.

[92] M.F.K. Ariffin, A. e Idris, Nanopartículas superparamagnéticas recubiertas de Fe_2O_3/quitosano que soportan la enzima lipasa de Candida Antarctica para la producción de biodiésel asistida por microondas, Energía Renov. 185 (2022) 1362−1375.

[93] D. Wan, C. Yan, Q. y Zhang, Síntesis fácil y rápida de nanoesferas de polidopamina mesoporosas magnéticas huecas con estructuras de poros sintonizables para la inmovilización de lipasa: producción verde de biodiesel, Investigación química industrial y de ingeniería 58 (36) (2019) 16358−16369.

[94] M. Shahedi, Z. Habibi, M. Yousefi, J. Brask, M. y Mohammadi, Mejora de la producción de biodiesel a partir de aceite de palma mediante coinmovilización de lipasa de Thermomyces lanuginosa y lipasa B de Candida antarctica: optimización utilizando metodología de superficie de respuesta, Revista internacional de macromoléculas biológicas 170 (2021) 490−502.

[95] D. Kumar, T. Das, B.S. Giri, B. Verma, Preparation and characterization of novel hybrid bio-support material immobilized from Pseudomonas cepacia lipase and its application to enhance biodiesel production, Renew. Energy 147 (2020a) 11−24.

[96] Y. Li, N. Zhong, L.Z. Cheong, J. Huang, H. Chen, S. y Lin, Inmovilización de Candida antarctica Lipase B en SBA-15 modificado orgánicamente para la producción eficiente de mono y diacilgliceroles a base de soja, Revista internacional de macromoléculas biológicas 120 (2018) 886−895.

[97] P. Priyanka, G.K. Kinsella, G.T. Henehan, B.J. Ryan, Isolation and characterization of a novel thermo-solvent-stable lipase from Pseudomonas brenneri and its application in biodiesel synthesis, Biocatal. Agricul. Biotechnol. 29 (2020) 101806.

[98] D. Kumar, T. Das, B.S. Giri, B. Verma, Optimization of biodiesel synthesis from nonedible oil using immobilized bio-support catalysts in jacketed packed bed bioreactor by response surface methodology, J. of Clean. Produc. 244 (2020b) 118700.

[99] V. Kumar, I.S. y Thakur, Producción de biodiesel a partir de la transesterificación de Serratia sp. Lípidos ISTD04 utilizando lipasa inmovilizada en materiales biocompuestos de productos biomineralizados de bacterias secuestradoras de dióxido de carbono, Tecnología de biorecursos 307 (2020) 123193.

[100] R.C. Rial, O.N. de Freitas, C.E.D. Nazário, L.H. Viana, Biodiesel from soybean oil using Porcine pancreas lipase immobilized on a new support: p-nitrobenzyl cellulose xanthate, Renew. Energy 149 (2020) 970−979.

[101] F. Esmi, T. Nematian, Z. Salehi, A.A. Khodadadi, A.K. y Dalai, Sílice mesoporosa funcionalizada con amina y aldehído en nanopartículas magnéticas para mejorar la inmovilización de la lipasa, la producción de biodiésel y la fácil separación, Combustible 291 (2021a) 120126.

[102] F. Esmi, T. Nematian, Z. Salehi, A.A. Khodadadi, A.K. Dalai, Amine and aldehyde functionalized mesoporous silica on magnetic nanoparticles for enhanced lipase immobilization, biodiesel production, and facile separation, Fuel 291 (2021b) 120126.

[103] L. Zhong, Y. Feng, H. Hu, J. Xu, Z. Wang, Y. Du, et al., Rendimiento enzimático mejorado de la lipasa inmovilizada en marcos orgánicos metálicos con recubrimiento superhidrofóbico para la producción de biodiesel, J. of Coll. Interf. Sci. 602 (2021) 426−436.

[104] P. Paitaid, A. y H-Kittikun, Mejora de la inmovilización de la lipasa ST11 de Aspergillus oryzae en la membrana nanofibrosa de poliacrilonitrilo por albúmina de suero bovino y su aplicación para la producción de biodiesel, Bioquímica preparativa y biotecnología 51 (6) (2021) 536−549.

[105] Z.W. Zhou, C.X. Cai, X. Xing, J. Li, Z.E. Hu, Z.B. Xie, et al., Magnetic COFs as satisfactory support for lipase immobilization and recovery to effectively achieve the production of biodiesel by maintenance of enzyme activity, Biotechnol. Biofuel. 14 (1) (2021) 1−12.

[106] M. Tian, J. Fu, Z. Wang, C. Miao, P. Lv, D. He, et al., Actividad y estabilidad mejoradas de la lipasa Rhizomucor miehei al mutar el sitio de glicosilación ligado a N y su aplicación en la producción de biodiesel, Combustible 304 (2021) 121514.

[107] H. Li, J. Watson, Y. Zhang, H. Lu, Z. Liu, Environment-enhancing process for algal wastewater treatment, heavy metal control and hydrothermal biofuel production: A critical review, Bioresource Technol. 298 (2020) 122421.

[108] Y.D. Wendy, M.N. Fauziah, Y.S. Baidurah, W.Y. Tong, C.K. Lee, Production and characterization of polyhydroxybutyrate (PHB) BY Burkholderia cepacia BPT1213 using waste glycerol as carbon source, Biocatal. Agricul. Biotechnol. 41 (2022) 102310.

[109] R.V. Quah, Y.H. Tan, N.M. Mubarak, J. Kansedo, M. Khalid, E.C. Abdullah, M.O. Abdullah, Magnetic biochar derived from waste palm kernel shell for biodiesel production via sulfonation, Waste Manage. 118 (2020) 626–636.

[110] J.C. Lee, B. Lee, Y.S. Ok, H. Lim, Preliminary techno-economic analysis of biodiesel production over solid-biochar, Bioresource Technol. 306 (2020) 123086.

[111] J. Vakros, Biochars y su uso como catalizadores de transesterificación para la producción de biodiesel: una breve revisión, Catalizadores 8 (11) (2018) 562.

[112] M. Qaim, K.T. Sibhatu, H. Siregar, I. Grass, Environmental, economic, and social consequences of the oil palm boom, Ann. Rev. Res. Econom. 12 (2020) 321–344.

[113] G.O. Ferrero, E.M.S. Faba, G.A. Eimer, Biodiesel production from alternative raw materials using a heterogeneous low ordered biosilicified enzyme as biocatalyst, Biotechnol. Biofuel. 14 (1) (2021) 1–11.

[114] T.M.I. Mahlia, Z.A.H.S. Syazmi, M. Mofijur, A.P. Abas, M.R. Bilad, H.C. Ong, A.S. Silitonga, Patent landscape review on biodiesel production: Technology updates, Renew. Sustain. Energy Rev. 118 (2020) 109526.

[115] N.M. Clauser, G. González, C.M. Mendieta, J. Kruyeniski, M.C. Area, M.E. Vallejos, Biomass waste as sustainable raw material for energy and fuels, Sustainability 13 (2) (2021) 794.

[116] F. Fernández-Tirado, C. Parra-López, M. Romero-Gámez, A multi-criteria sustainability assessment for biodiesel alternatives in Spain: Life cycle assessment normalization and weighting, Renew. Energy 164 (2021) 1195–1203.

[117] M. Arif, T. Jan, H. Munir, F. Rasul, M. Riaz, S. Fahad, et al., Climate-smart agriculture: assessment and adaptation strategies in changing climate, in: Global Climate Change and Environmental Policy, Springer, Singapore, 2020, pp. 351–377.

[118] K. Kurowska, R. Marks-Bielska, S. Bielski, H. Kryszk, A. Jasinskas, Food security in the context of liquid biofuels production, Energies 13 (23) (2020) 6247.

[119] A.P. Sinitsyn, O.A. Sinitsyna, Bioconversion of renewable plant biomass. Second-generation biofuels: raw materials, biomass pretreatment, enzymes, processes, and cost analysis, Biochemistry (Moscow) 86 (1) (2021) S166–S195.

[120] D. Kurczyński, P. Łagowski, G. Wcisło, Experimental study into the effect of the second-generation BBuE biofuel use on the diesel engine parameters and exhaust composition, Fuel 284 (2021) 118982.

[121] S. Foteinis, E. Chatzisymeon, A. Litinas, T. y Tsoutsos, Biodiésel de aceite de cocina usado: evaluación del ciclo de vida y comparación con biocombustibles de primera y tercera generación, Energía Renov. 153 (2020) 588–600.

[122] P. Sharma, M. Usman, E.S. Salama, M. Redina, N. Thakur, X. Li, Evaluation of various waste cooking oils for biodiesel production: A comprehensive analysis of feedstock, Waste Manage. 136 (2021) 219–229.

[123] H. Chowdhury, B. Loganathan, I. Mustary, F. Alam, S.M. Mobin, Algae for biofuels: The third generation of feedstock, in: Second and third generation of feedstocks, Elsevier, 2019, pp. 323–344.

[124] L.M. Guerrero Galindo, D. Díaz Mojica, Evaluación de la capacidad de biofijación de CO2 de un cultivo de microalgas en un proceso de combustión fija (Bachelor's thesis, Fundación Universidad de América), 2020.

[125] Z. Yin, L. Zhu, S. Li, T. Hu, R. Chu, F. Mo, et al., A comprehensive review on cultivation and harvesting of microalgae for biodiesel production: Environmental pollution control and future directions, Bioreso. Technol. 301 (2020) 122804.

[126] V. Godbole, M.K. Pal, P. Gautam, A critical perspective on the scope of interdisciplinary approaches used in fourth-generation biofuel production, Algal Res. 58 (2021) 102436.

[127] A. Syafiuddin, J.H. Chong, A. Yuniarto, T. Hadibarata, The current scenario and challenges of biodiesel production in Asian countries: A review, Biores. Technol. Rep. 12 (2020) 100608.

[128] M. Athar, S. Zaidi, A review of the feedstocks, catalysts, and intensification techniques for sustainable biodiesel production, J. of Environ. Chem. Eng. 8 (6) (2020) 104523.

[129] F. Esmi, V.B. Borugadda, A.K. Dalai, Heteropoly acids as supported solid acid catalysts for sustainable biodiesel production using vegetable oils: A Review, Catal. Today (2022).

[130] M.S. Habib, O. Asghar, A. Hussain, M. Imran, M.P. Mughal, B. Sarkar, A robust possibilistic programming approach toward animal fat-based biodiesel supply chain network design under uncertain environment, J. of Clean. Product. 278 (2021) 122403.

[131] L.T. Vargas-Ibáñez, J.J. Cano-Gómez, P. Zwolinski, D. Evrard, Environmental assessment of an animal fat based biodiesel: Defining goal, scope and life cycle inventory, Procedia CIRP 90 (2020) 215–219.

[132] N.S. Topare, R.I. Jogdand, H.P. Shinde, R.S. More, A. Khan, A.M. Asiri, A short review on approach for biodiesel production: Feedstock's, properties, process parameters and environmental sustainability, Mater. Today: Proceed. (2021).

[133] L.C. Pardo-Locarno, C.D. Dagua, M. Soto, Plagas asociadas a cultivos de palmas de importancia económica en colombia, in: 46° Congreso Socolen, 2019, July, p. 219.

[134] M. Mohadesi, B. Aghel, A. Gouran, M.H. Razmehgir, Transesterification of waste cooking oil using Clay/CaO as a solid base catalyst, Energy 242 (2022) 122536.

[135] A.S. Yusuff, A.O. Gbadamosi, L.T. Popoola, Biodiesel production from transesterified waste cooking oil by zinc-modified anthill catalyst: parametric optimization and biodiesel properties improvement, J. of Environ. Chem. Eng. 9 (2) (2021) 104955.

[136] J. Milano, A.H. Shamsuddin, A.S. Silitonga, A.H. Sebayang, M.A. Siregar, H.H. Masjuki, et al., Tribological study on the biodiesel produced from waste cooking oil, waste cooking oil blend with Calophyllum inophyllum and its diesel blends on lubricant oil, Energy Rep. 8 (2022) 1578–1590.

[137] H. Bai, J. Tian, D. Talifu, K. Okitsu, A. Abulizi, Process optimization of esterification for deacidification in waste cooking oil: RSM approach and for biodiesel production assisted with ultrasonic and solvent, Fuel 318 (2022) 123697.

[138] S.K. Nayak, A.T. Hoang, B. Nayak, P.C. Mishra, Influence of fish oil and waste cooking oil as post mixed binary biodiesel blends on performance improvement and emission reduction in diesel engine, Fuel 289 (2021) 119948.

[139] B. Karmakar, G. Halder, Accelerated conversion of waste cooking oil into biodiesel by injecting 2-propanol and methanol under superheated conditions: A novel approach, Energy Convers. Manage. 247 (2021) 114733.

[140] Q. Li, Y. Chen, S. Bai, X. Shao, L. Jiang, Q. y Li, Lipasa inmovilizada en marcos orgánicos de metal de base biológica construidos por mineralización biomimética: un biocatalizador sostenible para la síntesis de biodiesel, Coloid. y superficies B: Biointerfaces 188 (2020) 110812.

[141] M. Branco-Vieira, D.M. Costa, T.M. Mata, A.A. Martins, M.A. Freitas, N.S. Caetano, Environmental assessment of industrial production of microalgal biodiesel in central-south Chile, J. Clean. Product. 266 (2020) 121756.

[142] V. Makareviciene, E. Sendzikiene, Application of microalgae for the production of biodiesel fuel, in: Handbook of Algal Science, Technology and Medicine, Academic Press, 2020, pp. 353–365.

[143] A.A. Martins, T.M. Mata, N. de Sá Caetano, W.G.M. Junior, M. Gorgich, P.S. Corrêa, Microalgae for biotechnological applications: Cultivation, harvesting and biomass processing, 2020.

[144] L. Zhang, K.C. Loh, A. Kuroki, Y. Dai, Y.W. y Tong, Producción microbiana de biodiesel a partir de residuos orgánicos industriales por microorganismos oleaginosos: estado actual y perspectivas, Diario de materiales peligrosos 402 (2021) 123543.

[145] R. Raksasat, J.W. Lim, W. Kiatkittipong, K. Kiatkittipong, Y.C. Ho, M.K. Lam, et al., A review of organic waste enrichment for inducing palatability of black soldier fly larvae: Wastes to valuable resources, Environ. Pollut. 267 (2020) 115488.

[146] M. Kothri, M. Mavrommati, A.M. Elazzazy, M.N. Baeshen, T.A. Moussa, G. Aggelis, Microbial sources of polyunsaturated fatty acids (PUFAs) and the prospect of organic residues and wastes as growth media for PUFA-producing microorganisms, FEMS Microbiol. Lett. 367 (5) (2020) fnaa028.

[147] A. Patel, O. Sarkar, U. Rova, P. Christakopoulos, L. Matsakas, Valorization of volatile fatty acids derived from low-cost organic waste for lipogenesis in oleaginous microorganisms-a review, Biores. Technol. 321 (2021) 124457.

[148] H. Wang, X. Peng, H. Zhang, S. Yang, H. Li, Microorganisms-promoted biodiesel production from biomass: A review, Energy Convers. Manage.: X 12 (2021) 100137.

[149] T. Biswal, K.P. y Shadangi, Aplicación de la Nanotecnología en la Producción de Biocombustibles. Biocombustibles líquidos: fundamentos, caracterización y aplicaciones, 2021, pp. 487–515.

[150] A.P. Ingle, A.K. Chandel, R. Philippini, S.E. Martiniano, S.S. da Silva, Advances in nanocatalysts mediated biodiesel production: a critical appraisal, Symmetry 12 (2) (2020) 256.

[151] M. Shah (Ed.), Microbial Bioremediation & Biodegradation, Springer, Singapore, 2020.

[152] M. Shah (Ed.), Removal of Refractory Pollutants from Wastewater Treatment Plants, CRC Press, USA, 2021.

[153] M. Shah (Ed.), Removal of Emerging Contaminants through Microbial Processes, Springer, Singapore, 2021.

Further reading

[1] N. Li, L. Zhou, X. Jin, G. Owens, Z. Chen, Simultaneous removal of tetracycline and oxytetracycline antibiotics from wastewater using a ZIF-8 metal organic-framework, J. of Hazard. Mater. 366 (2019) 563–572.

[2] G.F. Mota, I.G. de Sousa, A.L.B. de Oliveira, A.L.G. Cavalcante, K. da Silva Moreira, F.T.T. Cavalcante, et al., Biodiesel production from microalgae using lipase-based catalysts: Current challenges and prospects, Algal Res. 62 (2022) 102616.

[3] Reem Shomal, Wei Du, Sulaiman Al-Zuhair, Immobilization of lipase on metal-organic frameworks for biodiesel production, J. of Environ. Chem. Eng.,

Index

Note: 'Page numbers followed by *f* indicate figures and *t* indicate tables.'

A

Acetogenesis, 441
Acetone–butanol–ethanol fermentation (ABE fermentation), 300–302
Acetyl CoA, 50
Acid catalysis, lignin conversion by, 215
Acid hydrolysis, 283
Acid pretreatment, 38, 228, 279–280
Acid/base transesterification, 285
　heterogeneous catalysis, 285
　homogenous catalysis, 285
Acidic culture, 95
Acidogens, 41–42
Acids, pretreatment using, 208
Acoustic cavitation and chemical and physical consequences, 243–244, 244f
Acoustic intensity effect, 250–252
Acrolein, 79–80
Actinastrum, 391
Additional biomass-based fuels, 131–132
Adenosine triphosphate (ATP), 447
Advanced biofuels, 203
Agaricus volvacea, 29
Agricultural crops, 89
Agricultural residues, 25, 28, 154, 463–464, 481
　properties of, 463–465
　technoeconomic analysis of pyrolysis process, 466–467
　thermochemical conversion process for, 465–466
Agricultural waste (AW), 326, 444, 450
　bioethanol production from biomass created by, 447
Alcohol, 89, 127
Alcoholic fermentation, 127, 449
Alcoholysis. *See* Transesterification
Algae, 46–47, 65, 123, 132, 168–169, 312–313, 315, 391, 444
　absorb carbon dioxide, 60–61
　advantages of algae as biofuel source, 413–414
　algae-based biodiesel, 123
　algae-based biofuel production, 46
　algae-derived biofuels, 123
　algae-to-biocrude process, 395
　benefits and limitations of using algae for bioenergy production purposes, 64–65
　biofuel, 8
　biomass
　　steps of bioethanol production from, 422
　　steps of biohydrogen production from, 424–425
　cultivation
　　conditions, 59–60
　　and effects on biodiesel production, 92–93
　　heterotrophy, 92–93
　　mixotrophy, 93
　　photoautotrophy, 92
　economic importance of algae as biofuel sources, 131–132
　physicochemical parameters for biofuel production from, 49–50
　　effect of light, 49
　　effect of nutrients, 49–50
　　effect of pH, 49
　　effect of salinity, 49
　　effect of temperature, 49
　potentiality of algae as biologic source of energy, 411
　production systems, 59
　steps of biodiesel production from algae biomass, 424–425
　steps of biohydrogen production from algae biomass, 424–425
Algal biodiesel, 130
Algal bioenergy
　algal biofuel, 413–414
　algal resources in Egypt, 425
　biofuel, 412–413
　climate impacts on mediterranean countries, 410
　climatic change consequences on energy, 409–410
　closed system, 417–418
　cultivation systems, 421–422
　difference between diesel and biodiesel, 420
　growth requirements for optimal microalgal growth, 416
　lipids yield of microalgae, 415
　macroalgal cultures, 419
　methodologies of production, 422–424
　　bioethanol recovery, 424
　　fermentation process, 423
　　pretreatment process, 422
　　steps of bioethanol production from algae biomass, 422
　microalga species for biodiesel, 421
　open system, 418–419
　potential strains, 411–412
　species, 411
　steps of biodiesel production from algae biomass, 424–425
　types of, 419–420
　　biodiesel production, 419–420
　　bioethanol production, 419
　types of algal bioenergy, 419–420
Algal biofuel, 413–414
　advantages of algae as biofuel source, 413–414
　artificial cultured biomass, 414
　different policies for algal biofuel commercialization, 7–8
　sources of algal biomass, 414
　types, 128–130
　　biodiesel, 128–130
　　bioethanol, 128
　　biogas, 128
Algal biomass, 54, 60–61, 63–64, 126, 168–170, 312
　algae, 46–47
　algal, 168–169
　anaerobic digestion of, 127
　bioconversion technologies of algal biomass to biofuel, 52–56
　　biodiesel, 53
　　bioelectricity, 54
　　bioethanol, 53
　　biogas, 54
　　biohydrogen production, 54
　　biomass production and phycoremediation, 54
　　value-added products, 55–56
　to biofuels, 100
　conversion techniques/methods of, 126–128
　biofuels obtained from biovalorization of, 174–176
　categories of biofuel generation, 168
　chemical composition, 60–61
　cultivation, 169–170
　economics of bioenergy production from, 65
　genetically engineered algae for biofuel production, 50
　harvesting of, 184–185
　methods for algal biomass conversion to bioenergy products, 63–64
　microalgal-based biofuel, 47
　opportunities and challenges in production of biofuels from, 176

491

492 Index

Algal biomass (*Continued*)
  physicochemical parameters for biofuel production from algae, 49—50
  sources of, 414
    natural standing crop, 414
  techniques for biofuel production from, 98—100
  techniques involved in transformation of third-generation biofuels from, 170—174
    anaerobic digestion of microalgal biomass, 173—174
    extraction of oil or lipid from algal biomass, 170—171
    lipids or oil transesterification, 171
    microbial fermentation of algal biomass, 173
    saccharification and pretreatment of algal biomass, 172
  technologies for macroalgal cultivation, 48
  technologies for microalgal cultivation, 47—48
Algal cultivation, 132
Algal fuel properties, 131
Algal genomes, 50
Algal oil, 55, 76, 98
Algal resources in Egypt, 425
Algal-based biofuel (ABB), 413
Aliphatic hydrocarbons, 471
Alkali pretreatment, 38, 280
Alkaline culture, 95
Alkaline method, pretreatment by, 208
Alkaline pretreatment, 228—229
Alloy based nanocatalyst, 192—193
Alternative Motor Fuels Act (AMFA), 3
Aluminum oxide (Al_2O_3), 18
Aluminum sulfate, 95
Amanita virgata, 29
American Society for Testing and Materials (ASTM), 192
Amino acids, 55
Ammonia, 388
  fiber explosion, 279
  pretreatment, 39
Ammonia fiber expansion (AFEX), 39, 209
Ammonia recycle percolation (ARP), 39
Ammonia solution (AS), 159—161
Anaerobic co-digestion (AcoD), 161
  process, 161
Anaerobic digestion (AD), 302, 305, 331, 440, 443, 446
Anaerobic lagoons, 446
Anaerobic methods, 483
Anaerobically and aerobically treated swine wastewater (AnATSW), 396
Animal fat for biodiesel, 482
Animal feedstuff, 77—78
Animal manure, 316
Anthropogenic processes, 474
Anti-gravity-based separation technique, 94
Antioxidants, 472
Aquatic crop, 326
Aquatic ecosystems, 123
Aquatic Species Program (ASP), 7—8

Aqueous coproducts (ACPs), 395
Aromatic compounds, 472—474
Arsenic (As), 374
Arthrospira platensis, 455
Artificial cultured biomass, 414
  microalgal cultures, 414
Artificial Neural Network model (ANN model), 148—149
Arundo donax, 443—444
Ash-free dry weight (afdw), 395
ASPEN Plus models, 147—148
Aspergillus
  A. awamori, 39, 95
  A. niger, 39, 82
Atmospheric and room temperature plasma (ARTP), 455—456
Atomic force microscopy (AFM), 193
Auto-flocculation process, 114, 185
Autoclaving, 279
Automotive diesel engine, 290
Autotrophic microorganisms, 343, 395—396
2,20-azinobis-(3)-ethyl-benzyl thiazoline-6-sulfonate (ABTS), 402

B

Bacillariophyceae, 440
Bacillariophyta, 45
Bacteria, 167
Bacterial biomass, 64
Bacterial nanocellulose (BC), 332
Bag photobioreactor, 112
Basolites, 476
Bead milling, 278
Bio-based products, 326
Bio-oil, 42
Bioactive chemicals, 60—61
Bioalcohol, 15—16, 434—435
Biobutanol, 300—302, 305
Biocatalyst, 289
Biochar (BC), 25, 175—176, 213, 332, 465—466, 479
Biochemical conversion process, 127—128. *See also* Thermochemical conversion process
  alcoholic fermentation, 127
  anaerobic digestion of algal biomass, 127
  photobiological hydrogen production, 128
Biochemicals, 60—61
Biodiesel, 1—2, 4, 14, 25, 35, 53, 63, 71—74, 83, 89, 107—108, 118, 128—130, 174—175, 191, 313, 332, 350, 393, 409, 471—479
  application, 74—75, 290—292
    biodiesel and human health, 75
    biodiesel improves engine operation, 75
    biodiesel reduces greenhouse gas emissions, 75
    biodiesel reduces tailpipe emissions, 75
    of microbial prospecting, 83—84
    of nanotechnologies in production of biodiesel, 483
  bioconversion of crude glycerol waste into value-added products, 78—83

biodiesel from microalgae oils, 482—483
biodiesel production by transesterification, 474—477
biofuel production using microalgae, 186
biomass (mahua seeds) for conversion to, 16—20
chemistry of biodiesel transesterification of oils, 72—73
crude glycerol waste from biodiesel as by-product, 77—78
different antioxidants as additives to improve stability and durability of, 473t
enzyme-catalyzed transesterification for biodiesel production, 478—479
extraction methods for, 317—318
microbial bioprospecting, 83
problems brought about by, 73—74
production, 75—76, 419—420, 440, 454
  from algae biomass, steps of, 424—425
  algae cultivation and effects on, 92—93
  difference between diesel and, 420
  improvement of performance, 17—18
  microalga species for, 421
  process and recent advances, 17—20
  reduction of exhaust emission, 18—20
  by transesterification, 474—477, 475f
production of biodiesel from oleaginous microorganisms with organic waste as raw materials, 483
prospective oleaginous microorganisms for biodiesel production, 272—274
purification, 286
qualitative determination of, 320
refinery, 20
sources, 76
  algal oil, 76
  vegetable oil, 76
  waste cooking oil, 76
sources for synthesis of biodiesel, 479—482
synthesis
  downstream processes for, 94—100
  upstream processes for, 90—93
transesterification, 118
transition to fuel, 73
from used cooking oil, 482
Bioelectricity, 54, 445
Bioenergy, 59, 336—342, 400, 443, 471
  biomass
    conversion techniques to produce electricity and heat, 445—447
    sources, 443—444
  considerations, 345
  enzymatic hydrolysis, 447—450
  fermentation, 447—450
  genetically engineered microorganisms for CO_2 fixation, 343
  key barriers to converting biomass into bioenergy, 344—345
  LCBs, 339
  microalgae, 341—342
  pectin biomass, 340—341
  production
    algae cultivation conditions, 59—60

algal biomass chemical composition, 60–61
benefits and limitations of using algae for bioenergy production purposes, 64–65
economics of bioenergy production from algal biomass, 65
macro and microalgae biomass production, 60
macro and microalgae for bioenergy production, 61–62
methods for algal biomass conversion to bioenergy products, 63–64
techniques, 445
starchy biomass, 340
Bioenergy production with Carbon Capture and Storage (BECCS), 174
Bioethanol, 5, 25, 35, 40, 53, 107–108, 128, 175, 300, 305, 312, 350, 434, 453–454
biofuel production using microalgae, 186
biomass (mahua flowers) for conversion to, 14–16
extraction methods for, 317–318
production, 108, 115–118, 400–404, 419, 437–439
from biomass created by agricultural waste, 447
enzyme saccharification, 117
ethanol production from lignocelluloses, 438–439
ethanol production from sugar and starch, 437–438
fermentation, 118
pretreatment, 116–117
pretreatment, 447
process, 15–16
rice straw potential for, 400
steps of bioethanol production from algae biomass, 422
systems, 15
techno-economic worthiness, 404
recovery, 424
refinery, 16
Bioflocculation method, 185, 289
Biofuel Decarbonization Credit (CBIO), 4
Biofuels, 1–2, 25, 35, 45, 59, 65, 71–72, 107, 123, 132, 153–154, 173, 191, 271–272, 311, 332, 335, 404, 412–413, 450
advantages and disadvantages of producing biofuels from rice straw and socioeconomic evaluation, 31
benefits of using microalgae for, 181
carbohydrate–protein–lipid composition, 182t
bioconversion technologies of algal biomass to, 52–56
biodiesel, 332
biofuels and economics, 433–434
biomass feedstocks in production of biofuel, 315–316
conversion of rice straw into, 40–42
conversion techniques/methods of algal biomass to, 126–128

cultivation and harvesting of microalgae for, 181–184
economy, 31
environmental benefits, 314–315
ethanol, 332
extraction and determination of different biofuels, 316–320
feedstock processing technologies, 437
feedstocks for biofuel production, 435–436
corn, 435–436
jatropha, 436
palm oil, 436
rapeseed/canola, 436
soybean plants, 436
sugarcane, 436
first-generation biofuel, 412
functions, 181
generation, 167, 203–204, 465
categories of, 168
of biofuels, 434
first-generation biofuels, 203
second-generation biofuels, 203–204
third-and fourth-generation biofuels, 204
genetic engineering for enhancing microalgae biomass and biofuel components, 457–458
genetically engineered algae for biofuel production, 50
genetic engineering to steer hydrogen production, 50
lipid augmentation by genetic alterations, 50
history of biofuel research, 411–412
HTL, 100
hydrogen production from biofuels obtained by thermochemical processes of microalgae biomass, 383–385
LCA and SWOT analysis for biofuels production from rice straw, 30–31
LCA and SWOT analysis for biofuels production from rice straw, 156–157
Madhuca species as green approach for, 14
makers, 84
mathematical models and tools to scrutinize sustainability of, 163–164
modification of growth and physiologic conditions for enhancing microalgae biomass and biofuel components, 454–455
obtained from biovalorization of algal biomass, 174–176
BC, 175–176
biodiesel, 174–175
bioethanol, 175
biomethane, 174
opportunities and challenges in production of biofuels from algal biomass, 176
plant biofuels, 434–435
policies for biofuel production and commercialization, 1–2
Brazil's biofuel national policies implemented for commercialization, 4
challenges with policies and future recommendations, 8–9

different policies for algal biofuel commercialization, 7–8
factors affecting commercialization of cellulosic biofuel, 2–3
latest policies undertaken by USA to commercialize cellulosic biofuels, 3–4
national policy of India to promote biofuel commercialization program, 6–7
recent progress in policies taken by government of China to drive biofuel generation, 5
potential source of, 46–47
macroalgae as source of biofuel, 46–47
microalgae as source of biofuel, 46
pretreatment of RS, 302–304
biological pretreatment, 304
chemical pretreatment, 303–304
physical pretreatment, 303
production strategies, 437–441
biodiesel production, 440
bioethanol production, 437–439
biomethane production, 440–441
process, 304–305
processes to improve biofuel yield from rice straw, 157–163
anaerobic codigestion process, 161
innovative process based on additives, 161–163
pretreatment processes, 157–161
production, 15, 29, 62, 391
biodiesel, 186
bioethanol, 186
biogas, 186
biohydrogen, 185–186
catalysts used so far for, 192–196
comparative highlights of catalytic efficiency of developed catalysts for, 196
different approaches for, 191–192
using microalgae, 185–186
physicochemical parameters for biofuel production from algae, 49–50
process, 29–30, 453–454
strategies, 437–441
pyrolysis, 99
renewable fuel standards II, 435
renewable hydrocarbon fuels, 332
from RS, 300–305, 301t
second-generation biofuel, 412
in situ catalytic transesterification, 99
in situ noncatalytic transesterification, 99
and sources, 312–314
sources
advantages of algae as, 413–414
economic importance of algae as, 131–132
switchgrass, 436–437
third-generation biofuel, 412–413
two-step catalytic transesterification, 98–99
types, 300–302
Biogas, 5, 25, 35, 41–42, 54, 128, 314, 320, 446, 465–466
biofuel production using microalgae, 186

Biogas (*Continued*)
　　extraction methods for, 317–318
　　production of, 29
Biogas power generation (off-grid) and thermal energy application (BPGTP), 6
Biohydrogen (BioH$_2$), 25, 35, 118, 223, 383
　　biofuel production using microalgae, 185–186
　　manufacturing process, 401
　　from microalgae, 118
　　production, 54, 421–422, 424
　　steps of biohydrogen production from algae biomass, 424–425
　　　　direct biophotolysis, 424–425
　　　　indirect biophotolysis, 425
　　　　light-independent reaction, 425
Biohydrogen yield (BHY), 223–224
Biologic source of energy, potentiality of algae as, 411
Biological conversion, 331
Biological molecules cellulose, 437
Biological pretreatment, 402, 447
　　of lignocellulosic biomass, 281
　　process, 159
Biomass, 13, 25, 35, 45, 48, 127, 137, 153–154, 191, 202, 326–329, 391, 443–444
　　advances, 333
　　applications of, 132
　　availability and abundance of, 202–203
　　bioethanol production from biomass created by agricultural waste, 447
　　biomass conversion techniques, 138
　　biomass-based catalysts, 194–196
　　biomass-based fuels, 434
　　challenges, 333–334
　　characteristics of biomass, 138
　　classification of biomass, 137
　　combustion process, 139
　　composition, 59
　　composition and energy content, 327–328
　　conversion techniques to produce electricity and heat, 445–447
　　　　bioethanol production from biomass created by agricultural waste, 447
　　　　fermentation, 446
　　　　gasification techniques, 446
　　　　pyrolysis, 446
　　conversion technologies, 329–331, 330t, 371t
　　conversion to energy, 137
　　cultivation, 454
　　energy, 137
　　environmental impacts, 333
　　feedstocks, 444
　　　　application, 315–316
　　　　in production of biofuel, 315–316
　　forms, 326–327
　　　　food crop as biomass, 326
　　　　nonfood biomass, 326–327
　　fractionation, 365
　　gasification, 148
　　implementation, 333
　　key barriers to converting biomass into bioenergy, 344–345
　　(mahua flowers) for conversion to bioethanol, 14–16
　　　　bioethanol production process, 15–16
　　　　bioethanol refinery, 16
　　(mahua seeds) for conversion to biodiesel, 16–20
　　　　biodiesel production process and recent advances, 17–20
　　　　biodiesel refinery, 20
　　　　seed oil and oil yield, 16
　　measuring and analysis, 329
　　modeling studies, 146–149
　　production and phycoremediation, 54
　　productivity, 90
　　products and applications, 332
　　sources, 443–444
　　　　biomass feedstock, 444
　　　　for integrated biorefineries, 336–342
　　thermochemical conversion, 138–146
Biomethane, 174, 302, 305
　　algal biofuel types, 128
　　production, 440–441
　　　　methane process description, 441
　　　　methanogenic environments, 440
Biomethane potentials (BMPs), 174
Biophotolysis, 185–186, 385, 424
Biopower, 443
Bioprospecting
　　microalgae for biofuel synthesis
　　　　genetic engineering for enhancing microalgae biomass and biofuel components, 457–458
　　　　modification of growth and physiologic conditions for enhancing carbohydrate content of microalgae, 456–457
　　　　modification of growth and physiologic conditions for enhancing lipid content of microalgae, 455–456
　　　　modification of growth and physiologic conditions for enhancing microalgae biomass and biofuel components, 454–455
　　process, 83
Bioreactors
　　cultivation techniques of, 236
　　used for culturing, 183–184
　　　　open ponds, 183
　　　　photobioreactor, 184
Biorefinery, 42, 107, 271–272, 289, 335, 369
　　biomass, 349, 349f
　　classification, 370
　　potential of rice residue in, 37–40
　　system, 55–56
Bioremediation capacity of microalgae, 372–375
　　environmental toxicant monitoring, 373–374
　　use of microalgae for wastewater treatment, 374–375
Biosensor, 238
Biosyngas, 25, 35

Biotechnology industry, 433
Black/dark earth. *See* Biochar
Bligh method, 97
Blue-Green 11 (BG-11), 393
Borax/phth test, 320
Borosilicate glass, 110
Botryococcus, 107
　　B. braunii, 125, 168–171, 181, 312, 393
Brake specific fuel consumption (BSFC), 18
Brake thermal efficiency (BTE), 17–18
Brake-specific fuel consumption (BSFC), 18
Brazil's biofuel national policies implemented for commercialization, 4
Brown algae (*Phaeophyta*), 46–47
Bubble column, 417–418
Bubbling Fluidized Bed (BFB), 139–140
Bulk harvesting, 361
　　flocculation, 361
　　flotation, 361
Bulk rice straw production, 26
Burnable gases, 142
By-product, crude glycerol waste from biodiesel as, 77–78

C

C-phycocyanin, 119
Calcium hydroxide (Ca(OH)$_2$), 402
Calcium oxide (CaO), 159–161, 195, 476
Caldicellulosiruptor saccharolyticus, 223–224
California Air Resources Board (CARB), 75
Calvin cycle, 185–186
Calvin–Benson–Bassham cycle (CBB cycle), 343
Candida antartica, 479
Canola, 440
Carbohydrates, 55, 107, 342, 453–454, 482
　　content, 456–457
　　　modification of growth and physiologic conditions for enhancing carbohydrate content of microalgae, 456–457
　　polymers, 3
　　techniques to identify and quantify, 365
Carbon (C), 60
Carbon capture, utilization, and storage (CCUS), 375
Carbon capture and storage (CCS), 375
Carbon capture and utilization (CCU), 375
Carbon cycle, 123
Carbon dioxide (CO$_2$), 47, 64, 124, 141, 183, 201, 438, 455–456
　　emissions, 13, 123
Carbon emission reduction, 3
Carbon materials, 196
Carbon monoxide (CO), 40, 153–154
Carbon neutral, 72
Carbonyl sulphide, 27
Carboxymethylcellulose (CMC), 230
Cascade extraction, 365
Castor oil, 73
Catalysis, lignin conversion through base, 215
Catalysts

comparative highlights of catalytic efficiency of catalysts for biofuel production, 196
recovery, 282
used for biofuel production, 192–196
biomass-based catalysts, 194–196
metal oxide nanoparticles, 194
metal/alloy based nanocatalyst, 192–193
Cavitation bubbles, 243–244
Cell culture, 117
Cell disruption techniques, 97
Cell dry weight (CDW), 172
Cell wall composition, 91
Cellular lipids, 90
Cellulolytic enzyme lignin (CEL), 212–213
Cellulolytic enzymes, 399–400
Cellulolytic strain culture and enzyme activity testing, 401
Cellulose, 36, 138, 155, 168, 226, 328, 339, 447
biomass, 400
in LB, 226–227
Cellulose, lignin, and hemicellulose (CHL), 223–224
Cellulose nanofibers (CNFs), 332
Cellulosic biofuels, 3
factors affecting commercialization of, 2–3
barriers, 3
drivers, 2–3
latest policies undertaken by USA to commercialize, 3–4
Cellulosic materials, 8–9
Cellulosic microfibrils, 154
Cellulosic nanocrystals (CNCs), 332
Centrifugal sedimentation, 362
Centrifugation process, 96, 115, 184
Ceriporiopsis subvermispora, 39
Cetane number (CN), 100–101
Chaetoceros calcitrans, 312
Challenges
in cultivation and harvesting, 115
in lignin conversion, 217–218
of microalgae-based biofuel synthesis, 103
Chemical flocculation process, 114
Chemical hydrolysis, 283
Chemical oxygen demand (COD), 393–394
Chemical pretreatments, 402, 447
Chemical techniques, 400–401
Chemical treatment of lignocellulosic biomass, 279–281
Chemosynthesis process, 107
Chlamydomonas, 391
C. reinhardtii, 49–50, 91, 125, 186
Chlorella sp., 91, 96, 107, 112, 124, 181, 312, 391
C. minutissima, 95
C. protothecoides, 48, 168–169, 313, 483
C. pyernoidosa, 49–50, 188, 393, 455–456, 483
C. saccharophila, 393
C. vulgaris, 48, 96, 98–99, 170–171
C. zofingiensis, 92–93
Chlorophyceae, 440
Chrome (Cr), 374

Chrysophyceae, 440
Circular bioeconomy, 349, 376
co-production of microalgae-derived products for circular bio-economy-based bio-refinery, 376
Circular economy, 335
Circular ponds, 419
Circulating Fluidized Bed (CFB), 139–140
Citric acid, 78, 82
Clean Air Act, 435
Clean Energy Act, 4
Climate change, 2, 45, 201–202, 223–224
Climate impacts on mediterranean countries, 410
pollution problems of coal energy and need for clean alternatives, 410
potentiality of algae as biologic source of energy, 411
size classification, 411
types of microalgae, 411
Climatic change consequences on energy, 409–410
Closed photoautotrophic production, 358
Closed photobioreactor system, photoautotrophic algal production, 124–125
Closed systems, 170, 417–418
flat-panel photobioreactor, 418
horizontal tube photobioreactor, 417
photobioreactor, 110–113
bag photobioreactor, 112
FP PBR, 112
hybrid photobioreactor, 113
pyramid photobioreactor, 112–113
STR, 112
tubular photobioreactor, 112
stirring photobioreactor, 418
vertical tube photobioreactor, 417–418
Clostridium
C. beijerinckii, 223–224
C. butyricum, 78
Cloud point (CP), 102
Coal, 411–412
pollution problems of coal energy and need for clean alternatives, 410
Cobalt (Co), 374
Cocultivation, 289–290
Cold atmospheric pressure plasma (CAPP), 455
Cold filter plugging point (CFPP), 102
Cold flow properties, 102
Column photobioreactor, 125
Combined heat and power (CHP), 6, 202–203
Combined pretreatment methods, 159–161, 403
Combustion, 250–251, 331, 465
process of thermochemical conversion, 139–140
reactors for combustion, 139–140
studies on combustion, 140
techniques, 445
Composite oxide, 217

Compression ignition engines (CI engines), 18
Computational Fluid Dynamics model (CFD model), 148
Coniophora puteana, 39
Consequential life cycle assessment (cLCA), 20–21
Consolidated bioprocessing (CBP), 224–225
Continuous hydrogen production (CHP), 81
Continuously stirred tank reactors (CSTR), 233–234
Control lipid biosynthesis, 90
Conventional extraction methods, 97
Conversion
methods, 326
reaction, 474
techniques/methods of algal biomass to biofuels, 126–128
biochemical conversion, 127–128
thermochemical conversion, 126–127
technologies, 138
Cooking oil, biodiesel from used, 482
Copper (Cu), 374
Corn, 435–436
COVID-19, 1
distribution, 4
pandemic, 435
Crallus domesticus, 195
Crop residues, 26, 36, 154, 223–224, 316
burning of, 28
emissions formed from crop residues and associated health hazard, 27
utilization of crop residues off-field, 28–29
composting of residues, 29
forage for livestock, 28
mushroom grown in rice straw, 29
production of biofuel and biogas, 29
roof thatching, 29
transportation packaging, 29
Crops, 25, 123, 156
production, 400
Crude glycerol, 71, 78–79, 83–84
bioconversion of crude glycerol waste into value-added products, 78–83
waste from biodiesel as by-product, 77–78
value-added opportunities for crude glycerol, 77–78
Crude oil, 191
Cultivation, 414
challenges in, 115
conditions, 132
of microalgae for biofuel production, 181–184
systems, 110–113, 169–170, 421–422
biohydrogen production, 421–422
closed systems, 170
closed-system photobioreactor, 110–113
open system, 110
open-air systems, 169–170
photoautotrophic versus heterotrophic, 421
techniques
of algae, 47
of bioreactor, 236

Culture system, 183
Cyanobacteria, 108, 167, 370, 425, 444, 481
Cyanophyceae, 440

D

Dark fermentation (DF), 223–224, 386
  comparative study with dark fermentation, 234
  factor affecting dark fermentation, 235–236
  lignocellulosic biomass for, 225
  strategies for, 232–234
    natural strategy relevant to dark fermentation, 233
    pretreatment followed by enzyme hydrolysis, 232–233
    separation, 233–234
Decarbonization, 381
Deconstruction, 437
  high-temperature deconstruction, 437
  low-temperature deconstruction, 437
Dense phase region, 147
Density, 102
Desorption, 257
Di-tert-butyl peroxide (DTBP), 18
Diacylglycerols (DG), 440, 471, 471f
2,4-dichlorophenoxyacetic acid, 458
Diesel, 76
  engines, 436
Dihydroxyacetone, 79–80
2,3-dihydroxypropionic acid, 82
Dilute region, 147
Dinoflagellates, 45
9,10-diphenylanthracene, 455–456
Direct bio-photolysis, 385–386, 424–425
Direct combustion process, 127, 383
Docosahexaenoic acid (DHA), 82–83
Downstream processes for biodiesel synthesis, 94–100. *See also* Upstream processes for biodiesel synthesis
  microalgae harvesting techniques, 94–96
  techniques for biofuel production from algal biomass, 98–100
  technologies for effective lipid extraction from algae, 96–98
Downstream processing of biomass, 283–286
  fermentation, 284
  hydrolysis/saccharification, 283
  transesterification, 285–286
Drilling fluids, 291
Drivers Pressure State Impact Response (DPSIR), 427
Drop-in fuels, 332
Dry milling process, 438
Dry weight (DW), 64
Dunaliella sp., 96, 107
  D. salina, 124
  D. tertiolecta, 393
Dyer method, 97

E

Economics, 433–434
Edible crops, 45
Eicosapentaenoic acids, 483
Electric field pretreatments, 116
Electrical efficiency, 445
Electricity
  biomass conversion techniques to produce electricity and heat, 445–447
  generation, 188
Electrochemical flocculation, 96
Electrolysis, 94, 381, 388
Emission control, 141
Endemic microalgae, 350
Endo-amylase hydrolyzes, 117
Energy, 153–154, 311, 443
  climatic change consequences on, 409–410
  consumption, 1, 59, 79, 167, 478
  content, 328
  crisis, 391
  crops, 202, 326, 443–444
  potential biomass feedstocks for energy production, 202
  potentiality of algae as biologic source of, 411
  resources, 89
  security, 2
  sources, 191, 472
Energy Independence and Security Act (EISA), 4, 435
Energy Information Administration of United States Department of Energy (DOE/EIA), 3
Energy Information Administration research (EIA), 409
Energy Policy and Conservation Act (EPCA), 3
Energy Returned on Energy Invested (EROEI), 96–97
Energy's Bioenergy Technologies Office (BETO), 7–8
Engine oil, 102
Environmental damage, 474
Environmental toxicant monitoring, 373–374
Enzymatic hydrolysis, 304, 447–450
  fermentation, 447–450
  of pretreated straw, 403
Enzymatic mild acidolysis lignin (EMAL), 212–213
Enzymatic pretreatment, 117
Enzymatic transesterification, 285–286
Enzyme monolignol transferase (PMT), 205–206
Enzymes, 437, 478
  cellulolytic strain culture and enzyme activity testing, 401
  enzyme-catalyzed transesterification for biodiesel production, 478–479
  hydrolysis, 283
  saccharification, 117
Equilibrium models, 146–147
Erosion processes, 409–410
Erucic acid, 436
Escherichia coli, 81–82
Ethane, 243
Ethanol, 3, 53, 175, 332, 399, 437–438
  blending, 434
  ethanol-biodiesel, 472
    from lignocelluloses, 438–439
    from sugar and starch, 437–438
Ethyl alcohol, 40
Ethyl tert-butyl ether (ETBE), 203
Eucalyptus globulus, 443–444
Euglena gracilis, 110
European Union (EU), 8
Exajoules (EJ), 202–203
Exhaust emission, reduction of, 18–20
Extracted lignin, elucidation and characterization of, 212–213
Extraction
  methods for biodiesel, bioethanol, and biogas, 317–318
  and purification of microalgal biomass, 126
Extractives, 328

F

Fast pyrolysis, 143, 316–319, 330–331
Fatty acid methyl esters (FAMEs), 63, 83–84, 171, 271–272, 313, 391, 471, 471f, 481
Fatty acids (FAs), 50, 72, 91–92
  profiling, 91–92
Feedstock
  for chemicals, 78
  processing technologies, 437
    deconstruction, 437
    upgrading, 437
[FeFe]-hydrogenase enzyme, 385
Fermentable sugars conversion, 400–401
Fermentation, 186, 284, 331, 441, 446
  biomass conversion techniques to produce electricity and heat, 446
    anaerobic digestion, 446
    enzymatic hydrolysis, 447–450
  process, 118, 423
  of rice straw hydrolysates, 403–404
  of sugars, 304–305
Ferredoxin (Fdrd), 386, 424–425
Ferric chloride, 95
Ferric sulfate, 95
Field burning, 316
Field management of rice straw, 155–156
  in-field rice straw management, 155
  off-field rice straw management, 156
Filamentous fungi, 273–274
Filtration, 362–363
  process, 96, 114, 184
First-generation biofuels, 45, 107, 168, 203, 311, 412
First-generation biorefineries, 349, 369–370
Fischer-Tropsch process, 167
Fixed Bed (FB), 139–140
Fixed carbon (FC), 464
Flash pyrolysis, 143
Flat-panel photobioreactor, 418
Flat-plate photobioreactor (FP PBR), 112
Flocculants, 114
Flocculation, 126, 361

auto-flocculation, 114
chemical flocculation, 114
process, 94–96, 114, 185
Flood-prone rice ecosystems, 298
Flotation process, 94, 115, 184, 361
Folch method, 97
Food crops, 275
as biomass, 326
Food industry, 449–450
Food safety, 472
Food security, 482
Food waste (FW), 161
generation, 340
Forest biomass feedstock, 444
Forest trees and residues, 275
Forest waste/residue (FR), 326
Forestry residue, 444
Formate hydrogen lyase (FHL), 386
Fossil energy, 71
Fossil fuels, 2, 13–14, 25, 27, 45, 107, 123, 137, 167, 325, 369, 433–435, 472
energy consumption, 223
reserves, 71
sources, 153–154
Fossil-derived fuel, 314
Fourth-generation biofuels, 107, 204
Free fatty acid (FFA), 14, 285
Freezing pretreatment, 117
Freshwater and marine microalgae, comparison between, 92
Fuel, 243, 311
transition to, 73
Fungi-assisted sedimentation, 115

G

Ganoderma lucidum (GL), 159
Gas chromatography, 394
Gaseous biofuel, 311–312
Gaseous fuel, 52, 191
Gaseous hydrogen, 388
Gasification, 40, 78, 204, 331, 384–385, 437, 465
process of thermochemical conversion, 141–143
reactors for gasification, 141
fixed bed, 141
fluidized bed gasification, 141
significant studies on gasification, 142–143
techniques, 446
thermochemical conversion process, 126
Gasoline, 435–436
gasoline-biodiesel, 472
Gastrointestinal tract, 440
Gene editing tools, 457–458
Generations of biofuels, 434
Genetic and metabolic engineering, 287–288
Genetic engineering
for enhancing microalgae biomass and biofuel components, 457–458
tools, 457–458
Genetically engineered microorganisms for CO_2 fixation, 343
Geographic Information System (GIS), 203
Global carbon dioxide, 13

Global energy consumption, 167
Global energy requirement, 137
Global fossil fuel, 453–454
Global petroleum usage, 409
Global population, 391
Global warming, 391, 409
Gluconobacter sp., 82
Glucose transformation, 402
Glucose-based sugars, 63–64
Glyceric acid (GA), 79–80, 82
Glycerides, 76
Glycerin, 475–476
Glycerol, 55–56, 79, 84, 118, 271–272, 478
Glycerophospholipids, 91
Glycolipids, 91
Glycolysis, 441
Gold silver core-shell nanoparticles (Au–Ag NPs), 192
Government of China to drive biofuel generation
recent progress in policies taken by, 5
Grain alcohol, 40
Gravimetric moisture measurement, 329
Gravity, 114, 362
separation process, 115
Green algae (*Chlorophyta*), 46–47, 168–169, 172
Green biomass, 142
Green economy, 409–410
Green hydrogen, 185, 381
Green sulfur bacteria, 386
Greenhouse emissions, 446
Greenhouse gas emissions (GHG emissions), 25, 27, 71, 107, 123, 153–154, 167, 201, 223, 297, 311, 335, 391, 399, 409, 433
biodiesel reduces greenhouse gas emissions, 75
Gross calorific value (GCV), 328
Growth requirements for optimal microalgal growth, 416
light, 416
mixing, 416
nutrient provision, 416
temperature, 416
Guaiacyl units (G), 205

H

Haematococcus, 107
H. pluvialis, 419
Harvesting
of algal biomass, 184–185
biologic method, 185
chemical methods, 185
physical methods, 184
challenges in, 115
method, 94
microalgae biomass, 94
of microalgae for biofuel production, 181–184
of microalgal biomass, 126
process, 113–115
centrifugation, 115
filtration, 114

flocculation, 114
flotation, 115
fungi-assisted sedimentation, 115
magnetic separation, 115
sedimentation, 114
Heat
biomass conversion techniques to produce electricity and, 445–447
energy, 138
and mass transfer, 257
Heavy metals, 60
Hemicellulose, 26, 138, 168, 226, 328
in LB, 226–227
pyrolysis, 27
Heterogeneous acid catalysis, 474
Heterogeneous acid catalysts, 196
Heterogeneous catalysis, 285, 475–476
Heterotrophic algal production, 125
Heterotrophic cultivation production, 358
Heterotrophy, 92–93
Heveochlorella sp., 95
Hexoses, 26
Heynigia, 391
Hibiscus cannabinus, 443–444
Hidden layers, 148–149
High hydrogen partial pressure (HPP), 235
High value-added by-product, 119
High-pressure homogenization, 278
High-temperature deconstruction, 437
Higher heating value (HHV), 102, 464–465
Homogenous catalysis, 285
Horizontal PBR, 112
Horizontal tube photobioreactor, 417
Hot water pretreatment, 209, 280–281
Human diet, 340
Human health, biodiesel and, 75
Husk, 316
Hybrid anaerobic baffled reactor (HABR), 394
Hybrid nanocatalysts, 193
Hybrid photobioreactors, 113, 419
Hybrid production systems, 125
Hydrocarbons (HCNs), 36
emissions, 409
fuels, 476–477
Hydrochloric acid (HCl), 475
Hydrodeoxygenation (HDO), 215–216
Hydrogen (H_2), 54, 78–81, 107–108, 188, 381
advantages and disadvantages of hydrogen production label colors, 382t
benefits and challenges of hydrogen production, 387–388
bonds, 399
discharging, 262–265
by microorganisms, 385–386
perspectives, 388–389
production, 143, 243
process, 425
production from biofuels obtained by thermochemical processes of microalgae biomass, 383–385
production from renewable energy, 381–383

Hydrogen (H_2) (*Continued*)
  storage, 259–262
Hydrogen peroxide, 38–39
Hydrogen yield (HY), 223–224
Hydrolysates, 37
Hydrolysis, 305, 403, 437, 441
  hydrolysis/saccharification, 283
Hydrothermal carbonization, 463
Hydrothermal gasification process, 30
Hydrothermal liquefaction (HTL), 40, 100, 145, 437
  processes, 146
Hydrothermal pretreatment (HTP), 116
3-hydroxyanthranilic acid (HA), 402
Hydroxymethylfurfural (HMF), 228
Hydroxyphenyl units (H), 205
2-hydroxypropanoic acid, 81
3-hydroxypropionaldehyde (3-HPA), 80–81
3-hydroxypropionic acid (3-HP), 81

I

Improved performance, 18
In situ catalytic transesterification, 99
In situ noncatalytic transesterification, 99
In-field rice straw management, 155
Inclined tubular PBR, 112
Indirect biophotolysis, 385–386, 425
Indirect injection diesel (IDI), 17
Indoleacetic acid (IAA), 454–455
Industrial process residues, 275
Industrial wastes, 76, 327
Inertial cavitation, 244
Inhalation, 420
Innovative process based on additives, 161–163
Inorganic flocculants, 95–96
Insects, 482
Insoluble polymers, 441
Instrumental methods, 329
Intergovernmental Panel on Climate Change (IPCC), 201
Intermediate pyrolysis, 331
Internal combustion engines (ICEs), 72
Internal rate of return (IRR), 463
International Energy Agency (IEA), 1–2, 13, 16, 107
Iodine value (IV), 100
Ionic liquids (ILs), 98, 208, 228
  hydrolysis, 283
Irpex lacteus, 39
Irrigated rice ecosystems, 298
Isochrysis galbana, 312, 454–455

J

Jatropha (*Jatropha curcas*), 14, 20, 45, 195, 436

K

Karanja (*Pongamia pinnata*), 20
Kinematic viscosity, 102
Kinetic models, 147
Kitchen waste (KW), 161

Klebsiella pneumoniae, 78
Kraft lignin via Kraft process, 210–211

L

Lactic acid, 81
  homofermentation, 449
Laminaria saccharina, 110
LaNi$_5$, 257
  analysis of absorption process, 259–262
  analysis of desorption process, 262–265
  model, 259
  physical description of unit, 258
  works, 258
Layered double hydroxides (LDHs), 215
Lead (Pb), 374
Life Cycle Assessment (LCA), 4, 13, 29–31, 156–157, 286
  of biofuels, 20–21
    biomass (mahua flowers) for conversion to bioethanol, 14–16
    biomass (mahua seeds) for conversion to biodiesel, 16–20
    coproduct, byproduct, residue, 21
    geographical and climatic resolution, 13
    goal, scope, and functional unit, 20
    impact assessment, 21
    inventory analysis, 21
    Madhuca species as green approach for biofuel production, 14
    system boundaries and reference system, 20–21
Life cycle GHG emissions, 433
Life cycle inventory (LCI), 21
Life-cycle assessment (LCA), 156
Light, 416
Light emitting diodes (LEDs), 454–455
Light-independent reaction, 425
Lignifications, 205–206
Lignin, 26, 138, 168, 226, 328, 339
  availability and abundance of biomass, 202–203
  challenges in lignin conversion, 217–218
  chemically, 204–206
    LCC, 206
    role of lignin in biological living world, 204–205
    synthesis of lignin in living system, 205–206
  climate change, 201–202
  content in LB, 226–227
  current energy status globally, 201
  elucidation and characterization of extracted lignin, 212–213
  generations of biofuel, 203–204
  isolation and extraction of lignin from lignocellulosic biomass, 206–212
  potential biomass feedstocks for energy production, 202
  routes of lignin conversion, 213–217
    lignin conversion by acid catalysis, 215
    lignin conversion through base catalysis, 215
    lignin conversion through reductive catalysis, 215–216

oxidative conversion of lignin, 216–217
pyrolysis method to convert lignin, 213
Lignin carbohydrate complex (LCC), 206
Lignin peroxidase (LiP), 402
LignoBoost, 210–211
Lignocellulose, 168, 271–272, 447
  biomass, 137, 297, 399
  ethanol production from, 438–439
Lignocellulosic biomass (LCBM), 8–9, 25, 42, 203, 223–224, 312, 316, 339, 401
  advancement, 287–290
  applications of biodiesel, 290–292
  as carbon source, 274–275
    food crops, 275
    forest trees and residues, 275
    industrial process residues, 275
    nonfood/energy crops, 275
  characteristics of lignocellulosic biomass, 226
  chemical components of different types of, 36
  commercially available lignin, 210–212
    sulfur-bearing process, 210–211
    sulfur-free process, 211–212
  comparative study with dark fermentation, 234
  compositional variation in cellulose, hemicellulose, and lignin content in LB, 226–227
  cultivation techniques of bioreactor, 236
  dark fermentation advantages and stoichiometry, 234–235
  environmental issues, 286
  factor affecting dark fermentation, 235–236
  future prospects, 292–293
  inhibitor formation and solution strategies to overcome inhibitor formation for darkfermentation, 231–232
  isolation and extraction of lignin from, 206–212
  lignocellulosic biomass for dark fermentation, 225
  opportunities and challenges, 286–287
  pretreatment, 277–282
    advantages and disadvantages, 281t–282t
    mechanical pretreatment, 277–279
    nonmechanical pretreatment, 279–281
  pretreatment methods for dark fermentation, 228–231
  pretreatment technologies to separate and depolymerize lignin, 206–210
  role of pretreatment, 227
  sources of lignocellulosic biomass, 225–226
  strategies for dark fermentation, 232–234
  structural features of lignocellulosic biomass and recalcitrance, 276–277
  transesterification, 285–286
  yield enhancement strategies, 236–238
Lignocellulosic feedstocks, 438
Lignocellulosic materials, 154
  biorefineries, 349
Lignocellulosic substrates, 271–272
Lignosulfonate lignin via sulfite pulping, 211

Lipase, 118, 478
  enzyme, 99, 285–286
Lipid(s), 129, 421, 481
  accumulation, 458
  from algal biomass, extraction of, 170–171
  augmentation by genetic alterations, 50
  content, 393, 455–456
    microalgae, modification of growth and physiologic conditions for enhancing of, 455–456
  extraction technique, 170–171
  induction technique, 48
  productivity, 90–91
  techniques to identify and quantify, 365
  technologies for effective lipid extraction from algae, 96–98
    cell disruption techniques, 97
    conventional extraction methods, 97
    ILs, 98
    liquid CO_2 technology, 98
    simultaneous distillation and extraction process, 97
    supercritical fluid technology, 97–98
  transesterification, 171
  yield of microalgae, 415
Liquefaction, 383–384
  of thermochemical conversion, 145–146
    reactors for liquefaction, 145
    significant studies on liquefaction, 145–146
  thermochemical conversion process, 126–127
Liquefied biomethane, 302
Liquid
  biofuel, 311–312
    production methods, 64
  fuel, 52, 191
  petroleum gas, 446
  shear forces, 277
Liquid CO_2 (L-CO_2), 98
  technology, 98
Liquid fractions of digestate (LFD), 159–161
Low bulk density, 25
Low-cost cellulosic ethanol, 156–157
Low-energy process, 114
Low-temperature deconstruction, 437
Lubricants, 188
Lyle Cummins, 73

M
Macaw oil (*Acrocomia aculeata*), 76
Macroalgae, 76, 108
  for bioenergy production, 61–62
  biomass production, 60
  as source of biofuel, 46–47
Macroalgae-derived biofuel, 62
Macroalgal
  biomass, 45
  cultures, 419
  technologies for macroalgal cultivation, 48
    filtration, 48–49
    flocculation, 48
    flotation, 48
    sedimentation, 48
Macromolecule cellulose, 26
Macronutrients, 395
Madhuca
  M. latifolia L., 13
  species as green approach for biofuel production, 14
Magnesium oxide (MgO), 476
Magnetic separation process, 115
Mahua (*Madhuca indica*), 13, 20
  biomass (mahua flowers) for conversion to bioethanol, 14–16
Mahua biodiesel (MU), 17–18
Mahua flowers, 14
Mahua methyl ester (MME), 20
Mahua-based biodiesel, 14
Maize straw, 154
Manganese peroxidase (MnP), 402
Mannitol, 175
Manure and sewage sludge, 326–327
Marine Biomass Program (MBP), 7–8
Marine microalgae, comparison between freshwater and, 92
Marine vessels, 290
Mathematical model, 258
Mechanical pretreatments for microalgal biomass, 117
Mediterranean countries, climate impacts on, 410
Mediterranean Network of Transmission System Operators (Med-TSO), 409–410
Mercury (Hg), 374
Meruliporia incrassate, 39
Mesoporous and macroporous catalysts (MCM), 145
Metabolic modeling, 289
Metal based nanocatalyst, 192–193
Metal hydride (MH), 257
Metal oxides, 18
  nanoparticles, 194
Metal-organic frameworks (MOFs), 476
Methane (CH_4), 27, 30, 62, 64, 167, 201, 243, 441
  fermentation, 128
  process description, 441
    acetogenesis, 441
    fermentation, 441
    hydrolysis, 441
    methanogenesis, 441
Methanogenesis, 441
Methanogenic environments, 440
Methanol, 83–84, 98, 467
  sono-conversion
    acoustic intensity effect, 250–252
    literature data, 245–246
    model, 246
    optimal gas conditions for H_2 production, 246–248
    sonochemical process, 243–245
    ultrasound frequency effect, 249–250
4-methoxybenzyl alcohol, 194
Meyerozyma caribbica, 16

Micractinium, 391
Micro-fibrils, 226
Microalga species for biodiesel, 421
Microalgae, 47, 76, 89–92, 108, 118, 123, 132, 167–168, 181, 274, 341–342, 350, 391, 453–454, 482
  advantages and disadvantages, 373t
  algal biofuel types, 128–130
  algal fuel properties, 131
  applications, 186–188
    aviation, 188
    lubricants, 188
    power/electricity generation, 188
    transport, 186
  applications of biomass, 132
  benefits of using microalgae for biofuel productions, 181
  biodiesel from microalgae oils, 482–483
  biohydrogen from, 118
  for bioenergy production, 61–62
  biofuel production using, 185–186
  biofuel production using microalgae, 185–186
  biomass, 107–108, 457–458
    and biofuel components, genetic engineering for enhancing, 457–458
    biofuel components, modification of growth and physiologic conditions for e and nhancing, 454–455
    production, 60
  bioremediation capacity of, 372–375
  co-production of microalgae-derived products for circular bio-economy-based bio-refinery, 376
  conversion techniques/methods of algal biomass to biofuels, 126–128
  cultivation, 130–131
    growth factors for, 182–183
    systems, 169
  cultivation and harvesting of microalgae for biofuel production, 181–184
    bioreactors used for culturing, 183–184
    growth factors for microalgae cultivation, 182–183
  cultures, 60
  current status of third-generation bio-refineries with, 370
  economic importance of algae as biofuel sources, 131–132
  extraction and purification of microalgal biomass, 126
  future perspectives, 376–377
  genetically modified algae, 130–131
  harvesting of algal biomass, 184–185
  harvesting of microalgal biomass, 126
  harvesting techniques, 90, 94–96
    centrifugation, 96
    filtration, 96
    flocculation, 94–96
    flotation, 94
  integration and reuse of industrial waste for microalgae cultivation, 375–376
  isolation, 352f
  lipids yield of, 415

Microalgae (*Continued*)
  liquid media references for growing microalgae species, 353t
  methodologies to extract and isolate, 350–352
  microalgal biomass production, 124–125
  mitigation capacity of CO_2, 375
  modification of growth and physiologic conditions for enhancing carbohydrate content of, 456–457
  modification of growth and physiologic conditions for enhancing lipid content of, 455–456
  optimal culture media for laboratory-level photobioreactor biomass production, depending on, 357–365
  photosynthesis diagram, 372f
  review of morphological microalgae identification, 352
  as source of biofuel, 46
  and sustainability, 370–372
  techniques to identify and quantify lipids and carbohydrates, 365
  technologic applications of, 108–119
    biodiesel, 118
    bioethanol production, 115–118
    biohydrogen from microalgae, 118
    challenges in cultivation and harvesting, 115
    cultivation system, 110–113
    harvesting, 113–115
    high value-added by-product, 119
    Spirulina platensis, 108–110
  types of, 411
  use of microalgae for wastewater treatment, 374–375
Microalgae-based biofuel synthesis
  challenges and opportunities, 103
  downstream processes for biodiesel synthesis, 90–92
  properties of microalgal biodiesel, 100–102
  upstream processes for biodiesel synthesis, 90–93
Microalgae-based third-generation biofuel, 65
Microalgal biodiesel
  production, 395
  properties of, 100–102
    CN, 100–101
    cold flow properties, 102
    density, 102
    HHV, 102
    IV, 100
    kinematic viscosity, 102
    OS, 101–102
    SV, 101
Microalgal biomass, 45, 108, 123, 181
  anaerobic digestion of, 173–174
  extraction and purification of, 126
  harvesting of, 126
  production, 124–125
    heterotrophic algal production, 125
    mixotrophic algal production, 125
    photoautotrophic algal production, 124–125
Microalgal cells, 95
Microalgal cultivation
  technologies for, 47–48
    lipid induction technique, 48
    open ponds, 47
    photobioreactors, 47
    two-stage hybrid system, 48
    wastewater used for, 125
Microalgal cultures, 89, 414
Microalgal lipids, 91, 123
Microalgal oils, 97–98
Microalgal species, 411
Microalgal strain, 396
Microalgal-based biofuel, 47
Microbes, 71–72
Microbial bioprospecting, 83
Microbial fermentation of algal biomass, 173
Microbial oils, 272
Microbial pretreatment process, 159
Microbial prospecting, application of, 83–84
Microjet, 244–245
Microorganisms, 50, 83, 437
  cocultivation of, 274
Microwave
  irradiation, 37, 278–279, 402
  pretreatment, 116
  pretreatment by, 209
  spectroscopy, 329
Microwave-assisted pyrolysis (MAP), 465–466
Microwave-facilitated extraction, 170–171
Milled wood lignin (MWL), 212–213
Ministry of New and Renewable Energy of India (MNRE), 333
Miscanthus, 45
  M. giganteus, 443–444
Mixing, 416
Mixotrophic algal production, 125
  wastewater used for microalgal cultivation, 125
Mixotrophic microalgae, 93
Mixotrophic production, 358
Mixotrophs, 168–169
Mixotrophy, 93
Modeling studies of biomass, 146–149
  ANN model, 148–149
  ASPEN Plus models, 147–148
  CFD model, 148
  equilibrium models, 146–147
  kinetic models, 147
Moisture content (MC), 464
  of biomass, 329
Moisture measurement, 329
Monoacyl glycerol (MAG), 440, 471, 471f
Monoglycerides, 79–80, 471
Monolignols, 205
Mucor circinelloides, 41–42
Mulch, retention of surface as, 28
Multi-walled carbon nanotubes (MWCNT), 18
Municipal solid waste (MSW), 202, 316
Mushroom grown in rice straw, 29

N

Nannochloropsis sp., 48, 91, 96
  N. oceanic, 457–458
  N. oceanica, 454–455
  N. oculata, 342, 483
  N. salina, 454–455
Nanocatalyst, 192, 483–484
  metal/alloy based nanocatalyst, 192–193
Nanocellulose, 332
Nanomaterial production, 339
Nanosecond pulse electric field (nsPEF), 170–171
"National Biofuel Fund", 6–7
National Biofuels Policy, 6
National Development and Reform Commission (NDRC), 5
National policy of India to promote biofuel commercialization program, 6–7
National Program for Biodiesel Production and Use (PNPB), 4
Natural gas, 223, 433
Natural sources, 474
Natural standing crop, 414
Natural strategy relevant to dark fermentation, 233
Naturalistic environments, 401
Near-infrared spectroscopy (NIR), 329
Neem (*Azadirachta indica*), 20
Net present value (NPV), 466
Neutral lipids, 91
Nickel (Ni), 40, 374
Nitric oxide (NO), 18
Nitrogen (N), 50, 60, 169
  restriction, 456–457
Nitrogen dioxide (NO_2), 183
Nitrogen oxide (NOx), 36, 75
Nitrous oxides (NOx), 153–154, 201
Non-lignocelluloses biomass, 137
Non-Timber Forest Produce (NTFP), 15
Noncatalytic transesterification, 99
Nonfood biomass, 326–327
Nonfood/energy crops, 275
Nonfuel coproducts, 2
Nonglucose-based sugars, 63–64
Nonrenewable energy sources, 13
NTDEs, 75
Nutrients, 49–50, 55, 60
  provision, 416
Nutritional deficiency, 416

O

Off-field rice straw management, 156
Off-road vehicles, 290
Oil palm trunk (OPT), 228
Oils
  chemistry of biodiesel transesterification of, 72
  crops, 45
  expeller, 277–278
  extraction of oil from algal biomass, 170–171
  oil-rich biomaterial, 316

transesterification, 171
  yield, 16
Oleaginous bacteria, 274
Oleaginous microorganisms (OMs), 271–272
  and lipid content, 273t
  with organic waste as raw materials, production of biodiesel from, 483
  prospective oleaginous microorganisms for biodiesel production, 272–274
Oleaginous yeast, 273–274
Oleic acid, 91
  content, 391
Oligoplasts, 481
Olive trees, 73
Omnibus Budget Reconciliation Act, 3
On-road vehicles, 290
Open ponds, 47
  bioreactors used for culturing, 183
  production system, 124
  system, 110
Open raceway pond, 183
Open system, 418–419
  circular ponds, 419
  hybrid photobioreactors, 419
  raceway pond, 418
  static pond, 418
Open-air systems, 47, 169–170
Optimal culture media for laboratory-level photobioreactor biomass production, depending on microalgae species, 357–365
  closed system production, 358
    closed photoautotrophic production, 358
    heterotrophic cultivation production, 358
    mixotrophic production, 358
    production of carbohydrates, 363–365
    production of lipids, 363
    recovery of microalgal biomass, 358–363
Optimal gas conditions for H_2 production, 246–248
Optimal microalgal growth, growth requirements for, 416
Organic carbon, 60, 396, 454–455
Organic dye, 454–455
Organic fraction of municipal solid wastes (OFMSW), 161, 316
Organic ligands, 476
Organic materials, 443
Organic molecules, 47
Organic polymers, 95
Organic solvent, pretreatment using, 208
Organic waste as raw materials, production of biodiesel from, oleaginous microorganisms with, 483
Organosolv lignin via solvent pulping, 212
Organosolv process, 229
Organotrophic acetoclastic methanogens, 441
Orthophosphate content, 393
Osmotic stress, 290
Oxidation, 449
Oxidation stability (OS), 101–102
Oxidative compounds method, 229
Oxidative pretreatment, 229

Oxidative stress, 455
Ozone (O_3), 201
  O_3-floatation method, 395

P

p-coumaric acid, 26
Palm oil, 436
Palmitoleic acid, 91
Panicum virgatum, 443–444
Particulate matter (PM), 28, 75
Pasteurization, 437–438
Pectin biomass, 340–341
Pelleted fungus, 115
Pentose, 231
Peracetic acid, 38–39
Phaeodactylum tricornutum, 65, 96, 454–455
Phalaris arundinacea, 443–444
Phanerochaete chrysosporium (PC), 39, 159
Phlebia subserialis, 39
Phosphoribosyl kinase (PRK), 343
Phosphorus (P), 50, 60
Photo-fermentation, 386
  ethanol, 186
Photoautotrophic algal production, 124–125
  closed photobioreactor system, 124–125
  hybrid production systems, 125
  open pond production system, 124
Photoautotrophic cultivation systems, 92, 454–455
Photoautotrophism, 92
Photoautotrophy, 92
Photobiological hydrogen production, 128
Photobioreactors (PBRs), 47, 170, 181–182, 418
  bag photobioreactor, 112
  bioreactors used for culturing, 184
  hybrid photobioreactor, 113
  pyramid photobioreactor, 112–113
  tubular photobioreactor, 112
Photoheterotrophs, 168–169
Photoinhibition process, 182–183
Photosynthesis process, 107, 371
Photosynthetic algae, 434
Photosynthetic efficiency (PE), 46
Photosynthetic organisms, 59
Photosystem II (PSII), 385, 416, 425
Phycocyanin (PC), 119
Physical pretreatments, 402, 447
Pig manure (PM), 161
Pigments, 55
Plant biofuels, 434–435
Plant oils, 73
Plant-based cooking oils, 72
Plasma gasification, 446
Plastoquinone pool (PQ), 385–386
Pleurotus eryngii, 39
Pleurotus ostreatus (PO), 39, 159
Polar lipids, 91
Policymakers, 2
Pollution problems of coal energy and need for clean alternatives, 410
Polycyclic aromatic hydrocarbons (PAHs), 153–154, 463

Polyethylene (PE), 110
Polyhedral crystalline system, 476–477
Polymethyl methacrylate, 110
Polysaccharides, 400, 441
Polyunsaturated fatty acids (PUFAs), 91
Polyvinyl chloride (PVC), 110
Porous materials, 483–484
Porphyidium cruentum, 168–169
Posthydrothermal liquefaction wastewater (PHWW), 395–396
Potassium hydroxide (KOH), 474
Potato dextrose agar (PDA), 401
Potential strains, 411–412
  history of biofuel research, 411–412
Pour point (PP), 102
Power generation, 188
Pressurized liquid extraction (PLE), 170–171
Pretreated straw, enzymatic hydrolysis of, 403
Pretreatment methods, 37, 226–227, 422, 447
  of algal biomass, 172
  bioethanol production, 116–117
    enzymatic pretreatment, 117
    freezing/thawing pretreatment, 117
    HTP, 116
    mechanical pretreatment, 117
    microwave pretreatment, 116
    pulse/electric field pretreatments, 116
    ultrasound pretreatment, 116–117
  biofuel yield from rice straw, 157–161
    biological pretreatment, 159
    chemical pretreatment, 158–159
    combined pretreatment methods, 159–161
    physical pretreatment, 157–158
  biological methods, 210
  biological pretreatment, 447
  chemical methods, 208
    pretreatment by alkaline method, 208
    pretreatment using acids, 208
    pretreatment using ILs, 208
    pretreatment using organic solvent, 208
  chemical pretreatment, 447
  for dark fermentation, 228–231
    biological pretreatment methods, 230
    chemical pretreatment methods for dark fermentation, 228–229
    combination of pretreatment methods, 230–231
    physical pretreatment method for dark fermentation, 228
  followed by enzyme hydrolysis, 232–233
  LB role of, 227
  with oxidizing agents, 38–39
  physical methods, 207
    pretreatment by mechanical extrusion, 207
    pretreatment by milling, 207
    pretreatment by sonication, 207
  physical pretreatment, 447
  physico-chemical methods, 208–209
    explosion by supercritical CO_2, 209

Pretreatment methods (*Continued*)
  pretreatment by AFEX, 209
  pretreatment by hot water, 209
  pretreatment by microwave, 209
  pretreatment by steam explosion, 208–209
  pretreatment by wet oxidation, 209
  techniques, 401–403
    biological pretreatment, 402
    chemical pretreatments, 402
    combined pretreatment, 403
    enzymatic hydrolysis of pretreated straw, 403
    physical pretreatments, 402
  technologies to separate and depolymerize lignin, 206–210
Primary biofuel, 168
Primary food crop, 326
1,2-propanediol, 79–80
1,3-propanediol (1,3-PDO), 78, 80–81
Proteins, 55, 60–61, 108–110, 328
Pseudomona cepacian, 479
Pulse field pretreatments, 116
Pure glycerol, 83–84
Pure water sonolysis, 243
Purification of biodiesel, 286
Purple nonsulfur bacteria, 386
Pyramid photobioreactor, 112–113
Pyrolysis process, 39, 64, 330–331, 385, 437, 446, 465–466
  of agricultural residue, technoeconomic analysis of, 466–467
  fast pyrolysis, 330–331
  intermediate pyrolysis, 331
  method to convert lignin, 213
  oil, 316–317
  slow pyrolysis, 331
  techniques for biofuel production from algal biomass, 99
  of thermochemical conversion, 143–145
    classification of pyrolysis process, 143
    reactors for pyrolysis, 144
    significant studies on pyrolysis, 144–145
    thermochemical conversion process, 127
Pyruvate formate lyase (PFL), 386

Q
Qualitative determination of biodiesel, 320
Quantitative determination of ethanol, 319

R
Raceway pond, 418
Rainfed lowland ecosystems, 298
Rapeseed biodiesel (RA biodiesel), 18
Rapeseed/canola, 436
Raw material, 447, 472
  production of biodiesel from, oleaginous microorganisms with organic waste as, 483
Reactors
  for combustion, 139–140
  for gasification, 141
  for liquefaction, 145
  for pyrolysis, 144
Red algae (*Rhodophyta*), 46–47
Renewable biofuels, 181
Renewable electricity, 443
Renewable Energy, 5, 59, 181, 191, 325
  development, 5
  hydrogen production from, 381–383
  sources, 137, 399
Renewable fuel standard program (RFS program), 203, 433
  RFS II, 435
Renewable hydrocarbon fuels, 332
Renewable resource, 225
  feedstocks, 71
Renewable sources, 107, 443–444
  of biofuels, 25
Renewal energies, 474
Research and development (R&D), 2
Residual biomass biorefineries, 335, 336t–338t
Reusable carbon sources, 444
Ribulose 1,5-biphosphate carboxylase/oxygenase (RuBisCO), 343
Rice (*Oryza sativa*L.), 153–154
  composition of rice by-products, 36
  crop residue, 26
  ecosystem, 297–298
  hull, 36, 42
  husks, 36, 154
  production in Asian countries, 35
  stalk diseases, 399
Rice residue in biorefineries, potential of, 37–40
  biological pretreatment, 39
  chemical pretreatment, 37–39
    acid pretreatment, 38
    alkali pretreatment, 38
    ammonia pretreatment, 39
    pretreatment with oxidizing agents, 38–39
  physical pretreatment, 37
  thermochemical pretreatment, 39–40
    gasification, 40
    hydrothermal liquefaction, 40
    pyrolysis, 39
    torrefaction, 39
Rice straw (RS), 35–37, 42, 154, 297–300, 316
  annual rice production in few Asian countries, 298t
  availability and management, 298–299
  bioethanol from production of, 400–404
    basic concepts of, 400–401
    bioethanol production's techno-economic worthiness, 404
    cellulolytic strain culture and enzyme activity testing, 401
    fermentation of rice straw hydrolysates, 403–404
    pretreatment techniques, 401–403
  for biofuel production
    crop residues, 154
    LCA and SWOT analysis for biofuels production from rice straw, 156–157
    mathematical models and tools to scrutinize sustainability of biofuels, 163–164
    processes to improve biofuel yield from rice straw, 157–163
  biofuel production from, 300–305, 301t
  biofuels from rice straw and future perspectives, 25
    advantages and disadvantages of producing biofuels from rice straw and socioeconomic evaluation, 31
    biofuel production and processes, 29–30
    composition and properties, 26–27
    emissions formed from crop residues and associated health hazard, 27
    LCA and SWOT analysis for biofuels production from rice straw, 30–31
    rice crop residue, 26
    utilization of crop residues off-field, 28–29
  concept of, 155–156
  composition of, 155
  field management of, 155–156
  constituents and characteristics, 299–300
  conversion of rice straw into biofuels, 40–42
    bio-oil, 42
    bioethanol, 40
    biogas, 41–42
    value-based products, 42
  economic evaluation of biofuel produced from, 305–306
  management of rice straw in field, 27–29
    burning of crop residue, 28
    incorporation of residue in soil, 28
    retention of surface as mulch, 28
  management techniques, 299t
  mushroom grown in, 29
  potential for bioethanol production, 400
  production of bioethanol from, 400–404
  rice straw potential for bioethanol production, 400
Rice straw mushroom (RSM), 156–157
rRNA gene sequencing, 393
Rural economic growth, 2–3

S
Saccharification, 447
  of algal biomass, 172
Saccharomyces cerevisiae, 107–108
Saponification value (SV), 101
Saturated fatty acids (SFAs), 91, 472
Scanning electron microscopy (SEM), 193
Scenedesmus, 91, 391
  S. dimorphus, 168–169
  S. incrassatulus, 100
  S. obliquus, 49–50, 125
Schizochytrium limacinum, 312
Scripps Institution of Oceanography (SIO), 8
Seawater, 48
Seaweed, 414
Second-generation biofuels, 45, 107, 203–204, 311, 412
  generation, 168

Second-generation biomass, 223–224
Second-generation biorefineries, 349, 369–370
Secondary building units (SBUs), 476
Sedimentation process, 114, 184
Seed oil, 16
Selenastrum capricornutum, 110
Separate hydrolysis and fermentation (SHF), 173, 305
Serpula lacrymans, 39
Sewage wastewater, 393
Shock waves, 244–245
Silicon dioxide (SiO_2), 18
Simultaneous distillation and extraction process, 97
Simultaneous saccharification and cofermentation (SSCF), 223–224, 305
Simultaneous saccharification and fermentation (SSF), 173, 204, 223–224, 305, 403
Single-cell oil (SCO), 271–272
Slow pyrolysis, 143, 331
Soda lignin via soda pulping, 211
Sodium chloride (NaCl), 49
Sodium hydroxide (NaOH), 210–211, 303, 402, 474
Soft rot fungi, 210
Soil, incorporation of residue in, 28
Soil organic carbon (SOC), 28, 155
Solar-driven membrane distillation (SDMD), 16
Solid biofuel, 311–312
Solid biomass, 437
Solid fuel, 52
Solid shear forces, 277
Solid-state fermentation (SSF), 284, 402
Solid-state hydrogen storage materials, 257
Solvents, 290
Sonochemical process, 243–245
Soybeans, 436
 plants, 436
Species, 411
Spirulina, 107, 124
 S. maxima, 168–169, 186
 S. platensis, 107–110, 125
Starch, 328, 340
 ethanol production from, 437–438
 sugar crop, 326
Starchy biomass, 340
Starchy feedstocks, 312
Starchy residues, 340
Static pond, 418
Stationary power generation, 291–292
Steam explosion, pretreatment by, 208–209
Steam pretreatment/steam explosion, 280
Steer hydrogen production, genetic engineering to, 50
Sterculia foetida, 45
Stirred-tank bioreactor (STR), 112
Stirring photobioreactor, 418
Straight vegetable oils (SVOs), 72
Straw, 316
Strengths, weaknesses, opportunities, threats (SWOT), 31
 analysis for biofuels production from rice straw, 30–31, 156–157
Streptomyces griseorubens JSD-1, 39
Submerged fermentation (SmF), 15, 284
Sugar, 328
 beets, 437–438
 ethanol production from, 437–438
 platform, 37
Sugarcane, 436
 bagasse, 154
 harvesting, 436
Suitable microalgae harvesting technology, 94
Sulfur oxide (SOx), 36
Sulfur-bearing process, 210–211
 Kraft lignin via Kraft process, 210–211
 lignosulfonate lignin via sulfite pulping, 211
Sulfur-free process, 211–212
 organosolv lignin via solvent pulping, 212
 soda lignin via soda pulping, 211
Sulfuric acid (H_2SO_4), 283, 402, 475
Supercritical alcohols, 99
Supercritical CO_2, explosion by, 209
Supercritical fluid carbon dioxide (Sc-CO_2), 97–98
Supercritical fluid extraction (SFE), 170–171
Supercritical fluid technology, 97–98
Sustainability, 2, 370–372
Sustainable biofuels, 167
Sustainable development goals (SDGs), 410
Sustainable production of biofuels
 chemical components of different types of lignocellulosic biomass, 36
 composition of rice by-products, 36
 conversion of rice straw into biofuels, 40–42
 potential of rice residue in biorefineries, 37–40
 rice production in Asian countries, 35
Switchgrass, 436–437
Synechococcus sp., 112
Syngas, 40
Synthesis gas platform, 37
Synthetic gas (syn gas), 465
 Crude glycerol, 78

T

Tank to wheel (TTW), 20
Technoeconomic analysis (TEA), 2, 463, 466
 of biofuel production
  properties of agricultural residues, 463–465
  technoeconomic analysis, 466
  technoeconomic analysis of pyrolysis process of agricultural residue, 466–467
  thermochemical conversion process for agricultural residue, 465–466
 of pyrolysis process of agricultural residue, 466–467
Technology push policies, 2
Temperature, 416
Temperature uniformity, 141
Tetradesmus, 91
 T. obliquus, 100
Thawing pretreatment, 117
Thermal conversion of biomass, 140
Thermal decomposition of biomass, 143
Thermal energy, 443
Thermal process, 39
Thermobaculum terrenum TreS (TtTreS), 81–82
Thermochemical conversion process, 126–127, 384–385, 463, 465–466. *See also* Biochemical conversion process
 for agricultural residue, 465–466
  combustion, 465
  gasification, 465
  pyrolysis, 465–466
 of biomass, 138–146
  combustion, 139–140
  gasification, 141–143
  liquefaction, 145–146
  pyrolysis, 143–145
 direct combustion, 127
 gasification, 126
 liquefaction, 126–127
 pyrolysis, 127
Thermodynamic equilibrium models, 146
Thermogravimetric analysis (TGA), 195
Thermomechanical pulp lignin (TMPL), 212–213
Thetotal organic carbon, 393
Thickened waste activated sludge (TWAS), 161
Thickening, 361–363
 filtration, 362–363
 gravity and centrifugal sedimentation, 362
 ultrasonic aggregation, 362
Third generation biodiesel, 89
Third-generation biofuels, 45, 107, 204, 311, 412–413
 feedstock, 123
 generation, 168
 from microalgae
  technologic applications of microalgae, 108–119
  third-generation biorefinery, 108
 techniques involved in transformation of third-generation biofuels from algal biomass, 170–174
Third-generation biorefinery, 108, 349, 369–370, 371f
 current status of third-generation bio-refineries with microalgae, 370
Torrefaction, 39, 331
Total suspended solids (TSS), 394
Traditional biomass, 168
Traditional fossil fuels, 167

Index

Transcription activator-like effector nucleases (TALEN), 457–458
Transesterification process, 72, 79, 98, 118, 129, 285–286, 319–320, 474
  acid/base transesterification, 285
  biodiesel production by, 474–477
  enzymatic transesterification, 285–286
  of microalgal oil, 99
  reaction, 440
Transformation, 434
Transmission electron microscopy (TEM), 192
Trehalose, 81–82
Trehalose synthase (TreS), 81–82
Triacetin, 79–80
Triacyclglycerols (TG), 471, 471f
Triacylglycerides (TAGs), 415
Triacylglycerols (TAGs), 50, 91, 272, 350, 440
Triglycerides (TAGs), 174–175
Triglycerides, 72
Tubular photobioreactor, 112, 125
Two-stage hybrid system, 48
Two-step catalytic transesterification, 98–99

U

U.S. Energy Information Administration (EIA), 191, 201
Ultrasonic aggregation, 362
Ultrasonic reactors, 245
Ultrasonication, 278
Ultrasound
  frequency effect, 249–250
  pretreatment, 116–117
  ultrasound-based extraction, 170–171
Ultraviolet irradiation (UV irradiation), 455–457
  mutagenesis, 459
Unburned hydrocarbon (UBHC), 18
United Nations Convention on Climate Change (UNFCC), 201
United States Department of Energy (DOE), 3, 8
United States of America (USA), 26
Unsaturated fatty acids (UFAs), 95
Upland ecosystems, 298
Upstream processes for biodiesel synthesis, 90–93. *See also* Downstream processes for biodiesel synthesis
  algae cultivation and effects on biodiesel production, 92–93
  strain selection, 90–92
    cell wall composition, 91
    comparison between freshwater and marine microalgae, 92
    fatty acid profiling, 91–92
    lipid productivity, 90–91
Urbanization, 409

V

Vaginata virgata, 29
Valuable resource, 84
Value-added opportunities for crude glycerol, 77–78
  animal feedstuff, 77–78
  chemicals produced through conventional catalytic conversions, 78
  feedstocks for chemicals, 78
Value-added products, 55–56
  bioconversion of crude glycerol waste into value-added products, 78–83
Variable compression ratiodiesel engine (VCRDE), 18
Vegetable oil, 76, 434
  for biodiesel, 481
Vegetables, 474
Veratryl alcohol (VA), 402
Vertical PBR, 112
Vertical tube photobioreactor, 417–418
Viscosity, 102
Volatile fatty acids (VFAs), 159, 305
Volatile matter (VM), 464
Volatile organic compounds (VOCs), 153–154
Volatile suspended solids (VSS), 393
Volume of biomass, 329
Volvariella volvacea, 29, 156–157

W

Waste cooking oil (WCO), 76
Waste generation, 463
Waste-to-energy technique (WtE technique), 463
Wastewater (WW), 65, 453–454
  algal biofuel production using, 391–396
  sludge, 316
  use of microalgae for wastewater treatment, 374–375
  used for microalgal cultivation, 125
Water
  pollution, 391
  resources, 454–455
  systems, 426
Weight of biomass, 329
Well to tank (WTT), 20
Well to wheel (WTW), 20
Wet milling process, 438
Wet oxidation, pretreatment by, 209
Wheat straw, 154
Woody biomass, 316
Woody debris, 444

X

X-ray diffraction (XRD), 192–193
X-ray spectroscopy, 329
Xylem tissues, 204–205
Xylose, 226, 231

Y

Yeast strain (*Pichia kudriavzevii*), 16

Z

Zero-Valent Iron (ZVI), 80–81
Zinc (Zn), 374
Zinc-finger nucleases (ZFN), 9
Zymomonas mobilis, 15

Printed in the United States
by Baker & Taylor Publisher Services